# Advanced Topics in Mathematical Analysis

# Advanced Topics in Mathematical Analysis

*Edited by*
Michael Ruzhansky
Hemen Dutta

CRC Press
Taylor & Francis Group
Boca Raton London New York

CRC Press is an imprint of the
Taylor & Francis Group, an **informa** business

A CHAPMAN & HALL BOOK

CRC Press
Taylor & Francis Group
6000 Broken Sound Parkway NW, Suite 300
Boca Raton, FL 33487-2742

First issued in paperback 2020

© 2019 by Taylor & Francis Group, LLC
CRC Press is an imprint of Taylor & Francis Group, an Informa business

No claim to original U.S. Government works

Version Date: 20181204

ISBN 13: 978-0-367-65676-8 (pbk)
ISBN 13: 978-0-8153-5087-3 (hbk)

**Library of Congress Cataloging-in-Publication Data**

Names: Ruzhansky, M. (Michael), author. | Dutta, Hemen, 1981- author.
Title: Advanced topics in mathematical analysis / Michael Ruzhansky and Hemen Dutta.
Description: Boca Raton, Florida : CRC Press, [2018] | Includes bibliographical references
and index.
Identifiers: LCCN 2018024670| ISBN 9780815350873 (hardback : alk. paper) |
ISBN 9781351142120 (ebook)
Subjects: LCSH: Mathematical analysis--Study and teaching.
Classification: LCC QA300 .R89 2018 | DDC 515--dc23
LC record available at https://lccn.loc.gov/2018024670

**Visit the Taylor & Francis Web site at
http://www.taylorandfrancis.com**

**and the CRC Press Web site at
http://www.crcpress.com**

# Contents

# Preface

We aim this book at researchers, graduate students, and educators with interest in mathematical analysis in particular, and in mathematics, in general. The book aims to present theory, methods, and applications of the chosen topics under several chapters that seem to have recent research importance and use. An emphasis is made to present the basic developments concerning an idea with reasonable details, and also contain the recent advances made in an area of study. The text is attempted to be presented in a self-contained manner, providing at least an idea of the proof of all results, and giving enough references to enable an interested reader to follow subsequent studies in a still developing field. There are seventeen chapters in this book which are organised as follows.

The chapter "Random Measures in Infinite-Dimensional Dynamics" is devoted to a family of problems from measurable dynamics in general settings where solutions depend on measure classes and yet take a relatively explicit form. The setting includes families of reversible processes, a new family of graph Laplacians, and measurable fields, including Gaussian fields. The measure classes include atomic vs non-atomic, measures from the theory of fractal, and it includes the study of relatively singular vs. relatively absolutely continuous pairs of measures. The emphasis is on the use of positive definite functions and associated Hilbert spaces, but now adapting to the measure theoretic framework, and therefore going beyond the traditional setting of Aronszajn kernels.

In the chapter "Extensions of Some Matrix Inequalities via Matrix Means", some matrix inequalities which are relating to the Furuta inequalities are investigated. Then extensions of these matrix inequalities via the Karcher mean are studied.

The chapter "Functional Equations on Affine Groups" presents some recent developments concerning functional equations on affine groups. Exponential polynomials are the basic building blocks of spectral synthesis. Recently it has turned out that the fundamental theorem of L. Schwartz about spectral synthesis on the reals can be generalized to several dimensions via some reasonable modification of the original setting. As affine groups play a basic role in this generalization, it is reasonable to study the functional equations characterizing the class of exponential polynomials and related functions on these objects.

The chapter "Locally Pseudoconvex Spaces and Algebras" shows that the topology of any topological linear space and any topological algebra are possible to define by a collection of $F$-seminorms. Main classes of locally pseudoconvex spaces and locally pseudoconvex algebras are described and several structural results for locally pseudoconvex algebras are proved. Gelfand-Mazur theorem, Arens-Michael theorem, and Akkar theorem are generalized in the case of locally pseudoconvex algebras.

The chapter "Applications of Singular Integral Operators and Commutators" aims to study some qualitative properties of solutions to elliptic, parabolic, and ultraparabolic partial differential equations with discontinuous coefficients, in nondivergence and divergence form. Specifically, it is aimed to obtain regularity results applying some estimates for singular integral operators and commutators. Some regularity results of solutions to equations whose coefficients belong to the Sarason class of function with vanishing mean oscillation are shown. The contents move from the classical $L^p$-theory to recent applications in the framework of Morrey-type spaces. Precisely, the chapter deals with classical and generalized Morrey spaces and the new functional class of mixed Morrey spaces. The last class is very useful in the study of parabolic equations. The hearth of several results shown in the chapter is the use of an explicit representation formula for some derivatives of solutions to the equations under consideration. These formulas involve singular integral operators of various types and for this reason it is very important to find useful estimates for specific integral operators that appear in the representation formulas above. In some cases are also stated continuity results.

The chapter "Composite Submeasures and Supermeasures with Applications to Classical Inequalities" provides some composite sub- and supermeasures that are naturally associated to classical inequalities such as Jensen's inequality, Hölder's inequality, Minkowski's inequality, Cauchy-Bunyakovsky-Schwarz's inequality, Čebyšev's inequality, Hermite-Hadamard's inequalities, and the definition of convexity property. Some refinements of the above inequalities are also obtained.

The chapter "Generalized Double Statistical Weighted Summability and Its Application to Korovkin Type Approximation Theorem" is divided into five sections. The first section provides a brief introductory overview of the developments of the subjects presented. Section 2 reviews some definitions and results on difference operators of double sequences, $(p, q)$-integers, $\alpha\beta$-statistical convergence, weighted statistical convergence and statistical weighted summability. Section 3 defines a double difference operator with respect to two integer orders based on $(p, q)$-integers and extends the notions of statistical summability and statistical convergence by

the weighted method; these are largely used to obtain interesting applications to approximation theorems. Section 4 studies some inclusion relations between newly proposed methods with an illustrative example. Section 5 proves a Korovkin-type approximation theorem for functions of two variables and also presents an example via $(p, q)$-analogue of generalized bivariate

Bleimann-Butzer-Hahn operators to show that the proposed method is stronger than its classical and statistical versions.

The chapter "Birkhoff-James Orthogonality and Its Application in the Study of Geometry of Banach Space" studies the relation between Birkhoff-James orthogonality in a finite dimensional real Banach space $X$ and the space of bounded linear operators $B(X)$ in terms of the norm attainment set of a bounded linear operator. Authors also obtain an elegant characterization of inner product spaces in terms of operator norm attainment. Various geometric and analytic properties of the space of bounded linear operators $B(X)$ have been studied using Birkhoff-James orthogonality. A complete characterization of Birkhoff-James orthogonality of bounded linear operators has been obtained for Hilbert spaces of any dimension, as well as finite dimensional Banach spaces. Finally, the symmetry of Birkhoff-James orthogonality of bounded linear operators has been discussed for Hilbert spaces, as well as Banach spaces.

The purpose of the chapter "Fixed Point of Mappings on Metric Space Endowed with Graphic Structure" is to construct the fixed-point results of multivalued graphic contraction mappings defined on the family of closed and bounded subsets of a metric space endowed with a graph. Authors also prove some coincidence and common fixed point results for multivalued generalized graphic contractive mappings defined on the family of sets endowed with a graph. These results are obtained without appealing to any form of continuity of mappings involved. Some examples are presented to support the results.

The chapter "Some Subclasses of Analytic Functions and Their Properties" investigates some subclasses of analytic functions in the open unit disk in the complex plane. Authors derive characteristic properties of the functions belonging to these classes and find upper bound estimates for some functions belonging to these subclasses. Also, for some other functions in these classes, several coefficient inequalities are given. Furthermore, two recent general $p$-valent integral operators in the unit disc $U$ are introduced and the properties of $p$-valent starlikeness and $p$-valent convexity of these integral operators of $p$-valent functions on some classes of $\beta$-uniformly $p$-valent starlike and $\beta$-uniformly $p$-valent convex functions of complex order and type $\alpha$ $(0 \leq \alpha < p)$ including theirs in some special cases are obtained.

The chapter "Free Stochastic Integrals for Weighted-Semicircular Motion Induced by Orthogonal Projections" constructs weighted-semicircular elements from mutually orthogonal integer-many projections, and then: (i) establishes weighted-semicircular processes induced by the weighted-semicircular elements; and (ii) studies the corresponding free stochastic integrals. In particular, the author considers free distributions of such integrals by understanding them as free-probabilistic objects. The main results not only provide interesting examples, and applications in free stochastic calculus, but also show how a family of orthogonal projections induces stochastic calculus models.

In the chapter "On Difference Operators and Their Applications", a new difference operator of fractional order and related sequence spaces has been introduced, and, subsequently, an application of this operator, the notion of the derivatives and the integrals of a function in the case of non-integer order has been generalized. In fact, the entire chapter is divided into five sections. In the beginning, some definitions and known results in the literature are presented, which is the basic part of Section 1. In Section 2, the authors define a generalized new fractional difference operator, which induces several difference operators in the literature and also provides some relations among such operators with different orders. In Section 3, the authors define related sequence spaces and study their topological and certain other properties. As an application of the proposed operator, in Section 4, authors determine the fractional derivatives of certain functions and study some results involving Leibniz and Chain rules. Finally, Section 5 deals with the investigation of unusual and non-uniform behaviour of the proposed operator and fractional calculus. It also interprets the unusual and non-uniform behaviours of fractional calculus geometrically.

The chapter "Composite Ostrowski and Trapezoid Type Inequalities for Riemann-Liouville Fractional Integrals of Functions with Bounded Variation" establishes some composite Ostrowski and generalized trapezoid type inequalities for the Riemann-Liouville fractional integrals of functions of bounded variation. Applications for composite mid-point and trapezoid inequalities are provided as well. They generalize the known results holding for the classical Riemann integral.

The chapter "*kn-2* Scalar Matrix and Its Functional Equations by Mathematical Modeling" defines *kn-2* scalar matrix and discusses its eigenvalues and eigenvectors. This chapter elucidates a proper method to model various types of functional equations like additive, quadratic and mixed type of additive and quadratic using eigenvalues and eigenvectors of newly established *kn-2* scalar matrices with suitable numerical examples.

In the chapter "Renorming $c_0$ and Fixed Point Property", under affinity assumption, it is shown first that can be renormed so that there exists a large class of non-weakly compact, closed, bounded, convex (c.b.c.) subsets of with fixed point property for nonexpansive mappings [FPP (n.e)]. Next, it is proved that there is a family of equivalent norms on that has FPP (n.e) if the mappings are affine.

The chapter "Steinhaus Type Theorems over Valued Fields: A Survey" presents a brief survey of the Steinhaus type theorems over fields such as a complete, non-trivially valued field or a complete, non-trivially valued, non-Archimedean field proved so far.

The chapter "The Interplay Between Topological Algebras Theory and Algebras of Holomorphic Functions" reviews classical results in the subject concerning interactions between the general theory of commutative topological algebras over $C$ with unity and that of the holomorphic functions of one or several complex variables, as well as lists some long-standing open problems. This chapter is divided into two parts. The first part reviews known results on the general theory of Banach algebras, Frechet algebras, $LB$ and $LF$ algebras, as well as the open problems in the field. The second part examines *concrete* algebras of holomorphic functions, again reviewing known results and open problems.

The editors are grateful to many people in the form of authors, reviewers, colleagues, and well-wishers for their valuable co-operation and contributions in some ways. We are hopeful that the book will serve its intended purposes.

**Michael Ruzhansky**, London, UK
**Hemen Dutta**, Guwahati, India

January 2018

# Contributors

**Mujahid Abbas**
Department of Mathematics,
Government College University
Lahore, Pakistan
and
Department of Mathematics,
King Abdulaziz University
Jeddah, Saudi Arabia
abbas.mujahid@gmail.com

**Mati Abel**
Institute of Mathematics and
Statistics, University of Tartu
Tartu, Estonia
mati.abel@ut.ee

**Pinakadhar Baliarsingh**
Department of Mathematics, School
of Applied Sciences, KIIT University
Bhubaneswar, India
pb.math10@gmail.com

**Ilwoo Cho**
Department of Mathematics, Saint
Ambrose University Davenport,
Iowa, USA
choilwoo@sau.edu

**Silvestru Sever Dragomir**
Mathematics, College of Engineering
& Science, Victoria University
Melbourne City, Australia
and
DST-NRF Centre of Excellence in
Mathematical and Statistical
Sciences, School of Computer Science
and Applied Mathematics,
University of the Witwatersrand,
Johannesburg, South Africa
sever.dragomir@vu.edu.au

**Hemen Dutta**
Department of Mathematics,
Gauhati University
Guwahati, India
hemen_dutta08@rediffmail.com

**Palle E.T. Jorgensen**
Department of Mathematics,
University of Iowa Iowa City,
Iowa, USA
palle-jorgensen@uiowa.edu

**Ugur Kadak**
Department of Mathematics,
Gazi University
Ankara, Turkey
ugurkadak@gmail.com

**Nizami Mustafa**
Faculty of Science and Letters,
Department of Mathematics, Kafkas
University Kars, Turkey
nizamimustafa@gmail.com

**Pasupathi Narasimman**
Department of Mathematics,
Thiruvalluvar University College of
Arts and Science,
Kariyampatti, Tirupattur,
Tamilnadu, India
drpnarasimman@gmail.com

**P.N. Natarajan**
R.A. Puram
Chennai, India
Formerly Head of the Department of
Mathematics, Ramakrishna
Mission Vivekananda College
Chennai, India
pinnangudinatarajan@gmail.com

**Talat Nazir**
Department of Mathematics,
COMSATS Institute of Information
Technology
Abbottabad, Pakistan
and
Department of Mathematics,
University of Jeddah,
Jeddah, Saudi Arabia
dr.talatnazir@gmail.com

**Vessel Nezir**
Faculty of Science and Letters,
Department of Mathematics,
Kafkas University
Kars, Turkey
veyselnezir@yahoo.com

**Kallol Paul** Department of
Mathematics, Jadavpur University
Kolkata,
West Bengal, India
kalloldada@gmail.com

**Debmalya Sain**
Department of Mathematics, Indian
Institute of Science
Bengaluru, Karnataka, India
saindebmalya@gmail.com

**Alberto Saracco** Dipartimento di
Scienze Matematiche, Fisiche e
Informatiche
Università degli Studi di Parma,
Italy
alberto.saracco@unipr.it

**Andrea Scapellato**
Department of Mathematics and
Computer Science, University of
Catania
Viale Andrea Doria, Catania, Italy
scapellato@dmi.unict.it

**László Székelyhidi**
Institute of Mathematics, University
of Debrecen
Debrecen, Hungary
lszekelyhidi@gmail.com

**Feng Tian**
Department of Mathematics,
Hampton University
Hampton, Virginia, USA
feng.tian@hamptonu.edu

**Takeaki Yamazaki**
Department of Electrical, Electronic
and Computer Engineering, Toyo
University
Kawagoe, Japan
t-yamazaki@toyo.jp

# Chapter 1

## Random Measures in Infinite-Dimensional Dynamics

**Palle E.T. Jorgensen**

*Department of Mathematics, The University of Iowa, Iowa City, U.S.A.*

**Feng Tian**

*Department of Mathematics, Hampton University, Hampton, U.S.A.*

## 1.1 Introduction, history, and motivation

We study a family of problems from measurable dynamics and their connection to the theory of positive definite functions. In particular, the aim of the present chapter is two-fold. One is an extension of the traditional setting for reproducing kernel Hilbert space (RKHS) theory. We also extend the traditional context of Aronszajn Aronszajn (1943, 1950) to better adapt it to a host of applications to probability theory, stochastic processes Alpay and Jorgensen (2012); Applebaum (2009); Jorgensen and Tian (2016); Mumford (2000), mathematical physics Haeseler et al. (2014); Konno et al. (2005); Osterwalder and Schrader (1973); Parussini et al. (2017); Rodgers et al. (2005); Tsong and Wu (2006) and measurable dynamics, specifically reversible processes (see, e.g., Bezuglyi et al. (2014); Chang et al. (2015); Dutkay and Jorgensen (2011); Hersonsky (2012); Roblin (2011); Skopenkov (2013); Tosiek and Brzykcy (2013), and also Bishop and Peres (2017); Lubetzky and Peres (2016); Peres et al. (2016); Peres and Sousi (2016)). For applications to random processes, a kernel in the sense of Aronszajn will typically represent a covariance kernel.

The second main theme in the chapter will be the study of reversible processes, and their applications to generalized graph Laplacians. See, e.g., Barrière et al. (2006); Burioni and Cassi (2005); Freschi (2011); Jorgensen and Pearse (2010, 2011, 2013); Strichartz (2010); Zhou et al. (2014).

Now in the standard approach to RKHSs of Aronszajn, one starts with a positive definite function (p.d), $K$ on $X \times X$ (often called a p.d. kernel) where $X$ is a given set. The term "reproducing" refers to the fact that for every $f$ in $\mathscr{H}$, the values $f(x)$ can be reproduced from the inner Hilbert-product in $\mathscr{H}$. With a standard construction, starting with $X$ and $K$, one then arrives at a Hilbert space $\mathscr{H}$ of functions on $X$, the so called reproducing kernel Hilbert space (RKHS). It depends on the pair $(X, K)$ of course; so is denoted $\mathscr{H}(K)$ when the kernel is not given from the context. *A priori*, the set $X$ is not given any additional structure, but a key point is that both $K$ and the functions $f$ in the RKHS $\mathscr{H}(K)$ are defined everywhere on $X$. If for example, $X$ is a complex domain, in interesting applications, then the functions in $\mathscr{H}(K)$ will be analytic, or in the case of the familiar RKHS of Bargmann (Theorem 1.5.4), the functions in $\mathscr{H}(K)$ will be entire analytic. If $X$ has a topology, and if $K$ is assumed continuous, then the functions in $\mathscr{H}(K)$ will also be continuous.

But up to now, many of the applications have focused on Hilbert spaces of regular functions. If for example, a kernel represents a Green's function for an elliptic partial differential operator (PDO), then the associated RKHS will consist of functions which have some degree of smoothness.

Turning to stochastic analysis, we note that the use of p.d. kernels, and associated RKHSs, has been especially successful in the context of Gaussian processes; mainly due to the following theorem: For every p.d. kernel $K$ on an arbitrary set, say $X$, there is an associated square integrable Gaussian process, say $W$, which is then indexed by $X$, i.e., $W$ has moments of second order, and such that $W$ has $K$ as its covariance kernel, i.e., $\mathbb{E}[W_x W_y] = K(x, y), \forall (x, y) \in X \times X$ (see Theorem A.1.12 in Appendix A at the end of this chapter for details.)

In the simplest cases, such as Brownian motion, the process (still called $W$) is indexed by time, so the set $X$ under discussion may then instead represent some kind of time-indexing. Now many processes are indexed differently, and the terminology "random field" is often used for this more general context. Many Gaussian processes are not continuous in their indexing, and so a modification of the standard approach to p.d. kernel theory is called for. And there are yet other applications which dictate such an extension.

But even with this particular area of applications, the Aronszajn approach has serious limitations: Often functions will be defined only almost everywhere with respect to some measure which is prescribed on the set $X$; for example, if $X$ represents time, in one or more dimensions, the prescribed measure $\lambda$ is often Lebesgue measure. For fractal random fields, $\lambda$ may be a fractal measure. For this reason, and others (to be outlined in this chapter), it is useful to instead let $X$ be a measure space, say $(X, \mathscr{B}, \lambda)$. And for important reasons (details below), it is appropriate to assume that $X$ is a locally compact Hausdorff space, that $\mathscr{B}$ is the corresponding sigma-algebra, and that $\lambda$ is a fixed positive measure, and assumed to be a regular measure on $(X, \mathscr{B})$. The modification in the resulting new definition of p.d. kernels $K$ in this context is subtle. Here we just mention that, for a p.d. system $(X, \mathscr{B}, \lambda)$ and kernel $K$ in the measurable category, the associated RKHS $\mathscr{H}$ will now instead be a Hilbert space of measurable functions on $X$, more precisely, measurable with respect to $\mathscr{B}$, and locally in $L^2(\lambda)$. We shall say that $\mathscr{H}$ is contained in $L^2_{loc}(\lambda)$. The p.d. kernel $K$ itself will be a random family of signed measures on $(X, \mathscr{B})$.

**Discussion of the literature.** The theory of RKHS and its applications is vast, and below we only make a selection. Readers will be able to find more cited within the selection. As for the general theory of RKHS in the pointwise category, we find useful Alpay et al. (1993); Alpay and Dym (1992, 1993); Lata et al. (2009); Paulsen and Raghupathi (2016). The applications include fractals (see e.g., Alpay et al. (2013); Aronszajn (1943); Bezuglyi and Handelman (2014)); probability theory Cortes et al. (2017); El Machkouri et al. (2017); Hsing and Eubank (2015); Jørsboe (1968); Muandet et al. (2016); Parussini et al. (2017); Saitoh (2016); and application to learning theory Jorgensen and Tian (2015); Smale and Zhou (2004, 2009a,b).

We now turn to the details. In the first half of the chapter, we shall focus on theory, and in the second half, applications.

Our approach is that of book chapter; and this viewpoint has shaped our presentation in two ways. First, in order to make the material more accessible to non-specialists, and to researchers from neighboring areas, we have added an appendix which should be useful for readers who encounter one theme or the other for the first time. Secondly, we have strived to avoid excessively technical details inside some of the proofs. As a result, some proofs below have been formulated in order to stress the underlying main ideas, the intuition, and the motivation. As for more in-depth technical fine-points, and some longer proofs, we have referred in a few instances instead to research papers, some currently in the works.

---

## 1.2 RKHSs in measurable category: new settings for positivity conditions

We begin with an outline making the comparison between the reproducing property for the RKHS, in the context of Aronszajn on the one hand, and the corresponding *reproducing property* valid in the measurable context. The details referred to in Table 1.1 will be made clear inside the section.

Setting.

- $X$: A locally compact Hausdorff space;

- $\mathscr{B}$: Borel $\sigma$-algebra of subsets of $X$;

- $\lambda$: A positive measure on $(X, \mathscr{B})$.

| Aronszajn | Measurable |
|---|---|
| $X \times X \xrightarrow{K} \mathbb{R}$ | $(X, \mathscr{B}, \lambda)$ fixed |
| $\mathscr{H}$ | $\mathscr{H} \subset L^2_{loc}(\lambda)$ |
| $f \in \mathscr{H} \colon f(x) = \langle f, K(x, \cdot) \rangle_{\mathscr{H}},$ $\forall x \in X$ | $f \in \mathscr{H} \colon \int \varphi f d\lambda = \langle f, K^{\varphi} \rangle_{\mathscr{H}},$ $\forall \varphi \in C_c(X)$ |

**TABLE 1.1**: Reproducing properties.

- Set $\mathscr{B}_{fin} = \{E \in \mathscr{B} \mid \lambda(B) < \infty\}$.

- $\mathscr{H} \subset L^2_{loc}(\lambda)$. That is, every $f \in \mathscr{H}$ is assumed to satisfy

$$\int_B |f|^2 \, d\lambda < \infty, \ \forall B \in \mathscr{B}_{fin}. \tag{1.1}$$

**Definition 1.2.1.** A pair $(\mathscr{H}, \lambda)$ as above is said to be *positive definite* (p.d.) if and only if, for all $B \in \mathscr{B}_{fin}$,

$$\sup_{f \in \mathscr{H}, \, \|f\|_{\mathscr{H}} \leq 1} \left| \int_B f d\lambda \right| < \infty. \tag{1.2}$$

Note that (1.2) makes sense since $f$ is assumed to be in $L^2_{loc}(\lambda)$.

*Remark.* In the proofs below we shall mostly restrict attention to real valued functions, but all arguments easily extend to the complex case.

**Lemma 1.2.2.** *The following are equivalent:*

$$\sup \left\{ \left| \int_A f d\lambda \right|, \, f \in \mathscr{H}, \, \|f\|_{\mathscr{H}} \leq 1 \right\} < \infty, \ \forall A \in \mathscr{B}_{fin}, \tag{1.3}$$

$$\Updownarrow$$

$$\sup \left\{ \left| \int \varphi f d\lambda \right|, \, f \in \mathscr{H}, \, \|f\|_{\mathscr{H}} \leq 1 \right\} < \infty, \ \forall \varphi \in C_c(X). \tag{1.4}$$

*Proof.* A use of approximation rules from measure theory: with simple functions, and with $C_c(X)$. $\qquad \square$

**Lemma 1.2.3.** *Suppose $(\mathscr{H}, \lambda)$ is positive definite.*

*(i) Then, for $\forall B \in \mathscr{B}_{fin}$, $\exists! K^B \in \mathscr{H}$ such that*

$$\int_B f d\lambda = \langle f, K^B \rangle_{\mathscr{H}}, \ \forall f \in \mathscr{H}. \tag{1.5}$$

(ii) Setting $K(\cdot, B) = K^B$, $B \in \mathscr{B}_{fin}$, note that $x \longmapsto K(x, \cdot)$ is then a measurable field of random signed measures on $(X, \mathscr{B})$, and that

$$\rho_{\mathscr{H}}(A, B) = \langle K^A, K^B \rangle_{\mathscr{H}} \tag{1.6}$$

$$= \int_A K(x, B)\, d\lambda(x) = \overline{\int_B K(y, A)\, d\lambda(y)}.$$

Then $\rho_{\mathscr{H}}$ is p.d. on $\mathscr{B}_{fin} \times \mathscr{B}_{fin}$. (See Definition A.1.1.)

*Proof.* (i) Apply Riesz to the linear functional $\mathscr{H} \ni f \longmapsto \int_B f d\lambda$. Part (ii) is immediate. $\qquad\square$

**Corollary 1.2.4.** *Let* $(\mathscr{H}, \lambda)$ *be as above; and let* $\{f_i\}_{i \in \mathbb{N}}$ *be a fixed ONB in* $\mathscr{H}$, *or a Parseval Frame, then, for all* $B \in \mathscr{B}_{fin}$, *we have:*

$$K^B(x) = \sum_{i=1}^{\infty} \left( \int_B f_i d\lambda \right) f_i(x); \tag{1.7}$$

*with convergence in the* $\mathscr{H}$-*norm, and*

$$\|K^B\|_{\mathscr{H}}^2 = \sum_{i=1}^{\infty} \left| \int_B f_i d\lambda \right|^2. \tag{1.8}$$

*Proof.* Use the reproducing property, $\langle f_i, K^B \rangle = \int_B f_i d\lambda$, for all $B \in \mathscr{B}_{fin}$. $\qquad\square$

*Remark* 1.2.5. Let $(\mathscr{H}, \lambda)$ be as specified in Definition 1.2.1. In this case, the corresponding reproducing property takes any one of the following equivalent forms:

1. For all $B \in \mathscr{B}_{fin}$, we have $\mathscr{H} \subset L^2_{loc}(\lambda)$, and

$$\int_B f d\lambda = \langle f(\cdot), K(\cdot, B) \rangle_{\mathscr{H}}, \quad \forall B \in \mathscr{B}_{fin}, \forall f \in \mathscr{H}. \tag{1.9}$$

2. Setting $K^{\varphi}(\cdot) = \int \varphi(y) K(\cdot, dy) (\in \mathscr{H})$, then we have

$$\int f\varphi d\lambda = \langle f, K^{\varphi} \rangle_{\mathscr{H}}, \quad \forall \varphi \in C_c(X), \forall f \in \mathscr{H}. \tag{1.10}$$

When $(\mathscr{H}, \lambda)$ is given as above, and $f \in \mathscr{H}$, we set $\mu_f$ to be the signed measure

$$\mu_f(B) := \langle f(\cdot), K(\cdot, B) \rangle_{\mathscr{H}}, \quad B \in \mathscr{B}_{fin}. \tag{1.11}$$

**Definition 1.2.6.** Let $(X, \lambda)$ be as above. Let $\rho_{\mathscr{H}}$ be the p.d. function on $\mathscr{B}_{fin} \times \mathscr{B}_{fin}$ as in (1.6). We denote $\mathscr{H}^{(Aronsz)}(\rho_{\mathscr{H}})$ the corresponding RKHS in the sense of Aronszajn's.

**Theorem 1.2.7.** *Let $(\mathscr{H}, \lambda)$ be as above, i.e., positive definite (p.d.). Then the following hold:*

*(i) For all $f \in \mathscr{H}$, let $\mu_f$ be the signed measure as in (1.11). Then $\mu_f \ll \lambda$, with the Radon-Nikodym derivative*

$$\frac{d\mu_f}{d\lambda} = f \in L^1_{loc}(\lambda). \tag{1.12}$$

*(ii) For all $f \in \mathscr{H}$, we have $\mu_f \in \mathscr{H}^{(Aronsz)}(\rho_{\mathscr{H}})$, and*

$$\left\| \mu_f \right\|_{\mathscr{H}^{(Aronsz)}(\rho_{\mathscr{H}})} = \|f\|_{\mathscr{H}}, \ \forall f \in \mathscr{H}. \tag{1.13}$$

*More precisely, the functions in $\mathscr{H}^{(Aronsz)}(\rho_{\mathscr{H}})$ are signed measures on $(X, \mathscr{B})$; and the map*

$$\mathscr{H} \ni f \xmapsto[\simeq]{T} \mu_f = f d\lambda \in \mathscr{H}^{(Aronsz)}(\rho_{\mathscr{H}}), \tag{1.14}$$

*is an isometric isomorphism from $\mathscr{H}$ onto $\mathscr{H}^{(Aronsz)}(\rho_{\mathscr{H}})$.*

*Proof.* For part (i), note that

$$\mu_f(B) = \left\langle f, K^B \right\rangle_{\mathscr{H}} = \int_B f d\lambda, \ \forall f \in \mathscr{H}, \ \forall B \in \mathscr{B}_{fin};$$

see (1.5), and (1.11).

Part (ii). Given $f \in \mathscr{H}$, we have

$$\left| \sum_i \xi_i \mu_f(A_i) \right|^2 = \left| \left\langle f, \sum_i \xi_i K^{A_i} \right\rangle_{\mathscr{H}} \right|^2$$
$$\leq \|f\|_{\mathscr{H}}^2 \sum_i \sum_j \xi_i \xi_j \rho_{\mathscr{H}}(A_i, A_j);$$

valid for $\forall (\xi_i)_{i=1}^n$ in $\mathbb{R}$, $\forall (A_i)_{i=1}^n$ in $\mathscr{B}_{fin}$, $\forall n \in \mathbb{N}$. Thus, $\mu_f \in \mathscr{H}^{(Aronsz)}(\rho_{\mathscr{H}})$, $\forall f \in \mathscr{H}$. (See Theorem A.1.4.)

Note also that $span\{K(\cdot, A)\}_{A \in \mathscr{B}_{fin}}$ and $span\{\rho_{\mathscr{H}}(\cdot, A)\}_{A \in \mathscr{B}_{fin}}$ are dense in $\mathscr{H}$ and $\mathscr{H}^{(Aronsz)}(\rho_{\mathscr{H}})$, respectively; and

$$\left\| \sum \xi_i K^{A_i} \right\|_{\mathscr{H}}^2 = \sum_i \sum_j \xi_i \xi_j \left\langle K^{A_i}, K^{A_j} \right\rangle_{\mathscr{H}}$$
$$= \sum_i \sum_j \xi_i \xi_j \rho_{\mathscr{H}}(A_i, A_j) = \left\| \sum \xi_i \rho_{\mathscr{H}}(\cdot, A_i) \right\|_{\mathscr{H}^{(Aronsz)}(\rho_{\mathscr{H}})}^2.$$

Thus, $T : K^A \longmapsto \rho_{\mathscr{H}}(\cdot, A)$ extends to an isometric isomorphism from $\mathscr{H}$ onto $\mathscr{H}^{(Aronsz)}(\rho_{\mathscr{H}})$. In particular, (1.13) holds. $\square$

*Remark* 1.2.8. Starting with a p.d. pair $(\mathscr{H}, \lambda)$, see Definition 1.2.1, we introduce $\rho_{\mathscr{H}} : \mathscr{B}_{fin} \times \mathscr{B}_{fin} \longrightarrow \mathbb{R}$, $\rho_{\mathscr{H}}(A, B) = \left\langle K^A, K^B \right\rangle_{\mathscr{H}}$. It is clear that $\rho_{\mathscr{H}}$ is then positive definite. Indeed, if $\{\xi_i\}_{i=1}^n$ in $\mathbb{R}$, $(A_i)_{i=1}^n$ in $\mathscr{B}_{fin}$, then

$$\sum_i \sum_j \xi_i \xi_j \rho_{\mathscr{H}}(A_i, A_j) = \left\| \sum_i \xi K^{A_i} \right\|_{\mathscr{H}}^2 \geq 0.$$

See also Definition A.1.1.

## 1.3   $L_{loc}^2(\lambda)$ vs $L^2(\lambda)$: Local versus global $L^2$-conditions

Given $(X, \mathscr{B}, \mathscr{H}, \lambda)$ as above, i.e., $(\mathscr{H}, \lambda)$ is p.d. (Definition 1.2.1). Let $\{K(\cdot, A)\}_{A \in \mathscr{B}_{fin}}$ be the corresponding random signed measures.

On the dense domain $C_c(X)$ in $L^2(\lambda)$, set

$$L\varphi := K^\varphi, \ \varphi \in C_c(X), \tag{1.15}$$

where $K^\varphi(\cdot) = \int \varphi(y) K(\cdot, dy) \in \mathscr{H}$. We shall consider the two operators $L$ and $L^*$ as follows:

$$L^2(\lambda) \ni \varphi \quad \overset{L}{\underset{L^*}{\rightleftarrows}} \quad K^\varphi \in \mathscr{H} \tag{1.16}$$

**Proposition 1.3.1.** *Let $L$ be as in (1.15), then $f \in dom(L^*)$ iff $\exists g \in L^2(\lambda)$ s.t. $L^* f = g$, and $f = g$, $\lambda$-a.e.*

*Proof.* Let $f \in dom(L^*)$, and set $g = L^* f$, then

$$\langle f, L\varphi \rangle_{\mathscr{H}} = \langle L^* f, \varphi \rangle_{L^2(\lambda)}$$
$$\Updownarrow$$
$$\langle f, K^\varphi \rangle_{\mathscr{H}} = \langle g, \varphi \rangle_{L^2(\lambda)}$$
$$\Updownarrow$$
$$\int \varphi f d\lambda = \int \varphi g d\lambda, \ \forall \varphi \in C_c(X).$$

Thus, $f = g$ w.r.t. $\lambda$. $\qquad\square$

*Remark* 1.3.2 (A non-closable operator). The question of when some $f \in \mathscr{H} \subset L_{loc}^2(\lambda)$ may not be in $L^2(\lambda)$ is related to "closable" for a certain unbounded operator.

**Lemma 1.3.3.** *The operator $L$ (see (1.15)-(1.16)) is not closable iff there exists $f \in \mathscr{H} \setminus \{0\}$ s.t. $f \notin L^2(\lambda)$.*

*Proof.* Recall that, by definition, $L$ is closed iff its graph, $G(L)$, is a closed subspace in $L^2(\lambda) \oplus \mathscr{H}$, where

$$G(L) = \left\{ \binom{\varphi}{K^\varphi} \mid \varphi \in C_c(X) \right\} \subset \begin{matrix} L^2(\lambda) \\ \oplus \\ \mathscr{H} \end{matrix}. \tag{1.17}$$

Hence $L$ is *not* closable iff $\exists \{\varphi_n\}$ in $C_c(X)$, $\varphi_n \to 0$ in $L^2(\lambda)$, but

$$K^{\varphi_n} \to f \neq 0, \ f \in \mathscr{H},$$

and so $f \in \mathscr{H} \setminus L^2(\lambda)$ iff $L$ is non-closable. $\qquad\square$

*Remark* 1.3.4. We showed that $dom\,(L^*) = \mathscr{H} \cap L^2\,(\lambda)$ (Proposition 1.3.1). So if $\mathscr{H} \cap L^2\,(\lambda)$ is a "small" subspace in $\mathscr{H}$, then $L$ is "very" non-closable. This also implies that $dom\,(L^*)$ is *not* dense in $\mathscr{H}$.

**Theorem 1.3.5.** *Given $(\mathscr{H}, \lambda)$ be p.d. as above, we then get two densely defined operators, $L$ and the adjoint $L^*$. For the domain of $L^*$, we have $dom\,(L^*) = \mathscr{H} \cap L^2\,(\lambda)$ which is generally not dense in $\mathscr{H}$, so $L$ will then be non-closable.*

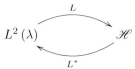

*where*

(i) *$dom\,(L) = C_c\,(X)$, and*

$$L\varphi = \int \varphi\,(y)\,K\,(x, dy)\,, \quad \forall \varphi \in C_c\,(X)\,. \tag{1.18}$$

(ii) *$L^* : \mathscr{H} \longrightarrow L^2\,(\lambda)$,*

$$dom\,(L^*) = \{f \in \mathscr{H} \mid \langle f, K\,(\cdot, dy)\rangle_{\mathscr{H}} = f\,(y)\,\lambda\,(dy)\}\,. \tag{1.19}$$

*Equivalently, $f \in dom\,(L^*)$ iff*

$$\mu_f := \langle f, K\,(\cdot, dy)\rangle_{\mathscr{H}} \ll \lambda \tag{1.20}$$

*with Radon-Nikodym derivative*

$$\frac{d\mu_f}{d\lambda}\,(y) = f\,(y) \in L^2_{loc}\,(\lambda)\,. \tag{1.21}$$

*Then $L^*f = f$, for all $f \in dom\,(L^*)$.*

*Remark.* The setting here is more restrictive than the most general case of $K\,(\cdot, dy)$, p.d. on $(X, \mathscr{B}, \lambda)$.

*Proof.* We shall only consider real-valued case. With formulas (1.18)-(1.21) and the respective domains of $L$ and $L^*$, we now prove

$$\langle L\varphi, f\rangle_{\mathscr{H}} = \langle \varphi, L^*f\rangle_{L^2(\lambda)}\,, \tag{1.22}$$

for all $\varphi \in dom\,(L)$, and all $f \in dom\,(L^*)$. Note that

$$
\begin{aligned}
\mathrm{LHS}_{(1.22)} &= \left\langle f, \int \varphi\,(y)\,K\,(\cdot, dy)\right\rangle_{\mathscr{H}} \\
&= \int \varphi\,(y)\,\langle f, K\,(\cdot, dy)\rangle_{\mathscr{H}} \qquad \text{since } \varphi \in C_c\,(X) \\
&= \int \varphi\,(y)\,f\,(y)\,d\lambda\,(y) \qquad \text{by } (1.21) \\
&= \langle \varphi, L^*f\rangle_{L^2(\lambda)} = \mathrm{RHS}_{(1.22)},
\end{aligned}
$$

and the desired conclusion holds: If $f \in \mathscr{H}$ satisfies $f \in dom\,(L^*)$, then (Radon-Nikodym derivative)

$$L^* f = \frac{d\mu_f}{d\lambda} = \frac{d\,\langle f, K\,(\cdot, dy)\rangle_{\mathscr{H}}}{d\lambda} = f;$$

see (1.20)-(1.20); so $L^*$ is non-closable if $L^2\,(\lambda) \cap \mathscr{H}$ is not dense in $\mathscr{H}$. From the general theory of unbounded operators, we know that an operator $L$ with dense domain is closable iff its adjoint $L^*$ has dense domain. □

*Remark* 1.3.6 (Two Variants). We have two equivalent versions of p.d.:

1. $\rho_{\mathscr{H}} : \mathscr{B}_{fin} \times \mathscr{B}_{fin} \longrightarrow \mathbb{C}$,

$$\rho_{\mathscr{H}}\,(A, B) = \left\langle K^A, K^B \right\rangle_{\mathscr{H}} = \int_A K\,(x, B)\,d\lambda\,(x)\,; \tag{1.23}$$

2. The map $C_c\,(X) \times C_c\,(X) \longrightarrow \mathbb{C}$,

$$(\varphi, \psi) \longmapsto \varphi K \psi = \int \varphi\,(x) \int \psi\,(y)\,K\,(x, dy)\,d\lambda\,(x)\,. \tag{1.24}$$

The function $\rho_{\mathscr{H}}$ in (1.23) is p.d. in the sense of Aronszajn. The corresponding RKHS $\mathscr{H}^{Aronsz}\,(\rho_{\mathscr{H}})$ (see Definition 1.2.6) is a Hilbert space of functions on $\mathscr{B}$; and the realization here are Hilbert spaces $\mathscr{H} = \mathscr{H}\,(K, \lambda)$ of measurable functions on $(X, \mathscr{B})$.

Thus, $\mathscr{H}^{(Aronsz)}\,(\rho_{\mathscr{H}}) \simeq \mathscr{H}\,(K, \lambda)$, via $\rho_{\mathscr{H}}\,(\cdot, A) \longmapsto K\,(\cdot, A)$, $\forall A \in \mathscr{B}_{fin}$.

The generalized RKHS $\mathscr{H}\,(K, \lambda)$ has the reproducing property as follows:

**Theorem 1.3.7.** *$\mathscr{H}\,(K, \lambda)$ is the Hilbert-completion of functions $\{K^\varphi \mid \varphi \in C_c\,(X)\}$ w.r.t.*

$$\left\langle K^\varphi, K^\psi \right\rangle_{\mathscr{H}(K,\lambda)} = \int \varphi\,(x) \int \psi\,(y)\,K\,(x, dy)\,d\lambda\,(x)\,. \tag{1.25}$$

*For all $B \in \mathscr{B}_{fin}$, and $f \in \mathscr{H}\,(K, \lambda)$, we have*

$$\mu_f\,(B) := \langle f\,(\cdot)\,, K\,(\cdot, B)\rangle_{\mathscr{H}(K,\lambda)} = \int_B f\,(y)\,d\lambda\,(y)\,; \tag{1.26}$$

*so $\mu_f \ll \lambda$, $d\mu_f/d\lambda = f$ in $L^1_{loc}\,(\lambda)$.*

*Remark.* We emphasize two important points:

1. For all $f \in \mathscr{H}\,(K, \lambda)$, $\exists\,\{\varphi_n\}$ in $C_c\,(X)$ s.t.

$$\|f - K^{\varphi_n}\|_{\mathscr{H}(K,\lambda)} \xrightarrow[n \to \infty]{} 0;$$

and

$$\langle K^\varphi, f\rangle_{\mathscr{H}} = \int \varphi\,(y)\,d\mu_f\,(y) = \int \varphi\,(y)\,f\,(y)\,d\lambda\,(y)\,, \; \forall \varphi \subset C_c\,(X)\,. \tag{1.27}$$

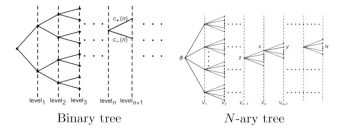

| Binary tree | $N$-ary tree |

**FIGURE 1.1**: Examples of discrete networks.

2. All conditions are purely local, i.e., (1.27) holds for all $\varphi \in C_c(X)$. Also, we assume $\lambda$ is positive and *regular*:

    (i) For all compact subset $C \subset X$, $\lambda(C) < \infty$;

    (i) For all open set $\mathscr{O} \subset X$, $\mathscr{O} \neq \emptyset$, we have $\lambda(\mathscr{O}) > 0$;

    (i) $\lambda$ is inner, and outer, regular; see Rudin (1987).

---

## 1.4 Continuous networks: Atomic versus non-atomic models

There is already a rich body of research on discrete network theory, both because of its relevance to discrete harmonic analysis, discrete graph Laplacians, and the study of reversible Markov chains; see the cited papers in the References below. As is known, the two topics are closely connected. Some key questions for these Markov models deal with infinite path analysis; for example, in a given model, decide whether it is transient or recurrent.

In more detail: The discrete framework is a given pair $(V, E)$, where $V$ is a countable discrete set of vertices such that each $x$ in $V$ has a finite number of nearest neighbors, but excluding loops $(xx)$. The set $E$ of edges consists of undirected lines $(xy)$ between such nearest neighbors. The pair $(V, E)$ is then the associated network, or graph. See Figures 1.1-1.2.

In principle, $V$ may be finite, but the questions we address here are of main interest when $V$ is assumed infinite. Furthermore, we usually assume that the network is connected; meaning that any two vertices (points in $V$), say $x_0$ and $y_0$, can be connected by a finite path of edges, so a finite set of "line segments" chosen from $E$, each such finite path starting at $x_0$ and ending at $y_0$. In the interesting models there will be many such choices of finite paths. We think of the two vertices $x_0$ and $y_0$ as distant.

In an electric network, one assumes in addition that there is a prescribed positive function $c$ on $E$ representing conduction. In electrical network models

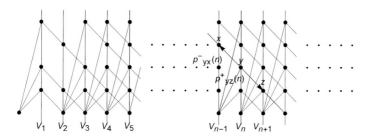

**FIGURE 1.2**: A Bratteli diagram with vertex-set $V = \{\emptyset\} \cup V_1 \cup V_2 \cup \cdots$ and transition between neighboring levels.

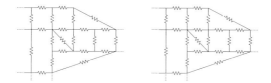

**FIGURE 1.3**: Examples of electric network.

$c$ is the reciprocal of resistance; so an electric network is a system of fixed resistors. Computations on $(V, E, c)$ are done with the rules from electricity: Kirchhoff's rule and Ohm's law. See Figure 1.3.

A typical question in electrical network models is this: Given a pair of distant vertices, say $x_0$ and $y_0$, what is the voltage drop from $x_0$ to $y_0$ when 1 Amp is injected at $x_0$, and again 1 Amp is extracted from $y_0$? In the papers on discrete networks, this is computable, and it is called the resistance distance, see Figure 1.4 and also, Cho and Jorgensen (2011); Jorgensen and Pearse (2010, 2011).

For the discrete models, calculations are done with the use of counting measure on $V$; hence "discrete" or atomic. In the other extreme (continuous, or perhaps better, non-atomic) the starting point is a measure space $(V, \mathscr{B}, \mu)$, where $\mathscr{B}$ is a prescribed sigma algebra and $\mu$ is a fixed measure defined on it. The measure $\mu$ will typically be infinite.

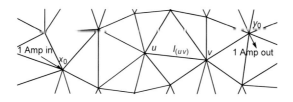

**FIGURE 1.4**: The convex set $W_{x_0 y_0}$. On edges $(u, v) \in E$ from $x_0$ to $y_0$ in $V$, the current is $I_{uv} = c_{uv} (f(u) - f(v))$, and $f$ denotes a voltage-distribution.

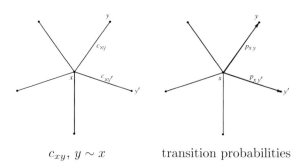

$c_{xy}$, $y \sim x$            transition probabilities

**FIGURE 1.5**: Neighbors of $x$.

Hence, when non-atomic, the measure $\mu$ here allows for the generalization: we replace counting measure on $V$ with an arbitrary separable measure space, subject to "mild" restrictions, see below. Interesting choices for $\mu$ will be families of fractal measures, for example Cantor measures. Recall that the Cantor measures are non-atomic; although of course singular.

In the detailed discussion below we shall refer to the two cases as old and new for simplicity. We refer to the atomic (or discrete) case as "old" and the general measure theoretic setting (non-atomic) as "new."

In the present section, we study a measure theoretic variant of the discrete networks (see Table 1.2). We formulate each question in such a way that there are both discrete and continuous variants. We first recall a natural setting in the case of a fixed continuous network, $(V, E, \rho, \mu)$. In the discrete case, $\mu$ is the counting measure, and $\rho$ is a given conductance function.

*Remark* 1.4.1. In discrete networks, $V$ will be the set of vertices for a given graph $G = (V, E)$ with edges $E$, and a conductance function $c$ defined on $E$ (Figure 1.5).

Specifically, $V$ is a countable discrete set, and let $E \subset V \times V \backslash (\text{diagonal})$ be a subset such that:

1. $(xy) \in E \Longleftrightarrow (yx) \in E$; $x, y \in V$.

2. For all $x \in V$, the set $E_x := \{y \in V \mid (xy) \in E\}$ satisfies $0 < \#E_x < \infty$.

3. $\exists o \in V$ s.t. for all $y \in V$ $\exists x_0, x_1, \ldots, x_n \in V$ with $x_0 = o$, $x_n = y$, $(x_{i-1}x_i) \in E$, $\forall i = 1, \ldots, n$.

4. $c : E \longrightarrow \mathbb{R}_+$, i.e., $c_{xy} > 0$, $\forall (xy) \in E$.

5. Set $c(x) = \sum_{y \sim x} c_{xy}$.

There is a large literature dealing with analysis on infinite graphs; see e.g., Jorgensen and Pearse (2010, 2011, 2013); see also Boyle et al. (2007); Cho and Jorgensen (2011); Okoudjou and Strichartz (2005).

The setting here is more general:

| **Discrete (old)** | **Continuous (new)** |
|---|---|
| 1. $G = (V, E, c)$, $V$ countable discrete; | 1' $(V, \mathscr{B}, \mu)$ measure space; |
| 2. $E \subset V \times V \setminus \{\text{diagonal}\}$; | 2' $E \subset V \times V \setminus \{\text{diagonal}\}$; |
| 3. conductance $c : E \longrightarrow \mathbb{R}_+$; | 3' $\rho$ on $E$, $\rho = \int \rho^{(x)} d\mu(x)$; |
| 4. $G$ is connected, and $$0 < \#E_x < \infty, \ \forall x \in V;$$ | 4' Defn. connectedness: $$c(x) = \rho^{(x)}(V) < \infty, \ \forall x \in V;$$ |
| 5. Laplacian $$\Delta f(x) = \sum_{y \sim x} c_{xy}(f(x) - f(y));$$ | 5' Laplacian $$\Delta f(x) = \int (f(x) - f(y)) d\rho^{(x)}(y);$$ |
| 6. Hilbert space $\mathscr{H} = l^2(V)$ $$\|f\|_{\mathscr{H}}^2 = \frac{1}{2}\sum_{x \sim y}\sum c_{xy}|f(x) - f(y)|^2;$$ | 6' Hilbert space $\mathscr{H} = L^2(V, \mu)$, $$\|f\|_{\mathscr{H}}^2 = \frac{1}{2}\iint_E |f(x) - f(y)|^2 d\rho(x,y);$$ |
| 7. $\delta_x \in \mathscr{H}$, $\|\delta_x\|_{\mathscr{H}}^2 = c(x)$, $$\|f\|_{\mathscr{H}}^2 = \langle f, \Delta f \rangle_{l^2} + \text{bdr term};$$ | 7' For all $\chi_A \in \mathscr{H}$, $\forall A \in \mathscr{B}_{fin}$, $$\|\chi_A\|_{\mathscr{H}}^2 = \int_A \rho^{(x)}(V \setminus A) d\mu(x);$$ |
| 8. For all $x \in V$, $f \in l^2(V)$, $$\Delta f(x) = \langle \delta_x, f \rangle_{\mathscr{H}}.$$ | 8' For all $A \in \mathscr{B}_{fin}$, and all $f \in \mathscr{H}$, $$\langle \chi_A, f \rangle_{\mathscr{H}} = \int_A (\Delta f) d\mu.$$ |

**TABLE 1.2**: Discrete versus continuous networks.

Let $(V, \mathscr{B}, \mu)$ be a measure space, where $V$ is locally compact and Hausdorff, $\mathscr{B}$ denotes the Borel $\sigma$-algebra of subsets in $V$, and $\mu$ is a positive measure on $(V, \mathscr{B})$.

Fix $E \subset V \times V \setminus \{\text{diagonal}\}$, and let $\rho$ be a positive measure on $E$ s.t. $\rho(x, y) = \rho(y, x)$. We further assume that $(\rho, \mu)$ is a *disintegration pair*, i.e., that $\exists \left\{ \rho^{(x)} \right\}_{x \in V}$, measures on $V$ s.t.

$$\rho = \int \rho^{(x)} d\mu(x). \tag{1.28}$$

**Definition 1.4.2.** A measure system $(V, \mathscr{B}, \mu)$ and $\rho$ on $V \times V$ is said to be a *measurable network-system* iff

(i) $\rho$ is symmetric;

(ii) There is a measurable partition $\{E_x\}$ for $\rho$ such that $supp\left(\rho^{(x)}\right) = E_x$, and $\rho = \int_V \rho^{(x)} d\mu(x)$; and

(iii) $\rho(A \times (V \setminus A)) < \infty$ holds for all $A \in \mathscr{B}_{fin}$.

See the terms in the right hand side column in Table 1.2.

**Lemma 1.4.3.** *Let $\mu$, $\rho$ be measures satisfying the conditions in Definition 1.4.2, then for all $A \in \mathscr{B}_{fin}$, we have $\|\chi_A\|_{\mathscr{H}}^2 = \rho(A \times (V \setminus A))$.*

*Proof.* A direct computation using

$$\|f\|_{\mathscr{H}}^2 = \frac{1}{2} \iint_E |f(x) - f(y)|^2 \, d\rho(x, y)$$

yields

$$\|\chi_A\|_{\mathscr{H}}^2 = \int_A \rho^{(x)}(V \setminus A) \, d\mu(x)$$
$$= \rho(A \times (V \setminus A)) = \rho((V \setminus A) \times A)$$

which is assumed finite by (iii) in Definition 1.4.2. $\square$

**Lemma 1.4.4.** *Let*

$$Harm = \{f \in \mathscr{H} \mid \Delta f = 0\}, \text{ and}$$
$$Fin = \text{closure in } \mathscr{H} \text{ of span} \{\chi_A \mid A \in \mathscr{B}_{fin}\}.$$

*Then*

$$\mathscr{H} = Fin \oplus Harm. \tag{1.29}$$

*Proof.* Suppose $f \in \mathscr{H} \ominus Fin$, then $\int_A \Delta f d\mu = 0$, $\forall A \in \mathscr{B}_{fin}$, thus $\Delta f = 0$. $\square$

By 8' above, we have

$$\langle \chi_A, f \rangle_{\mathscr{H}} = \int_A (\Delta f) \, d\mu, \ \forall A \in \mathscr{B}_{fin}, \ \forall f \in \mathscr{H}.$$

Set

$$\mu_f(A) := \langle \chi_A, f \rangle_{\mathscr{H}}, \ \forall A \in \mathscr{B}_{fin}, \tag{1.30}$$

then $\mu_f \ll \mu$ with local Radon-Nikodym derivative

$$\frac{d\mu_f}{d\mu} = \Delta f, \ \text{and} \ \Delta f \in L^1_{loc}(\mu). \tag{1.31}$$

Thus, we get a new p.d. function $\rho_{\mathscr{H}}$ on $\mathscr{B}_{fin} \times \mathscr{B}_{fin}$,

$$\rho_{\mathscr{H}}(A, B) = \langle \chi_A, \chi_B \rangle_{\mathscr{H}}, \ A, B \in \mathscr{B}_{fin}. \tag{1.32}$$

See Section 1.2 for details.

**Lemma 1.4.5** (Generalized Dipoles). *Let* $\left(V, \mathscr{B}, \mu, E, \rho, \rho^{(x)}, \mathscr{H}\right)$ *be as above. For all* $A, B \in \mathscr{B}_{fin}, \exists C = C_{AB} > 0$ *s.t.*

$$\left| \int_A f \, d\mu - \int_B f \, d\mu \right| \leq C \|f\|_{\mathscr{H}}, \ \forall f \in \mathscr{H};$$

*hence* $\exists! w_{AB} \in \mathscr{H}$ *s.t.*

$$\int_A f \, d\mu - \int_B f \, d\mu = \langle w_{AB}, f \rangle_{\mathscr{H}}, \ \forall f \in \mathscr{H}. \tag{1.33}$$

*Moreover,*

$$\Delta w_{AB} = \chi_A - \chi_B, \ A, B \in \mathscr{B}_{fin}. \tag{1.34}$$

*Proof.* An application of Riesz to $\mathscr{H}$. $\qquad\qquad\square$

**Theorem 1.4.6.** *Let* $\left(V, \mathscr{B}, \mu, E, \rho, \rho^{(x)}, \mathscr{H}\right)$ *be as above, then*

$$\rho_{\mathscr{H}}(A, B) := \langle \chi_A, \chi_B \rangle_{\mathscr{H}}$$
$$= \int_{A \cap B} c(x) \, d\mu(x) - \frac{1}{2} \left( \int_A \rho^{(x)}(B) \, d\mu(x) + \int_B \rho^{(x)}(A) \, d\mu(x) \right). \tag{1.35}$$

*In particular,*

$$\rho_{\mathscr{H}}(A, A) = \int_A \rho^{(x)}(V \backslash A) \, d\mu(x), \ A \in \mathscr{B}_{fin}. \tag{1.36}$$

*Proof.* Let $A \in \mathscr{B}_{fin}$, then it follows from 7', that

$$\|\chi_A\|_{\mathscr{H}}^2 = \frac{1}{2} \iint |\chi_A(x) - \chi_A(y)|^2 \, d\rho(x,y)$$

$$= \frac{1}{2} \iint (\chi_A(x) - 2\chi_A(x)\chi_A(y) + \chi_A(y)) \, d\rho(x,y)$$

$$= \frac{1}{2} \left[ 2 \int_A c(x) \, d\mu(x) - 2 \int_A \rho^{(x)}(A) \, d\mu(x) \right]$$

$$= \int_A \rho^{(x)}(V \backslash A) \, d\mu(x),$$

where $c(x) - \rho^{(x)}(A) = \rho^{(x)}(V \backslash A)$, and $c(x) := \rho^{(x)}(V)$.

Now for (1.35), we have

$$\langle \chi_A, \chi_B \rangle_{\mathscr{H}} = \frac{1}{2} \iint (\chi_A(x) - \chi_A(y))(\chi_B(x) - \chi_B(y)) \, d\rho(x,y)$$

$$= \int_{A \cap B} c(x) \, d\mu(x) - \frac{1}{2} \left( \int_A \rho^{(x)}(B) \, d\mu(x) + \int_B \rho^{(x)}(A) \, d\mu(x) \right),$$

using $\rho = \int_V \rho^{(x)} d\mu(x)$; see 3'. $\qquad\qquad\qquad\square$

*Remark 1.4.7.* Let $\left( V, \mathscr{B}, \mu, E, \rho, \rho^{(x)}, \mathscr{H} \right)$ be as above, where $\mathscr{H} = L^2(V, \mu)$ is the energy Hilbert space. By (1.35)-(1.36), $\rho_{\mathscr{H}} : \mathscr{B}_{fin} \times \mathscr{B}_{fin} \longrightarrow \mathbb{C}$ is p.d. in Aronszajn's sense, so we get an RKHS $\mathscr{H}^{(Aronsz)} \left( \rho_{\mathscr{H}} \right)$ which is isomorphic to $\mathscr{H}$. Also see Theorem 1.2.7 and Remark 1.3.6.

---

## 1.5    New correspondences

In the present section we make precise three equivalent formulations of the notion of positive definite system and the corresponding reproducing property. All three refer to a fixed measure space $(X, \mathscr{B}, \lambda)$. Each refers to pairs with the measure $\lambda$ fixed in the second place: They are: $(\mathscr{H}, \lambda)$, $(K, \lambda)$, and $(M, \lambda)$, where $\mathscr{H}$ is a Hilbert space as outlined in Definition 1.2.1, $K$ is the associated kernel (see Lemma 1.2.3), and $M$ is a measure on $X \times X$ which is defined from a certain factorization property; see item ③ below.

**Theorem 1.5.1.** *Let $X$ be a locally compact Hausdorff space with Borel $\sigma$-algebra $\mathscr{B}$, and let $\lambda$ be a fixed positive regular measure on $(X, \mathscr{B})$. Then a positive definite (p.d.) system is a pair $(\mathscr{H}, \lambda)$ with a Hilbert space $\mathscr{H} \subset L^2_{loc}(\lambda)$, and such that, for all $\varphi \in C_c(X)$, there is a $K^\varphi \in \mathscr{H}$ with $\int \varphi f d\lambda = \langle f, K^\varphi \rangle_{\mathscr{H}}, \forall f \in \mathscr{H}$; the reproducing property.*

*The following three characterizations, listed as ①,②, and ③, are equivalent:*

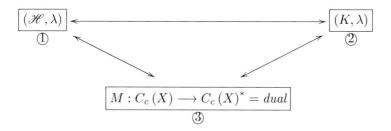

Notation: Below we work with real valued functions and real valued signed measures, but one can easily extend the results to the complex case as well.

## Axioms for ①:

(1a) $\mathscr{H} \subset L^2_{loc}(\lambda)$

(1b) For $\forall \varphi \in C_c(X)$, $\exists Const_\varphi < \infty$ s.t.

$$\int \varphi f d\lambda \leq Const_\varphi \|f\|_{\mathscr{H}}, \ \forall f \in \mathscr{H}. \tag{1.37}$$

From this, we get a kernel $K(\cdot, \cdot)$ as a random signed measure, and (1b) implies that $\exists K^\varphi$, where $K^\varphi = \int \varphi(y) K(\cdot, dy) \in \mathscr{H}$, with

$$\varphi K \varphi = \int \varphi(x) \int \varphi(y) K(y, dy) d\lambda(x) \tag{1.38}$$

satisfying

$$0 \leq \varphi K \varphi < \infty, \ \forall \varphi \in C_c(X). \tag{1.39}$$

## Axioms for ②:

Given $(K, \lambda)$, where $K : X \xrightarrow{\text{measurable}} C_c(X)^*$ is a random signed measure, i.e., we assume that (1.38)-(1.39) hold.

## Axioms for ③:

Here, we assume that given $M : C_c(X) \longrightarrow C_c(X)^*$ p.d. satisfying

$$(\varphi, M\varphi) = \int \varphi(x) d(M\varphi)(x)$$

and that

(3a) $0 \leq (\varphi, M\varphi) < \infty$, and

(3b) $dM$ has a disintegration,

$$(dM)(\cdot) = \int K(x, \cdot) \, d\lambda(x)$$

as a measure on $X \times X$, where $K(x, \cdot)$ is a measure, $\forall x \in X$, satisfying the conditions (1.38)-(1.39).

**Discussion of the signed measures $f d\lambda$.**

Starting with ①, recall that $\mathscr{H} \subset L^2_{loc}(\lambda)$. Since $f \in L^2_{loc}(\lambda) \subset L^1_{loc}(\lambda)$, so we get the estimate

$$\int \varphi f d\lambda \leq \max_x |\varphi(x)| \int_{supp(\varphi)} |f| \, d\lambda$$

$$\leq \max_x |\varphi(x)| \, \lambda \left(supp(\varphi)\right)^{\frac{1}{2}} \left( \int_{supp(\varphi)} |f|^2 \, d\lambda \right)^{\frac{1}{2}}, \; \forall \varphi \in C_c(X). \tag{1.40}$$

Hence $f d\lambda$ is well defined as a signed measure; see also Remark 1.5.2 below.

Consider $f d\lambda$ as a signed measure. The following properties hold: By (1b), for $\forall \varphi \in C_c(X) \; \exists! \; K^\varphi \in \mathscr{H}$ s.t.

$$\int \varphi f d\lambda = \langle f, K^\varphi \rangle_{\mathscr{H}}. \tag{1.41}$$

Fix $f \in \mathscr{H}$, then the LHS in (1.41) is a signed measure, and therefore so is RHS of (1.41), call it $d\mu_f$; we have

$$\int \varphi f d\lambda = \int \varphi d\mu_f; \text{ and so } d\mu_f = f d\lambda, \tag{1.42}$$

and $f$ is a local Radon-Nikodym derivative.

*Remark* 1.5.2. In the discussion, we make use of standard properties of signed measures, see, e.g., Rudin (1987). Recall the stated assumptions:

- $X$ is locally compact and Hausdorff, and $\mathscr{B}$ is the Borel $\sigma$-algebra;

- $\lambda$ is a fixed *regular positive* measure on $(X, \mathscr{B})$. In particular, $\lambda(E) < \infty$, for all compact set $E \subset X$, and so $\lambda(supp(\varphi)) < \infty$, $\forall \varphi \in C_c(X)$, which gives the finite constant in the estimate (1.40).

Using Equation (1.40) and Axiom (1b), we conclude that $\mu_f$ is a signed additive measure. Since $d\mu_f = f d\lambda$ holds locally, and $\lambda$ is regular (see Rudin (1987)), it follows that $|\mu_f(A)| < \infty$ for all $A \in \mathscr{B}_{fin}$, i.e., $A \in \mathscr{B}$ with $\lambda(A) < \infty$.

With $A \in \mathscr{B}_{fin}$, we get a constant $Const_A < \infty$ s.t.

$$\left| \mu_f (A) \right| \leq Const_A \left\| f \right\|_{\mathscr{H}}, \ \forall f \in \mathscr{H}. \tag{1.43}$$

By Reisz, $\exists! \ K(\cdot, A) \in \mathscr{H}$ s.t.

$$\mu_f (A) = \langle f(\cdot), K(\cdot, A) \rangle_{\mathscr{H}}, \forall A \in \mathscr{B}_{fin}. \tag{1.44}$$

Fix $\varphi \in C_c(X)$. Now use standard approximation of $\int \varphi d\mu_f$ with simple functions $\sum_i c_i \chi_{A_i}$, $A_i \in \mathscr{B}_{fin}$,

$$\begin{cases} \lim \sum_i c_i \chi_{A_i} & \longrightarrow \quad \varphi \ \text{(simple function)} \\ \lim \sum_i c_i \lambda(A_i) & \longrightarrow \quad \int \varphi f d\lambda = \int \varphi d\mu_f. \end{cases} \tag{1.45}$$

In (1.45), the limit is taken over measurable partitions $(A_i)$. We also used the (1.40).

Combining (1.41) and (1.44), we get $K^\varphi (\cdot) = \int \varphi(y) K(\cdot, dy)$; and (1.38)-(1.39) hold.

*Proof of Theorem 1.5.1.* Outline of the correspondences.

① $\longrightarrow$ ② Let $(\mathscr{H}, \lambda)$ be a p.d. system so that ① is as specified in (1a)-(1b). From Riesz applied to $\mathscr{H}$, we then get: for $\forall \varphi \in C_c(X) \ \exists! K^\varphi \in \mathscr{H}$ s.t.

$$\int \varphi f d\lambda = \langle f, K^\varphi \rangle_{\mathscr{H}} \tag{1.46}$$

For $\forall A \in \mathscr{B}_{fin}$, $\exists! K(\cdot, A) \in \mathscr{H}$ s.t. $\int_A f d\lambda = \langle f, K(\cdot, A) \rangle_{\mathscr{H}}$, with the properties we discussed. We conclude that

$$K^\varphi (\cdot) = \int \varphi(y) K(\cdot, dy). \tag{1.47}$$

Apply (1.46) to $f = K^\psi$, $\psi \in C_c(X)$; we get

$$\int \varphi(x) K^\psi (x) d\lambda(x) = \left\langle K^\psi, K^\varphi \right\rangle_{\mathscr{H}}, \tag{1.48}$$

and finally substitute (1.47) into (1.48). We get, by expanding (1.48),

$$\text{LHS}_{(1.48)} = \int \varphi(x) \int \psi(y) K(x, dy) d\lambda(x) = \left\langle K^\psi, K^\varphi \right\rangle_{\mathscr{H}}, \tag{1.49}$$

and, for $\psi = \psi$, we then get

$$0 \leq \int \varphi(x) \int \varphi(y) K(x, dy) d\lambda(x) = \left\| K^\varphi \right\|_{\mathscr{H}}^2 < \infty, \tag{1.50}$$

which is the desired property for $(K, \lambda)$, as outlined in ② from the list.

② $\longrightarrow$ ③ Assume a given pair $(K, \lambda)$ satisfying the conditions from ②.

From this, there is a standard way of assigning measure $M$ on $X \times X$, i.e., $M \in (C_c(X \times X))^*$ such that, for $\forall F \in C_c(X \times X)$, we have:

$$\iint F \, dM = \iint F(x, y) K(x, dy) \, d\lambda(x). \tag{1.51}$$

To do this, we note that $M$ is determined by

$$\iint (\varphi \otimes \psi) \, dM = \int \varphi(x) \int \psi(y) K(x, dy) \, d\lambda(x), \tag{1.52}$$

for $\varphi, \psi \in C_c(X)$; and so set

$$(M\varphi)(\cdot) = \int \varphi(x) K(x, \cdot) \, d\lambda(x), \tag{1.53}$$

and we get

$$\varphi M \varphi = \int \varphi d(M\varphi) \underset{\text{by } (1.53)}{=} \int \varphi(x) \int \varphi(y) K(x, dy) \, d\lambda(x)$$
$$= (\varphi, K\varphi);$$

see the axioms under ③. In particular (3a) and (3b) from the axioms ③ are satisfied. This is a direct application of the properties of $(K, \lambda)$, as listed in ②.

③ $\longrightarrow$ ① Assume that $M : C_c(X) \longrightarrow C_c(X)^*$ is given p.d. as in the axiom system ③, and we assume that $M$ has a disintegration w.r.t $\lambda$ where $\lambda$ is a given positive regular measure on $(X, \mathscr{B})$. Rewriting $\varphi M \varphi$ using $K$, we get,

$$(\varphi M \varphi) = (\varphi, K\varphi) = \int \varphi(x) \int \varphi(y) K(x, dy) \, d\lambda(x). \tag{1.54}$$

For $K^\varphi = \int \varphi(y) K(\cdot, dy)$, $K^\psi = \int \psi(y) K(\cdot, dy)$, set

$$\left\langle K^\varphi, K^\psi \right\rangle_{\mathscr{H}} := (\varphi M \psi), \tag{1.55}$$

and use (1.54), then (1.55) defines a pre-Hilbert space on the vector space of functions $\{K^\varphi\}_{\varphi \in C_c(X)}$, and we may do the usual Hilbert completion

$$\mathscr{H} := \left( K^\psi / \{(\varphi M \varphi) = 0\} \right)^\sim,$$

and we claim that the pair $(\mathscr{H}, \lambda)$ satisfies the axiom system from ①, i.e., (the real valued case)

$$\int (\varphi \otimes \psi) \, dM = \int \varphi(x) \left( \int \psi(y) K(x, dy) \right) d\lambda(x) \tag{1.56}$$

*Remark* 1.5.3. Measures $M$ on $X \times M$ of the above form are said to have a disintegration, i.e., for $F \in C_c(X \times X)$,

$$\iint F \, dM = \iint F(x, y) K(x, dy) \, d\lambda(x). \tag{1.57}$$

Other notations for (1.56) and (1.57):

$$M = \int_X K(x, \cdot) \, d\lambda(x). \tag{1.58}$$

The p.d. estimate for (1.56) is

$$0 \leq \iint \varphi(x) \varphi(y) K(x, dy) \, d\lambda(x) < \infty, \ \forall \varphi \in C_c(X).$$

Now, by definition $M(\varphi) \in C_c(X)^*$, $\forall \varphi \in C_c(X)$, and

$$\psi M \varphi = \int \psi dM(\varphi) \tag{1.59}$$

is p.d. As a result, we get for $\forall A \in \mathscr{B}_{fin}$,

$$\int_A \varphi(y) K(x, dy) \, d\lambda(x) = \int_X \varphi(x) K(x, A) \, d\lambda(x). \tag{1.60}$$

*Claim (Absolute continuity conclusion)*:

$$M(\varphi) \ll \lambda. \tag{1.61}$$

*Proof of* (1.61). Pick $A \in \mathscr{B}_{fin}$ s.t. $\lambda(A) = 0$. Then by (1.60), we get

$$\int_X \varphi(x) K(x, A) \, d\lambda(x) = 0,$$

and $M(\varphi)(A) = 0$ which is (1.61), $\forall \varphi \in C_c(X)$.

Since $f = K(\cdot, A)$, we get

$$f d\lambda = d\mu_f \tag{1.62}$$

But $span\{K(\cdot, A) \mid A \in \mathscr{B}_{fin}\}$ is dense in $\mathscr{H}(M) = \mathscr{H}(K)$, and we conclude that (1.62) holds for $\forall f \in \mathscr{H}$. And so $\mathscr{H} \subset L^2_{loc}(\lambda)$.

This ends the proof of ③ $\longrightarrow$ ①. $\qquad \square$

**Theorem 1.5.4.** *Consider $K(x, y) = e^{xy}$, $(x, y) \in \mathbb{R} \times \mathbb{R}$ as a p.d. kernel on $\mathbb{R}$, and let $\mathscr{H}$ be the corresponding RKHS (in the sense of Aronszajn). Then $\mathscr{H}$ is a Hilbert space of entire analytic functions on $\mathbb{R}$, i.e., absolutely convergent power-series on $\mathbb{R}$. Set, for $\forall (c_n) \in l^2(\mathbb{N}_0)$,*

$$T(c)(x) = \sum_{n=0}^{\infty} \frac{c_n x^n}{\sqrt{n!}};$$

*then $T$ is an isometric isomorphism of $l^2(\mathbb{N}_0)$ onto $\mathscr{H}$.*

**Review of properties of the RKHS $\mathscr{H}$ of $e^{xy}$.**

1. $\mathscr{H}$ consists of entire analytic functions on $\mathbb{R}$,

$$\mathscr{H} = \left\{ \sum_{n=0}^{\infty} c_n \frac{x^n}{\sqrt{n!}} \mid (c_n) \in l^2 \right\}. \tag{1.63}$$

That is, for all $f \in \mathscr{H}$, $\exists (c_n) \in l^2 (\mathbb{N}_0)$ s.t.

$$f(x) = \sum_{n=0}^{\infty} c_n \frac{x^n}{\sqrt{n!}}, \quad c_n = \frac{f^{(n)}(0)}{\sqrt{n!}} = \langle f, h_n \rangle_{\mathscr{H}}; \tag{1.64}$$

where $h_n(x) = \frac{x^n}{\sqrt{n!}}$; and

$$\|f\|_{\mathscr{H}} = \|(c_n)\|_{l^2}. \tag{1.65}$$

2. $\forall f \in \mathscr{H}$, $\forall x \in \mathbb{R}$, we have

$$|f(x)| \leq \|(c_n)\|_{l^2} e^{x^2/2} = \|f\|_{\mathscr{H}} e^{x^2/2}. \tag{1.66}$$

3. For $\forall \epsilon > 0$, we have

$$\int_{\mathbb{R}} |f(x)|^2 e^{-(1+\epsilon)x^2} dx \leq \|(c_n)\|_{l^2}^2 \int_{\mathbb{R}} e^{-\epsilon x^2} dx < \infty. \tag{1.67}$$

*Proof of Theorem 1.5.4 .* Use $e^{xy} = \sum_{n=0}^{\infty} \frac{x^n}{\sqrt{n!}} \frac{x^y}{\sqrt{n!}}$, so $h_n(x) = \frac{x^n}{\sqrt{n!}}$ is an ONB in $\mathscr{H}$. Hence functions $f$ in $\mathscr{H}$ have the form

$$f(x) = \underbrace{\sum_{n=0}^{\infty} c_n \frac{x^n}{\sqrt{n!}}}_{Tc} = \underbrace{\sum_{n=0}^{\infty} \frac{f^{(n)}(0)}{n!} x^n}_{T^{-1}f = \left( \frac{f^{(n)}(0)}{\sqrt{n!}} \right)}, \tag{1.68}$$

where $c_n = \langle f, h_n \rangle_{\mathscr{H}}$, and $(c_n) \in l^2$. It follows that

$$\sum_{n=0}^{\infty} \left| \frac{c_n x^n}{\sqrt{n!}} \right| \leq \left( \sum_{n=0}^{\infty} |c_n|^2 \right)^{\frac{1}{2}} \left( \sum_{n=0}^{\infty} \frac{x^{2n}}{n!} \right)^{\frac{1}{2}} = \|(c_n)\|_{l^2} e^{x^2/2}, \forall x \in \mathbb{R}.$$

Note that $f \in \mathscr{H} \Longleftrightarrow$ it has the representation in (1.68) $\Longleftrightarrow \left( \frac{f^{(n)}(0)}{\sqrt{n!}} \right) \in l^2$.

We check directly that the following reproducing property holds:

$$\langle e^{\cdot x}, f(\cdot) \rangle_{\mathscr{H}} = f(x), \forall f \in \mathscr{H}, \forall x \in \mathbb{R}. \tag{1.69}$$

Indeed,

$$\text{LHS}_{(1.69)} = \sum_{n=0}^{\infty} \frac{x^n}{\sqrt{n!}} \frac{f^{(n)}(0)}{\sqrt{n!}} = \sum_{n=0}^{\infty} \frac{f^{(n)}(0)}{n!} x^n = f(x) = \text{RHS}_{(1.69)}.$$

Since $f$ is entire analytic on $\mathbb{R}$, so $\text{RHS}_{(1.69)}$ is well defined and absolutely convergent. $\qquad \square$

**Corollary 1.5.5.** *There is no positive measure on* $(\mathbb{R}, \mathscr{B})$ *such that* $\mathscr{H} \xrightarrow[\text{(isom.)}]{} L^2(\mathbb{R}, \mu)$ *with an isometric embedding. Here* $\mathscr{H} := RKHS(e^{xy})$.

*Proof.* Indirect: Assume to the contrary that some measure $\mu$ exists defining an isometric embedding as stated. Then, since $h_n(x) = \frac{x^n}{\sqrt{n!}}$ is an ONB in $\mathscr{H}$, so we have

$$\int_{\mathbb{R}} h_n(x) h_m(x) \, d\mu(x) = \delta_{nm}, \ \forall n, m \in \mathbb{N}_0,$$

But using orthogonality, we have

$$0 = \langle h_0, h_2 \rangle_{\mathscr{H}} = \int_{\mathbb{R}} 1 \cdot \frac{x^2}{\sqrt{2!}} d\mu(x), \tag{1.70}$$

and

$$1 = \int_{\mathbb{R}} |h_1(x)|^2 \, d\mu(x) = \|h_1\|_{\mathscr{H}}^2 = \int x^2 d\mu(x), \tag{1.71}$$

which contradicts (1.70) satisfying that $\int_{\mathbb{R}} x^2 d\mu(x) = 0$. $\square$

**Corollary 1.5.6.** *Let* $(X, \mathscr{B}, \lambda)$ *be given as in Theorem 1.5.1. Let* $(\mathscr{H}, \lambda)$ *be a pair satisfying the axioms in part* ① *of the theorem, i.e.,*

$$\mathscr{H} \subset L_{loc}^2(\lambda), \tag{1.72}$$

*and for all* $\varphi \in C_c(X)$,

$$\sup_{f \in \mathscr{H}, \|f\| \leq 1} \left| \int_X \varphi f d\lambda \right| < \infty. \tag{1.73}$$

*Let* $\{h_n\}_{n \in \mathbb{N}_0}$, $\mathbb{N}_0 = \{0, 1, 2, \cdots\} = \{0\} \cup \mathbb{N}$, *be a fixed orthonormal basis (ONB) in* $\mathscr{H}$. *Let* $K$ *be the kernel from part* ② *in the theorem, i.e.,*

$$\langle K^{\varphi}, K^{\psi} \rangle_{\mathscr{H}} = \int_X \varphi(x) \int_X \psi(y) K(x, dy) \, d\lambda(x), \ \forall \varphi, \psi \in C_c(X). \tag{1.74}$$

(i) *Then the following representation for* $K$ *is well defined, and convergent,*

$$K(x, dy) = \sum_{n=0}^{\infty} h_n(x) h_n(y) \, d\lambda(y). \tag{1.75}$$

(ii) *For all* $f \in \mathscr{H}$, *the sequence* $\left( \int_X f h_n d\lambda \right)_{n \in \mathbb{N}_0} \in l^2(\mathbb{N}_0)$, *and for* $f = K^{\varphi}$, *set*

$$Tf = \left( \int_X \varphi h_n d\lambda \right)_n. \tag{1.76}$$

*Then* $T$ *defines an isometric isomorphism of* $\mathscr{H}$ *into* $l^2$, $T \cdot \mathscr{H} \longrightarrow l^2$.

*(iii) The adjoint operator* $T^* : l^2 \longrightarrow \mathscr{H}$ *is given by:*

$$T^* \left( (c_n)_{n \in \mathbb{N}_0} \right)(x) = \sum_{n=0}^{\infty} c_n h_n(x). \qquad (1.77)$$

*Proof.* Since $\{h_n\}_{n \in \mathbb{N}_0}$ is given to be an ONB in $\mathscr{H}$, we have, for all $f \in \mathscr{H}$, the Parseval formula:

$$\|f\|_{\mathscr{H}}^2 = \sum_{n=0}^{\infty} |\langle f, h_n \rangle_{\mathscr{H}}|^2,$$

and

$$\langle f, g \rangle_{\mathscr{H}} = \sum_{n=0}^{\infty} \langle f, h_n \rangle_{\mathscr{H}} \langle h_n, g \rangle_{\mathscr{H}}, \ \forall f, g \in \mathscr{H}.$$

Now apply these to $f = K^{\varphi}$, $g = K^{\psi}$, and substitute into (1.74). We get

$$\begin{aligned}
\mathrm{LHS}_{(1.74)} &= \sum_{n=0}^{\infty} \langle K^{\varphi}, h_n \rangle_{\mathscr{H}} \langle h_n, K^{\psi} \rangle_{\mathscr{H}} \\
&= \sum_{n=0}^{\infty} \int_X \varphi h_n d\lambda \int_X h_n \psi d\lambda \\
&= \int_X \varphi(x) \psi(y) \left( \sum_{n=0}^{\infty} h_n(x) h_n(y) \right) d\lambda(x) \, d\lambda(y).
\end{aligned}$$

Now substitute the formula (1.75) into the RHS of (1.74), and the desired conclusion follows.  □

*Acknowledgement.* The co-authors thank the following colleagues for helpful and enlightening discussions: Professors Daniel Alpay, Sergii Bezuglyi, Ilwoo Cho, Paul Muhly, Myung-Sin Song, Wayne Polyzou, and members in the Math Physics seminar at The University of Iowa.

---

## A.1   Applications of RKHSs: An overview

> "The simplicities of natural laws arise through the complexities
> of the language we use for their expression."
> — Eugene Wigner

A reproducing kernel Hilbert space (RKHS) is a Hilbert space $\mathscr{H}$ of functions on a prescribed set, say $V$, with the property that point-evaluation for functions $f \in \mathscr{H}$ is continuous with respect to the $\mathscr{H}$-norm. They are called kernel spaces, because, for every $x \in V$, the point-evaluation for functions $f \in \mathcal{H}$, $f(x)$ must then be given as a $\mathcal{H}$-inner product of $f$ and a vector $k_x$ in $\mathcal{H}$; called the kernel, i.e., $f(x) = \langle k_x, f \rangle_{\mathcal{H}}$, $\forall f \in \mathcal{H}$, $x \in V$.

The RKHSs have been studied extensively since the pioneering papers by Aronszajn Aronszajn (1943, 1950). They further play an important role in the theory of partial differential operators (PDOs); for example as Green's functions of second order elliptic PDOs Haeseler et al. (2014); Nelson (1957). Other applications include engineering, physics, machine-learning theory Cucker and Smale (2002); Kulkarni and Harman (2011); Smale and Zhou (2009b), stochastic processes Alpay et al. (1993); Alpay and Dym (1992, 1993); Alpay et al. (2013, 2014), numerical analysis, and more Ha Quang et al. (2010); Hedenmalm and Nieminen (2014); Lata and Paulsen (2011); Lin and Brown (2004); Schlkopf and Smola (2001); Schramm and Sheffield (2013); Shawe-Taylor and Cristianini (2004); Vuletić (2013); Zhang et al. (2012).

An illustration from *neural networks*: An extreme learning machine (ELM) is a neural network configuration in which a hidden layer of weights are randomly sampled Rasmussen and Williams (2006), and the object is then to determine analytically resulting output layer weights. Hence ELM may be thought of as an approximation to a network with infinite number of hidden units.

The literature so far has focused on the theory of kernel functions defined on continuous domains, either domains in Euclidean space, or complex domains in one or more variables. For these cases, the Dirac $\delta_x$ distributions do not have finite $\mathcal{H}$-norm. But for RKHSs over discrete point distributions, it is reasonable to expect that the Dirac $\delta_x$ functions will in fact have finite $\mathcal{H}$-norm (see, e.g., Jorgensen and Tian (2015, 2016)).

There is a related reproducing kernel notion called "relative:" This means that increments have kernel representations. In detail: Consider functions $f$ in $\mathcal{H}$, but suppose instead that, for every pair of points $x, y$ in $V$, each of the differences $f(x) - f(y)$ can be represented by a kernel from $\mathcal{H}$. We then say that $\mathcal{H}$ is a relative RKHS. The "relative" variant is of more recent vintage, and it is used in the study of electrical networks (voltage differences, and in analysis of Gaussian processes such as Gaussian fields).

In the theory of non-uniform sampling, one studies Hilbert spaces consisting of signals, understood in a very general sense. One then develops analytic tools and algorithms, allowing one to draw inference for an "entire" (or global) signal from partial information obtained from carefully chosen distributions of sample points. While the better known and classical sampling algorithms (Shannon and others) are based on interpolation, modern theories go beyond this. An early motivation is the work of Henry Landau. In this setting, it is possible to make precise the notion of "average sampling rates" in general configurations of sample points. Our present study turns the tables. We start with the general axiom system of positive definite kernels and their associated reproducing kernel Hilbert spaces (RKHSs), or *relative* RKHSs. With some use of metric geometry and of spectral theory for operators in Hilbert space, we are then able to obtain sampling theorems for a host of non-uniform point configurations. The modern theory of non-uniform sampling is vast, and it specializes into a variety of sub-areas. The following papers (and the literature cited there) will give an idea of the diversity of points of view: Aldroubi and Leonetti (2008); Khan et al. (2013); Landau (1967); Martinez et al. (2014); Zhou et al. (2014).

**Definition A.1.1.** Let $X$ be any set. A function $K : X \times X \to \mathbb{C}$ is *positive definite* iff

$$\sum_i \sum_j \overline{c_i} c_j K\left(x_i, x_j\right) \geq 0, \tag{A.1}$$

for all $\{x_i\}_{i=1}^n \subset X$, and all $(c_i)_{i=1}^n \in \mathbb{C}^n$.

**Theorem A.1.2** (Aronszajn Aronszajn (1950))**.** *Let $X$ be a set, and let $K : X \times X \longrightarrow \mathbb{C}$ be a positive definite function as in (A.1).*

*Then there is a Hilbert space $\mathscr{H} = \mathscr{H}(K)$ such that the functions*

$$K_x\left(\cdot\right) = K\left(\cdot, x\right), \quad x \in X \tag{A.2}$$

*span a dense subspace in $\mathscr{H}(K)$, called the reproducing kernel Hilbert space (RKHS), and for all $f \in \mathscr{H}(K)$, we have*

$$f\left(x\right) = \langle K_x, f \rangle_{\mathscr{H}}, \quad x \in X. \tag{A.3}$$

*Proof.* (Sketch) Let $\mathscr{H}(K)$ be the Hilbert-completion of the *span* $\{K_x : x \in X\}$, with respect to the inner product

$$\left\langle \sum c_x K_x, \sum d_y K_y \right\rangle_{\mathscr{H}(K)} := \sum \sum \overline{c_x} d_y K\left(x, y\right) \tag{A.4}$$

$\mathscr{H}(K)$ is then a RKHS, with the reproducing property:

$$\langle K_x, f \rangle_{\mathscr{H}(K)} = f\left(x\right), \forall x \in X, \forall f \in \mathscr{H}(K). \tag{A.5}$$

$\square$

*Remark* A.1.3. The summations in (A.4) are all finite. Starting with finitely supported summations in (A.4), the RKHS $\mathscr{H}(K)$ is then obtained by Hilbert space completion. We use physicists' convention, so that the inner product is conjugate linear in the first variable, and linear in the second variable.

**Theorem A.1.4.** *A function $f$ on $X$ is in $\mathscr{H}(K)$ if and only if there is a constant $C = C(f)$ such that for all $n$, $(c_j)_1^n$, $(x_j)_1^n$, as above, we have*

$$\left|\sum_j c_j f\left(x_j\right)\right|^2 \leq C \sum_i \sum_j \overline{c_i} c_j K\left(x_i, x_j\right). \tag{A.6}$$

*Remark* A.1.5. It follows that reproducing kernel Hilbert spaces (RKHSs) arise from a given positive definite kernel $K$, a corresponding pre-Hilbert form;

and then a Hilbert-completion. The question arises: What are the functions in the completion? The *a priori* estimate (A.6) above is an answer to the question. By contrast, the Hilbert space completions are subtle; they are classical Hilbert spaces of functions, not always transparent from the naked kernel $K$ itself. Examples of classical RKHSs include Hardy spaces or Bergman spaces (for complex domains), Sobolev spaces and Dirichlet spaces Okoudjou et al. (2013); Strichartz (2010); Strichartz and Teplyaev (2012) (for real domains, or for fractals), band-limited $L^2$ functions (from signal analysis), and Cameron-Martin Hilbert spaces from Gaussian processes (in continuous time domain).

## Application to Optimization

One of the more recent applications of kernels and the associated reproducing kernel Hilbert spaces (RKHS) is to optimization, also called kernel-optimization. In the context of machine learning, it refers to training-data and feature spaces. In the context of numerical analysis, a popular version of the method is used to produce splines from sample points; and to create best spline fits. In statistics, there are analogous optimization problems going by the names "least-square fitting," and "maximum-likelihood" estimation. In the latter instance, the object to be determined is a suitable probability distribution which makes most likely the occurrence of some data which arises from experiments, or from testing.

What these methods have in common is a minimization (or a max problem) involving a quadratic expression $Q$ with two terms. The first in $Q$ measures a suitable $L^2(\mu)$-square applied to a difference of a measurement and a best fit. The latter will then to be chosen from any one of a number of suitable reproducing kernel Hilbert spaces (RKHS). The choice of kernel and RKHS will serve to select desirable features. So we will minimize a quantity $Q$ which is the sum of two terms as follows: (i) a $L^2$-square applied to a difference, and (ii) a penalty term which is a RKHS norm-squared. (See equation (A.8).) The term in (ii) is often called the penalty term. In the application to determination of splines, the penalty term may be a suitable Sobolev normed-square; i.e., $L^2$ norm-squared applied to a chosen number of derivatives. Hence non-differentiable choices will be penalized.

In all of the cases, discussed above, there will be a good choice of (i) and (ii), and we show an explicit formula for the optimal solution; see equation (A.11) in Theorem A.1.6 below.

Let $X$ be a set, and let $K : X \times X \longrightarrow \mathbb{C}$ be a positive definite (p.d.) kernel. Let $\mathscr{H}(K)$ be the corresponding reproducing kernel Hilbert space (RKHS). Let $\mathscr{B}$ be a sigma-algebra of subsets of $X$, and let $\mu$ be a positive measure on the corresponding measure space $(X, \mathscr{B})$. We assume that $\mu$ is sigma-finite. We shall further assume that the associated operator $T$ given by

$$\mathscr{H}(K) \ni f \xrightarrow{T} (f(x))_{x \in X} \in L^2(\mu) \tag{A.7}$$

is densely defined and closable.

Fix $\beta > 0$, and $\psi \in L^2(\mu)$, and set

$$Q_{\psi,\beta}(f) = \|\psi - Tf\|^2_{L^2(\mu)} + \beta \|f\|^2_{\mathcal{H}(K)} \tag{A.8}$$

defined for $f \in \mathcal{H}(K)$, or in the dense subspace $dom(T)$ where $T$ is the operator in (A.7). Let

$$L^2(\mu) \xrightarrow{T^*} \mathcal{H}(K) \tag{A.9}$$

be the corresponding adjoint operator, i.e.,

$$\langle F, T^*\psi \rangle_{\mathcal{H}(K)} = \langle Tf, \psi \rangle_{L^2(\mu)} = \int_X \overline{f(s)} \psi(s) \, d\mu(s). \tag{A.10}$$

**Theorem A.1.6.** *Let $K$, $\mu$, $\psi$, $\beta$ be as specified above; then the optimization problem*

$$\inf_{f \in \mathcal{H}(K)} Q_{\psi,\beta}(f)$$

*has a unique solution $F$ in $\mathcal{H}(K)$, it is*

$$F = (\beta I + T^*T)^{-1} T^*\psi \tag{A.11}$$

*where the operator $T$ and $T^*$ are as specified in* (A.7)-(A.10).

*Proof.* (Sketch) We fix $F$, and assign $f_\varepsilon := F + \varepsilon h$ where $h$ varies in the dense domain $dom(T)$ from (A.7). For the derivative $\frac{d}{d\varepsilon}\big|_{\varepsilon=0}$ we then have:

$$\frac{d}{d\varepsilon}\Big|_{\varepsilon=0} Q_{\psi,\beta}(f_\varepsilon) = 2\Re \langle h, (\beta I + T^*T) F - T^*\psi \rangle_{\mathcal{H}(K)} = 0$$

for all $h$ in a dense subspace in $\mathcal{H}(K)$. The desired conclusion follows.

$\square$

### Application: Least-square Optimization

We now specialize the optimization formula from Theorem A.1.6 to the problem of minimize a quadratic quantity $Q$. It is still the sum of two individual terms: (i) a $L^2$-square applied to a difference, and (ii) a penalty term which is the RKHS norm-squared. But the least-square term in (i) will simply be a sum of a finite number of squares of differences; hence "least-squares." As an application, we then get an easy formula (Theorem A.1.7) for the optimal solution.

Let $K$ be a positive definite kernel on $X \times X$ where $X$ is an arbitrary set, and let $\mathcal{H}(K)$ be the corresponding reproducing kernel Hilbert space (RKHS). Let $m \in \mathbb{N}$, and consider sample points:

$\{t_j\}_{j=1}^m$ as a finite subset in $X$, and

$\{y_i\}_{i=1}^m$ as a finite subset in $\mathbb{R}$, or equivalently, a point in $\mathbb{R}^m$.

Fix $\beta > 0$, and consider $Q = Q_{(\beta,t,y)}$, defined by

$$Q(f) = \sum_{i=1}^m \underbrace{|f(t_i) - y_i|^2}_{\text{least square}} + \underbrace{\beta \|f\|^2_{\mathcal{H}(K)}}_{\text{penalty form}}, \quad f \in \mathcal{H}(K). \tag{A.12}$$

We introduce the associated dual pair of operators as follows:

$$T : \mathscr{H}(K) \longrightarrow \mathbb{R}^m \simeq l_m^2, \text{ and}$$
$$T^* : l_m^2 \longrightarrow \mathscr{H}(K) \tag{A.13}$$

where

$$Tf = (f(t_i))_{i=1}^m, \quad f \in \mathscr{H}(K); \text{ and} \tag{A.14}$$

$$T^* y = \sum_{i=1}^m y_i K(\cdot, t_i) \in \mathscr{H}(K), \tag{A.15}$$

for all $\vec{y} = (y_i) \in \mathbb{R}^m$.

Note that the duality then takes the following form:

$$\langle T^* y, f \rangle_{\mathscr{H}(K)} = \langle y, Tf \rangle_{l_m^2}, \quad \forall f \in \mathscr{H}(K), \forall y \in l_m^2; \tag{A.16}$$

consistent with (A.10).

Applying Theorem A.1.6 to the counting measure

$$\mu - \sum_{i=1}^m \delta_{t_i} - \delta_{\{t_i\}}$$

for the set of sample points $\{t_i\}_{i=1}^m$, we get the two formulas:

$$T^* T f = \sum_{i=1}^m f(t_i) K(\cdot, t_i) = \sum_{i=1}^m f(t_i) K_{t_i}, \text{ and} \tag{A.17}$$

$$T T^* y = K_m \vec{y} \tag{A.18}$$

where $K_m$ denotes the $m \times m$ matrix

$$K_m = (K(t_i, t_j))_{i,j=1}^m = \begin{pmatrix} K(t_1, t_1) & \cdots & \cdots & K(t_1, t_m) \\ K(t_2, t_1) & \cdots & \cdots & K(t_2, t_m) \\ \vdots & & & \\ K(t_m, t_1) & & & K(t_m, t_m) \end{pmatrix}. \tag{A.19}$$

**Theorem A.1.7.** *Let $K$, $X$, $\{t_i\}_{i=1}^m$, and $\{y_i\}_{i=1}^m$ be as above, and let $K_m$ be the induced sample matrix (A.19).*

*Fix $\beta > 0$; consider the optimization problem with*

$$Q_{\beta, \{t_i\}, \{y_i\}}(f) = \sum_{i=1}^m |y_i - f(t_i)|^2 + \beta \|f\|_{\mathscr{H}(K)}^2, \quad f \in \mathscr{H}(K). \tag{A.20}$$

*Then the unique solution to (A.20) is given by*

$$F(\cdot) = \sum_{i=1}^m (K_m + \beta I_m)_i^{-1} K(\cdot, t_i) \text{ on } X; \tag{A.21}$$

*i.e., $F = \arg \min Q$ on $\mathscr{H}(K)$.*

*Proof.* From Theorem A.1.6, we get that the unique solution $F \in \mathscr{H}(K)$ is given by:
$$\beta F + T^* T F = T^* y,$$
and by (A.17)-(A.18), we further get

$$\beta F(\cdot) = \sum_{i=1}^{m} (y_i - F(t_i)) K(\cdot, t_i) \qquad (A.22)$$

where the dot $\cdot$ refers to a free variable in $X$. An evaluation of (A.22) on the sample points yields:

$$\beta \vec{F} = K_m \left( \vec{y} - \vec{F} \right) \qquad (A.23)$$

where $\vec{F} := (F(t_i))_{i=1}^{m}$, and $\vec{y} = (y_i)_{i=1}^{m}$. Hence

$$\vec{F} = (\beta I_m + K_m)^{-1} K_m \vec{y}. \qquad (A.24)$$

Now substitute (A.24) into (A.23), and the desired conclusion in the theorem follows. We used the matrix identity

$$I_m - (\beta I_m + K_m)^{-1} K_m = \beta (\beta I_m + K_m)^{-1}.$$

$\square$

## A digression: Positive definite functions and stochastic processes

The interest in positive definite functions has at least three roots: (i) Fourier analysis, and harmonic analysis more generally, including the non-commutative variant where we study unitary representations of groups; (ii) optimization and approximation problems, involving for example spline approximations as envisioned by I. Schöenberg; and (iii) the study of stochastic (random) processes.

**Definition A.1.8.** By a *probability space*, we mean a triple $(\Omega, \mathscr{F}, \mathbb{P})$ where:

- $\Omega$ is a set,

- $\mathscr{F}$ is a $\sigma$-algebra of subsets of $\Omega$, and

- $\mathbb{P}$ is a probability measure defined on $(\Omega, \mathscr{F})$, i.e., $\mathbb{P}(\emptyset) = 0$, $\mathbb{P}(\Omega) = 1$, $\mathbb{P}(F) \geq 0$, $\forall F \in \mathscr{B}$, and if $\{F_i\}_{i \in \mathbb{N}} \subset \mathscr{B}$, $F_i \cap F_j = \emptyset$, $i \neq j$ in $\mathbb{N}$, then $\mathbb{P}(\cup_i F_i) = \sum_i \mathbb{P}(F_i)$.

**Definition A.1.9.** A *stochastic process* is an indexed family of random variables $\{X_s\}_{s \in S}$ based on a fixed probability space $(\Omega, \mathscr{F}, \mathbb{P})$.

In our present analysis, the processes will be indexed by some group $G$; for example, $G = \mathbb{R}$, or $G = \mathbb{Z}$ correspond to processes indexed by real time or discrete time. A main tool in the analysis of stochastic processes is an associated covariance function, see (A.25).

**FIGURE A.1**: Standard 2D Brownian motion starting at $(0,0)$ (five sample paths).

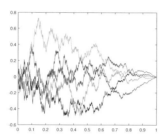

**FIGURE A.2**: Brownian bridge in 1D (five sample paths).

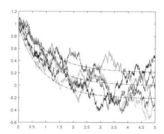

**FIGURE A.3**: The OU-process in 1 dimension. $dx_t = -\gamma x_t dt + \beta dW_t$, with $x_0 = 1$, $\gamma = 1$, $\beta = 0.3$, $0 \le t \le 5$.

A process $\{X_g \mid g \in G\}$ is called Gaussian if each random variable $X_g$ is Gaussian, i.e., its distribution is Gaussian. For Gaussian processes we only need two moments. So if we normalize, setting the mean equal to 0, then the process is determined by the covariance function. In general the covariance function is a function on $G \times G$, or on a subset, but if the process is stationary, the covariance function will in fact be a positive definite function defined on $G$, or a subset of $G$. Examples include Brownian motion, Brownian bridge, and the Ornstein-Uhlenbeck (OU) process etc, all Gaussian or Itō integrals. See Figures A.1 through A.3.

We outline a brief sketch of these facts below.

Let $G$ be a locally compact group, and let $(\Omega, \mathscr{F}, \mathbb{P})$ be a probability space, $\mathscr{F}$ a sigma-algebra, and $\mathbb{P}$ a probability measure defined on $\mathscr{F}$. A stochastic $L^2$-process is a system of random variables $\{X_g\}_{g \in G}$, $X_g \in L^2(\Omega, \mathscr{F}, \mathbb{P})$. The covariance function $c_X$ of the process is the function $G \times G \to \mathbb{C}$ given by

$$c_X(g_1, g_2) = \mathbb{E}\left(\overline{X}_{g_1} X_{g_2}\right), \ \forall (g_1, g_2) \in G \times G. \tag{A.25}$$

To simplify we will assume that $\mathbb{E}(X_g) = \int_\Omega X_g d\mathbb{P}(\omega) = 0$ for all $g \in G$.

**Definition A.1.10.** We say that $(X_g)$ is stationary iff

$$c_X(hg_1, hg_2) = c_X(g_1, g_2), \ \forall h \in G. \tag{A.26}$$

In this case $c_X$ is a function of $g_1^{-1} g_2$, i.e.,

$$\mathbb{E}(X_{g_1} X_{g_2}) = c_X\left(g_1^{-1} g_2\right), \ \forall g_1, g_2 \in G. \tag{A.27}$$

(Just take $h = g_1^{-1}$ in (A.26).)

We now recall the following theorem of Kolmogorov (see Parthasarathy and Schmidt (1975)). One direction is easy, and the other is more complex.

**Definition A.1.11.** A function $c$ defined on a subset of $G$ is said to be *positive definite* iff

$$\sum_i \sum_j \overline{\lambda_i} \lambda_j c\left(g_i^{-1} g_j\right) \geq 0$$

for all finite summation, where $\lambda_i \in \mathbb{C}$ and $g_i^{-1} g_j$ in the domain of $c$.

**Theorem A.1.12** (Kolmogorov). *A function $c : G \to \mathbb{C}$ is positive definite if and only if there is a stationary Gaussian process $(\Omega, \mathscr{F}, \mathbb{P}, X)$ with mean zero, such that $c = c_X$.*

*Proof.* To stress the idea, we include the easy part of the theorem, and we refer to Parthasarathy and Schmidt (1975) for the non-trivial direction:

Let $\lambda_1, \lambda_2, \ldots, \lambda_n \in \mathbb{C}$, and $\{g_i\}_{i=1}^{N} \subset G$, then for all finite summations, we have:

$$\sum_i \sum_j \overline{\lambda_i} \lambda_j c_X\left(g_i^{-1} g_j\right) = \mathbb{E}\left(\left|\sum_i \lambda_i X_{g_i}\right|^2\right) \geq 0.$$

$\square$

---

## A.2 Duality of representations, and some of their applications

Below we describe another measurable-factorization family.

**Definition A.2.1.** Fix a positive measure $\lambda$ on $(X, \mathscr{B})$, and set

$$\mathscr{K}(\lambda) := \left\{ \text{Hilbert space } \mathscr{H} \text{ on } (X, \mathscr{B}) \mid \text{For } \forall \varphi \in C_c(X) \right.$$

$$\left. \sup\left\{ \left| \int \varphi f d\lambda \right|, \|f\|_{\mathscr{H}} \leq 1 \right\} < \infty \right\}. \tag{A.28}$$

If $\mathscr{H}$ is a Hilbert space of measurable functions on $(X, \mathscr{B})$, set

$$\mathscr{L}(\mathscr{H}) := \left\{ \text{signed measures } \lambda \mid \sup\left\{ \left| \int \varphi f d\lambda \right|, \|f\|_{\mathscr{H}} \leq 1 \right\} < \infty \right\}, \tag{A.29}$$

and

$$\mathscr{L}_+(\mathscr{H}) = \{\lambda \in \mathscr{L}(\mathscr{H}) \mid \lambda \text{ is positive}\}. \tag{A.30}$$

**Definition A.2.2.** Given $(X, \mathscr{B}, \lambda)$ as above, let $\{K(x, \cdot)\}_{x \in X}$ be a family of random signed measures, i.e., for all $x \in X$, if $(B_i)_{i=1}^n \subset \mathscr{B}$, $B_i \cap B_j = \emptyset$, $i \neq j$, then

$$K(x, \cup_i B_i) = \sum_i K(x, B_i). \tag{A.31}$$

We say $\mathscr{H}$ is *admissible* if $\mathscr{H} \in \mathscr{K}(\lambda)$.

**Lemma A.2.3.** *Every admissible $\mathscr{H}$ induces a family of measures $\{\mu_f\}_{f \in \mathscr{H}}$, where $\mu_f$ is given by (1.11), i.e., $\mu_f(B) = \langle f, K(\cdot, B) \rangle_{\mathscr{H}}$, $B \in \mathscr{B}_{fin}$.*

*Proof.* Immediate from the definition. □

**Definition A.2.4.** We say a positive measure $\lambda$ on $(X, \mathscr{B})$ is *symmetric* with respect to $\mathscr{H}$ iff $\mu_f \ll \lambda$, $\forall f \in \mathscr{H}$, and

$$\mu_f(B) = \int_B f d\lambda, \quad \forall B \in \mathscr{B}_{fin},$$

and so $d\mu_f / d\lambda = f$ on $B$, i.e., $f \in L_{loc}^1(\lambda)$.

**Corollary A.2.5.** *The pair $(K, \lambda)$ in (A.2.2) yields a symmetric $\mathscr{H}$, so $\mathscr{H} \in \mathscr{K}(\lambda)$.*

*Proof.* We already proved that $(\mathscr{H}, \lambda)$ satisfies the symmetry condition. □

**Factorization.** Given $K$ as in (A.31), i.e., $x \to K(x, \cdot)$ is a random signed measure; set $K^B = K(\cdot, B)$, $K^\varphi(x) = \int \varphi(y) K(x, dy)$, $\forall \varphi \in C_c(X)$. Let $\mathscr{H}$ be admissible, and $f \to \mu_f$ is the indexed system of measures.

We have proved that:

**Theorem A.2.6.** $\mathscr{H} \in \mathscr{K}(\mathscr{L}(\mathscr{H}))$, *and* $\lambda \in \mathscr{L}(\mathscr{K}(\lambda))$.

*Proof.* See Section 1.2, especially Lemma 1.2.3 and Theorem 1.2.7. □

**Question.** *When do we have $\mathscr{L}(\mathscr{K}(\mathscr{L}(\mathscr{H}))) = \mathscr{L}(\mathscr{H})$? (The inclusion $\subseteq$ is immediate.)*

# Bibliography

Aldroubi A. and Leonetti C. Non-uniform sampling and reconstruction from sampling sets with unknown jitter. *Sampl. Theory Signal Image Process.*, 7(2):187–195, 2008.

Alpay D., Bolotnikov V., Dijksma A., and de Snoo H., 1993. On some operator colligations and associated reproducing kernel Hilbert spaces. In *Operator extensions, interpolation of functions and related topics*, volume 61 of *Oper. Theory Adv. Appl.*, pages 1–27. Birkhäuser, Basel.

Alpay D. and Dym H., 1992. On reproducing kernel spaces, the Schur algorithm, and interpolation in a general class of domains. In *Operator theory and complex analysis (Sapporo, 1991)*, volume 59 of *Oper. Theory Adv. Appl.*, pages 30–77. Birkhäuser, Basel.

Alpay D. and Dym H. On a new class of structured reproducing kernel spaces. *J. Funct. Anal.*, 111(1):1–28, 1993.

Alpay D., Jorgensen P., Seager R., and Volok D. On discrete analytic functions: products, rational functions and reproducing kernels. *J. Appl. Math. Comput.*, 41(1-2):393–426, 2013.

Alpay D., Jorgensen P., and Volok D. Relative reproducing kernel Hilbert spaces. *Proc. Amer. Math. Soc.*, 142(11):3889–3895, 2014.

Alpay D. and Jorgensen P. E. T. Stochastic processes induced by singular operators. *Numer. Funct. Anal. Optim.*, 33(7-9):708–735, 2012.

Applebaum D. *Lévy processes and stochastic calculus*, volume 116 of *Cambridge Studies in Advanced Mathematics*. Cambridge University Press, Cambridge, 2009, second edition.

Aronszajn N. La théorie des noyaux reproduisants et ses applications. I. *Proc. Cambridge Philos. Soc.*, 39:133–153, 1943.

Aronszajn N. Theory of reproducing kernels. *Trans. Amer. Math. Soc.*, 68:337–404, 1950.

Barrière L., Comellas F., and Dalfó C. Fractality and the small-world effect in Sierpinski graphs. *J. Phys. A*, 39(38):11739–11753, 2006.

Bezuglyi S. and Handelman D. Measures on Cantor sets: the good, the ugly, the bad. *Trans. Amer. Math. Soc.*, 366(12):6247–6311, 2014.

Bezuglyi S., Kwiatkowski J., and Yassawi R. Perfect orderings on finite rank Bratteli diagrams. *Canad. J. Math.*, 66(1):57–101, 2014.

Bishop C. J. and Peres Y. *Fractals in probability and analysis*, volume 162 of *Cambridge Studies in Advanced Mathematics*. Cambridge University Press, Cambridge, 2017.

Boyle B., Cekala K., Ferrone D., Rifkin N., and Teplyaev A. Electrical resistance of *N*-gasket fractal networks. *Pacific J. Math.*, 233(1):15–40, 2007.

Burioni R. and Cassi D. Random walks on graphs: ideas, techniques and results. *J. Phys. A*, 38(8):R45–R78, 2005.

Chang X., Xu H., and Yau S.-T. Spanning trees and random walks on weighted graphs. *Pacific J. Math.*, 273(1):241–255, 2015.

Cho I. and Jorgensen P. Free probability induced by electric resistance networks on energy Hilbert spaces. *Opuscula Math.*, 31(4):549–598, 2011.

Cortes R. X., Martins T. G., Prates M. O., and Silva B. A. Inference on dynamic models for non-Gaussian random fields using INLA. *Braz. J. Probab. Stat.*, 31(1):1–23, 2017.

Cucker F. and Smale S. On the mathematical foundations of learning. *Bull. Amer. Math. Soc. (N.S.)*, 39(1):1–49 (electronic), 2002.

Dutkay D. E. and Jorgensen P. E. T. Affine fractals as boundaries and their harmonic analysis. *Proc. Amer. Math. Soc.*, 139(9):3291–3305, 2011.

El Machkouri M., Es-Sebaiy K., and Ouassou I. On local linear regression for strongly mixing random fields. *J. Multivariate Anal.*, 156:103–115, 2017.

Freschi V. Improved biological network reconstruction using graph Laplacian regularization. *J. Comput. Biol.*, 18(8):987–996, 2011.

Ha Quang M., Kang S. H., and Le T. M. Image and video colorization using vector-valued reproducing kernel Hilbert spaces. *J. Math. Imaging Vision*, 37(1):49–65, 2010.

Haeseler S., Keller M., Lenz D., Masamune J., and Schmidt M. Global properties of Dirichlet forms in terms of Green's formula. *ArXiv e-prints*, 2014.

Hedenmalm H. and Nieminen P. J. The Gaussian free field and Hadamard's variational formula. *Probab. Theory Related Fields*, 159(1-2):61–73, 2014.

Hersonsky S. Boundary value problems on planar graphs and flat surfaces with integer cone singularities, I: The Dirichlet problem. *J. Reine Angew. Math.*, 670:65–92, 2012.

Hsing T. and Eubank R. *Theoretical foundations of functional data analysis, with an introduction to linear operators*. Wiley Series in Probability and Statistics. John Wiley & Sons, Ltd., Chichester, 2015.

Jø rsboe O. G. *Equivalence or singularity of Gaussian measures on function spaces*. Various Publications Series, No. 4. Matematisk Institut, Aarhus Universitet, Aarhus, 1968.

Jorgensen P. and Pearse E. P. A Hilbert space approach to effective resistance metric. *Complex Anal. Oper. Theory*, 4(4):975–1013, 2010.

Jorgensen P. and Pearse E. P., 2011. Resistance boundaries of infinite networks. In *Random walks, boundaries and spectra*, volume 64 of *Progr. Probab.*, pages 111–142. Birkhäuser/Springer Basel AG, Basel.

Jorgensen P. and Pearse E. P. A discrete Gauss-Green identity for unbounded Laplace operators, and the transience of random walks. *Israel J. Math.*, 196(1):113–160, 2013.

Jorgensen P. and Tian F. Discrete reproducing kernel Hilbert spaces: sampling and distribution of Dirac-masses. *J. Mach. Learn. Res.*, 16:3079–3114, 2015.

Jorgensen P. and Tian F. Graph Laplacians and discrete reproducing kernel Hilbert spaces from restrictions. *Stoch. Anal. Appl.*, 34(4):722–747, 2016.

Khan S., Goodall R. M., and Dixon R. Non-uniform sampling strategies for digital control. *Internat. J. Systems Sci.*, 44(12):2234–2254, 2013.

Konno N., Masuda N., Roy R., and Sarkar A. Rigorous results on the threshold network model. *J. Phys. A*, 38(28):6277–6291, 2005.

Kulkarni S. and Harman G. *An elementary introduction to statistical learning theory.* Wiley Series in Probability and Statistics. John Wiley & Sons, Inc., Hoboken, NJ, 2011.

Landau H. J. Necessary density conditions for sampling and interpolation of certain entire functions. *Acta Math.*, 117:37–52, 1967.

Lata S., Mittal M., and Paulsen V. I. An operator algebraic proof of Agler's factorization theorem. *Proc. Amer. Math. Soc.*, 137(11):3741–3748, 2009.

Lata S. and Paulsen V. The Feichtinger conjecture and reproducing kernel Hilbert spaces. *Indiana Univ. Math. J.*, 60(4):1303–1317, 2011.

Lin Y. and Brown L. D. Statistical properties of the method of regularization with periodic Gaussian reproducing kernel. *Ann. Statist.*, 32(4):1723–1743, 2004.

Lubetzky E. and Peres Y. Cutoff on all Ramanujan graphs. *Geom. Funct. Anal.*, 26(4):1190–1216, 2016.

Martinez A., Gelb A., and Gutierrez A. Edge detection from non-uniform Fourier data using the convolutional gridding algorithm. *J. Sci. Comput.*, 61(3):490–512, 2014.

Muandet K., Sriperumbudur B., Fukumizu K., Gretton A., and Schölkopf B. Kernel mean shrinkage estimators. *J. Mach. Learn. Res.*, 17:Paper No. 48, 41, 2016.

Mumford D., 2000. The dawning of the age of stochasticity. In *Mathematics: frontiers and perspectives*, pages 197–218. Amer. Math. Soc., Providence, RI.

Nelson E. Kernel functions and eigenfunction expansions. *Duke Math. J.*, 25:15–27, 1957.

Okoudjou K. A. and Strichartz R. S. Weak uncertainty principles on fractals. *J. Fourier Anal. Appl.*, 11(3):315–331, 2005.

Okoudjou K. A., Strichartz R. S., and Tuley E. K. Orthogonal polynomials on the Sierpinski gasket. *Constr. Approx.*, 37(3):311–340, 2013.

Osterwalder K. and Schrader R. Axioms for Euclidean Green's functions. *Comm. Math. Phys.*, 31:83–112, 1973.

Parthasarathy K. R. and Schmidt K. Stable positive definite functions. *Trans. Amer. Math. Soc.*, 203:161–174, 1975.

Parussini L., Venturi D., Perdikaris P., and Karniadakis G. E. Multi-fidelity Gaussian process regression for prediction of random fields. *J. Comput. Phys.*, 336:36–50, 2017.

Paulsen V. I. and Raghupathi M. *An introduction to the theory of reproducing kernel Hilbert spaces*, volume 152 of *Cambridge Studies in Advanced Mathematics*. Cambridge University Press, Cambridge, 2016.

Peres Y., Schapira B., and Sousi P. Martingale defocusing and transience of a self-interacting random walk. *Ann. Inst. Henri Poincaré Probab. Stat.*, 52(3):1009–1022, 2016.

Peres Y. and Sousi P. Dimension of fractional Brownian motion with variable drift. *Probab. Theory Related Fields*, 165(3-4):771–794, 2016.

Rasmussen C. E. and Williams C. K. I. *Gaussian processes for machine learning*. Adaptive Computation and Machine Learning. MIT Press, Cambridge, MA, 2006.

Roblin T. Comportement harmonique des densités conformes et frontière de Martin. *Bull. Soc. Math. France*, 139(1):97–128, 2011.

Rodgers G. J., Austin K., Kahng B., and Kim D. Eigenvalue spectra of complex networks. *J. Phys. A*, 38(43):9431–9437, 2005.

Rudin W. *Real and complex analysis*. McGraw-Hill Book Co., New York, 1987, third edition.

Saitoh S., 2016. A reproducing kernel theory with some general applications. In *Mathematical analysis, probability and applications—plenary lectures*, volume 177 of *Springer Proc. Math. Stat.*, pages 151–182. Springer, [Cham].

Schlkopf B. and Smola A. J. *Learning with Kernels: Support Vector Machines, Regularization, Optimization, and Beyond (Adaptive Computation and Machine Learning).* The MIT Press, 2001, 1st edition.

Schramm O. and Sheffield S. A contour line of the continuum Gaussian free field. *Probab. Theory Related Fields*, 157(1-2):47–80, 2013.

Shawe-Taylor J. and Cristianini N. *Kernel Methods for Pattern Analysis.* Cambridge University Press, 2004.

Skopenkov M. The boundary value problem for discrete analytic functions. *Adv. Math.*, 240:61–87, 2013.

Smale S. and Zhou D.-X. Shannon sampling and function reconstruction from point values. *Bull. Amer. Math. Soc. (N.S.)*, 41(3):279–305, 2004.

Smale S. and Zhou D.-X. Geometry on probability spaces. *Constr. Approx.*, 30(3):311–323, 2009a.

Smale S. and Zhou D.-X. Online learning with Markov sampling. *Anal. Appl. (Singap.)*, 7(1):87–113, 2009b.

Strichartz R. S. Transformation of spectra of graph Laplacians. *Rocky Mountain J. Math.*, 40(6):2037–2062, 2010.

Strichartz R. S. and Teplyaev A. Spectral analysis on infinite Sierpiński fractafolds. *J. Anal. Math.*, 116:255–297, 2012.

Tosiek J. and Brzykcy P. States in the Hilbert space formulation and in the phase space formulation of quantum mechanics. *Ann. Physics*, 332:1–15, 2013.

Tzeng W. J. and Wu F. Y. Theory of impedance networks: the two-point impedance and *LC* resonances. *J. Phys. A*, 39(27):8579–8591, 2006.

Vuletić M., 2013. The Gaussian free field and strict plane partitions. In *25th International Conference on Formal Power Series and Algebraic Combinatorics (FPSAC 2013)*, Discrete Math. Theor. Comput. Sci. Proc., AS, pages 1041–1052. Assoc. Discrete Math. Theor. Comput. Sci., Nancy.

Zhang H., Xu Y., and Zhang Q. Refinement of operator-valued reproducing kernels. *J. Mach. Learn. Res.*, 13:91–136, 2012.

Zhou L., Li X., and Pan F. Gradient-based iterative identification for Wiener nonlinear systems with non-uniform sampling. *Nonlinear Dynam.*, 76(1):627–634, 2014.

# Chapter 2

## Extensions of Some Matrix Inequalities via Matrix Means

**Takeaki Yamazaki**

*Department of Electrical, Electronic and Computer Engineering, Toyo University, Kawagoe, Japan*

## 2.1 Matrix monotone function

Let $\mathcal{M}_n$ be the set of all $n$–by–$n$ matrices on $\mathbb{C}^n$ with an identity matrix $I$. A matrix $A \in \mathcal{M}_n$ is said to be positive semi-definite if $\langle Ax, x \rangle \geq 0$ holds for all $x \in \mathbb{C}^n$. Moreover $A \in \mathcal{M}_n$ is called positive definite if $A$ is positive semi-definite and invertible. We use a notation $A \geq 0$ (resp. $A > 0$) for positive semi-definite (resp. positive definite) matrices. In this chapter, $\mathcal{PS}_n$ and $\mathcal{P}_n$ denote the sets of all positive semi-definite and positive definite matrices, respectively. For Hermitian $A, B \in \mathcal{M}_n$, $A \leq B$ is defined by $0 \leq B - A$. A real-valued function $f$ defined on a real interval $I$ is said to be matrix monotone if $f(A) \leq f(B)$ holds for all Hermitian $A, B \in \mathcal{M}_n$ such that $A \leq B$ and $\sigma(A), \sigma(B) \subset I$, where $\sigma(T)$ means the spectrum of a $T \in \mathcal{M}_n$. Here $f(A)$ is defined by

$$f(A) = \sum_{i=1}^{n} f(\lambda_i) P_i,$$

where $A = \sum_{i=1}^{n} \lambda_i P_i$ is the spectral decomposition of an Hermitian $A$ with the eigenvalues $\lambda_i$ of $A$ and the spectral projections $P_i$ $(i = 1, 2, ..., n)$.

Typical examples of matrix monotone functions are $f(x) = x^\alpha$ $(\alpha \in [0, 1])$ and $f(x) = \log x$. We remark that $f(x) = x^\alpha$ $(\alpha > 1)$ is not matrix monotone. In fact let $A = \begin{pmatrix} 3 & 1 \\ 1 & 2 \end{pmatrix}$ and $B = \begin{pmatrix} 2 & 0 \\ 0 & 1 \end{pmatrix}$. Then $A, B \in \mathcal{P}_n$ (i.e., $A$ and $B$ are Hermitian, and all eigenvalues of $A$ and $B$ are positive) and $A - B =$

$\begin{pmatrix} 1 & 1 \\ 1 & 1 \end{pmatrix} \geq 0$. But

$$A^2 - B^2 = \begin{pmatrix} 6 & 5 \\ 5 & 4 \end{pmatrix} \ngeq 0. \tag{2.1}$$

Assume that $f(x) = x^\alpha$ is matrix monotone for some $\alpha > 1$. Then we have

$$B \leq A \quad \Longrightarrow \quad B^\alpha \leq A^\alpha \quad \Longrightarrow \quad B^{\alpha^2} \leq A^{\alpha^2} \quad \Longrightarrow \cdots \Longrightarrow B^{\alpha^n} \leq A^{\alpha^n}$$

for any natural number $n$. Since $f(x) = x^\alpha$ is matrix monotone for all $\alpha \in [0,1]$, we have

$$B^2 = \left(B^{\alpha^n}\right)^{\frac{2}{\alpha^n}} \leq \left(A^{\alpha^n}\right)^{\frac{2}{\alpha^n}} = A^2.$$

It is a contradiction to (2.1). Hence $f(x) = x^\alpha$ is not matrix monotone for any $\alpha > 1$.

**Theorem 2.1.1** (Loewner-Heinz inequality Heinz (1951); Loewner (1934)). *Let $A, B \in \mathcal{PS}_n$ satisfying $B \leq A$. Then $B^\alpha \leq A^\alpha$ holds for all $\alpha \in [0,1]$.*

*Proof.* First of all, we only prove Theorem 2.1.1 in the case $A, B \in \mathcal{P}_n$. In fact, for $A, B \in \mathcal{PS}_n$ and $\varepsilon > 0$, put $A_\varepsilon = A + \varepsilon I$ and $B_\varepsilon = B + \varepsilon I$. Then $A_\varepsilon, B_\varepsilon \in \mathcal{P}_n$ and $B_\varepsilon \leq A_\varepsilon$ if $B \leq A$. If we can prove Theorem 2.1.1 for $A_\varepsilon$ and $B_\varepsilon$, then it holds for $A, B \in \mathcal{PS}_n$ by $\varepsilon \to 0$.

Next, for $A, B \in \mathcal{P}_n$, we note that

$$B \leq A \quad \Longleftrightarrow \quad A^{\frac{-1}{2}} B A^{\frac{-1}{2}} \leq I \quad \Longleftrightarrow \quad \|A^{\frac{-1}{2}} B A^{\frac{-1}{2}}\| \leq 1,$$

where $\|\cdot\|$ means the spectral norm. Hence, it is enough to show that

$$\|A^{\frac{-1}{2}} B A^{\frac{-1}{2}}\| \leq 1 \quad \Longrightarrow \quad \|A^{\frac{-\alpha}{2}} B^\alpha A^{\frac{-\alpha}{2}}\| \leq 1 \tag{2.2}$$

for all $\alpha \in [0,1]$. Define

$$D := \{\alpha \in [0,1] : (2.2) \text{ holds for all } A, B \in \mathcal{P}_n\}.$$

Then we shall show $[0,1] \subseteq D$. Since $0, 1 \in D$, $D$ is not an empty set, it is enough to prove that $D$ is a convex set. Assume $\alpha, \beta \in D$, if $\|A^{\frac{-1}{2}} B A^{\frac{-1}{2}}\| \leq 1$, then we have

$$\left\|A^{\frac{-(\alpha+\beta)}{4}} B^{\frac{\alpha+\beta}{2}} A^{\frac{-(\alpha+\beta)}{4}}\right\|^2 = r\left(B^{\frac{\alpha+\beta}{2}} A^{\frac{-(\alpha+\beta)}{2}}\right)^2$$

$$= r\left(A^{\frac{-\alpha}{2}} B^{\frac{\alpha}{2}} \cdot B^{\frac{\beta}{2}} A^{\frac{-\beta}{2}}\right)^2$$

$$\leq \left\|A^{\frac{-\alpha}{2}} B^{\frac{\alpha}{2}}\right\|^2 \cdot \left\|B^{\frac{\beta}{2}} A^{\frac{-\beta}{2}}\right\|^2$$

$$= \left\|A^{\frac{-\alpha}{2}} B^\alpha A^{\frac{-\alpha}{2}}\right\| \cdot \left\|A^{\frac{-\beta}{2}} B^\beta A^{\frac{-\beta}{2}}\right\| \leq 1,$$

where $r(T)$ means the spectral radius of $T$. Then $\frac{\alpha+\beta}{2} \in D$, and $D$ is a convex set since $A^\alpha$ is continuous on $\alpha \in \mathbb{R}$. Therefore the proof is completed. $\quad\square$

**Corollary 2.1.2.** $f(x) = \log x$ *is matrix monotone on* $x > 0$.

*Proof.* By Theorem 2.1.1, $B \leq A$ implies $B^\alpha \leq A^\alpha$ for all $\alpha \in [0, 1]$. Then

$$\frac{B^\alpha - I}{\alpha} \leq \frac{A^\alpha - I}{\alpha}$$

holds for all $\alpha \in (0, 1]$. By using the fact

$$\lim_{\alpha \to 0} \frac{X^\alpha - 1}{\alpha} = \log X \tag{2.3}$$

for any $X \in P_n$, we have Corollary 2.1.2. $\qquad\square$

The following important inequalities hold for a similar condition of Theorem 2.1.1.

**Theorem 2.1.3** (Hansen's inequality Hansen (7980)). *Let* $A \in PS_n$, *and* $X \in M_n$ *be a contraction (i.e.,* $\|X\| \leq 1$). *Then*

(i) $(X^*AX)^\alpha \leq X^*A^\alpha X$ *holds for all* $\alpha \in [1, 2]$;

(ii) $X^*A^\alpha X \leq (X^*AX)^\alpha$ *holds for all* $\alpha \in [0, 1]$.

To prove this, the following lemma is useful.

**Lemma 2.1.4** (Furuta (1995)). *For* $A \in P_n$ *and invertible* $X \in M_n$,

$$(X^*AX)^\alpha = X^*A^{\frac{1}{2}}(A^{\frac{1}{2}}XX^*A^{\frac{1}{2}})^{\alpha-1}A^{\frac{1}{2}}X$$

*holds for any* $\alpha \in \mathbb{R}$. *If* $\alpha \geq 1$, *then it holds for any* $A \in PS_n$ *and* $X \in M_n$.

*Proof.* Let $X^*A^{\frac{1}{2}} = UP$ be the polar decomposition such that $U$ is a partial isometry and $P \in P_n$. Then

$$
\begin{aligned}
(X^*AX^{\frac{1}{2}})^\alpha &= (UP^2U^*)^\alpha \\
&= UP^{2\alpha}U^* \\
&= UP \cdot P^{2(\alpha-1)}PU^* \\
&= X^*A^{\frac{1}{2}}\left|X^*A^{\frac{1}{2}}\right|^{2(\alpha-1)}A^{\frac{1}{2}}X^{\frac{1}{2}} \\
&= X^*A^{\frac{1}{2}}(A^{\frac{1}{2}}XX^*A^{\frac{1}{2}})^{\alpha-1}A^{\frac{1}{2}}X.
\end{aligned}
$$

$\qquad\sqcup$

*Proof of Theorem 2.1.3.* (i) We note that $XX^* \leq I$ holds. Firstly, let $\alpha \in [1, 2]$. Then by Lemma 2.1.4,

$$(X^*AX)^\alpha = X^*A^{\frac{1}{2}}(A^{\frac{1}{2}}XX^*A^{\frac{1}{2}})^{\alpha-1}A^{\frac{1}{2}}X \leq X^*A^\alpha X,$$

where the inequality holds by $XX^* \leq I$, $\alpha - 1 \in [0, 1]$ and Theorem 2.1.1.

(ii) By Theorem 2.1.1, (i) ensures $X^* A_1 X \le (X^* A_1^t X)^{\frac{1}{t}}$ for all $t \in [1, 2]$ and $A_1 \in \mathcal{P}S_n$. By putting $\alpha = \frac{1}{t}$ and $A = A_1^t$, we have

$$X^* A^\alpha X \le (X^* A X)^\alpha$$

for $\alpha \in [\frac{1}{2}, 1]$. Let $\alpha_1 \in [\frac{1}{4}, \frac{1}{2}]$. Since $2\alpha_1 \in [\frac{1}{2}, 1]$, we have

$$X^* A^{\alpha_1} X = X^* (A^{\frac{1}{2}})^{2\alpha_1} X \le (X^* A^{\frac{1}{2}} X)^{2\alpha_1} \le (X^* A X)^{\alpha_1}.$$

Repeating this way, we obtain (ii). □

A real valued function $f$ defined on an interval $I$ is called matrix convex if

$$f\left((1 - \lambda)A + \lambda B\right) \le (1 - \lambda)f(A) + \lambda f(B) \tag{2.4}$$

holds for all $\lambda \in [0, 1]$ and Hermitian $A, B \in M_n$, and if the function $f$ satisfies an opposite inequality to (2.4), we call $f$ a matrix concave function. As a consequence of Theorem 2.1.3, we have a following result.

**Corollary 2.1.5.** *Let $f(x) = x^\alpha$. Then*

  *(i) $f(x)$ is a matrix convex for all $\alpha \in [1, 2]$;*

  *(ii) $f(x)$ is a matrix concave for all $\alpha \in [0, 1]$.*

*Proof.* (i) Let $\lambda \in [0, 1]$ and $A, B \in \mathcal{P}S_n$. Define

$$\hat{X} = \begin{pmatrix} \sqrt{1 - \lambda}I & 0 \\ \sqrt{\lambda}I & 0 \end{pmatrix} \in M_{2n} \quad \text{and} \quad \hat{A} = \begin{pmatrix} A & 0 \\ 0 & B \end{pmatrix} \in \mathcal{P}S_{2n}.$$

Since $\|\hat{X}\| = 1$, $\hat{X}$ is a contraction. Then by (i) of Theorem 2.1.3, we have

$$\begin{pmatrix} ((1 - \lambda)A + \lambda B)^\alpha & 0 \\ 0 & 0 \end{pmatrix} = (\hat{X}^* \hat{A} \hat{X})^\alpha$$

$$\le \hat{X}^* \hat{A}^\alpha \hat{X} = \begin{pmatrix} (1 - \lambda)A^\alpha + \lambda B^\alpha & 0 \\ 0 & 0 \end{pmatrix}$$

for all $\alpha \in [1, 2]$, that is, $f(x) = x^\alpha$ is matrix convex for all $\alpha \in [1, 2]$. (ii) can be proven by the same way. □

By the same argument to the proof of Corollary 2.1.2, $f(x) = \log x$ is also a matrix concave function.

An an extension of Theorem 2.1.1, the following Furuta inequality is very famous.

**Theorem 2.1.6** (Furuta inequality Furuta (1987)).
*If $A \ge B \ge 0$, then for each $r \ge 0$, (i)*

$$(B^{\frac{r}{2}} A^p B^{\frac{r}{2}})^{\frac{1}{q}} \ge (B^{\frac{r}{2}} B^p B^{\frac{r}{2}})^{\frac{1}{q}}$$

*and*

(ii)

$$(A^{\frac{r}{2}}A^p A^{\frac{r}{2}})^{\frac{1}{q}} \geq (A^{\frac{r}{2}}B^p A^{\frac{r}{2}})^{\frac{1}{q}}$$

*hold for $p \geq 0$ and $q \geq 1$ with $(1+r)q \geq p+r$.*

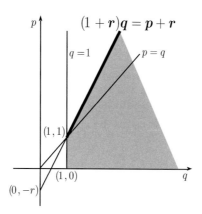

**FIGURE 2.1**: Domain for $p$, $q$ and $r$

Theorem 2.1.1 follows from Theorem 2.1.6 by putting $r = 0$ and $\alpha = \frac{p}{q}$. The domain for the parameters $p, q, r$ in Theorem 2.1.6 is the best possible Tanahashi (1996) (see Figure 2.1).

*Proof of Theorem 2.1.6.* By the same reason to the proof of Theorem 2.1.1, we may assume $A, B \in \mathcal{P}_n$. Moreover (i) and (ii) follow from each other. In fact, assume (i) holds. Then we can obtain (ii) as follows.

$$B \leq A \Longleftrightarrow A^{-1} \leq B^{-1}$$

$$\Longrightarrow \left[(A^{-1})^{\frac{r}{2}}(A^{-1})^p(A^{-1})^{\frac{r}{2}}\right]^{\frac{1}{q}} \leq \left[(A^{-1})^{\frac{r}{2}}(B^{-1})^p(A^{-1})^{\frac{r}{2}}\right]^{\frac{1}{q}} \quad \text{(by (i))}$$

$$\Longleftrightarrow \left(A^{\frac{r}{2}}B^p A^{\frac{r}{2}}\right)^{\frac{1}{q}} \leq \left(A^{\frac{r}{2}}A^p A^{\frac{r}{2}}\right)^{\frac{1}{q}}.$$

Therefore we shall only prove (i). Moreover by Theorem 2.1.1, we only consider the case $p \geq 1$ and $\frac{1}{q} = \frac{1+r}{p+r}$, i.e., we shall prove

$$B \leq A \implies B^{1+r} = (B^{\frac{r}{2}}B^p B^{\frac{r}{2}})^{\frac{1+r}{p+r}} \leq (B^{\frac{r}{2}}A^p B^{\frac{r}{2}})^{\frac{1+r}{p+r}} \quad \text{for } p > 1 \text{ and } r > 0,$$

$$\tag{2.5}$$

The case $r \in [0, 1]$.

$$(B^{\frac{r}{2}}A^p B^{\frac{r}{2}})^{\frac{1+r}{p+r}} = B^{\frac{r}{2}}A^{\frac{p}{2}}(A^{\frac{p}{2}}B^r A^{\frac{p}{2}})^{\frac{1-p}{p+r}}A^{\frac{p}{2}}B^{\frac{r}{2}} \quad \text{by Lemma 2.1.4}$$

$$\geq B^{\frac{r}{2}}A^{\frac{p}{2}}(A^{\frac{p}{2}}A^r A^{\frac{p}{2}})^{\frac{1-p}{p+r}}A^{\frac{p}{2}}B^{\frac{r}{2}}$$

$$= B^{\frac{r}{2}}AB^{\frac{r}{2}} \geq B^{1+r},$$

where the first inequality follows from Theorem 2.1.1 since $r \in [0,1]$ and $\frac{1-p}{p+r} \in [-1,0]$. Hence we have

$$B \leq A \implies B^{1+r} \leq (B^{\frac{r}{2}} A^p B^{\frac{r}{2}})^{\frac{1+r}{p+r}} \quad \text{for } p \geq 1 \text{ and } r \in [0,1]. \tag{2.6}$$

Next, we shall show (2.5) in the case $r \in [1,3]$. Let $B_1 := B^{1+r}$ and $A_1 := (B^{\frac{r}{2}} A^p B^{\frac{r}{2}})^{\frac{1+r}{p+r}}$. Then $B \leq A$ implies $B_1 \leq A_1$ for $p \geq 1$ and $r \in [0,1]$ by (2.6). By applying (2.6) to $A_1$ and $B_1$, we have

$$B_1^{1+r_1} \leq (B_1^{\frac{r_1}{2}} A_1^{p_1} B_1^{\frac{r_1}{2}})^{\frac{1+r_1}{p_1+r_1}}$$

for $p_1 \geq 1$ and $r_1 \in [0,1]$. Put $p_1 = \frac{p+r}{1+r} \geq 1$ and $r_1 = 1$. Then we have

$$B^{1+(1+2r)} \leq (B^{\frac{1+2r}{2}} A^p B^{\frac{1+2r}{2}})^{\frac{1+(1+2r)}{p+(1+2r)}}$$

for $p \geq 1$ and $r \in [0,1]$, that is,

$$B^{1+s} \leq (B^{\frac{s}{2}} A^p B^{\frac{s}{2}})^{\frac{1+s}{p+s}} \quad \text{holds for } p \geq 1 \text{ and } s \in [1,3] \tag{2.7}$$

by putting $s = 1 + 2r$. Combining (2.6) and (2.7), we have (2.5) for all $p \geq 1$ and $r \in [0,3]$. Repeating this method, we can get (i), and the proof is completed. □

As in the proof of Theorem 2.1.6, the essential part is the following form.

**Theorem 2.1.6′** For $0 \leq B \leq A$,

$$\text{(i) } B^{1+r} \leq (B^{\frac{r}{2}} A^p B^{\frac{r}{2}})^{\frac{1+r}{p+r}} \quad \text{and} \quad \text{(ii) } (A^{\frac{r}{2}} B^p A^{\frac{r}{2}})^{\frac{1+r}{p+r}} \leq A^{1+r}$$

hold for $p \geq 1$ and $r \geq 0$.

The next inequalities consider a weaker assumption than $B \leq A$.

**Theorem 2.1.7** (Ando (1987); Fujii et al. (1993); Furuta (1992); Uchiyama (1999)). *Let $A, B \in \mathcal{P}_n$. Then the following inequalities follow from each other.*

*(i)* $\log B \leq \log A$ *(this order is called the chaotic order);*

*(ii)* $(A^{\frac{r}{2}} B^r A^{\frac{r}{2}})^{\frac{1}{2}} \leq A^r$ *for all $r \geq 0$;*

*(iii)* $(A^{\frac{p}{2}} B^r A^{\frac{p}{2}})^{\frac{r}{p+r}} \leq A^r$ *for all $p, r \geq 0$.*

Before proving Theorem 2.1.7, we recall the following famous formula: For $X \in \mathcal{P}_n$,

$$\lim_{n \to \infty} \left( I + \frac{1}{n} \log X \right)^n = X. \tag{2.8}$$

*Proof of Theorem 2.1.7.* Proof of (i) $\implies$ (iii). Let $A_1 = I + \frac{1}{n}\log A$ and $B_1 = I + \frac{1}{n}\log B$. Then $A_1, B_1 \in \mathcal{P}_n$ for sufficiently large $n$, and $B_1 \leq A_1$ by (i). By (ii) of Theorem 2.1.6', we have

$$(A_1^{\frac{r}{2}} B_1^{p} A_1^{\frac{r}{2}})^{\frac{1+r}{p+r}} \leq A_1^{1+r}$$

for $p \geq 1$ and $r \geq 0$. Replacing $p$ and $r$ into $np$ and $nr$, respectively, we have

$$\left[\left(I + \frac{1}{n}\log A\right)^{\frac{nr}{2}}\left(I + \frac{1}{n}\log B\right)^{np}\left(I + \frac{1}{n}\log A\right)^{\frac{nr}{2}}\right]^{\frac{1/n+r}{p+r}} \leq \left(I + \frac{1}{n}\log A\right)^{n(\frac{1}{n}+r)}$$

for $r \geq 0$ and $p \geq \frac{1}{n}$. Hence we have

$$(A^{\frac{r}{2}} B^{p} A^{\frac{r}{2}})^{\frac{r}{p+r}} \leq A^r$$

for $r \geq 0$ and $p \geq 0$ by $n \to \infty$ and (2.8).

Proof of (iii) $\implies$ (ii) is obvious by putting $p = r$ in (iii).

Proof of (ii) $\implies$ (i). By (ii), we have

$$\frac{A^{\frac{r}{2}} B^r A^{\frac{r}{2}} - A^r + A^r - I}{r\{(A^{\frac{r}{2}} B^r A^{\frac{r}{2}})^{\frac{1}{2}} + I\}} = \frac{(A^{\frac{r}{2}} B^r A^{\frac{r}{2}})^{\frac{1}{2}} - I}{r} \leq \frac{A^r - I}{r}.$$

Hence by (2.3), we have

$$\frac{1}{2}(\log B + \log A) \leq \log A,$$

i.e., $\log B \leq \log A$.

Therefore the proof is completed.     $\square$

By using Theorem 2.1.6, we have an additional property of the inequality $(A^{\frac{r}{2}} B^p A^{\frac{r}{2}})^{\frac{r}{p+r}} \leq A^r$.

**Theorem 2.1.8.** *Let $A, B \in \mathcal{PS}_n$. If $(A^{\frac{r_0}{2}} B^{p_0} A^{\frac{r_0}{2}})^{\frac{r_0}{p_0+r_0}} \leq A^{r_0}$ holds for some $p_0, r_0 \geq 0$, then $(A^{\frac{r}{2}} B^p A^{\frac{r}{2}})^{\frac{r}{p+r}} \leq A^r$ holds for all $p_0 \leq p$ and $r_0 \leq r$.*

*Proof.* We may assume $A, B \in \mathcal{P}_n$. Let $B_1 = (A^{\frac{r_0}{2}} B^{p_0} A^{\frac{r_0}{2}})^{\frac{r_0}{p_0+r_0}}$ and $A_1 = A^{r_0}$. Then $B_1 \leq A_1$ by the assumption. By (ii) of Theorem 2.1.6', we have

$$(A_1^{\frac{s}{2}} B_1^{p} A_1^{\frac{s}{2}})^{\frac{1+s}{p+s}} \leq A_1^{1+s}$$

for $p \geq 1$ and $s \geq 0$. Put $p = \frac{p_0+r_0}{r_0} \geq 1$. We have

$$\left(A^{\frac{(1+s)r_0}{2}} B^{p_0} A^{\frac{(1+s)r_0}{2}}\right)^{\frac{(1+s)r_0}{p_0+(1+s)r_0}} \leq A^{(1+s)r_0}.$$

Put $r = (1+s)r_0 \geq r_0$. We have

$$(A^{\frac{r}{2}} B^{p_0} A^{\frac{r}{2}})^{\frac{r}{p_0+r}} \leq A^r \qquad (2.9)$$

for some $p_0 \geq 0$ and all $r \geq r_0$. Here using Lemma 2.1.4 to the left hand side,

$$(A^{\frac{r}{2}} B^{p_0} A^{\frac{r}{2}})^{\frac{r}{p_0+r}} = A^{\frac{r}{2}} B^{\frac{p_0}{2}} (B^{\frac{p_0}{2}} A^r B^{\frac{p_0}{2}})^{\frac{-p_0}{p_0+r}} B^{\frac{p_0}{2}} A^{\frac{r}{2}}$$

$$= A^{\frac{r}{2}} B^{\frac{p_0}{2}} \left( (B^{-1})^{\frac{p_0}{2}} (A^{-1})^r (B^{-1})^{\frac{p_0}{2}} \right)^{\frac{p_0}{p_0+r}} B^{\frac{p_0}{2}} A^{\frac{r}{2}}.$$

(2.9) is equivalent to

$$\left( (B^{-1})^{\frac{p_0}{2}} (A^{-1})^r (B^{-1})^{\frac{p_0}{2}} \right)^{\frac{p_0}{p_0+r}} \leq (B^{-1})^{p_0}.$$

By the same way to obtain (2.9),

$$\left( (B^{-1})^{\frac{p}{2}} (A^{-1})^r (B^{-1})^{\frac{p}{2}} \right)^{\frac{p}{p+r}} \leq (B^{-1})^p$$

holds for all $p \geq p_0$ and $r \geq r_0$. Again by using Lemma 2.1.4, it is equivalent to

$$(A^{\frac{r}{2}} B^p A^{\frac{r}{2}})^{\frac{r}{p+r}} \leq A^r$$

for all $p_0 \leq p$ and $r_0 \leq r$. It completes the proof. $\qquad\qquad\square$

---

## 2.2    Matrix means

It is easy to define the arithmetic mean of any two matrices $A$ and $B$ by $\frac{A+B}{2}$. However, it is not easy to define some other means of two matrices, for example, the geometric mean of two positive definite matrices can not be defined in the usual sense because matrices are not commuting under the product. In this section, we introduce how to consider means of two matrices. The axiom of matrix means is given as follows.

**Definition 2.2.1** (Matrix mean Kubo and Ando (7980)). Let $A, B \in \mathcal{PS}_n$. Then the binary operation $A\sigma B \in \mathcal{PS}_n$ is called a matrix mean if the following four conditions hold;

  (i) $A \leq C$ and $B \leq D$ ensures $A\sigma B \leq C\sigma D$;

  (ii) $X(A\sigma B)X \leq (XAX)\sigma(XBX)$ for any Hermitian $X \in \mathcal{M}_n$;

  (iii) if $A_n \searrow A$ and $B_n \searrow B$, then $A_n\sigma B_n \searrow A\sigma B$;

  (iv) $I\sigma I = I$.

Let $A, B \in \mathcal{PS}_n$. Then for any $\varepsilon > 0$, $A_1 = A + \varepsilon I$ and $B_1 = B + \varepsilon I$ are positive definite, and $A_1 \searrow A$ and $B_1 \searrow B$ as $\varepsilon \to 0$. Moreover, by the condition (iii), we have $A_1\sigma B_1 \searrow A\sigma B$ as $\varepsilon \to 0$. Hence it is enough to consider that $A, B \in \mathcal{P}_n$. The following characterization is very important.

**Theorem 2.2.2** (Kubo and Ando (7980)). *Let $\sigma$ be a matrix mean. Then there exists a matrix monotone function $f$ defined on $(0, \infty)$ such that $f(1) = 1$ and*

$$f(x)I = I\sigma(xI). \tag{2.10}$$

*Moreover,*

$$A\sigma B = A^{\frac{1}{2}} f(A^{\frac{-1}{2}} B A^{\frac{-1}{2}}) A^{\frac{1}{2}}$$

*holds for all $A \in \mathcal{P}_n$ and $B \in \mathcal{P}S_n$.*

A matrix monotone function which is defined by a matrix mean $\sigma$ is called a representing function of $\sigma$. To prove Theorem 2.2.2, we prepare the following lemma.

**Lemma 2.2.3.** *Assume that $\sigma$ is a matrix mean. If an Hermitian $X \in \mathcal{M}_n$ is invertible, then*

$$X(A\sigma B)X = (XAX)\sigma(XBX), \tag{2.11}$$

*and for every $\alpha \geq 0$,*

$$\alpha(A\sigma B) = (\alpha A)\sigma(\alpha B) \tag{2.12}$$

*holds.*

*Proof.* In the inequality (ii) of Definition 2.2.1, $A$ and $B$ are replaced by $X^{-1}AX^{-1}$ and $X^{-1}BX^{-1}$, respectively:

$$A\sigma B \geq X\left[(X^{-1}AX^{-1})\sigma(X^{-1}BX^{-1})\right]X.$$

Replacing $X$ with $X^{-1}$, we have

$$X(A\sigma B)X \geq (XAX)\sigma(XBX).$$

This and (ii) of Definition 2.2.1 yield equality.

When $\alpha > 0$, letting $C := \alpha^{\frac{1}{2}}I$ in (2.11) implies (2.12). When $\alpha = 0$, let $0 < \alpha_n \searrow 0$. Then $(\alpha_n I)\sigma(\alpha_n I) \searrow 0\sigma 0$ by (iii) above while $(\alpha_n I)\sigma(\alpha_n I) = \alpha_n(I\sigma I) \searrow 0$. Hence $0 = (0\sigma 0)$ which is (2.12) for $\alpha = 0$.   □

*Proof of Theorem 2.2.2.* Let $\sigma$ be a matrix mean. First we show that if an orthogonal projection $P$ commutes with $A$ and $B$, then $P$ commutes with $A\sigma B$ and

$$(AP)\sigma(BP)P = (A\sigma B)P. \tag{2.13}$$

Since $PAP = AP \leq A$ and $PBP = BP \leq B$, it follows from (ii) and (i) of Definition 2.2.1 that

$$P(A\sigma B)P \leq (PAP)\sigma(PBP) = (AP)\sigma(BP) \leq A\sigma B. \tag{2.14}$$

Hence $[A\sigma B - P(A\sigma B)P]^{\frac{1}{2}}$ exists so that

$$\left|[A\sigma B - P(A\sigma B)P]^{\frac{1}{2}} P\right|^2 = P[A\sigma B - P(A\sigma B)P]P = 0.$$

Therefore, $[A\sigma B - P(A\sigma B)P]^{\frac{1}{2}}P = 0$ and so $(A\sigma B)P = P(A\sigma B)P$. This implies that $P$ commutes with $A\sigma B$. Similarly, $P$ commutes with $(AP)\sigma(BP)$ as well, and (2.13) follows from (2.14). For every $x \geq 0$, since $I\sigma(xI)$ commutes with all orthogonal projections, it is a scalar multiple of $I$. Thus, we see that there is a function $f \geq 0$ on $[0, \infty)$ satisfying (2.10). The uniqueness of such a function $f$ is obvious, and it follows from (iii) of Definition 2.2.1 that right-continuous for $x \geq 0$. Since $x^{-1}f(x)I = (x^{-1}I)\sigma I$ for $x > 0$ thanks to (2.12), it follows from (iii) of Definition 2.2.1 again that $x^{-1}f(x)$ is left-continuous for $x > 0$ and so is $f(x)$. Hence $f$ is continuous on $[0, \infty)$.

To show the matrix monotonicity of $f$, let us prove that

$$f(A) = I\sigma A. \tag{2.15}$$

Let $A = \sum_{i=1}^{m}\alpha_i P_i$ be the spectral decomposition of $A$, where $\alpha_i > 0$ and $P_i$ are projections with $\sum_{i=1}^{m} P_i = I$. Since each $P_i$ commutes with $A$, using (2.13) twice we have

$$I\sigma A = \sum_{i=1}^{m}(I\sigma A)P_i = \sum_{i=1}^{m}\{P_i\sigma(AP_i)\}P_i = \sum_{i=1}^{m}\{P_i\sigma(\alpha_i P_i)\}P_i$$
$$= \sum_{i=1}^{m}\{I\sigma(\alpha_i I)\}P_i = \sum_{i=1}^{m}f(\alpha_i)P_i = f(A).$$

For general $A \geq 0$ choose a sequence $0 < A_n$ of the above form such that $A_n \searrow A$. By upper semi-continuity we have

$$I\sigma A = \lim_{n\to\infty} I\sigma A_n = \lim_{n\to\infty} f(A_n) = f(A).$$

So (2.15) is shown. Hence, if $0 \leq A \leq B$, then

$$f(A) = I\sigma A \leq I\sigma B = f(B)$$

and we conclude that $f$ is matrix monotone.

When $A$ is invertible, we can use (2.11):

$$A\sigma B = A^{\frac{1}{2}}\{I\sigma(A^{\frac{-1}{2}}BA^{\frac{-1}{2}})\}A^{\frac{1}{2}} = A^{\frac{1}{2}}f(A^{\frac{-1}{2}}BA^{\frac{-1}{2}})A^{\frac{1}{2}}$$

and the proof is completed. $\qquad\square$

A typical example of matrix means is the (weighted) geometric mean. Its representing function is

$$g_\lambda(x) = x^\lambda$$

for $\lambda \in [0, 1]$. By Theorem 2.1.1, $g_\lambda(x)$ is a matrix monotone function. For $A, B \in \mathcal{P}_n$, the geometric mean $A\sharp_\lambda B$ is defined by

$$A\sharp_\lambda B = A^{\frac{1}{2}}g_\lambda(A^{\frac{-1}{2}}BA^{\frac{-1}{2}})A^{\frac{1}{2}} = A^{\frac{1}{2}}(A^{\frac{-1}{2}}BA^{\frac{-1}{2}})^\lambda A^{\frac{1}{2}}.$$

The arithmetic $\nabla_\lambda$ and harmonic $!_\lambda$ means are also examples of matrix means. In fact for $A, B \in \mathcal{P}_n$, we have

$$A\nabla_\lambda B = A^{\frac{1}{2}}\left[(1-\lambda)I + \lambda(A^{\frac{-1}{2}}BA^{\frac{-1}{2}})\right]A^{\frac{1}{2}} = (1-\lambda)A + \lambda B,$$

$$A!_\lambda B = A^{\frac{1}{2}}\left[(1-\lambda)I + \lambda(A^{\frac{-1}{2}}BA^{\frac{-1}{2}})^{-1}\right]^{-1}A^{\frac{1}{2}} = \left[(1-\lambda)A^{-1} + \lambda B^{-1}\right]^{-1}.$$

If the representing functions $f$ and $g$ with respect to matrix means $\sigma_f$ and $\sigma_g$ which satisfy $f(x) \le g(x)$ for all $x > 0$, then it is equivalent to $A\sigma_f B \le A\sigma_g B$ for all $A, B \in \mathcal{P}_n$. Especially, we have arithmetic-geometric-harmonic mean inequality as follows.

$$A!_\lambda B \le A\sharp_\lambda B \le A\nabla_\lambda B \tag{2.16}$$

for all $\lambda \in [0,1]$ and $A, B \in \mathcal{P}_n$.

We remark that the weighted geometric mean $A\sharp_\lambda B$ has the following properties.

**Proposition 2.2.4.** *Let* $\lambda \in [0,1]$ *and* $A, B \in \mathcal{P}_n$. *Then*

(i) $A\sharp_\lambda B = B\sharp_{1-\lambda}A$;

(ii) $(1-\alpha)A\sharp_\lambda B + \alpha C\sharp_\lambda D \le [(1-\alpha)A + \alpha C]\sharp_\lambda[(1-\alpha)B + \alpha D]$.

*Proof.* (i) is easy by only using Lemma 2.1.4 to $A\sharp_\lambda B = A^{\frac{1}{2}}(A^{\frac{-1}{2}}BA^{\frac{-1}{2}})^\lambda A^{\frac{1}{2}}$. (ii) follows from (i) and Theorem 2.1.3 (ii).     □

Theorems 2.1.7 and 2.1.8 can be rewritten in the following forms.

**Theorem 2.2.5.** *Let* $A, B \in \mathcal{P}_n$. *Then the following inequalities follow from each other.*

(i) $\log A + \log B \le 0$;

(ii) $A^r\sharp_{\frac{1}{2}}B^r \le I$ *for all* $r \ge 0$;

(iii) $A^r\sharp_{\frac{r}{p+r}}B^p \le I$ *for all* $p, r \ge 0$.

*Proof.* By replacing $A$ with $A^{-1}$ in Theorem 2.1.7, we have Theorem 2.2.5.     □

**Theorem 2.2.6** (Ando-Hiai inequality Ando and Hiai (1994)). *Let* $A, B \in \mathcal{P}_n$ *such that* $A\sharp_\lambda B \le I$ *for a fixed* $\lambda \in [0,1]$. *Then* $A^r\sharp_\lambda B^r \le I$ *holds for all* $r \ge 1$.

*Proof.* The cases $\lambda = 0, 1$ are obvious. For each $\lambda \in (0,1)$, there exist $p_0, r_0 > 0$ such that $\frac{r_0}{p_0+r_0} = \lambda$. Then $A\sharp_\lambda B \le I$ is equivalent to

$$\left((A^{\frac{-1}{r_0}})^{\frac{r_0}{2}}(B^{\frac{1}{p_0}})^{p_0}(A^{\frac{-1}{r_0}})^{\frac{r_0}{2}}\right)^{\frac{r_0}{p_0+r_0}} \le (A^{\frac{-1}{r_0}})^{r_0}.$$

By Theorem 2.1.8,

$$\left((A^{\frac{-1}{r_0}})^{\frac{rr_0}{2}}(B^{\frac{1}{p_0}})^{rp_0}(A^{\frac{-1}{r_0}})^{\frac{rr_0}{2}}\right)^{\frac{rr_0}{rp_0+rr_0}} \le (A^{\frac{-1}{r_0}})^{rr_0}.$$

holds for all $r \ge 1$. It is equivalent to $A^r\sharp_\lambda B^r \le I$ for all $r \ge 1$.     □

Other matrix means also satisfy the property of Theorem 2.2.6. In fact, the property of Theorem 2.2.6 depends on the representing function of a matrix mean.

**Theorem 2.2.7** (Wada (2014)). *Let $\sigma$ be a matrix mean with the representing function $f$. Then the following are equivalent:*

(i) $f(x^r) \le f(x)^r$ *holds for all $x > 0$ and $r \ge 1$;*

(ii) *For $A, B \in \mathcal{P}_n$, $A\sigma B \le I$ implies $A^r \sigma B^r \le I$ for all $r \ge 1$.*

*Proof.* Proof of (i) $\implies$ (ii). Let us consider the case $1 \le r \le 2$. Set $C := A^{\frac{-1}{2}} B A^{\frac{-1}{2}}$ and $\varepsilon := 2 - r$. From $A\sigma B \le I$, we have $A \le \frac{1}{f(C)}$. Thus

$$
\begin{aligned}
A^r \sigma B^r &= A^{\frac{r}{2}} f(A^{-\frac{r}{2}+\frac{1}{2}} C A^{\frac{1}{2}} B^{-\varepsilon} A^{\frac{1}{2}} C A^{-\frac{r}{2}+\frac{1}{2}}) A^{\frac{r}{2}} \\
&= A^{\frac{r}{2}} f\left(A^{-\frac{r}{2}+\frac{1}{2}} C A^{\frac{1}{2}} (A^{\frac{-1}{2}} C^{-1} A^{\frac{-1}{2}})^\varepsilon A^{\frac{1}{2}} C A^{-\frac{r}{2}+\frac{1}{2}}\right) A^{\frac{r}{2}} \\
&= A^{\frac{1}{2}} A^{\frac{1-\varepsilon}{2}} f\left(A^{-\frac{1-\varepsilon}{2}} C (A\sharp_\varepsilon C^{-1}) C A^{-\frac{1-\varepsilon}{2}}\right) A^{\frac{1-\varepsilon}{2}} A^{\frac{1}{2}} \\
&= A^{\frac{1}{2}} \left[A^{1-\varepsilon} \sigma\{C(A\sharp_\varepsilon C^{-1})C\}\right] A^{\frac{1}{2}} \\
&\le A^{\frac{1}{2}} \left[\frac{1}{f(C)^{1-\varepsilon}} \sigma C\left\{\frac{1}{f(C)}\sharp_\varepsilon C^{-1}\right\} C\right] A^{\frac{1}{2}} \quad \left(\text{by } A \le \frac{1}{f(C)}\right) \\
&= A^{\frac{1}{2}} \left(\frac{f(C^{2-\varepsilon})}{f(C)^{1-\varepsilon}}\right) A^{\frac{1}{2}} \\
&= A^{\frac{1}{2}} \left(\frac{f(C^r)}{f(C)^{r-1}}\right) A^{\frac{1}{2}} \\
&\le A^{\frac{1}{2}} f(C) A^{\frac{1}{2}} = A\sigma B,
\end{aligned}
$$

where the last inequality follows from the condition (i).

When $r > 2$, there exists a positive integer $n$ and $1 \le r_0 \le 2$ such that $r = r_0^n$. Iterating the above result gives (ii).

Proof of (ii) $\implies$ (i). For $x > 0$, set $\alpha := 1\sigma x \ (= f(x))$. Then $1 = \frac{1}{\alpha}\sigma\frac{x}{\alpha}$. From the assumption, we have $1 \ge (\frac{1}{\alpha})^r \sigma(\frac{x}{\alpha})^r = \frac{1}{\alpha^r}\sigma\frac{x^r}{\alpha^r}$, which means $f(x)^r = \alpha^r \ge 1\sigma x^r = f(x^r)$. $\square$

For a representing function $f$ of a matrix mean $\sigma$, $f(x^{-1})^{-1}$ is also matrix monotone. For the function, we can define a matrix mean $\sigma^*$. Applying Theorem 2.2.7 to $\sigma^*$, we have the following theorem.

**Theorem 2.2.8** (Wada (2014)). *Let $\sigma$ be a matrix mean with the representing function $f$. Then the following are equivalent.*

(i) $f(x)^r \le f(x^r)$ *for all $x > 0$ and $r \ge 1$;*

(ii) *For $A, B \in \mathcal{P}_n$, $A\sigma B \ge I$ implies $A^r \sigma B^r \ge I$ for all $r \ge 1$.*

We introduce below a characterization of the representing functions of matrix means without the proof.

**Theorem 2.2.9** (Loewner (1934)). *Let $f$ be a positive matrix monotone function on $(0, \infty)$ satisfying $f(1) = 1$. Then there exists a probability measure $\mu$ on $[0, 1]$ such that*

$$f(t) = \int_0^1 [1 - \lambda + \lambda t^{-1}]^{-1} d\mu(\lambda).$$

Matrix means can be considered by the convex combination of the weighted harmonic means. Simplified proof of Theorem 2.2.9 appears in Hansen (2013), for example.

---

## 2.3 Matrix power mean

In the previous section, we treat the geometric mean of two positive definite matrices. It has desirable properties; however, it was a long standing problem to extend the geometric means of two matrices into geometric means of more than three variables. In this section, we shall introduce the power and geometric means of $m$-matrices. In what follows, $\Delta_m$ denotes the set of all probability vectors of order $m$, i.e.,

$$\Delta_m = \{\omega = (w_1, ..., w_m) \in (0, 1)^m : \sum_{i=1}^m w_i = 1\}.$$

For $\omega = (w_1, ..., w_m) \in (0, \infty)^m$, we define a map $\delta : (0, \infty)^m \to \Delta_m$ by $\delta(\omega) = \alpha(w_1, ..., w_m)$, where $\alpha = (\sum_{i=1}^m w_i)^{-1}$. For $\mathbb{A} = (A_1, ..., A_m) \in \mathcal{P}_n^m$, an invertible $M \in \mathcal{M}_n$, $\mathbf{a} = (a_1, ..., a_m) \in (0, \infty)^m$, $\omega = (w_1, ..., w_m) \in \Delta_m$, $p \in \mathbb{R}$, and for a permutation $\sigma$ in $m$ letters, we set

$$M\mathbb{A}M^* = (MA_1M^*, ..., MA_mM^*),$$
$$\mathbb{A}_\sigma = (A_{\sigma(1)}, ..., A_{\sigma(m)}), \quad \omega_\sigma = (w_{\sigma(1)}, ..., w_{\sigma(m)}),$$
$$\mathbb{A}^p = (A_1^p, ..., A_m^p), \quad \mathbf{a}^p = (a_1^p, ..., a_n^p),$$
$$\mathbf{a} \cdot \mathbb{A} = (a_1 A_1, ..., a_m A_m), \quad \omega \cdot \mathbf{a} = (w_1 a_1, ..., w_m a_m).$$

The Thompson metric on $\mathcal{P}_n$ is defined by $d(A, B) = \| \log(A^{\frac{-1}{2}} B A^{\frac{-1}{2}}) \|$ for $A, B \in \mathcal{P}_n$. It is known that $d$ is a complete metric in $\mathcal{P}_n$ in Thompson (1963). Let $d = d(A, B)$. Then

$$A \leq e^d B \quad \text{and} \quad B \leq e^d A. \tag{2.17}$$

In other words,

$$d(A, B) = \inf\{\alpha > 0 : \ A \le e^{\alpha} B \text{ and } B \le e^{\alpha} A\}. \tag{2.18}$$

**Lemma 2.3.1** (Lim and Pálfia (2012)). *The following hold:*

(i) $d(A, B) = d(A^{-1}, B^{-1}) = d(MAM^*, MBM^*)$ *for any invertible* $M \in \mathcal{M}_n$;

(ii) $d(A \sharp B, A) = d(A \sharp B, B) = \frac{1}{2} d(A, B)$;

(iii) $d(A \sharp_t B, C \sharp_t D) \le (1 - t) d(A, C) + t d(B, D)$ *for* $t \in [0, 1]$;

(iv) $d\left( \sum_{i=1}^{n} t_i A_i, \sum_{i=1}^{n} t_i B_i \right) \le \max_{1 \le i \le m} \{d(A_i, B_i)\}$ *for* $t_i > 0$, $i = 1, ..., m$.

*Proof.* (i) $d(A^{-1}, B^{-1}) = \| \log(A^{\frac{1}{2}} B^{-1} A^{\frac{1}{2}}) \| = \| - \log(A^{\frac{-1}{2}} B A^{\frac{-1}{2}}) \| = d(A, B)$. $d(A, B) = d(MAM^*, MBM^*)$ follows form (2.17), immediately.

(ii) By (2.17), $d(A, B) = d(B, A)$ holds. Hence

$$d(A \sharp B, A) = d(A, A \sharp B)$$
$$= d(I, I \sharp A^{\frac{-1}{2}} B A^{\frac{-1}{2}}) \quad \text{(by (i))}$$
$$= \| \log(A^{\frac{-1}{2}} B A^{\frac{-1}{2}})^{\frac{1}{2}} \| = \frac{1}{2} d(A, B).$$

$d(A \sharp B, B) = \frac{1}{2} d(A, B)$ can be proven by the same way.

(iii) Let $d_1 = d(A, C)$ and $d_2 = d(B, D)$. Then by (2.17) and the monotonicity of the geometric mean, we have

$$A \sharp_t B \le (e^{d_1} C) \sharp_t (e^{d_2} D) = e^{(1-t)d_1 + td_2} C \sharp_t D,$$
$$C \sharp_t D \le (e^{d_1} A) \sharp_t (e^{d_2} B) = e^{(1-t)d_1 + td_2} A \sharp_t B.$$

Hence $d(A \sharp_t B, C \sharp_t D) \le (1 - t) d_1 + t d_2$ by (2.18).

(iv) Let $d_i = d(A_i, B_i)$. Then by (2.17), we have

$$\sum_{i=1}^{m} t_i A_i \le \sum_{i=1}^{m} t_i e^{d_i} B_i \le \left( \max_{1 \le k \le m} e^{d_k} \right) \sum_{i=1}^{m} t_i B_i,$$
$$\sum_{i=1}^{m} t_i B_i \le \sum_{i=1}^{m} t_i e^{d_i} A_i \le \left( \max_{1 \le k \le m} e^{d_k} \right) \sum_{i=1}^{m} t_i A_i.$$

Hence we have (iv) by (2.18). $\square$

**Theorem 2.3.2** (Lim and Pálfia (2012)). *Let* $A_1, ..., A_m \in \mathcal{P}_n$ *and let* $\omega = (w_1, ..., w_m) \in \Delta_m$. *Then for each* $\mathbf{t} = (t_1, ..., t_m) \in (0, 1]^m$, *the following matrix equation has a unique positive definite solution:*

$$X = \sum_{i=1}^{m} w_i X \sharp_{t_i} A_i. \tag{2.19}$$

*Furthermore, the solution varies continuously over* $\mathbf{t} \in (0, 1]^m$.

*Proof.* We will show that the map $f : \mathcal{P}_n \to \mathcal{P}_n$ defined by $f(X) = \sum_{i=1}^{m} w_i X \natural_{t_i} A_i$ is a strict contraction with respect to the Thompson metric. Let $X, Y > 0$, By Lemma 2.3.1,

$$d(f(X), f(Y)) \leq \max_{1 \leq i \leq m} \{d(X \natural_{t_i} A_i, Y \natural_{t_i} A_i)\}$$

$$\leq \max_{1 \leq i \leq m} \{(1 - t_i)d(X, Y)\} = \max_{1 \leq i \leq m} \{1 - t_i\}d(X, Y).$$

Since $\max_{1 \leq i \leq m} \{1 - t_i\} \in [0, 1)$, $f$ is a strict contraction.

By the continuity of fixed points of strict contractions, the solution of (2.19) varies continuously over $\mathbf{t} \in (0, 1]^m$. □

**Definition 2.3.3** (Matrix power means Lim and Pálfia (2012)). Let $\mathbb{A} = (A_1, ..., A_m) \in \mathcal{P}_n^m$ and $\omega = (w_1, ..., w_m) \in \Delta_m$. For $t \in (0, 1]$, we define $P_t(\omega; \mathbb{A})$ the $\omega$ weighted power mean of order $t$ of $A_1, ..., A_m$ by a unique positive solution of the following matrix equation:

$$X = \sum_{i=1}^{m} w_i X \natural_t A_i. \tag{2.20}$$

To simplify the notation we write $P_t(\mathbb{A}) = P_t(\frac{1}{m}, ..., \frac{1}{m}; \mathbb{A})$.

We note that $P_1(\omega; \mathbb{A}) = \sum_{i=1}^{m} w_i A_i$ and $P_{-1}(\omega; \mathbb{A}) = (\sum_{i=1}^{m} w_i A_i^{-1})^{-1}$, the weighted arithmetic and harmonic means of $A_1, ..., A_m$, respectively. For $t \in [-1, 0)$, $P_t(\omega; \mathbb{A})$ is defined by $P_t(\omega; \mathbb{A}) := P_{-t}(\omega; \mathbb{A}^{-1})^{-1}$.

**Proposition 2.3.4** (Lim and Pálfia (2012)). *Let* $\mathbb{A} = (A_1, ..., A_m), \mathbb{B} = (B_1, ..., B_m) \in \mathcal{P}_n^m$, $\omega \in \Delta_m$, $\mathbf{a} = (a_1, ..., a_m) \in (0, \infty)^m$ *and let* $t \in [-1, 1] \setminus \{0\}$.

(i) $P_t(\omega; \mathbb{A}) = (\sum_{i=1}^{m} w_i A_i^t)^{\frac{1}{t}}$ *if* $A_i$'s *commute;*

(ii) $P_t(\omega; \mathbf{a} \cdot \mathbb{A}) = (\sum_{i=1}^{m} w_i a_i^t)^{\frac{1}{t}} P_t(\delta(\omega \cdot \mathbf{a}^t); \mathbb{A})$;

(iii) $P_t(\omega_\sigma; \mathbb{A}_\sigma) = P_t(\omega; \mathbb{A})$ *for any permutation* $\sigma$;

(iv) $P_t(\omega; \mathbb{A}) \leq P_t(\omega; \mathbb{B})$ *if* $A_i \leq B_i$ *for all* $i = 1, 2, ..., m$;

(v) $d(P_t(\omega; \mathbb{A}), P_t(\omega; \mathbb{B})) \leq \max_{i \leq 1 \leq m} \{d(A_i, B_i)\}$;

(vi) $(1 - \lambda)P_{|t|}(\omega; \mathbb{A}) + \lambda P_{|t|}(\omega; \mathbb{B}) \leq P_{|t|}(\omega; (1 - \lambda)\mathbb{A} + \lambda \mathbb{B})$ *for any* $\lambda \in [0, 1]$;

(vii) $P_t(\omega; M\mathbb{A}M^*) = M P_t(\omega; \mathbb{A})M^*$ *for any invertible* $M \in \mathcal{M}_n$;

(viii) $P_t(\omega; \mathbb{A}^{-1})^{-1} = P_{-t}(\omega; \mathbb{A})$;

(ix) $(\sum_{i=1}^{m} w_i A_i^{-1})^{-1} \leq P_t(\omega; \mathbb{A}) \leq \sum_{i=1}^{m} w_i A_i$.

*Proof.* (i) Suppose that the $A_i$'s commute. Let $t \in (0, 1]$ and $X = (\sum_{i=1}^{m} w_i A_i^t)^{\frac{1}{t}}$. Then $X \natural_t A_i = X^{1-t} A_i^t$ and

$$\sum_{i=1}^{m} w_i X \sharp_t A_i = \sum_{i=1}^{m} w_i X^{1-t} A_i^t = X^{1-t} \sum_{i=1}^{m} w_i A_i^t = X^{1-t} X^t = X.$$

By uniqueness, $(\sum_{i=1}^{m} w_i A_i^t)^{\frac{1}{t}} = X = P_t(\omega; \mathbb{A})$. Furthermore, $P_{-t}(\omega; \mathbb{A}) = P_t(\omega; \mathbb{A}^{-1})^{-1} = (\sum_{i=1}^{m} w_i A_i^{-t})^{\frac{1}{-t}}$.

(ii) Let $t \in (0,1]$. Set $\beta = (\sum_{i=1}^{m} w_i a_i^t)^{\frac{1}{t}}$, $\zeta = \delta(\omega \cdot \mathbf{a}^t) \in \Delta_m$ and $X = P_t(\delta(\omega \cdot \mathbf{a}^t); \mathbb{A})$. Then $\zeta_i = (\delta(\omega \cdot \mathbf{a}^t))_i = \frac{1}{\sum_{i=1}^{m} w_i a_i^t} w_i a_i^t$ and $X = \sum_{i=1}^{m} \zeta_i X \sharp_t A_i$. Therefore

$$\sum_{i=1}^{m} w_i (\beta X) \sharp_t (a_i A_i) = \sum_{i=1}^{m} \beta^{1-t} w_i a_i^t X \sharp_t A_i$$

$$= \sum_{i=1}^{m} \beta^{1-t} \zeta_i \beta^t X \sharp_t A_i = \beta \sum_{i=1}^{m} \zeta_i X \sharp_t A_i = \beta X.$$

By uniqueness, $(\sum_{i=1}^{m} w_i a_i^t)^{\frac{1}{t}} P_t(\delta(\omega \cdot \mathbf{a}^t); \mathbb{A}) = \beta X = P_t(\omega; \mathbf{a} \cdot \mathbb{A})$.

For $t \in [-1, 0)$, we have

$$P_t(\omega; \mathbf{a} \cdot \mathbb{A}) = P_{-t}(\omega; (\mathbf{a} \cdot \mathbb{A})^{-1})^{-1}$$

$$= \left[ \left( \sum_{i=1}^{m} w_i a_i^t \right)^{\frac{-1}{t}} P_{-t}(\delta(\omega \cdot \mathbf{a}^t); \mathbb{A}^{-1}) \right]^{-1}$$

$$= \left( \sum_{i=1}^{m} w_i a_i^t \right)^{\frac{1}{t}} P_{-t}(\delta(\omega \cdot \mathbf{a}^t); \mathbb{A}^{-1})^{-1}$$

$$= \left( \sum_{i=1}^{m} w_i a_i^t \right)^{\frac{1}{t}} P_t(\delta(\omega \cdot \mathbf{a}^t); \mathbb{A}).$$

(iii) Follows from (2.20).

(iv) Suppose that $A_i \leq B_i$ for all $i = 1, 2, ..., m$. Let $t \in (0, 1]$. Define

$$f(X) = \sum_{i=1}^{m} w_i X \sharp_t A_i \quad \text{and} \quad g(X) = \sum_{i=1}^{m} w_i X \sharp_t B_i.$$

Then $P_t(\omega; \mathbb{A}) = \lim_{k \to \infty} f^{\circ k}(X)$ and $P_t(\omega; \mathbb{B}) = \lim_{k \to \infty} g^{\circ k}(X)$ for any $X \in \mathcal{P}_n$, by the Banach fixed point theorem. By the monotonicity of geometric mean, $f(X) \leq g(X)$ for all $X \in \mathcal{P}_n$, and $f(X) \leq f(Y), g(X) \leq g(Y)$ whenever $X \leq Y$. Let $X_0 > 0$. Then $f(X_0) \leq g(X_0)$ and $f^{\circ 2}(X_0) = f(f(X_0)) \leq g(f(X_0)) \leq g^{\circ 2}(X_0)$. Inductively, we have $f^{\circ k}(X_0) \leq g^{\circ k}(X_0)$ for all $k \in \mathbb{N}$. Therefore, $P_t(\omega; \mathbb{A}) = \lim_{k \to \infty} f^{\circ k}(X_0) \leq \lim_{k \to \infty} g^{\circ k}(X_0) = P_t(\omega; \mathbb{B})$.

Let $t \in [-1, 0)$. Then $\mathbb{A}^{-1} \geq \mathbb{B}^{-1}$ and thus $P_{-t}(\omega; \mathbb{A}^{-1}) \geq P_{-t}(\omega; \mathbb{B}^{-1})$. Therefore, $P_t(\omega; \mathbb{A}) = P_{-t}(\omega; \mathbb{A}^{-1})^{-1} \leq P_{-t}(\omega; \mathbb{B}^{-1})^{-1} = P_t(\omega; \mathbb{B})$.

(v) Let $t \in (0,1]$. Let $X = P_t(\omega; \mathbb{A})$ and $Y = P_t(\omega; \mathbb{B})$. Then by Lemma 2.3.1,

$$
\begin{aligned}
d(X,Y) &= d\left(\sum_{i=1}^{m} w_i X \natural_t A_i, \sum_{i=1}^{m} w_i Y \natural_t B_i\right) \\
&\leq \max_{1 \leq i \leq m} \{d(X \natural_t A_i, Y \natural_t B_i)\} \\
&\leq \max_{1 \leq i \leq m} \{(1-t)d(X,Y) + td(A_i, B_i)\} \\
&= (1-t)d(X,Y) + t \max_{1 \leq i \leq m} \{d(A_i, B_i)\},
\end{aligned}
$$

which implies that $d(X,Y) \leq \max_{1 \leq i \leq m}\{d(A_i, B_i)\}$. By (i) of Lemma 2.3.1, we also have

$$
\begin{aligned}
d(P_t(\omega; \mathbb{A}), P_t(\omega; \mathbb{B})) &= d(P_{-t}(\omega; \mathbb{A}^{-1})^{-1}, P_{-t}(\omega; \mathbb{B}^{-1})^{-1}) \\
&= d(P_{-t}(\omega; \mathbb{A}^{-1}), P_{-t}(\omega; \mathbb{B}^{-1})) \\
&\leq \max_{1 \leq i \leq m} \{d(A_i^{-1}, B_i^{-1})\} = \max_{1 \leq i \leq m} \{d(A_i, B_i)\}.
\end{aligned}
$$

(vi) Let $t \in (0,1]$. Let $X = P_t(\omega; \mathbb{A})$ and $Y = P_t(\omega; \mathbb{B})$. For $\lambda \in [0,1]$, we set $Z_\lambda = (1-\lambda)X + \lambda Y$. Let $f(Z) = \sum_{i=1}^{m} w_i Z \natural_t ((1-\lambda)A_i + \lambda B_i)$. Then by Proposition 2.2.4 (ii), we have

$$
\begin{aligned}
Z_\lambda &= (1-\lambda)X + \lambda Y \\
&= \sum_{i=1}^{m} w_i[(1-\lambda)X \natural_t A_i + \lambda Y \natural_t B_i] \\
&\leq \sum_{i=1}^{m} w_i((1-\lambda)X + \lambda Y) \natural_t ((1-\lambda)A_i + \lambda B_i)) = f(Z_\lambda).
\end{aligned}
$$

Inductively, $Z_\lambda \leq f^{\circ k}(Z_\lambda)$ for all $k \in \mathbb{N}$. Therefore, $(1-\lambda)P_t(\omega; \mathbb{A}) + \lambda P_t(\omega; \mathbb{B}) = Z_\lambda \leq P_t(\omega; (1-\lambda)\mathbb{A} + \lambda\mathbb{B})$.

(vii) Follows from (2.20) and the uniqueness of the positive definite solution.

(viii) True by definition.

(ix) Let $t \in (0,1]$. Let $X = P_t(\omega; \mathbb{A})$. By using (2.16), we obtain

$$
X = \sum_{i=1}^{m} w_i X \natural_t A_i \leq \sum_{i=1}^{m} w_i[(1-t)X + tA_i] = (1-t)X + t\sum_{i=1}^{m} w_i A_i,
$$

which implies that $X \leq \sum_{i=1}^{m} w_i A_i$. Similarly,

$$
X = \sum_{i=1}^{m} w_i X \natural_t A_i \geq \left[\sum_{i=1}^{m} w_i (X \natural_t A_i)^{-1}\right]^{-1} = \left[\sum_{i=1}^{m} w_i X^{-1} \natural_t A_i^{-1}\right]^{-1}.
$$

Taking inverses of both sides leads to

$$X^{-1} \le \sum_{i=1}^{m} w_i X^{-1} \sharp_t A_i^{-1} \le \sum_{i=1}^{m} w_i[(1-t)X^{-1} + tA_i^{-1}] = (1-t)X^{-1} + t\sum_{i=1}^{m} w_i A_i^{-1},$$

which implies that $X \ge (\sum_{i=1}^{m} w_i A_i^{-1})^{-1}$.

The case $t \in [-1, 0)$ holds by duality. $\qquad\qquad\qquad\qquad\qquad\square$

**Theorem 2.3.5** (Lim and Yamazaki (2013)). *Let* $\mathbb{A} = (A_1, \ldots, A_m) \in \mathcal{P}_n^m$ *and* $\omega = (w_1, \ldots, w_m) \in \Delta_m$. *Then*

(i) *for* $t \in (0, 1]$, $\sum_{i=1}^{m} w_i A_i^t \le I$ *implies* $P_t(\omega; \mathbb{A}) \le I$;

(ii) *for* $t \in [-1, 0)$, $\sum_{i=1}^{m} w_i A_i^t \le I$ *implies* $P_t(\omega; \mathbb{A}) \ge I$.

*Proof.* (i) Let $t \in (0, 1]$. Suppose that $\sum_{i=1}^{m} w_i A_i^t \le I$, equivalently $\sum_{i=1}^{m} \frac{w_i}{2} A_i^t \le \frac{1}{2}I$. Then one can find $X \ge I$ such that

$$\sum_{i=1}^{m} \frac{w_i}{2} A_i^t + \frac{1}{2}X^t = I.$$

From $A^t = I\#_t A$,

$$I = \sum_{i=1}^{m} \frac{w_i}{2} A_i^t + \frac{1}{2}X^t = \sum_{i=1}^{m} w_i'(I\#_t A_i) + w_{m+1}'(I\#_t X),$$

where $\omega' = (w_1', \ldots, w_{m+1}') := (\frac{w_1}{2}, \ldots, \frac{w_m}{2}, \frac{1}{2}) \in \Delta_{m+1}$. By Definition 2.3.3,

$$P_t(\omega'; A_1, \ldots, A_m, X) = I.$$

Proposition 2.3.4 (iv) ensures that

$$P_t(\omega'; A_1, \ldots, A_m, I) \le P_t(\omega'; A_1, \ldots, A_m, X) = I.$$

Let $X_k := P_t(\omega'; A_1, \ldots, A_m, X_{k-1})$ and $X_0 = I$. Then

$$I = X_0 \ge X_1 \ge X_2 \ge \cdots \ge X_k \ge \cdots > 0,$$

and $\{X_k\}$ converges to $X \in \mathcal{P}_n$. It satisfies $P_t(\omega'; A_1, \ldots, A_m, X) = X$. Here by Definition 2.3.3, we have $X = P_t(\omega; \mathbb{A})$, and $X = P_t(\omega; \mathbb{A}) \le I$.

(ii) Let $t \in [-1, 0)$. Suppose that $\sum_{i=1}^{m} w_i A_i^t \le I$. Set $s = -t \in (0, 1]$ and $B_i = A_i^{-1}, i = 1, \ldots, m$. Then

$$\sum_{i=1}^{m} w_i B_i^s = \sum_{i=1}^{m} w_i A_i^t \le I$$

and by (i), $P_s(\omega; B_1, \ldots, B_m) \le I$. By the order reverting property of inversion,

$$P_t(\omega; \mathbb{A}) = P_{-t}(\omega; \mathbb{A}^{-1})^{-1} = P_s(\omega; B_1, \ldots, B_m)^{-1} \ge I.$$

$\qquad\qquad\qquad\qquad\qquad\qquad\qquad\qquad\qquad\qquad\qquad\qquad\qquad\square$

We obtain a variant of the Ando-Hiai inequality of Theorems 2.2.6 to 2.2.8 on power means.

**Corollary 2.3.6** (Ando Hiai inequality Lim and Yamazaki (2013)). *Let $t \in (0,1]$. Then*

(i) $P_t(\omega; \mathbb{A}) \leq I$ implies $P_{\frac{t}{p}}(\omega; \mathbb{A}^p) \leq I$ for all $p \geq 1$,

(ii) $P_{-t}(\omega; \mathbb{A}) \geq I$ implies $P_{-\frac{t}{p}}(\omega; \mathbb{A}^p) \geq I$ for all $p \geq 1$.

*Proof of Corollary 2.3.6.* Let $t \in (0,1]$. Suppose that $X := P_t(\omega; \mathbb{A}) \leq I$. By (vii) of Proposition 2.3.4 and (i) of Theorem 2.1.3, for $p \in [1,2]$, we have

$$I = \sum_{i=1}^{n} w_i (X^{-\frac{1}{2}} A_i X^{-\frac{1}{2}})^t = \sum_{i=1}^{m} w_i (X^{-\frac{1}{2}} A_i X^{-\frac{1}{2}})^{p \cdot \frac{t}{p}}$$

$$= \sum_{i=1}^{m} w_i (X^{\frac{1}{2}} A_i^{-1} X^{\frac{1}{2}})^{p \cdot (\frac{-t}{p})}$$

$$\geq \sum_{i=1}^{m} w_i (X^{\frac{1}{2}} A_i^{-p} X^{\frac{1}{2}})^{\frac{-t}{p}} = \sum_{i=1}^{m} w_i (X^{-\frac{1}{2}} A_i^p X^{-\frac{1}{2}})^{\frac{t}{p}}.$$

By (i) of Theorem 2.3.5,

$$P_{\frac{t}{p}}(\omega; X^{-\frac{1}{2}} \mathbb{A}^p X^{-\frac{1}{2}}) \leq I$$

and by Proposition 2.3.4 (vii),

$$P_{\frac{t}{p}}(\omega; \mathbb{A}^p) \leq X = P_t(\omega; \mathbb{A}) \leq I$$

for $p \in [1,2]$. Repeating inductively this procedure to $P_{\frac{t}{p}}(\omega; \mathbb{A}^p) \leq I$ yields $P_{\frac{t}{p}}(\omega; \mathbb{A}^p) \leq I$ for all $p \geq 1$.

By the order reverting property of inversion and $P_t(\omega; \mathbb{A}) = P_{-t}(\omega; \mathbb{A}^{-1})^{-1}$, the second assertion (ii) follows by (i). $\square$

The next result gives us a concrete formula of the power mean of two matrices.

**Proposition 2.3.7.** *For $t \in [-1,1] \setminus \{0\}$, $\lambda \in (0,1)$ and $A, B \in \mathcal{P}_n$,*

$$F_t(1 - \lambda, \lambda; A, B) = A^{\frac{1}{2}}[(1 - \lambda)I + \lambda(A^{\frac{-1}{2}} B A^{\frac{-1}{2}})^t]^{\frac{1}{t}} A^{\frac{1}{2}},$$

*i.e., $p_t(x) = [1 - \lambda + \lambda x^t]^{\frac{1}{t}}$ is matrix monotone on $(0, \infty)$ for $t \in [-1,1] \setminus \{0\}$ and $\lambda \in [0,1]$.*

*Proof.* For $t \in (0,1]$, let $X = P_t(1 - \lambda, \lambda; A, B)$. Then by (2.20), $X = (1 - \lambda)X \sharp_t A + \lambda X \sharp_t B$. Setting $U = A^{\frac{-1}{2}} X A^{\frac{-1}{2}}$ and $Z = A^{\frac{-1}{2}} B A^{\frac{-1}{2}}$ yields $U =$

$(1-\lambda)U^{1-t} + \lambda U \natural_t Z$, which is equivalent to $I = (1-\lambda)U^{-t} + \lambda(U^{\frac{-1}{2}}ZU^{\frac{-1}{2}})^t$, that is, $Z = (\frac{U^t - (1-\lambda)I}{\lambda})^{\frac{1}{t}}$. This implies that $U = [(1-\lambda)I + \lambda Z^t]^{\frac{1}{t}}$ and

$$X = A^{\frac{1}{2}}UA^{\frac{1}{2}} = A^{\frac{1}{2}}[(1-\lambda)I + \lambda Z^t]^{\frac{1}{t}}A^{\frac{1}{2}} = A^{\frac{1}{2}}[(1-\lambda)I + \lambda(A^{\frac{-1}{2}}BA^{\frac{-1}{2}})^t]^{\frac{1}{t}}A^{\frac{1}{2}}.$$

For $t \in [-1, 0)$,

$$P_t(1-\lambda, \lambda; A, B) = P_{-t}(1-\lambda, \lambda; A^{-1}, B^{-1})^{-1}$$

$$= A^{\frac{1}{2}}\left[(1-\lambda)I + \lambda(A^{\frac{-1}{2}}BA^{\frac{-1}{2}})^t\right]^{\frac{1}{t}}A^{\frac{1}{2}}.$$

$\square$

We observe that for any $(1-\lambda, \lambda) \in \Delta_2$,

$$\lim_{t \to 0} P_t(1-\lambda, \lambda; A, B) = A \natural_\lambda B.$$

Indeed, setting $Z = A^{\frac{-1}{2}}BA^{\frac{-1}{2}}$, we have for $t > 0$

$$A^{\frac{-1}{2}}P_t(1-\lambda, \lambda; A, B)A^{\frac{-1}{2}} = [(1-\lambda)I + \lambda Z^t]^{\frac{1}{t}} \to Z^\lambda.$$

That is, $P_t(1-\lambda, \lambda; A, B) \to A^{\frac{1}{2}}Z^\lambda A^{\frac{1}{2}} = A^{\frac{1}{2}}(A^{\frac{-1}{2}}BA^{\frac{-1}{2}})^\lambda A^{\frac{1}{2}} = A \natural_\lambda B$. This further implies that

$$\lim_{t \to 0-} P_t(1-\lambda, \lambda; A, B) = \lim_{t \to 0-} P_{-t}(1-\lambda, \lambda; A^{-1}, B^{-1})^{-1}$$

$$= (A^{-1} \natural_\lambda B^{-1})^{-1} = A \natural_\lambda B.$$

## 2.4   Karcher mean

As a limit $t \to 0$ of the power mean can be considered as a geometric mean in the two matrices case. But this fact can be extended to more than three matrices.

**Definition 2.4.1** (Karcher mean Bhatia and Holbrook (2006); Lawson and Lim (2014); Lim and Pálfia (2012); Moakher (2005))**.** Let $\mathbb{A} = (A_1, ..., A_m) \in \mathcal{P}_n^m$, $\omega = (w_1, ..., w_m) \in \Delta_m$. The $\omega$-weighted Karcher mean $\Lambda(\omega; \mathbb{A})$ of $\mathbb{A}$ is defined to be the unique positive definite solution of the equation (it is called the Karcher equation)

$$\sum_{i=1}^{m} w_i \log(X^{\frac{-1}{2}}A_i X^{\frac{-1}{2}}) = 0. \tag{2.21}$$

To simplify the notation we write $\Lambda(\mathbb{A}) = \Lambda(\frac{1}{m}, ..., \frac{1}{m}; \mathbb{A})$.

We notice that by (2.21),

$$\Lambda(\omega; \mathbb{A}^{-1})^{-1} = \Lambda(\omega; \mathbb{A}) \tag{2.22}$$

holds.

**Theorem 2.4.2.**
$$\lim_{t\to 0} P_t(\omega; \mathbb{A}) = \Lambda(\omega; \mathbb{A}).$$

*Proof.* By Proposition 2.3.4 (viii) and (2.22), it suffices to show that $\lim_{t\to 0+} P_t(\omega; \mathbb{A}) = \Lambda(\omega; \mathbb{A})$. Set $X_t = P_t(\omega; \mathbb{A})$. Since $X_t = \sum_{i=1}^m w_i X_t \natural_t A_i$,

$$I = \sum_{i=1}^m w_i X_t^{\frac{-1}{2}} (X_t \natural_t A_i) X_t^{\frac{-1}{2}} = \sum_{i=1}^m w_i (X_t^{\frac{-1}{2}} A_i X_t^{\frac{-1}{2}})^t.$$

In particular for all $t \in (0, 1]$,

$$0 = \sum_{i=1}^m w_i \left[ \frac{(X_t^{\frac{-1}{2}} A_i X_t^{\frac{-1}{2}})^t - I}{t} \right]. \tag{2.23}$$

Let $\{t_k\}_{k=1}^\infty$ be a sequence in $(0, 1]$ converging to $0$. Since $X_t$ lies in the order interval determined by the $\omega$-weighted harmonic mean and arithmetic mean, which is compact, the sequence $\{X_{t_k}\}$ has at least one limit point. Suppose that $X_0$ is a limit point of $\{X_{t_k}\}$. We will show that $X_0 = \Lambda(\omega; \mathbb{A})$. Passing to a subsequence, we may assume that $X_{t_k} \to X_0$, as $k \to \infty$. Then $X_{t_k}^{\frac{-1}{2}} A_i X_{t_k}^{\frac{-1}{2}} \to X_0^{\frac{-1}{2}} A_i X_0^{\frac{-1}{2}}$ for all $i$. Setting $Y_{t_k} = X_{t_k}^{\frac{-1}{2}} A_i X_{t_k}^{\frac{-1}{2}}$ and $Y_0 = X_0^{\frac{-1}{2}} A_i X_0^{\frac{-1}{2}}$. Let $U_{t_k}$ be a unitary matrix such that $U_{t_k} Y_{t_k} U_{t_k}^* := D_{t_k}$ is a diagonal matrix. Since $Y_{t_k}$ converges to $Y_0$ and the unitary group is compact, we may assume that $U_{t_k} \to U_0$ and $D_{t_k} \to D_0$ for some unitary matrix $U_0$ and a diagonal matrix $D_0$. Indeed, first consider a subsequence of $U_{t_k}$ converging to a unitary matrix $U_0$, second the corresponding subsequence of $Y_{t_k}$, which always converges to $Y_0$, and then finally consider the corresponding subsequence of $D_{t_k}$, which converges to $D_0 := U_0 Y_0 U_0^*$. By (2.3), $\lim_{k\to\infty} \frac{D_{t_k}^{t_k} - I}{t_k} = \log D_0$. This implies that

$$\lim_{k\to\infty} \left[ \frac{(X_{t_k}^{\frac{-1}{2}} A_i X_{t_k}^{\frac{-1}{2}})^{t_k} - I}{t_k} \right] = \lim_{k\to\infty} \left[ \frac{Y_{t_k}^{t_k} - I}{t_k} \right]$$

$$= \lim_{k\to\infty} \left[ \frac{U_{t_k}^* D_{t_k}^{t_k} U_{t_k} - I}{t_k} \right]$$

$$= \lim_{k\to\infty} U_{t_k}^* \left[ \frac{D_{t_k}^{t_k} - I}{t_k} \right] U_{t_k}$$

$$= U_0^* (\log D_0) U_0$$

$$= \log Y_0 = \log(X_0^{\frac{-1}{2}} A_i X_0^{\frac{-1}{2}}).$$

This together with (2.23) yields $0 = \sum_{i=1}^n w_i \log(X_0^{\frac{-1}{2}} A_i X_0^{\frac{-1}{2}})$, that is, $X_0 = \Lambda(\omega; \mathbb{A})$. □

For $G, H : \Delta_m \times \mathcal{P}_n^m \to \mathcal{P}_n$, we define $G \leq H$ if $G(\omega; \mathbb{A}) \leq H(\omega; \mathbb{A})$ for all $\omega \in \Delta_m$ and $\mathbb{A} \in \mathcal{P}_n^m$. We note that $H \leq A$, the arithmetic-harmonic mean inequality.

**Corollary 2.4.3.** *For* $0 < s \leq t \leq 1$,

$$H = P_{-1} \leq P_{-t} \leq P_{-s} \leq \cdots \leq P_s \leq P_t \leq P_1 = A.$$

*Proof.* By order reverting property of inversion, it is enough to show that $P_s \leq P_t$ for any $0 < s \leq t$. Let $\mathbb{A} = (A_1, \ldots, A_m) \in \mathcal{P}_n^m$ and $\omega = (w_1, \ldots, w_m) \in \Delta_m$.

By (i) of Corollary 2.1.5, we have

$$\left[ \sum_{i=1}^m w_i A_i \right]^r \leq \sum_{i=1}^m w_i A_i^r$$

for $r \in [1, 2]$.

Let $0 < s \leq t \leq 1$ and $X = P_t(\omega; \mathbb{A})$. Suppose that $t \leq 2s$. Putting $r = \frac{t}{s} \in [1, 2]$ and replacing $A_i$ into $A_i^s$ yield

$$\left[ \sum_{i=1}^m w_i A_i^s \right]^{\frac{t}{s}} \leq \sum_{i=1}^m w_i A_i^t.$$

By Definition 2.3.3, we have

$$I = \sum_{i=1}^m w_i (X^{-\frac{1}{2}} A_i X^{-\frac{1}{2}})^t \geq \left[ \sum_{i=1}^m w_i (X^{-\frac{1}{2}} A_i X^{-\frac{1}{2}})^s \right]^{\frac{t}{s}},$$

that is, $I \geq \sum_{i=1}^m w_i (X^{-\frac{1}{2}} A_i X^{-\frac{1}{2}})^s$. It then follows by (i) of Theorem 2.3.5 that

$$P_s(\omega; X^{-\frac{1}{2}} \mathbb{A} X^{-\frac{1}{2}}) \leq I$$

and by Proposition 2.3.4 (vii)

$$P_s(\omega; \mathbb{A}) \leq X = P_t(\omega; \mathbb{A}).$$

This shows that $P_s \leq P_t$ for all pair $(s, t)$ satisfying $0 < s \leq t \leq 1$ and $t \leq 2s$. Applying this procedure inductively yields that $P_s \leq P_t$ for all $0 < s \leq t \leq 1$. $P_{-t} \leq P_{-s}$ for all $0 < s \leq t \leq 1$ can be shown, easily. □

The monotone family of power means $\{P_t\}_{t \in [-1,1]}$ interpolates between the harmonic mean $H$ and the arithmetic mean $A$ with the "center" $P_0 := \lim_{t \to 0} P_t = \Lambda$.

**Corollary 2.4.4.** *The Karcher mean satisfies the following properties:*

(i) $\Lambda(\omega; \mathbb{A}) = A_1^{w_1} \cdots A_n^{w_m}$ *if $A_i$'s commute;*

(ii) $\Lambda(\omega; \mathbf{a} \cdot \mathbb{A}) = a_1^{w_1} \cdots a_m^{w_m} \Lambda(\omega; \mathbb{A})$;

(iii) $\Lambda(\omega_\sigma; \mathbb{A}_\sigma) = \Lambda(\omega; \mathbb{A})$;

(iv) *If $B_i \leq A_i$ for all $1 \leq i \leq m$, then $\Lambda(\omega; \mathbb{B}) \leq \Lambda(\omega; \mathbb{A})$;*

(v) $d(\Lambda(\omega; \mathbb{A}), \Lambda(\omega; \mathbb{B})) \leq \max_{1 \leq i \leq m}\{d(A_i, B_i)\}$;

(vi) $\Lambda(\omega; M^* \mathbb{A} M) = M^* \Lambda(\omega; \mathbb{A}) M$ *for any invertible $M \in \mathcal{M}_n$;*

(vii) $\Lambda(\omega; (1 - \lambda)\mathbb{A} + \lambda \mathbb{B}) \geq (1 - \lambda)\Lambda(\omega; \mathbb{A}) + \lambda\Lambda(\omega; \mathbb{B})$ *for $0 \leq \lambda \leq 1$;*

(viii) $\Lambda(\omega; \mathbb{A}^{-1})^{-1} = \Lambda(\omega; \mathbb{A})$;

(ix) $(\sum_{i=1}^m w_i A_i^{-1})^{-1} \leq \Lambda(\omega; \mathbb{A}) \leq \sum_{i=1}^m w_i A_i$.

*Proof.* By Proposition 2.3.4, Theorem 2.4.2 and (2.21), Corollary 2.4.4 is immediate. ⊔

The following results are natural extensions of Theorems 2.2.5 and 2.2.6.

**Theorem 2.4.5** (Yamazaki (2012)). *Let $\mathbb{A} = (A_1, ..., A_m) \in \mathcal{P}_n^m$ and $\omega = (w_1, ..., w_m) \in \Delta_m$. If $\sum_{i=1}^m w_i \log A_i \leq 0$, then $\Lambda(\omega; \mathbb{A}) \leq I$.*

**Theorem 2.4.6** (Ando-Hiai inequality Yamazaki (2012)). *Let $\mathbb{A} \in \mathcal{P}_n^m$ and $\omega \in \Delta_m$. If $\Lambda(\omega; \mathbb{A}) \leq I$, then $\Lambda(\omega; \mathbb{A}^p) \leq I$ holds for all $p \geq 1$.*

The proofs of Theorems 2.4.5 and 2.4.6 are easy by letting $t \to 0$ in Theorem 2.3.5, Corollary 2.3.6, and using (2.3) and Theorem 2.4.2.

Let $p_1, ..., p_m$ be positive numbers. For $i = 1, 2, ..., m$, $\prod_{j \neq i} p_j$ denotes $p_{\neq i}$.

**Theorem 2.4.7** (Yamazaki (2012)). *Let $\mathbb{A} = (A_1, ..., A_m) \in \mathcal{P}_n^m$. Then the following inequalities follow from each other.*

(i) $\sum_{i=1}^m \log A_i \leq 0$;

(ii) $\Lambda(\mathbb{A}^p) \leq I$ *for all $p > 0$;*

(iii) $\Lambda(\delta(\omega); A_1^{p_1}, ..., A_m^{p_m}) \leq I$ *for all $p_i > 0$, $i = 1, 2, ..., m$,*

*where $\omega = (p_{\neq 1}, ..., p_{\neq m})$.*

To prove Theorem 2.4.7, we need the following well-known formula: Let $\mathbb{A} = (A_1, ..., A_m) \in \mathcal{P}_n^m$. Then

$$\lim_{p \to 0} \left(\frac{1}{m} \sum_{i=1}^m A_i^p\right)^{\frac{1}{p}} = \exp\left(\frac{1}{m} \sum_{i=1}^m \log A_i\right). \tag{2.24}$$

*Proof of Theorem 2.4.7.* Proof of (i) $\implies$ (iii). If $\sum_{i=1}^{m} \log A_i \leq 0$, then we have

$$\prod_{j=1}^{m} p_j \left( \sum_{i=1}^{m} \log A_i \right) \leq 0,$$

i.e.,

$$\sum_{i=1}^{m} p_{\neq i} \log A_i^{p_i} \leq 0.$$

Hence by Theorem 2.4.5, we have $\Lambda(\delta(\omega); A_1^{p_1}, ..., A_m^{p_m}) \leq I$ for all $p_i > 0$, $i = 1, 2, ..., m$.

Proof of (iii) $\implies$ (ii) is easy by putting $p_1 = \cdots = p_m = p$.

Proof of (ii) $\implies$ (i). By Corollary 2.4.4 (ix), we have

$$I \geq \Lambda(\mathbb{A}^p) \geq \left( \frac{1}{m} \sum_{i=1}^{m} A_i^{-p} \right)^{-1}.$$

By (2.24), we have

$$I \geq \lim_{p \to 0} \left( \frac{1}{m} \sum_{i=1}^{m} A_i^{-p} \right)^{\frac{-1}{p}} = \left[ \exp \left( \frac{1}{m} \sum_{i=1}^{m} \log A_i^{-1} \right) \right]^{-1} = \exp \left( \frac{1}{m} \sum_{i=1}^{m} \log A_i \right).$$

Hence we have (i). $\qquad \square$

The properties of Theorems 2.4.5 and 2.4.6 characterize the Karcher mean as follows.

**Theorem 2.4.8** (Lim and Pálfia (2012); Yamazaki (2012))**.** *Let $\mathbb{A} \in \mathcal{P}_n^m$, $\omega \in \Delta_m$, and $G(\omega; \mathbb{A})$ be a weighted geometric mean satisfying properties of (vi) and (viii) in Corollary 2.4.4. If the weighted geometric mean satisfies Theorem 2.4.5, then $G(\omega; \mathbb{A})$ should be the Karcher mean.*

*Proof.* Let $\mathbb{A} = (A_1, ..., A_m)$ and $\omega = (w_1, ..., w_m)$. If $G(\omega; \mathbb{A})$ satisfies Theorem 2.4.5, we have

$$\sum_{i=1}^{m} w_i \log A_i \geq 0 \Longleftrightarrow \sum_{i=1}^{m} w_i \log A_i^{-1} \leq 0$$

$$\implies G(\omega; \mathbb{A}^{-1}) \leq I$$

$$\Longleftrightarrow G(\omega; \mathbb{A}) \geq I \quad \text{(by (viii))}.$$

Hence we obtain

$$\sum_{i=1}^{m} w_i \log A_i = 0 \quad \implies \quad G(\omega; \mathbb{A}) = I. \tag{2.25}$$

Let $X = \Lambda(\omega; \mathbb{A})$, i.e., the Karcher mean. Then by the Karcher equation, we have

$$\sum_{i=1}^{m} w_i \log X^{\frac{-1}{2}} A_i X^{\frac{-1}{2}} = 0,$$

and by (2.25) and Corollary 2.4.4 (vi),

$$G(\omega; X^{\frac{-1}{2}} \mathbb{A} X^{\frac{-1}{2}}) = I \iff G(\omega; \mathbb{A}) = X = \Lambda(\omega; \mathbb{A}).$$

This completes the proof. $\qquad\square$

**Theorem 2.4.9** (Lim and Pálfia (2012); Yamazaki (2012)). *Let $\mathbb{A} \in \mathcal{P}_n^m$, $\omega \in \Delta_m$, and $G(\omega; \mathbb{A})$ be a weighted geometric mean satisfying properties of (vi), (viii) and (ix) in Corollary 2.4.4. If the weighted geometric mean satisfies Theorem 2.4.6, then $G(\omega; \mathbb{A})$ should be the Karcher mean.*

*Proof.* Let $\mathbb{A} = (A_1, ..., A_m)$ and $\omega = (w_1, ..., w_m)$. If $\sum_{i=1}^{m} w_i \log A_i \leq 0$ is satisfied, then by (ix), we have

$$I \geq \sum_{i=1}^{m} w_i \left( I + \frac{\log A_i}{k} \right) \geq G \left( \omega; I + \frac{\log A_1}{k}, ..., I + \frac{\log A_m}{k} \right)$$

for sufficiently large $k$. Since the weighted geometric mean $G$ satisfies Theorem 2.4.6, we have

$$G \left( \omega; \left( I + \frac{\log A_1}{k} \right)^k, ..., \left( I + \frac{\log A_m}{k} \right)^k \right) \leq I.$$

By (2.8), we have $G(\omega; \mathbb{A}) \leq I$, i.e., $G(\omega; \mathbb{A})$ satisfies Theorem 2.4.5. Hence by Theorem 2.4.8, $G(\omega; \mathbb{A})$ should coincide with the Karcher mean. $\qquad\square$

In the rest of this chapter, we can extend Theorem 2.1.6 (the Furuta inequality) via the Karcher mean by the following argument. The next result is an extension of Theorem 2.4.6

**Theorem 2.4.10** (Ando-Hiai inequality Ito (2012)). *Let $\mathbb{A} = (A_1, ..., A_m) \in \mathcal{P}_n^m$ and $\omega = (w_1, ..., w_m) \in \Delta_m$. If $\Lambda(\omega; \mathbb{A}) \leq I$, then*

$$\Lambda(\delta(\omega'); A_1^{p_1}, ..., A_m^{p_m}) \leq \Lambda(\omega; \mathbb{A}) \leq I$$

*for $p_1, ..., p_m \geq 1$, where $\omega' = (\frac{w_1}{p_1}, ..., \frac{w_m}{p_m}) \in (0, \infty)^m$.*

*Proof.* Let $X = \Lambda(\omega; \mathbb{A}) \leq I$. Then for each $p_1, ..., p_m \in [1, 2]$, by (2.21) and (i) of Theorem 2.1.3, we have

$$0 = \sum_{i=1}^{m} w_i \log X^{\frac{1}{2}} A_i^{-1} X^{\frac{1}{2}} = \sum_{i=1}^{m} \frac{w_i}{p_i} \log(X^{\frac{1}{2}} A_i^{-1} X^{\frac{1}{2}})^{p_i} \leq \sum_{i=1}^{m} \frac{w_i}{p_i} \log X^{\frac{1}{2}} A_i^{-p_i} X^{\frac{1}{2}},$$

that is, $\sum_{i=1}^{m} \frac{w_i}{p_i} \log X^{\frac{-1}{2}} A_i^{p_i} X^{\frac{-1}{2}} \leq 0$. By applying Theorem 2.4.5,

$$\lambda(\delta(\omega'); X^{\frac{-1}{2}} A_1^{p_1} X^{\frac{-1}{2}}, ..., X^{\frac{-1}{2}} A_n^{p_m} X^{\frac{-1}{2}}) \leq I.$$

Therefore we have that

$$X \leq I \text{ implies } \Lambda(\delta(\omega'); A_1^{p_1}, ..., A_m^{p_m}) \leq X \leq I \text{ for } p_1, ..., p_m \in [1, 2]. \quad (2.26)$$

Put $Y = \Lambda(\delta(\omega'); A_1^{p_1}, ..., A_m^{p_m}) \leq I$. Then by (2.26), we get

$$\Lambda(\delta(\omega''); A_1^{p_1 p_1'}, ..., A_n^{p_m p_m'}) \leq Y \leq X \leq I$$

for $p_1', ..., p_m' \in [1, 2]$, where $\omega'' = (\frac{w_1}{p_1 p_1'}, ..., \frac{w_n}{p_m p_m'}) \in (0, \infty)$. Therefore, by putting $q_i = p_i p_i'$ for $i = 1, ..., n$, we have that

$$X \leq I \text{ implies } \Lambda(\delta(\omega''); A_1^{q_1}, ..., A_m^{q_m}) \leq X \leq I \text{ for } q_1, ..., q_m \in [1, 4]. \quad (2.27)$$

By repeating the same steps from (2.26) to (2.27), we have the conclusion. $\square$

**Theorem 2.4.11** (Ito (2012)). *Let* $\mathbb{A} = (A_1, ..., A_m) \in \mathcal{P}_n^m$ *and* $\omega = (w_1, ..., w_m) \in \Delta_m$. *For each* $i = 1, ..., n$ *and* $q \in \mathbb{R}$, *if*

$$\Lambda(\omega; A_1^{p_1}, ..., A_m^{p_m}) \leq A_i^q \text{ for } p_1, ..., p_m \in \mathbb{R} \text{ with } p_i > q,$$

*then*

$$\Lambda(\delta(\omega'); A_1^{p_1}, ..., A_{i-1}^{p_{i-1}}, A_i^{p_i'}, A_{i+1}^{p_{i+1}}, ..., A_m^{p_m})$$
$$\leq \Lambda(\omega; A_1^{p_1}, ..., A_{i-1}^{p_{i-1}}, A_i^{p_i}, A_{i+1}^{p_{i+1}}, ..., A_m^{p_m}) \leq A_i^q$$

*for* $p_i' \geq p_i$, *where* $\omega' = (w_1, ..., w_{i-1}, \frac{p_i - q}{p_i' - q} w_i, w_{i+1}, ..., w_m) \in (0, \infty)^m$.

*Proof.* We may assume $i = 1$ by Corollary 2.4.4 (iii).

For $p_1, ..., p_m \in \mathbb{R}$ with $p_1 \geq q$, $\Lambda(\omega; A_1^{p_1}, ..., A_m^{p_m}) \leq A_1^q$ if and only if

$$\Lambda(\omega; A_1^{p_1 - q}, A_1^{\frac{-q}{2}} A_2^{p_2} A_1^{\frac{-q}{2}}, ..., A_1^{\frac{-q}{2}} A_m^{p_m} A_1^{\frac{-q}{2}}) \leq I.$$

By applying Theorem 2.4.10,

$$\Lambda(\delta(\omega'); A_1^{p_1' - q}, A_1^{\frac{-q}{2}} A_2^{p_2} A_1^{\frac{-q}{2}}, ..., A_1^{\frac{-q}{2}} A_m^{p_m} A_1^{\frac{-q}{2}})$$
$$\leq \Lambda(\delta(\omega'); A_1^{p_1 - q}, A_1^{\frac{-q}{2}} A_2^{p_2} A_1^{\frac{-q}{2}}, ..., A_1^{\frac{-q}{2}} A_m^{p_m} A_1^{\frac{-q}{2}}) \leq I$$

holds for $\frac{p_1' - q}{p_1 - q} \geq 1$, where $\omega' = (\frac{p_1' - q}{p_1 - q} w_1, w_2, ..., w_m) \in (0, \infty)^m$. Therefore

$$\Lambda(\delta(\omega'); A_1^{p_1'}, A_2^{p_2}, ..., A_m^{p_m}) \leq \Lambda(\delta(\omega'); A_1^{p_1}, A_2^{p_2}, ..., A_m^{p_m}) \leq A_1^q$$

holds for $p_1' \geq p_1$. $\square$

**Theorem 2.4.12** (Ito (2012)). *Let* $\mathbb{A} = (A_1, ..., A_m) \in \mathcal{P}_n^m$ *and* $w_1, ..., w_m > 0$. *If*

$$A_i^{q_i} \geq A_n^{q_m} > 0 \tag{2.28}$$

*and*

$$\sum_{i=1}^{m} \frac{w_i}{p_i - q_i} \log A_m^{\frac{-q_n}{2}} A_i^{p_i} A_m^{\frac{-q_m}{2}} \leq 0 \tag{2.29}$$

*hold for* $q_i \in \mathbb{R}$, $p_i > q_i$ *and* $i = 1, ..., n$, *then*

$$\Lambda(\delta(\omega'); A_1^{p_1'}, ..., A_m^{p_m'}) \leq \Lambda(\delta(\omega); A_1^{p_1}, ..., A_m^{p_m}) \leq A_m^{q_m}$$

*for all* $p_i' \geq p_i$ *and* $i = 1, ..., n$, *where* $\omega = (\frac{w_1}{p_1 - q_1}, ..., \frac{w_m}{p_m - q_m}) \in (0, \infty)^m$ *and* $\omega' = (\frac{w_1}{p_1' - q_1}, ..., \frac{w_m}{p_m' - q_m}) \in (0, \infty)^m$.

*Proof.* Applying Theorem 2.4.5 to (2.29), we have

$$\Lambda(\delta(\omega); A_m^{\frac{-q_m}{2}} A_1 A_m^{\frac{-q_m}{2}}, ..., A_m^{\frac{-q_m}{2}} A_{n-1} A_m^{\frac{-q_m}{2}}, A_m^{p_m - q_m}) \leq I,$$

so that by (2.28),

$$X_0 := \Lambda(\delta(\omega); A_1^{p_1}, ..., A_{m-1}^{p_{m-1}}, A_m^{p_m}) \leq A_m^{q_m} \leq A_1^{q_1}. \tag{2.30}$$

By applying Theorem 2.4.11 to (2.30) and by (2.28),

$$X_1 := \Lambda(\delta(\omega_1); A_1^{p_1'}, A_2^{p_2}, ..., A_m^{p_m}) \leq X_0 \leq A_m^{q_m} \leq A_2^{q_2}. \tag{2.31}$$

*for* $p_1' \geq p_1$, *where* $\omega_1 = (\frac{w_1}{p_1' - q_1}, \frac{w_2}{p_2 - q_2}, ..., \frac{w_m}{p_m - q_m}) \in (0, \infty)^m$. By applying Theorem 2.4.11 to (2.31) and by (2.28),

$$X_2 := \Lambda(\delta(\omega_2); A_1^{p_1'}, A_2^{p_2'}, A_3^{p_3}, ..., A_m^{p_m}) \leq X_1 \leq X_0 \leq A_m^{q_m} \leq A_3^{q_3}$$

*for* $p_1 \geq p_1$ *and* $p_2' \geq p_2$, *where* $\omega_2 = (\frac{w_1}{p_1' - q_1}, \frac{w_2}{p_2' - q_2}, \frac{w_3}{p_3 - q_2}, ..., \frac{w_m}{p_m - q_m}) \in (0, \infty)^m$. By repeating this argument, we can get

$$X_m := \Lambda(\delta(\omega'); A_1^{p_1'}, ..., A_m^{p_m'}) \leq X_{m-1} \leq \cdots \leq X_0 \leq A_m^{q_m}$$

*for* $p_i' \geq p_i$ *for* $i = 1, ..., n$, *where* $\omega' = (\frac{w_1}{p_1' - q_1}, ..., \frac{w_m}{p_m' - q_m}) \in (0, \infty)^m$. $\square$

**Theorem 2.4.13** (Furuta inequality Ito (2012)). *Let* $\mathbb{A} = (A_1, ..., A_m) \in \mathcal{P}_n^m$ *and* $q > 0$. *Then* $A_i^q \geq A_m > 0$ *for* $i = 1, ..., m - 1$ *implies*

$$\Lambda(\delta(\omega); A_1^{-p_1}, ..., A_{m-1}^{p_{m-1}}, A_m^{p_m}) \leq A_m^q \leq A_i^q \tag{2.32}$$

*for all* $p_i \geq 0$, $i = 1, ..., m - 1$ *and* $p_m > q$, *where* $\omega = (\frac{1}{p_1 + q}, ..., \frac{1}{p_{m-1} + q}, \frac{m-1}{p_m - q}) \in (0, \infty)^m$.

*Proof.* Assume that $A_i^q \geq A_m^q > 0$ for $q > 0$ and $i = 1, ..., m - 1$. Then $A_i^q \geq A_m^q > 0$ implies $\log A_i \geq \log A_m$. By (i) $\implies$ (iii) in Theorem 2.2.5. $\log A_i \geq \log A_m$ implies $A_i^{-p_i} \natural_{\frac{p_i}{q+p_i}} A_m^q \leq I$ for all $p_i \geq 0$. This is equivalent to $A_m^{-q} \natural_{\frac{q}{q+p_i}} A_i^{p_i} \geq I$, that is, $(A_m^{\frac{q}{2}} A_i^{p_i} A_m^{\frac{q}{2}})^{\frac{q}{p_i+q}} \geq A_m^q$. By taking the logarithm, we have $\frac{1}{p_i+q} \log A_m^{\frac{q}{2}} A_i A_m^{\frac{q}{2}} \geq \frac{1}{p_m-q} \log A_m^{p_m-q}$, that is,

$$\frac{1}{p_i + q} \log A_m^{\frac{-q}{2}} (A_i^{-1})^{p_i} A_m^{\frac{-q}{2}} + \frac{1}{p_m - q} \log A_m^{p_m - q} \leq 0 \qquad (2.33)$$

for all $p_i \geq 0$, $i = 1, ..., m-1$ and $p_m > q$. Summing up (2.33) for $i = 1, .., m-1$, we have

$$\sum_{i=1}^{m-1} \frac{1}{p_i + q} \log A_m^{\frac{-q}{2}} (A_i^{-1})^{p_i} A_m^{\frac{-q}{2}} + \frac{m-1}{p_m - q} \log A_m^{p_m - 1} \leq 0. \qquad (2.34)$$

By applying Theorem 2.4.12 to $(A_i^{-1})^{-q} \geq A_n^q > 0$ and (2.34), we can obtain

$$\Lambda(\omega; A_1^{-p_1}, ..., A_{m-1}^{-p_{m-1}}, A_m^{p_m}) \leq A_m^q \leq A_i^q$$

for all $p_i \geq 0 > -q$, $i = 1, ..., m-1$ and $p_m > q$. $\qquad \square$

*Alternative proof of Theorem 2.1.6.* Put $m = 2$, $p_1 = r$ and $p_2 = p$ in Theorem 2.4.13. Then since $\omega = (\frac{1}{r+q}, \frac{1}{p-q}) \in (0, \infty)^2$, $\delta(\omega) = (\frac{p-q}{p+r}, \frac{q+r}{p+r})$. Therefore we obtain the desired result. $\qquad \square$

**Notes and References.** The aim of this chapter is to introduce the Furuta inequality and its extension via the Karcher mean. So we mainly introduce the Furuta inequality and geometric means of two matrices in the first and second sections. But we have more general results for some theorems in these sections, for example, Theorems 2.1.1, 2.1.3, Corollary 2.1.5. These generalizations are introduced in Bhatia (1996, 2007); Hiai and Petz (2014). There are a lot of papers for the Furuta inequality Fujii (1990); Fujii and Kamei (2006); Furuta (1989, 1995, 2001); Kamei (1988). Through the series of the papers, we have a lot of ideas for studying the Karcher mean, for example, Theorems 2.4.7 and 2.4.10. However, we have not obtained the Furuta inequality for the power mean, yet. The generalization of Furuta inequality has been shown in Furuta (1995), and an extension of this via the Karcher mean has been shown in Ito (2014).

It was a long-standing problem to extend the geometric mean of two matrices to more than three matrices. For the problem, Ando, Li and Mathias provided a definition of a geometric mean of $m$-matrices in Ando et al. (2004) that satisfied Corollary 2.4.4. Before that, some papers defined geometric means of $m$-matrices but none satisfied some important properties in Corollary 2.4.4, for example, (iii) and (iv). Since Ando et al. (2004) appeared, many researchers have studied to give a better definition of the geometric mean of $m$-matrices in Bini et al. (2010); Izumino and Nakamura (2009); Jung et al. (2010); Lee et al. (2011); Yamazaki (2006). But since Theorems 2.4.5, 2.4.6 and 2.4.7 (Yamazaki (2012)), the Karcher mean can be considered as a natural extension of the geometric mean of two matrices. Then a lot of papers that discuss the Karcher mean appeared. Especially, the monotonicity property of the Karcher mean gives us another characterization of the Karcher mean in Lawson and Lim (2011). This characterization gives us an algorithm for computing the Karcher mean Holbrook (2012); Lim and Pálfia (2014); Sturm (2003). There are a lot of papers that discuss power and Karcher means Bhatia and Karandikar (2012); Lawson and Lim (2014); Yamazaki (2013).

Now, the Karcher mean has been generalized. We can consider any matrix means of $m$-matrices via the generalized Karcher equation Pálfia (2016). Moreover, we can generalize the Karcher equation by using integration of matrix connections instead of summation in Pálfia (2016).

---

# Bibliography

Ando T. On some operator inequalities. *Math. Ann.*, 279:157–159, 1987.

Ando T. and Hiai F. Log majorization and complementary Golden-Thompson type inequalities. *Linear Algebra Appl.*, 197/198:113–131, 1994.

Ando T., Li C.-K., and Mathias R. Geometric means. *Linear Algebra Appl.*, 385:305–334, 2004.

Bhatia R. *Matrix Analysis, Graduate Texts in Mathematics.* Springer, 1996.

Bhatia R. *Positive Definite Matrices, Princeton Series in Applied Mathematics.* Princeton University Press, Princeton, NJ, 2007.

Bhatia R. and Holbrook J. Riemannian geometry and matrix geometric means. *Linear Algebra Appl.*, 413:594–618, 2006.

Bhatia R. and Karandikar R. Monotonicity of the matrix geometric mean. *Math. Ann.*, 353:1453–1467, 2012.

Bini D. A., Meini B., and Poloni F. An effective matrix geometric mean satisfying the Ando-Li-Mathias properties. *Math. Comp.*, 79:437–452, 2010.

Fujii M. Furuta's inequality and its mean theoretic approach. *J. Operator Theory*, 23:67–72, 1990.

Fujii M., Furuta T., and Kamei E. Furuta's inequality and its application to Ando's theorem. *Linear Algebra Appl.*, 179:161–169, 1993.

Fujii M. and Kamei E. Ando-Hiai inequality and Furuta inequality. *Linear Algebra Appl.*, 416:541–545, 2006.

Furuta T. $A \geq B \geq 0$ assures $(B^r A^p B^r)^{1/q} \geq B^{(p+2r)/q}$ for $r \geq 0$, $p \geq 0$, $q \geq 1$ with $(1 + 2r)q \geq p + 2r$. *Proc. Amer. Math. Soc.*, 101:85–88, 1987.

Furuta T. An elementary proof of an order preserving inequality. *Proc. Japan Acad. Ser. A Math. Sci.*, 65:126, 1989.

Furuta T. Applications of order preserving operator inequalities. *Oper. Theory Adv. Appl.*, 59:180–190, 1992.

Furuta T. Extension of the Furuta inequality and Ando-Hiai log-majorization. *Linear Algebra Appl.*, 219:139–155, 1995.

Furuta T. *Invitation to Linear Operators.* Taylor & Francis, London, 2001.

Hansen F. An operator inequality. *Math. Ann.*, 246:249–250, 1979/80.

Hansen F. The fast track to Löwner's theorem. *Linear Algebra Appl.*, 438:4557–4571, 2013.

Heinz E. Beiträge zur Störungstheorie der Spektralzerlegung. *Math. Ann.*, 123:415–438, 1951.

Hiai F. and Petz D. *Introduction to matrix analysis and applications, Universitext.* Springer, New Delhi, 2014.

Holbrook J. No dice: a deterministic approach to the Cartan centroid. *J. Ramanujan Math. Soc.*, 27:509–521, 2012.

Ito M. Matrix inequalities including Furuta inequality via Riemannian mean of $n$-matrices. *J. Math. Inequal.*, 6:481–491, 2012.

Ito M. Matrix inequalities including grand Furuta inequality via Karcher mean. *J. Math. Inequal.*, 8:279–285, 2014.

Izumino S. and Nakamura N. Geometric means of positive operators II. *Sci. Math. Jpn.*, 69:35–44, 2009.

Jung C., Lee H., Lim Y., and Yamazaki T. Weighted geometric mean of $n$-operators with $n$ parameters. *Linear Algebra Appl.*, 432:1515–1530, 2010.

Kamei E. A satellite to Furuta's inequality. *Math. Japon.*, 33:883–886, 1988.

Kubo F. and Ando T. Means of positive linear operators. *Math. Ann.*, 246:205–224, 1979/80.

Lawson J. and Lim Y. Monotonic properties of the least squares mean. *Math. Ann.*, 351:267–279, 2011.

Lawson J. and Lim Y. Karcher means and Karcher equations of positive definite operators. *Trans. Amer. Math. Soc. Ser. B*, 1:1–22, 2014.

Lee H., Lim Y., and Yamazaki T. Multi-variable weighted geometric means of positive definite matrices. *Linear Algebra Appl.*, 435:307–322, 2011.

Lim Y. and Pálfia M. The matrix power means and the Karcher mean. *J. Funct. Anal.*, 262:1498–1514, 2012.

Lim Y. and Pálfia M. Weighted deterministic walks for the squares mean on Hadamard spaces. *Bull. Lond. Math. Soc.*, 46:561–570, 2014.

Lim Y. and Yamazaki T. On some inequalities for the matrix power and Karcher means. *Linear Algebra Appl.*, 438:1293–1304, 2013.

Loewner K. Über monotone Matrixfunktionen. *Math. Z.*, 38:177–216, 1934.

Moakher M. A differential geometric approach to the geometric mean of symmetric positive-definite matrices. *SIAM J. Matrix Anal. Appl.*, 26:735–747, 2005.

Pálfia M. Operator means of probability measures and generalized Karcher equations. *Adv. Math.*, 289:951–1007, 2016.

Sturm K.-T. Probability measures on metric spaces of nonpositive curvature. *Contemp. Math.*, 338:357–390, 2003.

Tanahashi T. Best possibility of the Furuta inequality. *Proc. Amer. Math. Soc.*, 124:141–146, 1996.

Thompson A. On certain contraction mappings in a partially ordered vector space. *Proc. Amer. Math. Soc.*, 14:438–443, 1963.

Uchiyama M. Some exponential operator inequalities. *Math. Inequal. Appl.*, 2:469–471, 1999.

Wada S. Some ways of constructing Furuta-type inequalities. *Linear Algebra Appl.*, 457:276–286, 2014.

Yamazaki T. An extension of Kantorovich inequality to *n*-operators via the geometric mean by Ando-Li-Mathias. *Linear Algebra Appl.*, 416:688–695, 2006.

Yamazaki T. The Riemannian mean and matrix inequalities related to the Ando-Hiai inequality and Chaotic order. *Oper. Matrices*, 6:577–588, 2012.

Yamazaki T. An elementary proof of arithmetic-geometric mean inequality of the weighted Riemannian mean of positive definite matrices. *Linear Algebra Appl.*, 438:1564–1569, 2013.

# Chapter 3

## Functional Equations on Affine Groups

László Székelyhidi

*Institute of Mathematics, University of Debrecen – Department of Mathematics and statistical Sciences, BIUST, Botswana*

## 3.1 Introduction

Spectral analysis and spectral synthesis deal with the description of translation invariant function spaces on topological groups. The basic building blocks of this description are the exponential functions and exponential monomials. The classical theorem of Laurent Schwartz completely explains the situation on the reals (see Schwartz (1947)): in every translation invariant linear space of complex valued continuous functions on the real line, which is also closed with respect to uniform convergence on compact sets the exponential monomials of the form $x \mapsto x^n e^{\lambda x}$ ($n$ is a natural number, $\lambda$ is a complex number) span a dense subspace. This property is called *spectral synthesis* of the function space in question. In the past years several new results have been published on this subject on discrete Abelian groups. For a detailed discussion of discrete spectral synthesis see the monograph Székelyhidi (2006) and also the research papers Lefranc (1958); Laczkovich and Székelyhidi (2005, 2007); Székelyhidi and Wilkens (2016, 2017); Székelyhidi (2018).

The case of non-discrete Abelian groups is more sophisticated. In fact, some counterexamples of D. I. Gurevich in Gurevič (1975) show that a direct extension of Schwartz's result is not possible even for functions in several real variables. To overcome this difficulty in Székelyhidi (2017) we proposed a modification of the classical spectral synthesis setting: instead of translation invariance we consider invariance with respect to some group of linear operators. On $\mathbb{R}^n$ we can take the affine group of $SO(n)$: the group of proper Euclidean motions which coincides with the the group of translations in the case $n = 1$. In this way we arrive at a proper extension of Schwartz's result for functions in several variables. In addition, this approach provides a starting point for non-commutative generalizations of spectral analysis and spectral synthesis: spherical spectral analysis and synthesis on affine groups. In this situation we need to study some new function classes which are the substitutes of exponential functions and exponential monomials. It turns out that on classical linear groups spherical functions and other special functions come into the picture together with new classes of functional equations (see Székelyhidi (2017)). The purpose of this work is to provide a survey on some problems and their solutions related to these questions: the study of functional equations on affine groups.

## 3.2   Semidirect products

Given a normed space $V$ isometries are fundamental from the point of view of the geometry of the space. By Theorem 11 on p. 144 in Aczél and Dhombres (1989), every surjective isometry between two real normed spaces is an *affine isometry*, i.e. the composition of a homogeneous linear isometry and of a translation. Affine isometries obviously form a group with respect to composition. Indeed, let

$$\varphi(x) = k \cdot x + u, \quad \psi(x) = l \cdot x + v,$$

where $k, l$ are linear isometries of $V$ and $u, v$ are vectors in $V$. Then the composition of $\varphi$ and $\psi$ has the form

$$\varphi \circ \psi(x) = (k \circ l) \cdot x + (k \cdot v + u),$$

hence it is an affine isometry, too. Clearly, the identity mapping

$$\iota(x) = id \cdot x + 0,$$

and the inverse of $\varphi$

$$\varphi^{-1}(x) = k^{-1} \cdot x - k^{-1} \cdot u$$

are affine isometries, too. More generally, all nonzero affine mappings of the form

$$\Phi(x) = k \cdot x + u$$

form a group with respect to composition, if $k$ is a linear automorphism of the vector space $V$ and $u$ is an element of $V$. We can generalize this construction in the following way. Let $N$ be a group and $H$ a subgroup of the automorphism group Aut $N$ of $N$. We introduce the dot operation $\cdot$ on $H \times N$ in the following manner:

$$(h, n) \cdot (h', n') = (h \circ h', (h \cdot n')n),$$

where $h, h'$ are in $H$ and $n, n'$ are in $N$. Clearly, this operation imitates the composition of the affine mappings

$$\varphi(x) = (h \cdot x)n, \quad \text{and} \quad \psi(x) = (h' \cdot x)n',$$

where $\circ$ stands for the composition of automorphisms, $\cdot$ denotes the action of automorphisms on elements of $N$, and juxtaposition means the operation in $N$. The composition of the two mappings $\varphi, \psi$ is given by the pair $(h \circ h', (h \cdot n')n)$. The identity mapping is described by $(id, e)$, where $id$ denotes the identity automorphism in $H$, and $e$ is the identity element in $N$, and the inverse of the mapping $\varphi$ is described by

$$(h, n)^{-1} = (h^{-1}, h^{-1}(n^{-1})).$$

Here we have to point out the exact meaning of $h^{-1}(n^{-1})$: indeed, $h^{-1}$ denotes the inverse mapping of $h$, which is an automorphism as well, hence we have

$$h^{-1}(n^{-1}) = [h^{-1}(n)]^{-1}$$

which is different from $h(n)$: the inner $^{-1}$ refers to the inverse mapping of $h$, and the outer $^{-1}$ refers to the inverse element of $h^{-1}(n)$ in the group $N$.

In this way we equip $H \times N$ with a group operation and this group will be called the *semidirect product* of $H$ and $N$, denoted by $H \ltimes N$.

It is easy to see that the set $\{(h, e) : h \in H\}$ is a subgroup in $H \ltimes N$ and the mapping $h \mapsto (h, e)$ is an isomorphism of $H$ onto this subgroup. We shall identify this subgroup with $H$ and we write $h$ instead of $(h, e)$. Similarly, it is easy to see that the set $\{(id, n) : n \in N\}$ is a normal subgroup in $H \ltimes N$ and the mapping $n \mapsto (id, n)$ is an isomorphism of $N$ onto this subgroup. Also, we shall identify this subgroup with $N$ and we write $n$ instead of $(id, n)$.

A special case of this semidirect product can be obtained if the automorphisms in $H$ are inner automorphisms, i.e. *conjugations*. More exactly, we may have the following situation: let $G$ be a group, and let $I, N$ be subgroups of $G$ with the property that $N$ is a normal subgroup, $I \cap N = \{e\}$, the identity element, and $G = I N$, i.e. every element of $G$ can be represented in the form of $h n$ with $h$ in $I$ and $n$ in $N$. In this case this representation is unique, as it is easy to see. Now we consider the set $H$ of inner automorphisms of $G$ corresponding to the elements of $I$:

$$H = \{h_i : i \in I, h_i(g) = i^{-1}gi \text{ for } g \in G\}.$$

Then the semidirect product $H \ltimes N$ is called the semidirect product of $I$ and $N$ and we write $G = I \ltimes N$. Hence in this case we have

$$(i, n) \cdot (i', n') = (i\, i', i^{-1}n'in),$$

or

$$i\, n \cdot i'\, n' = i\, i'\, i^{-1}n'in$$

as the new group operation in $G$. In particular, if the elements of $I$ and $N$ commute, then we have $(i, n) \cdot (i', n') = (i\, i', i^{-1}n'in) = (i\, i', n'\, n)$, hence the semidirect product of $I$ and $N$ is (isomorphic to) the direct product of $I$ and $N$. This consideration shows that the semidirect product is a kind of generalization of the direct product. Hence, in any case if $H$ and $N$ are subgroups of a group $G$ in which $N$ is normal, the semidirect product of $H$ and $N$ will be understood in this way, considering the elements of $H$ as inner automorphisms of $N$. We underline that the normality of $N$ is necessary for its invariance under the inner automorphisms corresponding to the elements of $H$. On the other hand, if both $H$ and $N$ are normal subgroups, then necessarily the elements of $H$ and $N$ commute, hence in this case the semidirect product reduces to the direct product.

We note that the mapping $(h, n) \mapsto h$ is a surjective homomorphism of $H \ltimes N$ onto $H$ with kernel $N$, hence $H \ltimes N / N \cong H$.

For instance, the *Euclidean group* of all rigid motions (isometries) of the plane $\mathbb{R}^2$ is isomorphic to the semidirect product of the Abelian group $\mathbb{R}^2$ and the orthogonal group $O(2)$. Here $\mathbb{R}^2$ describes translations by elements of the plane and $O(2)$ describes rotations and reflections, i.e. isometries which leave the origin fixed.

---

## 3.3 Linear groups

Let $V$ be an $n$-dimensional $\mathbb{K}$-normed space, where $\mathbb{K}$ denotes either the reals, or the complex numbers: $\mathbb{K} = \mathbb{R}$ or $\mathbb{C}$. We denote by $GL(V)$ the *general linear group* of $V$, that is, the group of all invertible linear operators on $V$. Using the standard basis in $\mathbb{K}$ we always identify $GL(V)$ with the corresponding matrix group. The matrix of a linear operator $k$ will be denoted by $[k]$. Also, identifying $GL(V)$ with a subset of $\mathbb{K}^{n^2}$, we equip it with the Euclidean topology, hence it is considered as a locally compact topological group. In fact, $GL(V)$ is a Lie group, and so are their closed subgroups, hence they also have a differentiable manifold structure.

We let $GL(V)$ act on functions $f : V \to \mathbb{C}$ defined on $V$ in the following manner: we consider the elements of $V$ as row vectors and $n \times n$ matrices from

$GL(V)$ act on such vectors as *right multiplication* which results in the rule

$$k \cdot f(v) = f(v \cdot [k])$$

for each $f : V \to \mathbb{C}$, $k$ in $GL(V)$ and $v$ in $V$. Given a subset $H \subseteq GL(V)$ we call $f$ $H$-*invariant* if $k \cdot f = f$ holds for each $k$ in $H$.

The linear groups are closed subgroups of $GL(V)$. One of them is the *special linear group* $SL(V)$, the set of all transformations in $GL(V)$ with determinant $+1$. By simple linear algebra, it is easy to see that this determinant condition is independent of the choice of the basis in $V$.

The *orthogonal group* $O(n)$, or the *rotation group* is the group of all isometries of the real vector space $V$ leaving the origin fixed. In terms of matrices $O(n)$ is the group of all $n \times n$ matrices with real entries and with determinant $\pm 1$. It is easy to see that $O(n)$ acts on the unit sphere of $V$ *transitively*: if $u, v$ are on the unit sphere in $V$, then there is a $k$ in $O(n)$ such that $k \cdot u = v$. It follows that $O(n)$-invariant functions are exactly the *radial* ones: $f$ is radial if it depends on the norm only, i.e. there is a function $\varphi : \mathbb{R} \to \mathbb{C}$ such that $f(u) = \varphi(\|u\|)$ holds for each $u$ in $V$. Obviously, the same holds for the *special orthogonal group* $SO(n)$ which is the subgroup of $O(n)$ consisting of *proper rotations*: those with determinant $+1$. Clearly, $SO(n) = SL(V) \cap O(n)$ and $SO(n)$ also acts transitively on the unit sphere in $V$, hence $SO(n)$-invariant functions are also the radial ones.

The orthogonal transformations can be characterized by the property that they leave the bilinear form

$$B(u, v) = \sum_{j=1}^{n} u_j v_j$$

fixed: $B(k \cdot u, k \cdot v) = B(u, v)$. We define the *generalized orthogonal group* as the group $O(p, q)$ of those linear transformations which leave the *indefinite bilinear form*

$$B(u, v) = \sum_{j=1}^{p} u_j v_j - \sum_{j=p+1}^{p+q} u_j v_j$$

with signature $(p, q)$ fixed. Here $p, q$ are natural numbers with $p + q = n$. Clearly, $O(p, 0)$ and $O(0, q)$ are isomorphic to $O(n)$. An important special case is the *Lorentz group* $O(1, 3)$: the group of those linear isometries of the *Minkowski spacetime* which leave the origin fixed. The Minkowski spacetime is the vector space $\mathbb{R} \oplus \mathbb{R}^3$ equipped with the indefinite inner product

$$B(u, v) = u_1 v_1 - u_2 v_2 - u_3 v_3 - u_4 v_4.$$

Traditionally we write $u = (t, x, y, z)$ and the indefinite quadratic form corresponding to the above inner product is

$$Q(t, x, y, z) = t^2 - x^2 - y^2 - z^2,$$

the *metric tensor* of the Minkowski spacetime.

The *proper Lorentz group* $SO(1,3)$ is obtained by collecting all elements in $O(1,3)$ with determinant $+1$.

Clearly, the matrices of the linear operators in the orthogonal group $O(n)$ can be characterized by the property that $M^t I_n M = I_n$, where $M^t$ is the transpose of $M$ and $I_n$ is the $n \times n$ identity matrix. The characterization of the matrices corresponding to elements of $O(p,q)$ is the following generalization: let $I_{p,q}$ denote the diagonal matrix with diagonal elements $+1$ in the first $p$ rows while the other $q$ elements are equal to $-1$. More visually,

$$I_{p,q} = \begin{pmatrix} 1 & 0 & 0 & \cdots & 0 & 0 & 0 & 0 \\ 0 & 1 & 0 & \cdots & 0 & 0 & 0 & 0 \\ \cdot & \cdot & \cdot & \cdots & \cdot & \cdot & \cdot & \cdot \\ \cdot & \cdot & \cdot & \cdots & \cdot & \cdot & \cdot & \cdot \\ 0 & 0 & 0 & \cdots & 1 & 0 & 0 & 0 \\ 0 & 0 & 0 & \cdots & 0 & -1 & 0 & 0 \\ \cdot & \cdot & \cdot & \cdots & 0 & 0 & -1 & 0 \\ \cdot & \cdot & \cdot & \cdots & \cdot & \cdot & \cdot & \cdot \\ 0 & 0 & 0 & \cdots 0 & 0 & 0 & 0 & -1 \end{pmatrix}.$$

Then the $n \times n$ matrix $M$ is the matrix of an element in $O(p,q)$ if and only if

$$M^t I_{p,q} M = I_{p,q}$$

holds.

We can generalize this setting as follows. Let $B : V \times V \to \mathbb{K}$ be a bilinear form and let $O(B)$ denote the set of all linear mappings $k$ in $GL(V)$ such that $B(k \cdot u, k \cdot v) = B(u,v)$ holds for each $u, v$ in $V$. Let $[B]$ denote the matrix of $B$. Hence the entries of $[B]$ are defined by

$$B(e_i, e_j) = \sum_{i=1}^{n} \sum_{j=1}^{n} [B]_{i,j} e_i e_j,$$

where $\{e_1, e_2, \ldots, e_n\}$ is the standard basis of $V$. Now the necessary and sufficient condition for the linear transformation $k$ in $GL(V)$ to be in $O(B)$ is

$$[k]^t [B][k] = [B],$$

where $[k]$ denotes the matrix of $k$.

Now let $V$ be a $2n$ dimensional $\mathbb{K}$-vector space and we define the block matrix

$$J_{2n} = \begin{pmatrix} 0 & I_n \\ -I_n & 0 \end{pmatrix},$$

where $I_n$ is the $n \times n$ identity matrix. The *symplectic group* $Sp(n)$ of rank $n$ is the set of all linear operators $k$ in $GL(V)$ such that their matrix satisfies

$[k]^t J_{2n}[k] = J_{2n}$. In other words, the linear operators in $Sp(n)$ are exactly those in $GL(V)$ which leave the bilinear form

$$J(u, v) = \sum_{j=1}^{n} (u_j v_{n+j} - u_{n+j} v_j)$$

fixed.

In the previous examples $V$ was a real vector space. However, most of these linear groups have their counterpart in the complex case. Let now $V$ be a complex vector space. The *unitary group* $U(n)$ is the group of all linear transformations $k$ in $GL(V)$ whose matrix is unitary: $[k]^*[k] = I_n$, where $[k]^*$ is the *adjoint* of $[k]$: the transpose of its complex conjugate. The *special unitary group* $SU(n)$ is the set of those unitary transformations with determinant $+1$, i.e. $SU(n) = SL(V) \cap U(n)$.

Obviously, the transformations in $U(n)$ are exactly those leaving the *Hermitian bilinear form*

$$B(u, v) = \sum_{j=1}^{n} u_j \bar{v}_j$$

fixed. The generalization of $U(n)$ to the *generalized unitary group* $U(p, q)$ is straightforward using the indefinite Hermitian form with signature $(p, q)$

$$B(u, v) = \sum_{j=1}^{p} u_j \bar{v}_j - \sum_{j=1}^{q} u_{p+j} \bar{v}_{p+j},$$

where $p + q = n$.

---

## 3.4 Affine groups

The affine groups we shall consider are related to the different linear groups listed above. Roughly speaking, the linear groups are *homogeneous transformation groups*, acting on the finite dimensional vector space $V$ and leaving the origin fixed. On the other hand, if we include translations then we arrive at affine transformations, which form *inhomogeneous transformation groups*.

The affine groups are all closed subgroups of the following semidirect product:

$$\text{Aff}\, GL(V) = GL(V) \ltimes V.$$

Here $V$ is equipped with the locally compact Euclidean, or unitary topology depending on whether $\mathbb{K} = \mathbb{R}$ or $\mathbb{C}$, and $GL(V)$ is considered as a subset of $\mathbb{R}^{n^2}$ or $\mathbb{C}^{n^2}$, respectively. Clearly, the product $GL(V) \times V$ is equipped with the locally compact product topology, hence the semidirect product $\text{Aff}\, GL(V)$ is a locally compact topological group – in fact, it is a Lie group. We recall that the elements $(k, u)$ of Aff $GL(V)$ can be considered as affine mappings on $V$ acting by the rule

$$(k, u) \cdot x = k \cdot x + u$$

for each $k$ in $GL(V)$ and $u, x$ in $V$. The operation in Aff $GL(V)$ corresponds to the composition of these affine mappings:

$$(k, u) \cdot (l, v) = (k \circ l, k \cdot v + u),$$

while the identity is $(id, 0)$, and the inverse of $(k, u)$ is given by $(k^{-1}, -k^{-1} \cdot u)$.

If we take a closed subgroup $H$ of $GL(V)$ then the semidirect product $H \ltimes V$ is called the *affine group of $H$ on $V$*, denoted by Aff $H$. For instance, in the trivial case $H = \{id\}$ the affine group Aff $H$ is isomorphic to $V$, this is the canonical representation of $V$ on itself by translations. On the other hand, in the case $H = SL(V)$ we obtain the *special affine group*: Aff $SL(V) = SL(V) \ltimes V$.

In the subsequent paragraphs we shall consider the two standard cases $V = \mathbb{R}^n$ and $V = \mathbb{C}^n$. First we suppose that $V = \mathbb{R}^n$ and we let $H = O(n)$. The resulting affine group Aff $O(n) = O(n) \ltimes \mathbb{R}^n$ is the group of *rigid Euclidean motions*. The elements of $O(n)$ represent rotations and the elements of $\mathbb{R}^n$ represent translations. As we have seen above $O(n)$ and $\mathbb{R}^n$ are embedded in this affine group: $O(n)$ is isomorphic to the subgroup $\{(k, 0) : k \in O(n)\}$ and $\mathbb{R}^n$ is isomorphic to the normal subgroup $\{(id, u) : u \in \mathbb{R}^n\}$. In fact, these isomorphisms are topological in the sense that they are homeomorphisms with respect to the topological group structures. In particular, $O(n)$ is isomorphic to a compact subgroup of Aff $O(n)$. Clearly, $O(n)$ is disconnected, e.g. $O(1) \cong \mathbb{Z}_2$, the two element cyclic group.

The next example is the affine group of $SO(n)$ on $\mathbb{R}^n$, the semidirect product Aff $SO(n) = SO(n) \ltimes \mathbb{R}^n$, the group of *proper Euclidean motions*. Clearly, $SO(n)$ is topologically isomorphic to a compact connected subgroup of Aff $SO(n)$. We note that $SO(1) \cong \{id\}$, hence Aff $SO(1)$ is identified with $\mathbb{R}$.

The following example is important in physics. First we consider the *Poincaré group*, which is the affine group of the proper Lorentz group on the Minkowski spacetime Aff $SO(1, 3) = SO(1, 3) \ltimes (\mathbb{R} \oplus \mathbb{R}^3)$. This is a special case of the affine group of the generalized special orthogonal group on $\mathbb{R}^p \oplus \mathbb{R}^q$, i.e. of the affine group Aff $S(p, q) = S(p, q) \ltimes \mathbb{R}^p \oplus \mathbb{R}^q$. Here $S(p, q)$ stands for the subgroup of $O(p, q)$ consisting of transformations with determinant $+1$. Of course, we may form the full affine group of $O(p, q)$ on $\mathbb{R}^p \oplus \mathbb{R}^q$. The action of Aff $O(p, q)$ on $\mathbb{R}^p \oplus \mathbb{R}^q$ can be described in the following way: on the vector $(x, y)$ in $\mathbb{R}^p \oplus \mathbb{R}^q$ the element $(k, u, v)$ in Aff $O(p, q)$ acts as

$$(k, u, v) \cdot (x, y) = k \cdot (x, y) + (u, v).$$

We note that neither $O(p, q)$ nor $SO(p, q)$ is compact if $pq \neq 0$. Nevertheless, by the Cartan–Iwasawa–Malcev Theorem (see e.g. Iwasawa (1949); Malcev

(1945); Stroppel (2006)), every connected Lie group (in fact, every connected locally compact group) has maximal compact subgroups which are conjugate to each other. Hence maximal compact subgroups are in some sense unique. For instance, a maximal compact subgroup in $O(p, q)$ is isomorphic to $O(p) \times O(q)$, and similarly, a maximal compact subgroup in $SO(p, q)$ is isomorphic to $SO(p) \times SO(q)$. Here $O(p) \times O(q)$, respectively, $SO(p) \times SO(q)$ acts on $\mathbb{R}^p \oplus \mathbb{R}^q$ "componentwise": the $O(p)$ part, respectively the $SO(p)$ part acts on $\mathbb{R}^p$, and the $O(q)$ part, respectively the $SO(q)$ part acts on $\mathbb{R}^q$. In particular, a maximal compact subgroup of the proper Lorentz group is isomorphic to $SO(1) \times SO(3) \cong SO(3)$, which acts on $\mathbb{R} \oplus \mathbb{R}^3$ by the rule: for $k$ in $SO(3)$ we have

$$k \cdot (t, x, y, z) = (t, [k] \cdot \begin{pmatrix} x \\ y \\ z \end{pmatrix}),$$

i.e. $k$ acts on the spatial part $(x, y, z)$ of the spacetime vector $(t, x, y, z)$ leaving the temporal part $t$ fixed.

Our last example is the *Heisenberg group* in three dimensions. This is the group of all real or complex matrices of the form

$$[h] = \begin{pmatrix} 1 & a & c \\ 0 & 1 & b \\ 0 & 0 & 1 \end{pmatrix}.$$

The identity of this group is the identity matrix and the inverse of $[h]$ is

$$[h]^{-1} = \begin{pmatrix} 1 & -a & ab - c \\ 0 & 1 & -b \\ 0 & 0 & 1 \end{pmatrix}.$$

In the real case the matrix $[h]$ can be considered as the matrix of the affine mapping $h : \mathbb{R}^2 \to \mathbb{R}^2$ defined by

$$h \cdot \begin{pmatrix} x \\ y \end{pmatrix} = \begin{pmatrix} 1 & a \\ 0 & 1 \end{pmatrix} \cdot \begin{pmatrix} x \\ y \end{pmatrix} + \begin{pmatrix} c \\ b \end{pmatrix}.$$

It follows that the Heisenberg group can be identified with a subgroup of the affine group Aff $GL(2) = GL(2) \ltimes \mathbb{R}^2$ of $GL(2)$ on $\mathbb{R}^2$. That subgroup corresponds to the linear group of transformations on $\mathbb{R}^2$ with matrices of the form

$$[A] = \begin{pmatrix} 1 & a \\ 0 & 1 \end{pmatrix}$$

with $a$ as a real or complex number. In fact, this group is isomorphic to $\mathbb{R}$.

The matrix multiplication on the Heisenberg group can be described on $\mathbb{R}^3$ if we identify the matrix $[h]$ above with the vector $(a, b, c)$. Then we have that the Heisenberg group is isomorphic to $\mathbb{R}^3$ equipped with the operation $\circ$ defined as

$$(a, b, c) \circ (a', b', c') = (a + a', b + b', c + c' + ab').$$

## 3.5 Invariant functions and measures

Let $N$ be a group and $H$ a subgroup of the automorphism group Aut $N$ of $N$. We consider the semidirect product $G = H \ltimes N$. The function $f : G \to \mathbb{C}$ is called $H$-*invariant* if it is constant on the double cosets of $H$. Formally, this is equivalent to the following property:

$$f\big((h', e) \cdot (h, n) \cdot (h'', e)\big) = f(h, n)$$

holds for each $h, h', h''$ in $H$ and $n$ in $N$. Indeed, it is obvious that the elements $(h, n)$ and $(h', e) \cdot (h, n) \cdot (h'', e)$ belong to the same double coset of $H$. In other words, $f : G \to \mathbb{C}$ is $H$-invariant if and only if it has the form

$$f = \tilde{f} \circ \Phi,$$

where $\tilde{f} : G//H \to \mathbb{C}$ is a function on the *double coset space* with respect to $H$, and $\Phi : G \to G//H$ is the canonical mapping of $G$ onto the double coset space $G//H$. We recall that the double coset space $G//H$ is the set of all *double cosets* $HgH$, where $g$ is in $G$ and

$$HgH = \{h'gh'' : h', h'' \in H\}.$$

All double cosets with respect to $H$ form a partition on $G$, and the factor set with respect to the corresponding equivalence relation is the double coset space $G//H$. Accordingly, the canonical mapping $\Phi : G \to G//H$ is defined by

$$\Phi(g) = HgH$$

for each $g$ in $G$. Consequently, $f : G \to \mathbb{C}$ is $H$-invariant if and only if we have $f(h, n) = \tilde{f}(H(h, n)H)$ for each $h$ in $H$ and $n$ in $N$. We can formulate this property in the following manner:

$$f(h, n) = f\big(h' \circ h \circ h'', h'(n)\big)$$

for each $h, h', h''$ in $H$ and $n$ in $n$. Given any $l$ in $H$ and choosing $h' = l$ further $h'' = h^{-1} \circ l^{-1}$ in the above equation we obtain

$$f(h, n) = f\big(id, l(n)\big).$$

This means that $f$ is $H$-invariant if and only if it depends only on the second component, i.e. it can be identified with a function on $N$: $f(h, n) = \varphi(n)$, and it is $H$-invariant on $N$: $\varphi(n) = \varphi(h(n))$ holds for each $h$ in $H$ and $n$ in $N$. In the subsequent paragraphs we shall utilize this identification: $H$-invariant function on $G$ simply means an $H$-invariant function on $N$, which is considered as a function on $G$ being independent of the first component.

The set of all $H$-invariant functions on $G$ form a locally convex topological vector space when equipped with the pointwise linear operations and with the topology of pointwise convergence. We denote this function space by $\mathcal{C}(G//H)$. In the case $H = \{e\}$ this coincides with the locally convex topological vector space of all complex valued functions on $G$ equipped with the pointwise linear operations and with the topology of pointwise convergence. In particular, by the above considerations, $\mathcal{C}(G//H)$ can be identified with a closed subspace of $\mathcal{C}(N)$, the subspace $\mathcal{C}_H(N)$ of $H$-invariant functions.

It is known (see e.g. Székelyhidi (2017)) that for any group $G$ the topological dual of $\mathcal{C}(G)$ can be identified with the space $\mathcal{M}_c(G)$ of all finitely supported complex measures on $G$ which is a (topological) algebra with the convolution of measures

$$\mu * \nu(f) = \sum_{x \in G} \sum_{y \in G} f(xy)\mu(x)\nu(y)$$

whenever $f$ is in $\mathcal{C}(G)$ and $\mu, \nu$ are in $\mathcal{M}_c(G)$. In particular, in our above situation the topological dual of the space $\mathcal{C}(G//H)$ of $H$-invariant functions can be identified with a closed subalgebra of $\mathcal{M}_c(N)$, the algebra of $H$-*invariant measures* which will be denoted by $\mathcal{M}_H(N)$ (see (Székelyhidi, 2017, Theorem 2, p. 4)).

In the case of affine groups we slightly modify the above concepts. Let $V$ be a finite dimensional real or complex vector space and suppose that $H$ is a closed subgroup of $GL(V)$. Then, with the notation $G = \operatorname{Aff} H = H \ltimes V$, $\mathcal{C}(G)$ denotes the space of all continuous complex valued functions on $G$ equipped with the pointwise linear operations and with the topology of uniform convergence on compact sets. Accordingly, $\mathcal{C}_H(V)$ denotes the space of continuous $H$-invariant functions on $G$, which is identified with the space of continuous $H$-invariant functions on $V$. The dual of this space is identified with the space $\mathcal{M}_H(V)$ of compactly supported complex Borel measures on $V$, which is also identified with the space of compactly supported $H$-invariant complex Borel measures on $G$. It is known (see Székelyhidi (2017)) that with the convolution $\mathcal{M}_H(V)$ is a commutative topological algebra.

The convolution of invariant measures in $\mathcal{M}_H(V)$ can be extended to a convolution between measures in $\mathcal{M}_H(V)$ and functions in $\mathcal{C}_H(V)$:

$$\mu * f(u) = \int_V f(u - v) \, d\mu(v)$$

which makes $\mathcal{C}_H(V)$ a topological module over $\mathcal{M}_H(V)$. Closed submodules of this module are called $H$-*varieties*. Given a function $f$ in $\mathcal{C}_H(V)$ the intersection of all $H$-varieties containing $f$ is obviously a $H$-variety, which is called the $H$-variety *generated by* $f$.

Now we suppose that $K$ is a compact subgroup of $GL(V)$ with the normalized Haar measure $\omega$. For each $f$ in $\mathcal{C}(G)$ we define its *projection* $f^\# : G \to \mathbb{C}$

as the function

$$f^{\#}(k, u) = \int_K \int_K f\Big((k', 0)(k, u)(k'', 0)\Big)\, d\omega(k')\, d\omega(k'') =$$

$$\int_K \int_K f(k' \circ k \circ k'', k' \cdot u)\, d\omega(k')\, d\omega(k'') = \int_K \int_K f(k'', k' \cdot u)\, d\omega(k')\, d\omega(k'')$$

whenever $(k, u)$ is in $G$. Then $f^{\#}$ is $K$-invariant. Indeed, we have

$$f^{\#}\Big((l', 0)(k, u)(l'', 0)\Big) = f^{\#}(l' \circ k \circ l'', l' \cdot u) =$$

$$\int_K \int_K f(k', k'' \cdot (l' \cdot u))\, d\omega(k')\, d\omega(k'') = \int_K \int_K f(k', (k'' \circ l') \cdot u))\, d\omega(k')\, d\omega(k'') =$$

$$\int_K \int_K f(k', k'' \cdot u)\, d\omega(k')\, d\omega(k'') = f^{\#}(k, u),$$

by the invariance of the Haar measure. Consequently, we can write, somewhat loosely

$$f^{\#}(k, u) = f^{\#}(u).$$

The mapping $f \mapsto f^{\#}$ is a continuous homomorphism of $\mathcal{C}(G)$ onto $\mathcal{C}_K(V)$ and $f$ in $\mathcal{C}(G)$ is $K$-invariant if and only if $f = f^{\#}$.

Similarly, we introduce the projection of measures in $\mathcal{M}_c(G)$. For each $\mu$ in $\mathcal{M}_c(G)$ we define its *projection* $\mu^{\#}$ by the equation

$$\langle \mu^{\#}, f \rangle = \langle \mu, f^{\#} \rangle$$

for each $f$ in $\mathcal{C}(G)$. Then $\mu^{\#}$ is a $K$-invariant measure in $\mathcal{M}_c(G)$ (see Székelyhidi (2017)). The mapping $\mu \mapsto \mu^{\#}$ is a continuous algebra homomorphism of $\mathcal{M}_c(G)$ onto $\mathcal{M}_K(V)$ and $\mu$ in $\mathcal{M}_c(G)$ is $K$-invariant if and only if $\mu = \mu^{\#}$.

An important role is played by the $K$-invariant measures $\delta^{\#}_{(k,u)}$, where $\delta_{(k,u)}$ is the *point mass* at the point $(k, u)$ in $G$. We have

$$\langle \delta^{\#}_{(k,u)}, f \rangle = \langle \delta_{(k,u)}, f^{\#} \rangle = f^{\#}(k, u) = \int_K \int_K f(k', k'' \cdot u)\, d\omega(k')\, d\omega(k'') = f^{\#}(u).$$

In particular, if $f$ is $K$-invariant, then we have

$$\langle \delta^{\#}_{(k,u)}, f \rangle = f(u),$$

the ordinary $\delta$-measure at $u$. We note that instead of $\delta^{\#}_{(k,u)}$ we can write $\delta^{\#}_u$, as $\delta^{\#}_{(k,u)}$ is $K$-invariant. In fact, $\delta^{\#}_{(k,u)} = \delta^{\#}_{(l,u)}$ for each $k, l$ in $K$ and $u$ in $V$.

## 3.6 Spherical functions and functional equations

In this section we suppose that $K$ is a compact subgroup of $GL(V)$. Using the projections of $\delta$-measures we define $K$-translation operators on $\mathcal{C}_K(V)$ in the following way:

$$\tau_v f(u) = \delta^{\#}_{-v} * f(u) = \int_V f(u - z) \, d\delta^{\#}_{-v}(z) = \int_K f(u + k \cdot v) \, d\omega(k).$$

The $K$-translation operators are continuous linear operators on $\mathcal{C}_K(V)$, and they form a commuting family:

$$(\tau_z \tau_v) f(u) = \int_K \tau_v f(u + k \cdot z) \, d\omega(k) = \int_K \int_K f(u + k \cdot z + l \cdot v) \, d\omega(k) \, d\omega(l),$$

which is symmetric in $z$ and $v$. This property is of basic importance and it is equivalent to the fact that $(G, K)$ forms a *Gelfand pair*: the measure algebra $\mathcal{M}_K(V)$ is commutative (see Székelyhidi (2017)).

The $K$-translation operators can be used to characterize $K$-varieties: in fact, a closed linear subspace in $\mathcal{C}_K(V)$ is a $K$-variety if and only if it is invariant with respect to all $K$-translations (see Székelyhidi (2017)). We remark that in the trivial case when $K = \{id\}$ we have $G \cong V$, every continuous function on $V$ is $K$-invariant, and the $K$-translation operators are exactly the ordinary translation operators on $\mathcal{C}(V)$. Consequently, $K$-varieties are nothing but the ordinary varieties: closed translation invariant linear subspaces of $\mathcal{C}(V)$. Clearly, the $K$-variety generated by a function $f$ in $\mathcal{C}_K(V)$ is the smallest closed linear subspace which contains all $K$-translates $\tau_v f$ of $f$.

The presence of translation operators makes it possible to study classical functional equations on affine groups. We start with the fundamental equation of the common eigenfunctions of all $K$-translations. The normalized common $K$-invariant eigenfunctions of all $K$-translation operators are called *K-spherical functions*. In other words, the nonzero $K$-invariant continuous function $s : V \to \mathbb{C}$ is a $K$-spherical function if and only if it satisfies

$$\int_K s(x + k \cdot y) \, d\omega(k) = s(x)s(y) \tag{3.1}$$

for each $x, y$ in $V$. We note that in the classical theory of spherical functions only the bounded solutions of the above equation are called spherical functions (see e.g. Dieudonné (1978); Helgason (1962); van Dijk (2009)). Here $K$-spherical functions are the generalizations of *exponential functions*, or *generalized characters*, which play a basic role in the theory of spectral analysis and spectral synthesis. In the subsequent paragraphs we shall study different basic classes of functions which are also fundamental from the point of view of spectral synthesis. In fact, the basic problems of spherical spectral analysis and synthesis were formulated in Székelyhidi (2017), and in some cases they also were proved. Our point here is to present descriptions of classes of exponential polynomials on affine groups with different choices of the compact group $K$.

Given the $K$-variety $X$ in $\mathcal{C}_K(V)$ we say that $K$-*spectral analysis* holds for $X$ if every nonzero subvariety of $X$ contains a $K$-spherical function. We say that $K$-*spectral analysis* holds on the affine group $G = K \ltimes V$ if $K$-spectral analysis holds for every $K$-variety on in $\mathcal{C}_K(V)$. In the case of $k = \{id\}$ this concept coincides with the ordinary spectral analysis. Moreover, if $K$ is a normal subgroup in $G$, then $K$-spectral analysis coincides with ordinary spectral analysis on the factor group $G/K$.

For the formulation of the spectral synthesis problems we need to define a reasonable concept of exponential polynomials. For this purpose we introduce the modified $K$-differences. Let $s$ be a $K$-spherical function and let $v$ be in $V$. The *modified $K$-difference* associated with $s$ and with increment $v$ is the $K$-invariant measure

$$\Delta_{s;v} = \delta^{\#}_{-v} - s(v)\delta_0,$$

where $\delta_0$ is the point mass at $0$ on $V$. Convolution with $\Delta_{s;v}$ is given by

$$\Delta_{s;v} * f(u) = \int_K f(u + k \cdot v)\, d\omega(k) - s(v)f(u)$$

whenever $u$ is in $V$ and $f$ is in $\mathcal{C}_K(V)$. Obviously, the intersection of the kernels of all convolution operators corresponding to modified $K$-differences associated with the $K$-spherical function $s$ is the one dimensional $K$-variety consisting all constant multiples of $s$. In the case $s \equiv 1$ we write $\Delta_v$ instead of $\Delta_{s;v}$ and we simply call it $K$-*difference*.

The *higher order $K$-differences* are the iterates of $\Delta_{s;v}$: we write for each natural number $n$ and for every $v_1, v_2, \ldots, v_{n+1}$

$$\Delta_{s;v_1,v_2,\ldots,v_{n+1}} = \Pi^{n+1}_{j=1}\Delta_{s;v_j},$$

where the product means convolution. In the case $v = v_1 = v_2 = \cdots = v_{n+1}$ we can write

$$\Delta^{n+1}_{s;v} = \Delta_{s;v_1,v_2,\ldots,v_{n+1}}.$$

We define $K$-*monomials* as the elements of the intersection of the kernels of higher order modified $K$-differences for a fixed $s$. More exactly, let $n$ be a natural number. The continuous $K$-invariant function $f$ is called a $K$-*monomial* if there exists a $K$-spherical function $s$ such that

$$\Delta_{s;v_1,v_2,\ldots,v_{n+1}} * f = 0$$

holds for each $v_1, v_2, \ldots, v_{n+1}$ in $V$. If $f$ is nonzero, then $s$ is unique (see Székelyhidi (2017)) and we call $f$ a *generalized $s$-monomial*, and the smallest $n$

with the above property is its *degree*. A generalized *s*-monomial is called an *s-monomial* if it generates a finite dimensional $K$-variety. The zero function is an *s*-monomial for each *s* and we consider its degree zero. For instance, constant multiples of *s* are exactly the *s*-monomials of degree zero. The generalized *s*-monomials of degree 1 are called *s-sine functions*, they are the solutions of the functional equation

$$\int_K \int_K f(x + k \cdot y + l \cdot z) \, d\omega(k) \, d\omega(l) + s(y)s(z)f(x) = \qquad (3.2)$$

$$s(y) \int_K f(x + k \cdot z) \, d\omega(k) + s(z) \int_K f(x + k \cdot y) \, d\omega(k)$$

for each $x, y, z$ in $V$. The following simple theorem shows that *s*-sine functions are *s*-monomials.

**Theorem 3.6.1.** *Given the K-spherical function* $s : V \to \mathbb{C}$ *the function* $f : V \to \mathbb{C}$ *in* $\mathcal{C}_K(V)$ *is a generalized s-monomial of degree at most one if and only if it satisfies*

$$\int_K f(x + k \cdot y) \, d\omega(k) + s(x)s(y)f(0) = s(x)f(y) + s(y)f(x) \qquad (3.3)$$

*for each* $x, y$ *in* $V$.

*Proof.* Let $x = 0$ in (3.2) then we have

$$\int_K \int_K f(k \cdot y + l \cdot z) \, d\omega(k) \, d\omega(l) + s(y)s(z)f(0) = s(y)f(z) + s(z)f(y),$$

as $f$ is $K$-invariant. The first term on the left can be written in the form

$$\int_K \int_K f(k \cdot y + l \cdot z) \, d\omega(k) \, d\omega(l) = \int_K \int_K f\big(k \cdot (y + k^{-1} \circ l \cdot z)\big) \, d\omega(k) \, d\omega(l) =$$

$$\int_K \int_K f\big(y + k^{-1} \circ l \cdot z\big) \, d\omega(l) = \int_K f(y + l \cdot z) \, d\omega(l),$$

by the $K$-invariance of $f$ and by the translation invariance of the Haar measure $\omega$. This proves the necessity of equation (3.3).

Conversely, suppose that $f$ satisfies (3.3). Substituting $y + l \cdot z$ for $y$ in (3.3) and then integrating with respect to $l$ over $K$ we get

$$\int_K \int_K f(x + k \cdot y + k \circ l \cdot z) \, d\omega(k) \, d\omega(l) + s(x)s(y)s(z)f(0) = \qquad (3.4)$$

$$s(x) \int_K f(y + l \cdot z) \, d\omega(l) + s(y)s(z)f(x) =$$

$$s(x)s(y)f(z) + s(x)s(z)f(y) + s(y)s(z)f(x) - s(x)s(y)s(z)f(0),$$

which can be written as

$$\int_K \int_K f(x + k \cdot y + k \circ l \cdot z)\, d\omega(k)\, d\omega(l) + s(x)s(y)f(z) =$$

$$s(y)s(z)f(x) + s(x)s(z)f(y) + 2s(x)s(y)f(z) - 2s(x)s(y)s(z)f(0)$$

for each $x, y, z$ in $V$. On the other hand, from (3.3) we have the equations

$$s(x)\int_K f(y + k \cdot z)\, d\omega(k) = s(x)s(y)f(z) + s(x)s(z)f(y) - s(x)s(y)s(z)f(0),$$

$$(3.5)$$

and

$$s(y)\int_K f(x + k \cdot z)\, d\omega(k) = s(y)s(z)f(x) + s(y)s(x)f(z) - s(x)s(y)s(z)f(0).$$

$$(3.6)$$

Adding (3.5) and (3.6) we get the statement. $\qquad\square$

**Corollary 3.6.1.** *Every s-sine function is an s-monomial.*

*Proof.* Indeed, equation (3.3) shows that every $K$-translate of $f$ is a linear combination of $f$ and $s$, hence the $K$-variety generated by $f$ is at most two dimensional. $\qquad\square$

In the case $s \equiv 1$ the $s$-monomials which vanish at the origin satisfy the equation

$$\int_K f(x + k \cdot y)\, d\omega(k) = f(x) + f(y) \tag{3.7}$$

and they are called $K$-*additive functions*.

Another basic function class related to $K$-monomials is the class of $K$-*moment functions*. Let $N$ be a natural number. We say that the sequence of $K$-invariant continuous functions $f_0, f_1, \ldots, f_N$ forms a $K$-*moment sequence* if we have

$$\int_K f_j(x + k \cdot y)\, d\omega(k) = \sum_{i=0}^{j} \binom{j}{i} f_i(x) f_{j-i}(y) \tag{3.8}$$

holds for each $j = 0, 1, \ldots, N$. We say that $f_j$ is a $K$-*moment function of order* $j$. Clearly, $K$-moment functions of order 0 are exactly the $K$-spherical functions and those of order 1 are the $s$-sine functions with $f_0 = s$. More generally, we have the following result.

**Theorem 3.6.2.** *Let $f_0, f_1, \ldots, f_N$ be a $K$-moment sequence. Then $f_j$ is an $f_0$-monomial of degree at most $j$ for $j = 0, 1, \ldots, N$.*

*Proof.* Let $s = f_0$. First we prove that $f_j$ is a generalized $f_0$-monomial of degree at most $j$ for $j = 0, 1, \ldots, N$. The statement holds for $j = 0$. We suppose that it holds for $j \geq 0$ and we prove it for $j + 1$. We have

$$\Delta_{s; y_1, y_2, \ldots, y_{j+1}, y_{j+2}} * f_j(x) =$$

$$\Delta_{s;y_1,y_2,\dots,y_{j+1}} * \left[ \int_K f_j(x + k \cdot y_{j+2}) \, d\omega(k) - s(y_{j+2}) f(x) \right] =$$

$$\Delta_{s;y_1,y_2,\dots,y_{j+1}} * \left[ \sum_{i=1}^{j} \binom{j}{i} f_i(y_{j+2}) f_{j-i}(x) \right] =$$

$$\sum_{i=1}^{j} \binom{j}{i} f_i(y_{j+2}) \Delta_{s;y_1,y_2,\dots,y_{j+1}} * f_{j-i}(x) = 0,$$

as $f_{j-i}$ is an $s$-monomial of degree at most $j-i$, by assumption, and $j-i \leq j-1$ for $j = 1, 2, \dots, j$.

Finally, equation (3.8) shows that the $K$-variety generated by $f_j$ is spanned by the functions $f_0, f_1, \dots, f_j$, hence it is finite dimensional. The proof is complete. $\qquad\square$

## 3.7   Functional equations on Euclidean motion groups

In this section we first consider the affine group of Euclidean motions $E(n)$, sometimes also denoted by $ISO(n)$. As we have seen above a *Euclidean motion* is an affine transformation on $\mathbb{R}^n$ whose linear part is an orthogonal transformation. In other words, $E(n) = \operatorname{Aff} O(n) = O(n) \ltimes \mathbb{R}^n$. It is clear that $O(n)$ is a compact subgroup of $GL(\mathbb{R}^n)$. In fact, it is a maximal compact subgroup as it is shown by the following theorem.

**Theorem 3.7.1.** *If $K \subseteq GL(\mathbb{R}^n)$ is a compact subgroup containing $O(n)$, then we have $K = O(n)$.*

*Proof.* Let $k$ be in $K$ and let $k = o \circ p$ be its polar decomposition, where $o$ is in $O(n)$ and $p$ is a positive definite symmetric matrix. We have $p = o^{-1}k$ is in $K$, as $o^{-1}$ is in $O(n)$. It follows that $p^m$ is in $K$ for each natural number $m$, and $K$ is compact, which implies that $p$ is the identity matrix and $K = O(n)$. $\quad\square$

First we consider the simplest case of $n = 1$. Now $K = O(1) \cong \mathbb{Z}_2$, the two element group which is identified with $\{-1, 1\}$ with multiplication. Let $G = K \ltimes \mathbb{R}$, then the continuous function $f : \mathbb{R} \to \mathbb{C}$ is $K$-invariant if and only if $f(x) = f(|x|)$, e.g. $f$ is even. Further the measure $\mu$ in $\mathcal{M}_c(\mathbb{R})$ is $K$-invariant if and only if

$$\int_{\mathbb{R}} f(|x|) \, d\mu(x) = \int_{\mathbb{R}} f(x) \, d\mu(x)$$

holds for each continuous function $f : \mathbb{R} \to \mathbb{C}$.

Clearly, the $K$-projection of the continuous function $f : K \times \mathbb{R} \to \mathbb{C}$ is given by

$$f^{\#}(x) = \frac{1}{2}[f(-1, -x) + f(1, x)] \tag{3.9}$$

for each $x$ in $\mathbb{R}$. We also have

$$\tau_y f(x) = \delta_{-y}^{\#} * f(x) = \int_K f(x + k \cdot y)\, d\omega(k) = \frac{1}{2}\big[f(x + y) + f(x - y)\big]$$

for every $K$-invariant function $f$ and for each $x, y$ in $\mathbb{R}$. This is the $K$-translation with $y$ in $\mathbb{R}$. The closed linear subspace $V$ in $\mathcal{C}_K(\mathbb{R})$ is a $K$-variety if and only for each $y$ in $\mathbb{R}$ it contains the function $x \mapsto \varphi(x + y) + \varphi(x - y)$ whenever $\varphi$ is in $V$. In particular, $V$ is a one dimensional $K$-variety if and only if it consists of all constant multiples of a nonzero continuous even function $\varphi : \mathbb{R} \to \mathbb{C}$ satisfying

$$\varphi(x + y) + \varphi(x - y) = 2\psi(y)\varphi(x) \tag{3.10}$$

for each $x, y$ in $\mathbb{R}$, where $\psi : \mathbb{R} \to \mathbb{C}$ is continuous and nonzero. Clearly, we have $\varphi(0) \neq 0$, and $\varphi(x) = \varphi(0)\psi(x)$. It follows that $\psi : \mathbb{R} \to \mathbb{C}$ is even, too, and it satisfies d'Alembert's functional equation

$$\psi(x + y) + \psi(x - y) = 2\psi(y)\psi(x) \tag{3.11}$$

for each $x, y$ in $\mathbb{R}$. In particular, the nonzero continuous even function $\varphi : \mathbb{R} \to \mathbb{C}$ is a $K$-spherical function if and only if it is a solution of the functional equation (3.11) for each $x, y$ in $\mathbb{R}$. We conclude that $\psi(x) = \cosh \lambda x$ with some complex number $\lambda$, and these are the generating functions of one dimensional $K$-varieties.

Now we consider the general case of $K = O(n)$ where $n \geq 2$ is an integer. As $O(n)$ acts transitively on the unit sphere of $\mathbb{R}^n$ it follows that $K$-invariant functions are again those depending on the norm only. Continuous $K$-invariant functions will be called $K$-*radial functions*. On the other hand, $K$-invariant measures are those measures $\mu$ in $\mathcal{M}_c(\mathbb{R}^n)$ satisfying

$$\int_{\mathbb{R}^n} f(x)\, d\mu(x) = \int_{\mathbb{R}^n} f(k \cdot x)\, d\mu(x)$$

for each continuous function $f : \mathbb{R}^n \to \mathbb{C}$ and for each real orthogonal $n \times n$ matrix $k$. These are called $K$-*radial measures*. The following theorem is fundamental (see Székelyhidi (2017); van Dijk (2009)).

**Theorem 3.7.2.** *Let $n \geq 2$. The $K$-radial function $\varphi : \mathbb{R}^n \to \mathbb{C}$ is a $K$-spherical function if and only if it is $\mathcal{C}^{\infty}$, it is an eigenfunction of the Laplacian, and $\varphi(0) = 1$.*

It follows that $s(x) = \varphi(\|x\|)$ holds for each $K$-spherical function $s$ on $\mathbb{R}^n$, where $\varphi : \mathbb{R} \to \mathbb{C}$ is an even $\mathcal{C}^{\infty}$ function with $\varphi(0) = 1$ such that

$\Delta s(x) = \lambda s(x)$ holds for each $x$ in $\mathbb{R}^n$ with some complex number $\lambda$. Using the radial form of the Laplacian in $\mathbb{R}^n$ we have that $\varphi$ is a solution of the Bessel differential equation

$$\frac{d^2}{dr^2}\varphi(r) + \frac{n-1}{r}\frac{d}{dr}\varphi(r) = \lambda\varphi(r) \tag{3.12}$$

for $r > 0$ such that $\varphi$ is regular at $0$ and $\varphi(0) = 1$. The unique $\varphi$ satisfying these conditions is

$$\varphi(r) = J_\lambda(r) = \Gamma\left(\frac{n}{2}\right)\sum_{k=0}^{\infty}\frac{\lambda^k}{k!\,\Gamma\left(k+\frac{n}{2}\right)}\left(\frac{r}{2}\right)^{2k}.$$

Finally, we obtain $s_\lambda(x) = J_\lambda(\|x\|)$ for each $x$ in $\mathbb{R}^n$ and $\lambda$ in $\mathbb{C}$.

To describe spherical monomials on $\mathrm{Aff}\,O(n)$ we can use the same idea as in (Székelyhidi, 2017, Corollary 2). In fact, we have the following theorem.

**Theorem 3.7.3.** *Let $m$ be a natural number, $\lambda$ a complex number. Given the $K$-spherical function $s_\lambda$ on the affine group $\mathrm{Aff}\,O(n)$ $s_\lambda$-monomials are exactly the linear combinations of the functions $\frac{\partial^j}{\partial\lambda^j}s_\lambda$ $(j = 0, 1, \dots, m)$.*

In particular, $K$-additive functions are constant multiples of

$$\frac{d}{d\lambda}J_\lambda(\|x\|) = \Gamma\left(\frac{n}{2}\right)\sum_{k=1}^{\infty}\frac{k\lambda^{k-1}}{k!\,\Gamma\left(k+\frac{n}{2}\right)}\left(\frac{\|x\|}{2}\right)^{2k} =$$

$$\Gamma\left(\frac{n}{2}\right)\sum_{k=0}^{\infty}\frac{\lambda^k}{k!\,\Gamma\left(k+1+\frac{n}{2}\right)}\left(\frac{\|x\|}{2}\right)^{2(k+1)}$$

at $\lambda = 0$, hence we have the general form of $K$-additive functions as

$$a(x) = constant \cdot \Gamma\left(\frac{n}{2}\right)\frac{1}{\Gamma\left(1+\frac{n}{2}\right)}\left(\frac{\|x\|}{2}\right)^2 = c\|x\|^2$$

with some complex number $c$. We can check easily the equation

$$\int_{O(n)} a(x + k\cdot y)\,d\omega(k) = \int_{O(n)}\|x + k\cdot y\|^2\,d\omega(k) =$$

$$\|x\|^2 + \|y\|^2 + \int_{O(n)}\langle x, k\cdot y\rangle\,d\omega(k) = \|x\|^2 + \|y\|^2 = a(x) + a(y),$$

as

$$\int_{O(n)}\langle x, k\cdot y\rangle\,d\omega(k) = 0.$$

Here $\langle,\rangle$ denotes the Euclidean inner product on $\mathbb{R}^n$. Indeed, it is clear that the mapping $x \mapsto \int_{O(n)} \langle x, k \cdot y \rangle \, d\omega(k)$ is linear in $x$ for each $y$ in $\mathbb{R}^n$, and it is also $O(n)$-invariant. It follows that

$$\int_{O(n)} \langle x, k \cdot y \rangle \, d\omega(k) = \int_{O(n)} \langle -x, k \cdot y \rangle \, d\omega(k),$$

which implies our statement.

As another example, we let $K = SO(n)$, the special orthogonal group and its affine group over $\mathbb{R}^n$, the group $G = \mathrm{Aff}\, SO(n) = SO(n) \ltimes \mathbb{R}^n$ which is called the *group of Euclidean motions* (see also van Dijk (2009)). Clearly, $SO(1) \cong \{id\}$, the trivial group, hence $\mathrm{Aff}\, SO(1)$ is identical with $\mathbb{R}$. Therefore we assume $n \geq 2$. Obviously, $SO(n)$ acts transitively on the unite sphere of $\mathbb{R}^n$, consequently the above considerations about $O(n)$ remain unchanged in this case. Namely, $K$-invariant continuous functions and $K$-spherical functions are the same $s_\lambda$'s as above, and so are the $s_\lambda$-monomials for each complex number $\lambda$ (see also Székelyhidi (2017)).

---

## 3.8   Affine groups of generalized orthogonal groups

In this section we consider functional equations on the affine group of the indefinite orthogonal group $O(p, q)$, i.e. $\mathrm{Aff}\, O(p, q) = O(p, q) \ltimes (\mathbb{R}^p \oplus \mathbb{R}^q)$ where $p, q$ are positive integers. We note that for $p = 0$ or $q = 0$ the group $O(p, q)$ is isomorphic to $O(p + q)$.

As we pointed out above $O(p, q)$ is non-compact, and all maximal compact subgroups of it are conjugate to $K = O(p) \times O(q)$. Hence it is reasonable to consider the affine group $G = \mathrm{Aff}\, K = \left( O(p) \times O(q) \right) \ltimes (\mathbb{R}^p \oplus \mathbb{R}^q)$. According to our above considerations $K$-invariant continuous functions on $\mathrm{Aff}\, K$ are those functions $f$ in $\mathcal{C}(\mathbb{R}^p \oplus \mathbb{R}^q)$ satisfying

$$f(k \cdot x, l \cdot y) = f(x, y)$$

for each $x$ in $\mathbb{R}^p$, $y$ in $\mathbb{R}^q$, $k$ in $O(p)$ and $l$ in $O(q)$. The $K$-translation $\tau_{(u,v)}$ with increment $(u, v)$ in $\mathbb{R}^p \oplus \mathbb{R}^q$ acts in the following way:

$$\tau_{(u,v)} f(x, y) = \int_{O(p)} \int_{O(q)} f(x + k \cdot u, y + l \cdot v) \, d\omega_p(k) \, d\omega_q(l)$$

for each $f$ in $\mathcal{C}_K(\mathbb{R}^p \oplus \mathbb{R}^q)$ and $(x, y)$ in $\mathbb{R}^p \oplus \mathbb{R}^q$, where $\omega_p$, respectively $\omega_q$ is the normalized Haar measure on $O(p)$, respectively on $O(q)$. Accordingly, the $K$-spherical functions are those functions $s$ in $\mathcal{C}_K(\mathbb{R}^p \oplus \mathbb{R}^q)$ satisfying

$$\int_{O(p)} \int_{O(q)} s(x + k \cdot u, y + l \cdot v) \, d\omega_p(k) \, d\omega_q(l) = s(x, y) s(u, v), \quad s(0, 0) = 1$$

$$(3.13)$$

for each $x, u$ in $\mathbb{R}^p$ and $y, v$ in $\mathbb{R}^q$. We have the following result.

**Theorem 3.8.1.** *The function $s : \mathbb{R}^p \oplus \mathbb{R}^q \to \mathbb{C}$ is a $K$-spherical function if and only if it has the form*

$$s_{\lambda,\mu}(x,y) = J_\lambda(\|x\|) J_\mu(\|y\|) \tag{3.14}$$

*for each $y, v$ in $\mathbb{R}^q$, with some complex numbers $\lambda, \mu$.*

*Proof.* Putting $y = v = 0$ in (3.13) we have

$$\int_{O(p)} s(x + k \cdot u, 0)\, d\omega_p(k) = s(x,0)s(u,0) \tag{3.15}$$

whenever $x, u$ is in $\mathbb{R}^p$. It follows that $x \mapsto s(x,0)$ is an $O(p)$-spherical function on $\mathbb{R}^p$, hence it has the form

$$s(x,0) = s_\lambda(x) = J_\lambda(\|x\|)$$

for each $x$ in $\mathbb{R}^p$ with some complex number $\lambda$. Similarly, we have that

$$s(0,y) = s_\mu(y) = J_\mu(\|y\|)$$

for each $y$ in $\mathbb{R}^q$ with some complex number $\mu$. Now we substitute $u = 0$, $y = 0$ in (3.13) to obtain

$$\int_{O(q)} s(x, l \cdot v)\, d\omega_q(l) = s(x,v) = s(x,0)s(0,v) \tag{3.16}$$

for each $x$ in $\mathbb{R}^p$ and $v$ in $\mathbb{R}^q$, which proves our statement. □

As a corollary we obtain the following result.

**Corollary 3.8.1.** *Given the natural number $m$ and the complex numbers $\lambda, \mu$ the $s_{\lambda,\mu}$-monomials of degree at most $m$ are exactly the linear combinations of the functions*

$$(x,y) \mapsto \partial_\lambda^i \partial_\mu^j s_{\lambda,\mu}(x,y)$$

*for $0 \le i + j \le m$.*

For instance, the $K$-polynomials can be obtained with the substitution $\lambda = \mu = 0$, as the following corollary shows.

**Corollary 3.8.2.** *Given the natural number $m$ the $K$-polynomials of degree at most $m$ are exactly the polynomials in $\|x\|^2$ and $\|y\|^2$ of degree at most $m$:*

$$p(x,y) = \sum_{0 \le i + j \le m} c_{ij} \|x\|^{2i} \|y\|^{2j}$$

*with some complex numbers $c_{ij}$.*

It follows that $K$-additive functions have the following form

$$a(x, y) = c\|x\|^2 + d\|y\|^2$$

with some complex numbers $c, d$.

An important special case is the Poincaré group, i.e. the affine group of the proper Lorentz group Aff $SO(1,3) = SO(1,3) \ltimes (\mathbb{R} \oplus \mathbb{R}^3)$. As we have seen, $SO(1,3)$ is not compact, but its maximal compact subgroups are conjugate to $K = SO(1) \times SO(3) \cong SO(3)$. Hence we consider the affine group $G =$ Aff $SO(3) \ltimes (\mathbb{R} \oplus \mathbb{R}^3)$, and we recall that $SO(3)$ acts on $\mathbb{R} \oplus \mathbb{R}^3$ by the rule

$$k \cdot (t, x) = (t, k \cdot x)$$

for each $k$ in $SO(3)$, $t$ in $\mathbb{R}$ and $x$ in $\mathbb{R}^3$. The $K$-translation in this case is given by

$$\tau_{(u,v)} f(t, x) = \int_{SO(3)} f(t + u, x + k \cdot v) \, d\omega(k),$$

hence the $K$-spherical function $s : \mathbb{R} \oplus \mathbb{R}^3 \to \mathbb{C}$ is characterized by the functional equations

$$\int_{SO(3)} s(t + u, x + k \cdot v) \, d\omega(k) = s(t, x)s(u, v), \quad s(0, 0) = 1 \qquad (3.17)$$

for each $t, u$ in $\mathbb{R}$ and $x, v$ in $\mathbb{R}^3$. We have the following theorem.

**Theorem 3.8.2.** *The function $s : \mathbb{R} \oplus \mathbb{R}^3 \to \mathbb{C}$ is a $K$-spherical function if and only if it has the form*

$$s(t, x) = s_{\lambda,\mu}(t, x) = e^{\lambda t} J_\mu(\|x\|)$$

*for each $t$ in $\mathbb{R}$ and $x$ in $\mathbb{R}^3$ where $\lambda, \mu$ are arbitrary complex numbers.*

*Proof.* Plugging $u = 0$ and $x = 0$ in (3.17) we obtain

$$s(t, v) = s(t, 0)s(0, v).$$

The substitution $x = v = 0$ in (3.17) gives

$$s(t + u, 0) = s(t, 0)s(u, 0), \quad s(0, 0) = 1,$$

consequently $s(t, 0) = e^{\lambda t}$ with some complex number $\lambda$. On the other hand, the substitution $t = u = 0$ in (3.17) gives

$$\int_{SO(3)} s(0, x + k \cdot v) \, d\omega(k) = s(0, x)s(0, v), \quad s(0, 0) = 1,$$

hence $x \mapsto s(0, x)$ is an $SO(3)$-spherical function on $\mathbb{R}^n$, consequently

$$s(0, x) = J_\mu(\|x\|)$$

holds with some complex number $\mu$. The proof is complete. $\qquad \square$

We note that in the case $n = 3$ the function $J_\mu(\|x\|)$ can be written in the form

$$J_\mu(\|x\|) = \Gamma\left(\frac{3}{2}\right) \sum_{k=0}^{\infty} \frac{\mu^k}{k!\Gamma\left(k + \frac{3}{2}\right)} \left(\frac{\|x\|}{2}\right)^{2k} = \frac{\sinh\mu\|x\|}{\mu\|x\|},$$

hence

$$s_{\lambda,\mu}(t, x) = \frac{\sinh\mu\|x\|}{\mu\|x\|} e^{\lambda t}$$

holds for each $t$ in $\mathbb{R}$ and $x$ in $\mathbb{R}^3$. In this case the general form of $K$-additive functions is

$$a(t, x) = ct + d\|x\|^2$$

with some complex numbers $c, d$.

## Acknowledgment

Research was supported by OTKA Grant No. K111651.

## Bibliography

Aczél J. and Dhombres J. *Functional equations in several variables.* volume 31 of *Encyclopedia of Mathematics and its Applications.* Cambridge University Press, Cambridge, 1989.

Dieudonné J. *Treatise on analysis. Vol. VI.* Academic Press, New York, 1978.

Gurevič D. I. Counterexamples to a problem of L. Schwartz, *Funkcional. Anal. i Priložen.*, 9(2):29–35, 1975.

Helgason S. *Differential geometry and symmetric spaces.* Academic Press, London, 1962.

Iwasawa K. On some types of topological groups, *Ann. of Math. (2)*, 50:507–558, 1949.

Lefranc M. L'analyse harmonique dans $Z_n$, *C. R. Acad. Sci. Paris*, 246:1951–1953, 1958.

Laczkovich M. and Székelyhidi G. Harmonic analysis on discrete Abelian groups, *Proc. Amer. Math. Soc.*, 133(6):1581–1586, 2005.

Laczkovich M. and Székelyhidi L. Spectral synthesis on discrete Abelian groups, *Math. Proc. Cambridge Philos. Soc.*, 143(1):103–120, 2007.

Malcev A. On the theory of the Lie groups in the large, *Rec. Math. [Mat. Sbornik] N. S.*, 16(58):163–190, 1945.

Schwartz L. Théorie générale des fonctions moyenne-périodiques, *Ann. of Math. (2)*, 48:857–929, 1947.

Stroppel M. *Locally compact groups, EMS Textbooks in Mathematics*, European Mathematical Society (EMS), Zürich, 2006.

Székelyhidi L. *Discrete spectral synthesis and its applications*, Springer Monographs in Mathematics. Springer, Dordrecht, 2006.

Székelyhidi L. and Wilkens B. Spectral analysis and synthesis on varieties, *J. Math. Anal. Appl.*, 433(2):1329–1332, 2016.

Székelyhidi L. On spectral synthesis in several variables, *Adv. Operator Theory*, 2(2):179–191, 2017.

Székelyhidi L. Spherical spectral synthesis, *Acta Math. Hung.*, 1–23, 2017.

Székelyhidi L. and Wilkens B. Spectral synthesis and residually finite-dimensional algebras. *Jour. of Algebra and Its Appl.*, 16(10): 1–10, 2017.

Székelyhidi L. Spectral Synthesis and Its Applications, Chapter 7 in Mathematical Analysis and Applications, Wiley, 60 pages, 2018.

van Dijk G. *Introduction to harmonic analysis and generalized Gelfand pairs*, de Gruyter Studies in Mathematics. Walter de Gruyter & Co., Berlin, 2009.

# Chapter 4

## Locally Pseudoconvex Spaces and Algebras

**Mati Abel**

*J. Liivi 2-615, Institute of Mathematics and Statistics, University of Tartu, Tartu, Estonia*

## 4.1   Brief introduction

The structure and properties of locally convex spaces ($LCS$, in short) are described in numerous papers and in many books. By contrast, the properties of locally pseudoconvex spaces ($LPS$, in short) have been considered not so much: only in seveal papers and, partly, in books Adasch et al. (1978), Balachandran (2000), Bayoumi (2003), Jarchow (1981), Kohn (1979), Rolewicz (1985), Waelbroeck (1971) and Wells (1968). Properties of locally $m$-convex algebras ($LmCA$, in short) are relatively well studied in numerous papers and in books Balachandran (2000), Beckenstein et al. (1977), Kızmaz (1981), Mallios (1986), Michael (1952), Page (1988), Waelbroeck (1967) and Zelazko

(1965)–Zelazko (1973) and locally bounded algebras ($LBA$, in short) in several papers and mostly in books Balachandran (2000), Zelazko (1965) and Zelazko (1973). Herewith, properties of locally pseudoconvex algebras ($LPA$, in short) have been studied mostly in papers and in part in Balachandran (2000), Turpin (1966b), Waelbroeck (1966) and Waelbroeck (1971).

Some results about the structure of $LPA$ (as the triviality of division $LPA$ and the representations of $LPA$ by the projective and inductive limits by Banach algebras, locally bounded algebras ($LBA$, in short) and $F$-algebras ($FA$, in short)) are given in this chapter. For this, several ideas from Abel (2004c), Abel (2007a), Abel (2010), Abel (2018), Abel (2015), Jarchow (1981), Khan (2013) and Et and Colak (1995 ) have been used.

Next we recall the main concepts and results we will use later on.

### 4.1.1   Linear spaces

In this section we recall some basic definitions about linear spaces.

a) Let $\mathbb{K}$ denote one of the fields $\mathbb{R}$ of real numbers or $\mathbb{C}$ of complex numbers and $X$ be a linear space over $\mathbb{K}$. Later on we will use the following subsets of $X$.

A subset $U$ of $X$ is called

*absorbing* or *absorbent* if, for each $x \in X$, there is a number $k = k(x) > 0$ such that $x \in \lambda U$ for all $\lambda \in \mathbb{K}$ with $|\lambda| \geqslant k$ (or, equivalently, there exists a number $l = l(x) > 0$ such that $\lambda x \in U$ for all $\lambda \in \mathbb{K}$ with $|\lambda| \leqslant l$);

*balanced* if $\lambda x \in U$ for all $x \in U$ and $\lambda \in \mathbb{K}$ with $|\lambda| \leqslant 1$;

*convex* if $\lambda x + \mu y \in U$ for all $x, y \in U$ and $\lambda, \mu \in [0, 1]$ such that $\lambda + \mu = 1$;

*absolutely convex* if $\lambda x + \mu y \in U$ for all $x, y \in U$ and $\lambda, \mu \in \mathbb{K}$ such that $|\lambda| + |\mu| \leqslant 1$;

*k-convex* with $k \in (0, 1]$ if $\lambda x + \mu y \in U$ for all $x, y \in U$ and $\lambda, \mu \in [0, 1]$ such that $\lambda^k + \mu^k = 1$;

*absolutely k-convex* with $k \in (0, 1]$ if $\lambda x + \mu y \in U$ for all $x, y \in U$ and $\lambda, \mu \in \mathbb{K}$ such that $|\lambda|^k + |\mu|^k \leqslant 1$;

*pseudoconvex* if there exists a number $\lambda > 0$ such that $U + U \subseteq \lambda U$;

*absolutely pseudoconvex* if $U$ is absolutely $k$-convex for some $k \in (0, 1]$.

Hereby, every convex set $U$ satisfies the property $U + U \subseteq 2U$. When $U$ is balanced, then we can assume in the definition of a pseudoconvex set, that $\lambda \geqslant 2$, otherwise, from

$$2U \subseteq U + U \subseteq \lambda U = 2\left(\frac{\lambda}{2}\right)U \subseteq 2U$$

follows that $U + U = 2U$ what is not correct. It is more convenient to write the number $\lambda$ in the definition of a pseudoconvex set on the form $\lambda = 2^{\frac{1}{k}}$. Then $k = \frac{\log 2}{\log \lambda}$. In case of balanced pseudoconvex set $U$, the number $k \in (0, 1]$.

Let again $k \in (0, 1]$. The set

$$\Gamma_k(U) = \left\{ \sum_{v=1}^{n} \lambda_v u_v : n \in \mathbb{N}, u_1, \ldots, u_n \in U, \lambda_1, \ldots, \lambda_n \in \mathbb{K}, \sum_{v=1}^{n} |\lambda_v|^k \leqslant 1 \right\}$$

is called an *absolutely k-convex hull* of $U$. Then $U \subseteq \Gamma_k(U)$ for every set $U$ and $\Gamma_k(U)$ is a balanced an absolutely $k$-convex set. Moreover, a set $U$ is absolutely convex if and only if $U = \Gamma_1(U)$, absolutely $k$-convex if and only if $U = \Gamma_k(U)$ and absolutely pseudoconvex if and only if $U = \Gamma_k(U)$ for some $k \in (0, 1]$.

b) A sequence $\mathcal{U} = (U_n)$ with $n \in \mathbb{N}' = \mathbb{N} \cup \{0\}$ of sets in a linear space $X$ over $\mathbb{K}$ is called a *string* in $X$ if every set $U_n \in \mathcal{U}$ is absorbing, balanced and *summative* (it means that $U_{n+1} + U_{n+1} \subseteq U_n$ for all $n \in \mathbb{N}_0$). The first member $U_0$ of this sequence $(U_n)$ is called the *beginning* of the string $\mathcal{U}$ and the set $U_n$ is called the $n$-th *knot* of $\mathcal{U}$. For every string $\mathcal{U}$ the set

$$N(\mathcal{U}) = \bigcap_{n \in \mathbb{N}'} U_n$$

is called the *kernel* of $\mathcal{U}$. In addition, for strings $\mathcal{U} = (U_n)$ and $\mathcal{V} = (V_n)$ we write

($\alpha$) $\lambda \mathcal{U} = (\lambda U_n)$ for each $\lambda \in \mathbb{K}$;
($\beta$) $\mathcal{U} + \mathcal{V} = (U_n + V_n)$ (the *sum of strings*);
($\gamma$) $\mathcal{U} \cap \mathcal{V} = (U_n \cap V_n)$ (the *intersection of strings*).

Moreover, we write $\mathcal{U} \subseteq \mathcal{V}$ if $U_n \subseteq V_n$ for all $n \in \mathbb{N}'$.

c) Let $k \in (0, 1]$. A function $p : X \to \mathbb{R}$ is called a *k-homogeneous seminorm* (in case, when $k = 1$, a *seminorm*) on $X$ if

(1) $p(x) \geqslant 0$ for all $x \in X$;
(2) $p(x) = 0$ if $x = \theta_A$ (the zero element of $X$);
(3) $p(x + y) \leqslant p(x) + p(y)$ for each $x, y \in X$ (the subadditivity);
(4) $p(\lambda x) = |\lambda|^k p(x)$ for all $x \in X$ and $\lambda \in \mathbb{K}$ (the $k$-homogenity);

and it is called an *F-seminorm* on $X$ if $p$ satisfies, in addition the conditions (1)–(3), also the conditions,

(5) $p(\lambda x) \leqslant p(x)$ for all $x \in X$ and $\lambda \in \mathbb{K}$ with $|\lambda| \leqslant 1$;
(6) if $(\lambda_n) \to 0$, then $(p(\lambda_n x)) \to 0$ for all $x \in X$
(7) if $(x_n) \to \theta_X$ (*the zero element of* $X$), then $(p(\lambda x_n)) \to 0$ for all $\lambda \in \mathbb{K}$.

In particular case, when from $p(x) = 0$ follows that $x = \theta_X$, the $k$-seminorm $p$ is called a *k-norm* (if $k = 1$, a *norm*) and the $F$-seminorm $p$ is called an *F-norm*. In this case, the map $d$, defined by $d(x, y) = p(x - y)$ for each $x, y \in X$, is a metric on $X$, which has the property $d(x+z, y+z) = d(x, y)$ for each $x, y, z \in X$.

### 4.1.2    Topological spaces

In this section we recall some basic definitions for topological spaces and main results, used later on.

A *topology* on a nonempty set $X$ is a collection $\tau$ of subsets of $X$ such that

(a) $X$ and $\emptyset$ belong to $\tau$;
(b) the union of an arbitrary number of sets in $\tau$ belongs to $\tau$,
(c) the intersection of a finite number of sets in $\tau$ belongs to $\tau$.

In this case the pair $(X, \tau)$ is called a *topological space*, each set $U$ in $\tau$ is called an *open set* and the complement $X \setminus U$ of an open set $U$ is called a *closed set*. The *closure* of a set $U$ in $(X, \tau)$, denoted by $\mathrm{cl}_X(U)$, is defined as the intersection of all closed sets in the topology $\tau$ containing $U$, the *interior* of a set $U$ in $(X, \tau)$, denoted by $\mathrm{int}_X(U)$, is defined as the union of all open sets in the topology $\tau$ contained in $U$, and $U$ is called dense in $(X, \tau)$, if $\mathrm{cl}_X(U) = X$. Thus, the set $U$ in $(X, \tau)$ is open if and only if $\mathrm{int}_X(U) = U$ and is closed in $(X, \tau)$ if and only if $\mathrm{cl}_X(U) = U$. Moreover, for any $x \in X$, a subset $O$ of $X$ is called a *neighborhood* of $x$, if there exists an open set $U$ in $X$ such that $x \in U \subseteq O$. Hence, every open set $O$ containing $x$, is an open neighborhood of $x$. The collection $\mathcal{N}(x)$ of all neighborhoods of $x$ is called a *neighborhood system* of $x$. A *base of neighborhoods* of $x$ is a subcollection $\mathcal{B}(x)$ of $\mathcal{N}(x)$ such that, for each $O \in \mathcal{N}(x)$, there exists some $U \in \mathcal{B}(x)$ such that $U \subseteq O$, and the elements of $\mathcal{B}(x)$ are called the *basic neighborhoods* of $x$.

**Theorem 4.1.1.** *Let $(X, \tau)$ be a topological space and $\mathcal{B}(x)$ a base of neighborhoods of $x$ for any $x \in X$. Then*

$(NB_1)$ $\mathcal{B}(x)$ *is not empty;*
$(NB_2)$ *if $U \in \mathcal{B}(x)$, then $x \in U$;*
$(NB_3)$ *if $U, V \in \mathcal{N}(x)$, there is some $W \in \mathcal{B}(x)$ such that $W \subseteq U \cap V$;*
$(NB_4)$ *if $V \in \mathcal{B}(x)$, there is some $O \in \mathcal{B}(x)$ such that if $y \in O$, there is some $W \in \mathcal{B}(y)$ with $W \subseteq V$ (that is, $V$ contains a basic neighborhood of each $y \in O$).*

*Conversely, if for each $x \in X$, the collection $\mathcal{B}(x)$ satisfies the conditions $(NB_1)$–$(NB_4)$ for every $x \in X$, then there is a topology*

$$\tau = \{U \subseteq X : \text{ for each } x \in X \text{ there is a } V \in \mathcal{B}(x) \text{ such that } V \subseteq U\}$$

*on $X$ such that $\mathcal{B}(x)$ is a base of neighborhoods of $x$ for every $x \in X$ in the topology $\tau$.*

*Proof.* It is clear that $\mathcal{B}(x)$ satisfies the conditions $(NB_1)$ and $(NB_2)$. If $U, V \in \mathcal{B}(x)$, then $U \cap V$ is a neighborhood of $x$. Therefore, there is a $W \in \mathcal{B}(x)$ such that $W \subseteq U \cap V$. Let now $V \in \mathcal{B}(x)$. Then there is an open set $O \in \mathcal{B}(x)$ such that $O \subseteq V$. If now $y \in O$, then $O$ is a neighborhood of $y$. Therefore, there is a $W \in \mathcal{B}(y)$ such that $W \subseteq O \subseteq V$. Hence, $\mathcal{B}(x)$ satisfies also $(NB_3)$ and $(NB_4)$.

Conversely, it is easy to see that $\tau$ is a topology on $X$ by the conditions

$(NB_1)$–$(NB_4)$ for every $x \in X$ and $\mathcal{B}(x)$ is a base of neighborhoods of $x$ in the topology $\tau$.   $\square$

The next result helps us to verify when for two given topologies $\tau_1$ and $\tau_2$ on a nonempty set $X$ holds $\tau_1 \subseteq \tau_2$.

**Lemma 4.1.2.** *Let $\tau_1$ and $\tau_2$ be two topologies on a nonempty set $X$ and for every $x \in X$ let $\mathcal{B}_1(x)$ and $\mathcal{B}_2(x)$ be bases of neighborhoods of $x$ in the topologies $\tau_1$ and $\tau_2$, respectively. Then $\tau_1 \subseteq \tau_2$ if and only if for every $U \in \mathcal{B}_1(x)$ there exists a $V \in \mathcal{B}_2(x)$ such that $V \subseteq U$.*

*Proof.* Let $\tau_1 \subseteq \tau_2$, $x \in X$ and $O \in \mathcal{B}_1(x)$. Then $O$ is a neighborhood of $x$ in the topology $\tau_1$ and, by the definition, there exists a $U \in \tau_1$ such that $x \in U \subseteq O$. Because $U \in \tau_2$, then $U$ is also a neighborhood of $x$ in the topology $\tau_2$. Therefore, there exists a $V \in \mathcal{B}_2(x)$ such that $V \subseteq U$.

Let now $O \in \tau_1$ and $x \in O$. Then $O$ is a neighborhood of $x$ in the topology $\tau_1$ and there exists a neighborhood $U \in \mathcal{B}_1(x)$ such that $x \in U \subseteq O$. By the assumption, there is now a $V \in \mathcal{B}_2(x)$ such that $x \in V \subseteq U$. Consequently, $O \in \tau_2$.   $\square$

Let $(X, \tau)$ and $(Y, \tau')$ be topological spaces. A map $f : X \to Y$ is called *continuous at* $x_0 \in X$, if for any neighborhood $O(f(x_0))$ of $f(x_0)$ in $(Y, \tau')$, there is a neighborhood $O(x_0)$ of $x_0$ in $(X, \tau)$ such that $f(O(x_0)) \subseteq O(f(x_0))$, and $f$ is called *continuous on $X$*, when $f$ is continuous at every $x \in X$.

In the particular case, when $f$ is a bijection, then $f$ has the *inverse map* $f^{-1} : Y \to X$, defined by $f^{-1}(f(x)) = x$. In this case $f$ is called a *homeomorphism* if $f$ and $f^{-1}$ are continuous.

---

## 4.2   Topological linear spaces

Let $X$ be a linear space over $\mathbb{K}$ and $\tau$ a topology on $X$. The pair $(X, \tau)$ is called a *topological linear space* (*TLS*, in short) if the following conditions hold:

$(TLS_1)$ the operation of addition $(x, y) \to x + y$ of $X \times X \to X$ is jointly continuous (that is, for any given $x, y \in X$ and any neighborhood $O$ of $x + y$ in $(X, \tau)$, there exist neighborhoods $U$ of $x$ and $V$ of $y$ in $(X, \tau)$ such that $U + V \subseteq O$; here, and later on, $A + B = \{a + b : a \in A \text{ and } b \in B\}$);

$(TLS_2)$ the operation of multiplication by scalars $(\lambda, x) \to \lambda x$ of $\mathbb{K} \times X \to X$ is jointly continuous (that is, for any given $x \in X$, $\lambda \in \mathbb{K}$ and neighborhood $O$ of $\lambda x$ in $(X, \tau)$, there exist a neighborhood $U$ of $x$ in $(X, \tau)$ and a neighborhood $K$ of $\lambda$ in $\mathbb{K}$ in the usual topology of $\mathbb{K}$ such that $KU \subseteq O$; here, and later on, $KB = \{\lambda b : \lambda \in K \text{ and } b \in B\}$).

**Lemma 4.2.1.** *Let $(X, \tau)$ be a TLS over $\mathbb{K}$. Then the following statements hold:*

(a) *for any fixed $a \in X$, the translation map $f_a : X \to X$, defined by $f_a(x) = x + a$ for each $x \in X$, is a homeomorphism of $(X, \tau)$ into itself;*

(b) *for any fixed $\lambda \in \mathbb{K}$, $\lambda \neq 0$, the multiplication (by scalars) map $g_\lambda : X \to X$, defined by $g_\lambda(x) = \lambda x$ for each $x \in X$, is a homeomorphism of $(X, \tau)$ into itself.*

*Proof.* It is easy to see that $f_a$ and $g_\lambda$ are continuous bijections, and $f_a^{-1} = f_{-a}$ and $g_\lambda^{-1} = g_{\frac{1}{\lambda}}$ are continuous. Hence, $f_a$ and $g_\lambda$ are homeomorphisms.    □

**Proposition 4.2.2.** *Let $(X, \tau)$ be a TLS, $x \in X$, $\lambda \in \mathbb{K}$ and $\lambda \neq 0$. Then*

(a) *if $O$ is an open (closed) set in $(X, \tau)$, then $a + O$ and $\lambda O$ are open (respectively, closed) sets in $X$;*

(b) *$O$ is a neighborhood of zero in $(X, \tau)$ if and only if $x + O$ is a neighborhood of $x$ in $(X, \tau)$;*

(c) *$O$ is a neighborhood of zero in $(X, \tau)$ if and only if $\lambda O$ is a neighborhood of zero in $(X, \tau)$;*

(d) *every neighborhood of zero in $(X, \tau)$ is absorbing;*

(e) *every neighborhood $O(x)$ of a point $x \in X$ in the topology $\tau$ has the form $O(x) = x + O$ for some neighborhood $O$ of zero in $(X, \tau)$;*

(f) *a collection $\mathcal{B}$ of subsets of $X$ is a base of neighborhoods of zero in $(X, \tau)$ if and only if the collection*

$$x + \mathcal{B} = \{x + O : O \in \mathcal{B}\}$$

*is a base of neighborhoods of $x$ in $(X, \tau)$.*

*Proof.* (a) By Lemma 5, the translation map $f_x$ and multiplication map $g_\lambda$ are homeomorphisms. Therefore, these maps are open and closed. Hence, $f_x(O) = x + O$ and $g_\lambda(O) = \lambda O$ are open (respectively, closed) sets in $X$.

(b) and (c) The statements (b) and (c) follow from (a).

(d) Let $O$ be a neighborhood of zero in $(X, \tau)$ and $x_0 \in X$ an arbitrary point. By $(TLS_2)$, the map $f_{x_0} : \mathbb{K} \to X$, defined by $f_{x_0}(\lambda) = \lambda x_0$ for each $\lambda \in \mathbb{K}$, is continuous at 0. So, there is a number $\varepsilon > 0$ such that $f_{x_0}(\overline{O}_\varepsilon) \subseteq O$, where $\overline{O}_\varepsilon = \{\lambda \in \mathbb{K} : |\lambda| \leqslant \varepsilon\}$. Hence, $\lambda x_0 \in O$, whenever $|\lambda| \leqslant \varepsilon$. It means that $O$ is absorbing.

(e) Let $x \in X$ and $O(x)$ be a neighborhood of $x \in X$ in the topology $\tau$. Then $O = O(x) - x = f_{-x}(O(x))$ is a neighborhood of zero in $X$, because $f_{-x}$ is a homeomorphism by Lemma 5 (a).

(f) By the statement (b), $O$ is a neighborhood of zer0 in $(X, \tau)$ if and only if $x + O$ is a neighborhood of $x$ in $(X, \tau)$. Taking this into account, $\mathcal{B}$ is a base of neighborhoods of zero in $(X, \tau)$ if and only if $x + \mathcal{B}$ is a base of neighborhoods of $x$ in $(X, \tau)$.    □

The next result shows that every $TLS$ has a base $\mathcal{B}$ of neighborhoods of zero, consisting of closed, balanced and absorbing sets.

**Theorem 4.2.3.** *Let $(X, \tau)$ be a TLS and $\mathcal{B}$ a base of neighborhoods of zero in $X$ in the topology $\tau$. Then*

$(B_1)$ *for every $U, V \in \mathcal{B}$, there exists $W \in \mathcal{B}$ such that $W \subseteq U \cap V$;*

$(B_2)$ *for each $U \in \mathcal{B}$, there exists $V \in \mathcal{B}$ such that $V + V \subseteq U$;*

$(B_3)$ *for each $U \in \mathcal{B}$, there exists a balanced neighborhood $W$ of zero in $(X, \tau)$ such that $W \subseteq U$;*

$(B_4)$ *for each $U \in \mathcal{B}$, there exists a closed neighborhood $W$ of zero in $(X, \tau)$ such that $W \subseteq U$.*

$(B_5)$ *for each $U \in \mathcal{B}$, there exists an open (closed) balanced neighborhood $W$ of zero in $(X, \tau)$ such that $W \subseteq U$.*

*Conversely, if $\mathcal{B}$ is a nonempty collection of absorbing subsets of $X$ which has the properties $(B_1)$–$(B_3)$, then there exists a topology $\tau$ on $X$ which makes $(X, \tau)$ a TLS with a base $\mathcal{B}$ of neighborhoods of zero.*

*Proof.* $(B_1)$ This statement holds by the property $(NB_3)$.

$(B_2)$ Let $U \in \mathcal{B}$. By $(TLS_1)$, the operation $l : (x, y) \to x + y$ is continuous at $(\theta_X, \theta_X)$. As $l(\theta_X, \theta_X) = \theta_X$, then there exist neighborhoods $O_1$ and $O_2$ of zero in $(X, \tau)$ such that $l(O_1 \times O_2) = O_1 + O_2 \subset U$. Now, there is, by $(B_1)$, a neighborhood $V \in \mathcal{B}$ such that $V \subseteq O_1 \cap O_2$. Therefore, $V + V \subseteq O_1 + O_2 \subseteq U$.

$(B_3)$ Let $U \in \mathcal{B}$. The map $g : \mathbb{K} \times X \to X$, defined by $g(\lambda, x) = \lambda x$ for each $\lambda \in \mathbb{K}$ and $x \in X$, is jointly continuous by $(TLS_2)$. Because $g(0, \theta_A) = \theta_A$, there exist a number $\varepsilon > 0$ and a neighborhood $V$ of zero in $(X, \tau)$ such that

$$W = \{\lambda \in \mathbb{K} : |\lambda| \leqslant \varepsilon\} V = \bigcup_{|\mu| \leqslant \varepsilon} \mu V \subseteq U.$$

Then $W$ is a balanced neighborhood of zero in $(X, \tau)$, because $\varepsilon V \subseteq W$.

$(B_4)$ Let $U \in \mathcal{B}$. By $(B_2)$ and $(B_3)$, we can find a balanced $W \in \mathcal{B}$ such that $W + W \subseteq U$. To show that $\mathrm{cl}_X(W) \subseteq U$, let $x \in \mathrm{cl}_X(W)$. Since $x + W$ is a neighborhood of $x$ by Proposition 4.2.2 (b), then $(x + W) \cap W \neq \emptyset$. Take $y \in (x + W) \cap W$. Then $y = x + v = w$, where $v, w \in W$. Therefore,

$$x = w - v \in W + (-W) = W + W \subseteq U.$$

Hence, $\mathrm{cl}_X(W) \subseteq U$.

$(B_5)$ Similarly, as in the proof of the statement $(B_3)$, there exist a number $\varepsilon > 0$ and $(B_4)$ by the definition of neighborhood of an element and by the statement $(B_4)$ an open (closed) neighborhood $V$ of zero in $(X, \tau)$ such that

$$W = \bigcup_{|\lambda| \leqslant \varepsilon} \lambda V \subseteq U.$$

If $V$ is open, let

$$W_0 = \bigcup_{0 < |\lambda| \leqslant \varepsilon} \lambda V.$$

Then $W_0$ is open (because every set $\lambda V$ is open by Lemma 5 (b)), balanced and $W_0 = W \subseteq U$. If $V$ is closed, then, by the statement $(B_3)$, there exists a balanced neighborhood $W'$ of zero in $(X, \tau)$ such that $W' \subseteq V$. Therefore,

$$\mathrm{cl}_X(\varepsilon W') \subset \mathrm{cl}_X(\varepsilon V) = \varepsilon V \subseteq U$$

and

$$\mu \mathrm{cl}_X(\varepsilon W') = \mathrm{cl}_X(\varepsilon(\mu W')) \subseteq \mathrm{cl}_X(\varepsilon W')$$

for each $|\mu| \leqslant 1$, because $\varepsilon W'$ is balanced and the map $g_\mu$, defined by $g_\mu(x) = \mu x$ on $X$, is a homeomorphism by Lemma 5 (b). Hence, $\mathrm{cl}_X(\varepsilon W')$ is closed and balanced.

Conversely, let

$$\mathcal{E} = \{V \subseteq X : V \text{ contains some } U \in \mathcal{B}\}.$$

Then $\mathcal{B} \subseteq \mathcal{E}$. First, we show that $\mathcal{C}(x) = x + \mathcal{E}$ satisfies the conditions $(NB_1)$–$(NB_4)$ for each $x \in X$. It is clear that $\mathcal{C}(x)$ satisfies the conditions $(NB_1)$ and $(NB_2)$ for each $x$. Let $V, W \in \mathcal{E}$. Then there are $U_V, U_W \in \mathcal{B}$ such that $U_V \subseteq V$ and $U_W \subseteq W$. By $(B_1)$, there is a set $U \in \mathcal{B}$ such that $U \subseteq U_V \cap U_W$. Taking this into account,

$$x + U \subseteq x + U_V \cap U_W \subseteq x + V \cap W \subseteq (x + V) \cap (x + W).$$

Thus, $\mathcal{C}(x)$ satisfies the condition $(NB_3)$.

Let $V \in \mathcal{E}$. Then there is a $U \in \mathcal{B}$ such that $U \subseteq V$ and, by $(B_2)$, a $W \in \mathcal{B}$ such that $W + W \subseteq U$. Now, if $y \in x + W$, then

$$y + W \subseteq x + W + W \subseteq x + V.$$

Hence, $\mathcal{C}(x)$ satisfies the condition $(NB_4)$. Consequently, by Theorem 4.1.1, there is a topology $\tau$ on $X$ is such that $\mathcal{C}(x)$ is a base of neighborhoods of $x$ in the topology $\tau$.

Now, $\mathcal{B} = \mathcal{C}(\theta_X)$ is a base of neighborhoods of zero in $(X, \tau)$. We show that $(X, \tau)$ is a *TLS*. First, we show that the map $(x, y) \to x + y$ is jointly continuous. For this, let $x_0, y_0 \in X$ and $U \in \mathcal{B}$ be fixed. Since $x_0 + y_0 + U$ is a neighborhood of $x_0 + y_0$, by Proposition 4.2.2 (b), then, by $(B_2)$, there is a $V \in \mathcal{B}$ such that $V + V \subseteq U$. Here $x_0 + V$ is a neighborhood of $x_0$ and $y_0 + V$ is a neighborhood of $y_0$, by Proposition 4.2.2 (b), and

$$(x_0 + V) + (y_0 + V) = x_0 + y_0 + V + V \subseteq x_0 + y_0 + U.$$

Therefore, the map $(x, y) \to x + y$ is continuous at any point $(x_0, y_0) \in X \times X$. Hence, it is continuous on $X \times X$. Consequently, $(X, \tau)$ satisfies the condition $(TLS_1)$.

To show that the map $(\lambda, y) \to \lambda y$ is jointly continuous, let $\lambda_0 \in \mathbb{K}$, $n \in \mathbb{N}$ be such that $|\lambda_0| \leqslant n$, $x_0 \in X$ and $W \in \mathcal{B}$ be balanced. Then $\lambda_0 x_0 + W$ is a neighborhood of $\lambda_0 x_0$ in $(X, \tau)$. By $(B_2)$ and $(B_3)$, there is a balanced $V \in \mathcal{B}$ such that

$$V_{n+2} = \underbrace{V + \cdots + V}_{n+2 \text{ summans}} \subseteq W.$$

Hence, $nV + V + V \subseteq V_{n+2} \subseteq W$. Since $V$ is an absorbing set by the assumption, then there is an $r \geqslant 1$ such that $x_0 \in rV$.

Let $\lambda \in O(\lambda_0, \frac{1}{r}) = \{\lambda \in \mathbb{K} : |\lambda - \lambda_0| \leqslant \frac{1}{r}\}$ and $x \in x_0 + V$. Since

$$(\lambda - \lambda_0)(x - x_0) \in \frac{1}{r} V \subseteq V$$

and $\lambda_0(x - x_0) \in nV$, then

$$\lambda x - \lambda_0 x_0 = (\lambda - \lambda_0)x_0 + (\lambda - \lambda_0)(x - x_0) + \lambda_0(x - x_0) \in V + V + nV \subseteq W$$

for each $\lambda \in O(\lambda_0, \frac{1}{r})$ and $x \in x_0 + V$. Consequently, the map $(\lambda, x) \to \lambda x$ is continuous at any point $(\lambda_0, x_0)$. So, $(X, \tau)$ satisfies the condition $(TLS_2)$ also. Hence, $(X, \tau)$ is a $TLS$. $\qquad\square$

### 4.2.1 Topological algebras

An algebra over $\mathbb{K}$ is called a *topological algebra* ($TA$, in short) if $A$ is a $TLS$ in which the multiplication of elements $(a, b) \to ab$ from $A \times A$ into $A$ is separately continuous. That is, $(A, \tau)$ is a $TA$ if

$(TA_1)$ $(A, \tau)$ is a $TLS$;

$(TA_2)$ $a \to ab$ and $b \to ab$ from $A \times A$ into $A$ are continuous maps.

In particular, when $(A, \tau)$ satisfies the conditions $(TA_1)$ and

$(TA_3)$, the map $(a, b) \to ab$ from $A \times A$ into $A$ is continuous,

then $(A, \tau)$ is called a $TA$ *with jointly continuous multiplication*.

Herewith, every $TA$ with jointly continuous multiplication is a $TA$ and every complete and metrizable $TA$ is a $TA$ with jointly continuous multiplication (see, for example, Zelazko (1971), Theorem 15).

---

## 4.3 Topology of a linear space and an algebra, defined by a collection of $F$-seminorms

Let $X$ be a linear space over $\mathbb{K}$, $\mathcal{P} = \{p_\lambda : \lambda \in \Lambda\}$ a nonempty collection of $F$-seminorms on $X$ and $\tau_\mathcal{P}$ the *initial topology* on $X$, defined by the collection

$\mathcal{P}$. That is, $\tau_{\mathcal{P}}$ is the weakest topology on $X$, in which all $F$-seminorms $p_\lambda$ are continuous. In this topology, the collection

$$\{O_{\lambda\varepsilon} : \lambda \in \Lambda, \varepsilon > 0\},$$

where $O_{\lambda\varepsilon} = \{x \in X : p_\lambda(x) < \varepsilon\}$ forms in $(X, \tau_{\mathcal{P}})$ a subbase of neigbourhoods of zero.

To show that the topology of every $TLS$ and $TA$ we can define by some collection of $F$-seminorms, we prove first

**Proposition 4.3.1.** *Let $X$ be a linear space over $\mathbb{K}$, $\mathcal{P} = \{p_\lambda : \lambda \in \Lambda\}$ a nonempty collection of $F$-seminorms on $X$ and $\tau_{\mathcal{P}}$ the initial topology on $X$, defined by the collection $\mathcal{P}$. Then $(X, \tau_{\mathcal{P}})$ is a $TLS$. In particular case, when $X$ is an algebra over $\mathbb{K}$, then $(X, \tau_{\mathcal{P}})$ is a $TA$, if $\mathcal{P}$ satisfies the condition*

*(a) for each fixed $x \in X$ and any $\varepsilon > 0$ and $\lambda \in \Lambda$, there exist $\delta_x > 0$ and $\lambda_x \in \Lambda$ such that $p_\lambda(xy) < \varepsilon$ and $p_\lambda(yx) < \varepsilon$, whenever $p_{\lambda_x}(y) < \delta_x$.*

*Moreover, $(X, \tau_{\mathcal{P}})$ is a $TA$ with jointly continuous multiplication, if $\mathcal{P}$ satisfies the condition*

*(b) for any $\varepsilon > 0$ and any $\lambda \in \Lambda$, there exist $\delta > 0$ and $\lambda' \in \Lambda$ such that $p_\lambda(xy) < \varepsilon$, whenever $p_{\lambda'}(x) < \delta$ and $p_{\lambda'}(y) < \delta$.*

*Proof.* It is easy to see that, for each $\varepsilon > 0$ and each $\lambda \in \Lambda$, every set $O_{\lambda\varepsilon}$ in the subbase $\{O_{\lambda\varepsilon} : \lambda \in \Lambda, \varepsilon > 0\}$ of neighborhoods of zero in $(X, \tau_{\mathcal{P}})$ is balanced by the condition (5) of an $F$-seminorm. Moreover, $O_{\lambda\varepsilon}$ is also absorbing by the condition (6) of an $F$-seminorm. Indeed, $p_\lambda(\frac{x}{n}) \to 0$ for each $x \in X$ by (6). Therefore, there is an $n_0 \in \mathbb{N}$ such that $\frac{x}{n} \in O_{\lambda\varepsilon}$, if $n > n_0$. Hence, $x \in n_1 O_{\lambda\varepsilon}$ for some $n_1 > n_0$.

To show that the addition in $(X, \tau_{\mathcal{P}})$ is jointly continuous, let $x_0, y_0 \in X$ and $O(x_0 + y_0)$ be a neighborhood of $x_0 + y_0$ in $(X, \tau_{\mathcal{P}})$. Then $O(x_0 + y_0) = x_0 + y_0 + O$ for some neighborhood $O$ of zero in $(X, \tau_{\mathcal{P}})$ by Proposition 4.2.2 (e), which defines $n \in \mathbb{N}$, $\varepsilon > 0$ and $\lambda_1, \ldots, \lambda_n \in \Lambda$ such that

$$\bigcap_{k=1}^{n} O_{\lambda_k \varepsilon} \subseteq O. \tag{4.1}$$

Let

$$V = \bigcap_{k=1}^{n} O_{\lambda_k \frac{\varepsilon}{2}},$$

$O(x_0) = x_0 + V$ and $O(y_0) = y_0 + V$. Since, $p_{\lambda_k}$ satisfies, for every $k$, the condition (3) of a $F$-seminorm, then from

$$p_{\lambda_k}[(x + y) - (x_0 + y_0)] \leqslant p_{\lambda_k}(x - x_0) + p_{\lambda_k}(y - y_0) < \frac{\varepsilon}{2} + \frac{\varepsilon}{2} = \varepsilon$$

for every $k$ follows that $x + y \in O(x_0 + y_0)$ for each $(x, y) \in O(x_0) \times O(y_0)$.

Therefore, the addition in $(X, \tau_{\mathcal{P}})$ is jointly continuous in any point $(x_0, y_0) \in X \times X$.

To show that the map $(\mu, x) \to \mu x$ from $\mathbb{K} \times X \to X$ is jointly continuous, let $\mu_0 \in \mathbb{K}$ and $x_0 \in X$ be fixed and $O(\mu_0 x_0)$ be a neighborhood of $\mu_0 x_0$ in $(X, \tau_{\mathcal{P}})$. Then $O(\mu_0 x_0) = \mu_0 x_0 + O$, where $O$ is a neighborhood of zero in $(X, \tau_{\mathcal{P}})$. Now there are again $n \in \mathbb{N}$, $\varepsilon > 0$ and $\lambda_1, \ldots, \lambda_n \in \Lambda$ such that the inclusion (1) holds. For every $m \in \mathbb{N}$, let

$$V_m = \bigcap_{k=1}^{n} O_{\lambda_k \frac{\varepsilon}{2^m}},$$

$v \in \mathbb{N}$ be such that $|\mu_0| \leqslant v$, $\rho \in (0, 1]$ be such that $\rho x_0 \in V_{v+2}$ (such $\rho$ exists, because $V_{v+2}$ is a neighborhood of zero and therefore absorbing by Proposition 4.2.2 (d)), $O(\mu_0) = \mu_0 + O_\rho$ (here $O_\rho = \{\lambda \in \mathbb{K} : |\lambda| \leqslant \rho\}$) and $O(x_0) = x_0 + V_{v+2}$. Let $\mu \in O(\mu_0)$. Since

$$\mu x - \mu_0 x_0 = (\mu - \mu_0)(x - x_0) + (\mu - \mu_0)x_0 + \mu_0(x - x_0),$$

then

$$p_{\lambda_k}(\mu x - \mu_0 x_0) \leqslant p_{\lambda_k}(x - x_0) + p_{\lambda_k}((\mu - \mu_0)x_0) + p_{\lambda_k}(\mu_0(x - x_0)) \quad (4.2)$$

by the conditions (3) and (5) of an $F$-seminorm. Moreover,

$$(\mu - \mu_0)x_0 \in \frac{\mu - \mu_0}{\rho} V_{v+2} \subseteq V_{v+2},$$

because $|\frac{\mu - \mu_0}{\rho}| \leqslant 1$ and $V_{v+2}$ is balanced, and $\mu_0(x - x_0) \in v V_{v+2}$. Therefore, for every $k$, by (2), we have that

$$p_{\lambda_k}(\mu x - \mu_0 x_0) \leqslant \frac{\varepsilon}{2^{v+2}} + \frac{\varepsilon}{2^{v+2}} + \frac{v\varepsilon}{2^{v+2}} = \frac{v+2}{2^{v+2}}\varepsilon < \varepsilon$$

or $\mu x - \mu_0 x_0 \in O_{\lambda_k \varepsilon}$ for each $k$ if $\mu \in O(\mu_0)$ and $x \in O(x_0)$. Hence, $\mu x \in O(\mu_0 x_0)$, if $\mu \in O(\mu_0)$ and $x \in O(x_0)$. So we have proved that the multiplication by scalar is jointly continuous in $(X, \tau_{\mathcal{P}})$. Hence, $(X, \tau_{\mathcal{P}})$ is a TLS.

If now $\mathcal{P}$ satisfies the condition (a), then the multiplication $(x, y) \to xy$ in $(X, \tau_{\mathcal{P}})$ is separately continuous. To show that, let $O$ be an arbitrary neighborhood of zero in $(X, \tau)$. Then there exist $\varepsilon > 0$, $n \in \mathbb{N}$ and $\lambda_1, \ldots, \lambda_n \in \Lambda$ such that the inclusion (1) holds. For each fixed $x \in X$ and each $k \in \{1, \ldots, n\}$ there are, by the condition (a), a number $\delta_x(k) > 0$ and an index $\lambda_x(k) \in \Lambda$ such that $p_{\lambda_k}(xy) < \varepsilon$ and $p_{\lambda_k}(yx) < \varepsilon$ whenever $p_{\lambda_x(k)}(y) < \delta_x(k)$. Let now

$$\delta_x = \min\{\delta_x(1), \ldots, \delta_x(n)\}$$

and

$$V_x = \bigcap_{k=1}^{n} O_{\lambda_x(k)\delta_x}.$$

Then $V_x$ is a neighborhood of zero in $X$ in the topology $\tau_\mathcal{P}$ and

$$xV_x \cup V_x x \subset \bigcap_{k=1}^{n} [xO_{\lambda_x(k)\delta_x(k)} \cup O_{\lambda_x(k)\delta_x(k)}x] \subset \bigcap_{k=1}^{n} O_{\lambda_k \varepsilon} \subset O.$$

Hence, the multiplication in $(X, \tau_\mathcal{P})$ is separately continuous. Consequently, $(X, \tau_\mathcal{P})$ is a $TA$.

If $\mathcal{P}$ satisfies the condition (b), then the multiplication $(x, y) \to xy$ is jointly continuous in $(X, \tau_\mathcal{P})$. To show that, let again $O$ be an arbitrary neighborhood of zero in the topology $\tau_\mathcal{P}$ on $X$. Then there are $\varepsilon > 0$, $n \in \mathbb{N}$ and $\lambda_1, \ldots, \lambda_n \in \Lambda$ such that holds (1). Now, for each $k$, there are, by the condition (b), a number $\delta_k > 0$ and an index $\lambda'_k \in \Lambda$ such that $p_{\lambda_k}(xy) < \varepsilon$, whenever $p_{\lambda'_k}(x) < \delta_k$ and $p_{\lambda'_k}(y) < \delta_k$. Let now $\delta = \min\{\delta_1, \ldots, \delta_n\}$ and

$$V = \bigcap_{k=1}^{n} O_{\lambda'_k \delta}.$$

Then $V$ is a neighborhood of zero in $(X, \tau_\mathcal{P})$ and

$$VV \subset \bigcap_{k=1}^{n} O_{\lambda'_k \delta_k} O_{\lambda'_k \delta_k} \subset \bigcap_{k=1}^{n} O_{\lambda_k \varepsilon} \subset O.$$

It means that the multiplication in $(X, \tau_\mathcal{P})$ is jointly continuous. $\qquad\square$

Next result shows that the topology of every $TLS$ and evry $TA$ can be given by a collection of $F$-seminorms.

**Theorem 4.3.2.** *Every $TLS$ $(X, \tau)$ defines a collection $\mathcal{P}$ of $F$-seminorms on $X$ such that $\tau = \tau_\mathcal{P}$. If, in the partcular case, $(X, \tau)$ is a $TA$, then $(X, \tau_\mathcal{P})$ is a $TA$ (in case, when $(X, \tau)$ is a $TA$ with jointly continuous multiplication, then $(X, \tau_\mathcal{P})$ is also a $TA$ with jointly continuous multiplication)*

*Proof.* Let $M$ be the dense subset of $\mathbb{R}^+$ which consists of all non-negative rational numbers, having a finite dyadic expansions, i.e, we may write every such number $\rho \in M$ in the form

$$\rho = \sum_{n=0}^{\infty} \delta_n(\rho) \cdot 2^{-n},$$

where $\delta_0(\rho) \in \mathbb{N}'$, $\delta_n(\rho) \in \{0, 1\}$ for each $n \in \mathbb{N}$ and $\delta_n(\rho) = 0$ for $n$ sufficiently large.

Let $\mathcal{L}_{(X, \tau)}$ be a base of neighborhoods of zero in $(X, \tau)$, consisting of closed balanced sets (such base exists by the condition $(B_5)$ of Theore 4.2.3), and $\mathcal{S} = \{S_\lambda : \lambda \in \Lambda\}$ be the set of all strings $\mathcal{S}_\lambda = (U_n(\lambda))$ in $\mathcal{L}_{(X, \tau)}$, that is, for every fixed $\lambda \in \Lambda$, the knot $U_n(\lambda) \in \mathcal{L}_{(X, \tau)}$ and $U_{n+1}(\lambda) + U_{n+1}(\lambda) \subseteq U_n(\lambda)$ for all $n \in \mathbb{N}'$.

For each $\lambda \in \Lambda$ and $\rho \in M$, let

$$V_\lambda(\rho) = \underbrace{U_0(\lambda) + \cdots + U_0(\lambda)}_{\delta_0(\rho) \text{ summands}} + \sum_{n=1}^{\infty} \delta_n(\rho) \cdot U_n(\lambda) \tag{4.3}$$

and

$$p_\lambda(x) = \inf\{\rho \in M : x \in V_\lambda(\rho)\} \tag{4.4}$$

for each $x \in X$ and $\lambda \in \Lambda$. To show that every $p_\lambda$ is an $F$-seminorm on $X$, we first prove that

$$V_\lambda(\rho) + V_\lambda(\sigma) \subseteq V_\lambda(\rho + \sigma) \tag{4.5}$$

for all $\rho, \sigma \in M$. This assertion holds, if $\rho, \sigma \in \mathbb{N}'$. Suppose that, for some $N \in \mathbb{N}'$, the inclusion (5) holds for all $\rho, \sigma \in M$ such that $\delta_n(\rho) = \delta_n(\sigma) = 0$ for every $n > N$. To use here the mathematical induction, let $\rho, \sigma \in M$ be such that $\delta_n(\rho) = \delta_n(\sigma) = 0$ for each $n > N + 1$. Let $\rho' = \rho - \frac{1}{2^{N+1}}\delta_{N+1}(\rho)$ and $\sigma' = \sigma - \frac{1}{2^{N+1}}\delta_{N+1}(\sigma)$. Then $\rho', \sigma' \in M$ and $\delta_n(\rho') = \delta_n(\sigma') = 0$ for each $n > N$. Hence,

$$V_\lambda(\rho') + V_\lambda(\sigma') \subseteq V_\lambda(\rho' + \sigma'),$$

by the assumption. Taking this into account,

$$V_\lambda(\rho) + V_\lambda(\sigma) = V_\lambda(\rho') + \delta_{N+1}(\rho)U_{N+1}(\lambda) + V_\lambda(\sigma') + \delta_{N+1}(\sigma)U_{N+1}(\lambda)$$
$$\subseteq V_\lambda(\rho' + \sigma') + \delta_{N+1}(\rho)U_{N+1}(\lambda) + \delta_{N+1}(\sigma)U_{N+1}(\lambda).$$

If $\delta_{N+1}(\rho) = \delta_{N+1}(\sigma) = 0$, then (5) holds. Let now $\delta_{N+1}(\rho) = 1$ and $\delta_{N+1}(\sigma) = 0$. Then, by the definition of the set $V_\lambda(\rho)$, we have

$$V_\lambda(\rho) + V_\lambda(\sigma) \subseteq V_\lambda(\rho' + \sigma') + U_{N+1}(\lambda) =$$
$$= V_\lambda\left(\rho' + \sigma' + \frac{1}{2^{N+1}}\right) = V_\lambda(\rho + \sigma).$$

In the same way, it is easy to show that the inclusion (5) holds, when $\delta_{N+1}(\rho) = 0$ and $\delta_{N+1}(\sigma) = 1$.

Finally, if $\delta_{N+1}(\rho) = \delta_{N+1}(\sigma) = 1$, we have

$$V_\lambda(\rho) + V_\lambda(\sigma) \subseteq V_\lambda(\rho' + \sigma') + U_{N+1}(\lambda) + U_{N+1}(\lambda) \subseteq V_\lambda(\rho' + \sigma') + U_N(\lambda)$$
$$= V_\lambda\left(\rho' + \sigma' + \frac{1}{2^N}\right) = V_\lambda(\rho + \sigma)$$

Hence, the inclusion (5) holds for every $\rho, \sigma \in M$.

Let now $\rho \in M \setminus \{0\}$. Since every $U_n(\lambda)$ is a neighborhood of zero in $(X, \tau)$, then $V_\lambda(\rho)$ is also a neighborhood of zero in $(X, \tau)$ (thus, an absorbing set in $X$, by Proposition 4.2.2 (d)) for each $\lambda \in \Lambda$ and $\rho \in M \setminus \{0\}$. Hence, for each given $x \in X$, there exists $n_0 \in \mathbb{N}$ such that $x \in n_0 V_\lambda(\rho)$. Therefore, $x \in V_\lambda(n_0\rho)$, by the inclusion (5). Taking this into account, $0 \leqslant p_\lambda(x) \leqslant n_0\rho < \infty$ for each $x \in X$. Thus, $p_\lambda$ is a map from $X$ to $\mathbb{R}^+$. Moreover, $p_\lambda(x) = 0$, if $x = \theta_A$

(from $\theta_A \in V_\lambda(\frac{1}{2^n})$ for each $n \in \mathbb{N}$ follows that $p_\lambda(x) \leqslant \frac{1}{2^n}$ for each $n$) and $p_\lambda(\mu x) \leqslant p_\lambda(x)$, if $|\mu| \leqslant 1$. Indeed, the condition $p_\lambda(x) < \rho$ means that $x \in V_\lambda(\rho)$. Since $V_\lambda(\rho)$ is balanced, then $\mu x \in V_\lambda(\rho)$ for all $\mu \in \mathbb{K}$ with $|\mu| \leqslant 1$. Therefore, from $p_\lambda(x) < \rho$ follows that $p_\lambda(\mu x) < \rho$, whenever $|\mu| \leqslant 1$. So, $p_\lambda$ satisfies the conditions (1), (2) and (5) of an $F$-seminorm.

To show that $p_\lambda$ satisfies the condition (4) of an $F$-seminorm, fix $x, y \in X$ and $\varepsilon > 0$. Let $\rho, \sigma \in M$ be such that $x \in V_\lambda(\rho)$, $y \in V_\lambda(\sigma)$, $\rho \leqslant p_\lambda(x) + \frac{1}{2}\varepsilon$ and $\sigma \leqslant p_\lambda(y) + \frac{1}{2}\varepsilon$. Now, $V_\lambda(\rho) + V_\lambda(\sigma) \subseteq V_\lambda(\rho + \sigma)$ by the inclusion (4). Therefore, $p_\lambda(x + y) \leqslant \rho + \sigma \leqslant p_\lambda(x) + p_\lambda(y) + \varepsilon$. Because $\varepsilon$ is arbitrary, we have shown that the condition (4) of an $F$-seminorm holds.

Next, let $(\mu_n)$ be a sequence in $\mathbb{R}$, which converges to 0. Then there is a number $n_1 \in \mathbb{N}$ such that $\mu_n \leqslant \frac{1}{n}$ whenever $n > n_1$. Above we showed that, for every $x \in X$ and $\rho \in M$ with $\rho \neq 0$, there exists a number $n_0 \in \mathbb{N}$ such that $x \in V(n_0\rho)$ (also for very small $\rho$). Let $n_2 \geqslant n_0$ and $n_2 \geqslant n_1$. Since

$$p_\lambda(\mu_n x) \leqslant p_\lambda\left(\frac{1}{n}x\right) \leqslant p_\lambda\left(\frac{1}{n_0}x\right) \leqslant \rho$$

if $n > n_2$, by the condition (5) of an $F$-seminorm, then

$$\lim_{n \to \infty} p_\lambda(\mu_n x) = 0.$$

If now $(x_n)$ is a sequence in $X$ such that $(x_n) \to \theta_X$, then $(\mu x_n) \to \theta_X$ for each $\mu \in \mathbb{K}$. Let $\varepsilon > 0$ be an arbitrary small number. Fix an $n_0 \in \mathbb{N}$ such that $\frac{1}{2^{n_0}} < \varepsilon$. Since $V_\lambda(\frac{1}{2^{n_0}})$ is a neighborhood of zero in $(X, \tau)$, then there is an $n_1 \in \mathbb{N}$ such that $\mu x_n \in V_\lambda(\frac{1}{2^{n_0}})$ whenever $n \geqslant n_1$. Therefore, from $p_\lambda(\mu x_n) \leqslant \frac{1}{2^{n_0}} < \varepsilon$, whenever $n \geqslant n_1$, follows that

$$\lim_{n \to \infty} p_\lambda(\mu x_n) = 0$$

for each $\mu \in \mathbb{K}$. Consequently, every $p_\lambda$ satisfies also the conditions (6) and (7) of an $F$-seminorm. So, $\mathcal{S}$ defines a collection $\mathcal{P} = \{p_\lambda : \lambda \in \Lambda\}$ of $F$-seminorms on $X$.

Next, we show that $\tau = \tau_\mathcal{P}$. For this, let $U \in \mathcal{L}_{(X,\tau)}$, $S_{\lambda_0} = (U_n(\lambda_0))$ be a fixed string in $\mathcal{L}_{(X,\tau)}$ for which $U_0(\lambda_0) = U$ and let

$$u \in O_{\lambda_0 1} = \{x \in X : p_{\lambda_0}(x) < 1\}.$$

Then $u \in V_{\lambda_0}(1) = U_0(\lambda_0) = U$. It means that $O_{\lambda_0 1} \subset U$. Since $O_{\lambda_0 1}$ belongs to the base of neigbourhoods of zero in $(X, \tau_\mathcal{P})$, then $\tau \subseteq \tau_\mathcal{P}$ by Lemma 4.1.2.

Let now $O \in \mathcal{L}_{(X,\tau_\mathcal{P})}$. Then there are $\varepsilon > 0$, $n \in \mathbb{N}$ and $\lambda_1, \ldots, \lambda_n \in \Lambda$ (with this we fix $n$ strings $S_{\lambda_1} = (U_n(\lambda_1)), \ldots, S_{\lambda_n} = (U_n(\lambda_n))$ in $\mathcal{L}_{(X,\tau)}$) such that the inclusion (1) holds. Let $n_\varepsilon \in \mathbb{N}$ be such that $\frac{1}{2^{n_\varepsilon}} < \varepsilon$. Then

$$U = \bigcap_{k=1}^{n} U_{n_\varepsilon}(\lambda_k)$$

is a neighborhood of zero in $(X, \tau_{\mathcal{P}})$. Since

$$U_{n_\varepsilon}(\lambda_k) = V_{\lambda_k}\left(\frac{1}{2^{n_\varepsilon}}\right)$$

for each $k = 1, \ldots, n$, then from $u \in U$ follows that $p_{\lambda_k}(u) \leqslant \frac{1}{2^{n_\varepsilon}} < \varepsilon$ for each $k = 1, \ldots, n$. Hence, $U \subset O$. It means that $\tau_{\mathcal{P}} \subseteq \tau$, by Lemma 4.1.2. Consequently, $\tau = \tau_{\mathcal{P}}$.

To show that $\mathcal{P}$ satisfies the condition (a) of Proposition 4.3.1, let $x \in X$, $\lambda \in \Lambda$ (by this we fix a string $S_\lambda = (U_n(\lambda))$ in $\mathcal{L}_{(X,\tau)}$) and $\varepsilon > 0$. Then there exists a number $n_\varepsilon \in \mathbb{N}$ such that $\frac{1}{2^{n_\varepsilon}} < \varepsilon$. Since the multiplication $(x, y) \to xy$ in $(X, \tau)$ is separately continuous, then there exists a neighborhood $V_x \in \mathcal{L}_{(X,\tau)}$ such that $xV_x \cup V_x x \subseteq U_{n_\varepsilon}(\lambda)$. Let $(U_n)$ be a string in $\mathcal{L}_{(X,\tau)}$, which is generated by $V_x$, that is, $U_0 = V_x$ and other members of this string are defined by $V_x$. Hence, there exists an index $\lambda_x \in \Lambda$ such that $S_{\lambda_x} = (U_n(\lambda_x))$, where $U_0(\lambda_x) = V_x$. Now, $V_x = V_{\lambda_x}(1)$ (or $p_{\lambda_x}(y) < 1$ if $y \in V_x$) and

$$xV_x \cup V_x x \subset U_{n_\varepsilon}(\lambda) = V_\lambda\left(\frac{1}{2^{n_\varepsilon}}\right).$$

Therefore, for every fixed $x \in X$, any $\lambda \in \Lambda$ and $\varepsilon > 0$, there exists an index $\lambda_x \in \Lambda$ such that $p_\lambda(xy) \leqslant 2^{-n_\varepsilon} < \varepsilon$ and $p_\lambda(yx) \leqslant 2^{-n_\varepsilon} < \varepsilon$, whenever $p_{\lambda_x}(y) < 1$. Hence, the collection $\mathcal{P}$ satisfies the condition (a) of the Proposition 4.3.1, because of which $(X, \tau_{\mathcal{P}})$ is a $TA$.

In particular, when $(X, \tau)$ is a $TA$ with jointly continuous multiplication, then the multiplication in $(X, \tau_{\mathcal{P}})$ is also jointly continuous. Indeed, let $\lambda \in \Lambda$ (by this we fix again a string $S_\lambda = (U_n(\lambda))$ in $\mathcal{L}_{(X,\tau)}$),0 and $\varepsilon > 0$. Then there is again a number $n_\varepsilon \in \mathbb{N}$ such that $\frac{1}{2^{n_\varepsilon}} < \varepsilon$. Since the multiplication $(x, y) \to xy$ in $(X, \tau)$ is jointly continuous, then there exists an element $V \in \mathcal{L}_{(X,\tau)}$ such that $VV \subset U_{n_\varepsilon}(\lambda)$. Let $(U_n)$ be a string in $\mathcal{L}_{(X,\tau)}$ for which $U_0 = V$. Then there exists an index $\lambda' \in \Lambda$ such that $S_{\lambda'} = (U_n(\lambda'))$, where $U_0(\lambda') = V$. If $x, y \in V = V_{\lambda'}(1)$, then $p_{\lambda'}(x) < 1$, $p_{\lambda'}(y) < 1$ and from $xy \in VV \subset U_{n_\varepsilon}(\lambda) = V_\lambda(\frac{1}{2^{n_\varepsilon}})$ follows that $p_\lambda(xy) \leqslant \frac{1}{2^{n_\varepsilon}} < \varepsilon$. Hence, for each $\lambda \in \Lambda$ and $\varepsilon > 0$, there exists $\lambda' \in \Lambda$ such that $p_\lambda(xy) < \varepsilon$, whenever $p_{\lambda'}(x) < 1$ and $p_{\lambda'}(y) < 1$. This shows that $\mathcal{P}$ satisfies the condition (b) of Proposition 4.3.1. Consequently, in this case, $(X, \tau_{\mathcal{P}})$ is a TA with jointly continuous multiplication.  □

## 4.4    Locally pseudoconvex spaces and algebras

A *TLS* $(X, \tau)$ is called a *LPS*, if $(X, \tau)$ has a base of neighborhoods of zero consisting of balanced pseudoconvex sets and a *TLS* $(X, \tau)$ is called a *locally k-convex space* with $k \in (0,1]$ (*LkPS*, in short) if $(X,\tau)$ has a base of neighborhoods of zero consisting of balanced $k$-convex sets. Moreover, a $TA$ is called a *locally pseudoconvex algebra* (*LPA*, in short) if the underlying $TLS$ of $(X,\tau)$ is a *LPS* and a a $TA$ is called *locally k-convex algebra* with $k \in (0,1]$ (*LkPA*, in short) if the underlying $TLS$ of $(X,\tau)$ is a *LkPS*. Instead of terms ,,locally 1-connvex space" and ,,locally 1-convex algebra" the terms *locally convex space* and *locally convex algebra* are used.

To show that $(X,\tau)$ has a base of neighborhoods of zero consisting of absolutely pseudoconvex sets, we need the following results.

**Proposition 4.4.1.** *Let $(X,\tau)$ be a LPS and $U$ a neighborhood of zero in $(X,\tau)$, satisfying the condition $U + U \subseteq 2^{\frac{1}{k}}U$ for some $k \in (0,1]$. Then*

$$\sum_{i=1}^{n} 2^{-\frac{k_i}{k}} U \subseteq U \tag{4.6}$$

*for each $n \geq 2$, where $k_i \in \mathbb{N}$ are such that $\sum_{i=1}^{n} \frac{1}{2^{k_i}} \leq 1$.*

*Proof.* We can assume that $k_i \in \mathbb{N}$ are such that

$$\sum_{i=1}^{n} \frac{1}{2^{k_i}} = 1, \tag{4.7}$$

(otherwise we add to this sum several fractions with the same powers). Then at least two numbers of $k_1, \ldots, k_n$ coincide. We show that, if

$$1 \leq k_1 < k_2 < \cdots < k_n,$$

then

$$\sum_{i=1}^{n} \frac{1}{2^{k_i}} < 1 \tag{4.8}$$

for every $n \geq 2$. To show this, we use the mathematical induction. First we put $n = 2$. Then $k_2 = k_1 + m$ for some $m \in \mathbb{N}$ and

$$\frac{1}{2^{k_1}} + \frac{1}{2^{k_2}} = \frac{2^{k_2-k_1}+1}{2^{k_2}} = \frac{2^m+1}{2^{k_1+m}} = \frac{1+2^{-m}}{2^{k_1}} < 2^{1-k_1} \leq 1.$$

Assume now that (8) holds for $n = N$. Next we put $n = N + 1$ and $m_i = k_{i+1} - k_1$ for each $i \in \{1, \ldots, N\}$. Then $1 \leq m_1 < \ldots < m_N$. Therefore

$$\sum_{i=1}^{N} \frac{1}{2^{m_i}} < 1.$$

Hence,

$$\sum_{i=1}^{N+1} \frac{1}{2^{k_i}} = \frac{1}{2^{k_1}}\left(1 + \sum_{i=1}^{N} \frac{1}{2^{m_i}}\right) < 2^{1-k_1} \leq 1.$$

So, the inequality (8) holds for every $n \geqslant 2$.

To prove the inclusion (6), we put first $n = 2$ and $k_1 = k_2 = 1$. Then from $U + U \subseteq 2^{\frac{1}{k}}U$ follows that $2^{-\frac{1}{k}}U + 2^{-\frac{1}{k}}U \subseteq U$. Assume that the inclusion (6) holds for $n = N$, that is, we consider numbers $k_1, \ldots, k_{N+1}$ such that (7) holds for $n = N + 1$. Since at least two of powers $k_1, \ldots, k_{N+1}$ coincide, we can fixe that $k_N = k_{N+1}$ (otherwise we rename the indexes), and put $m_i = k_i$ for $i \in \{1, \ldots, N-1\}$ and $m_N = k_N - 1$. Then

$$\sum_{i=1}^{N} \frac{1}{2^{m_i}} = 1.$$

Therefore,

$$\sum_{i=1}^{N+1} 2^{-\frac{k_i}{k}} U = \sum_{i=1}^{N-1} 2^{-\frac{k_i}{k}} U + 2^{-\frac{k_N}{k}}(U + U) \subseteq \sum_{i=1}^{N} 2^{-\frac{m_i}{k}} U \subseteq U.$$

Hence, the inclusion (6) holds for all $n \geqslant 2$ and $k_i \in \mathbb{N}$ satisfying the condition (6). $\qquad\square$

**Proposition 4.4.2.** *Let $(X, \tau)$ be a TLS and $U$ a balanced neighborhood of zero in $(X, \tau)$ such that $U + U \subseteq 2^{\frac{1}{k}}U$ for some $k \in (0, 1]$. Then*

$$\Gamma_k(U) \subseteq 2^{\frac{1}{k}}U. \tag{4.9}$$

*Proof.* Let $x \in \Gamma_k(U)$. Then there are $n \in \mathbb{N}$, $u_1, \ldots, u_n \in U$ and $\mu_1, \ldots, \mu_n \in \mathbb{K}$ with

$$\sum_{i=1}^{n} |\mu_i|^k \leqslant 1$$

such that

$$x = \sum_{i=1}^{n} \mu_i u_i.$$

Moreover, let $i \in \{1, \ldots, n\}$ and $k_i = -\log_2 |\mu_i|^k + 1$. Then

$$k_i - 1 = -\log_2 |\mu_i|^k < k_i,$$

because of which

$$2^{k_i-1} = 2^{-\log_2 |\mu_i|^k} < 2^{k_i}.$$

Since

$$2^{-\log_2 |\mu_i|^k} = \frac{1}{|\mu_i|^k},$$

then $2^{-k_i} < |\mu_i|^k = 2^{1-k_i}$. Hence,

$$\sum_{i=1}^{n} \frac{1}{2^{k_i}} < \sum_{i=1}^{n} |\mu_i|^k \leqslant 1.$$

Therefore, the inclusion (6) holds, by Proposition 4.4.1. Because $|2^{\frac{k_i-1}{k}}\mu_i| = 1$ and $U$ is balanced, then

$$x = \sum_{i=1}^{n}\mu_i u_i = 2^{\frac{1}{k}}\sum_{i=1}^{n}2^{-\frac{k_i}{k}}\left(2^{\frac{k_i}{k}}2^{-\frac{1}{k}}\mu_i\right)u_i \subseteq 2^{\frac{1}{k}}\sum_{i=1}^{n}2^{-\frac{k_i}{k}}U \subseteq 2^{\frac{1}{k}}U. \qquad \square$$

**Theorem 4.4.3.** *Absolutely pseudoconvex neighborhoods of zero form a base of neighborhoods of zero in every LPS.*

*Proof.* Let $(X,\tau)$ be a *LPS*. Then every neighborhood $O$ of zero in $(X,\tau)$ contains a balanced and $k$-convex neighborhood $U$ of zero for some $k \in (0,1]$. Then

$$2^{-\frac{1}{k}}\Gamma_k(U) \subseteq U \subseteq O,$$

by Proposition 4.4.2. Since $2^{-\frac{1}{k}}\Gamma_k(U)$ is a balanced absolutely $k$-convex neighborhood of zero, then the statement of Theorem 4.4.3 is true. $\qquad \square$

**Proposition 4.4.4.** *Let $(X,\tau)$ be a TLS and $U$ an absorbing absolutely $k$-convex set for some $k \in (0,1]$. Then the $k$-homogeneous Minkovsky functional $p_U$ of $U$, defined by*

$$p_U(x) = \inf\{\lambda^k : \lambda > 0 \text{ and } x \in \lambda U\}$$

*for each $x \in X$, satisfies the following conditions:*

(a) $p_U(x) \in [0,\infty)$ *for each $x \in X$;*
(b) $p_U(\mu x) = |\mu|^k p_U(x)$ *for each $x \in X$ and $\mu \in \mathbb{K}$ ($k$-homogeneity);*
(c) $p_U(x+y) \leqslant p_U(x) + p_U(y)$ *for each $x,y \in X$ (subadditivity).*

*Proof.* a) It is clear that $p_U(x) \geqslant 0$ for all $x \in X$. Since $U$ is an absorbing set, then for each $x \in X$ there exists a number $\varepsilon_x > 0$ such that $x \in \lambda U$ for all $\lambda \in \mathbb{K}$ with $|\lambda| \geqslant \varepsilon_x$. Hence, $p_U(x) \leqslant \varepsilon_x^k < \infty$.

b) If $\mu = 0$, then the equality (b) holds, because $p_U(\theta_A) = 0$. Let now $\mu = \xi\sigma \neq 0$, where $\xi = |\mu|$ and $|\sigma| = 1$.

If $x \in X$ and $\lambda > 0$ are such that $\lambda^k > p_U(x)$, then $x \in \lambda U$. Since $\lambda U$ is balanced, then $\sigma x \in \lambda U$. Hence, $p_U(\sigma x) \leqslant \lambda^k$. It means that the set $K = \{\lambda : \lambda^k > p_U(x)\}$ is bounded from below by $p_U(\sigma x)$. Hence $p_U(\sigma x) \leqslant p_U(x)$ and $p_U(x) = p_U(\frac{1}{\sigma}\sigma x) \leqslant p_U(\sigma x)$, because $|\frac{1}{\sigma}| = 1$. Consequently, $p_U(\sigma x) = p_U(x)$ for each $x \in X$ and $\sigma \in \mathbb{K}$ with $|\sigma| = 1$. Since $\xi > 0$ and $x \in \lambda U$, then from $\xi x \in \xi\lambda U$ follows that $p_U(\xi x) \leqslant \xi^k \lambda^k$, that is, the set $K$ is bounded from below by $p_U(\xi x)\frac{1}{\xi^k}$. Therefore, $p_U(\xi x) \leqslant \xi^k p_U(x)$ and

$$p_U(x) = p_U\left(\frac{1}{\xi}\xi x\right) \leqslant \frac{1}{\xi^k}p_U(\xi x),$$

because $\xi^{-1} > 0$. Thus, $p_U(\xi x) = \xi^k p_U(x)$.

Taking this into account,

$$p_U(\mu x) = p_U(\xi(\sigma x)) = \xi^k p_U(\sigma x) = \xi^k p_U(x) = |\mu|^k p_U(x)$$

for each $x \in X$ and $\mu \in \mathbb{K}$.

c) Let $x, y \in X$ and $\lambda, \mu \in \mathbb{R}$ be such that $\lambda^k > p_U(x)$ and $\mu^k > p_U(y)$. Then $x \in \lambda U$ and $y \in \mu U$. Therefore, we put $x = \lambda u$ and $y = \mu v$ for some $u, v \in U$. Since $U$ is absolutely $k$-convex and

$$\left| \frac{\lambda}{(\lambda^k + \mu^k)^{\frac{1}{k}}} \right|^k + \left| \frac{\mu}{(\lambda^k + \mu^k)^{\frac{1}{k}}} \right|^k = 1,$$

then

$$\frac{x+y}{(\lambda^k + \mu^k)^{\frac{1}{k}}} = \frac{\lambda}{(\lambda^k + \mu^k)^{\frac{1}{k}}} u + \frac{\mu}{(\lambda^k + \mu^k)^{\frac{1}{k}}} v \in \Gamma_k(U) = U$$

or $x + y \in (\lambda^k + \mu^k)^{\frac{1}{k}} U$. Hence, $p_U(x+y) \leqslant \lambda^k + \mu^k$. Fixing now, one of $\lambda$ or $\mu$, taking the infimum with respect to it, and then taking the infimum with respect of other, we have that the condition (c) holds. $\qquad \square$

**Corollary 4.4.5.** *Let* $k \in (0, 1]$ *and* $U$ *be an absorbing absolutely $k$-convex set in a TLS* $(X, \tau)$. *Then*

$$\{x \in X : p_U(x) < 1\} \subset U \subset \{x \in X : p_U(x) \leqslant 1\}.$$

**Theorem 4.4.6.** *Every LPS* $(X, \tau)$ *defines a collection* $\mathcal{P}$ *of nonhomogeneous seminorms on* $X$ *such that* $\tau = \tau_{\mathcal{P}}$. *If, in the particular case,* $(X, \tau)$ *is a LPA, then* $(X, \tau_{\mathcal{P}})$ *is a LPA (in case, when* $(X, \tau)$ *is a LPA with jointly continuous multiplication, then* $(X, \tau_{\mathcal{P}})$ *is also a LPA with jointly continuous multiplication).*

*Proof.* Let $(X, \tau)$ be a *LPS*. Then $(X, \tau)$ has a base $\mathcal{L}_X$ of absolutely pseudoconvex neighborhoods of zero by Theorem 4.4.3. As every neighborhood of zero in $(X, \tau)$ is absorbing, by Proposition 4.2.2 (d), then every neighborhood $U \in \mathcal{L}_X$ defines a $k \in (0, 1]$ and a $k$-homogeneous seminorm $p_U$ on $X$, by Proposition 4.4.4.

Let $\mathcal{P} = \{p_U : U \in \mathcal{L}_X\}$ and $\tau_{\mathcal{P}}$ be the initial topology on $X$, defined by the collection $\mathcal{P}$. To show that $\tau = \tau_{\mathcal{P}}$, let $O$ be a base neighborhood of zero in the topology $\tau$. Since $(X, \tau)$ is a *LPS*, then there is a $U \in \mathcal{L}_X$ such that $U \subseteq O$. Take $x \in X$ such that $p_U(x) < \frac{1}{2^{k^2}}$. Then $x \in \frac{1}{2^k} U \subseteq \frac{1}{2^k} \Gamma_k(U) \subseteq U$ by Proposition 4.4.2. Hence,

$$V = \{x \in X : p_U(x) < \frac{1}{2^{k^2}}\} \subset O.$$

Because $V \in \tau_{\mathcal{P}}$, then $\tau \subseteq \tau_{\mathcal{P}}$ by Lemma 4.1.2.

To show that $\tau_{\mathcal{P}} \subseteq \tau$, let $O \in \tau_{\mathcal{P}}$. Then there is $n \in \mathbb{N}$, positive numbers $\varepsilon_1, \ldots, \varepsilon_n$ and $U_1, \ldots, U_n \in \mathcal{L}_X$ such that

$$O = \bigcap_{i=1}^{n} \{x \in X : p_{U_k}(x) < \varepsilon_k\}.$$

Since every $p_{U_k}$ is continuous, then $O \in \tau$. It means, that $\tau_{\mathcal{P}} \subseteq \tau$ by Lemma 4.1.2. Consequently, $\tau = \tau_{\mathcal{P}}$. $\qquad \square$

Next, we show how $F$-seminorms and homogeneous seminorms on every $LCS$ are connected with one another. Let $O$ be an absolutely convex neighborhood of zero in a $LCS$ $(X, \tau)$ and $S = (\frac{O}{2^n})$. Since

$$\frac{O}{2^{n+1}} + \frac{O}{2^{n+1}} \subseteq \frac{1}{2^n}\left(\frac{O}{2} + \frac{O}{2}\right) \subseteq \frac{\Gamma(O)}{2^n} = \frac{O}{2^n},$$

then $S$ is a string. Let now $p_S$ be the $F$-seminorm on $X$, defined by the string $S$, and $p_O$ the $k$-homogeneous seminorm on $X$, defined by $O$. Since

$$\rho O = \sum_{n=0}^{\infty} \delta_n(\rho)\frac{O}{2^n} = V(\rho)$$

for every $\rho \in M$, then

$$p_S(x) = \inf\{\rho \in M : x \in V(\rho)\} = \inf\{\rho > 0 : x \in \rho O\} = p_O(x)$$

for each $x \in X$. Thus, $p_S$ is a homogeneous seminorm.

### 4.4.1 Saturation of a collection of seminorms

Let $(X, \tau)$ be a $LPS$ and $\mathcal{P} = \{p_\lambda : \lambda \in \Lambda\}$ a collection of $k_\lambda$-homogenous seminorms $p_\lambda$ on $X$ with $k_\lambda \in (0, 1]$ for each $\lambda \in \Lambda$. In addition, let $n \in \mathbb{N}$, $\lambda_1, \ldots, \lambda_n \in \Lambda$, $\overline{\lambda} = \{(\lambda_1, \ldots, \lambda_n)\}$, $k = \min\{k_{\lambda_1}, \ldots, k_{\lambda_n}\}$ and

$$p_{\overline{\lambda}}(x) = \max\{p_{\lambda_1}(x)^{\frac{k}{k_{\lambda_1}}}, \ldots, p_{\lambda_n}(x)^{\frac{k}{k_{\lambda_n}}}\} \tag{4.10}$$

for each $x \in X$. Then $p_{\overline{\lambda}}$ is a $k$-homogeneous seminorm on $X$.

Indeed, $0 \leqslant p_{\overline{\lambda}}(x) < \infty$ for each $x \in X$. Since

$$p_{\lambda_i}(\mu x)^{\frac{k}{k_{\lambda_i}}} = (|\mu|^{k_{\lambda_i}}p_{\lambda_i}(x))^{\frac{k}{k_{\lambda_i}}} = |\mu|^k p_{\lambda_i}(x)^{\frac{k}{k_{\lambda_i}}}$$

for each $i \in \{i, \ldots, n\}$, $x \in X$ and $\mu \in \mathbb{K}$, then $p_{\overline{\lambda}}(\mu x) = |\mu|^k p_{\overline{\lambda}}(x)$ for each $x \in X$ and $\mu \in \mathbb{K}$. It is known (see, for example, Narici and Beckenstein (2011), Theorem 4.6.1) that $(a + b)^\alpha \leqslant a^\alpha + b^\alpha$ for all nonnegative numbers $a$ and $b$ and for any $\alpha \in (0, 1]$. Using this inequality,

$$p_{\lambda_i}(x + y)^{\frac{k}{k_{\lambda_i}}} \leqslant (p_{\lambda_i}(x) + p_{\lambda_i}(y))^{\frac{k}{k_{\lambda_i}}} \leqslant p_{\lambda_i}(x)^{\frac{k}{k_{\lambda_i}}} + p_{\lambda_i}(y)^{\frac{k}{k_{\lambda_i}}} \leqslant p_{\overline{\lambda}}(x) + p_{\overline{\lambda}}(y)$$

for each $x, y \in X$ and $i \in \{i, \ldots, n\}$, because $\frac{k}{k_{\lambda_i}} \in (0, 1]$. Therefore, we have that $p_{\overline{\lambda}}(x + y) \leqslant p_{\overline{\lambda}}(x) + p_{\overline{\lambda}}(x)$ for each $x, y \in X$. Consequently, every $p_{\overline{\lambda}}$ is a $k$-homogeneous seminorm on $X$.

For every collection $\mathcal{P} = \{p_\lambda : \lambda \in \Lambda\}$ of (homogeneous or nonhomogeneous) seminorms on $X$, the collection

$$\overline{\mathcal{P}} = \{p_{\{(\lambda_1, \ldots, \lambda_n)\}} : n \in \mathbb{N}, \lambda_1, \ldots, \lambda_n \in \Lambda\} \tag{4.11}$$

of seminorms on $X$ is called a *saturation* for $\mathcal{P}$ and $\mathcal{P}$ is called *saturated* if $\mathcal{P} = \overline{\mathcal{P}}$. It means that the collection $\mathcal{P}$ is saturated if $\mathcal{P}$ contains all seminorms on the form (11), defined by a finite number of seminorms from $\mathcal{P}$.

**Proposition 4.4.7.** *Let $(X, \tau)$ be a LPS, $\mathcal{P} = \{p_\lambda : \lambda \in \Lambda\}$ a collection of $k_\lambda$-homogeneous seminorms with $k_\lambda \in (0, 1]$ for each $\lambda \in \Lambda$, which defines the topology $\tau$, and $\overline{\mathcal{P}}$ the saturated collection for $\mathcal{P}$. Then $\tau_{\mathcal{P}} = \tau_{\overline{\mathcal{P}}}$.*

*Proof.* To show that $\tau_{\mathcal{P}} \subseteq \tau_{\overline{\mathcal{P}}}$, let $O$ be a neighborhood of zero in the topology $\tau_{\mathcal{P}}$. Then there are $n \in \mathbb{N}$, $\varepsilon_1, \ldots, \varepsilon_n > 0$ (we can assume that every $\varepsilon_i < 1$ otherwise we take smaller one ) and $\lambda_1, \ldots, \lambda_n \in \Lambda$ such that

$$\bigcap_{i=1}^{n} \{x \in X : p_{\lambda_i}(x) < \varepsilon_i\} \subseteq O.$$

Let $\varepsilon = \min\{\varepsilon_1, \ldots, \varepsilon_n\}$, $k = \min\{k_{\lambda_1}, \ldots, k_{\lambda_n}\}$, $m = \max\{\frac{k}{k_{\lambda_1}}, \ldots, \frac{k}{k_{\lambda_n}}\}$ and

$$U = \{x \in X : p_{\{(\lambda_1, \ldots, \lambda_n)\}}(x) < \varepsilon^m\}.$$

Then $U \in \tau_{\overline{\mathcal{P}}}$. Since by

$$p_{\lambda_i}(x)^{\frac{k}{k_{\lambda_i}}} \leqslant p_{\{(\lambda_1, \ldots, \lambda_n)\}}(x) < \varepsilon^m \leqslant \varepsilon_i^m \leqslant \varepsilon_i^{\frac{k}{k_{\lambda_i}}}$$

it is clear that $p_{\lambda_i}(x) < \varepsilon_i$ for each $x \in X$ and $i \in \{1, \ldots, n\}$, then $U \subseteq O$. It means that $\tau_{\mathcal{P}} \subseteq \tau_{\overline{\mathcal{P}}}$, by Lemma 4.1.2. To show that $\tau_{\overline{\mathcal{P}}} \subseteq \tau_{\mathcal{P}}$, let $O$ be a neighborhood of zero in the topology $\tau_{\overline{\mathcal{P}}}$. Then there are $n \in \mathbb{N}$, $\varepsilon_1, \ldots, \varepsilon_n > 0$ and $\overline{\lambda}_1 = \{\lambda_{11}, \ldots, \lambda_{1m_1}\}, \ldots, \overline{\lambda}_n = \{(\lambda_{n1}, \ldots, \lambda_{nm_n})\}$ for some $\lambda_{11}, \ldots, \lambda_{nm_n} \in \Lambda$ such that

$$\bigcap_{i=1}^{n} \{x \in X : p_{\overline{\lambda}_i}(x) < \varepsilon_i\} \subseteq O.$$

Let $\varepsilon = \min\{\varepsilon_1, \ldots, \varepsilon_m\}$, $k_{\lambda_{iv}} \in (0, 1]$ the homogeneous power of seminorm $p_{\lambda_{iv}}$, $k_i = \min\{k_{\lambda_i 1}, \ldots, k_{\lambda_i m_i}\}$ and

$$U = \bigcap_{i=1}^{n} \bigcap_{v=1}^{m_i} \left\{x \in X : p_{\lambda_{iv}}(x) < \varepsilon^{\frac{k_{\lambda_i v}}{k_i}}\right\}.$$

Then $U \in \tau_{\mathcal{P}}$. From

$$p_{\lambda_{iv}}(x) < \varepsilon^{\frac{k_{\lambda_i v}}{k_i}} \leqslant \varepsilon_i^{\frac{k_{\lambda_i v}}{k_i}}$$

for each $i$ and $v$ follows that $p_{\overline{\lambda}_i}(x) < \varepsilon_i$ for each $i$. Therefore $U \subseteq O$. It means that $\tau_{\overline{\mathcal{P}}} \subseteq \tau_{\mathcal{P}}$, by Lemma 4.1.2. Consequently, $\tau_{\mathcal{P}} = \tau_{\overline{\mathcal{P}}}$. □

**Corollary 4.4.8.** *Let $(X, \tau)$ be a LPS, the topology $\tau$ of which is given by a saturated collection $\mathcal{P} = \{p_\lambda : \lambda \in \Lambda\}$ of (homogeneous or nonhomogeneous) seminorms $p_\lambda$. Then the set*

$$\{\{x \in X : p_\lambda(x) < \varepsilon\} : \lambda \in \Lambda, \varepsilon > 0\}$$

*forms a base of neighborhoods of zero in $(X, \tau)$.*

## 4.5   Locally bounded space and algebra

Let $(X, \tau)$ be a *TLS* over $\mathbb{K}$. A subset $U$ of $X$ is a *bounded* in $(X, \tau)$, if for every neighborhood $O$ of zero there is a number $\mu = \mu(O) > 0$ such that $U \subseteq \mu O$. A *TLS* $(X, \tau)$ is called a *locally bounded space* (*LBS*, in short), if $(X, \tau)$ has a bounded neighborhood of zero and a *TA* $(A, \tau)$ is a *locally bounded algebra*, if the underlying *TLS* of $(A, \tau)$ is a *LBS*.

To describe the topology $\tau$ of *LBS* and *LBA* $(X, \tau)$, we need the following results:

**Proposition 4.5.1.** *The following statements hold:*

a) *if $U$ is a bounded set in a TLS $(X, \tau)$, then $U + U$ is also bounded in $(X, \tau)$;*

b) *every bounded neighborhood of zero in a TLS $(X, \tau)$ is pseudoconvex;*

c) *every LBS $(X, \tau)$ has a balanced and bounded neighborhood $U$ of zero and*

$$\left\{\frac{1}{n}U : n \in \mathbb{N}\right\}$$

*is a base of balanced and pseudoconvex neighborhoods of zero in $(X, \tau)$.*

*Proof.* a) Let $U$ be a bounded subset and $O$ a neighborhood of zero in a *TLS* $(X, \tau)$. Then there is another neighborhood $O'$ of zero in $(X, \tau)$ such that $O' + O' \subseteq O$ by the statement $(B_2)$ of Theorem 4.2.3. Since $U$ is bounded, then there is a number $\mu > 0$ such that $U \subseteq \mu O'$. Hence, $U + U \subseteq \mu(O' + O') \subseteq \mu O$. It means that $U + U$ is bounded.

b) Let $U$ be a bounded neighborhood of zero in a *TLS* $(X, \tau)$. Then there is a $\mu > 0$ such that $U + U \subseteq \mu U$ by the statement a). Hence, $U$ is pseudoconvex.

c) Let $O$ be an arbitrary neighborhood of zero and $V$ the bounded neighborhood of zero in $(X, \tau)$. Then, by the statement $(B_3)$ of Theorem 4.2.3, there is a balanced and bounded neighborhood $U$ of zero in $(X, \tau)$ such that $U \subseteq V$ and a number $\mu > 0$ such that $U \subseteq \mu O$, i.e. $\frac{1}{\mu}U \subseteq O$. Let $n_0 \in \mathbb{N}$ be such that $n_0 > \mu$. Then

$$\frac{1}{n_0}U \subset \frac{1}{\mu}\left(\frac{\mu}{n_0}U\right) \subseteq \frac{1}{\mu}U \subseteq O.$$

It means that $\left\{\frac{1}{n}U : n \in \mathbb{N}\right\}$ is a base of balanced and pseudoconvex neighborhoods of zero in $(X, \tau)$. $\qquad\square$

**Corollary 4.5.2.** *Every Hausdorff LBS is metrizable.*

*Proof.* By Proposition 4.5.1, every *LBS* $(X, \tau)$ has a countable base of neighborhoods of zero. Since $(X, \tau)$ is a Hausdorff space, then $(X, \tau)$ is metrizable (see Schaefer (1966), result I.6.2). $\qquad\square$

Let $(X, \tau)$ be a *LBS* and $V$ a bounded neighborhood of zero in it. Then $V$ is pseudoconvex by the statement b) of Proposition 4.5.1. Therefore, there is a number $k \in (0, 1]$ such that $V + V \subseteq 2^{\frac{1}{k}} V$. Let now

$$p(x) = \inf\{\mu^k : \mu > 0, x \in \mu \Gamma_k(V)\} \qquad (4.12)$$

and

$$p_n(x) = \inf\left\{\mu^k : \mu > 0, x \in \mu \Gamma_k\left(\frac{1}{n} V\right)\right\}$$

for each $x \in X$ and $n \in \mathbb{N} \setminus \{1\}$. Then $p$ and $p_n$ are $k$-homogeneous seminorms on $(X, \tau)$, by Proposition 4.4.4, because $\Gamma_k(V)$ and $\Gamma_k(\frac{1}{n} V)$ are absorbing and absolutely $k$-convex sets. Moreover,

$$p_n(x) = n^k p(x)$$

for each $x \in X$ and $n \geqslant 2$. To show that, let $x \in X$ and $\lambda > 0$ be such that $x \in \lambda \Gamma_k(V)$. Then $\frac{1}{n} x \in \frac{\lambda}{n} \Gamma_k(V) = \lambda \Gamma_k(\frac{1}{n} V)$ for every $n \in \mathbb{N}$. Hence, $p_n(\frac{1}{n} x) \leqslant \lambda^k$. Going to infimum according to $\lambda^k$, we have that $p_n(\frac{1}{n} x) \leqslant p(x)$ or $p_n(x) \leqslant n^k p(x)$ for each $x \in X$ and $n \in \mathbb{N}$. Let now $x \in X$ and $\lambda > 0$ be such that

$$x \in \lambda \Gamma_k\left(\frac{1}{n} x\right) = \frac{\lambda}{n} \Gamma_k(V).$$

Then $p(nx) \leqslant \lambda^k$ and, similarly as above, we have that

$$n^k p(x) = p(nx) \leqslant p_n(x)$$

for eaxh $x \in X$ and $n \in \mathbb{N}$. Consequently, $p_n(x) = n^k p(x)$ for each $x \in X$ and $n \in \mathbb{N}$. Taking this into account, we have

**Proposition 4.5.3.** *Every LBS $(X, \tau)$ defines a number $k \in (0, 1]$ and a $k$-homogeneous seminorm $p$ on $X$ such that $\tau$ coincides with the topology $\tau_p$, defined by $p$.*

*Proof.* Let $(X, \tau)$ be a *LBS*. Then $(X, \tau)$ has a bounded neighborhood $V$ of zero which defines a number $k \in (0, 1]$ and a $k$-homogeneous seminorm $p$ on $X$ in the form (4.12). Let $\tau_p$ denote the topology on $X$, defined by $p$, that is, the topology in which sets $O_\varepsilon = \{x \in X : p(x) < \varepsilon\}$ with $\varepsilon > 0$ form a base of neighborhoods of zero.

To show that $\tau \subseteq \tau_p$, let $O$ be a neighborhood of zero in $(X, \tau)$. Since sets $\frac{1}{n} V$ form a base of neighborhoods of zero in $(X, \tau)$ by Proposition 4.5.1 (c), there is a $n_0 \in \mathbb{N}$ such that $\frac{1}{n_0} V \subseteq O$. Moreover, $\Gamma_k(V) \subseteq 2^{\frac{1}{k}} V$, by Proposition 4.4.2. Then

$$\frac{\Gamma_k\left(\frac{V}{n_0}\right)}{2^{\frac{1}{k}}} = \frac{\Gamma_k(V)}{n_0 2^{\frac{1}{k}}} \subseteq \frac{V}{n_0} \subseteq O$$

and $n_0^k p(x) = p_{n_0}(x) \leqslant \frac{1}{2}$, if

$$x \in \frac{\Gamma_k\left(\frac{V}{n_0}\right)}{2^{\frac{1}{k}}}.$$

Now,

$$U_0 = \left\{ x \in X : p(x) \leqslant \frac{1}{2n_0^k} \right\}$$

is a neighborhood of zero in $(X, \tau_p)$ and $U_0 \subseteq O$. It means that $\tau \subseteq \tau_p$.

Let now $O$ be a neighborhood of zero in $(X, \tau_p)$ and $n_0 \in \mathbb{N}$ with $n_0 \geqslant 2$. Then $O_\varepsilon = \{ x \in X : p(x) < \varepsilon \} \subset O$ for some $\varepsilon > 0$. Since

$$U_{n_0} = \{ x \in X : p_{n_0}(x) < n_0^k \varepsilon \}$$

is a neighborhood of zero in $(X, \tau)$ and $U_{n_0} \subseteq O_\varepsilon$, then $\tau_p \subseteq \tau$. Consequently, $\tau = \tau_p$.      $\square$

---

## 4.6    Main types of locally pseudoconvex algebras

Let $(A, \tau)$ be a *LPA* over $\mathbb{K}$ and $\mathcal{P} = \{ p_\lambda : \lambda \in \Lambda \}$ the collection of $k_\lambda$-homogeneous seminorms on $A$ with $k_\lambda \in (0, 1]$ for each $\lambda \in \Lambda$, which defines the topology $\tau$. If

$$k = \inf\{ k_\lambda : \lambda \in \Lambda \} > 0,$$

then $(A, \tau)$ is a *LkCA*. To see this, let $q_\lambda = p_\lambda^{\frac{k}{k_\lambda}}$ for each $\lambda \in \Lambda$. Since $\frac{k}{k_\lambda} \in (0, 1]$, then

$$p_\lambda(x + y)^{\frac{k}{k_\lambda}} \leqslant (p_\lambda(x) + p_\lambda(y))^{\frac{k}{k_\lambda}} \leqslant p_\lambda(x)^{\frac{k}{k_\lambda}} + p_\lambda(y)^{\frac{k}{k_\lambda}}$$

(by the inequality $(a + b)^\alpha \leqslant a^\alpha + b^\alpha$ for all nonnegative numbers $a$ and $b$ and any $\alpha \in (0, 1]$, given above). Hence, $q_\lambda(a + b) \leqslant q_\lambda(a) + q_\lambda(b)$ for each $a, b \in A$ and $\lambda \in \Lambda$. Moreover,

$$q_\lambda(\mu a) = p_\lambda(\mu a)^{\frac{k}{k_\lambda}} = |\mu|^k q_\lambda(a)$$

for each $a \in A$ and $\lambda \in \Lambda$. Hence, the collection $\mathcal{P}$ defines on $A$ a collection $\mathcal{Q}$ of $k$-homogeneous seminorms such that $\tau_{\mathcal{P}} = \tau_{\mathcal{Q}}$, because

$$\{ a \in A : p_\lambda(a) < \varepsilon \} = \{ a \in A : q_\lambda(a) < \varepsilon^{\frac{k}{k_\lambda}} \}$$

and

$$\{ a \in A : q_\lambda(a) < \varepsilon \} = \{ a \in A : p_\lambda(a) < \varepsilon^{\frac{k_\lambda}{k}} \}$$

for each $\lambda \in \Lambda$. But there exists also *LPA* with $k = 0$. Such algebra is, for example, an algebra (with point-wise algebraic operation) of all functions $f$ on a set $\mathbb{R}$ with seminorms

$$p_n(f) = \sup_{t \in [-n, n]} \sqrt[n]{|f(t)|}$$

for each $n \in \mathbb{N}$. Here, $p_n(\mu f) = |\mu|^{\frac{1}{n}} p_n(f)$ for each $\mu \in \mathbb{K}$ and $n \in \mathbb{N}$, that is, in the present case, $k_n = \frac{1}{n}$ for each $n \in \mathbb{N}$ and $k = 0$.

Let $(A, \tau)$ be $LPA$ and $\mathcal{P} = \{p_\lambda : \lambda \in \Lambda\}$ a collection of $k_\lambda$-homogeneous seminorms $p_\lambda$, with $k_\lambda \in (0, 1]$ for each $\lambda \in \Lambda$, which defines the topology $\tau$. If every seminorm $p_\lambda$ in $\mathcal{P}$ is *submultiplicative*, that is,

$$p_\lambda(ab) \leqslant p_\lambda(a)p_\lambda(b)$$

for each $a, b \in A$, then $(A, \tau)$ is called a *locally multiplicatively pseudoconvex algebra* or *locally m-pseudoconvex algebra* ($LmPA$, in short), and, if for every $a \in A$ and $\lambda \in \Lambda$, there exists a number $M(a, \lambda) > 0$ such that

$$p_\lambda(ab) \leqslant M(a, \lambda)p_\lambda(b) \quad \text{and} \quad p_\lambda(ba) \leqslant M(a, \lambda)p_\lambda(b) \qquad (4.13)$$

for each $b \in A$, then $(A, \tau)$ is called a *locally absorbingly pseudoconvex algebra* or *locally A-pseudoconvex algebra* ($LAPA$, in short). In the language of neighborhoods of zero, it means that a $LPA$ $(A, \tau)$ is a $LmPA$ if and only if $(A, \tau)$ has a base of neighborhoods $U$ of zero which are absolutely pseudoconvex and *idempotent sets* (that is, $UU \subseteq U$), and $(A, \tau)$ is a $LAPA$ if and only if $(A, \tau)$ has a base of *absorbingly pseudoconvex neighborhoods $U$* of zero. It means that every neighborhood in this base is an absolutely pseudoconvex set, which absorbs the set $aU \cup Ua$ for each $a \in A$ (that is, for any $a \in A$, there exists a number $M(a, U) > 0$ such that $aU \cup Ua \subseteq M(a, U)U$). So, every $LmPA$ (that is, the case when $M(a, \lambda) = p_\lambda(a)$) is a $LAPA$. In pariicular case, when $k > 0$, we speak about *locally m-(k-convex) algebra* ($LmkCA$, in short) and about *locally A-(k-convex) algebra* ($LAkCA$, in short). In case, when $k = 1$, we speak about *locally m-convex algebra* and *locally A-convex algebra* and, when the collection $\mathcal{P}$ consists of one seminorm, then about *A-seminormed algebra* and *A-normed algebra*.

**Theorem 4.6.1.** *For every commutative $LmPA$ $(A, \tau)$, there exists a topology $\tau'$ on $A$, weaker than the topology $\tau$, such that $(A, \tau')$ is a commutative $LmCA$.*

*Proof.* Let $(A, \tau)$ be a unital commutative $LmPA$, topology of which has been defined by a collection $\{p_\lambda : \lambda \in \Lambda\}$ of $k_\lambda$-homogeneous submultiplicative seminorms $p_\lambda$, with $k_\lambda \in (0, 1]$ for each $\lambda \in \Lambda$. Let $\{\nu_\lambda : \lambda \in \Lambda\}$ be the collection of maps $\nu_\lambda$, defined by

$$\nu_\lambda(a) = \lim_{n \to \infty} \sqrt[n]{p_\lambda(a^n)}$$

for each $a \in A$. Since every $p_\lambda$ is a $\mathbb{R}$ valued (non negative) submultiplicative function on $A$, the limit $\nu_\lambda(a)$ exists in $(A, \tau)$ by Lemma 3.3.6 from Balachandran (2000), p. 119. Moreover, $\nu_\lambda(ab) \leqslant \nu_\lambda(a)\nu_\lambda(b)$ and $\nu_\lambda(e_A) = 1$ by Proposition 3.3.7 from Balachandran (2000), p. 120, for each $\lambda \in \Lambda$ and $a, b \in A$.

For every $\lambda \in \Lambda$, let

$$\overline{B}_\lambda = \{a \in A : \nu_\lambda(a) \leqslant 1\}.$$

Similarly as in Balachandran (2000) (see the proof of Theorem 4.8.6, p. 218), we show that $\overline{B}_\lambda$ is a convex subset in $A$ (using the result that a closed subset is convex if and only if it is midpoint convex (see, for example, Skarakis (2008), Theorem 2)). Since every $\overline{B}_\lambda$ is, moreover, an absorbing, balanced and idempotent subset in $A$, then $p_{\overline{B}_\lambda}$ (the Minkowski functional of $\overline{B}_\lambda$), defined by

$$p_{\overline{B}_\lambda}(a) = \inf\{\mu > 0 : a \in \mu\overline{B}_\lambda\}$$

for each $a \in A$, is a submultiplicative $m$-convex seminorm on $A$ and

$$p_{\overline{B}_\lambda}(a) = \nu_\lambda(a)^{\frac{1}{k_\lambda}}$$

for each $\lambda \in \Lambda$ and all $a \in A$. To see this, let $\mu \in \mathbb{R}^+$ be such that $p_{\overline{B}_\lambda}(a) < \mu$. Then $a \in \mu\overline{B}_\lambda$. Therefore, $\nu_\lambda(\frac{a}{\mu}) \leqslant 1$ or $\nu_\lambda(a)^{\frac{1}{k_\lambda}} \leqslant \mu$. Going here to infimum according to $\mu$, we have that $\nu_\lambda(a)^{\frac{1}{k_\lambda}} \leqslant p_{\overline{B}_\lambda}(a)$ for each $a \in A$.

Let now $\mu \in \mathbb{R}$ and $\mu > 0$ be such that $\nu_\lambda(a) < \mu$. Then $\nu_\lambda\left(\frac{a}{\mu^{\frac{1}{k_\lambda}}}\right) < 1$, i.e. $a \in \mu^{\frac{1}{k_\lambda}}\overline{B}_\lambda$. Hence, going to infimum according to $\mu$ in $p_{\overline{B}_\lambda}(a)^{k_\lambda} < \mu$, we have that

$$p_{\overline{B}_\lambda}(a) \leqslant \nu_\lambda(a)^{\frac{1}{k_\lambda}}$$

for each $\lambda \in \Lambda$ and $a \in A$.

Now, the collection $\{p_{\overline{B}_\lambda} : \lambda \in \Lambda\}$ defines on $A$ a topology $\tau'$ such that $(A, \tau')$ is a commutative $LmCA$. To show that $\tau' \subseteq \tau$, let $O$ be a neighborhood of zero in $(A, \tau')$. Then there is $\lambda_0 \in \Lambda$ and $\varepsilon > 0$ such that

$$O_0 = \{a \in A : p_{\overline{B}_{\lambda_0}}(a) < \varepsilon\} \subset O$$

(we can assume that the topology $\tau$ is saturated). Now

$$U_0 = \{a \in A : p_{\lambda_0}(a) < \varepsilon^{k_{\lambda_0}}\}$$

is a neigbourhood of zero in $(A, \tau)$. To show that $U_0 \subseteq O_0$, let $a \in U_0$. Then $\nu_{\lambda_0}(a) \leqslant p_{\lambda_0}(a) < \varepsilon^{k_{\lambda_0}}$, because $p_{\lambda_0}$ is submultiplicative. Therefore,

$$p_{\overline{B}_{\lambda_0}}(a) = \nu_{\lambda_0}(a)^{\frac{1}{k_{\lambda_0}}} < \varepsilon.$$

Hence, $a \in O_0$. So, by Lemma 4.1.2, $\tau' \subseteq \tau$. $\qquad\square$

To prove the next result, we need

**Lemma 4.6.2.** *Let $(A, \tau)$ be a LAPA and $\mathcal{L}$ a collection of absorbing and absorbingly pseudoconvex sets in $(A, \tau)$. Then*

a) *for every $U \in \mathcal{L}$ the set $U' = \{x \in U : xU \cup Ux \subseteq U\}$ is an absorbing, idempotent and absolutely pseudoconvex set in $(A, \tau)$;*

b) *the intersection of finite number of absorbing, idempotent and absolutely pseudoconvex sets is an absorbing, idempotent and absolutely pseudoconvex set in $(A, \tau)$.*

*Proof.* a) Let $U \in \mathcal{L}$. Then $U'U \cup UU' \subseteq U$ and there exists $k \in (0, 1]$ such that $U$ is an absolutely $k$-convex set. Since

$$(U'U')U \cup U(U'U') = U'(U'U) \cup (UU')U' \subseteq U'U \cup UU' \subseteq U,$$

then $U'U' \subseteq U'$. Let now $x, y \in U'$ and $\alpha, \beta \in \mathbb{K}$ be such that $|\alpha|^k + |\beta|^k \leqslant 1$. Because $xU, Ux, yU, Uy \subseteq U$ and $U$ is absolutely $k$-convex, then

$$(\alpha x + \beta y)U \cup U(\alpha x + \beta y) \subseteq (\alpha(xU) + \beta(yU)) \cup (\alpha(Ux) + \beta(Uy)) \subseteq \alpha U + \beta U \subseteq U.$$

Therefore, $\alpha x + \beta y \in U'$. Hence, $U'$ is absolutely $k$-convex. To show that $U'$ is absorbing, let $a \in A$. Because $U$ is an absorbing set in $(A, \tau)$, then there is a number $\nu_0 > 0$ such that $a \in \lambda U$, whenever $|\lambda| \geqslant \nu_0$, by Proposition 4 (d). Therefore,

$$\frac{a}{\lambda_0} U \cup U \frac{a}{\lambda_0} \subset aU \cup Ua \subseteq U$$

for every $\lambda_0 > \nu_0$, because $U$ is balanced and we can take $\lambda_0 > 1$. Hence, $a \in \lambda_0 U'$.

b) It is enough to show that the intersection of two absorbing, idempotent and absolutely pseudoconvex sets is an absorbing, idempotent and absolutely pseudoconvex set in $(A, \tau)$. For this, let $U$ and $V$ be two absorbing, idempotent and absolutely pseudoconvex sets in $(A, \tau)$. Then $W = U \cap V$ is an absorbing set in $(A, \tau)$. To show this, let $w \in W$. Since $U$ and $V$ are absorbing sets, then there exist numbers $k > 0$ and $r > 0$ such that $w \in \mu U$ when $|\mu| \geqslant k$ and $w \in \mu V$ when $|\mu| \geqslant r$. We can take $\lambda_w > \max\{k, r\}$. Then

$$w \in kU \cap rV = \lambda_w \left[ \left( \frac{k}{\lambda_w} \right) U \cap \left( \frac{r}{\lambda_w} \right) V \right] \subseteq \lambda_w W.$$

Moreover, if $a_1, a_2 \in W$, then $a_1 a_2 \in UU \cap VV \subseteq W$. Hence, $W$ is an idempotent set. If $U$ is an absolutely $k_1$-convex set and $V$ is an absolutely $k_2$-convex set with $k_1, k_2 \in (0, 1]$, then $W$ is an absolutely $k$-convex set for $k = \min\{k_1, k_2\}$. To show that, let $w_1, w_2 \in W$ and $\alpha, \beta \in \mathbb{K}$ with $|\alpha|^k + |\beta|^k \leqslant 1$. Since $|\lambda|^{k_1} + |\beta|^{k_1} \leqslant |\alpha|^k + |\beta|^k$ and $|\lambda|^{k_2} + |\beta|^{k_2} \leqslant |\alpha|^k + |\beta|^k$, then $|\lambda|^{k_1} + |\beta|^{k_1} \leqslant 1$ and $|\lambda|^{k_2} + |\beta|^{k_2} \leqslant 1$, if $|\lambda|^k + |\beta|^k \leqslant 1$. Since $U$ is absolutely $k_1$-convex and $V$ is $k_2$-convex, then $\alpha w_1 + \beta w_2 \in U \cap V = W$. Therefore, $W$ is absolutely $k$-convex. $\square$

**Theorem 4.6.3.** *Let $(A, \tau)$ be a LAPA and $\mathcal{L} = \{U_\lambda : \lambda \in \Lambda\}$ the collection of all absorbing and absorbingly pseudoconvex sets in $(A, \tau)$ and*

$$\mathcal{B}' = \{\varepsilon U_\lambda' : \varepsilon \in (0, 1], U_\lambda \in \mathcal{L}\},$$

*where $U_\lambda' = \{a \in U_\lambda : aU_\lambda \cup U_\lambda x \subseteq U_\lambda\}$ for each $\lambda \in \Lambda$. Then every $U_\lambda'$ is absolutely $k_\lambda$-convex, if $U_\lambda$ is absolutely $k_\lambda$-convex, and $\mathcal{B}'$ forms a subbase of the neighborhoods of zero for a locally $m$-pseudoconvex topology $\tau'$ on $A$ such that $\tau \subseteq \tau'$. In particular, when $(A, \tau)$ is a LmPA, then $\tau' = \tau$.*

*Proof.* Let $\mathcal{E}$ be the family of all finite intersections of sets in $\mathcal{B}'$ and $E$ and $F$ arbitrary elements in $\mathcal{E}$. Then there exist numbers $n, m \in \mathbb{N}$ and $\varepsilon_1, \ldots, \varepsilon_n, \delta_1, \ldots, \delta_m \in (0, 1]$ and neighborhoods of zero $U_1, \ldots, U_n, V_1, \ldots, V_m \in \mathcal{L}$ such that

$$E = \bigcap_{v=1}^{n} \varepsilon_v U_v' \quad \text{and} \quad F = \bigcap_{s=1}^{m} \delta_s V_s'. \tag{4.14}$$

Then

$$S = \bigcap_{v=1}^{n} U_v' \cap \bigcap_{s=1}^{m} V_s'$$

is an absorbing, idempotent and absolutely pseudoconvex set in $(A, \tau)$ by Lemma 1.3 b). Let $\nu = \min\{\varepsilon_1, \ldots, \varepsilon_n, \delta_1, \ldots, \delta_m\}$. Then $\nu \in (0, 1]$ and

$$\nu S' = \nu\{a \in S : aS \cup Sa \subseteq S\} \subseteq \nu S = \bigcap_{v=1}^{n} \varepsilon_v \left(\frac{\nu}{\varepsilon_v} U_v'\right) \cap \bigcap_{s=1}^{m} \delta_s \left(\frac{\nu}{\delta_s} V_s'\right) \subseteq E \cap F,$$

because $U_v'$ and $V_s'$ are balanced by Lemma 6 a). Since every idempotent and absolutely pseudoconvex set is absorbingy pseudoconvex, then $S \in \mathcal{L}$. Hence, $\mathcal{E}$ satisfies the condition $(B_1)$ of Theorem 4.2.3.

Let now $E \in \mathcal{E}$ be as above (where every $U_v'$ is absolutely $k_v$-convex set for some $k_v \in (0, 1]$) and

$$F = \bigcap_{v=1}^{n} \gamma_v U_v',$$

where $\gamma_v = \varepsilon_v 2^{-\frac{1}{k_v}} \in (0, 1]$ for each $v$. Then $F \in \mathcal{E}$. If $a_1, a_2 \in F$, then, for each fixed $v$, there are elements $b_v^1, b_v^2 \in U_v'$ such that $a_1 = \gamma_v b_v^1$ and $a_2 = \gamma_v b_v^2$. Now,

$$(a_1 + a_2)U_v \cup U_v(a_1 + a_2) \subseteq \gamma_v[(U_v'U_v + U_v'U_v) \cup (U_v U_v' + U_v U_v')]$$
$$\subseteq \gamma_v(U_v + U_v)$$
$$\subseteq \gamma_v 2^{\frac{1}{k_v}} U_v = \varepsilon_v U_v.$$

Hence,

$$\frac{1}{\varepsilon_v}(a_1 + a_2)U_v \cup \frac{1}{\varepsilon_v} U_v(a_1 + a_2) \subset U_v$$

or $a_1 + a_2 \in \varepsilon_v U_v'$ for each $v$. Thus, $F + F \subseteq E$. So, $\mathcal{E}$ satisfies also the condition $(B_2)$ of Theorem 4.2.3. Moreover, since every set $U_v'$ is balanced, by Lemma 1.3 a), and $\varepsilon U_v'$ is balanced (hence every finite intersection of balanced sets is also balanced), then every $E \in \mathcal{E}$ is a balanced set. Therefore, $\mathcal{E}$ satisfies also the condition $(B_3)$ of Theorem 4.2.3. Consequently, by Theorem 4.2.3, there exists a topology $\tau'$ on $A$, which makes $(A, \tau')$ a TLS in which $\mathcal{E}$ is a base of neighborhoods of zero.

To show that every set in $\mathcal{E}$ is idempotent and absolutely pseudoconvex, let

$E \in \mathcal{E}$ and $a_1, a_2 \in E$. Then there are numbers $n \in \mathbb{N}$, $\varepsilon_1, \ldots, \varepsilon_n \in (0,1]$ and $U_1, \ldots U_n \in \mathcal{L}$ such that (4.14) holds and $\frac{a_1}{\varepsilon_v}, \frac{a_2}{\varepsilon_v} \in U_v'$ for each $v$. Therefore, from

$$\frac{a_1}{\varepsilon_v} \frac{a_2}{\varepsilon_v} U \cup U \frac{a_1}{\varepsilon_v} \frac{a_2}{\varepsilon_v} \subseteq U_v' U_v' U_v \cup U_v U_v' U_v' \subseteq U_v' U_v \cup U_v U_v' \subseteq U$$

follows that $a_1 a_2 \in \varepsilon_v^2 U_v' \subset \varepsilon_v U_v'$ for each $v$. Hence, $EE \subseteq E$. Since every set $\varepsilon_v U_v'$ in $E$ is absolutely pseudoconvex, the intersection of such sets is also absolutely pseudoconvex, by Lemma 6 b). Consequently, $(A, \tau')$ is a $LmPA$. Moreover, $\tau \subseteq \tau'$. Indeed, since $(A, \tau)$ is a $LAPA$, then $(A, \tau)$ has a base $\mathcal{L}_A$ of absorbingly pseudoconvex neighborhoods of zero and every neighborhood of zero is an absorbing set, by Proposition 4.2.2 (d), then $\mathcal{L}_A \subseteq \mathcal{L}$. Every $U \in \mathcal{L}_A$ defines $U' \in \mathcal{E}$ such that $U' \subset U$. Therefore, $\tau \subseteq \tau'$, by Lemma 4.1.2.

In case $(A, \tau)$ is a $LmCA$, every $U' = U$. Hence, in this case, $\tau = \tau'$. $\square$

---

## 4.7 Generalizations of Gelfand-Mazur theorem for locally pseudoconvex division algebras

One of the fundamental question in the general theory of $TAs$ is the following: *when a given topological division algebra over $\mathbb{C}$ is topologically isomorphic to $\mathbb{C}$?* This problem has been studied by many mathematitians.

In 1938, S. Mazur in Mazur (1938) stated (the first proof of this result has been published in the book Zelazko (1973)) that every normed division algebra over the field $\mathbb{R}$ of real numbers is one of the fields $\mathbb{R}$ and $\mathbb{C}$ or the skew field $\mathbb{H}$ of real quaternions. In 1939, I. Gelfand, in Gelfand (1939), Theorem 3, stated without proof (the proof of this result was published in 1941 in Gelfand (19XX)) that every complex Banach field is isometrically isomorphic to $\mathbb{C}$. Nowadays, these results of Mazur and Gelfand are known as the *Gelfand-Mazur Theorem*. In the complex case it is formulated in the form: *every normed division algebra over $\mathbb{C}$ is isometrically isomorphic to $\mathbb{C}$*.

In 1947, R. Arens in Arens (1947), Theorem 1, generalized the Gelfand-Mazur Theorem to the case of Hausdorff division $LCA$ with continuous inversion and, in 1952, in Michael (1952), Proposition 2.9, E. A. Michael generalized to the case of Hausdorff division $LmCA$.

In 1954, J. H. Williamson in Wiljamson (1954), p. 371, showed, that it is possible to endowe the field $C(t)$ (of all rational functions in the indeterminate $t$ over $\mathbb{C}$) with such metrizable topology, in which the addition, both multiplications and the inversion are continuous, but $\mathbb{C} \subsetneq C(t)$. It means that, there exists a metrizable division algebra with continuous multiplication and continuous inversion, which is not topologically isomorphic to $\mathbb{C}$.

In 1960, W. Żelazko in Zelazko (1960b), Theorem 11, generalized the

Gelfand-Mazur Theorem to the case of Hausdorff division $LBA$; in the same paper (Theorem 22) to the case of Hausdorff division $LmPA$ and in Zelazko (1960a) to the case of complete metrizable $LCA$. Moreover, in 1965, G. R. Allan in Allan (1965), Corollary 3.10, generalized the the Gelfand-Mazur Theorem to the case of Hausdorff division $LCA$ with bounded elements; in 1966, P. Turpin in Turpin (1966a), Theorem 3 (see also Turpin (1966b)) to the case of Hausdorff division $LPA$ with bounded elements and, in the same year, L. Waelbroeck announced in Waelbroeck (1966), p. 153, (see also Turpin and Waelbroeck (1968), p. 195) that the Gelfand-Mazur Theorem holds in case of Hausdorff division $LPA$ with continuous inversion (up to now it is not clear whether the proof of this statement is correct or not).

In addition to these, in 1973 A. C. Cochran Carleson (1962), Theorem 2.1, generalized the Gelfand-Mazur Theorem to the case of Hausdorff division $LACA$ and, in 1979, V. Kaushik in Kaushik (1979), to the case of Hausdorff division $LAPA$.

To describe the main types of topological division algebras, topologically isomorphic to $\mathbb{C}$, we need the following concepts.

A unital TA $(A, \tau)$ is called

a) a *Q-algebra*, if the set Inv$A$ of all invertible elements of $A$ is open in $(A, \tau)$;

b) a *Waelbroeck algebra*, if $(A, \tau)$ is a $Q$-algebra, in which the inversion $a \to a^{-1}$ is continuous;

c) an *exponentially galbed algebra*, if for each neighborhood $O$ of zero in $(A, \tau)$, there exists another neighborhood $O'$ of zero such that

$$\left\{ \sum_{v=1}^{n} \frac{a_v}{2^v} : a_1, \ldots, a_n \in O' \right\} \subseteq O$$

for each $n \in \mathbb{N}$.

Herewith, Banach algebras and many other TA are Waelbroeck algebras (see, for example, Balachandran (2000), Kızmaz (1981), Mallios (1986), Michael (1952), Zelazko (1965) and Zelazko (1973)). It is easy to see that every $LPA$ $(A, \tau)$ is exponentially galbed. Indeed, let $\mathcal{P} = \{p_\lambda : \lambda \in \Lambda\}$ be a collection of $k_\lambda$-homogeneous seminorms, with $k_\lambda \in (0, 1]$ for each $\lambda \in \Lambda$, which defines the topology $\tau$. By Proposition 4.4.7, we can assume that the collection $\mathcal{P}$ is saturated. Then every neighborhood $O$ of zero in $(A, \tau)$ defines a $\lambda_0 \in \Lambda$ and a $\varepsilon > 0$ such that

$$O_{\lambda_0 \varepsilon} = \{a \in A : p_{\lambda_0}(a) < \varepsilon\} \subseteq O,$$

by Corollary 4.4.8. If $\delta \in (0, (1 - 2^{-k_{\lambda_0}})\varepsilon)$, $n \in \mathbb{N}$ and $a_1, \ldots, a_n \in O_{\lambda_0 \delta}$, then

$$p_{\lambda_0}\left( \sum_{v=1}^{n} \frac{a_v}{2^v} \right) \leqslant \sum_{v=1}^{n} \frac{p_{\lambda_0}(a_v)}{2^{v k_{\lambda_0}}} < \delta \sum_{v=0}^{\infty} \frac{1}{2^{v k_{\lambda_0}}} = \delta \frac{1}{1 - 2^{-k_{\lambda_0}}} < \varepsilon.$$

Hence,

$$\sum_{v=1}^{n} \frac{a_v}{2^v} \subseteq O,$$

whenever $n \in \mathbb{N}$ and $a_1, \ldots, a_n \in O_{\lambda_0 \delta}$. Hence, every $LPA$ is an exponentially galbed algebra.

Moreover, an element $a$ of a $TA$ $(A, \tau)$ is *bounded* in $(A, \tau)$, if there exists a number $\lambda > 0$ such that the set

$$S(a, \lambda) = \left\{ \left( \frac{a}{\lambda} \right)^n : n \in \mathbb{N} \right\}$$

is bounded in $(A, \tau)$. The *spectrum* $\mathrm{sp}_A(a)$ of $a$ in any unital algebra $A$ over $\mathbb{C}$ with the unit element $e_A$ is defined by

$$\mathrm{sp}_A(a) = \{ \lambda \in \mathbb{C} : a - \lambda e_A \notin \mathrm{Inv} A \}$$

and the *resolvent* $R_a$ of an element $a \in A$ is defined by

$$R_a(\lambda) = (a - \lambda e_A)^{-1}$$

for each $\lambda \in \mathbb{C} \setminus \mathrm{sp}_A(a)$. It is well known that the spectrum of every element $a$ in a $Q$-algebra $(A, \tau)$ is closed and

$$\lim_{|\lambda| \to \infty} R_a(\lambda) = - \lim_{|\lambda| \to \infty} \frac{1}{\lambda} \left( e_A - \frac{a}{\lambda} \right)^{-1} = \theta_A, \qquad (4.15)$$

if, in addition, the inversion $a \to a^{-1}$ in $(A, \tau)$ is continuous.

**Proposition 4.7.1.** *Let $(A, \tau)$ be a $TA$ over $\mathbb{C}$ with bounded elements. Then the resolvent $R_a$ of $a \in A$ is an $A$-valued analytic map on $\mathbb{C} \cup \{\infty\} \setminus \mathrm{cl}(\mathrm{sp}_A(a))$. If, in addition, $(A, \tau)$ is an exponentially galbed algebra, then $R_a$ satisfies the condition (15).*

*Proof.* Let $a \in A$, $\mathbb{C}_a = \mathbb{C} \setminus \mathrm{cl}(\mathrm{sp}_A(a))$, $\lambda_0 \in \mathbb{C}_a$ and $\lambda_1 > 0$ such that the set $S(R_a(\lambda_0), \lambda_1)$ is bounded in $(A, \tau)$. Since $\mathbb{C}_a$ is open in $\mathbb{C}$, there is a number $\varepsilon > 0$ such that $\lambda_0 + \lambda \in \mathbb{C}_a$, whenever $|\lambda| < \varepsilon$. It is easy to show that

$$R_a(\lambda_0 + \lambda) = R_a(\lambda_0) + \lambda R_a(\lambda_0) R_a(\lambda_0 + \lambda).$$

Putting now on the right side of this equation $R_a(\lambda_0) + \lambda R_a(\lambda_0) R_a(\lambda_0 + \lambda)$ in place of $R_a(\lambda_0 + \lambda)$ and repeating this procedure $n$ times, we have that

$$R_a(\lambda_0 + \lambda) = \sum_{v=1}^{n} R_a(\lambda_0)^{v+1} \lambda^v + \lambda^{n+1} R_a(\lambda_0)^{n+1} R_a(\lambda_0 + \lambda)$$

for each $n \in \mathbb{N}$ and $\lambda \in \mathbb{C}_a$. Since

$$(\lambda R_a(\lambda_0))^{n+1} = (\lambda \lambda_1)^{n+1} \left( \frac{R_a(\lambda_0)}{\lambda_1} \right)^{n+1}$$

for each $n \in \mathbb{N}$ and $\lambda \in \mathbb{C}_a$, the sequence

$$\left( \left( \frac{R_a(\lambda_0)}{\lambda_1} \right)^{n+1} \right)$$

is bounded and the sequence $(\lambda^n)$ vanishes, then

$$\lim_{n \to \infty} \lambda^{n+1} R_a(\lambda_o)^{n+1} R_a(\lambda_0 + \lambda) = \theta_A,$$

whenever $|\lambda| < \delta = \min\{\varepsilon, |\lambda_1|^{-1}\}$. Hence,

$$R_a(\lambda_0 + \lambda) = \sum_{v=1}^{\infty} R_a(\lambda_0)^{v+1} \lambda^v,$$

whenever $|\lambda| < \delta$.

Let now $\lambda_2 > 0$ be such that the set $S(a, \lambda_2)$ is bounded in $(A, \tau)$. It is easy to show that

$$R_a(\lambda) = -\frac{e_A}{\lambda} + \frac{a}{\lambda} R_a(\lambda).$$

Putting now on the right side of this equation $-\frac{e_A}{\lambda} + \frac{a}{\lambda} R_a(\lambda)$ in place of $R_a(\lambda)$ and repeating this procedure $n$ times, we have that

$$R_a(\lambda) = -\sum_{v=0}^{n} \frac{a^v}{\lambda^{v+1}} + \left( \frac{a}{\lambda} \right)^{n+1} R_a(\lambda)$$

for each $n \in \mathbb{N}$ and $\lambda \neq 0$. Since

$$\left( \frac{a}{\lambda} \right)^{n+1} = \left( \frac{a}{\lambda_2} \right)^{n+1} \left( \frac{\lambda_2}{\lambda} \right)^{n+1}$$

for each $n \in \mathbb{N}$ and $\lambda \neq 0$, the sequence $((\frac{a}{\lambda_2})^n)$ is bounded and the sequence $((\frac{\lambda_2}{\lambda})^n)$ vanishes, when $|\lambda| > \lambda_2$, then

$$\lim_{n \to \infty} \left( \frac{a}{\lambda} \right)^{n+1} R_a(\lambda) = \theta_A,$$

whenever $|\lambda| > \lambda_2$. Hence,

$$R_a(\lambda) = -\sum_{v=0}^{\infty} \frac{a^v}{\lambda^{v+1}},$$

whenever $|\lambda| > \lambda_2$. Thus, $R_a$ is an $A$-valued analytic map on

$$\mathbb{C} \cup \{\infty\} \setminus \mathrm{cl}(\mathrm{sp}_A(a)).$$

Let now $(A, \tau)$ be an exponentially galbed algebra, $\lambda_0 > 0$ a number such that the set $S(a, \lambda_0)$ is bounded in $(A, \tau)$ and $O$ a neighborhood of zero in $(A, \tau)$. Then, by Theorem 4.2.3, there is a closed neighborhood $O_1$ of zero in

$(A, \tau)$ such that $O_1 \subseteq O$. Since $(A, \tau)$ is an exponentially galbed algebra, then $O_1$ defines another balanced neighborhood $O_2$ in $(A, \tau)$ such that

$$\sum_{v=0}^{n} \frac{a_v}{2^v} \in O_1$$

for every $n \in \mathbb{N}$ and $a_1, \ldots, a_n \in O_2$. Let $\mu \geqslant 1$ be such that $\left(\frac{a}{\lambda_0}\right)^n \subset \mu O_2$ for each $n \in \mathbb{N}$, and let $r = \max\{\mu, 2\lambda_0\}$. Since

$$a_v = -\frac{(2a)^v}{\lambda^{v+1}} = -\frac{1}{\mu}\left(\frac{a}{\lambda_0}\right)^v \left(\frac{2\lambda_0}{\lambda}\right)^v \frac{\mu}{\lambda} \in O_2$$

for each $v \in \mathbb{N}$, whenever $|\lambda| > r$, because $O_2$ is balanced. Hence

$$-\sum_{v=0}^{n} \frac{a^v}{\lambda^{v+1}} = \sum_{v=0}^{n} \frac{a_v}{2^v} \in O_1$$

for each $n \in \mathbb{N}$, whenever $|\lambda| > r$. Therefore,

$$R_a(\lambda) = -\sum_{v=0}^{\infty} \frac{a^v}{\lambda^{v+1}} \in O_1,$$

whenever $|\lambda| > r$. It means that (4.15) holds. $\qquad\square$

**Proposition 4.7.2.** *Let $(A, \tau)$ be a unital Waelbroeck algebra over $\mathbb{C}$ and let $A^*$ be the topological dual of $(A, \tau)$. If $A^*$ has nonzero functionals, then every element $a \in A$ is weakly bounded (it means that there is a number $\lambda = \lambda(\varphi)$ such that the set $S(\varphi(a), \lambda)$ is bounded in $\mathbb{C}$ for every $\varphi \in A^*$).*

*Proof.* Let $(A, \tau)$ be a unital Waelbroeck algebra over $\mathbb{C}$. Then $A$ is a $Q$-algebra in which the inversion $a \to a^{-1}$ is continuous. Therefore, the set $\text{Inv}A$ is open in $(A, \tau)$. Hence, there exists a neighborhood $O$ of zero in $(A, \tau)$, such that $e_A + O \subset \text{Inv}A$. Moreover, for every $a \in A$, there exists an open neighborhood

$$O_{\varepsilon_a} = \{\lambda \in \mathbb{C} : |\lambda| < \varepsilon_a\}$$

of zero in $\mathbb{C}$ with $\varepsilon_a > 0$ such that $O_{\varepsilon_a} a \subset O$. So, for every $a \in A$ the map $f_a$, defined by

$$f_a(\lambda) = (e_A - \lambda a)^{-1}$$

for each $\lambda \in O_{\varepsilon_a}$, is continuous on $O_{\varepsilon_a}$. Since $A^*$ of $(A, \tau)$ contain nonzero functionals, we fix one of such $\varphi \in A^*$. Then $\varphi \circ f_a$ is a continuous $\mathbb{C}$-valued map on $O_{\varepsilon_a}$. To show that $\varphi \circ f_a$ is infinitely many times differentiable at every point of $O_{\varepsilon_a}$, let $\lambda_0 \in O_{\varepsilon_a}$ be an arbitrary fixed number and $\lambda \in \mathbb{C}$ be such that $\lambda_0 + \lambda \in O_{\varepsilon_a}$. Since $f_a$ and $\varphi$ are continuous maps, the multiplication in $(A, \tau)$ is separately continuous,

$$f_a(\lambda_0 + \lambda) - f_a(\lambda_0) = \lambda f_a(\lambda_0 + \lambda) a f_a(\lambda_0)$$

for each $\lambda \in O_{\varepsilon_a}$ and $af_a(\lambda_0) = f_a(\lambda_0)a$ for each $a \in A$, then

$$
\begin{aligned}
(\varphi \circ f_a)'(\lambda_0) &= \lim_{\lambda \to 0} \frac{(\varphi \circ f_a)(\lambda_0 + \lambda) - (\varphi \circ f_a)(\lambda_0)}{\lambda} \\
&= \varphi \left[ \lim_{\lambda \to 0} \frac{f_a(\lambda_0 + \lambda) - f_a(\lambda_0)}{\lambda} \right] \\
&= \varphi \left[ \lim_{\lambda \to 0} f_a(\lambda_0 + \lambda) a f_a(\lambda_0) \right] \\
&= \varphi(f_a(\lambda_0) a f_a(\lambda_0)) = \varphi(a f_a(\lambda_0)^2).
\end{aligned}
$$

By Lemmaa 3 inAbel (2015) we have that

$$
(\varphi \circ f_a)^{(n)}(\lambda_0) = n! \varphi(a^n f_a(\lambda_0)^{n+1})
$$

and $(\varphi \circ f_a)^{(n)}(0) = n! \varphi(a^n)$ for every $n \in \mathbb{N}$ (here $f_a^{(n)}(\lambda)$ denotes the $n$-th complex derivative of $f_a(\lambda)$ at $\lambda$). It means that $\varphi \circ f_a$ has derivatives of every order at every point $\lambda \in O_{\varepsilon_a}$. Therefore, the Taylor's series of $\varphi \circ f_a$ at 0 has the form

$$
(\varphi \circ f_a)(0) = \sum_{k=0}^{\infty} \frac{(\varphi \circ f_a)^{(k)}(0)}{k!} \lambda^n = \sum_{k=0}^{\infty} \varphi(\lambda a)^n
$$

in the neighborhood $O_{\varepsilon_a}$. Hence, for every $\varphi \in A^*$, the series $\sum_{k=0}^{\infty} \varphi(\lambda a)^n$ converges at every point $\lambda \in O_{\varepsilon_a}$. This means, that for every $\lambda \in O_{\varepsilon_a} \setminus \{0\}$, the set $S(\varphi(a), \frac{1}{\lambda})$ is weakly bounded. $\qquad \square$

**Corollary 4.7.3.** *Let $(A, \tau)$ be a unital Waelbroeck algebra over $\mathbb{C}$ such that*

    *a) the topological dual space $A^*$ of $(A, \tau)$ has nonzero functionals;*

    *b) for every $a \in A$, there exists a neighborhood $O_{\varepsilon_a}$ of zero in $\mathbb{C}$ such that, for some $\lambda \in O_{\varepsilon_a}$, from the weak boundedness of $\{(\lambda a)^n : n \in \mathbb{N}\}$ in $(A, \tau)$ follows the boundedness of this set in $(A, \tau)$.*

    *Then every element in $(A, \tau)$ is bounded.*

**Theorem 4.7.4.** *Let $(A, \tau)$ be a unital Waelbroeck LCA over $\mathbb{C}$. Then every element in $(A, \tau)$ is bounded.*

*Proof.* By assumption, $(A, \tau)$ is a $LCA$. Therefore $A^*$ has nonzero functionals (see, for example, (Balachandran, 2000, Proposition 4.7.6)) and every weakly bounded set in $(A, \tau)$ is bounded (see (Rolewicz, 1985, Theorem 3.18)). Therefore, the set $\{(\lambda a)^n : n \in \mathbb{N}\}$ is bounded in $(A, \tau)$ at least for one $\lambda$ and each $a \in A$. Hence, every element in $(A, \tau)$ is bounded, by Corollary 4.7.3. $\qquad \square$

    The next lemma gives a necessary condition for a unital division TA over $\mathbb{C}$ to be topologically isomorphic to $\mathbb{C}$.

**Lemma 4.7.5.** *Let $(A, \tau)$ be a division $TA$ over $\mathbb{C}$ with unit element $e_A$. If $(A, \tau)$ is topologically isomorphic to $\mathbb{C}$, then every element in $(A, \tau)$ is bounded.*

*Proof.* If $(A, \tau)$ is topologically isomorphic to $\mathbb{C}$, then every $a \in A$ has the form $a = \lambda_a e_A$ for some $\lambda_a \in \mathbb{C}$ and $S(a, \lambda_a) = \{e_A\}$ is bounded in $(A, \tau)$. $\square$

**Theorem 4.7.6.** *Let $(A, \tau)$ be one of the following Hausdorff TA over $\mathbb{C}$:*

a) *a division LPA in which all elements are bounded;*

b) *a division LCA in which the inversion $a \to a^{-1}$ is continuous;*

c) *a division LmCA;*

d) *a division complete metrizable LCA;*

e) *a division LmPA;*

f) *a division LAPA;*

g) *a division complete LBA.*

*Then $(A, \tau)$ is topologically isomorphic to $\mathbb{C}$ (in its usual topology).*

*Proof.* a) Let $(A, \tau)$ be a division $LPA$. Then $(A, \tau)$ is an exponentially galbed algebra, as has been showed above. Assume, that there is an element $a \in A$ with empty spectrum $\mathrm{sp}_A(a)$. Since all elements in $(A, \tau)$ are bounded, the resolvent $R_a$ of $a$ is an $A$-valued analytic map on $\mathbb{C} \cup \{\infty\}$, by Proposition 4.7.1. Therefore, $R_a$ is a constant map, by Turpin result (see Turpin (1975)). Since $(A, \tau)$ satisfies the condition (15), by Proposition 4.7.1, then $R_a(\lambda) \equiv \theta_A$. Hence, $a^{-1} = R_a(0) = \theta_A$ or $e_A = \theta_A$. But this is not possible. Consequently, the spectrum $\mathrm{sp}_A(a)$ of every element $a \in A$ is not empty. Since $A$ is a division algebra, then every nonzero element of $A$ is invertible. Hence, for each $a \in A$, there is a $\lambda_a \in \mathbb{C}$ such that $a = \lambda_a e_A$. Therefore, $A = \mathbb{C} e_A$.

b) Every Hausdorff division TA is a $Q$-algebra. Hence, $(A, \tau)$ is a Waelbroeck $LCA$. Terefore, by Hahn-Banach theorem, the topological dual space $A^*$ of $(A, \tau)$, has nonzero functionals. Assume, that there is an element $a \in A$ such that $\mathrm{sp}_A(a)$ is empty. Then $\varphi \circ R_a$ is diferentiable on all the space $\mathbb{C}$ for each $\varphi \in A^*$ and

$$\lim_{|\lambda| \to \infty} (\varphi \circ R_a)(\lambda) = 0$$

by (4.15). Hence, $\varphi \circ R_a$ is differentiable on $\mathbb{C}$ and bounded on the extended plane $\mathbb{C} \cup \{\infty\}$ for each $\varphi \in A^*$. Therefore, by the Liouville's theorem, $\varphi \circ R_a$ is a constant function on $\mathbb{C}$. Because $(A, \tau)$ is a $LCS$, then $A^*$ separates the points of $A$, by Hahn-Banach theorem. Taking this into account, $R_a$ is a constant map on $\mathbb{C}$. Thus, $R_a(\lambda) = \theta_A$ for each $\lambda \in \mathbb{C}$ by (4.15). Similarly, as above, $e_A = \theta_A$. But this is not possible. Consequently, the spectrum $\mathrm{sp}_A(a)$ is not empty for every $a \in A$ and therefore, $A = \mathbb{C} e_A$ (similarly, as in the proof of the statement a) of Theorem 4.7.6).

c) Since the inversion $a \to a^{-1}$ is continuous in every unital $LmCA$ (see, for example, Zelazko (1971), Theorem 12.4), then $A = \mathbb{C} e_A$ (similarly, as in the proof of the statement b) of Theorem 4.7.6).

d) Since the inversion $a \to a^{-1}$ is continuous in every division $F$-algebra (see Zelazko (1971), Corollary 1.9), then $A = \mathbb{C} e_A$ (similarly, as in the proof of the statement b) of Theorem 4.7.6).

e) Let $(A, \tau)$ be a division $LmPA$, $a$ an arbitrary element in $A$ and $A_a$ the commutative subalgebra of $A$, generated by $a$ and $e_A$. Then there exists the maximal commutative subalgebra $B$ of $A$ such that $A_a \subseteq B$. Now, $B$ is a commutative division $LmPA$ (because in this case $\mathrm{Inv} B = B \cap \mathrm{Inv} A$) in the subset topology. By Theorem 4.6.1, there exists a topology $\tau'$ on $B$ such that $(B, \tau')$ is a commutative $LmCA$ over $\mathbb{C}$. Hence, similarly as in the proof of the statement c) of Theorem 4.7.6, there exists a complex number $\mu_a$ such that $a = \mu_a e_A$. Consequently, $A = \mathbb{C} e_A$.

f) Since every $LAPA$ $(A, \tau)$ has a topology $\tau'$ such that $(A, \tau')$ is a $LmPA$, by Theorem 4.6.3, then $A = \mathbb{C} e_A$ (similarly, as in the proof of the statement e) of Theorem 4.7.6).

g) Let $(A, \tau)$ be a division complete $LBA$ and $p$ a $k$-homogeneous seminorm on $A$ for some $k \in (0, 1)$, which defines the topology $\tau$. Since $(A, \tau)$ is complete, then we can assume that $p$ is submultiplicative, otherwise we can define on $A$ a submultiplicative seminorm, which gives the same topology (see Zelazko (1971), Theorem 2.3). Hence, $(A, \tau)$ is a $LMPA$ and, therefore, $A = \mathbb{C} e_A$ (similarly, as in the proof of the statement c) of Theorem 4.7.6).

Let now $\rho$ be a map, defined by $\rho(a) = \lambda_a$ for each $a \in A$. Then $\rho$ is an isomorphism from $A$ onto $\mathbb{C}$, whose inverse map $\lambda_a \to \lambda_a e_A$ is continuous. To show the continuity of $\rho$, let $O$ be a neighborhood of zero in $\mathbb{C}$. Then there is a number $\varepsilon > 0$ such that $O_\varepsilon \subseteq O$. Let $\lambda_0 \in O_\varepsilon \setminus \{0\}$. Then there exists a balanced neigbourhood $V$ of zero of $(A, \tau)$ such that $\lambda_0 e_A \notin V$, because $(A, \tau)$ is a Hausdorff space. If now $|\lambda_a| \geqslant |\lambda_0|$, then $\left|\frac{\lambda_0}{\lambda_a}\right| \leqslant 1$. Therefore, $\lambda_0 e_A = \left(\frac{\lambda_0}{\lambda_a}\right) a \in V$ for each $a \in V$. As this is not possible, then $\lambda_a \in O$ for each $a \in V$. Consequently, $\rho$ is continuous. $\qquad \square$

**Remark.** Theorem 4.7.6 takes together several known results and some generalizations of these (in some cases with new proofs) of R. Arens in Arens (1947), E. A. Michael in Michael (1952), W. Żelazko in Zelazko (1960a) and in Zelazko (1960b), G. R. Allan in Allan (1965), A C. Cochran in Carleson (1962) , P. Turpin in Turpin (1966b) and Turpin (1966a), V. Kaushik in Kaushik (1979), M. Abel and A. Kokk in Abel and Kokk (1988), M. Abel in Abel (2018) and others. L. Waelbroeck announced in Waelbroeck (1966), p. 153, (see also Turpin and Waelbroeck (1968), p. 195) that the Gelfand-Mazur Theorem holds in case of Hausdorff division $LPA$ with continuous inversions (up to now we have not seen any correct proof for this statement) and in Waelbroeck (1971), see Proposition 1, that the Gelfand-Mazur Theorem holds also in case when a division TA has a nontrivial continuous $k$-homogeneous seminorm for some $k \in (0, 1]$ (the proof of this result has a gap, he suggested to use the ,,small bornology" but it is not clear how to use that).

The first generalization of the Gelfand-Mazur Theorem to an exponetially galbed algebra (that is, the case, when the topology of the algebra is (not neccesarily) defined by (homogeneous or not homogeneous) seminorms) gave P. Turpin in Turpin (1975) and later M. Abel in Abel (2005a). They showed that every exponentially galbed Hausdorff division algebra over $\mathbb{C}$ with bounded elements is topologically isomorphic to $\mathbb{C}$. Other generalizations for more general cases (using bornological properties of TA) you can find from Abel (1990), ?, Abel (2004c), Abel (2005a), Abel (2005b), Abel (2007b), Abel (2008), Abel (2013), Allan et al. (1971) and Anjidani (2014). Applications of Theorem 1.9 for the description of all closed maximal (one-sided and two-sided) regular ideals in subalgebras of some $TA$ of $A$-valued ($A$ is a $LPA$) continuous functions on a topological space, are given in Kilbas et al. (2006), Blutzer and Torvik (1996) and Dreisigmeyer and Young (2003).

## 4.8 Generalizations of Arens-Michael theorem for locally pseudoconvex algebras

It is well-known (see Michael (1952), p. 17) that every complete Hausdorff $LmCA$ is topologically isomorphic to the projective limit of Banach algebras. This result has been generalized to the case of complete Hausdorff $LmkCA$ for some $k \in (0,1]$ in Abel (1989), Theorem 5; to the case of complete Hausdorff $LACA$ in Akkar et al. (1989), Theorem 2.2, and to the case of complete Hausdorff $LmPA$ in Balachandran (2000), pp. 202–204. Similar representations of TA with jointly continuous multiplication by projective limits of TA with more simple structure, are given in Michael (1952), Theorem 1, that is, *every complete Hausdorff $TA$ with jointly continuous multiplication is topologically isomorphic to the projective limit of $F$-algebras* and *every complete Hausdorff $LCA$ with jointly continuous multiplication is topologically isomorphic to the projective limit of locally convex $F$-algebras* see. Michael (1952). Next, we give a detailed proof for Müldner's (as well Arens-Michael's) result.

To do this, first of all, we recall the definition of the projective limit of $TA$s.

Let $(A, \tau)$ be a $TA$, $(I, \prec)$ a directed set (that is, an ordered set for which for each $i, j \in I$ there exists a $k \in I$ such that $i \prec k$ and $j \prec k$) and $\{(A_i, \tau_i) : i \in I\}$ a collection of TAs. If for each $i, j \in I$ with $i \prec j$ has been given continuous homomorphisms $h_{ij}$ from $(A_j, \tau_j)$ into $(A_i, \tau_i)$ such that $h_{ii}$ is the identity map on $A_i$ for each $i \in I$ and $h_{ij} = h_{ik} \circ h_{kj}$ for each $i, j, k \in I$ with $i \prec k \prec j$, then the collection $\{(A_i, \tau_i); h_{ij}; I\}$ is called a *projective system* or *inverse system* of $(A_i, \tau_i)$ and the set

$$\varprojlim A_i = \left\{ (a_i)_{i \in I} \in \prod_{i \in I} A_i : h_{ij}(a_j) = a_i \text{ for each } i, j \in I \text{ with } i \prec j \right\}$$

is called the *projective* (or *inverse*) *limit* of that projective system. All algebraic operations in $\varprojlim A_i$ are defined point-wise as in the product $\prod_{i \in I} A_i$ and $\varprojlim A_i$ is provided with the subset topology, defined by the product topology of $\prod_{i \in I} A_i$. Then $\varprojlim A_i$ is a $TA$.

Next, let $(A, \tau)$ be a $TA$ with jointly continuous multiplication, $\mathcal{L}$ a base of neighborhoods of zero in $(A, \tau)$, consisting of closed balanced sets, and $\mathcal{S} = \{S_\lambda : \lambda \in \Lambda\}$ the set of all *algebraic strings* $S_\lambda = (U_n^\lambda)$ in $\mathcal{L}$, that is, $U_n^\lambda \in \mathcal{L}$,

$$U_{n+1}^\lambda + U_{n+1}^\lambda \subseteq U_n^\lambda$$

and

$$U_{n+1}^\lambda U_{n+1}^\lambda \subseteq U_n^\lambda$$

for each $n \in \mathbb{N}' = \mathbb{N} \cup \{0\}$. For each $\lambda \in \Lambda$, $S_\lambda = (U_n^\lambda) \in \mathcal{S}$ and $\rho \in M$ (the dense subset of $\mathbb{R}^+$, consisting of all non-negative rational numbers with finite dyadic expansions) let hold (4.3) and (4.4) for each $a \in A$ and $\lambda \in \Lambda$. Then every $p_\lambda$ is an $F$-seminorm on $A$ (see the proof of Theorem 4.3.2) and

$$\ker p_\lambda = \bigcap_{n=0}^\infty U_n^\lambda.$$

Intead, if $a \in \ker p_\lambda$, then $0 = p_\lambda(a) < 2^{-n}$, hence $a \in U_n^\lambda$ for each $n \in \mathbb{N}'$. Let now $a \in U_n^\lambda = V_\lambda(\frac{1}{2^n})$ for each $n \in \mathbb{N}'$. Then $p_\lambda(a) \leqslant 2^{-n}$ for each $n \in \mathbb{N}'$. Therefore, $p_\lambda(a) = 0$ or $a \in \ker p_\lambda$.

Let $\mathcal{P} = \{p_\lambda : \lambda \in \Lambda\}$ and $\tau_\mathcal{P}$ be the initial topology on $A$, defined by the collection $\mathcal{P}$. By Theorem 4.3.2, $(A, \tau_\mathcal{P})$ is a $TA$ with jointly continuous multiplication and $\tau = \tau_\mathcal{P}$.

To give a new proof for the Müldner's result, we need

**Lemma 4.8.1.** *Let $(A, \tau)$ be a $TA$ with jointly continuous multiplication, $\mathcal{L}$ the base of all closed and balanced neighborhoods of zero in $(A, \tau)$ and $S_A = (U_n)$ an algebraic string in $\mathcal{L}$. If $(A, \tau)$ is a $T_1$-space, then the kernel*

$$N(S_A) = \bigcap_{n=1}^\infty U_n$$

*of $S_A$ is a closed two-sided ideal in $(A, \tau)$.*

*Proof.* When $N(S_A) = \{\theta_A\}$, then $N(S_A)$ is a closed two-sided ideal in $A$. Suppose now that $N(S_A) \neq \{\theta_A\}$. Take any elements $a, b \in N(S_A) \backslash \{\theta_A\}$. Let $n \in \mathbb{N}$ be an arbitrary fixed number. Since $N(S_A) \subset U_{n+1}$ and $U_{n+1} + U_{n+1} \subset U_n$, then $a + b \in U_n$ for each $n \in \mathbb{N}$. Hence, $a + b \in N(S_A)$.

Let $\lambda$ be a real or complex number and $a \in N(S_A)$. Then $a \in U_n$ for each $n \in \mathbb{N}$. If $|\lambda| \leqslant 1$, then $\lambda a \in U_n$ for each $n \in \mathbb{N}$, because $U_n$ is balanced. If $|\lambda| > 1$, let $n_0 \in \mathbb{N}$ be a natural number such that $[|\lambda|] + 1 \leqslant 2^{n_0}$ (here $[r]$ denotes the entire part of a real number $r$) and $n$ an arbitrary fixed natural number. Since $a \in U_{n+n_0}$ and

$$\lambda a = [|\lambda|] \frac{\lambda}{|\lambda|} a + (|\lambda| - [|\lambda|]) \frac{\lambda}{|\lambda|} a \in \underbrace{U_{n+n_0} + \cdots + U_{n+n_0}}_{[|\lambda|]+1 \text{ summands}} \subset U_n,$$

because $\left|\frac{\lambda}{|\lambda|}\right| = 1$, $\left|(|\lambda| - [|\lambda|])\frac{\lambda}{|\lambda|}\right| < 1$ and every $U_n$ is balanced, then $\lambda a \in U_n$ for each $n \in \mathbb{N}$. Thus, $\lambda a \in N(S_A)$ for all $\lambda \in \mathbb{K}$ and $a \in N(S_A)$.

Let now $a \in A$, $b \in N(S_A)$, $n \in \mathbb{N}$ and $m = n + 1$. Then there exists a positive number $\varepsilon_m$ such that $a \in \varepsilon_m U_m$ (because every neighborhood of zero absorbs points). If $|\varepsilon_m| \leqslant 1$, then $\varepsilon_m U_m \subseteq U_m$, as $U_m$ is balanced, and if $|\varepsilon_m| > 1$, then, from $\varepsilon_m b \subseteq \varepsilon_m N(S_A) \subseteq N(S_A) \subseteq U_m$ follows that

$$ab \in (\varepsilon_m U_m)(\varepsilon_m^{-1} U_m) \subset U_m U_m \subset U_n.$$

Hence, $ab \in N(S_A)$. Similarly, we can show that $ba \in N(S_A)$. Consequently, $N(S_A)$ is a two-sided ideal in $A$. $\qquad\square$

**Theorem 4.8.2.** a) [Müldner Theorem] *Every Hausdorff TA $(A, \tau)$ with jointly continuous multiplication defines a projective system $\{\tilde{A}_\lambda; \tilde{h}_{\lambda\mu}, \Lambda\}$ of F-algebras and continuous homomorphisms $\tilde{h}_{\lambda\mu}$ from $\tilde{A}_\mu$ to $\tilde{A}_\lambda$ (whenever $\lambda \prec \mu$) such that $(A, \tau)$ is topologically isomorphic to a dense subalgebra of the projective limit $\varprojlim \tilde{A}_\lambda$ of this system. (Here $\tilde{A}$ denotes the completion of A).*

*Moreover, if $(A, \tau)$ is a Hausdorff LPA (in particular, a LCA) with jointly continuous multiplication, then $(A, \tau)$ is topologically isomorphic to a dense subalgebra of the projective limit $\varprojlim \tilde{A}_\lambda$ of complete and metrizable LPA (respectively, LCA). In both cases, when $(A, \tau)$ is complete, then $(A, \tau)$ and $\varprojlim \tilde{A}_\lambda$ are topologically isomorphic.*

b) [Generalized Arens-Michael Theorem] *Every Hausdorff LmPA (in particular, LmCA) defines a projective system $\{\tilde{A}_\lambda; \tilde{h}_{\lambda\mu}, \Lambda\}$ of complete LBA (respectively, Banach algebras) and continuous homomorphisms $\tilde{h}_{\lambda\mu}$ from $\tilde{A}_\mu$ to $\tilde{A}_\lambda$ (whenever $\lambda \prec \mu$) such that $(A, \tau)$ is topologically isomorphic to a dense subalgebra of the projective limit $\varprojlim \tilde{A}_\lambda$ of this system. In case, when $(A, \tau)$ is complete, $(A, \tau)$ and $\varprojlim \tilde{A}_\lambda$ are topologically isomorphic.*

*Proof.* a) Let $(A, \tau)$ be a Hausdorff $TA$ with jointly continuous multiplication, $\mathcal{L}$ the base of closed and balanced neighborhoods of zero in $(A, \tau)$ and $\mathcal{S} = \{S_\lambda : \lambda \in \Lambda\}$ the collection of all algebraic strings in $\mathcal{L}$. That is, every $S_\lambda \in \mathcal{S}$ is a sequence $(O_n^\lambda)$ in $\mathcal{L}$, knots $O_n^\lambda$ of which satisfy the conditions

$$O_{n+1}^\lambda + O_{n+1}^\lambda \subseteq O_n^\lambda$$

and

$$O_{n+1}^\lambda O_{n+1}^\lambda \subseteq U_n^\lambda$$

for each $n \in \mathbb{N}'$. We define the ordering $\prec$ in $\Lambda$ in the following way: we say that $\lambda \prec \mu$ in $\Lambda$ if and only if $S_\mu \subseteq S_\lambda$, that is, if $S_\lambda = (O_n^\lambda)$ and $S_\mu = (O_n^\mu)$, then $O_n^\mu \subseteq O_n^\lambda$ for each $n \in \mathbb{N}'$. It is easy to see that $(\Lambda, \prec)$ is a partially ordered set. To show that $(\Lambda, \prec)$ is a directed set, let $S_{\lambda_1} = (O_n^{\lambda_1})$ and $S_{\lambda_2} = (O_n^{\lambda_2})$ be arbitrary fixed algebraic strings in $\mathcal{S}$ and let $S_\mu = (O_n^\mu)$ be the algebraic string in $\mathcal{L}$, which we define in the following way: let $O_0^\mu \in \mathcal{L}$ be such that

$O_0^\mu \subseteq O_0^{\lambda_1} \cap O_0^{\lambda_2}$ and for each $n \geqslant 0$, let $U_{n+1}$ be a neighborhood of zero in $\mathcal{L}$ such that $U_{n+1} + U_{n+1} \subseteq O_n^\mu$ and $U_{n+1}U_{n+1} \subseteq O_n^\mu$ (such neighborhood of zero exists by the joint continuity of the addition and multiplication in $(A, \tau)$). Let now $O_{n+1}^\mu$ be a neighborhood in $\mathcal{L}$ such that

$$O_{n+1}^\mu \subseteq U_{n+1} \cap O_{n+1}^{\lambda_1} \cap O_{n+1}^{\lambda_2}.$$

Then

$$O_{n+1}^\mu + O_{n+1}^\mu \subseteq U_{n+1} \cap O_{n+1}^{\lambda_1} \cap O_{n+1}^{\lambda_2} + U_{n+1} \cap O_{n+1}^{\lambda_1} \cap O_{n+1}^{\lambda_2} \subset U_{n+1} + U_{n+1} \subseteq O_n^\mu$$

and

$$O_{n+1}^\mu O_{n+1}^\mu \subseteq (U_{n+1} \cap O_{n+1}^{\lambda_1} \cap O_{n+1}^{\lambda_2})^2 \subset U_{n+1}U_{n+1} \subseteq O_n^\mu$$

for each $n \in \mathbb{N}$. Since $S_\mu \in \mathcal{S}$ and $O_n^\mu \subseteq O_n^{\lambda_1} \cap O_n^{\lambda_2}$ for each $n \in \mathbb{N}_0$, then $\lambda_1 \prec \mu$ and $\lambda_2 \prec \mu$. It means that $(\Lambda, \prec)$ is a directed set.

Let $p_\lambda$ be the $F$-seminorm on $A$, which is defined by the string $S_\lambda = (O_n^\lambda)$ for each $\lambda \in \Lambda$ and let $\mathcal{P}_A = \{p_\lambda : \lambda \in \Lambda\}$. Then $\ker p_\lambda$ is a closed two-sided ideal in $A$, by Lemma 1.5. For each $\lambda \in \Lambda$, let $A_\lambda = A/\ker p_\lambda$ and $\pi_\lambda$ be the canonical homomorphism from $A$ onto $A_\lambda$. Moreover, let $\overline{p}_\lambda(\pi_\lambda(a)) = p_\lambda(a)$ for each $a \in A$. Then $\overline{p}_\lambda$ is an $F$-norm on $A_\lambda$ and the multiplication in $A_\lambda$ in the topology $\tau_{\overline{p}_\lambda}$, defined by $\overline{p}_\lambda$, is jointly continuous. To show this, let $O$ be an arbitrary neighborhood of zero in $(A_\lambda, \tau_{\overline{p}_\lambda})$. Then there is an $\varepsilon > 0$ such that $\{x \in A_\lambda : \overline{p}_\lambda(x) < \varepsilon\} \subseteq O$. Now, there exists $n_\varepsilon \in \mathbb{N}$ such that $2^{-n_\varepsilon} < \varepsilon$. Since

$$U = \left\{ x \in A_\lambda : \overline{p}_\lambda(x) < \frac{1}{2^{n_\varepsilon+1}} \right\}$$

is a neighborhood of zero in $(A_\lambda, \tau_{\overline{p}_\lambda})$ and from $x, y \in U$ follows that

$$xy \in V_\lambda \left( \frac{1}{2^{n_\varepsilon+1}} \right)^2 = (U_{n_\varepsilon+1}^\lambda)^2 \subseteq U_{n_\varepsilon}^\lambda = V_\lambda \left( \frac{1}{2^{n_\varepsilon}} \right),$$

then $\overline{p}_\lambda(xy) \leqslant 2^{-n_\varepsilon} < \varepsilon$ for each $x, y \in U$. It means that the multiplication in $(A_\lambda, \tau_{\overline{p}_\lambda})$ is continuous at $(\theta_{A_\lambda}, \theta_{A_\lambda})$. Consequently, the multiplication in $A_\lambda$ (as a map $l : A_\lambda \times A_\lambda \to A_\lambda$, defined by $l(x, y) = xy$ for each $x, y \in A_\lambda$) is continuous (see Jarchow (1981), Proposition 3, p. 88).

Let $\tilde{A}_\lambda$ be the completion of $A_\lambda$ (here $\tilde{A}_\lambda$ is a $TA$, because the multiplication in $(A_\lambda, \tau_{\overline{p}_\lambda})$ is jointly continuous), $\nu_\lambda$ the topological homeomorphism from $\tilde{A}_\lambda$ onto a dense subalgebra of $\tilde{A}_\lambda$ (defined by the completion of $A_\lambda$), $\tilde{p}_\lambda$ the extension of $\overline{p}_\lambda \circ \nu_\lambda^{-1}$ to $\tilde{A}_\lambda$ and $\tilde{\tau}_\lambda$ the topology on $\tilde{A}_\lambda$, defined by $\tilde{p}_\lambda$. Then

$$\tilde{p}_\lambda[(\nu_\lambda \circ \pi_\lambda)(a)] = \overline{p}_\lambda(\pi_\lambda(a)) = p_\lambda(a)$$

for each $a \in A$. Therefore, $\tilde{p}_\lambda$ is an $F$-norm on $\tilde{A}_\lambda$, because of which $(\tilde{A}_\lambda, \tilde{\tau}_\lambda)$ is metrizable. Hence, $(\tilde{A}_\lambda, \tilde{\tau}_\lambda)$ is a $F$-algebra fo each $\lambda \in \Lambda$.

For each $\lambda, \mu \in \Lambda$, with $\lambda \prec \mu$, we define the map $h_{\lambda\mu}$ by $h_{\lambda\mu}(\pi_\mu(a)) = \pi_\lambda(a)$ for each $a \in A$. Then $h_{\lambda\mu}$ is a continuous homomorphism from $A_\mu$ onto

$A_\lambda$, $h_{\lambda\lambda}$ is the identity mapping on $A_\lambda$ for each $\lambda \in \Lambda$ and $h_{\lambda\mu} \circ h_{\mu\gamma} = h_{\lambda\gamma}$ for each $\lambda, \mu, \gamma \in \Lambda$ with $\lambda \prec \mu \prec \gamma$. Since $\nu_\lambda \circ h_{\lambda\mu} \circ \nu_\mu^{-1}$ is a continuous homomorphism from $\nu_\mu(A_\mu)$ into $\tilde{A}_\lambda$, then there exists a continuous extension $\tilde{h}_{\lambda\mu}$ from $\tilde{A}_\mu$ into $\tilde{A}_\lambda$, which is linear (by Proposition 5 in Horváth (1966)) and submultiplicative, by the continuity of multiplication in $\tilde{A}_\mu$ (similarly as in the proof of Proposition 1, pp. 4–5, in Kohn (1979) or in the proof of Proposition 3 in Aglić Aljinović (2014)). Moreover,

$$\tilde{h}_{\lambda\mu}[\nu_\mu(\pi_\mu(a))] = \nu_\lambda[h_{\lambda\mu}(\pi_\mu(a))] = \nu_\lambda[\pi_\lambda(a)]$$

for each $a \in A$ and $\lambda, \mu \in \Lambda$ with $\lambda \prec \mu$. Since $\tilde{h}_{\lambda\lambda}$ is the identity map on $\tilde{A}_\lambda$ for each $\lambda \in \Lambda$ and $\tilde{h}_{\lambda\mu} \circ \tilde{h}_{\mu\gamma} = \tilde{h}_{\lambda\gamma}$, whenever $\lambda, \mu, \gamma \in \Lambda$ and $\lambda \prec \mu \prec \gamma$, then $\{\tilde{A}_\lambda; \tilde{h}_{\lambda\mu}; \Lambda\}$ is a projective system of $F$-algebras $(\tilde{A}_\lambda, \tilde{\tau}_\lambda)$ with continuous homomorphisms $\tilde{h}_{\lambda\mu}$ from $\tilde{A}_\mu$ into $\tilde{A}_\lambda$ and

$$\varprojlim \tilde{A}_\lambda = \{(\nu_\lambda[\pi_\lambda(a)])_{\lambda\in\Lambda} \in \prod_{\lambda\in\Lambda} \tilde{A}_\lambda : \tilde{h}_{\lambda\mu}[\nu_\mu(\pi_\mu(a))] = \nu_\lambda(\pi_\lambda(a)) \text{ if } \lambda \prec \mu\}$$

is the projective limit of this system.

Let $\tilde{e}$ from $A$ into $\prod_{\mu\in\Lambda} \tilde{A}_\mu$ be defined by $\tilde{e}(a) = (\nu_\lambda[\pi_\lambda(a)])_{\lambda\in\Lambda}$ for each $a \in A$, and $\mathrm{pr}_\lambda$ the projection of $\prod_{\mu\in\Lambda} \tilde{A}_\mu$ onto $\tilde{A}_\lambda$ for each $\lambda \in \Lambda$. Since $\mathrm{pr}_\lambda(\tilde{e}(a)) = \nu_\lambda[\pi_\lambda(a)]$ for each $a \in A$ and $\lambda \in \Lambda$ and $\nu_\lambda \circ \pi_\lambda$ is continuous for each $\lambda \in \Lambda$, then $\tilde{e}$ is a continuous map from $A$ into $\prod_{\mu\in\Lambda} \tilde{A}_\mu$ (see, for example, Willard (1970), Theorem 8.8). Moreover, if $a, b \in A$ and $\tilde{e}(a) = \tilde{e}(b)$, then $(\nu_\lambda \circ \pi_\lambda)(a) = (\nu_\lambda \circ \pi_\lambda)(b)$ for each $\lambda \in \Lambda$. As $\nu_\lambda$ is an injection, then $\pi_\lambda(a) = \pi_\lambda(b)$ or $a - b \in \ker p_\lambda$ for each $\lambda \in \Lambda$. Therefore,

$$a - b \in \bigcap_{\lambda\in\Lambda} \ker p_\lambda = \bigcap_{O\in\mathcal{L}} O = \theta_A,$$

because $(A, \tau)$ is a Hausdorff space and $\mathcal{S}$ is the collection of all algebraic strings in $\mathcal{L}$. It means that $a = b$. Hence, $\tilde{e}$ is a one-to-one map.

Let now $O$ be an open subset in $A$, $o$ a point in $O$, $\alpha$ a fixed index in $\Lambda$ and

$$U = \left[\prod_{\lambda\in\Lambda} U_\lambda\right] \cap \tilde{e}(A),$$

where $U_\alpha = (\nu_\alpha \circ \pi_\alpha)(O)$ and $U_\lambda = \tilde{A}_\lambda$, if $\lambda \neq \alpha$. Then $U$ is a neighborhood of $\tilde{e}(o)$ in $\tilde{e}(A)$. Since

$$\mathrm{pr}_\alpha(U) \subset (\nu_\alpha \circ \pi_\alpha)(O) = \mathrm{pr}_\alpha(\tilde{e}(O))$$

and $\alpha$ is arbitrary, then $U \subset \tilde{e}(O)$. Hence, $\tilde{e}$ is an open map. Taking this into account, $\tilde{e}$ is a topological isomorphism from $(A, \tau)$ into $\prod_{\lambda\in\Lambda}(\tilde{A}_\lambda, \tilde{\tau}_\lambda)$.

To show that $\tilde{e}(A)$ is dense in $\varprojlim \tilde{A}_\lambda$, let $(\tilde{a}_\lambda)_{\lambda\in\Lambda} \in \varprojlim \tilde{A}_\lambda$ be an arbitrary element and $O$ an arbitrary neighborhood of $(\tilde{a}_\lambda)_{\lambda\in\Lambda}$ in $\varprojlim \tilde{A}_\lambda$. Then there is a neighborhood $U$ of $(\tilde{a}_\lambda)_{\lambda \in \Lambda}$ in $\prod_{\lambda \in \Lambda} \tilde{A}_\lambda$ such that $O = U \cap \varprojlim \tilde{A}_\lambda$. Now, there is a finite subset $H \subset \Lambda$ such that $\prod_{\lambda \in \Lambda} U_\lambda \subset U$, where $U_\lambda$ is a neighborhood of $\tilde{a}_\lambda$ in $(\tilde{A}_\lambda, \tilde{\tau}_\lambda)$, if $\lambda \in H$, and $U_\lambda = \tilde{A}_\lambda$, if $\lambda \in \Lambda \setminus H$. Let $\mu \in \Lambda$ be such that $\lambda \prec \mu$ for every $\lambda \in H$ and

$$V = \bigcap_{\lambda \in H} \tilde{h}_{\lambda\mu}^{-1}(U_\lambda).$$

Then $V$ is a neighborhood of $\tilde{a}_\mu$ in $\tilde{A}_\mu$. Take an element $a \in (\nu_\mu \circ \pi_\mu)^{-1}(V)$. Then $(\nu_\mu \circ \pi_\mu)(a) \in V$. Therefore, $(\nu_\lambda \circ \pi_\lambda)(a) = \tilde{h}_{\lambda\mu}((\nu_\mu \circ \pi_\mu)(a)) \in U_\lambda$ for each $\lambda \in H$. It means that $\tilde{e}(a) \in U \cap \tilde{e}(A)$. Consequently, $\tilde{e}(A)$ is dense in $\varprojlim \tilde{A}_\lambda$.

If now $A$ is a Hausdorff $LPA$ (in particular, $LCA$) with jointly continuous multiplication, then every $\tilde{A}_\lambda$ is a complete $LPA$ (respectively, $LCA$) with jointly continuous multiplication. Therefore, $(A, \tau)$ is topologically isomorphic to a dense subalgebra of the projective limit $\varprojlim \tilde{A}_\lambda$ of complete $LPA$s (respectively, $LCA$s). In both cases, if $(A, \tau)$ is complete, then $(A, \tau)$ and $\varprojlim \tilde{A}_\lambda$ are topologically isomorphic.

b) Let $A(, \tau)$ be a Hausdorff $LmPA$ ( $LmCA$) and $\mathcal{P} = \{p_\lambda; \lambda \in \Lambda\}$ the collection of $k_\lambda$-homogeneous (respectively, homogeneous) seminorms on $A$, with $k_\lambda \in (0, 1]$ for each $\lambda \in \Lambda$, which defines the topology $\tau$. Then the kernel $\ker p_\lambda$ of every seminorm $p_\lambda$ is a two-sided ideal in $A$ for each $\lambda \in \Lambda$. Similarly, as in the part a), we define the algebras $\tilde{A}_\lambda$ everyone of which is a complete $LBA$ (respectively, a Banach algebra). Therefore, in the present case, $(A, \tau)$ is topologically isomorphic to a dense subalgebra of the projective limit $\varprojlim \tilde{A}_\lambda$ of complete $LBA$ (respectively, Banach algebras) and when $(A, \tau)$ is complete, then $\tilde{A}_\lambda$ and $\varprojlim \tilde{A}_\lambda$ are topologically isomorphic.          □

The structure of Banach algebras and complete $LBA$s is well studied. Therefore, by Theorem 4.8.2, it is possible to describe the properties of $LmCA$ and of complete metrizable $LmPA$.

## 4.9   Generalization of Akkar teorem for locally pseudo-convex algebras

M. Akkar showed in ?, p. 943, that every Hausdorff $LmCA$ $(A, \tau)$ is a bornological inductive limit (with continuous canonical injections) of metrizable locally $m$-convex subalgebras of $(A, \tau)$ and every complete $LmCA$ is a bornological inductive limit (with continuous canonical injections) of complete and metrizable locally $m$-convex subalgebras of $(A, \tau)$. Later on, M. Oudadess in Oudadess (1997), Theorem 6.2, (see also M. (1982) and Oudadess (1983)) and M. Akkar, A. Beddaa and

M. Oudadess in Akkar et al. (1996), Theorem 6.2, gave a similar result in case when instead of completeness of $(A, \tau)$ was considered weaker kinds of completenesses, so called, Mackey completeness and advertible completness of $(A, \tau)$, respectively. All these results are true also in case of $LmkCA$ with $k \in (0, 1]$.

To generalize these results to locally pseudoconvex (or more general) case, we need the following concepts.

1. A topological algebra $(A, \tau)$ is called a *locally idempotent algebra* ($LIA$, in short) if $(A, \tau)$ has a base of neighborhoods of zero, consisting of idempotent sets. This class of $TA$s has been introduced in Zelazko (1965), p. 31. So, every $LmPA$ is a $LIA$.

For any topological algebra $(A, \tau)$, let $\mathcal{B}_A$ denote the *von Neumann bornology* on $A$, that is, $\mathcal{B}_A$ is the set of all bounded subsets of $(A, \tau)$. If, for any $B \in \mathcal{B}_A$, there exists a number $k_B \in (0, 1]$, such that $\Gamma_{k_B}(B) \in \mathcal{B}_A$, then $\mathcal{B}_A$ is *pseudoconvex* (see, Hogbe-Nlend (1971), p. 101, or ?, p. A1058). In particular, if the number $k$ does not depend on sets $B$ (that is, when $k_B = k$ for all $B \in \mathcal{B}_A$), then $\mathcal{B}_A$ is *k-convex* (see Wells (1968)), and a $k$-convex bornology is *convex*, if $k = 1$. It is known (see Wells (1968), Proposition 1.2.15) that the von Neumann bornology on any $LkCS$ is $k$-convex and there exists a locally non-convex space with convex von Neumann bornology (see Wells (1968), Example 1.2.7). Moreover (see ?, Theorems 1 and 2, Metzler (1967) and Hogbe-Nlend (1971), p. 102–103), the von Neumann bornology $\mathcal{B}$ on a $LPS$ $(X, \tau)$ is pseudoconvex, if $\mathcal{B}$ has a countable base, and every metrizable $LTS$ is $LkCS$ for some $k \in (0, 1]$, if $\mathcal{B}$ is pseudoconvex.

2. A net $(a_\lambda)_{\lambda \in \Lambda}$ in a $TA$ $(A, \tau)$ is called a *Mackey-Cauchy net*, if there exists a balanced set $B \in \mathcal{B}_A$ and for every $\varepsilon > 0$ a number $\lambda_\varepsilon \in \Lambda$ such that $a_\lambda - a_\mu \in \varepsilon B$, whenever $\mu > \lambda > \lambda_\varepsilon$. Moreover, a net $(a_\lambda)_{\lambda \in \Lambda}$ in a $TA$ $(A, \tau)$ is called *advertibly convergent*, if there is an element $a \in A$ such that $(a \circ a_\lambda)_{\lambda \in \Lambda}$ and $(a_\lambda \circ a)_{\lambda \in \Lambda}$ converge to zero (in particular case, when $A$ is a unital algebra with the unit element $e_A$, then $(aa_\lambda)_{\lambda \in \Lambda}$ and $(a_\lambda a)_{\lambda \in \Lambda}$ converge to $e_A$ (here $a \circ b = a + b - ab$ for each $a, b \in A$)). A $TA$ is called *Mackey complete*, if every Mackey-Cauchy net in $(A, \tau)$ converges in $(A, \tau)$ and *advertibly complete*, if every advertibly convergent Cauchy net in $(A, \tau)$ converges in $(A, \tau)$.

It is easy to see that every Mackey-Cauchy net is a Cauchy net. The converse statement is false in general (see Hogbe-Nlend (1977), p. 122) but it is true in case of metrizable $TA$ (see Hogbe-Nlend (1977), p. 27, or Wells (1968), Proposition 1.2.4). We say that a TA $(A, \tau)$ is *sequentially Mackey complete* and *sequentially advertibly complete*, if instead of nets we consider only sequences. Consequently, every sequentially complete algebra (and every complete algebra) is sequentially Mackey complete. Moreover, every complete $TA$ is advertibly complete (see, Mallios (1986), p. 60).

3. Let $(I, \prec)$ be an upward directed set, $(A, \tau)$ a $TA$ and $\{(A_i, \tau_i) : i \in I\}$ a collection of subalgebras $(A_i, \tau_i)$ of $(A, \tau)$. For each $i, j \subset I$ with $i \prec j$ let $f_{ji}$ be a continuous homomorphism from $(A_i, \tau_i)$ into $(A_j, \tau_j)$ such that $f_{ii}$ is

the identity map on $A_i$ for each $i \in I$, and $f_{ki} = f_{kj} \circ f_{ji}$ for each $i, j, k \in I$ with $i \prec j \prec k$. Then $\{(A_i, \tau_i); f_{ji}; I\}$ is an *inductive system* or *direct system* of subalgebras $(A_i, \tau_i)$ of $(A, \tau)$, defined by the homomorphisms $f_{ji}$ and

$$\varinjlim \tilde{A}_i = \bigcup_{i \in I} A_i$$

is the *inductive (direct) limit* of subalgebras $(A_i, \tau_i)$. Herewith, $\varinjlim \tilde{A}_i$ is called a *regular inductive limit* (see Jarchow (1981), p. 83), if for every bounded set $B$ in $(A, \tau)$ there exists an index $i_0 \in I$ such that $B \subset A_{i_0}$ and is bounded there, and a *bornological inductive limit* (see Hogbe-Nlend (1977), p. 43), if every subset $B$ in $A$ is bounded if and only if $B$ is bounded in some $(A_i, \tau_i)$.

Next, we present an analogue of Akkar's result for $LIA$ and $LmPA$.

**Proposition 4.9.1.** 1) *Let* $(A, \tau)$ *be a Hausdorff* $LIA$ *with pseudoconvex von Neumann bornology* $\mathcal{B}_A$. *Then every basis* $\beta_A$ *of* $\mathcal{B}_A$ *defines an inductive system* $\{A_B : B \in \beta_A\}$ *of metrizable* $Lmk_BCAs$ *with* $k_B \in (0, 1]$ *such that* $(A, \tau)$ *is a regular inductive limit of this system.*

2) *Let* $(A, \tau)$ *be a Hausdorff* $LmPA$ *with pseudoconvex von Neumann bornology* $\mathcal{B}_A$. *Then every basis* $\beta_A$ *of* $\mathcal{B}_A$ *defines an inductive system* $\{A_B : B \in \beta_A\}$ *of metrizable* $Lmk_BCAs$ *with* $k_B \in (0, 1]$ *such that* $(A, \tau)$ *is a bornological inductive limit of this system.*

*Herewith, when* $(A, \tau)$ *is, in addition, a sequentially Mackey complete, every subalgebra* $A_B$ *is a complete and metrizable* $Lmk_BCA$, *and when* $(A, \tau)$ *is sequentially advertibly complete, then every subalgebra* $A_B$ *is an advertibly complete metrizable* $Lmk_BCA$.

*Proof.* 1) Let $(A, \tau)$ be a $LIA$, $\mathcal{B}_A$ the von Neumann bornology on $A$, $\beta_A$ a basis of pseudoconvex sets in $\mathcal{B}_A$ and $\mathfrak{L}_A$ a base of idempotent and balanced neighborhoods of zero in $(A, \tau)$. Then every $B \in \beta_A$ defines a number $k_B \in (0, 1]$ such that $\Gamma_{k_B}(B) \in \mathcal{B}_A$. For each $n \in \mathbb{N}$ and $B \in \beta_A$, let

$$\mathfrak{L}_n^B = \{O \in \mathfrak{L}_A : \Gamma_{k_B}(B) \subseteq nO\}.$$

If now some of sets $\mathfrak{L}_n^B$ are empty, then we omit them, receiving a new sequence of sets $(\mathfrak{L}_{v_n}^B)$, in which all members $\mathfrak{L}_{v_n}^B$ are nonempty. Next, we put

$$\mathfrak{O}_n^B = \cap\{O : O \in \mathfrak{L}_{v_n}^B\}.$$

Since every $\mathfrak{O}_n^B$ is an idempotent subset in $A$, then $C_n^B(k_B) = \mathrm{cl}_A(\Gamma_{k_B}(\mathfrak{O}_n^B))$ is a closed, idempotent and absolutely $k_B$-convex subset of $A$ (see Jarchow (1981), p. 103, and Michael (1952), Lemma 4). Therefore, there is a countable set of $k_B$-homogeneous submultiplicative seminorms $p_n^B$ on

$$A_B = \{a \in A : C_n^B(k_B) \text{ absorbs } a \text{ for each } n \in \mathbb{N}\},$$

defined by

$$p_n^B(a) = \inf\{\mu^{k_B} : a \in \mu C_n^B(k_B)\}$$

for each $a \in A_B$. It is not difficult to verify that $B \subseteq A_B$ for each $B \in \beta_A$ (because $B \subseteq v_n C_n^B(k_B)$ for each $n \in \mathbb{N}$) and $A_B$ is a subalgebra of $A$. Indeed, if $a, b \in A_B$, then there exist $\gamma_1 > 0$ and $\gamma_2 > 0$ such that $a \in \gamma_1 C$ and $b \in \gamma_2 C$, where $C = C_n^B(k_B)$. Therefore, $a = \gamma_1 c_1$ and $b = \gamma_2 c_2$ for some $c_1, c_2 \in C$,

$$ab \in \gamma_1 \gamma_2 CC \subseteq \gamma_1 \gamma_2 C,$$

because $C$ is idempotent, and

$$a + b = \mu \left[ \left( \frac{\gamma_1}{\mu} \right) c_1 + \left( \frac{\gamma_2}{\mu} \right) c_2 \right] \in \mu \Gamma_{k_B}(C) = \mu C$$

for $\mu > (\gamma_1^{k_B} + \gamma_2^{k_B})^{\frac{1}{k_B}}$, because $C$ is absolutely $k_B$-convex. Now

$$A = \bigcup_{B \in \beta_A} A_B, \tag{4.16}$$

because, for every $a \in A$, there exists a $B \in \beta_A$ such that $\{a\} \subset B$, and

$$\mathfrak{L}_A = \bigcup_{n \in \mathbb{N}} \mathfrak{L}_{v_n}^B \tag{4.17}$$

for each $B \in \beta_A$, because, for every $O \in \mathcal{L}_A$, there is an $n \in \mathbb{N}$ such that $\Gamma_{k_B}(B) \subset v_n O$. Moreover, every $U \in \mathcal{B}_A$ defines a set $B_0 \in \beta_A$ such that $U \subseteq B_0 \subseteq \Gamma_{k_{B_0}}(B_0)$. Since

$$\frac{1}{v_n} U \subseteq \mathfrak{O}_n^{B_0} \subseteq \Gamma_{k_{B_0}}(\mathfrak{O}_n^{B_0}) \subseteq C_n^{B_0}(k_{B_0})$$

for each $n \in \mathbb{N}$, then $C_n^{B_0}(k_{B_0})$ absorbs $U$ for each $n \in \mathbb{N}$. Hence, $U \subseteq A_{B_0}$ and $p_n^{B_0}(u) \leqslant v_n^{k_{B_0}}$ for each $u \in U$ and fixed $n \in \mathbb{N}$. It means that $U$ is bounded in $A_{B_0}$ in the subset topology. Consequently, every bounded subset of $(A, \tau)$ is bounded in some subalgebra $A_B$ of $A$ with $B \in \beta_B$.

Let now $n \in \mathbb{N}$ be fixed and $B, B' \in \beta_A$. We define the order on $\beta_A$ by inclusion: $B \prec B'$ if and only if $B \subset B'$. Since $\beta_A$ is a basis of $\mathcal{B}_A$, then, for any $B, B' \in \beta_A$, there exists a $B'' \in \beta_A$ such that $B \cup B' \subseteq B''$. Hence, $(\beta_A, \prec)$ is a directed set. Now, for any $B, B' \in \beta_A$ with $B \prec B'$, it is true that $\mathfrak{L}_{v_n}^{B'} \subseteq \mathfrak{L}_{v_n}^B$, $\mathfrak{O}_n^B \subseteq \mathfrak{O}_n^{B'}$, $C_n^B(k_B) \subseteq C_n^{B'}(k_{B'})$, $A_B \subset A_{B'}$ and

$$p_n^{B'}(a)^{k_B} \leqslant p_n^B(a)^{k_{B'}} \tag{4.18}$$

for each $n \in \mathbb{N}$ and $a \in A_B$. To show that the inequality (18) holds, let $a \in A_B$ and $\mu \in \mathbb{R}$ be such that $p_n^B(a) < \mu$. Then

$$a \in \mu^{\frac{1}{k_B}} C_n^B(k_B) \subseteq \mu^{\frac{1}{k_B}} C_n^{B'}(k_{B'}).$$

Hence,

$$p_n^{B'}(a) \leqslant \mu^{\frac{k_{B'}}{k_B}}.$$

It means that the set of numbers $\mu$ is bounded below by $p_n^{B'}(a)^{\frac{k_b}{k_{B'}}}$. Taking the infimum of values of $\mu$, we have the inequality (4.18).

For each $B, B' \in \beta_A$ with $B \prec B'$, let $i_{B'B}$ denote the canonical injection from $A_B$ into $A_{B'}$ and, for each $B \in \beta_A$, let $i_B$ denote the canonical injection from $A_B$ into $A$. Then

$$p_n^{B'}(i_{B'B}(a))^{k_B} \leqslant p_n^B(a)^{k_{B'}}$$

for each $n \in \mathbb{N}$ and $a \in A_B$ by (4.18). Taking this into account, $\{A_B; i_{B'B}; \beta_A\}$ is an inductive system (with continuous canonical injections $i_{B'B}$) of metrizable locally $m$-($k_B$-convex) subalgebras $A_B$ of $A$ and $A$ is, by (4.16), the inductive limit of this system with (not necessarily continuous) injections $i_B$.

2) Let $(A, \tau)$ be a Hausdorff $LmPA$ with pseudoconvex von Neumann bornology $\mathcal{B}_A$. Then the injection $i_B$ from $A_B$ into $A$ is continuous for each $B \in \beta_B$. To show that, let $B \in \beta_A$ and $O$ be an arbitrary neighborhood of zero in $(A, \tau)$. Since $(A, \tau)$ is a $LmPA$, then $(A, \tau)$ has a base $\mathfrak{L}_A$ of neighborhoods of zero, consisting of idempotent and absolutely pseudoconvex sets. Hence, there are a number $k \in (0, 1]$ and a closed idempotent and absolutely $k$-convex neighborhood $O_0$ of zero in $\mathfrak{L}_A$ such that $O_0 \subseteq O$. Moreover, there exists a number $k_B \in (0, 1]$ such that $\Gamma_{k_B}(B) \in \mathcal{B}_B$, because $\mathcal{B}_B$ is pseudoconvex. We can assume that $k \leqslant k_B$ (otherwise we take, instead of $k_B$, a number $k'_B \geqslant k$, because, in case $k_B < k'_B$, every absolutely $k_B$-convex set is absolutely $k'_B$-convex). Now, $O_0$ defines a number $n_0 \in \mathbb{N}$ such that $O_0 \in \mathfrak{L}_{v_{n_0}}^B$ by (4.17). Hence, $\mathfrak{O}_{v_{n_0}}^B \subseteq O_0$. Therefore, from

$$O_{n_0}^B \subseteq C_{n_0}^B(k_B) = \mathrm{cl}_A(\Gamma_{k_B}(\mathfrak{O}_{v_{n_0}}^B)) \subseteq \mathrm{cl}_A(\Gamma_k(\mathfrak{O}_{v_{n_0}}^B)) \subseteq \mathrm{cl}_A(\Gamma_k(O_0)) = O_0 \subseteq O$$

follows that $i_B(O_{n_0}^B) \subseteq O$, where

$$O_{n_0}^B = \{a \in A_B : p_{n_0}^B(a) < 1\}$$

is a neighborhood of zero in $A_B$ (in the topology, defined by $p_{n_0}^B$) for each fixed $B \in \beta_A$. Hence, $i_B$ is continuous.

Next, let $U$ be a bounded subset in $A_B$. Then for any $n \in \mathbb{N}$ there is a positive number $M_n$ such that $p_n^B(u) \leqslant M_n^{k_B}$ for all $u \in U$. Hence, $O$ defines an $n \in \mathbb{N}$ such that

$$U \subseteq M_n C_n^B(k_B) = M_n \mathrm{cl}_A(\Gamma_k(\mathfrak{O}_n^B)) \subseteq M_n \mathrm{cl}_A(\Gamma_k(O_0)) = M_n O_0 \subseteq M_n O.$$

That is, $U \in \mathcal{B}_A$. Consequently, every Hausdorff $LmPA$ $(A, \tau)$ with pseudoconvex von Neumann bornology is a bornological inductive limit of metrizable $m$-($k_B$-convex) subalgebras $A_B$ with continuous canonical injections from $A_B$ into $A$.

Let now $(A, \tau)$ be a sequentially Mackey complete Hausdorff $LIA$ with pseudoconvex von Neumann bornology $\mathcal{B}_A$, $B \in \beta_A$, $(a_m)$ a Cauchy sequence in $A_B$ and

$$V_B = \{a_k - a_l : k, l \in \mathbb{N}\}$$

and

$$O_{n\nu}^B = \{a \in A_B : p_n^B(a) < \nu\}$$

for each $n \in \mathbb{N}$ and $\nu > 0$.

Then $V_B$ is bounded in $A_B$, $O_{n\nu}^B$ is a neighborhood of zero in $A_B$ and

$$O_{n\nu}^B = \nu^{\frac{1}{k_B}} O_{n1}^B$$

for each $n \in \mathbb{N}$ and $\nu > 0$. Hence, for each $n \in \mathbb{N}$, there exists a number $\mu_n > 0$ such that $V_B \subset \mu_n O_{n1}^B$. Now, let $\epsilon > 0$, $(\alpha_n)$ be a sequence of strictly positive numbers, which converges to 0, $\lambda_n = \frac{\mu_n}{\alpha_n}$ for each $n \in \mathbb{N}$ and

$$U = \bigcap_{n \in \mathbb{N}} \lambda_n O_{n1}^B. \tag{4.19}$$

Then $U$ is a bounded and balanced set in $A_B$, $\frac{\lambda_n}{\mu_n} = \frac{1}{\alpha_n}$ tends to $\infty$, if $n \to \infty$, and there is a number $s \in \mathbb{N}$ such that $\frac{\lambda_n}{\mu_n} \geq \frac{1}{\epsilon}$ for each $n > s$. Hence $\mu_n \leqslant \epsilon\lambda_n$ and $V_B \subset \mu_n O_{n1}^B \subseteq \epsilon\lambda_n O_{n1}^B$ for each $n > s$. Since

$$W_B = \bigcap_{n \leqslant s} \epsilon\lambda_n O_{n1}^B$$

is a neighborhood of zero in $A_B$ (in the topology, defined by $p_n^B$), there exists $l \in \mathbb{N}$ and $\alpha > 0$ such that $O_{l\alpha}^B \subseteq W_B$. Thus,

$$V_B \cap O_{l\alpha}^B \subseteq \left(\bigcap_{n>s} \epsilon\lambda_n O_{n1}^B\right)\bigcap\left(\bigcap_{n\leqslant s} \epsilon\lambda_n O_{n1}^B\right) = \bigcap_{n \in \mathbb{N}} \epsilon\lambda_n O_{n1}^B = \epsilon U. \tag{4.20}$$

Since $(a_m)$ is a Cauchy sequence in $A_B$, then there is a number $r \in \mathbb{N}$ such that $a_s - a_t \in O_{l\alpha}^B$, whenever $s > t > r$. Taking this into account, it is clear, by (20), that $a_s - a_t \in V_B \cap O_{l\alpha}^B \subseteq \epsilon U$, whenever $s > t > r$. Consequently, $(a_m)$ is a Mackey-Cauchy sequence in $A_B$. Since the canonical injection $i_B$ from $A_B$ into $(A, \tau)$ is continuous, then $U$ is bounded also in $A$ and $(a_m)$ is a Cauchy-Mackey sequence also in $(A, \tau)$. Hence, $(a_m)$ converges in $(A, \tau)$, say to $a_0$.

As $(a_m)$ is a bounded sequence in $A_B$, then, for each fixed $n \in \mathbb{N}$, there exists a number $M_n > 0$ such that

$$p_n^B(a_m) < M_n^{k_B}$$

for all $m \in \mathbb{N}$. Hence, $a_m \in M_n C_n^B(k_B)$ for each $n \in \mathbb{N}$ and all $m \in \mathbb{N}$. It is easy to see that $M_n C_n^B(k_B)$ is a closed and balanced subset of $(A, \tau)$. Therefore,

$$a_0 \in M_n C_n^B(k_B) = \mu\left(\frac{M_n}{\mu}\right)C_n^B(k_B) \subseteq \mu C_n^B(k_B),$$

whenever $|\mu| \geqslant M_n$, because $C_n^B(k_B)$ is balanced. Consequently, $C_n^B(k_B)$

absorbs $a_0$ for each $n \in \mathbb{N}$. Hence, $a_0 \in A_B$. Since $(a_n)$ is a Cauchy sequence in $A_B$, then for every $\varepsilon > 0$ there exist $\delta \in (0, \varepsilon)$ and $r_\delta \in \mathbb{N}$ such that $p_n^B(a_s - a_t) < \delta$ for each $s > t > r_\delta$. Taking this into account, $p_n^B(a_0 - a_t) \leq \delta < \epsilon$ for each $t > r$, because $p_n^B$ is continuous on $A_B$. Consequently, $(a_n)$ converges to $a_0$ in $A_B$. It means that $A_B$ is a complete and metrizable $LmkCA$.

Let now $(A, \tau)$ be a sequentially advertibly complete Hausdorff $LmPA$ with pseudoconvex von Neumann bornology $\mathcal{B}_A$, $\beta_A$ a basis of $\mathcal{B}_A$ and let $B \in \mathcal{B}_A$. Then the canonical injection $i_B$ from $A_B$ into $A$ is continuous (as it was shown above). Therefore, the topology $\tau_{A_B}$ on $A_B$, defined by the system $\{p_n^B : n \in \mathbb{N}\}$ of seminorms, is stronger than the subspace topology $\tau|_{A_B}$ on $A_B$. If $(a_n)$ is an advertibly convergent Cauchy sequence in $A_B$, then there is an element $a \in A_B$ such that $(a \circ a_n)$ and $(a_n \circ a)$ converge to zero in the topology $\tau_{A_B}$. Since it is stronger that $\tau|_{A_B}$, then $(a_n)$ is a Cauchy sequence in $A$, which advertibly converges in the topology $\tau$ as well. Hence, $(a_n)$ converges in $(A, \tau)$, because $(A, \tau)$ is sequentially advertibly complete.

Let $a_0$ be the limit of $(a_n)$ in $(A, \tau)$. It is easy to see that $a_0$ is the quasi-inverse of $a$ in $A$. Since every Cauchy sequence is bounded, then, similarly as above, $C_n^B(k_B)$ absorbs $a_0$ for all $n \in \mathbb{N}$. Thus, $a_0 \in A_B$. Since $(a_n) = (a_0 \circ (a \circ a_n))$ converges to $a_0 \circ \theta_A = a_0$, then $A_B$ is an advertibly complete metrizable $Lmk_BCA$ with $k_B \in (0, 1]$ for each $B \in \mathcal{B}_A$.  $\square$

# Bibliography

Abel M., Projective limits of topological algebras. *Ül. Toimetised 836, 3–27 (Russian)*, 1989.

Abel M., On the Gelfand-Mazur theorem for exponentially galbed algebras. *Ül. Toimetised 899, 65–70*, 1990.

Abel M., *Structure of Gelfand-Mazur algebras*. Dissertation, University of Tartu, Tartu, 2003. Dissertationes Mathematicae Universitatis Tartuensis 31. Tartu University Press, Tartu, 2003.

Abel M., Description of all closed maximal regular ideals in subalgebras of the algebra C(X,A;σ). In *Topological algebras and their applications, Contemp. Math.* **341**, *Amer. Math. Soc., Providence, RI*, 1–15, 2004a.

Abel M., Description of closed maximal regular one-sided ideals in Gelfand-Mazur algebras without a unit. *Acta Univ. Oulu. Ser. A Sci. Rerum Natur.* **408** *(2004), 9–24*, 2004b.

Abel M., Galbed Gelfand-Mazur algebras. In *Topological vector spaces, algebras and related areas. Pitman Research Notes in Math. Series 316, Longman Group Ltd., Harlow, 116–129*, 2004c.

Abel M., 2005a. Galb algebras with jointly continuous mutiplication. In *General Sem. Math. Patras Univ., 52, 97–101*.

Abel M., Topological algebras with pseudoconvexly bounded elements. In *Topological algebras, their applications, and related topics, Banach Center Publ., 67, Polish Acad. Sci., Warsaw*, 21–33, 2005b.

Abel M., Structure of locally idempotent algebras. *Banach J. Math. Anal. 1, no, 2, 195–207*, 2007a.

Abel M., Topological algebras with idempotently pseudoconvex von Neumann bornology. *Topological algebras and applications, Contemp. Math., 427, Amer. Math. Soc. Providence, RI*, 15–29, 2007b.

Abel M., Topological algebras with galbed von Neumann bornology. *Inter. Conf. on Topol. Alg. and Appl., ICTAA2008, Math. Stud. (Tartu) 4, Est. Math. Soc., Tartu*, 18–28, 2008.

Abel M., Representations of topological algebras by projective limits. *Ann. Funct. Anal. 1, no.1, 144–157*, 2010.

Abel M., Liouville's theorem for vector-valued functions. *Bul. Acad. Ştiinţe Repub. Mold. Mat., no. 2–3, 5–16*, 2013.

Abel M., 2015. Topological algebras with all elements bounded. In *Proc. ICTA 2015 (accepted)*.

Abel M., 2018. Topological division algebras which are topologically isomorphic to the field of complex numbers. In *Proc. ICTAA 2014 (accepted)*.

Abel M., and Abel M. Pairs of topological algebras. *Rocky Mountain J. Math. 37, no. 1, 1–16*, 2007.

Abel M., and Kokk A. Locally pseudoconvex Gelfand-Mazur algebras. *Akad. Toimetised, Füüs.-Mat. 37, 377–386 (Russian)*, 1988.

Adasch N., Ernst B., and Keim D. *Topological vector spaces. The theory without convexity conditions*, Lecture Notes in Math. 639, Springer-Verlag, Berlin-New York, 1978.

Akkar M., Beddaa A., and Oudadess M. Sur une classe d'algèbres topologiques. *Bull. Belg. Math. Soc. Simon Stevin 3, no. 1, 13–24.*, 1996.

Akkar M., Chcikh O. H., and Oudadess M. Sur la structure des algèbres localement A-convexes. *Bull. Polish Acad. Sci. Math. 37, no. 7–12, 567–570*, 1989.

Allan G. R., A spectral theory for locally convex algebras. *Proc. Amer. Math. Soc.15, no. 3, 399–421*, 1965.

Allan G. R., Dales H. G., and Maclure J. P. Pseudo-Banach algebras. *Studia Math. 40, 55–69*, 1971.

Anjidani E., On Gelfand-Mazur theorem on a class of $F$-algebras. *Topol. Algebra Appl. 2, no. 2, 19–23*, 2014.

Arens R., Linear topological division algebras. *Bull. Amer. Math. Soc. 53, 623–630*, 1947.

Balachandran V. K., *Topological Algebras*. North-Holland Math. Studies 185, Elsevier, 2000.

Bayoumi A. *Fundations of complex analysis in non locally convex spaces. Function theory without convexity condition.* North-Holland Mat. Studies **193**, Elsevier Sci: B. V., Amsterdam, 2003.

Beckenstain E., Narici L., and C. S. *Topological algebras*. North-Holland Math. Studies 24, North-Holland Publ. Co., Amsterdam-New York-Oxford, 1977.

Cochran A. C. Representation of $A$-convex algebras. *Proc. Amer. Math. Soc. 41, no. 2, 473–479*, 1973.

Fragoulopoulou M. *Topological algebras with involution*. North-Holland Mathematics Studies 200. Elsevier Science B.V., Amsterdam, 2005.

Gelfand I. M. Normmierte Ringe. *Rec. Math. [Mat. Sbornik] N. S. 9(51), 3–24*.

Gelfand I. M. Normed Rings. *C. R. (Doklady) Acad. Sci. URSS (N.S.) 23, 430–432 (Russian)*, 1939.

Hogbe-Nlend H. *Théories des bornologies et applications*. Lecture Notes in Math. 213, Springer-Verlag, Berlin, 1971.

Hogbe-Nlend H. *Bornologies and Functional Analysis*. North-Holland Math. Stud. 26, North-Holland, Amsterdam, 1977.

Horváth J. *Topological vector spaces and distributions I*. Addison-Wesley Publ. Co., Reading, Mass., 1966.

Jarchow H. *Locally Convex Spaces*. B.G. Teubren, Stuttgart, 1981.

Kaushik V. Projective and inductive limits of $m$-convex algebras and tensor-products of topological algebras. *Math. Sem. Notes Kobo Univ. 7, no. 1, 49–72.*, 1979.

Khan L. A. *Linear topological spaces of continuous vector-valued functions.* Acad. Publ., Jeddah, 2013.

Kokk A. Description of the homomorphisms of topological module-algebras. *Eesti ENSV Tead. Akad. Toimetised Füüs.-Mat. 36, no. 1, 1–7 (Russian).*

Köthe G. *Topological vector spaces I.* Die Grundlehren der Mathematischen Wissenschaft 159, Springen-Verlag New York Inc., New York, 1969.

M. O. Unité et semi-normes dans les algèbres localement convexes. *Rev. Colombiana Mat. 16, 141–150,* 1982.

Mallios A. *Topological Algebras. Selected Topics.* North-Holland Math. Studies 124, North-Holland Publ. Co., Amsterdam, 1986.

Mazur S. Sur les anneaux linèaires. *C. R. Acad. Sci., Paris 207, 1025–107.,* 1938.

Metzler R. C. A remark on bounded sets in linear topological spaces. *Bull. Acad. Polon. Sci. Sér. Sci. Math. Astronom. Phys. 15, 317–318,* 1967.

Michael E. A., *Locally multiplicatively-convex topologial algebras.* Mem. Amer. Math. Soc., 1952.

Narici L. and Beckenstein E., *Topological Vector spaces. Second Edition.* Pure and Applied Math. **296**, CRC Press, Boca Raton, FL, 2011.

Oudadess M., Thérèmes de structures et propetiété fondamentals des algèbres localement uniformément *A*-convexes. *C. R. Acad. Sci. Paris Sér. I Math. 296, no. 20, 851–853,* 1983.

Oudadess M., Functional boundedness of some *M*-complete *m*-convex algebras. *Bull Greek Math. Soc. 39, 17–20,* 1997.

Page W., *Topological uniform structures. Revised reprint of the 1978 original.* Dover Publ., Inc., New York, 1988.

Rolewicz S., *Metric linear spaces. Second edition.* Math. and its Appl, 20, D. Reidel Publ. Co., Dordrecht; PWN–Polish Sciendific Publ., Warsawa, 1985.

Schaefer H. H., *Topological vector spaces.* The Macmillan Co., New York, Collier-Macmillan Ltd, London, 1966.

Skarakis C., *Convex optimisation theory and practice.* Univ. York, UK, 2008.

Turpin P., Sur un classe d'algèbre topologique. *C. R. Acad. Sc. Paris 263, Ser. A , 436–439,* 1966a.

Turpin P., 1966b. *Sur une classe d'algèbres topologiques generalisant les algèbres localement borneesl.* PhD thesis, Ph. D. Thesis, Faculty of sciences, University of Grenoble.

Turpin P., Espaces et opératuers exponentiellemnt algébres. *Math. Sem. Notes Kobo Univ. 7, no. 1, 49–72.*, 1975.

Turpin P., and Waelbroeck L. Algèbres localement pseudo-convex à inverse continu. *C. R. Acad.Sc. Paris A 267, A194–A195*, 1968.

Waelbroeck L., 1966. Continuous inverse locally pseudo-convex algebras. summer school on topological algebra theory. In *Eesti NSV Tead. Akad. Toimetised, Füüs.-Mat. 37, 377–386 (Russian)*. Bruges, 128–185.

Waelbroeck L., *Théorie des algèbres de Banach et des algèbres localement convexes. Séminaire de Mathérmatiques Supérieures 2.* Le Presses de l'Universté de Montréal, Montreal, 1967.

Waelbroeck L., *Topological vector spaces and algebras.* Lecture Notes in Math. 230, Springer-Verlag, Berlin-New York, 1971.

Waelbroeck L., *Bornological quotients. With the collaboration of Guy Noël.* Acad. Royale Belg. Classe de Sciences, Brussels, 2005.

Wiljamson J. H., On topologizing the field $C(t)$. *Proc. Amer. Math. Soc. 5, 729–734*, 1954.

Willard S., *General topology.* Addison-Wesley Publ. Company, Reading-Ontario, 1970.

Zelazko W., A theorem on $B_0$-division algebras. *Bull. Acad. Polon. Sci. Sér. Sci. Math. Astr. Phys. 8, 373–375*, 1960a.

Zelazko W., On locally bounded algebras and $m$-convex topological algebras. *Studia Math.19, 333–356*, 1960b.

Zelazko W., *Metric generalizations of Banach algebras.* Rozprawy Mat. 47, 1965.

Zelazko W., *Selected topics in topological algebras.* Lectures 1969/1970. Lect. Notes Ser. 31, Aarhus, 1971.

Zelazko W., *Banach algebras.* Amsterdam-London-New York; PWN–Polish Sci. Publ., Warsawa, 1973.

# Chapter 5

# Applications of Singular Integral Operators and Commutators

**Andrea Scapellato**

*Department of Mathematics and Computer Science, University of Catania, Viale Andrea Doria 6, Catania, Italy*

## 5.1   Introduction

Aim of this chapter is to study some qualitative properties of solutions to elliptic, parabolic and ultraparabolic partial differential equations with discontinuous coefficients, in nondivergence and divergence form. Specifically, our goal is to obtain regularity results applying some estimates for singular integral operators and commutators. Some regularity results of solutions to equations whose coefficients belong to the Sarason class of functions with vanishing mean oscillation are shown.

## 5.2 BMO class

BMO, the space of functions having bounded mean oscillation, was introduced by Fritz John and Louis Nirenberg in 1961 (see John and Nirenberg (1961)) in view of the mathematics community's interest in real analysis and in studies of partial differential equations.

**Definition 5.2.1** (John and Nirenberg (1961)). We say that $f \in L^1_{\mathrm{loc}}(\mathbb{R}^n)$ belongs to the space BMO (*bounded mean oscillation*) if

$$\|f\|_* \equiv \sup_B \frac{1}{|B|} \int_B |f(x) - f_B| \mathrm{d}x < +\infty$$

where $B$ ranges in the class of the balls of $\mathbb{R}^n$ and

$$f_B \equiv \frac{1}{|B|} \int_B f(x) \mathrm{d}x$$

is the integral average of $f$ in $B$.

Note that $\|f\|_*$ is a norm in BMO modulo constant functions under which BMO is a Banach space (see Neri (1977)).

The construction of the $\mathrm{BMO}\,(\Omega)$ space, where $\Omega$ is a subset of $\mathbb{R}^n$ having a smooth boundary, is similar to the previous one: in this case we consider all functions $f$ which are integrable in $\Omega$ and, in order to define $\|\cdot\|_*$, we take the supremum over all spheres $B$ contained in $\Omega$.

## 5.3 VMO class

In 1975 D. Sarason in Sarason (1975) defined a new class of functions.

**Definition 5.3.1** (Sarason (1975)). Let $f \in \mathrm{BMO}$ and $r > 0$. We set

$$\eta(r) \equiv \sup_{\rho \leq r} \frac{1}{|B|} \int_B |f(x) - f_B| \mathrm{d}x$$

where $B$ ranges in the class of the balls of $\mathbb{R}^n$ having radius $\rho$ less than or equal to $r$.

We say that $f \in \mathrm{BMO}$ belongs to the class VMO (*vanishing mean oscillation*) if

$$\lim_{r \to 0^+} \eta(r) = 0.$$

We say that the function $\eta(r)$ is the *VMO modulus* of $f$.

Bounded uniformly continuous (BUC) functions belong to VMO.

We underline that all functions in the Sobolev Spaces $W^{1,n}$ belong to VMO; this fact follows directly from the Poincaré inequality. Indeed,

$$\frac{1}{|B|} \int_B |f(x) - f_B| \mathrm{d}x \le c(n) \left( \int_B |\nabla f(x)|^n \mathrm{d}x \right)^{1/n}.$$

The following Sarason's characterization of VMO holds.

**Theorem 5.3.2** (Sarason (1975)). *Let $f \in \mathrm{BMO}$. The following conditions are equivalent:*

1. *$f$ belongs to VMO;*

2. *$f$ belongs to the BMO closure of BUC;*

3. $\lim_{h \to 0} \| f(\cdot - h) - f(\cdot) \|_* = 0.$

Let us observe that condition 3 implies the good behavior of VMO functions. Precisely, if $f \in \mathrm{VMO}$ and $\eta_f(r)$ is its VMO modulus, we can find a sequence $\{f_h\}_{h \in \mathbb{N}} \in C^\infty(\mathbb{R}^n)$ of functions with VMO modules $\eta_{f_h}(r)$, such that $f_h \to f$ in BMO as $h \to +\infty$ and $\eta_{f_h}(r) \le \eta_f(r)$ for all integers $h$.

---

## 5.4 Morrey-Campanato spaces

In this section we present some other kind of spaces that historically play an important role in the study of the boundedness of certain type of singular integral operators. We also recall some relations between them and the VMO class.

Let us begin with the definition of *Morrey Space $L^{p,\lambda}$*.

**Definition 5.4.1** (Chiarenza (1994)). *Let $p \in [1, +\infty[$ and $\lambda \in ]0, n[$. We say that a function $f \in L^p_{\mathrm{loc}}(\mathbb{R}^n)$ belongs to the space $L^{p,\lambda}$ if*

$$\|f\|_{p,\lambda} \equiv \sup_{r>0} \left( \frac{1}{r^\lambda} \int_{B_r} |f(x)|^p \mathrm{d}x \right)^{1/p} < +\infty,$$

where $B_r$ ranges in the class of the balls of $\mathbb{R}^n$.

We also define $\mathrm{VL}^{p,\lambda}(\mathbb{R}^n)$, being $p$ and $\lambda$ as above, as the subspace of functions $f \in L^{p,\lambda}(\mathbb{R}^n)$ such that

$$\vartheta(r) \equiv \sup_{\rho \le r} \left( \frac{1}{\rho^\lambda} \int_{B_\rho} |f(x)|^p \mathrm{d}x \right)^{1/p}$$

vanishes as $r$ approaches zero.

It is immediately seen that functions whose gradient belongs to $L^{1,n-1}$ are functions of BMO class and functions whose gradient belongs to $\mathrm{VL}^{1,n-1}$ are functions of VMO class. This shows that there exist VMO functions which are discontinuous; for instance one can take $f(x) = |\log|x||^\alpha$, $0 < \alpha < 1$.

It easy to see that $W^{\theta,n/\theta}(\mathbb{R}^n)$, $0 < \theta < 1$, is contained in VMO. Indeed:

$$
\begin{aligned}
\frac{1}{|B|} \int_B |f(x) - f_B| \mathrm{d}x &= \left( \frac{1}{|B|} \int_B |f(x) - f_B|^{n/\theta} \right)^{\theta/n} \\
&= \left( \frac{1}{|B|} \int_B \left| \frac{1}{|B|} \int_B (f(x) - f(y)) \mathrm{d}y \right|^{n/\theta} \mathrm{d}x \right)^{\theta/n} \\
&\le \left( \frac{1}{|B|} \int_B \frac{1}{|B|} \int_B |f(x) - f(y)|^{n/\theta} \mathrm{d}x \mathrm{d}y \right)^{\theta/n} \\
&\le c_n \left( \int_B \int_B \frac{|f(x) - f(y)|^{n/\theta}}{|x - y|^{2n}} \mathrm{d}x \mathrm{d}y \right)^{\theta/n}.
\end{aligned}
$$

Because of $f \in W^{\theta,n/\theta}(\mathbb{R}^n)$ it follows that

$$\left( \int_{\mathbb{R}^n} \int_{\mathbb{R}^n} \frac{|f(x) - f(y)|^{n/\theta}}{|x - y|^{2n}} \mathrm{d}x \mathrm{d}y \right)^{\theta/n} < +\infty,$$

thanks to the absolute continuity of the integral, we obtain the conclusion.

**Definition 5.4.2** (Campanato (1965)). *Let $p \in [1, +\infty[$ and $\lambda \in ]0, n + p]$. We say that a function $f \in L^p_{\mathrm{loc}}(\mathbb{R}^n)$ belongs to the space $\mathscr{L}^{p,\lambda}(\mathbb{R}^n)$ if*

$$\|f\|_{p,\lambda} \equiv \sup_{r>0} \left( \frac{1}{r^\lambda} \int_{B_r} |f(x) - f_{B_r}|^p \mathrm{d}x \right)^{1/p} < +\infty.$$

The family of spaces $\mathscr{L}^{p,\lambda}(\mathbb{R}^n)$ contains $C^{0,\alpha}$ e $\mathrm{BMO} = \mathscr{L}^{1,n}(\mathbb{R}^n)$. Indeed, the following theorem is valid:

**Theorem 5.4.3** (Campanato (1963a), Meyers (1964)). *Let $p \in [1, +\infty[$ and $\lambda \in ]n, n + p[$. Then $\mathscr{L}^{p,\lambda}(\mathbb{R}^n)$ coincides with the space of the Hölder-continuous functions $C^{0,\alpha}$, with $\alpha = (\lambda - n)/p$.*

We would to point out that in literature sometimes the authors define the previous spaces on a set $\Omega$ instead of $\mathbb{R}^n$ (see, e.g. Pick et al. (2013)). For a bounded domain $\Omega \subset \mathbb{R}^n$, let us denote

$$\delta = \operatorname{diam} \Omega = \sup\{|x - y| : x, y \in \Omega\}.$$

The bounded domain $\Omega \subset \mathbb{R}^n$ is said to be *of type A* if there exists a constant $A > 0$ such that for every $x \in \overline{\Omega}$ and all $\rho \in ]0, \delta[$,

$$|\Omega(x, \rho)| \geq A\rho^n,$$

where, as usual,

$$\Omega(x, r) = \{y \in \Omega : |x - y| < \rho\}.$$

Later in this section we shall consider only domains of type $A$.

In order to present some embedding result, it is convenient to fix the notation used to indicate the embeddings. In general, if $X$ and $Y$ are two normed linear spaces and there exists a continuous embedding from $X$ into $Y$, as usual, we write $X \hookrightarrow Y$. If simultaneously $X \hookrightarrow Y$ and $Y \hookrightarrow X$, then we shall write $X \rightleftarrows Y$.

The next theorems show the connections between Morrey, Campanato and Lebesgue spaces as well as their embedding properties.

**Theorem 5.4.4.** *Let $\Omega$ be a bounded domain of $\mathbb{R}^n$.*

1. *Let $p \in ]1, +\infty[$. Then*

$$L^{p,0}(\Omega) \rightleftarrows L^p(\Omega).$$

2. *Let $p \in ]1, +\infty[$. Then*

$$L^{p,n}(\Omega) \rightleftarrows L^\infty(\Omega).$$

3. *Let $1 \leq p \leq q < +\infty$, let $\lambda$ and $\nu$ be nonnegative numbers. If*

$$\frac{\lambda - n}{p} \leq \frac{\nu - n}{q},$$

*then*

$$L^{q,\nu}(\Omega) \hookrightarrow L^{p,\lambda}(\Omega).$$

*Remark.* It is easy to see that $L^{p,\lambda}(\Omega) = \{0\}$ for $\lambda > n$. Further, it follows from Theorem 5.4.4 that the collection $\{L^{p,\lambda}(\Omega)\}_{\lambda \in [0,n]}$ for fixed $p \in [1, +\infty[$ generates a "scale of spaces" between $L^p(\Omega)$ and $L^\infty(\Omega)$.

**Theorem 5.4.5.** *Let $\Omega$ be a bounded domain of $\mathbb{R}^n$.*

1. *Let $p \in ]1, +\infty[$. Then*

$$\mathscr{L}^{p,0}(\Omega) \rightleftarrows L^p(\Omega).$$

2. *Let $1 \leq p \leq q < +\infty$, let $\lambda$ and $\nu$ be nonnegative numbers. If*

$$\frac{\lambda - n}{p} \leq \frac{\nu - n}{q},$$

*then*

$$\mathscr{L}^{q,\nu}(\Omega) \hookrightarrow \mathscr{L}^{p,\lambda}(\Omega).$$

One easily realizes that functions of VMO class are Hölder-continuous if

$$\eta(r) \leq cr^{\alpha}.$$

A natural question arises: *when are functions in VMO continuous?*

The answer has been given by Spanne (in Spanne (1965)), who perhaps really invented VMO class about ten years before Sarason. Indeed, while studying some generalizations of the $\mathscr{L}^{1,\lambda}$ spaces, Spanne introduced in Spanne (1965) the $\mathscr{L}_\phi$ spaces.

Namely:

**Definition 5.4.6** (Spanne (1965)). Let $\phi : [0, +\infty[ \to [0, +\infty[$ a non-decreasing function and $f \in L^1_{\text{loc}}(\mathbb{R}^n)$. We say that the function $f$ belongs to $\mathscr{L}_\phi$ if

$$\|f\|_{\mathscr{L}_\phi} \equiv \sup \frac{1}{\phi(|B|^{\frac{1}{n}})} \frac{1}{|B|} \int_B |f(x) - f_B| \mathrm{d}x < +\infty.$$

It is clear that if in Definition 5.4.6 we set $\phi(t) = t^{\lambda-n}$, $\lambda > n$, we recover the Campanato spaces $\mathscr{L}^{1,\lambda}(\mathbb{R}^n)$. Also, if $f \in \mathscr{L}_\phi$ and $B_\rho \subseteq B_r$, we have

$$\frac{1}{|B_\rho|} \int_{B_\rho} |f(x) - f_{B_\rho}| \mathrm{d}x \leq \|f\|_{\mathscr{L}_\phi} \cdot \phi(c_n\rho) \leq \|f\|_{\mathscr{L}_\phi} \cdot \phi(c_n r).$$

The above inequality shows that $\mathscr{L}_\phi \subseteq$ VMO if $\phi(r)$ vanishes as $r$ approaches zero and that, as VMO modulus for $f \in \mathscr{L}_\phi$, we can take

$$\eta(r) = \|f\|_{\mathscr{L}_\phi} \cdot \phi(c_n r).$$

Spanne proved (Spanne (1965)) that if $\phi(t)$ is *Dini continuous*, i.e.

$$\exists \delta > 0 : \int_0^\delta \frac{\phi(t)}{t} \mathrm{d}t < +\infty,$$

then $\mathscr{L}_\phi \subset C^0$.

## 5.5 Generalized Morrey spaces

In the last years a lot of authors studied the *generalized Morrey spaces* and the boundedness of integral operators in such spaces. Mizuhara (Mizuhara

(1991)) and Nakai (Nakai (1994)) introduced generalized Morrey spaces. Later, Guliyev (Guliyev (2013)) defined the generalized Morrey spaces with normalized norm.

It is convenient to define the generalized Morrey spaces in the following form.

**Definition 5.5.1** (Guliyev et al. (2018)). Let $\varphi(x, r)$ be a positive measurable function on $\mathbb{R}^n \times (0, \infty)$ and $1 \leq p < \infty$. We denote by $M^{p,\varphi} \equiv M^{p,\varphi}(\mathbb{R}^n)$ the generalized Morrey space that is the space of all functions $f \in L^p_{\text{loc}}(\mathbb{R}^n)$ having finite quasinorm

$$\|f\|_{M^{p,\varphi}} = \sup_{x \in \mathbb{R}^n, r>0} \varphi(x,r)^{-1} |B(x,r)|^{-\frac{1}{p}} \|f\|_{L^p(B(x,r))}.$$

Also, by $WM^{p,\varphi} \equiv WM^{p,\varphi}(\mathbb{R}^n)$ we denote the weak generalized Morrey space of all functions $f \in WL^p_{\text{loc}}(\mathbb{R}^n)$ for which

$$\|f\|_{WM^{p,\varphi}} = \sup_{x \in \mathbb{R}^n, r>0} \varphi(x,r)^{-1} |B(x,r)|^{-\frac{1}{p}} \|f\|_{WL^p(B(x,r))} < \infty,$$

where $WL^p(B(x,r))$ denotes the weak $L^p$-space consisting of all measurable functions $f$ for which

$$\|f\|_{WL^p(B(x,r))} \equiv \|f\chi_{B(x,r)}\|_{WL^p(\mathbb{R}^n)} < \infty.$$

According to this definition we recover, for $0 \leq \lambda < n$, the Morrey space $M^{p,\lambda}$ and weak Morrey space $WM^{p,\lambda}$ under the choice $\varphi(x, r) = r^{\frac{\lambda-n}{p}}$:

$$M^{p,\lambda} = M^{p,\varphi}\Big|_{\varphi(x,r)=r^{\frac{\lambda-n}{p}}}, \qquad WM^{p,\lambda} = WM^{p,\varphi}\Big|_{\varphi(x,r)=r^{\frac{\lambda-n}{p}}}.$$

In Guliyev et al. (2018) the authors considered the local version of the previous spaces and studied the behaviour of Hardy-Littlewood maximal operator and singular integral operators in the framework of such generalized local Morrey spaces.

We set the following definitions.

**Definition 5.5.2** (Guliyev et al. (2018)). Let $\varphi(x, r)$ be a positive measurable function on $\mathbb{R}^n \times (0, \infty)$ and $1 \leq p < \infty$. We denote by $LM^{p,\varphi} \equiv LM^{p,\varphi}(\mathbb{R}^n)$ the local generalized Morrey space, the space of all functions $f \in L^p_{\text{loc}}(\mathbb{R}^n)$ having finite quasinorm

$$\|f\|_{LM^{p,\varphi}} = \sup_{r>0} \varphi(0,r)^{-1} |B(0,r)|^{-\frac{1}{p}} \|f\|_{L^p(B(0,r))}.$$

Also by $WLM^{p,\varphi} \equiv WLM^{p,\varphi}(\mathbb{R}^n)$ we denote the weak generalized Morrey space of all functions $f \in WL^p_{\text{loc}}(\mathbb{R}^n)$ for which

$$\|f\|_{WLM^{p,\varphi}} = \sup_{r>0} \varphi(0,r)^{-1} |B(0,r)|^{-\frac{1}{p}} \|f\|_{WL^p(B(0,r))} < \infty.$$

**Definition 5.5.3** (Guliyev et al. (2018)). Let $\varphi(x, r)$ be a positive measurable function on $\mathbb{R}^n \times (0, \infty)$ and $1 \leq p < \infty$. For any fixed $x_0 \in \mathbb{R}^n$ we denote by $LM_{\{x_0\}}^{p,\varphi} \equiv LM_{\{x_0\}}^{p,\varphi}(\mathbb{R}^n)$, the *local generalized Morrey space*, as the space of all functions $f \in L_{\text{loc}}^p(\mathbb{R}^n)$ having finite quasinorm

$$\|f\|_{LM_{\{x_0\}}^{p,\varphi}} = \|f(x_0 + \cdot)\|_{LM^{p,\varphi}}.$$

Also by $WLM_{\{x_0\}}^{p,\varphi} \equiv WLM_{\{x_0\}}^{p,\varphi}(\mathbb{R}^n)$ we denote the weak generalized Morrey space of all functions $f \in WL_{\text{loc}}^p(\mathbb{R}^n)$ for which

$$\|f\|_{WLM_{\{x_0\}}^{p,\varphi}} = \|f(x_0 + \cdot)\|_{WLM^{p,\varphi}} < \infty.$$

According to this definition we recover, for $0 \leq \lambda < n$, the local Morrey space $LM_{\{x_0\}}^{p,\lambda}$ and weak local Morrey space $WLM_{\{x_0\}}^{p,\lambda}$ under the choice $\varphi(x_0, r) = r^{\frac{\lambda-n}{p}}$:

$$LM_{\{x_0\}}^{p,\lambda} = LM_{\{x_0\}}^{p,\varphi}\Big|_{\varphi(x_0,r)=r^{\frac{\lambda-n}{p}}}, \qquad WLM_{\{x_0\}}^{p,\lambda} = WLM_{\{x_0\}}^{p,\varphi}\Big|_{\varphi(x_0,r)=r^{\frac{\lambda-n}{p}}}.$$

Useful tool in the sequel is the following definition.

**Definition 5.5.4** (Chiarenza et al. (1991)). Let $k : \mathbb{R}^n \setminus \{0\} \to \mathbb{R}$. We say that $k$ is a **Calderón-Zygmund kernel** (CZ) if:

1. $k \in C^\infty(\mathbb{R}^n \setminus \{0\})$;

2. $k(x)$ is homogeneous of degree $-n$;

3. $\int_\Sigma k(x) \, d\sigma_x = 0$ where $\Sigma \equiv \{x \in \mathbb{R}^n : |x| = 1\}$.

The following theorem was proved by Guliyev in Guliyev (2013).

**Theorem 5.5.5.** *Let $x_0 \in \mathbb{R}^n$, $1 \leq q < \infty$, $K$ be a Calderón-Zygmund singular integral operator and the functions $\varphi_1, \varphi_2$ satisfy the condition*

$$\int_r^\infty \frac{\operatorname*{ess\,inf}_{t<\tau<\infty} \varphi_1(x_0, \tau)\, \tau^{\frac{n}{q}}}{t^{\frac{n}{q}+1}} \, dt \leq C\, \varphi_2(x_0, r),$$

*where $C$ does not depend on $r$. Then, for $1 < q < \infty$ the operator $K$ is bounded from $LM_{\{x_0\}}^{q,\varphi_1}(\mathbb{R}^n)$ to $LM_{\{x_0\}}^{q,\varphi_2}(\mathbb{R}^n)$ and for $1 \leq q < \infty$ the operator $K$ is bounded from $LM_{\{x_0\}}^{q,\varphi_1}(\mathbb{R}^n)$ to $WLM_{\{x_0\}}^{q,\varphi_2}(\mathbb{R}^n)$. Moreover, for $1 < q < \infty$*

$$\|Kf\|_{LM_{\{x_0\}}^{q,\varphi_2}} \leq c\|f\|_{LM_{\{x_0\}}^{q,\varphi_1}} \leq c\|f^\sharp\|_{LM_{\{x_0\}}^{q,\varphi_1}},$$

*where $c$ does not depend on $x_0$ and $f$ and for $1 \leq q < \infty$*

$$\|Kf\|_{WLM_{\{x_0\}}^{q,\varphi_2}} \leq c\|f\|_{LM_{\{x_0\}}^{q,\varphi_1}} \leq c\|f^\sharp\|_{LM_{\{x_0\}}^{q,\varphi_1}},$$

*where $c$ does not depend on $x_0$ and $f$.*

**Theorem 5.5.6** (Guliyev et al. (2018)). *Let $x_0 \in \mathbb{R}^n$, $1 < q < s < p < +\infty$, $K$ be a Calderón-Zygmund singular integral operator and the function $\varphi$ satisfy the conditions*

$$\sup_{r<t<\infty} \frac{\operatorname*{ess\,inf}_{t<\tau<\infty} \varphi(x_0,\tau)\,\tau^{\frac{nq}{p}}}{t^{\frac{nq}{p}}} \leq C\,\varphi(x_0,r),$$

$$\sup_{r<t<\infty} \frac{\operatorname*{ess\,inf}_{t<\tau<\infty} \varphi(x_0,\tau)\,\tau^{\frac{ns}{p}}}{t^{\frac{ns}{p}}} \leq C\,\varphi(x_0,r)$$

*and*

$$\int_r^\infty \frac{\operatorname*{ess\,inf}_{t<\tau<\infty} \varphi(x_0,\tau)\,\tau^{\frac{n}{p}}}{t^{\frac{n}{p}+1}}\,dt \leq C\,\varphi(x_0,r),$$

*where $C$ does not depend on $r$.*

*If $a \in BMO(\mathbb{R}^n)$ then, the commutator*

$$[a,K](f) = a\,Kf - K\,(af)$$

*is a bounded operator from $LM_{\{x_0\}}^{p,\varphi}(\mathbb{R}^n)$ in itself. Precisely, $\forall f \in LM_{\{x_0\}}^{p,\varphi}(\mathbb{R}^n)$, we have*

$$\|[a,K](f)\|_{LM_{\{x_0\}}^{p,\varphi}} \leq c\,\|a\|_* \,\|f\|_{LM_{\{x_0\}}^{p,\varphi}} \leq c\,\|a\|_* \,\|f^\sharp\|_{LM_{\{x_0\}}^{q,\varphi}},$$

*for some constant $c \geq 0$ independent on $a$ and $f$.*

**Corollary 5.5.7** (Guliyev et al. (2018)). *Let $x_0 \in \mathbb{R}^n$, $1 < p < +\infty$, $K$ be a Calderón-Zygmund singular integral operator and the function $\varphi(x_0,\cdot) : (0,\infty) \to (0,\infty)$ be a decreasing function. Assume that the mapping $r \mapsto \varphi(x_0,r)\,r^{\frac{n}{p}}$ is almost increasing (there exists a constant $c$ such that for $s < r$ we have $\varphi(x_0,s)\,s^{\frac{n}{p}} \leq c\varphi(x_0,r)\,r^{\frac{n}{p}}$). Let also be true the following inequality*

$$\int_r^\infty \varphi(x_0,t)\,\frac{dt}{t} \leq C\,\varphi(x_0,r),$$

*where $C$ does not depend on $r$.*

*If $a \in BMO(\mathbb{R}^n)$, then the commutator $[a,K]$ is a bounded operator from $LM_{\{x_0\}}^{p,\varphi}(\mathbb{R}^n)$ in itself.*

## 5.6 Mixed Morrey spaces

Recently, in Ragusa and Scapellato (2017) the authors defined a new class of functions. These spaces generalize Morrey spaces and give a refinement of the classical Lebesgue spaces. The authors also proved some embeddings between these new classes and, as application, they obtain some regularity results for solutions to partial differential equations of parabolic type in non-divergence form. Preparatory to achieving these results is the study of the behavior of Hardy-Littlewood maximal function, Riesz potential, sharp and fractional maximal functions, singular integral operators with Calderón-Zygmund kernel and commutators.

In the sequel let $T > 0$ and $\Omega$ be a bounded open set of $\mathbb{R}^n$ such that $\exists A > 0 : \forall x \in \Omega$ and $0 \leq \rho \leq \operatorname{diam}(\Omega)$, $|Q(x, \rho) \cap \Omega| \geq A \rho^n$, being $Q(x, \rho)$ a cube centered in $x$, having edges parallel to the coordinate axes and lenght $2\rho$.

**Definition 5.6.1** (Ragusa and Scapellato (2017)). Let $1 < p, q < +\infty$, $0 < \lambda, \mu < n$. We define the set $L^{q,\mu}(0, T, L^{p,\lambda}(\Omega))$ as the class of functions $f$ such that the following norm

$$\|f\|_{L^{q,\mu}(0,T,L^{p,\lambda}(\Omega))} = \left( \sup_{\substack{t_0, t \in (0,T) \\ \rho > 0}} \frac{1}{\rho^{\mu}} \int_{(0,T) \cap (t_0 - \rho, t_0 + \rho)} \left( \sup_{\substack{x \in \Omega \\ \rho > 0}} \frac{1}{\rho^{\lambda}} \int_{\Omega \cap B_{\rho}(x)} |f(y, t)|^p \, dy \right)^{\frac{q}{p}} dt \right)^{\frac{1}{q}}$$

is finite.

The same definition holds if $\Omega = \mathbb{R}^n$.

---

## 5.7 Applications to partial differential equations

In this section we present some important applications of VMO class to the theory of partial differential equations. The most interesting aspect of the exposition is that the hypothesis of *vanishing mean oscillation* allows many authors to study several type of equations, taking into account both equations in nondivergence and divergence form and systems. Precisely, we will deal with elliptic, parabolic and ultraparabolic equations.

### 5.7.1 Elliptic equations

#### 5.7.1.1 Preliminary tools

It is well known that it is possible to associate to a CZ kernel a bounded operator in $L^p$. Precisely, the following result holds:

**Theorem 5.7.1** (Calderón and Zygmund (1952)). *Let $k(x)$ be a CZ kernel and $\epsilon$ a positive number, $f \in L^p(\mathbb{R}^n)$ and $p \in ]1, +\infty[$. We set*

$$K_\epsilon f(x) = \int_{|x-y| > \epsilon} k(x - y) f(y) \mathrm{d}y.$$

*Then, for any $f \in L^p(\mathbb{R}^n)$, there exists $Kf \in L^p(\mathbb{R}^n)$ such that*

$$\lim_{\epsilon \to 0} \|K_\epsilon f - Kf\|_{L^p(\mathbb{R}^n)} = 0.$$

*Also, there exists a constant $c = c(n,p)$ such that*

$$\|Kf\|_{L^p(\mathbb{R}^n)} \leq c \left( \int_\Sigma k^2 \, d\sigma \right)^{1/2} \|f\|_{L^p(\mathbb{R}^n)}, \qquad \forall f \in L^p(\mathbb{R}^n),$$

*where $\Sigma = \{x \in \mathbb{R}^n : |x| = 1\}$.*

We say $K$ a *Calderón-Zygmund singular integral operator*, using the following notation

$$Kf(x) = \text{P.V.} \, k * f(x) = \text{P.V.} \int_{\mathbb{R}^n} k(x-y)f(y) \, dy = \lim_{\epsilon \to 0} \int_{|x-y|>\epsilon} k(x-y)f(y) \, dy.$$

We now recall another definition useful in the sequel:

**Definition 5.7.2** (Coifman et al. (1976)). Let $\varphi \in \text{BMO}$, $k$ be a CZ kernel and $f \in L^p(\mathbb{R}^n)$, $1 < p < +\infty$. We define **commutator**, $Tf$, the principal value

$$Tf = C[\varphi, f] = \varphi(\text{P.V.} \, k * f) - \text{P.V.} \, (k * (\varphi f)).$$

The following theorem is valid:

**Theorem 5.7.3** (Coifman et al. (1976)). *Let $\varphi \in \text{BMO}$ and $k$ be a CZ kernel. Then, $C[\varphi, f]$ is well defined for $f \in L^p(\mathbb{R}^n)$, $\forall p \in ]1, +\infty[$. Moreover, $C[\varphi, f]$ is a bounded operator in $L^p(\mathbb{R}^n)$, i.e. there exists a constant $c = c(n, p, \|k\|_{L^2(\Sigma)})$ such that*

$$\|C[\varphi, f]\|_{L^p(\mathbb{R}^n)} \leq c\|\varphi\|_* \|f\|_{L^p(\mathbb{R}^n)}.$$

For sake of completeness, we now recall another important result from the literature which is a complement to Theorem 5.7.3.

**Theorem 5.7.4** (Uchiyama (1978)). *The commutator $Tf$ in Definition 5.7.2 is a compact operator from $L^p(\mathbb{R}^n)$ in itself if and only if $\varphi$ belongs to the BMO closure of $C_0^\infty(\mathbb{R}^n)$.*

We have the following theorem which is crucial in the proof of the interior estimates for solution of partial differential equations.

**Theorem 5.7.5** (Chiarenza et al. (1991)). *Let $\Omega$ be an open subset of $\mathbb{R}^n$. Let $k : \Omega \times (\mathbb{R}^n \setminus \{0\}) \to \mathbb{R}$ be a function verifying:*

1. *$k(x, \cdot)$ is a CZ kernel, for a.a. $x \in \Omega$;*

2. *$\displaystyle \max_{|j| \leq 2n} \left\| \frac{\partial^j}{\partial z^j} k(x, z) \right\|_{L^\infty(\Omega \times \Sigma)} = M < +\infty.$*

*For $f \in L^p(\Omega)$, $1 < p < +\infty$, $\varphi \in L^\infty(\mathbb{R}^n)$ and $x \in \Omega$, let us set*

$$K_\epsilon f(x) = \int_{\substack{|x-y|>\epsilon \\ y \in \Omega}} k(x, x-y)f(y)\, dy$$

*and*

$$C[\varphi, f](x) = \varphi(x)K_\epsilon f(x) - K_\epsilon(\varphi f)(x) = \int_{\substack{|x-y|>\epsilon \\ y \in \Omega}} k(x, x-y)[\varphi(x)-\varphi(y)]f(y)\, dy.$$

*Then, for any $f \in L^p(\Omega)$ there exist $Kf$, $C[\varphi, f] \in L^p(\Omega)$ such that*

$$\lim_{\epsilon \to 0} \|K_\epsilon f - Kf\|_{L^p(\Omega)} = \lim_{\epsilon \to 0} \|C_\epsilon[\varphi, f] - C[\varphi, f]\|_{L^p(\Omega)} = 0.$$

*Moreover, there exists a constant $c \equiv c(n, p, M)$ such that*

$$\|Kf\|_{L^p(\Omega)} \le c\|f\|_{L^p(\Omega)}, \qquad \|C[\varphi, f]\|_{L^p(\Omega)} \le c\|\varphi\|_*\|f\|_{L^p(\Omega)}.$$

From the boundedness of the commutator stated in Theorem 5.7.5 we can easily deduce the following localized estimate.

**Corollary 5.7.6** (Chiarenza et al. (1991)). *Let $k$ be as in Theorem 5.7.5, $a \in$ VMO $\cap L^\infty(\mathbb{R}^n)$ and $\eta$ the VMO modulus of $\varphi$. Then, for any $\epsilon > 0$ there exists $\rho_0 = \rho_0(\epsilon, \eta)$ such that, for $r \in ]0, \rho_0[$ and every sphere $B_r \subseteq \Omega$, we have*

$$\|C[a, f]\|_{L^p(B_r)} \le c\epsilon\|f\|_{L^p(B_r)}, \qquad \forall f \in L^p(B_r).$$

In order to achieve global regularity results, we need the next results. Let $\mathbb{R}_+^n = \{x = (x', x_n) \in \mathbb{R}^n : x' \in \mathbb{R}^{n-1}, x_n > 0\}$ and $\tilde{x} = (x', -x_n)$.

**Theorem 5.7.7** (Chiarenza et al. (1993)). *Let $f \in L^p(\mathbb{R}_+^n)$, $1 < p < +\infty$. For $x \in \mathbb{R}_+^n$, let us set*

$$\tilde{K}f(x) = \int_{\mathbb{R}_+^n} \frac{f(y)}{|\tilde{x} - y|^n}\, dy.$$

*Then, there exists a constant $c = c(n, p)$ such that*

$$\|\tilde{K}f\|_{L^p(\mathbb{R}_+^n)} \le c\|f\|_{L^p(\mathbb{R}_+^n)}.$$

**Theorem 5.7.8** (Chiarenza et al. (1993)). *Let $f \in L^p(\mathbb{R}_+^n)$, $1 < p < +\infty$ and $a \in$ VMO $\cap L^\infty(\mathbb{R}_+^n)$. For $x \in \mathbb{R}_+^n$, let us set*

$$\tilde{C}[a, f]f(x) = \int_{\mathbb{R}_+^n} \frac{a(x) - a(y)}{|\tilde{x} - y|^n} f(y)\, dy.$$

*Then, there exists a constant $c = c(n, p)$ such that*

$$\|\tilde{C}[a, f]\|_{L^p(\mathbb{R}_+^n)} \le c\|a\|_*\|f\|_{L^p(\mathbb{R}_+^n)}.$$

## 5.7.1.2   Nondivergence form elliptic equations

Let us start considering the following problem.

*Let $\Omega$ be an open subset, $\Omega \subseteq \mathbb{R}^n$, and let us consider a second order linear elliptic equation in nondivergence form*

$$Lu \equiv \sum_{i,j=1}^{n} a_{ij} u_{x_i x_j} = f, \quad \text{a.e. in } \Omega \qquad (5.1)$$

*what hypotheses ensure that, for all $p \in ]1, +\infty[$, the Dirichlet problem*

$$\begin{cases} Lu = f & \text{a.e. in } \Omega \\ u = 0 & \text{on } \partial\Omega \\ u \in W^{2,p}(\Omega) \cap W_0^{1,p}(\Omega), & 1 < p < +\infty \end{cases} \qquad (5.2)$$

*is well-posed?*

This problem was deeply studied in Chiarenza's pioneering work Chiarenza et al. (1991). The regularizing properties of $L$ in Hölder spaces, i.e. $Lu \subset C^\alpha(\overline{\Omega})$ implies $u \in C^{2+\alpha}(\overline{\Omega})$, was studied in the case of Hölder continuous coefficients $a_{ij}(x)$; also, unique classical solvability of the Dirichlet problem for (15.1) was studied in this case (Gilbarg and Trudinger (2001), Chapter 6).

If the coefficients $a_{ij}$ are uniformly continuous, Agmon, Douglis and Nirenberg in Agmon et al. (1959) developed an $L^p$−Schauder theory for the differential operator $L$. In particular, those authors proved that $Lu \in L^p(\Omega)$ implies that the strong solutions to (15.1) belong to the Sobolev space $W^{2,p}(\Omega)$, for each $p \in ]1, \infty[$.

However, the situation becomes rather difficult if we try to weaken the assumption on the coefficients. In general, it is well known that arbitrary discontinuity of $a_{ij}$'s breaks down as the $L^p$−theory of $L$, as the strong solvability of the Dirichlet problem associated to (15.1).

Only in the case of the two-dimensional domain, $\Omega \subseteq \mathbb{R}^2$, and $p = 2$, Talenti in Talenti (1966) assuming the coefficients $a_{ij}$ only bounded and measurable, obtained isomorphic properties of $L$ considered as a mapping from $W^{2,2}(\Omega) \cap W_0^{1,2}(\Omega)$ into $L^2(\Omega)$.

In the multidimensional case, $n \geq 3$, if $a_{ij} \in W^{1,n}$ (Miranda, Miranda (1963)) or if the difference between the largest and the smallest eigenvalues of $\{a_{ij}(x)\}$ is small enough (the *Cordes condition*, Campanato (1967)), then $Lu \in L^2(\Omega)$ yields $u \in W^{2,2}(\Omega)$ and these results can be extended to $W^{2,p}(\Omega)$ for $p \in ]2 - \epsilon, 2 + \epsilon[$ for sufficiently small $\epsilon$.

The Sarason class $VMO$ was employed by Chiarenza, Frasca and Longo in Chiarenza et al. (1991) and Chiarenza et al. (1993) in order to obtain local and global Sobolev regularity of the strong solutions to (15.1). Precisely, in a series of papers by Chiarenza, Frasca, Franciosi and Longo, it was proved that if we assume that $a_{ij} \in \text{VMO} \cap L^\infty(\Omega)$ and $Lu \in L^p(\Omega)$, then $u \in W^{2,p}(\Omega)$ for *each* value of $p$ in $]1, \infty[$.

Also, the well posedness of the Dirichlet problem for $Lu = f$ in $W^{2,p}(\Omega) \cap W_0^{1,p}(\Omega)$ was studied.

The technique employed in the scientific notes Chiarenza et al. (1991), Chiarenza et al. (1993) is based on the representation formula for second derivatives of the solutions in terms of singular integral operators and commutators with Calderón-Zygmund kernels.

### 5.7.1.3    Representation formula

In Chiarenza et al. (1991) the authors obtained a representation formula of the second derivatives of a solution to (15.1). Let us introduce below the argument.

Let $B$ a sphere in $\mathbb{R}^n$, $a_{ij} \in \text{VMO} \cap L^\infty(\mathbb{R}^n)$, $i,j = 1, ..., n$ and

1. $a_{ij}(x) = a_{ji}(x)$ a.e. in $B$, $\forall i, j = 1, ..., n$;

2. $\exists \lambda > 0 : \lambda^{-1} \leq \sum_{i,j=1}^n a_{ij}(x)\xi_i\xi_j \leq \lambda|\xi|^2$ a.e. in $B$, $\forall \xi \in \mathbb{R}^n$.

Denote by $\tilde{B}$ the subset of $B$ where both conditions 1 and 2 hold. In the sequel we set

$$\Gamma(x,t) = \frac{1}{(n-2)\omega_n(\det a_{ij}(x))^{1/2}} \left( \sum_{i,j=1}^n A_{ij}(x)t_it_j \right)^{(2-n)/2} \tag{5.3}$$

for a.a. $x \in B$ and all $t \in \mathbb{R} \setminus \{0\}$, where we denote by $A_{ij}(x)$ the entries of the inverse matrix of the matrix $(a_{ij})_{i,j=1,...,n}$ and $\omega_n$ is the measure of the unit sphere in $\mathbb{R}^n$.

For any fixed $x_0 \in \tilde{B}$, $\Gamma(x_0, t)$ is a fundamental solution for the operator $L_0 u(x) \equiv \sum_{i,j=1}^n a_{ij}(x_0)u_{x_ix_j}(x)$.

Also, we set

$$\Gamma_i(x,t) = \frac{\partial}{\partial t_i}\Gamma(x,t) \quad \text{and} \quad \Gamma_{ij}(x,t) = \frac{\partial^2}{\partial t_i \partial t_j}\Gamma(x,t). \tag{5.4}$$

It is known that $\Gamma_{ij}$ are CZ kernels in the $t$ variable. Indeed, they are the first derivatives of a homogeneous function of degree $1 - n$.

Taking into account all the notations above, the following fundamental theorem holds.

**Theorem 5.7.9** (Chiarenza et al. (1991)). *Let $n \geq 3$, $B$ and $(a_{ij})_{i,j=1,...,n}$ as above and $u \in W_0^{2,p}(B)$. Also set*

$$Lu(x) \equiv \sum_{i,j=1}^n a_{ij}(x)u_{x_ix_j}(x).$$

*Then, for a.a. $x \in B$,*

$$u_{x_i x_j} = \text{P.V.} \int_B \Gamma_{ij}(x, x-y) \left[ \sum_{h,k=1}^n (a_{hk}(x) - a_{hk}(y)) u_{x_h x_k} + Lu(y) \right] dy +$$

$$+ Lu(x) \int_{|t|=1} \Gamma_i(x,t) t_j \, d\sigma_t.$$

### 5.7.1.4 Regularity

In Chiarenza et al. (1991), the authors used the stated representation formula in the study of the $L^p$–regularity of solutions of (15.1). We make the following assumptions to which we collectively refer as *Assumption (A)*:

$$(\Lambda) \quad \begin{cases} 1. & a_{ij}(x) \in \text{VMO} \cap L^\infty(\mathbb{R}^n), \quad \forall i,j = 1, ..., n \\ 2. & a_{ij}(x) = a_{ji}(x) \text{ a.e. in } \Omega, \forall i, j = 1, ..., n \\ 3. & \exists \nu > 0 : \nu^{-1} \le \sum_{i,j=1}^n a_{ij}(x) \xi_i \xi_j \le \nu |\xi|^2 \text{ a.e. in } \Omega, \forall \xi \in \mathbb{R}^n. \end{cases}$$

*Furthermore, call $\eta_{ij}(r)$ the function corresponding to each coefficient $a_{ij}$ as in Definition 5.3.1 and set*

$$\eta(r) = \left( \sum_{i,j=1}^n \eta_{ij}^2(r) \right)^{1/2}.$$

*Finally, recalling the functions $\Gamma_{ij}$ as in (5.4), let us set*

$$\max_{i,j=1,...,n} \max_{|\alpha| \le 2n} \left\| \frac{\partial^\alpha \Gamma_{ij}(x,t)}{\partial t^\alpha} \right\|_{L^\infty(\Omega \times \Sigma)} = M.$$

We have the following result.

**Lemma 5.7.10** (Chiarenza et al. (1991)). *Let Assumption (A) be fulfilled. Then, for each $p \in ]1, +\infty[$ there exist positive numbers $c = c(n, p, M)$ and $\rho_0 = \rho_0(c, \eta)$ such that, for any ball $B_r \Subset \Omega$, $r < \rho_0$ and any $u \in W_0^{2,p}(B_r)$, we have*

$$\|u_{x_i x_j}\|_{L^p(B_r)} \le c \|Lu\|_{L^p(B_r)}, \quad \forall i, j = 1, ..., n.$$

**Theorem 5.7.11** (Chiarenza et al. (1991)). *Let Assumption (A) be fulfilled. Then, for all $p, q$ such that $1 < q \le p < +\infty$ and all $u \in W_{\text{loc}}^{2,q}(\Omega)$ such that $Lu \in L_{\text{loc}}^p(\Omega)$ we have $u \in W_{\text{loc}}^{2,p}(\Omega)$. Moreover, given $\Omega' \Subset \Omega'' \Subset \Omega$, where $\Omega'$ and $\Omega''$ are open sets, there exists a constant $c = c(n, \nu, p, M, \text{dist}(\Omega', \partial\Omega''), \eta)$ such that*

$$\|u\|_{W^{2,p}(\Omega')} \le c \left\{ \|u\|_{L^p(\Omega'')} + \|Lu\|_{L^p(\Omega'')} \right\}.$$

The standard techniques used in the regularity results contained in this section consist in obtaining suitable interior and boundary estimates for the solution of problem (15.5) and then, from these, an a priori estimate for the solutions of (15.5). Both interior and boundary estimates are consequences of explicit representation formulas for the solution of (15.5) and the boundedness in $L^p$ of some integral operators appearing in those formulas. Finally, for the uniqueness, the VMO assumption again played a crucial role, assuring that some operators in $L^p$ are contractions on this space.

Now, we make the following assumptions and we refer to them collectively as *Assumption (B)*:

$(B)$ 
$$\begin{cases} \text{Let } n \geq 3, \Omega \text{ be an open bounded subset of } \mathbb{R}^n, \text{ with } \partial\Omega \in C^{1,1} \text{ and let} \\ \qquad L = \sum_{i,j=1}^{n} a_{ij}(x) \frac{\partial^2}{\partial x_i \partial x_j} \\ \text{where} \\ 1. \quad a_{ij}(x) \in \text{VMO} \cap L^\infty(\mathbb{R}^n), \quad \forall i,j = 1, ..., n \\ 2. \quad a_{ij}(x) = a_{ji}(x) \text{ a.e. in } \Omega, \forall i,j = 1, ..., n \\ 3. \quad \exists \lambda > 0 : \lambda^{-1} \leq \sum_{i,j=1}^{n} a_{ij}(x)\xi_i\xi_j \leq \lambda|\xi|^2 \text{ a.e. in } \Omega, \forall \xi \in \mathbb{R}^n \end{cases}$$

*Furthermore, call $\eta_{ij}(r)$ the VMO modulus of $a_{ij}$ $(i,j = 1, ..., n)$ and set*

$$\eta(r) = \left( \sum_{i,j=1}^{n} \eta_{ij}^2(r) \right)^{1/2}.$$

*Finally, let $\Gamma$, $\Gamma_i$, $\Gamma_{ij}(x,t)$ as in (5.4) and set*

$$\max_{i,j=1,...,n} \max_{|\alpha| \leq 2n} \left\| \frac{\partial^\alpha \Gamma_{ij}(x,t)}{\partial t^\alpha} \right\|_{L^\infty(\Omega \times \Sigma)} = M. \qquad (5.5)$$

**Lemma 5.7.12** (Chiarenza et al. (1993)). *Assume $(B)$. Let $p, q \in ]1, +\infty[$, $q \leq p$, $f \in L^p(\Omega)$, $u \in W^{2,q}(\Omega) \cap W_0^{1,q}(\Omega)$ and $Lu = f$ a.e. in $\Omega$. Then, $u \in W_{\text{loc}}^{2,p}(\Omega)$. Moreover, given $\Omega' \Subset \Omega$, $\Omega'$ open, there exists a constant $c = c(n, p, M, \text{dist}(\Omega', \partial\Omega'), \lambda, \eta)$ such that*

$$\|u\|_{W^{2,p}(\Omega')} \leq c \left\{ \|u\|_{L^p(\Omega)} + \|f\|_{L^p(\Omega)} \right\}.$$

**Theorem 5.7.13** (Chiarenza et al. (1993)). *Assume $(B)$. Let $p, q \in ]1, +\infty[$, $q \leq p$, $f \in L^p(\Omega)$, $u \in W^{2,q}(\Omega) \cap W_0^{1,q}(\Omega)$ and $Lu = f$ a.e. in $\Omega$. Then, $u \in W^{2,p}(\Omega)$ and there exists a constant $c = c(n, p, M, \partial\Omega, \lambda, \eta)$ such that*

$$\|u\|_{W^{2,p}(\Omega)} \leq c \left\{ \|u\|_{L^p(\Omega)} + \|f\|_{L^p(\Omega)} \right\}.$$

Now we present two results; the first one deals with the uniqueness and the second with the existence of the solution to the Dirichlet problem under consideration.

**Theorem 5.7.14** (Uniqueness, Chiarenza et al. (1993)). *Assume (B) and let* $1 < p < +\infty$. *Then, the solution of the following Dirichlet problem*

$$\begin{cases} Lu = 0 & \text{a.e. in } \Omega \\ u \in W^{2,p}(\Omega) \cap W_0^{1,p}(\Omega) \end{cases}$$

*is* $u = 0$ *in* $\Omega$.

**Theorem 5.7.15** (Existence, Chiarenza et al. (1993)). *Assume (B). Let* $f \in L^p(\Omega)$, $p \in ]1, +\infty[$. *Then, the following Dirichlet problem*

$$\begin{cases} Lu = f & \text{a.e. in } \Omega \\ u \in W^{2,p}(\Omega) \cap W_0^{1,p}(\Omega) \end{cases}$$

*has a unique solution* $u$.
*Furthermore, there exists a positive constant* $c = c(n, p, M, \partial\Omega, \lambda, \eta)$ *such that*

$$\|u\|_{W^{2,p}(\Omega)} \le c\|f\|_{L^p(\Omega)}.$$

The results obtained by Chiarenza, Frasca and Longo in Chiarenza et al. (1991), Chiarenza et al. (1993) has been extended by Di Fazio and Ragusa in Morrey spaces in Di Fazio and Ragusa (1993), Di Fazio et al. (1999). Let us begin recalling the results contained in Di Fazio and Ragusa (1993).

Assume *Assumption (A)*: as a consequence of the representation formula and the estimate in Morrey space for singular integral operators and commutators, the authors obtain similar estimates to those obtained in Theorem 5.7.5 and in Corollary 5.7.6.

**Theorem 5.7.16** (Di Fazio and Ragusa (1993)). *Assume (A). Let* $p \in ]1, +\infty[$ *and* $\lambda \in ]0, n[$. *Then there exists a positive number* $C = C(n, p, \lambda, M)$ *and* $\rho_0 = \rho_0(C, n)$ *such that, for any ball* $B_r \Subset \Omega$, $r < \rho_0$ *and any* $u \in W_0^{2,p}(B_r)$ *such that* $u_{x_i x_j} \in L^{p,\lambda}(B_r)$ *we have*

$$\|u_{x_i x_j}\|_{L^{p,\lambda}(B_r)} \le C\|Lu\|_{L^{p,\lambda}(B_r)}, \qquad \forall i, j = 1, ..., n.$$

Furthermore, in Di Fazio and Ragusa (1993) the authors showed two regularity results.

The first result deals with regularity in Morrey spaces:

**Theorem 5.7.17** (Di Fazio and Ragusa (1993)). *Assume Assumption (A).* *Let* $1 < p < +\infty$, $0 < \lambda < n$. *Then, if* $u \in W^{2,p}(\Omega)$, $Lu \in L^{p,\lambda}(\Omega)$ *and given* $\Omega'$ *and* $\Omega''$ *such that* $\Omega' \Subset \Omega'' \Subset \Omega$ *we have* $D^2u \in L^{p,\lambda}(\Omega')$ *and there exists a positive constant* $C = C(n, p, \lambda, \nu, \mathrm{dist}(\Omega', \Omega''), \eta)$ *such that*

$$\|D^2u\|_{L^{p,\lambda}(\Omega')} \le C\left\{\|u\|_{L^{p,\lambda}(\Omega'')} + \|Lu\|_{L^{p,\lambda}(\Omega'')}\right\}.$$

The second result is a consequence of the previous one and deals with the $C^{0,\alpha}$-regularity:

**Theorem 5.7.18** (Di Fazio and Ragusa (1993)). *Let us assume Assumption (A). Let $1 < p < +\infty$, $n - p < \lambda < n$. Then, if $u \in W^{2,p}(\Omega)$, $Lu \in L^{p,\lambda}(\Omega)$, given $\Omega'$ and $\Omega''$ such that $\Omega' \Subset \Omega'' \Subset \Omega$, we have $Du \in C^{0,\alpha}(\overline{\Omega'})$, where $\alpha = 1 - \frac{n}{p} + \frac{\lambda}{p}$, and moreover*

$$\|Du\|_{C^{0,\alpha}(\overline{\Omega'})} \leq C \left\{ \|u\|_{L^{p,\lambda}(\Omega'')} + \|Lu\|_{L^{p,\lambda}(\Omega'')} \right\}.$$

*The constant $C$ has the same meaning as in Theorem 5.7.17.*

In Di Fazio et al. (1999) the authors studied the global Morrey regularity of strong solutions to the Dirichlet Problem for elliptic equations with discontinuous coefficients. Namely, Di Fazio, Palagachev and Ragusa in Di Fazio et al. (1999) extended the local results obtained in Di Fazio and Ragusa (1993) to global ones.

Let $\Omega$ be an open bounded subset of $\mathbb{R}^n$, $n \geq 2$, $1 < p < +\infty$, $0 < \lambda < n$, $a_{ij} \in \text{VMO} \cap L^\infty(\Omega)$, $\forall i, j = 1, ..., n$, $f \in L^{p,\lambda}(\Omega)$. Let us also consider the following Dirichlet problem

$$\begin{cases} Lu = f(x) & \text{a.e. in } \Omega \\ u \in W^{2,p,\lambda}(\Omega) \cap W_0^{1,p}(\Omega), \quad p \in ]1, +\infty[ \end{cases}, \tag{5.6}$$

where $W^{k,p,\lambda}(\Omega)$ is the Banach space of functions belonging to $W^{k,p}(\Omega)$ having $k^{\text{th}}$ order derivatives lying in the Morrey space $L^{p,\lambda}(\Omega)$. A natural norm in that space is:

$$\|f\|_{W^{k,p,\lambda}(\Omega)} = \|f\|_{L^p(\Omega)} + \|D^k f\|_{L^{p,\lambda}(\Omega)}.$$

As we have seen, problem (5.6), with $u \in W^{2,p}(\Omega) \cap W_0^{1,p}(\Omega)$, has been already studied in Chiarenza et al. (1993), then the regularity and existence results in $W^{2,p}(\Omega)$ are known. In Di Fazio et al. (1999) the authors have shown that finer regularity of the right-hand side $f(x)$ increases the regularity of the second derivatives of the solutions. Precisely, if $u \in W^{2,p}(\Omega)$, $1 < p < +\infty$, is a strong solution to (5.6), being $f \in L^{p,\lambda}(\Omega)$, then $u \in W^{2,p,\lambda}$.

The first result contained in Di Fazio et al. (1999) asserts boundary Morrey regularity for the second derivatives of the strong $W^{2,p}$−solution.

Let $\Omega$ be an open bounded subset of $\mathbb{R}^n$ with a $C^{1,1}$ smooth boundary and consider the second order differential operator

$$L = a_{ij}(x)D_{ij}, \qquad D_{ij} = \frac{\partial^2}{\partial x_i \partial x_j},$$

where we adopt the usual-summation convention on repeated indices.

In the sequel, we assume the following regularity and ellipticity assumption on the coefficients of $L$:

$$\begin{cases} a_{ij} \in \text{VMO}(\Omega) \\ a_{ij}(x) = a_{ji}(x) \\ \exists k > 0 : k^{-1}|\xi|^2 \leq a_{ij}(x)\xi_i\xi_j \leq k|\xi|^2, \qquad \forall \xi \in \mathbb{R}^n, \text{ a.e. in } \Omega. \end{cases} \tag{5.7}$$

Let $C_{\gamma_0}$ be the space of functions $u$ that are the restriction to

$$B_r^+ = \{x = (x_1, ..., x_n) \equiv (x', x_n) \in \mathbb{R}^n : x_n > 0\}$$

of functions belonging to $C_0^\infty(B_r)$, with $u(x', 0) = 0$.

Now, we define $W_{\gamma_0}^{2,p}(B_r^+)$ to be the closure of $C_{\gamma_0}$ in $W^{2,p}$. The following theorem is valid.

**Theorem 5.7.19** (Di Fazio et al. (1999)). *Let (5.7) be true. Let $1 < p < +\infty$ and $0 < \lambda < n$. There exist constants $c = c(n, k, p, \lambda, M, \partial\Omega)$, $\rho_0 \in ]0, r[$ such that, for every $\rho < \rho_0$ and every $u \in W_{\gamma_0}^{2,p}(B_r^+)$ satisfying $Lu \in L^{p,\lambda}(B_r^+)$ and $D_{ij}u \in L^{p,\lambda}(B_r^+)$, we have*

$$\|D_{ij}u\|_{L^{p,\lambda}(B_\rho^+)} \le c\|Lu\|_{L^{p,\lambda}(B_\rho^+)}.$$

The previous result can be refined, removing the assumption $D_{ij}u \in L^{p,\lambda}$.

**Theorem 5.7.20** (Di Fazio et al. (1999)). *Let (5.7) be true. Let $1 < p < +\infty$ and $0 < \lambda < n$. Assume further that $f \in L^{p,\lambda}(\Omega)$ and $u \in W^{2,p} \cap W_0^{1,p}(\Omega)$ are such that $Lu = f$ q.o. in $\Omega$. Then, $D_{ij} \in L^{p,\lambda}(\Omega)$ and exists a constant $c = c(n, k, p, \lambda, M, \partial\Omega)$ such that*

$$\|D_{ij}u\|_{L^{p,\lambda}(\Omega)} \le c(\|u\|_{L^{p,\lambda}(\Omega)} + \|f\|_{L^{p,\lambda}(B_\rho^+)}).$$

Using the previous theorems, the authors proved well-posedness of the Dirichlet problem (5.6) in the Morrey space $W^{2,p,\lambda}(\Omega)$.

**Theorem 5.7.21** (Di Fazio et al. (1999)). *Let (5.7) be true. Let $1 < p < +\infty$ and $0 < \lambda < n$. Then, for every $f \in L^{p,\lambda}(\Omega)$ there exists a unique solution to the Dirichlet problem*

$$\begin{cases} Lu = f(x) & \text{q.o. in } \Omega \\ u \in W^{2,p,\lambda}(\Omega) \cap W_0^{1,p}(\Omega) \end{cases} \tag{5.8}$$

*Moreover, there is a constant $c = c(n, k, p, \lambda, M, \eta, \partial\Omega)$ such that*

$$\|D_{ij}u\|_{L^{p,\lambda}(\Omega)} \le c\|f\|_{L^{p,\lambda}(\Omega)}.$$

The results obtained in Di Fazio et al. (1999) can be also applied in the study of the nonhomogeneous problems (5.8). Indeed, let $f \in L^{p,\lambda}(\Omega)$ and $\varphi \in W^{2,p,\lambda}(\Omega)$ and consider the Dirichlet problem

$$\begin{cases} Lu = f(x) & \text{a.e. in } \Omega \\ u = \varphi & \text{on } \partial\Omega, \quad u - \varphi \in W_0^{1,p}(\Omega) \end{cases} \tag{5.9}$$

We have $L\varphi \in L^{p,\lambda}(\Omega)$ and, therefore, the difference $u(x) - \varphi(x)$ solves the homogeneous Dirichlet problem associated to (5.9). This way, the strong solutions to (5.9) belong to the space $W^{2,p,\lambda}(\Omega)$.

As a consequence of Theorem 5.7.20 and the known properties of Morrey spaces for suitable $p$ and $\lambda$ (Campanato (1963b)), we obtain the global Hölder regularity for the gradient $Du$ of the strong solutions to (5.9):

**Corollary 5.7.22** (Di Fazio et al. (1999)). *Let $u \in W^{2,p}(\Omega)$ be a strong solution to (5.9), where $f \in L^{p,\lambda}(\Omega)$ and $\varphi \in W^{2,p,\lambda}(\Omega)$. Suppose $n - p < \lambda < n$. Then, the gradient $Du$ is a Hölder-continuous function on $\overline{\Omega}$ with exponent $\alpha = 1 - \frac{n-\lambda}{p}$.*

It is worth pointing out that recently Scapellato obtained regularity results for solutions to linear nondivergence form elliptic equations in the framework of generalized Morrey spaces (see Scapellato (2017a), Scapellato (2017b)).

### 5.7.1.5   Divergence form elliptic equations

In this section we examine some regularity results about elliptic equations in divergence form with discontinuous coefficients.

Ragusa in Ragusa (2000b) studies the following divergence form elliptic equation

$$Lu \equiv -\sum_{i,j=1}^{n} (a_{ij}(x)u_{x_i})_{x_j} = \operatorname{div} f(x), \qquad \text{for almost all } x \in \Omega, \qquad (5.10)$$

in a bounded open subset of $\Omega \subset \mathbb{R}^n$, $n \geq 3$.

Let us assume $L$ to be a linear elliptic operator with coefficients $a_{ij}(x) \in$ VMO $(\Omega)$, for all $i, j = 1, ..., n$ and $f = (f_1, ..., f_n)$ such that $f_i \in L^{p,\lambda}$, $1 < p < \infty$, $0 < \lambda < n$, $i = 1, ..., n$.

In Ragusa (2000b) the author obtain local regularity in Morrey spaces $L^{p,\lambda}$ of the first derivatives of the solutions to the equation (5.10). As a consequence of the main result, Ragusa proved a $C^{0,\alpha}_{loc}$-regularity result.

The argument rests on an integral representation formula for the first derivatives of the solutions of equation (5.10) and the boundedness in $L^{p,\lambda}$ of some integral operators and commutators appearing in this formula.

We assume that:

$$\exists p \in ]1, +\infty[, \lambda \in ]0, n[ \text{ such that } f = (f_1, ..., f_n) \in [L^{p,\lambda}(\Omega)]^n; \qquad (5.11)$$

$$\exists a_{ij}(x) \in \text{VMO} \cap L^{\infty}(\Omega), \qquad \forall i, j = 1, ..., n; \qquad (5.12)$$

$$\exists a_{ij}(x) = a_{ji}(x), \qquad \forall i, j = 1, ..., n, \quad \text{for a.a. } x \in \Omega; \qquad (5.13)$$

$$\exists \sigma > 0 : \sigma^{-1}|\xi|^2 \leq a_{ij}(x)\xi_i\xi_j \leq \sigma|\xi|^2, \quad \forall \xi \in \mathbb{R}^n, \text{ for a.a. } x \in \Omega. \qquad (5.14)$$

We say that $u$ is a solution to (5.10) if $u, \partial_{x_i} u \in L^p(\Omega)$, for all $i = 1, ..., n$ and for some $1 < p < +\infty$

$$\int_{\Omega} a_{ij} u_{x_i} \phi_{x_j} \, \mathrm{d}x = -\int_{\Omega} f_i \phi_{x_i}, \qquad \forall \phi \in C_0^{\infty}(\Omega).$$

Let us recall the definitions (5.3), (5.4), (5.5).

Let $r, R \in \mathbb{R}^+$, $r < R$ and $\phi \in C^{\infty}(\mathbb{R})$ be a standard cut-off function such that, for every $B_R \subset \Omega$, we have

$$\phi(x) = 1 \quad \text{in } B_R, \qquad \phi(x) = 0, \quad \forall x \notin B_R.$$

Then, if $u$ is a solution to (5.10) and $v = u\phi$, we have

$$L(v) = \operatorname{div} G + g, \tag{5.15}$$

where we set

$$G = \phi f + uAD\phi, \tag{5.16}$$
$$g = \langle ADu, D\phi \rangle - \langle f, D\phi \rangle. \tag{5.17}$$

In the following fundamental lemma, we can find an integral representation formula for the first derivatives of a solution $u$ of (5.10).

**Lemma 5.7.23** (Manfredini and Polidoro (1998)). *Let, for all $i = 1, ..., n$, $a_{ij} \in \mathrm{VMO} \cap L^\infty(\mathbb{R}^n)$ satisfy (5.13) and (5.14). Let also be $u$ a solution of (5.10) and $\phi, g$ and $G$ be defined as above.*
*Then, for every $i = 1, ..., n$, we have*

$$\partial_{x_i}(\phi u)(x) = \sum_{h,j=1}^{n} P.V. \int_{B_R} \Gamma_{ij}(x, x - y)\{(a_{jh}(x) - a_{jh}(y))\partial_{x_h}(\phi u)(y) - G_j(y)\}\mathrm{dy} -$$

$$- \int_{B_R} \Gamma_i(x, x - y)g(y)\mathrm{dy} + \sum_{h=1}^{n} c_{ih}(x)G_h(x), \quad \forall x \in B_R,$$

*setting $c_{ih} = \int_{|t|=1} \Gamma_i(x, t)t_h \, \mathrm{d}\sigma_t$.*

In Ragusa (2000b), as a consequence of Lemma 5.7.23, the author obtained an a priori estimate for the first derivatives of a solution of equation (5.10).

**Theorem 5.7.24** (Ragusa (2000b)). *Let the assumptions of Lemma 5.7.23 be fulfilled, $u, \partial_{x_i} u \in L^p(\Omega)$, for all $i = 1, ..., n$, $f \in [L^{p,\lambda}(\Omega)]^n$, $0 < \lambda < n$, $1 < p < +\infty$ and $K$ be a compact contained in $\Omega$.*
*Then, there exists a constant $c = c(n, p, \lambda, \sigma, \eta, \operatorname{dist}(K, \partial\Omega))$ such that:*

*(i) $\partial_{x_i} u \in L^{p,\lambda}(K)$, for all $i = 1, ..., n$;*

*(ii) $\|\partial_{x_i} u\|_{L^{p,\lambda}(K)} \leq c \left( \|u\|_{L^{p,\lambda}(\Omega)} + \|f\|_{L^{p,\lambda}(\Omega)} + \|\partial_{x_i} u\|_{L^{q,\mu}(\Omega)} \right)$,*

*where $q = p_*$ is such that $\frac{1}{p_*} = \frac{1}{p} + \frac{1}{n}$ and $\mu = \frac{\lambda p_*}{p}$.*

**Theorem 5.7.25** (Ragusa (2000b)). *Let $a_{ij}(x) \in \mathrm{VMO} \cap L^\infty(\mathbb{R}^n)$, for all $i, j - 1, ..., n$, such that (5.13) and (5.14) hold true, $f \in [L^{p,\lambda}(\Omega)]^n$, $0 < \lambda < n$, $1 < p < +\infty$ and let $u \in L^p(\Omega)$ the solution of $Lu = \operatorname{div} f$, with $\partial_{x_i} u \in L^p(\Omega)$, for all $i = 1, ..., n$. Then, for any compact set $K$ contained in $\Omega$, there exists a constant $c = c(n, p, \lambda, \sigma, \eta, K, \Omega)$ such that:*

$$\|\partial_{x_i} u\|_{L^{p,\lambda}(K)} \leq c \left( \|u\|_{L^2(\Omega)} + \|f\|_{L^{p,\lambda}(\Omega)} \right),$$

*for every $i = 1, ..., n$, $0 < \lambda < n$, $2 < p < +\infty$.*

Finally, the following theorem expresses the Hölder regularity of a solution $u$ of $Lu = \operatorname{div} f$.

**Theorem 5.7.26** (Ragusa (2000b)). *Let $0 < \lambda < n$, $2 < p < +\infty$ such that $p > n - \lambda$ and let $a_{ij}(x) \in \text{VMO} \cap L^\infty(\mathbb{R}^n)$, for all $i, j = 1, .., n$, satisfying (5.13) and (5.14). Then, if $u \in L^p(\Omega)$ is solution of $Lu = \operatorname{div} f$, with $\partial_{x_i} u \in L^p(\Omega)$ for all $i = 1, ..., n$ and $f \in [L^{p,\lambda}(\Omega)]^n$, we have that for any compact set $K$ contained in $\Omega$, $u \in C^{0,\alpha}(K)$ and*

$$\|u\|_{C^{0,\alpha}(K)} \le c \left( \|f\|_{L^{p,\lambda}(\Omega)} + \|u\|_{L^2(\Omega)} \right),$$

*where $\alpha = 1 - \frac{n}{p} + \frac{\lambda}{p}$ and $c$ has the same meaning as Theorem 5.7.25.*

It is interesting to study the well posedness of a Dirichlet problem associated to (5.10).

Let us define $H_0^{1,p,\lambda}(\Omega)$ as the subset of $u \in H_0^{1,p}$ such that its first order derivatives lie in the Morrey space $L^{p,\lambda}(\Omega)$, so:

$$H_0^{1,p,\lambda}(\Omega) = \{u \in H_0^{1,p}(\Omega) : Du = (\partial_{x_1} u, ..., \partial_{x_n} u) \text{ is such that}$$
$$\partial_{x_i} u \in L^{p,\lambda}(\Omega), \forall i = 1, ..., n\}.$$

In Ragusa (2000a) is studied the well posedness of the following Dirichlet problem:

$$\begin{cases} Lu = \operatorname{div} f(x) & \text{a.e. in } \Omega \\ u \in H_0^{1,p,\lambda}(\Omega), & 1 < p < +\infty, \ 0 < \lambda < n \end{cases}, \qquad (5.18)$$

where $L$ is in the form (5.10).

Let us assume that $f = (f_1, ..., f_n)$ is such that, for all $i = 1, ..., n$, $f_i$ belongs to the space $L^{p,\lambda}(\Omega)$, with $1 < p < +\infty$ and $0 < \lambda < n$, and that, for all $i, j = 1, ..., n$, $a_{ij} \in \text{VMO} \cap L^\infty(\Omega)$. Let also us assume that $\Omega \subset \mathbb{R}^n$ is an open bounded set such that $\partial\Omega \in C^{1,1}$ and that for the equation $Lu = \operatorname{div} f(x)$, a.e. in $\Omega$, the assumptions (5.12), (5.13), (5.14) hold.

The following result holds.

**Theorem 5.7.27** (Ragusa (2000a)). *Let $a_{ij}(x)$, $i, j = 1, ..., n$ such that (5.12), (5.13) and (5.14) are true in $B_\gamma^+$ for some $\gamma > 0$, $f \in [L^{p,\lambda}(B_\gamma^+)]^n$, $1 < p < +\infty$, $0 < \lambda < n$ and $u \in W_\gamma^{1,p}(B_\gamma^+)$, $u$ solution of $L(\phi u) = \operatorname{div} G + g$ in $B_\gamma^+$. Then,*

*(i)* $\partial_{x_i} u \in L^{p,\lambda}\left(B_{\frac{\gamma}{2}}^+\right)$, $i = 1, ..., n$;

*(ii) exists a constant $c = c(n, p, \lambda, \sigma, M, \partial\Omega, \eta_{ij})$ such that*

$$\|\partial_{x_i} u\|_{L^{p,\lambda}(B_{\frac{\gamma}{2}}^+)} \le c \left( \|u\|_{L^{p,\lambda}(B_\gamma^+)} + \|f\|_{L^{p,\lambda}(B_\gamma^+)} + \|\partial_{x_i} u\|_{L^{q,\mu}(B_\gamma^+)} \right),$$

*where $q = p_*$ is such that $\frac{1}{p_*} = \frac{1}{p} + \frac{1}{n}$ and $\mu = \frac{\lambda p_*}{p}$.*

From Theorem 5.7.27 the following theorem results.

**Theorem 5.7.28** (Ragusa (2000a)). *Let $a_{ij}(x) \in \text{VMO} \cap L^\infty(\Omega)$ such that (5.13) and (5.14) are true, $1 < p < +\infty$ and $0 < \lambda < n$. Then, for every $f \in [L^{p,\lambda}(\Omega)]^n$ there exists a unique solution to the Dirichlet problem*

$$\begin{cases} Lu = \text{div } f(x) & a.e. \text{ in } \Omega \\ u \in H_0^{1,p,\lambda}(\Omega) \end{cases}.$$

*Moreover, there exists a constant $c = c(n, \sigma, p, \lambda, M, \eta, \Omega)$ such that for every $2 < p < +\infty$*

$$\|\nabla u\|_{L^{p,\lambda}(\Omega)} \leq c\|f\|_{L^{p,\lambda}(\Omega)}.$$

From the previous theorem, follows the Hölder-regularity result below.

**Theorem 5.7.29** (Ragusa (2000a)). *Let $0 < \lambda < n$, $2 < p < +\infty$ such that $p > n - \lambda$, $a_{ij} \in \text{VMO} \cap L^\infty(\Omega)$ satisfying (5.13) and (5.14). Then, if $f \in [L^{p,\lambda}]^n$ and $u$ is the solution of the Dirichlet problem (5.18), there exists a constant $c = c(n, p, \lambda, o, M, \partial\Omega, \eta)$ such that*

$$\|u\|_{C^{0,\alpha}(\overline{\Omega})} \leq c\|f\|_{L^{p,\lambda}(\Omega)},$$

*where $\alpha = 1 - \frac{n}{p} + \frac{\lambda}{p}$.*

In Ragusa (1999), Ragusa studies another elliptic boundary value problem, but this time it is associated to an elliptic equation with lower order terms. Precisely, the elliptic equation under consideration in Ragusa (1999) is

$$Lu + b_i u_{x_i} - (d_i u)_{x_i} + cu = (f_j)_{x_j} \tag{5.19}$$

in an open bounded set $\Omega \subset \mathbb{R}^n$, $n \geq 3$, where we assume $L$ to be the elliptic second order operator in the divergence form

$$L \equiv -\frac{\partial}{\partial x_j}\left(a_{ij}\frac{\partial}{\partial x_i}\right),$$

with discontinuous coefficients $a_{ij}$ which belongs to the Sarason class VMO and the lower order terms $b_i$, $c$, $d_i$ belong to suitable Lebesgue spaces $L^s(\Omega)$.

Ragusa in Ragusa (1999) proves the well-posedness of the following Dirichlet problem

$$\begin{cases} Lu + b_i u_{x_i} - (d_i u)_{x_i} + cu = (f_j)_{x_j} & a.e. \text{ in } \Omega \\ u = 0 & \text{on } \partial\Omega \end{cases} \tag{5.20}$$

in the class of weak solutions $u \in H_0^{1,p}(\Omega)$, for all $1 < p < +\infty$.

Let us assume that the following *Hypothesis (K)* holds:

$$(K) \quad \begin{cases} 1. & a_{ij}(x) \in \text{VMO} \cap L^\infty(\mathbb{R}^n), \quad \forall i, j = 1, ..., n \\ 2. & a_{ij}(x) = a_{ji}(x) \text{ a.e. in } \Omega, \, \forall i, j = 1, ..., n \\ 3. & \exists \tau > 0 : \tau^{-1}|\xi|^2 \leq a_{ij}(x)\xi_i\xi_j \leq \tau|\xi|^2 \text{ a.e. in } \Omega, \, \forall \xi \in \mathbb{R}^n. \end{cases}$$

*and*

$$b_i, d_i \in L^r(\Omega), \, \forall i = 1, ..., n \text{ with } \begin{cases} r = n & \text{if } 1 < p < n \\ r > n & \text{if } p = n \\ r = p & \text{if } p > n \end{cases},$$

$$c \in L^{\frac{r}{2}}(\Omega), \quad \text{with } r \text{ is defined as above.}$$

*Let us also make the following assumption*

$$c - (d_j)_{x_j} \geq c_0 > 0.$$

The well posedness of the Dirichlet problem (5.20) is proved in the class $H_0^{1,p}(\Omega)$ for all $1 < p < \infty$ and, as a consequence, Ragusa obtained the Hölder regularity of the solution $u$.

**Theorem 5.7.30** (Ragusa (1999)). *Let $a_{ij}$, $b_i$, $c$, $d_i$ verify Hypothesis $(K)$, $f \in [L^p(\Omega)]^n$, $1 < p < +\infty$ and $\partial\Omega \in C^{1,1}$.*

*Then, the Dirichlet problem (5.20) has a unique solution and there exists a constant $k$ independent on $u$ and $f$ such that*

$$\|\nabla u\|_{L^p(\Omega)} \leq k\|f\|_{L^p(\Omega)}.$$

**Theorem 5.7.31** (Ragusa (1999)). *Let $a_{ij}$, $b_i$, $c$, $d_i$ verify Hypothesis $(K)$, $f \in [L^p(\Omega)]^n$, $p > n$ and $\partial\Omega \in C^{1,1}$.*

*The solution of the Dirichlet problem (5.20) is Hölder regular in $\overline{\Omega}$ and there exists a constant $k$ independent on $u$ and $f$ such that*

$$\|u\|_{C^{0,\alpha}(\overline{\Omega})} \leq k\|f\|_{L^p(\Omega)}.$$

We conclude the applications to elliptic equations, showing the results obtained by Palagachev, Ragusa and Softova in Palagachev et al. (2000).

Let $\Omega \subset \mathbb{R}^n$, $n \geq 3$, be a bounded domain with sufficiently smooth boundary $\partial\Omega$. Consider the unit vector field $\ell(x) = (\ell_1(x), ..., \ell_n(x))$ defined on $\partial\Omega$ and the first-order boundary operator

$$\mathscr{B} \equiv \frac{\partial}{\partial\ell(x)} + \sigma(x), \qquad x \in \partial\Omega.$$

In $\Omega$ we consider the second order uniformly elliptic operator

$$\mathscr{L} \equiv a^{ij}(x)D_{ij}$$

where the usual summation convention on repeated indices is accepted and

$D_{ij} \equiv \frac{\partial^2}{\partial x_i \partial x_j}$. In Palagachev et al. (2000) the authors investigate the global regularity, in the framework of Morrey spaces, for the next oblique derivative problem:

$$\begin{aligned} \mathscr{L}u &= f(x) && \text{for almost all } x \in \Omega, \\ \mathscr{B}u &= \varphi(x) && \text{in the trace sense on } \partial\Omega. \end{aligned} \tag{5.21}$$

Now, we list the assumptions considered in Palagachev et al. (2000).

- The operator $\mathscr{L}$ is uniformly elliptic with VMO coefficients. That is,

$$\begin{aligned} &\exists \kappa > 0 : \kappa^{-1}|\xi|^2 \leq a^{ij}(x)\xi_i\xi_j \leq \kappa|\xi|^2, && \forall \xi \in \mathbb{R}^n, \text{ a.a. in } \Omega, \\ &a^{ij}(x) \in \text{VMO}(\Omega), \quad a^{ij}(x) = a^{ji}(x), && \forall i, j = 1, ..., n. \end{aligned} \tag{5.22}$$

As usual, we denote by $\eta_{ij}(r)$ the VMO modulus of the function $a^{ij}(x)$, for all $i, j = 1, ..., n$ and let $\eta(r) = \left( \sum_{i,j=1}^n \eta_{ij}^2(r) \right)^{1/2}$.

- The boundary operator $\mathscr{B}$ is such that:

$$\begin{aligned} &\ell_i(x), \sigma(x) \in C^{0,1}(\partial\Omega), && \partial\Omega \in C^{1,1}, \\ &\ell(x) \cdot \nu(x) = \ell_i(x)\nu_i(x) > 0, \quad \sigma(x) < 0, && \text{for each } x \in \partial\Omega, \end{aligned} \tag{5.23}$$

with $\nu(x) = (\nu_1(x), ..., \nu_n(x))$ being the unit inward normal to $\partial\Omega$. The obvious geometric meaning of (5.23) is that the field $\ell(x)$ is nowhere tangential to $\partial\Omega$, that is, (5.21) is a *regular oblique derivative problem*.

To interpret the boundary condition in (5.21) in the trace sense on $\partial\Omega$, we use the space of function defined on $\partial\Omega$ which are traces of functions lying in $W^{1,p,\lambda}(\Omega)$. This functional class is well studied by Campanato. Thus, we define $W^{(p,\lambda)}(\partial\Omega)$ to be the Banach space containing the functions $\varphi$ defined on $\partial\Omega$ and having the finite norm

$$\|\varphi\|_{W^{(p,\lambda)}(\Omega)} = \left( \sup_{\substack{\rho>0 \\ z' \in \partial\Omega}} \rho^{-\bar{\lambda}} \int_{B_\rho(z')\cap\partial\Omega} |\varphi(x')|^p \, d\sigma_{x'} \right)^{\frac{1}{p}}$$

$$+ \left( \sup_{\substack{\rho>0 \\ z',\bar{z}' \in \partial\Omega}} \rho^{-\bar{\lambda}} \int_{B_\rho(z')\cap\partial\Omega} \int_{B_\rho(\bar{z}')\cap\partial\Omega} \frac{|\varphi(x') - \varphi(\bar{x}')|^p}{|x' - \bar{x}'|^{p+n-2}} \, d\sigma_{x'} d\sigma_{\bar{x}'} \right)^{\frac{1}{p}}.$$

with $\bar{\lambda} = \max\{\lambda - 1, 0\}$.

The main results proved in Palagachev et al. (2000) are listed below.

**Theorem 5.7.32** (Palagachev et al. (2000)). *Let* (5.22) *and* (5.23) *be true,* $p \in ]1, +\infty[$ *and* $\lambda \in ]0, n[$. *Assume further that* $u \in W^{2,p}(\Omega)$ *solves the problem* (5.21) *with* $f \in L^{p,\lambda}(\Omega)$ *and* $\varphi \in W^{(p,\lambda)}(\partial\Omega)$.

*Then,* $D_{ij}u \in L^{p,\lambda}(\Omega)$ *and there exists a constant* $C = C(n, p, \lambda, \kappa, \eta, \ell, \sigma, \partial\Omega)$ *such that*

$$\|u\|_{W^{2,p,\lambda}(\Omega)} \le C \left( \|u\|_{L^{p,\lambda}(\Omega)} + \|f\|_{L^{p,\lambda}(\Omega)} + \|\varphi\|_{W^{(p,\lambda)}(\partial\Omega)} \right).$$

**Theorem 5.7.33** (Palagachev et al. (2000)). *Let* (5.22) *and* (5.23) *be true,* $p \in ]1, +\infty[$ *and* $\lambda \in ]0, n[$.

*Then, for every* $f \in L^{p,\lambda}(\Omega)$ *and* $\varphi \in W^{(p,\lambda)}(\partial\Omega)$, *there exists a unique solution of the oblique derivative problem* (5.21). *Moreover, there exists a constant* $C = C(n, p, \lambda, \kappa, \eta, \ell, \sigma, \partial\Omega)$ *such that*

$$\|u\|_{W^{2,p,\lambda}(\Omega)} \le C \left( \|f\|_{L^{p,\lambda}(\Omega)} + \|\varphi\|_{W^{(p,\lambda)}(\partial\Omega)} \right).$$

Finally, an immediate consequence of Theorem 5.7.32 and the imbedding properties of the Morrey spaces for suitable values of $p$ and $\lambda$ (see Campanato (1963b)) is the next global Hölder regularity for the gradient $Du$ of the strong solutions to (5.21):

**Corollary 5.7.34** (Palagachev et al. (2000)). *Let* $u \in W^{2,p}(\Omega)$ *be a strong solution to* (5.21), *where* $f \in L^{p,\lambda}(\Omega)$ *and* $\varphi \in W^{(p,\lambda)}(\partial\Omega)$.

*Then, if* $n - p < \lambda < n$, *the gradient* $Du$ *is a Hölder continuous function on* $\overline{\Omega}$ *with exponent* $\alpha = 1 - \frac{n-\lambda}{p}$ *and*

$$\|Du\|_{C^{0,\alpha}(\overline{\Omega})} = \sup_{x,y \in \Omega} \frac{|Du(x) - Du(y)|}{|x - y|^\alpha} \le C \left( \|f\|_{L^{p,\lambda}(\Omega)} + \|\varphi\|_{W^{(p,\lambda)}(\partial\Omega)} \right).$$

Let us specify that only the assumptions $f \in L^p(\Omega)$ and $\varphi \in W^{1-1/p,p}(\partial\Omega)$ imply $u \in W^{2,p}(\Omega)$ (Di Fazio and Palagachev (1996)). Thus, if $p > n$, the Sobolev imbedding theorem ensures $Du \in C^\beta(\overline{\Omega})$ with $\beta = 1 - \frac{n}{p}$. On the other hand, Corollary 5.7.34 yields the Hölder continuity of the gradient *also* for $p \in ]1, n]$, assuming finer regularity of the data because they belong to $f \in L^{p,\lambda}(\Omega)$ with $\lambda \in ]n - p, n[$.

## 5.7.2   Parabolic equations

### 5.7.2.1   Preliminary tools

In this section we show the main results of Palagachev et al. (2003) and Ragusa (2004). In both papers the VMO hypothesis on the coefficients of the differential operator under consideration is crucial.

Let $\Omega \subset \mathbb{R}^n$ an open bounded set, $n \ge 1$ (in Palagachev et al. (2003)) and $n \ge 3$ (in Ragusa (2004)).

In the sequel $Q_T$ stands for the *cylinder* $\Omega \times ]0, T[$, with $T > 0$.

The *lateral surface* of $Q_T$ is denoted by

$$S_T = \partial\Omega \times ]0, T[$$

and the *parabolic boundary* of $Q_T$ by

$$\partial Q_T = \Omega \cup S_T.$$

Often, as usual, we adopt the standard summation convention on repeated upper and lower indices.

For semplicity, we denote the set of the *parabolic variables* by

$$x = (x', t) = (x_1, ..., x_n, t) \in \mathbb{R}^{n+1}$$

and we set

$$D_i u = \frac{\partial u}{\partial x_i}, \qquad D_{ij} u = \frac{\partial^2 u}{\partial x_i \partial x_j}, \qquad u_t = D_t u = \frac{\partial u}{\partial t}.$$

Moreover,

$$D_{x'} u = (D_1 u, \quad , D_n u)$$

stands for the *spatial gradient* of $u$.

Finally, we set

$$\mathbb{R}^{n+1}_+ = \mathbb{R}^n \times \mathbb{R}_+, \qquad \mathbb{D}^{n+1}_+ = \mathbb{R}^n_+ \times \mathbb{R}_+ = \{x' \in \mathbb{R}^n : x_n > 0\} \times \{t > 0\}.$$

In Palagachev et al. (2003), the authors assume that $a^{ij}(x) \in \text{VMO}$, for all $i, j = 1, ..., n$.

Before proceeding with the results showed in Palagachev et al. (2003), it is useful to recall some definitions.

A *parabolic operator* $\mathscr{P}$ is an operator of the form

$$\mathscr{P} = D_t - \sum_{i,j=1}^{n} a^{ij}(x) D_{ij}.$$

**Definition 5.7.35** (Palagachev et al. (2003)). We say that $\mathscr{P}$ is a **uniformly parabolic operator** if there exists a constant $\Lambda > 0$ such that

$$\Lambda^{-1}|\xi|^2 \le a^{ij}(x)\xi_i\xi_j \le \Lambda|\xi|^2, \qquad \text{a.a. in } Q_T, \ \forall \xi \in \mathbb{R}^n. \tag{5.24}$$

In the sequel we assume that the coefficient matrix $\mathbf{a} = \{a^{ij}\}_{i,j=1}^n$ is symmetric; this ensures the essential boundedness of $a^{ij}$'s.

Let us denote by $\mathscr{P}_0$ a linear parabolic operator with constant coefficients $a^{ij}_0$, which satisfies (5.24). The fundamental solution of the operator $\mathscr{P}_0$ with pole at the origin is given by the formula

$$\Gamma^0(y) = \Gamma^0(y', \tau) = \begin{cases} \dfrac{1}{(4\pi\tau)^{n/2}\sqrt{\det \mathbf{a}_0}} \exp\left\{-\dfrac{A_0^{ij} y_i y_j}{4\tau}\right\} & \text{as } \tau > 0 \\ 0 & \text{as } \tau < 0 \end{cases},$$

where $\mathbf{a}_0 = \{a_0^{ij}\}$ is the coefficients matrix of $\mathscr{P}_0$ and $\mathbf{A}_0 = \{A_0^{ij}\} = \mathbf{a}_0^{-1}$.

In the problem that we wish to analyze, the coefficients of the operator $\mathscr{P}$ depend on $x$ and it reflects also on the fundamental solution. To express this dependence we define

$$\Gamma(x, y) = \begin{cases} \dfrac{1}{(4\pi\tau)^{n/2}\sqrt{\det \mathbf{a}(x)}} \exp\left\{ -\dfrac{A^{ij}(x)y_i y_j}{4\tau} \right\} & \text{as } \tau > 0 \\ 0 & \text{as } \tau < 0 \end{cases}, \qquad (5.25)$$

where $\mathbf{a}(x) = \{a^{ij}\}$ and $\mathbf{a}(x) = \{A^{ij}(x)\} = \mathbf{a}(x)^{-1}$. Set also

$$\Gamma_i = \frac{\partial \Gamma(x; y', \tau)}{\partial y_i}, \qquad \Gamma_{ij} = \frac{\partial^2 \Gamma(x; y', \tau)}{\partial y_i \partial y_j}, \qquad i, j = 1, ..., n.$$

Besides the *standard parabolic metric*

$$\tilde{\rho}(x) = \max\{|x'|, |t|^{1/2}\}, \quad |x'| = \left(\sum_{i=1}^n x_i^2\right)^{1/2}, \qquad \tilde{d}(x, y) = \tilde{\rho}(x - y),$$

we are going to use the one introduced by Fabes and Riviére (Fabes and Rivière (1966))

$$\rho(x) = \sqrt{\frac{|x'|^2 + \sqrt{|x'|^4 + 4t^2}}{2}}, \qquad d(x, y) = \rho(x - y).$$

The topology induced by $d$ is defined through open ellipsoids centered at zero and of radius $r$:

$$\mathscr{E}_r(0) = \left\{ x \in \mathbb{R}^{n+1} : \frac{|x'|^2}{r^2} + \frac{t^2}{r^4} < 1 \right\}.$$

Obviously, the unit sphere with respect to that metric coincides with the unit Euclidean sphere in $\mathbb{R}^{n+1}$, i.e.

$$\partial \mathscr{E}_1(0) \equiv \Sigma_{n+1} = \left\{ x \in \mathbb{R}^{n+1} : |x| = \left(\sum_{i=1}^n x_i^2 + t^2\right)^{1/2} = 1 \right\}$$

and $\overline{x} = \frac{x}{\rho(x)} \in \Sigma_{n+1}$. It is easy to see that for any ellipsoid $\mathscr{E}_r$, there exist cylinders $\underline{I}$ and $\overline{I}$ (that are balls with respect to the metric $\tilde{\rho}$) with measures comparable to $r^{n+2}$ and such that $\underline{I} \subset \mathscr{E}_r \subset \overline{I}$. It is clear that the relation $\underline{I} \subset \mathscr{E}_r \subset \overline{I}$ gives an equivalence of the metrics $\rho$ and $\tilde{\rho}$ and the induced by them topologies.

**Definition 5.7.36** (Palagachev et al. (2003)). A function $k(y) : \mathbb{R}^{n+1} \setminus \{0\} \to \mathbb{R}$ is said to be a **constant parabolic Calderón-Zygmund kernel** (PCZ kernel) if

1. $k(y)$ is smooth on $\mathbb{R}^{n+1} \setminus \{0\}$;

2. *(Homogeneity condition)* $k(ry', r^2 t) = r^{-(n+2)} k(y', t)$ for each $r > 0$;

3. *(Cancellation property on ellipsoids)* $\int_{\rho(y)=r} k(y) \, d\sigma_y = 0$ for each $r > 0$.

A function $k(x; y) : \mathbb{R}^{n+1} \times (\mathbb{R}^{n+1} \setminus \{0\}) \to \mathbb{R}$ is a **variable Calderón-Zygmund kernel** if for any fixed $x \in \mathbb{R}^{n+1}$, $k(x; \cdot)$ is a parabolic PCZ kernel and

$$\sup_{\rho(y)=1} \left| \left( \frac{\partial}{\partial y} \right)^{\beta} k(x; y) \right| \leq C(\beta),$$

for every multiindex $\beta$, independently of $x$.

**Definition 5.7.37** (Ragusa (2005)). The **parabolic cube** of center $x = (x', t)$ and radius $r$ is the set

$$Q = Q_r(y) = \{y = (y', s) \in \mathbb{R}^{n+1} : |x' - y'| < r, |t - s| < r^2\}.$$

For sake of completeness, we recall here some definitions already given in the elliptic case, adapting them to the parabolic case.

**Definition 5.7.38** (Ragusa (2005)). Let $f \in L^1_{\mathrm{loc}}(\mathbb{R}^{n+1})$. We say that $f$ belongs to the parabolic BMO $(\mathbb{R}^{n+1})$ space if

$$\sup_{Q \subset \mathbb{R}^{n+1}} \frac{1}{|Q|} \int_Q |f(y) - f_Q| dy < +\infty,$$

where $Q$ ranges in the class of the parabolic cubes in $\mathbb{R}^{n+1}$ and $f_Q = \frac{1}{|Q|} \int_Q f(y) dy$.

Let $f \in \mathrm{BMO}(\mathbb{R}^{n+1})$ and $r > 0$. Let us define the VMO modulus of $f$ the function

$$\eta(r) = \sup_{\rho \leq r} \frac{1}{|Q_\rho|} \int_{Q_\rho} |f(y) - f_{Q_\rho}| dy,$$

where $Q_\rho$ is a parabolic cube in $\mathbb{R}^{n+1}$ of radius $\rho$, $\rho \leq r$.

**Definition 5.7.39** (Ragusa (2005)). We say that a function $f \in \mathrm{BMO}$ belongs to the Sarason class VMO $(\mathbb{R}^{n+1})$ if

$$\lim_{r \to 0^+} \eta(r) = 0.$$

The spaces BMO $(Q_T)$ and VMO $(Q_T)$ of functions given on $Q_T$, can be defined as in Definition 5.7.38, taking $Q \cap Q_T$ instead of $Q$ above.

**Definition 5.7.40** (Palagachev et al. (2003))**.** A measurable function $f \in L^1_{\text{loc}}(\mathbb{R}^{n+1})$ belongs to the **parabolic Morrey space** $L^{p,\lambda}(\mathbb{R}^{n+1})$ with $p \in ]1, +\infty[$ and $\lambda \in ]0, n+2[$ if the following norm is finite

$$\|f\|_{p,\lambda} = \left( \sup_{r>0} \frac{1}{r^{\lambda}} \int_{Q_r} |f(y)|^p \mathrm{d}y \right)^{1/p},$$

where $Q_r$ is any cylinder of radius $r$.

Similarly, a measurable function $f \in L^1_{\text{loc}}(\mathbb{R}^{n+1})$ is said to belong to the **parabolic Morrey space** $L^{p,\lambda}(Q_T)$ if the following norm is finite

$$\|f\|_{p,\lambda;Q_T} = \left( \sup_{r>0} \frac{1}{r^{\lambda}} \int_{Q_T \cap Q_r} |f(y)|^p \mathrm{d}y \right)^{1/p}.$$

**Definition 5.7.41** (Palagachev et al. (2003))**.** We say that the function $u(x)$ lies in $W^{2,1}_{p,\lambda}(Q_T)$, $1 < p < +\infty$, $0 < \lambda < n+2$, if it is weakly differentiable and belongs to $L^{p,\lambda}(Q_T)$ along with all its derivatives $D^r_t D^s_{x'} u$, with $0 \le 2r+s \le 2$.

The quantity

$$\|u\|_{W^{2,1}_{p,\lambda}(Q_T)} = \|u\|_{p,\lambda;Q_T} + \|D^2_{x'} u\|_{p,\lambda;Q_T} + \|D_t u\|_{p,\lambda;Q_T}$$

defines a norm under which $W^{2,1}_{p,\lambda}(Q_T)$ is a Banach space.

**Definition 5.7.42** (Ragusa (2005))**.** We define $L^{p,q}(Q_T)$, $1 \le p, q \le +\infty$ as the space of all measurable functions $h(x,t)$ such that

$$\|h\|_{L^{p,q}(Q_T)} \equiv \left( \int_0^T \left( \int_{\Omega} |h(x,t)|^p \mathrm{d}x \right)^{q/p} \mathrm{d}t \right)^{1/p} < +\infty$$

with obvious modifications if $p$ or $q$ or both the exponents are infinite.

If $p = q$, we simply write $L^p(Q_T)$ instead of $L^{p,p}(Q_T)$.

Moreover, if $h \in L^{p,q}(Q_T)$, we set

$$\omega(\sigma) = \sup_{|D| \le \sigma} \|h\|_{L^{p,q}(D)}$$

where $D \subset \mathbb{R}^{n+1}$ is Lebesgue measurable and $|D|$ is its Lebesgue measure.

The function $\omega(\sigma)$ is decreasing in $]0, |Q_T|[$ and is such that

$$\lim_{\sigma \to 0} \omega(\sigma) = 0.$$

We refer to $\omega(\sigma)$ as the *AC modulus* of $|h|$.

Let us define $W^{2,1}_p(Q_T)$ as the space of functions $u(x,t)$ with

$p$−summability in $Q_T$ such that the derivative with respect to $t$ belongs to $L^p(Q_T)$ and as functions of $t$ with values in $L^p(\Omega)$, belong to $L^p(0, T, W_p^2(\Omega))$. Let us assume in $W_p^{2,1}(Q_T)$ as a norm the quantity

$$\|u\|_{W_p^{2,1}(Q_T)} = \left( \int_{Q_T} \left( u_t^2 + \sum_{\alpha=1}^{2} |D^\alpha u|^2 \right)^{p/2} \mathrm{d}x\mathrm{d}t \right)^{1/p}.$$

**Definition 5.7.43** (Ragusa (2004)). Let $\Omega$ be an open bounded subset of $\mathbb{R}^n$. The space $H_p^{1-\frac{1}{p}}(-T, 0, L^p(\Omega, \mathbb{R}))$ consists of the functions $u$ in $L^p(Q_T)$ for which

$$[u]_{H_p^{1-\frac{1}{p}}(Q_T)} = \left\{ \int_{-T}^{0} \left( \int_{-T}^{0} \left( \int_\Omega \frac{|u(x,t) - u(x,\xi)|^p}{|t - \xi|^p} \mathrm{d}x \right) \mathrm{d}\xi \right) \mathrm{d}t \right\}^{\frac{1}{p}}$$

is finite.

The space $H_p^{1-\frac{1}{p}}$ equipped with the norm given by

$$\|u\|_{H_p^{1-\frac{1}{p}}(Q_T)} = \left( \|u\|_{L^p(Q_T)} + [u]_{H_p^{1-\frac{1}{p}}(Q_T)} \right)^{\frac{1}{p}},$$

where

$$\|u\|_{L^p(Q_T)} = \left( \int_{Q_T} |u(y)|^p \mathrm{d}y \right)^{\frac{1}{p}},$$

is a Banach space.

**Definition 5.7.44** (Ragusa (2004)). A **weak solution** of the equation

$$u_t - \sum_{i,j=1}^{n} (a_{ij}(x)u_{x_i})_{x_j} = \operatorname{div} f$$

is a function $u : Q_T \to \mathbb{R}$ such that $u, u_{x_j} \in L^2_{\mathrm{loc}}(Q_T)$, for all $j = 1, ..., n$ and

$$\int_{Q_T} \sum_{i,j=1}^{n} a_{ij}(x,t)u_{x_j}\phi_{x_j}\mathrm{d}x\mathrm{d}t - \int_{Q_T} u(x,t)\frac{\partial\phi(x,t)}{\partial t}\mathrm{d}x\mathrm{d}t$$

$$= -\int_{Q_T} \sum_{i=1}^{n} f_i(x,t)\phi_{x_i}(x,t)\mathrm{d}x\mathrm{d}t, \quad \forall\phi \in C_0^\infty(Q_T).$$

### 5.7.2.2 Nondivergence form parabolic equations

In Palagachev et al. (2003) the authors study qualitative properties in the framework of the parabolic Morrey spaces of the Cauchy-Dirichlet problem

$$\begin{cases} \mathscr{P}u \equiv u_t - \sum_{i,j=1}^{n} a^{ij}(x)D_{ij}u = f(x) & \text{a.e. in } Q_T \\ u(x) = 0 & \text{on } \partial\Omega \end{cases} \tag{5.26}$$

in the case of uniformly parabolic operator $\mathscr{P}$ (see Definition 5.7.35) with discontinuous coefficients. Assume that $\Omega \subset \mathbb{R}^n$ is a bounded and $C^{1,1}-$smooth domain, $n \geq 1$.

The following results, obtained in $W^{2,1}_{p,\lambda}(Q_T)$, hold.

**Theorem 5.7.45** (Palagachev et al. (2003)). *Suppose* $a^{ij} \in \mathrm{VMO}(Q_T)$, (5.24), $\partial\Omega \in C^{1,1}$ *and let* $u \in W^{2,1}_{p,\lambda}(Q_T)$, $p \in ]1, +\infty[$, $\lambda \in ]0, n+2[$, *be a strong solution to* (5.26).

*Then, there exists a constant* $C = C(n, p, \lambda, \Lambda, T, \eta_{ij}, \partial\Omega)$ *such that*

$$\|u\|_{W^{2,1}_{p,\lambda}(Q_T)} \leq C\|f\|_{p,\lambda;Q_T}.$$

In the proof of the previous theorem, play an important role the interior representation formula for second spatial derivatives (see Bramanti and Cerutti (1993)). This representation formula expresses $D_{ij}u$ in terms of singular integral operators and their commutators with kernels $\Gamma_{ij}(x; x-y)$ (the derivatives of the fundamental solution (5.25) with respect to the second variable).

Now, we present a result dealing with the existence of a unique strong solution to the Cauchy-Dirichlet problem (5.26).

**Theorem 5.7.46** (Palagachev et al. (2003)). *Assume* (5.24), $\partial\Omega \in C^{1,1}$ *and* $a^{ij} \in \mathrm{VMO}(Q_T)$. *Then, the problem* (5.26) *admits a unique strong solution* $u \in W^{2,1}_{p,\lambda}(Q_T)$ *with* $p \in ]1, +\infty[$, $\lambda \in ]0, n+2[$, *for every* $f \in L^{p,\lambda}(Q_T)$.

An immediate consequence of the last result is Hölder continuity of the strong solution $u$ to (5.26) or its spatial gradient $D_{x'}u$ for suitable values of $p$ and $\lambda$. Precisely, define

$$[u]_{\alpha;Q_T} = \sup_{\substack{(x',t),(y',\tau)\in Q_T \\ (x',t)\neq(y',\tau)}} \frac{|u(x',t) - u(y',\tau)|}{(|x'-y'|^2 + |t-\tau|)^{\alpha/2}}, \qquad 0 < \alpha < 1$$

and set $C^{0,\alpha}(\overline{Q_T})$ for the space of all functions $u : \overline{Q_T} \to \mathbb{R}$ of finite norm

$$\|u\|_{0,\alpha;Q_T} = \|u\|_{\infty;Q_T} + [u]_{\alpha;Q_T}.$$

**Corollary 5.7.47** (Palagachev et al. (2003)). *Assume* $a^{ij} \in \mathrm{VMO}(Q_T)$, $\partial\Omega \in C^{1,1}$, (5.24), $f \in L^{p,\lambda}(Q_T)$ *with* $p \in ]1, +\infty[$ *and* $\lambda \in ]0, n+2[$ *and let* $u \in W^{2,1}_{p,\lambda}(Q_T)$ *be the unique strong solution of the problem* (5.26). *Then:*

1. $u \in C^{0,\alpha}(\overline{Q_T})$ *and*

$$\|u\|_{0,\alpha;Q_T} \leq C\|f\|_{p,\lambda;Q_T}$$

*with* $\alpha = \frac{1}{n+1} + \frac{\lambda-(n+2)}{p}$ *if* $\lambda > \max\left\{0, n+2-\frac{p}{n+1}\right\}$;

2. $D_{x'}u \in C^{0,\alpha}(\overline{Q_T})$ *and*

$$\|D_{x'}u\|_{0,\alpha;Q_T} \leq C\|f\|_{p,\lambda;Q_T}$$

*with* $\alpha = 1 + \frac{\lambda-(n+2)}{p}$ *if* $\lambda > \max\{0, n+2-p\}$.

Another application to parabolic equations in nondivergence form is contained in Ragusa (2005), where Ragusa considers a linear parabolic operator $\mathscr{L}$ of the form

$$\mathscr{L}u \equiv u_t - \sum_{i,j=1}^{n} a_{ij} u_{x_i' x_j'} + \sum_{i=1}^{n} b_i(x) u_{x_i'} + cu \tag{5.27}$$

where $x = (x', t') = (x_1', ..., x_n', t) \in \mathbb{R}^{n+1}$.

Let $\Omega \subset \mathbb{R}^n$ be a bounded $C^{1,1}$ domain.

The Cauchy-Dirichlet studied in Ragusa (2005) is the following:

$$\begin{cases} \mathscr{L}u = f & \text{a.e. in } Q_T \\ u = 0 & \text{on } \partial\Omega \times ]0, T[ \\ u(x', 0) = 0 & \text{in } \Omega \end{cases} \tag{5.28}$$

where $f \in L^p(Q_T)$, $1 < p < +\infty$. The coefficients $a_{ij}(x)$ are discontinuous; precisely, they are in the space VMO. Moreover, the principal part of the operator under consideration is supposed to be symmetric and uniformly elliptic with ellipticity constant $\tau$. Let us assume also the following hypotheses:

(i) for every $i = 1, ..., n$, $b_i \in L^{t,\bar{t}}(Q_T)$, where

$$\begin{array}{lll} t = \bar{t} = p & & \text{if } p > n + 2, \\ t = p, \quad \bar{t} = \frac{p\rho}{\rho - p} & \text{or} \quad t = \frac{p\rho}{\rho - p}, \quad \bar{t} = p & \text{if } p = n + 2, \\ t = \frac{p\rho}{\rho - p}, \quad \bar{t} = \frac{p}{\theta} & & \text{if } 1 < p < n + 2, \end{array}$$

(ii) if $p < n + 2$, we set

$$\rho = \frac{np}{n - p + 2\theta}, \qquad \max\left\{0, \frac{p - n}{2}\right\} \le \theta \le \min\left\{1, \frac{p}{2}\right\},$$

(iii) $c \le 0$ a.e. in $Q_T$ and $c \in L^{s,\bar{s}}(Q_T)$ where

$$\begin{array}{lll} s = \bar{s} = p & & \text{if } p > \frac{n+2}{2}, \\ s = p, \quad \bar{s} = \frac{p\rho}{\rho - p} & \text{or} \quad s = \frac{p\rho}{\rho - p}, \quad \bar{s} = p, \quad \forall \rho > p & \text{if } p = \frac{n+2}{2}, \\ s = \frac{p\rho}{\rho - p}, \quad \bar{s} = \frac{p}{\theta} & & \text{if } 1 < p < \frac{n+2}{2}, \end{array}$$

(iv) if $p < \frac{n+2}{2}$, we set

$$\rho = \frac{np}{n - 2p + 2\theta}, \qquad \max\left\{0, \frac{2p - n}{2}\right\} \le \theta \le \min\{1, p\}.$$

**Theorem 5.7.48** (Ragusa (2005)). *Let us assume that the coefficients $b_i$ and $c$ satisfy (i), (ii), (iii), (iv), for every $i = 1, ..., n$. Then, for every $u \in W_p^{2,1}(Q_T)$, we have*

$$b_i Du, \ cu \in L^p(Q_T), \qquad \text{for every } 1 < p < +\infty$$

*and it results*

$$\|b_i Du\|_{L^p(Q_T)} + \|cu\|_{L^p(Q_T)} \leq C\|u\|_{W_p^{2,1}(Q_T)},$$

*where $C$ depends on $T$, $p$, $n$, $\Omega$, $\|c\|_{L^{s,\bar{s}}(Q_T)}$, $\|b_i\|_{L^{t,\bar{t}}(Q_T)}$ and the Sobolev constant.*

The following a priori estimate for the solution of the problem (5.28) holds.

**Theorem 5.7.49** (Ragusa (2005)). *Let $f \in L^p(Q_T)$, $1 < p < +\infty$, the coefficients $b_i$ and $c$ satisfying the assumptions (i), (ii), (iii), (iv), for every $i = 1, ..., n$, $a_{ij} \in \text{VMO} \cap L^\infty(Q_T)$ are symmetric and uniformly elliptic. Then, there exists a constant $k$ such that for any solution $u$ of the Cauchy-Dirichlet problem (5.28), we have*

$$\|u\|_{W_p^{2,1}(Q_T)} \leq k \left( \|\mathscr{L}u\|_{L^p(Q_T)} + \|u\|_{L^p(Q_T)} \right),$$

*where the constant $k$ depends on $n$, $p$, $\tau$, $\eta$, $|\Omega|$, $\partial\Omega$, $T$, $\|b_i\|_{L^{t,\bar{t}}}$, for every $i = 1, ..., n$, $\|c\|_{L^{s,\bar{s}}}$ and the AC moduli of $b_i$ and $c$.*

Finally, in Ragusa (2005) the author showed the well-posedness of the Cauchy-Dirichlet problem (5.28):

**Theorem 5.7.50** (Ragusa (2005)). *Let $f \in L^p(Q_T)$, $1 < p < +\infty$, the coefficients $b_i$ and $c$ satisfying the assumptions (i), (ii), (iii), (iv), for every $i = 1, ..., n$. Then, the Cauchy-Dirichlet problem (5.28) has a unique solution $u \in W_p^{2,1}(Q_T)$ and there exists a constant $C_0$ such that*

$$\|u\|_{W_p^{2,1}(Q_T)} \leq C_0 \|f\|_{L^p(Q_T)}.$$

*The constant $C_0$ depends on $n$, $p$, $\tau$, $\eta$, $|\Omega|$, $\partial\Omega$, $T$, $\|b_i\|_{L^{t,\bar{t}}}$, for every $i = 1, ..., n$, $\|c\|_{L^{s,\bar{s}}}$ and the AC moduli of $b_i$ and $c$.*

We conclude this section with the recent results obtained by Ragusa and Scapellato in Ragusa and Scapellato (2017).

Let $k(x, y)$ be a variable Calderón-Zygmund kernel for a.e. $x \in \mathbb{R}^{n+1}$, $f \in L^{q,\mu}(0, T, L^{p,\lambda}(\mathbb{R}^n))$ with $1 < p, q < \infty$ $0 < \lambda, \mu < n$, $a \in BMO(\mathbb{R}^{n+1})$. For $\varepsilon > 0$ let us define the operator $K_\varepsilon$ and the commutator $C_\varepsilon[a, f]$, as follows

$$K_\varepsilon f(x) = \int_{\rho(x-y)>\varepsilon} k(x, x - y) f(y) dy$$

$$\begin{aligned} C_\varepsilon[a, f] &= K_\varepsilon(af)(x) - a(x) K_\varepsilon f(x) \\ &= \int_{\rho(x-y)>\varepsilon} k(x, x - y)[a(x) - a(y)] f(y) dy. \end{aligned}$$

The authors in Ragusa and Scapellato (2017), prove that $K_\varepsilon f$ and $C_\varepsilon[a, f]$

are, uniformly in $\varepsilon$, bounded from $L^{q,\mu}(0,T,L^{p,\lambda}(\mathbb{R}^n))$ into itself. This fact allows us to let $\varepsilon \to 0$ obtaining as limits in $L^{q,\mu}(0,T,L^{p,\lambda}(\mathbb{R}^n))$ the following singular integral and commutator

$$Kf(x) = P.V. \int_{\mathbb{R}^n} k(x, x-y)f(y)dy = \lim_{\varepsilon \to 0} K_\varepsilon f(x)$$

$$C[a,f](x) = P.V. \int_{\mathbb{R}^n} k(x, x-y)[a(x) - a(y)]f(y)dy = \lim_{\varepsilon \to 0} C_\varepsilon[a,f](x)$$

These operators are bounded in the class $L^{q,\mu}(0,T,L^{p,\lambda}(\mathbb{R}^n))$.

**Theorem 5.7.51** (Ragusa and Scapellato (2017)). *Let $k(x,y)$ be a variable Calderón-Zygmund kernel, for a.e. $x \in \mathbb{R}^{n+1}$, $1 < p, q < \infty$, $0 < \lambda, \mu < n$ and $a \in VMO(\mathbb{R}^{n+1})$.*

*For any $f \in L^{q,\mu}(0,T,L^{p,\lambda}(\mathbb{R}^n))$ the singular integrals $Kf$, $C[a,f] \in L^{q,\mu}(0,T,L^{p,\lambda}(\mathbb{R}^n))$ exist as limits in $L^{q,\mu}(0,T,L^{p,\lambda}(\mathbb{R}^n))$, for $\varepsilon \to 0$, of $K_\varepsilon f$ and $C_\varepsilon[a,f]$, respectively. Then, the operators*

$$Kf,\; C[a,f] : L^{q,\mu}(0,T,L^{p,\lambda}(\mathbb{R}^n)) \to L^{q,\mu}(0,T,L^{p,\lambda}(\mathbb{R}^n))$$

*are bounded and satisfy the following inequalities*

$$\|Kf\|_{L^{q,\mu}(0,T,L^{p,\lambda}(\mathbb{R}^n))} \le c\|f\|_{L^{q,\mu}(0,T,L^{p,\lambda}(\mathbb{R}^n))}$$

$$\|C[a,f]\|_{L^{q,\mu}(0,T,L^{p,\lambda}(\mathbb{R}^n))} \le c\|a\|_*\|f\|_{L^{q,\mu}(0,T,L^{p,\lambda}(\mathbb{R}^n))}$$

*where $c = c(n, p, \lambda, \alpha, K)$.*

*Moreover, for every $\epsilon > 0$ there exists $\rho_0 > 0$ such that, if $B_r$ is a ball with radius $r$ such that $0 < r < \rho_0$, $k(x,y)$ satisfies the above assumptions and $f \in L^{q,\mu}(0,T,L^{p,\lambda}(B_r))$, we have*

$$\|C[a,f]\|_{L^{q,\mu}(0,T,L^{p,\lambda}(B_r))} \le c\,\epsilon\,\|f\|_{L^{q,\mu}(0,T,L^{p,\lambda}(B_r))} \tag{5.29}$$

*for some constant $c$ independent of $f$.*

As application of the previous results we obtain a regularity result for strong solutions to the nondivergence form parabolic equations.

Precisely, let $n \geq 3$, $Q_T = \Omega' \times (0,T)$ be a cylinder of $\mathbb{R}^{n+1}$ of base $\Omega' \subset \mathbb{R}^n$. In the sequel let us set $x = (x',t) = (x'_1, x_2, \ldots, x'_n, t)$ a generic point in $Q_T$, $f \in L^{q,\mu}(0,T,L^{p,\lambda}(\Omega'))$, $1 < p, q < \infty$, $0 < \lambda, \mu < n$ and

$$Lu = u_t - \sum_{i,j=1}^{n} a_{ij}(x',t)\frac{\partial^2 u}{\partial x'_i \partial x'_j}$$

where

$$a_{ij}(x',t) = a_{ji}(x',t), \qquad \forall i,j = 1, \ldots, n, \qquad \text{a.e. in } Q_T,$$

$$\exists \nu > 0 : \nu^{-1}|\xi|^2 \le \sum_{i,j=1}^{n} a_{ij}(x', t)\,\xi_i \xi_j \le \nu|\xi|^2, \quad \text{a.e. in } Q_T, \forall \xi \in \mathbb{R}^n,$$

$$a_{ij}(x', t) \in VMO(Q_T) \cap L^\infty(Q_T), \qquad \forall i, j = 1, \dots, n.$$

Let us consider

$$Lu(x', t) = f(x', t). \tag{5.30}$$

A strong solution to (5.30) is a function $u(x) \in L^{q,\mu}(0, T, L^{p,\lambda}(\Omega'))$ with all its weak derivatives $D_{x'_i} u$, $D_{x'_i x'_j} u$, $i, j = 1, \dots, n$ and $D_t u$, satisfying (5.30), $\forall x \in Q_T$.

Let us now fix the coefficient $x_0 = (x'_0, t_0) \in Q_T$. and consider the fundamental solution of $L_0 = L(x_0)$, is given, for $\tau > 0$, by

$$\Gamma(x_0; \theta) = \Gamma(x'_0, t_0; \zeta, \tau) = \frac{(4\pi\tau)^{\frac{1-n}{2}}}{\sqrt{a^{ij}(x_0)}} \exp\left(-\frac{A^{ij}(x_0)\zeta_i\zeta_j}{4\tau}\right)$$

that is equal to zero if $\tau \le 0$, being $A^{ij}(x_0)$ the entries of the inverse matrix $\{a^{ij}(x_0)\}^{-1}$.

The second order derivatives with respect to $\zeta_i$ and $\zeta_j$, denoted by $\Gamma_{ij}(x_0, t_0; \zeta, \tau)$, $i, j = 1, \dots, n$, and $\Gamma_{ij}(x; \theta)$, are kernels of mixed homogeneity.

**Theorem 5.7.52** (Ragusa and Scapellato (2017)). *Let* $n \ge 3, a_{ij} \in VMO(Q_T) \cap L^\infty(Q_T)$, $B_r \Subset \Omega'$ *a ball in* $\mathbb{R}^n$ *Then, for every* $u$ *having compact support in* $B_r \times (0, T)$, *solution of* $Lu = f$ *such that* $D_{x'_i x'_j} u \in L^{q,\mu}(0, T, L^{p,\lambda}(B_r))$ $\forall i, j = 1, \dots, n$, *there exists* $r_0 = r_0(n, p, \nu, \eta)$ *such that, if* $r < r_0$,

$$\|D_{x'_i x'_j} u\|_{L^{q,\mu}(0,T,L^{p,\lambda}(B_r))} \le C\|Lu\|_{L^{q,\mu}(0,T,L^{p,\lambda}(B_r))}, \quad i, j = 1, \dots, n,$$

$$\|u_t\|_{L^{q,\mu}(0,T,L^{p,\lambda}(B_r))} \le C\|Lu\|_{L^{q,\mu}(0,T,L^{p,\lambda}(B_r))},$$

### 5.7.2.3 Divergence form parabolic equations

In this section we are concerned with the results obtained by Ragusa in Ragusa (2004).

In Ragusa (2004) the author studied the well posedness of the Dirichlet problem for the following divergence form parabolic equation

$$\mathscr{L}u \equiv u_t - \sum_{i,j=1}^{n} (a_{ij}(x)u_{x'_i})_{x'_j} = \operatorname{div} f$$

in the cylinder $Q_T$, where $\Omega \subset \mathbb{R}^n$ is a bounded open set with sufficiently smooth boundary, $n \ge 3$ and $T > 0$.

We assume that the parabolic operator $\mathscr{L}$ satisfies the following assumptions:

1. $a_{ij}(x) = a_{ji}(x)$,     $\forall x \in Q_T$,   $\forall i, j = 1, ..., n$;

2. $\exists \mu > 0 : \mu^{-1}|\xi|^2 \leq a_{ij}(x)\xi_i\xi_j \leq \mu|\xi|^2$,   $\forall \xi \in \mathbb{R}^n$,  a.e. in $Q_T$.

In Ragusa (2004), Ragusa revealed the existence and uniqueness of the solution of the following Cauchy-Dirichlet problem:

$$\begin{cases} \mathscr{L}u = \operatorname{div} f & \text{in } Q_T \\ u = 0 & \text{on } \partial\Omega\times] - T, 0[ \\ u(x', -T) = 0 & \text{in } \Omega \end{cases} \tag{5.31}$$

where $f \in L^p(Q_T)$, $1 < p < +\infty$. The coefficients $a_{ij}$ of the principal part of $\mathscr{L}$ will be discontinuous. Finally, in Ragusa (2004), the author derived estimates in $H_p^{1-\frac{1}{p}}$ for every $1 < p < +\infty$.

The following theorem is valid.

**Theorem 5.7.53** (Ragusa (2004)). *Let $a_{ij} \in \text{VMO} \cap L^\infty(\mathbb{R}^{n+1})$, for every $i, j = 1, ..., n$, be symmetric and uniformly elliptic. Suppose that $f \in L^p(Q_T)$ where*

$1 < p < +\infty$.

*Then, the Dirichlet problem (5.31) has a unique solution $u$.*
*Moreover, if $u$ satisfies the above parabolic equation, we have that*

$$u \in H_p^{1-\frac{1}{p}}(-T, 0, L^p(\Omega, \mathbb{R}))$$

*and there exists a constant $C$ independent of $f$ such that*

$$\int_{-T}^0 dt \int_{-T}^0 d\xi \int_\Omega \frac{|u(x,t) - u(x,\xi)|^p}{|t - \xi|^p} dx \leq C \left\{ \int_{Q_T} |f|^p dx dt + \int_{Q_T} |u|^p dx dt \right\}.$$

## 5.7.3   Ultraparabolic equations

### 5.7.3.1   Preliminary tools

In this section we present some Sobolev-Morrey regularity results for ultraparabolic equations. It is important to underline that the coefficients of the principal parts of the differential operators under consideration will belong to a modified version of the classical Sarason class VMO. Precisely, they will lie in the class VMO $_L$, naturally associated with the group's structures that we will define in this section.

**Definition 5.7.54** (Polidoro and Ragusa (2001)). For every $(x, t), (\xi, t) \in \mathbb{R}^{n+1}$, we set

$$(x, t) \circ (\xi, t) = (\xi + E(\tau)x, t + \tau), \quad E(t) = \exp(-tB^T)$$

and

$$D(\lambda) = \operatorname{diag}\left(\lambda I_{m_0}, \lambda^3 I_{m_1}, ..., \lambda^{2r+1} I_{m_r}\right),$$

where $I_{m_j}$ is the $m_j \times m_j$ identity matrix.

We say that $(\mathbb{R}^{n+1}, \circ)$ is the **translation group associated to** $L$ and that $(D(\lambda), \lambda^2)_{\lambda > 0}$ is the **dilation group associated to** $L$.

We call **homogeneous dimension** of $\mathbb{R}^{n+1}$ the integer $Q + 2$, where

$$Q = m_0 + 3m_1 + \dots + (2r+1)m_r.$$

Let us point out that the zero of the group $(\mathbb{R}^{n+1}, \circ)$ is $(0,0)$ and that $(x,t)^{-1} = (-E(-t)x, -t)$.

**Definition 5.7.55** (Polidoro and Ragusa (2001)). Let $\alpha_1, ..., \alpha_n$ be the positive integers such that

$$D(\lambda) = \operatorname{diag}(\lambda^{\alpha_1}, ..., \lambda^{\alpha_n}).$$

If $z = 0$, we set $\|z\| = 0$ while, if $z \in \mathbb{R}^{n+1} \setminus \{0\}$, we define $\|z\| = \rho$ where $\rho$ is the unique positive solution to the equation

$$\frac{x_1^2}{\rho^{2\alpha_1}} + \frac{x_2^2}{\rho^{2\alpha_2}} + \dots + \frac{x_n^n}{\rho^{2\alpha_n}} + \frac{t^2}{\rho^4} = 1.$$

With the norm we associate the **quasidistance**

$$d(z,w) = \|w^{-1} \circ z\|$$

and we denote by $\mathscr{B}_r(z)$ the $d$–sphere centered in $z$ and with radius $r$.

Now, let us define the function spaces $\mathrm{BMO}_L$ and $\mathrm{VMO}_L$; these spaces extend respectively the John-Nirenberg class (see John and Nirenberg (1961)) and Sarason class (see Sarason (1975)).

As usual, let $u \in L^1_{\mathrm{loc}}(\mathbb{R}^{n+1})$ and

$$\fint_{\mathscr{B}} u = \frac{1}{|\mathscr{B}|} \int_{\mathscr{B}} u, \qquad u_{\mathscr{B}} = \fint_{\mathscr{B}} u.$$

**Definition 5.7.56** (Polidoro and Ragusa (2001)). For every function $u \in L^1_{\mathrm{loc}}(\mathbb{R}^{n+1})$ we set

$$\|u\|_* = \sup_{\mathscr{B}} \fint_{\mathscr{B}} |u(z) - u_{\mathscr{B}}| \, dz, \qquad \eta_u(r) = \sup_{\rho \leq r} \fint_{\mathscr{B}_\rho} |u(z) - u_{\mathscr{B}_\rho}| \, dz.$$

Then we define

$$\mathrm{BMO}_L = \{ u \in L^1_{\mathrm{loc}}(\mathbb{R}^{n+1}) : \|u\|_* < +\infty \},$$

$$\mathrm{VMO}_L = \left\{ u \in \mathrm{BMO}_L : \lim_{r \to 0} \eta_u(r) = 0 \right\}.$$

In the following definition, we define the Morrey spaces $L^{p,\lambda}(L, \Omega)$ associated with the group's structures of Definition 5.7.54.

**Definition 5.7.57** (Polidoro and Ragusa (2001)). Let $\Omega$ be an open subset of $\mathbb{R}^{n+1}$, $p \in ]1, +\infty[$ and $\lambda \in [0, Q+2]$. We set

$$\|f\|_{L^{p,\lambda}(L,\Omega)} = \left( \sup_{\substack{r>0 \\ z \in \Omega}} \frac{1}{r^{\lambda}} \int\limits_{\Omega \cap \mathcal{B}_r(z)} |f(w)|^p \, dw \right)^{1/p}$$

and we define

$$L^{p,\lambda}(L,\Omega) = \left\{ f \in L^p_{\text{loc}}(\mathbb{R}^{n+1}) : \|f\|_{L^{p,\lambda}(L,\Omega)} < +\infty \right\}.$$

*Remark.* Let $\Omega \subset \mathbb{R}^{n+1}$. As an immediate consequence of Hölder inequality we obtain that the space $L^{q,\nu}(L,\Omega)$ is continuously imbedded in every $L^{p,\lambda}(L,\Omega)$, with $p$ and $\lambda$ are such that $p \leq q$ and $\lambda \leq \frac{p}{q}\nu + (Q+2)(1 - \frac{p}{q})$.

**Definition 5.7.58** (Polidoro and Ragusa (1998)). Let $\Omega$ be an open subset of $\mathbb{R}^{n+1}$, $p \in ]1, +\infty[$ and $\lambda \in [0, Q+2]$. We denote by $S^{p,\lambda}(L,\Omega)$ the set of functions $f \in L^{p,\lambda}(L,\Omega)$ having weak derivatives $\partial_{x_j} f, \partial_{x_j,x_k} f \in L^{p,\lambda}(L,\Omega)$, $Yf \in L^{p,\lambda}(L,\Omega)$, for $j, k = 1, ..., m_0$.

The norm in this space is defined by

$$\|f\|_{S^{p,\lambda}(L,\Omega)} = \left( \|f\|^p_{L^{p,\lambda}(L,\Omega)} + \sum_{j=1}^{m_0} \|\partial_{x_j} f\|^p_{L^{p,\lambda}(L,\Omega)} + \right.$$

$$\left. + \sum_{j,k=1}^{m_0} \|\partial_{x_j,x_k} f\|^p_{L^{p,\lambda}(L,\Omega)} + \|Yf\|^p_{L^{p,\lambda}(L,\Omega)} \right)^{\frac{1}{p}}.$$

### 5.7.3.2   Nondivergence form ultraparabolic equations

In this paragraph we will present the results contained in Polidoro and Ragusa (1998), where Polidoro and Ragusa considered the class of hypoelliptic operators

$$Lu \equiv \sum_{i,j=1}^{m_0} a_{ij}(z)\partial_{x_i x_j} u + \sum_{i,j=1}^{n} b_{ij} x_i \partial_{x_j} u - \partial_t u \qquad (5.32)$$

where $z = (x,t) \in \mathbb{R}^{n+1}$, $0 < m_0 \leq n$.

In Polidoro and Ragusa (1998), the authors proved that a strong solution to the differential equation $Lu = f$, with $f$ in the Morrey space $L^{p,\lambda}$, belongs to a suitable Sobolev-Morrey space $S^{p,\lambda}$. Then, Polidoro and Ragusa prove some Morrey-type imbedding results that give a local Hölder continuity of the solution $u$.

Let us suppose that the following Hypothesis holds:

## HYPOTHESIS (R).

(i) The matrix $A_0(z) = (a_{ij}(z))_{i,j=1,\dots,m_0}$ is symmetric and $\Lambda > 0$ such that

$$\Lambda^{-1}|\xi|^2 \leq \langle A_0(z)\xi, \xi \rangle \leq \Lambda|\xi|^2, \quad \forall z \in \mathbb{R}^{n+1}, \ \forall \xi \in \mathbb{R}^{m_0}.$$

(ii) The matrix $B = (b_{ij})_{i,j=1,\dots,n}$ has constant entries and takes the following form

$$B(z) = \begin{pmatrix} 0 & B_1 & 0 & \cdots & 0 \\ 0 & 0 & B_2 & \cdots & 0 \\ \vdots & \vdots & \vdots & \ddots & \vdots \\ 0 & 0 & 0 & \cdots & B_r \\ 0 & 0 & 0 & \cdots & 0 \end{pmatrix},$$

where each $B_j$ is a $m_{j-1} \times m_j$ block matrix of rank $m_j$, with $j = 1, \dots, r$, and $m_0 \geq m_1 \geq \dots \geq m_r \geq 1$ and $m_0 + m_1 + \dots + m_r = n$.

We observe that the Lebesgue measure in $\mathbb{R}^{n+1}$ is invariant with respect to the translation group, because $\det E(t) = e^{ttrB} = 1$. Since $\det D(\lambda) = \lambda^Q$, we also have

$$|\mathscr{B}_r(0)| = r^{Q+2}|\mathscr{B}_1(0)|,$$

where $|\mathscr{B}_1(0)| = \omega_{n+1}$ is the Lebesgue measure of the Euclidean unit ball of $\mathbb{R}^{n+1}$.

Let us point out that the generic operator (5.32) can be written in the following form:

$$Lu \equiv \sum_{i,j=1}^{m_0} a_{ij}(z)\partial_{x_i x_j}u + Yu,$$

where $z = (x, t) \in \mathbb{R}^{n+1}$, $0 < m_0 \leq n$ and

$$Yu = \sum_{i,j=1}^{n} b_{ij}x_i\partial_{x_j}u - \partial_t u.$$

In Polidoro and Ragusa (1998) the authors are interested in local regularity problem for strong solutions to the differential equation $Lu = f$ where the coefficients $a_{ij}$ are weakly-continuous. In the next definition we specify the concept of *strong solution*.

**Definition 5.7.59** (Polidoro and Ragusa (1998)). We say that $u$ is a **strong solution** of $Lu = f$ in the open set $\Omega \subset \mathbb{R}^{n+1}$ if $u$ and its weak derivatives $\partial_{x_i}u$, $\partial_{x_i x_j}u$ $(i, j = 1, \dots, m_0)$ and $Yu$ belong to $L^q_{\text{loc}}(\Omega)$, for some $q > 1$, and the differential equation $Lu = f$ is verified almost everywhere in $\Omega$.

The main regularity results contained in Polidoro and Ragusa (1998) are listed below:

**Theorem 5.7.60** (Polidoro and Ragusa (1998)). *Let $\Omega$ be an open subset of $\mathbb{R}^{n+1}$ and let $u$ be a strong solution in $\Omega$ to the equation*

$$\sum_{i,j=1}^{m_0} a_{ij}(z)\partial_{x_i x_j} u + Yu = f.$$

*Suppose that $L$ satisfies the Hypothesis $(R)$, $a_{ij} \in \text{VMO}_L$, per $i,j = 1,...,m_0$, $u \in L^p(\Omega)$, $f \in L^{p,\lambda}(L,\Omega)$, with $p \in ]1,+\infty[$ and $\lambda \in ]0, Q+2[$.*
    *Then $\partial_{x_i x_j} \in L^{p,\lambda}_{\text{loc}}(L,\Omega)$, for $i,j = 1,...,m_0$, $Yu \in L^{p,\lambda}_{\text{loc}}(L,\Omega)$ and for every open set $\Omega' \Subset \Omega$ there exists a positive constant $c_1$, depending only on $p$, $\lambda$, $\Omega'$, $\Omega$, $L$, such that*

$$\|\partial_{x_i x_j} u\|_{L^{p,\lambda}(L,\Omega')} \leq c_1 \|f\|_{L^{p,\lambda}(L,\Omega)} + \|u\|_{L^p(L,\Omega)},$$
$$\|Yu\|_{L^{p,\lambda}(L,\Omega')} \leq c_1 \|f\|_{L^{p,\lambda}(L,\Omega)} + \|u\|_{L^p(L,\Omega)},$$

*for every $i,j = 1,...,m_0$.*

**Theorem 5.7.61** (Polidoro and Ragusa (1998)). *Let $\Omega$ be a bounded open set in $\mathbb{R}^{n+1}$ and let $u$ be a strong solution in $\Omega$ to the equation*

$$\sum_{i,j=1}^{m_0} a_{ij}(z)\partial_{x_i x_j} u + Yu = f.$$

*Suppose that $L$ satisfies $(R)$, $a_{ij} \in \text{VMO}_L$, for $i,j = 1,...,m_0$, $u \in L^p(\Omega)$, $f \in L^{p,\lambda}(L,\Omega)$, where $p \in ]1,+\infty[$ and $\lambda \in ]0, Q+2[$.*

- *If $2p + \lambda > Q + 2$ then, letting $\alpha = \min\{1, \frac{2p+\lambda-(Q+2)}{p}\}$, for any open set $\Omega' \Subset \Omega$ there exists a positive constant $c_2$, depending only on $p$, $\lambda$, $\Omega'$, $\Omega$ and the operator $L$, such that*

$$\frac{|u(z) - u(\zeta)|}{\|\zeta^{-1} \circ z\|^\alpha} \leq c_2 \left( \|f\|_{L^{p,\lambda}(L,\Omega)} + \|u\|_{L^p(\Omega)} \right),$$

    *for every $z, \zeta \in \Omega'$, $z \neq \zeta$.*

- *If $p + \lambda > Q + 2$ then, letting $\beta = \frac{p+\lambda-(Q+2)}{p}$, for any open set $\Omega' \Subset \Omega$ there exists a positive constant $c_3$, depending only on $p$, $\lambda$, $\Omega'$, $\Omega$ and the operator $L$, such that*

$$\frac{|\partial_{x_j} u(z) - \partial_{x_j} u(\zeta)|}{\|\zeta^{-1} \circ z\|^\beta} \leq c_3 \left( \|f\|_{L^{p,\lambda}(L,\Omega)} + \|u\|_{L^p(\Omega)} \right),$$

    *for every $z, \zeta \in \Omega'$, $z \neq \zeta$ and $j = 1,...,m_0$.*

### 5.7.3.3 Divergence form ultraparabolic equations

In Polidoro and Ragusa (2001) Polidoro and Ragusa studied interior regularity for weak solutions of the following divergence form ultraparabolic equation

$$Lu \equiv \sum_{i,j=1}^{m_0} \partial_{x_i}(a_{ij}(z)\partial_{x_j}u) + \sum_{i,j=1}^{n} b_{ij}x_i\partial_{x_j}u - \partial_t u = \sum_{j=1}^{m_0} \partial_{x_j}F_j(z) \qquad (5.33)$$

where $z = (x,t) \in \mathbb{R}^{n+1}$, $0 < m_0 \leq n$ and the functions $F_j$ belong to some $L^p$ spaces. We can equivalently rewrite (5.33) as follows

$$\operatorname{div}(A(z)Du) + Yu = \operatorname{div}(F),$$

where $F = (F_1, ..., F_{m_0}, 0, ..., 0)$, $Yu = \langle x, BDu \rangle - \partial_t u$ and $A(z)$ is the $n \times n$ matrix

$$A(z) = \begin{pmatrix} A_0(z) & 0 \\ 0 & 0 \end{pmatrix}.$$

We say that a function $u$ is a *weak solution* of (5.33) in the open set $\Omega \subset \mathbb{R}^{n+1}$ if $u$, $\partial_{x_1}u$, ..., $\partial_{x_{m_0}}u$, $Yu$ belong to $L_{\text{loc}}^q(\Omega)$ for some $q > 1$ and

$$\int_\Omega \langle A(z)Du(z), D\varphi(z) \rangle \mathrm{d}z - \int_\Omega Yu(z)\varphi(z)\,\mathrm{d}z$$
$$= \int_\Omega \langle F(z), D\varphi(z) \rangle \mathrm{d}z, \qquad \forall \varphi \in C_0^\infty(\Omega).$$

Now we list the main results contained in Polidoro and Ragusa (2001).

**Theorem 5.7.62** (Polidoro and Ragusa (2001)). *Let $\Omega$ be a bounded open set in $\mathbb{R}^{n+1}$ and $u$ a weak solution in $\Omega$ of equation (5.33):*

$$\operatorname{div}(A(x,t)Du) + Yu = \operatorname{div}(F).$$

*Suppose that the operator $L$ satisfies Hypothesis $(R)$, $a_{ij} \in \text{VMO}_L$, $i,j = 1, ..., m_0$, $u \in L^p(\Omega)$ and $F_j \in L^{p,\lambda}(\Omega)$ $\forall j = 1, ..., m_0$, $0 \leq \lambda < Q + 2$ and $1 < p < +\infty$.*

*Then, for any compact set $K \subset \Omega$ we have that $\partial_{x_j}u \in L^{p,\lambda}(K)$, $\forall j = 1, ..., m_0$, for every $1 < p < +\infty$, $0 \leq \lambda < Q + 2$.*

*Moreover, there exists a positive constant $c$ depending on $p$, $\lambda$, $K$, $\Omega$ and $L$ such that*

$$\|\partial_{x_j}u\|_{L^{p,\lambda}(L,K)} \leq c \left( \sum_{k=1}^{m_0} \|F_k\|_{L^{p,\lambda}(L,K)} + \|u\|_{L^p(\Omega)} \right), \qquad \forall j = 1, ..., m_0.$$

**Theorem 5.7.63** (Polidoro and Ragusa (2001)). *Let $\Omega$ be a bounded open set in $\mathbb{R}^{n+1}$ and $u$ a weak solution in $\Omega$ of equation (5.33):*

$$\operatorname{div}(A(x,t)Du) + Yu = \operatorname{div}(F).$$

*Suppose that the operator $L$ satisfies Hypothesis $(R)$, $a_{ij} \in \text{VMO}_L$, $i, j = 1, ..., m_0$, $u \in L^p(\Omega)$ and $F_j \in L^{p,\lambda}(\Omega)$ $\forall j = 1, ..., m_0$, $0 \le \lambda < Q + 2$ and $p > Q + 2 - \lambda$.*

*Then, for any compact $K \subset \Omega$ there exists a positive constant $c$ depending on $p$, $\lambda$, $K$, $\Omega$ and $L$ such that*

$$\frac{|u(z) - u(\zeta)|}{\|\zeta^{-1} \circ z\|^{1 - \frac{Q+2}{p} + \frac{\lambda}{p}}} \le c \left( \sum_{k=1}^{m_0} \|F_k\|_{L^{p,\lambda}(L,K)} + \|u\|_{L^p(\Omega)} \right),$$

*for all $z, \zeta \in \Omega'$, $z \neq \zeta$.*

---

# Bibliography

Agmon S., Douglis A., and Nirenberg L. Estimates near the boundary for solutions of elliptic partial differential equations satisfying general boundary conditions. I. *Comm. Pure Appl. Math.*, 12:623–727, 1959.

Bramanti M. and Cerutti M. C. $W_p^{1,2}$ solvability for the Cauchy-Dirichlet problem for parabolic equations with VMO coefficients. *Comm. Partial Differential Equations*, 18(9-10):1735–1763, 1993.

Calderón A. P. and Zygmund A. On the existence of certain singular integrals. *Acta Math.*, 88:85–139, 1952.

Campanato S. Proprietà di hölderianità di alcune classi di funzioni. *Ann. Scuola Norm. Sup. Pisa (3)*, 17:175–188, 1963a.

Campanato S. Proprietà di inclusione per spazi di Morrey. *Ricerche Mat.*, 12:67–86, 1963b.

Campanato S. Equazioni ellittiche del II° ordine e spazi $\mathcal{L}^{(2,\lambda)}$ . *Annali di Matematica Pura ed Applicata*, 69(1):321–381, 1965.

Campanato S. Un risultato relativo ad equazioni ellittiche del secondo ordine di tipo non variazionale. *Ann. Scuola Norm. Sup. Pisa (3)*, 21:701–707, 1967.

Chiarenza F., 1994. $L^p$ regularity for systems of PDEs, with coefficients in VMO. In *Nonlinear analysis, function spaces and applications, Vol. 5*, pages 1–32. Prometheus, Prague.

Chiarenza F., Frasca M., and Longo P. Interior $W^{2,p}$ estimates for nondivergence elliptic equations with discontinuous coefficients. *Ricerche Mat.*, 40(1):149–168, 1991.

Chiarenza F., Frasca M., and Longo P. $W^{2,p}-$ Solvability of the Dirichlet Problem for Nondivergence Elliptic Equations with VMO Coefficients. *Transactions of the American Mathematical Society*, 336, 1993.

Coifman R. R., Rochberg R., and Weiss G. Factorization theorems for Hardy spaces in several variables. *Ann. of Math. (2)*, 103(3):611–635, 1976.

Di Fazio G. and Palagachev D. K. Oblique derivative problem for elliptic equations in non-divergence form with VMO coefficients. *Comment. Math. Univ. Carolin.*, 37(3):537–556, 1996.

Di Fazio G., Palagachev D. K., and Ragusa M. A. Global Morrey regularity of strong solutions to the Dirichlet problem for elliptic equations with discontinuous coefficients. *J. Funct. Anal.*, 166(2):179–196, 1999.

Di Fazio G. and Ragusa M. A. Interior estimates in Morrey spaces for strong solutions to nondivergence form equations with discontinuous coefficients. *J. Funct. Anal.*, 112(2):241–256, 1993.

Fabes E. B. and Rivière N. M. Singular integrals with mixed homogeneity. *Studia Math.*, 27:19–38, 1966.

Gilbarg D. and Trudinger N. S. *Elliptic partial differential equations of second order*. Classics in Mathematics. Springer, Berlin, 2001. Reprint of the 1998 edition.

Guliyev V. S. Local generalized Morrey spaces and singular integrals with rough kernel. *Azerb. J. Math.*, 3(2):79–94, 2013.

Guliyev V. S., Omarova M. N., Ragusa M. A., and Scapellato A. Commutators and generalized local Morrey spaces. *J. Math. Anal. Appl.*, 457(2):1388–1402, 2018.

John F. and Nirenberg L. On functions of bounded mean oscillation. *Comm. Pure Appl. Math.*, 14:415–426, 1961.

Manfredini M. and Polidoro S. Interior regularity for weak solutions of ultraparabolic equations in divergence form with discontinuous coefficients. *Boll. Unione Mat. Ital. Sez. B Artic. Ric. Mat. (8)*, 1(3):651–675, 1998.

Meyers N. G. Mean oscillation over cubes and Hölder continuity. *Proc. Amer. Math. Soc.*, 15:717–721, 1964.

Miranda C. Sulle equazioni ellittiche del secondo ordine di tipo non variazionale, a coefficienti discontinui. *Ann. Mat. Pura Appl. (4)*, 63:353–386, 1963.

Mizuhara T., 1991. Boundedness of some classical operators on generalized Morrey spaces. In *Harmonic analysis (Sendai, 1990)*, ICM-90 Satell. Conf. Proc., pages 183–189. Springer, Tokyo.

Nakai E. Hardy-Littlewood maximal operator, singular integral operators and the Riesz potentials on generalized Morrey spaces. *Math. Nachr.*, 166:95–103, 1994.

Neri U. Some properties of functions with bounded mean oscillation. *Studia Math.*, 61(1):63–75, 1977.

Palagachev D. K., Ragusa M. A., and Softova L. G. Regular oblique derivative problem in Morrey spaces. *Electron. J. Differential Equations*, pages No. 39, 17, 2000.

Palagachev D. K., Ragusa M. A., and Softova L. G. Cauchy-Dirichlet problem in Morrey spaces for parabolic equations with discontinuous coefficients. *Boll. Unione Mat. Ital. Sez. B Artic. Ric. Mat. (8)*, 6(3):667–683, 2003.

Pick L., Kufner A., John O., and Fučík S. *Function spaces. Vol. 1*, volume 14 of De Gruyter Series in Nonlinear Analysis and Applications. Walter de Gruyter & Co., Berlin, 2013, extended edition.

Polidoro S. and Ragusa M. A. Sobolev-Morrey spaces related to an ultra parabolic equation. *Manuscripta Math.*, 96(3):371–392, 1998.

Polidoro S. and Ragusa M. A. Hölder regularity for solutions of ultraparabolic equations in divergence form. *Potential Anal.*, 14(4):341–350, 2001.

Ragusa M. A. Elliptic boundary value problem in vanishing mean oscillation hypothesis. *Comment. Math. Univ. Carolin.*, 40(4):651–663, 1999.

Ragusa M. A. $C^{(0,\alpha)}$-regularity of the solutions of Dirichlet problem for elliptic equations in divergence form. *Int. J. Differ. Equ. Appl.*, 1(1):113–126, 2000a.

Ragusa M. A. Regularity of solutions of divergence form elliptic equations. *Proc. Amer. Math. Soc.*, 128(2):533–540, 2000b.

Ragusa M. A. Cauchy-Dirichlet problem associated to divergence form parabolic equations. *Commun. Contemp. Math.*, 6(3):377–393, 2004.

Ragusa M. A. The Cauchy-Dirichlet problem for parabolic equations with VMO coefficients. *Math. Comput. Modelling*, 42(11-12):1245–1254, 2005.

Ragusa M. A. and Scapellato A. Mixed Morrey spaces and their applications to partial differential equations. *Nonlinear Anal.*, 151:51–65, 2017.

Sarason D. Functions of vanishing mean oscillation. *Trans. Amer. Math. Soc.*, 207:391–405, 1975.

Scapellato A., 2017a. On some qualitative results for the solution to a Dirichlet problem in local generalized Morrey spaces. In *AIP Conference Proceedings*, volume 1798.

Scapellato A., 2017b. Some properties of integral operators on generalized Morrey spaces. In *AIP Conference Proceedings*, volume 1863.

Spanne S. Some function spaces defined using the mean oscillation over cubes. *Ann. Scuola Norm. Sup. Pisa (3)*, 19:593–608, 1965.

Talenti G. Equazioni lineari ellittiche in due variabili. *Matematiche (Catania)*, 21:339–376, 1966.

Uchiyama A. On the compactness of operators of Hankel type. *Tôhoku Math. J. (2)*, 30(1):163–171, 1978.

# Chapter 6

## Composite Submeasures and Supermeasures with Applications to Classical Inequalities

**Silvestru Sever Dragomir**

*Mathematics, College of Engineering & Science, Victoria University, Melbourne City, Australia & DST-NRF Centre of Excellence, in the Mathematical and Statistical Sciences, School of Computer Science and Applied Mathematics, University of the Witwatersrand, Johannesburg, South Africa*

## 6.1 Introduction

Let $\Omega$ be a nonempty set. A subset $\mathcal{A}$ of the power set $2^{\Omega}$ is called an *algebra* if the following conditions are satisfied:

(i) $\Omega$ is in $\mathcal{A}$;

(ii) $\mathcal{A}$ is closed under complementation, namely, if $A \in \mathcal{A}$ then $\Omega \setminus A \in \mathcal{A}$;

(iii) $\mathcal{A}$ is closed under union, i.e. if $A, B \in \mathcal{A}$ then $A \cup B \in \mathcal{A}$.

By applying *de Morgan*'s laws it follows that $\mathcal{A}$ is closed under intersection, namely if $A, B \in \mathcal{A}$ then $A \cap B \in \mathcal{A}$. It also follows that the empty set $\emptyset$ belongs to $\mathcal{A}$. Elements of the algebra are called measurable sets. An ordered pair $(\Omega, \mathcal{A})$, where $\Omega$ is a set and $\mathcal{A}$ is a algebra over $\Omega$, is called a *measurable space*.

The function $\mu : \mathcal{A} \to [0, \infty)$ is called a *measure* [*submeasure* (*supermeasure*)] on $\mathcal{A}$ if

(a) For all $A \in \mathcal{A}$ we have $\mu(A) \geq 0$ (nonnegativity);

(aa) We have $\mu(\emptyset) = 0$ (null empty set);

(aaa) For any $A, B \in \mathcal{A}$ with $A \cap B = \emptyset$ we have

$$\mu(A \cup B) = [\leq (\geq)] \mu(A) + \mu(B), \qquad (6.1)$$

i.e., $\mu$ is *additive* [*subadditive* (*superadditive*)] on $\mathcal{A}$.

For $\mu$ as above we denote by

$$\mathcal{A}_\mu := \{A \in \mathcal{A} \mid \mu(A) > 0\}.$$

If $\mathcal{A}_\mu = \mathcal{A} \setminus \{\emptyset\}$ then we say that $\mu$ is *positive* on $\mathcal{A}$.

Let $A, B \in \mathcal{A}$ with $A \subset B$, then $B = A \cup (B \setminus A)$, $A \cap (B \setminus A) = \emptyset$ and $B \setminus A \in \mathcal{A}$. If $\mu$ is additive or superadditive on $\mathcal{A}$, then

$$\mu(B) = \mu(A \cup (B \setminus A)) = (\geq) \mu(A) + \mu(B \setminus A) \geq \mu(A)$$

showing that $\mu$ is *monotonic nondecreasing* on $\mathcal{A}$.

In this chapter, by the use of some measures and *submeasures* (*supermeasures*) on $\mathcal{A}$ we show that we can construct some natural composite functionals that are in their turn *submeasures* (*supermeasures*) on $\mathcal{A}$. We also provide some examples of supermeasures that can be naturally associated to classical inequalities such as Jensen's inequality, Hölder's inequality, Minkowski's inequality, Cauchy-Bunyakovsky-Schwarz's inequality, Čebyšev's inequality, Hermite-Hadamard's inequalities and the definition of convexity property. As a consequence of monotonic nondecreasing property of these supermeasures, some refinements of the above inequalities are also obtained.

---

## 6.2 Some composite functionals

We have:

**Theorem 6.2.1.** *Let $(\Omega, \mathcal{A})$ be a measurable space and $\mu : \mathcal{A} \to [0, \infty)$ a positive measure on $\mathcal{A} \setminus \{\emptyset\}$. If $\delta : \mathcal{A} \to [0, \infty)$ is a supermeasure (submeasure) on $\mathcal{A}$ and $p \geq 1$ $(0 < p < 1)$ then the functional*

$$\eta_p : \mathcal{A} \to [0, \infty), \eta_p(A) = \delta(A) \mu^{1 - \frac{1}{p}}(A) \qquad (6.2)$$

*is also a supermeasure (submeasure) on $\mathcal{A}$.*

*Proof.* First, we observe that the following elementary inequality holds:

$$(\alpha + \beta)^p \geq (\leq) \alpha^p + \beta^p \qquad (6.3)$$

for any $\alpha, \beta \geq 0$ and $p \geq 1$ $(0 < p < 1)$.

Indeed, if we consider the function

$$f_p : [0, \infty) \to \mathbb{R}, \ f_p(t) = (t+1)^p - t^p,$$

then we have

$$f_p'(t) = p\left[(t+1)^{p-1} - t^{p-1}\right].$$

Observe that for $p > 1$ and $t > 0$ we have that $f_p'(t) > 0$ showing that $f_p$ is strictly increasing on the interval $[0, \infty)$. Now for $t = \frac{\alpha}{\beta}$ $(\beta > 0, \alpha \geq 0)$ we have $f_p(t) > f_p(0)$ giving that

$$\left(\frac{\alpha}{\beta} + 1\right)^p - \left(\frac{\alpha}{\beta}\right)^p > 1,$$

i.e., the desired inequality (13.7).

For $p \in (0, 1)$ we have that $f_p$ is strictly decreasing on $[0, \infty)$, which proves the second case in (13.7).

Let $A, B \in \mathcal{A} \backslash \{\emptyset\}$ with $A \cap B = \emptyset$.

Now, if $\delta$ is superadditive (subadditive) and $p \geq 1$ $(0 < p < 1)$, then we have by (13.7) that

$$\delta^p (A \cup B) \geq (\leq) [\delta(A) + \delta(B)]^p \geq (\leq) \delta^p(A) + \delta^p(B). \qquad (6.4)$$

Utilising (13.9) and the additivity property of $\mu$ we have that

$$\frac{\delta^p(A \cup B)}{\mu(A \cup B)} \geq (\leq) \frac{\delta^p(A) + \delta^p(B)}{\mu(A) + \mu(B)} \qquad (6.5)$$

$$= \frac{\mu(A) \cdot \frac{\delta^p(A)}{\mu(A)} + \mu(B) \cdot \frac{\delta^p(B)}{\mu(B)}}{\mu(A) + \mu(B)}$$

$$= \frac{\mu(A) \cdot \left[\frac{\delta(A)}{\mu^{1/p}(A)}\right]^p + \mu(B) \cdot \left[\frac{\delta(B)}{\mu^{1/p}(B)}\right]^p}{\mu(A) + \mu(B)} =: I.$$

Since for $p \geq 1$ $(0 < p < 1)$ the power function $g(t) = t^p$ is convex (concave), then

$$I \geq (\leq) \left[\frac{\mu(A) \cdot \frac{\delta(A)}{\mu^{1/p}(A)} + \mu(B) \cdot \frac{\delta(B)}{\mu^{1/p}(B)}}{\mu(A) + \mu(B)}\right]^p \qquad (6.6)$$

$$= \left[\frac{\delta(A) \mu^{1-1/p}(A) + \delta(B) \mu^{1-1/p}(B)}{\mu(A \cup B)}\right]^p.$$

By combining (13.10) with (13.11) we get

$$\frac{\delta^p (A \cup B)}{\mu (A \cup B)} \geq (\leq) \left[ \frac{\delta (A) \mu^{1-1/p} (A) + \delta (B) \mu^{1-1/p} (B)}{\mu (A \cup B)} \right]^p,$$

which is equivalent to

$$\frac{\delta (A \cup B)}{\mu^{1/p} (A \cup B)} \geq (\leq) \frac{\delta (A) \mu^{1-1/p} (A) + \delta (B) \mu^{1-1/p} (B)}{\mu (A \cup B)}$$

i.e., by multiplying with $\mu (A \cup B)$,

$$\eta_p (A \cup B) \geq (\leq) \eta_p (A) + \eta_p (B)$$

and the proof is complete. $\qquad\square$

**Corollary 6.2.2.** *With the assumptions of Theorem 6.2.1 for $(\Omega, \mathcal{A})$, $\mu$ and if $\delta : \mathcal{A} \to [0, \infty)$ is a supermeasure (submeasure) on $\mathcal{A}$ while $p, q \geq 1$ $(0 < p, q < 1)$ then the two parameter functional*

$$\eta_{p,q} : \mathcal{A} \to [0, \infty), \eta_{p,q} (A) = \delta^q (A) \mu^{q\left(1-\frac{1}{p}\right)} (A) \qquad (6.7)$$

*is also a supermeasure (submeasure) on $\mathcal{A}$.*

*Proof.* Let $A, B \in \mathcal{A} \backslash \{\emptyset\}$ with $A \cap B = \emptyset$. Observe that $\eta_{p,q} (A) = \left[ \eta_p (A) \right]^q$ for $A \in \mathcal{A}$. Therefore, by Theorem 6.2.1 and the inequality (13.7) for $q \geq 1$ $(0 < q < 1)$ we have that

$$\begin{aligned} \eta_{p,q} (A \cup B) &= \left[ \eta_p (A \cup B) \right]^q \geq (\leq) \left[ \eta_p (A) + \eta_p (B) \right]^q \\ &\geq (\leq) \left[ \eta_p (A) \right]^q + \left[ \eta_p (B) \right]^q = \eta_{p,q} (A) + \eta_{p,q} (B) . \end{aligned}$$

$\qquad\square$

*Remark* 6.2.3. If, for $q = p$ in the above corollary, we consider the functional $\psi_p : \mathcal{A} \to [0, \infty)$ defined by

$$\psi_p (A) := \delta^p (A) \mu^{p-1} (A) \qquad (6.8)$$

for $p \geq 1$ $(0 < p < 1)$ and $\delta : \mathcal{A} \to [0, \infty)$ is a supermeasure (submeasure) on $\mathcal{A}$, then the functional $\psi_p$ is also a supermeasure (submeasure) on $\mathcal{A}$.

We have:

**Theorem 6.2.4.** *Let $(\Omega, \mathcal{A})$ a measurable space and $\mu : \mathcal{A} \to [0, \infty)$ a positive measure on $\mathcal{A} \backslash \{\emptyset\}$. If $\delta : \mathcal{A} \to (0, \infty)$ is a supermeasure on $\mathcal{A} \backslash \{\emptyset\}$ and $0 < p < 1$ then the functional*

$$\varphi_p : \mathcal{A} \backslash \{\emptyset\} \to [0, \infty), \eta_p (A) = \frac{\mu^{1-\frac{1}{p}} (A)}{\delta (A)} \qquad (6.9)$$

*is a submeasure on $\mathcal{A} \backslash \{\emptyset\}$.*

*Proof.* Let $s := -p \in [-1, 0)$. For $s < 0$ we have the following inequality

$$(\alpha + \beta)^s \leq \alpha^s + \beta^s \qquad (6.10)$$

for any $\alpha, \beta > 0$.

Indeed, by the convexity of the function $f_s(t) = t^s$ on $(0, \infty)$ with $s < 0$ we have that

$$(\alpha + \beta)^s \leq 2^{s-1} (\alpha^s + \beta^s)$$

for any $\alpha, \beta > 0$ and since, obviously, $2^{s-1} (\alpha^s + \beta^s) \leq \alpha^s + \beta^s$, then (13.13) holds true.

Let $A, B \in \mathcal{A} \setminus \{\emptyset\}$ with $A \cap B = \emptyset$. Taking into account that $\delta$ is super-additive, then by (13.13) we have

$$\delta^s (A \cup B) \leq [\delta(A) + \delta(B)]^s \leq \delta^s(A) + \delta^s(B) \qquad (6.11)$$

for any $A, B \in \mathcal{A} \setminus \{\emptyset\}$ with $A \cap B = \emptyset$.

Since $\mu$ is additive, then by (13.13) we have that

$$\frac{\delta^s (A \cup B)}{\mu (A \cup B)} \leq \frac{\delta^s(A) + \delta^s(B)}{\mu(A) + \mu(B)} \qquad (6.12)$$

$$= \frac{\mu(A) \cdot \left[\frac{\delta(A)}{\mu^{1/s}(A)}\right]^s + \mu(B) \cdot \left[\frac{\delta(B)}{\mu^{1/s}(B)}\right]^s}{\mu(A) + \mu(B)}$$

$$= \frac{\mu(A) \cdot \left[\frac{\mu^{1/s}(A)}{\delta(A)}\right]^{-s} + \mu(B) \cdot \left[\frac{\mu^{1/s}(B)}{\delta(B)}\right]^{-s}}{\mu(A) + \mu(B)} =: J.$$

By the concavity of the function $g(t) = t^{-s}$ with $s \in [-1, 0)$ we also have

$$J \leq \left[\frac{\mu(A) \cdot \frac{\mu^{1/s}(A)}{\delta(A)} + \mu(B) \cdot \frac{\mu^{1/s}(B)}{\delta(B)}}{\mu(A) + \mu(B)}\right]^{-s}. \qquad (6.13)$$

Making use of (13.18) and (13.19) we get

$$\frac{\delta^s (A \cup B)}{\mu (A \cup B)} \leq \left[\frac{\mu(A) \cdot \frac{\mu^{1/s}(A)}{\delta(A)} + \mu(B) \cdot \frac{\mu^{1/s}(B)}{\delta(B)}}{\mu(A) + \mu(B)}\right]^{-s}$$

for any $A, B \in \mathcal{A}$, which is equivalent to

$$\frac{\delta^{-1} (A \cup B)}{\mu^{-1/s} (A \cup B)} \leq \frac{\frac{\mu^{1+1/s}(A)}{\delta(A)} + \frac{\mu^{1+1/s}(B)}{\delta(B)}}{\mu(A) + \mu(B)}$$

and, since $\mu(A \cup B) = \mu(A) + \mu(B)$, with

$$\frac{\mu^{1+1/s} (A + B)}{\delta (A + B)} \leq \frac{\mu^{1+1/s}(A)}{\delta(A)} + \frac{\mu^{1+1/s}(B)}{\delta(B)}$$

for any $A, B \in \mathcal{A} \setminus \{\emptyset\}$ with $A \cap B = \emptyset$.

This completes the proof. $\qquad \square$

The following result may be stated as well:

**Corollary 6.2.5.** *Assume that $(\Omega, \mathcal{A})$ and $\mu$ are as in Theorem 6.2.4. If $\delta : \mathcal{A} \to (0, \infty)$ is a supermeasure on $\mathcal{A} \backslash \{\emptyset\}$ and $0 < p, q < 1$ then the two parameter functional*

$$\varphi_{p,q} : \mathcal{A} \backslash \{\emptyset\} \to [0, \infty), \varphi_{p,q}(A) = \frac{\mu^{q\left(1 - \frac{1}{p}\right)}(A)}{\delta^q(A)} \qquad (6.14)$$

*is a submeasure on $\mathcal{A} \backslash \{\emptyset\}$.*

*Proof.* Observe that $\varphi_{p,q}(A) = [\varphi_p(A)]^q$ for $A \in \mathcal{A}$. Therefore, by Theorem 6.2.4 and the inequality (13.7) for $0 < q < 1$ we have that

$$\begin{aligned} \varphi_{p,q}(A \cup B) &= [\varphi_p(A \cup B)]^q \leq [\varphi_p(A) + \varphi_p(B)]^q \\ &\leq [\varphi_p(A)]^q + [\varphi_p(B)]^q = \varphi_{p,q}(A) + \varphi_{p,q}(B) \end{aligned}$$

for any $A, B \in \mathcal{A} \backslash \{\emptyset\}$ with $A \cap B = \emptyset$ and the statement is proved. $\square$

*Remark* 6.2.6. If we consider the functional for $0 < p < 1$ defined by

$$\sigma_p(A) := \frac{\mu^{p-1}(A)}{\delta^p(A)},$$

where $\mu : \mathcal{A} \to (0, \infty)$ is a measure on $\mathcal{A} \backslash \{\emptyset\}$ and $\delta : \mathcal{A} \backslash \{\emptyset\} \to (0, \infty)$ is a supermeasure on $\mathcal{A} \backslash \{\emptyset\}$, then the functional $\sigma_p$ is a submeasure on $\mathcal{A} \backslash \{\emptyset\}$.

---

## 6.3 Applications for Jensen's inequality

Let $(\Omega, \mathcal{A}, \nu)$ be a measurable space consisting of a set $\Omega$, a $\sigma$-algebra $\mathcal{A}$ of parts of $\Omega$ and a countably additive and positive measure $\nu$ on $\mathcal{A}$ with values in $\mathbb{R} \cup \{\infty\}$. For a $\nu$-measurable function $w : \Omega \to \mathbb{R}$, with $w(x) \geq 0$ for $\nu$-a.e. (almost every) $x \in \Omega$, consider the *Lebesgue space*

$$L_w(\Omega, \nu) := \{f : \Omega \to \mathbb{R}, \ f \text{ is } \nu\text{-measurable and } \int_\Omega w(x) |f(x)| \, d\nu(x) < \infty\}.$$

For simplicity of notation we write everywhere in the sequel $\int_\Omega w d\nu$ instead of $\int_\Omega w(x) \, d\nu(x)$.

Let also

$$\mathcal{A}_\nu := \{A \in \mathcal{A} \mid \nu(A) > 0\}.$$

For a $\nu$-measurable function $w : \Omega \to \mathbb{R}$, with $w(x) > 0$ for $\nu$-a.e. $x \in \Omega$, we consider the functional $J(\cdot, w; \Phi, f) : \mathcal{A}_\nu \to [0, \infty)$ defined by

$$J(A, w; \Phi, f) := \int_A w(\Phi \circ f) \, d\nu - \Phi\left(\frac{\int_A wf d\nu}{\int_A w d\nu}\right) \int_A w d\nu \geq 0, \qquad (6.15)$$

where $\Phi : [m, M] \to \mathbb{R}$ is a continuous convex function on the interval of real numbers $[m, M]$, $f : \Omega \to [m, M]$ is $\nu$-measurable and such that $f$, $\Phi \circ f \in L_w(\Omega, \nu)$.

We use the following result, see Dragomir (2015):

**Lemma 6.3.1.** *Let* $\Phi : [m, M] \to \mathbb{R}$ *be a continuous convex function on the interval of real numbers* $[m, M]$, $f : \Omega \to [m, M]$ *is* $\nu$*-measurable and such that* $f$, $\Phi \circ f \in L_w(\Omega, \nu)$. *Then the functional* $J(\cdot, w; \Phi, f)$ *defined by (6.15) is a supermeasure on* $\mathcal{A}_\nu$.

For some Jensen's inequality related functionals and their properties see Abramovich (2013), Aldaz (2012), Bari et al. (2009), Dragomir (1995), Dragomir et al. (1996), Dragomir (2006), Dragomir (2010) and Dragomir et al. (2011).

For $p, q \geq 1$ consider the two parameter functional $J_{p,q}(\cdot, w; \Phi, f) : \mathcal{A}_\nu \to [0, \infty)$,

$$J_{p,q}(A, w; \Phi, f) = J^q(A, w; \Phi, f) \left( \int_A w \, d\nu \right)^{q\left(1 - \frac{1}{p}\right)}. \tag{6.16}$$

Observe that

$$J_{p,q}(A, w; \Phi, f) = \left( \int_A w \, d\nu \right)^{-\frac{q}{p}} \left[ \frac{\int_A w(\Phi \circ f) \, d\nu}{\int_A w \, d\nu} - \Phi\left( \frac{\int_A wf \, d\nu}{\int_A w \, d\nu} \right) \right]^q. \tag{6.17}$$

In particular, for $q = p$ we have

$$J_p(A, w; \Phi, f) := \left( \int_A w \, d\nu \right)^{-1} \left[ \frac{\int_A w(\Phi \circ f) \, d\nu}{\int_A w \, d\nu} - \Phi\left( \frac{\int_A wf \, d\nu}{\int_A w \, d\nu} \right) \right]^p. \tag{6.18}$$

**Theorem 6.3.2.** *Let* $\Phi : [m, M] \to \mathbb{R}$ *be a continuous convex function on the interval of real numbers* $[m, M]$, $f : \Omega \to [m, M]$ *is* $\nu$*-measurable and such that* $f$, $\Phi \circ f \in L_w(\Omega, \nu)$. *Then the functional* $J_{p,q}(\cdot, w; \Phi, f)$ *defined by (13.41) is a supermeasure on* $\mathcal{A}_\nu$ *for any* $p, q \geq 1$.

The proof follows by Corollary 6.2.2 for $p, q \geq 1$, the measure $\mu : \mathcal{A}_\nu \to (0, \infty)$, $\mu(A) := \int_A w \, d\nu$ and the supermeasure $\delta : \mathcal{A}_\nu \to [0, \infty)$, $\delta(A) := J(A, w; \Phi, f)$.

If $\Phi : [m, M] \to \mathbb{R}$ is a continuous strictly convex function on $[m, M]$ then Jensen's inequality is strict, namely

$$J(A, w; \Phi, f) := \int_A w(\Phi \circ f) \, d\nu - \Phi\left( \frac{\int_A wf \, d\nu}{\int_A w \, d\nu} \right) \int_A w \, d\nu > 0$$

for any $A \in \mathcal{A}_\nu$.

In this situation we can also define the two parameters of functional $\widetilde{J}_{p,q}(A, w; \Phi, f) : \mathcal{A}_\nu \to [0, \infty)$,

$$\widetilde{J}_{p,q}(A, w; \Phi, f) = \frac{\left( \int_A w \, d\nu \right)^{q\left(1 - \frac{1}{p}\right)}}{J^q(A, w; \Phi, f)} \tag{6.19}$$

where $0 < p, q < 1$.

Observe that

$$\widetilde{J}_{p,q}\left(A, w; \Phi, f\right) = \frac{1}{\left(\int_A w d\nu\right)^{\frac{q}{p}} \left[\frac{\int_A w(\Phi \circ f) d\nu}{\int_A w d\nu} - \Phi\left(\frac{\int_A w f d\nu}{\int_A w d\nu}\right)\right]^q}$$

and

$$\widetilde{J}_p\left(A, w; \Phi, f\right) := \frac{1}{\left[\frac{\int_A w(\Phi \circ f) d\nu}{\int_A w d\nu} - \Phi\left(\frac{\int_A w f d\nu}{\int_A w d\nu}\right)\right]^p \int_A w d\nu}.$$

**Theorem 6.3.3.** *Let $\Phi : [m, M] \to \mathbb{R}$ be a continuous strictly convex function on the interval of real numbers $[m, M]$, $f : \Omega \to [m, M]$ is $\nu$-measurable and such that $f$, $\Phi \circ f \in L_w(\Omega, \nu)$. Then the functional $\widetilde{J}_{p,q}(\cdot, w; \Phi, f)$ defined by (13.44) is a submeasure on $\mathcal{A}_\nu$ for any $0 < p, q < 1$.*

The proof follows by Corollary 6.2.5for $0 < p, q < 1$, the measure $\mu : \mathcal{A}_\nu \to (0, \infty)$, $\mu(A) := \int_A w d\nu$ and the supermeasure $\delta : \mathcal{A}_\nu \to [0, \infty)$, $\delta(A) := J(A, w; \Phi, f)$.

Let $\Phi : [m, M] \to \mathbb{R}$ be a continuous convex function on the interval of real numbers $[m, M]$, $x = (x_i)_{i \in \mathbb{N}}$ a sequence of real numbers with $x_i \in [m, M]$, $i \in \mathbb{N}$, and $w = (w_i)_{i \in \mathbb{N}}$ a sequence of positive real numbers.

Let $\Omega = \mathbb{N}$ and $\mathcal{P}_f(\mathbb{N})$ be the algebra of finite parts of natural numbers $\mathbb{N}$. By the monotonicity property of supermeasure on $\mathcal{P}_f(\mathbb{N})$ we have from the above results that the sequence

$$J_{n,p}(w; \Phi, x) := \left(\sum_{i=0}^{n} w_i\right)^{-\frac{1}{p}} \left[\frac{\sum_{i=0}^{n} w_i \Phi(x_i)}{\sum_{i=0}^{n} w_i} - \Phi\left(\frac{\sum_{i=0}^{n} w_i x_i}{\sum_{i=0}^{n} w_i}\right)\right],$$

where $p \geq 1$ is monotonic nondecreasing, namely

$$\left(\sum_{i=0}^{n+1} w_i\right)^{-\frac{1}{p}} \left[\frac{\sum_{i=0}^{n+1} w_i \Phi(x_i)}{\sum_{i=0}^{n+1} w_i} - \Phi\left(\frac{\sum_{i=0}^{n+1} w_i x_i}{\sum_{i=0}^{n+1} w_i}\right)\right] \quad (6.20)$$

$$\geq \left(\sum_{i=0}^{n} w_i\right)^{-\frac{1}{p}} \left[\frac{\sum_{i=0}^{n} w_i \Phi(x_i)}{\sum_{i=0}^{n} w_i} - \Phi\left(\frac{\sum_{i=0}^{n} w_i x_i}{\sum_{i=0}^{n} w_i}\right)\right]$$

for any $n \in \mathbb{N}$ and

$$J_{n,p}(w; \Phi, x) \quad (6.21)$$

$$\geq \max_{0 \leq i \neq j \leq n} \left\{(w_i + w_j)^{-\frac{1}{p}} \left[\frac{w_i \Phi(x_i) + w_j \Phi(x_j)}{w_i + w_j} - \Phi\left(\frac{w_i x_i + w_j x_j}{w_i + w_j}\right)\right]\right\}.$$

We also have for $n \geq 1$ that

$$
\left( \sum_{i=0}^{2n} w_i \right)^{-\frac{q}{p}} \left[ \frac{\sum_{i=0}^{2n} w_i \Phi(x_i)}{\sum_{i=0}^{2n} w_i} - \Phi \left( \frac{\sum_{i=0}^{2n} w_i x_i}{\sum_{i=0}^{2n} w_i} \right) \right]^q \tag{6.22}
$$

$$
\geq \left( \sum_{i=0}^{n} w_{2i} \right)^{-\frac{q}{p}} \left[ \frac{\sum_{i=0}^{n} w_{2i} \Phi(x_{2i})}{\sum_{i=0}^{n} w_{2i}} - \Phi \left( \frac{\sum_{i=0}^{n} w_{2i} x_{2i}}{\sum_{i=0}^{n} w_{2i}} \right) \right]^q
$$

$$
+ \left( \sum_{i=0}^{n-1} w_{2i+1} \right)^{-\frac{q}{p}} \left[ \frac{\sum_{i=0}^{n-1} w_{2i+1} \Phi(x_{2i+1})}{\sum_{i=0}^{n-1} w_{2i+1}} - \Phi \left( \frac{\sum_{i=0}^{n-1} w_{2i+1} x_{2i+1}}{\sum_{i=0}^{n-1} w_{2i+1}} \right) \right]^q
$$

and

$$
\left( \sum_{i=0}^{2n+1} w_i \right)^{-\frac{q}{p}} \left[ \frac{\sum_{i=0}^{2n+1} w_i \Phi(x_i)}{\sum_{i=0}^{2n} w_i} - \Phi \left( \frac{\sum_{i=0}^{2n+1} w_i x_i}{\sum_{i=0}^{2n} w_i} \right) \right]^q \tag{6.23}
$$

$$
\geq \left( \sum_{i=0}^{n} w_{2i} \right)^{-\frac{q}{p}} \left[ \frac{\sum_{i=0}^{n} w_{2i} \Phi(x_{2i})}{\sum_{i=0}^{n} w_{2i}} - \Phi \left( \frac{\sum_{i=0}^{n} w_{2i} x_{2i}}{\sum_{i=0}^{n} w_{2i}} \right) \right]^q
$$

$$
+ \left( \sum_{i=0}^{n} w_{2i+1} \right)^{-\frac{q}{p}} \left[ \frac{\sum_{i=0}^{n} w_{2i+1} \Phi(x_{2i+1})}{\sum_{i=0}^{n} w_{2i+1}} - \Phi \left( \frac{\sum_{i=0}^{n} w_{2i+1} x_{2i+1}}{\sum_{i=0}^{n} w_{2i+1}} \right) \right]^q .
$$

## 6.4  Applications for Hölder's inequality

Let $(\Omega, \mathcal{A}, \nu)$ be a measurable space consisting of a set $\Omega$, a $\sigma$-algebra $\mathcal{A}$ of parts of $\Omega$ and a countably additive and positive measure $\nu$ on $\mathcal{A}$ with values in $\mathbb{R} \cup \{\infty\}$. For a $\nu$-measurable function $w : \Omega \to \mathbb{C}$, with $w(x) \geq 0$ for $\nu$-a.e. (almost every) $x \in \Omega$, consider the $\alpha$-Lebesgue space

$$
L_w^{\alpha}(\Omega, \nu) := \{ f : \Omega \to \mathbb{C}, \ f \text{ is } \nu\text{-measurable and } \int_{\Omega} w |f|^{\alpha} \, d\nu < \infty \},
$$

for $\alpha \geq 1$.

The following inequality is well known in the literature as *Hölder's inequality*

$$
\left| \int_{\Omega} w f g d\nu \right| \leq \left( \int_{\Omega} w |f|^{\alpha} \, d\nu \right)^{1/\alpha} \left( \int_{\Omega} w |g|^{\beta} \, d\nu \right)^{1/\beta} \tag{6.24}
$$

where $\alpha, \beta > 1$ with $\frac{1}{\alpha} + \frac{1}{\beta} = 1$ and $f \in L_w^{\alpha}(\Omega, \nu)$, $g \in L_w^{\beta}(\Omega, \nu)$.

We consider the functional $H_{\alpha,\beta}(\cdot, w; f, g) : \mathcal{A}_{\nu} \to [0, \infty)$ defined by

$$
H_{\alpha,\beta}(A, w; f, g) = \left( \int_A w |f|^{\alpha} \, d\nu \right)^{1/\alpha} \left( \int_A w |g|^{\beta} \, d\nu \right)^{1/\beta} \left| \int_A w f g d\nu \right| . \tag{6.25}
$$

We use the following result, see Dragomir (2015):

**Lemma 6.4.1.** *Let* $f \in L_w^{\alpha}(\Omega, \nu)$, $g \in L_w^{\beta}(\Omega, \nu)$ *where* $\alpha, \beta > 1$ *with* $\frac{1}{\alpha} + \frac{1}{\beta} = 1$. *Then the functional* $H_{\alpha, \beta}(\cdot, w; f, g) : \mathcal{A}_\nu \to [0, \infty)$ *defined by (13.58) is a supermeasure.*

For some Hölder's inequality related functionals and their properties see Agarwal and Dragomir (1998 ), Dragomir (2009) and Everitt (1961).

Consider the two parameter functional $H_{\alpha, \beta, p, q}(\cdot, w; f, g) : \mathcal{A} \to [0, \infty)$ defined by

$$H_{\alpha, \beta, p, q}(A, w; f, g) \tag{6.26}$$

$$:= \left[ \left( \frac{\int_A w |f|^{\alpha} d\nu}{\int_A w d\nu} \right)^{1/\alpha} \left( \frac{\int_A w |g|^{\beta} d\nu}{\int_A w d\nu} \right)^{1/\beta} - \left| \frac{\int_A w f g d\nu}{\int_A w d\nu} \right| \right]^q$$

$$\times \left( \int_A w d\nu \right)^{q \left( 2 - \frac{1}{p} \right)},$$

where $p, q \geq 1$.

Also, define

$$H_{\alpha, \beta, p}(A, w; f, g) \tag{6.27}$$

$$:= \left[ \left( \frac{\int_A w |f|^{\alpha} d\nu}{\int_A w d\nu} \right)^{1/\alpha} \left( \frac{\int_A w |g|^{\beta} d\nu}{\int_A w d\nu} \right)^{1/\beta} - \left| \frac{\int_A w f g d\nu}{\int_A w d\nu} \right| \right]^p$$

$$\times \left( \int_A w d\nu \right)^{2p-1},$$

where $p \geq 1$.

**Theorem 6.4.2.** *Let* $f \in L_w^{\alpha}(\Omega, \nu)$, $g \in L_w^{\beta}(\Omega, \nu)$ *where* $\alpha, \beta > 1$ *with* $\frac{1}{\alpha} + \frac{1}{\beta} = 1$. *Then the functional* $H_{\alpha, \beta, p, q}(\cdot, w; f, g) : \mathcal{A}_\nu \to [0, \infty)$ *with* $p, q \geq 1$, *defined by (13.61) is a supermeasure.*

The proof follows by Corollary 6.2.2 for $p, q \geq 1$, the measure $\mu : \mathcal{A}_\nu \to (0, \infty)$, $\mu(A) := \int_A w d\nu$ and the supermeasure $\delta : \mathcal{A}_\nu \to [0, \infty)$, $\delta(A) := H_{\alpha, \beta}(A, w; f, g)$.

Let $x = (x_i)_{i \in \mathbb{N}}$ and $x = (y_i)_{i \in \mathbb{N}}$ be sequences of complex numbers and $w = (w_i)_{i \in \mathbb{N}}$ a sequence of positive real numbers. Assume that $\alpha, \beta > 1$ with $\frac{1}{\alpha} + \frac{1}{\beta} = 1$.

Let $\Omega = \mathbb{N}$ and $\mathcal{P}_f(\mathbb{N})$ be the algebra of finite parts of natural numbers $\mathbb{N}$. By the monotonicity property of supermeasure on $\mathcal{P}_f(\mathbb{N})$ we have from the above results that the sequence

$$H_{n,\alpha,\beta,p}\left(w;x,y\right) := \left[\left(\sum_{i=0}^{n} w_i \left|x_i\right|^{\alpha}\right)^{1/\alpha} \left(\sum_{i=0}^{n} w_i \left|y_i\right|^{\beta}\right)^{1/\beta} - \left|\sum_{i=0}^{n} w_i x_i y_i\right|\right]$$

$$\times \left(\sum_{i=0}^{n} w_i\right)^{1-\frac{1}{p}},$$

(6.28)

is monotonic nondecreasing and

$$H_{n,\alpha,\beta,p}\left(w;x,y\right)$$

(6.29)

$$\geq \max_{0 \leq i \neq j \leq n} \left\{ \left[\left(w_i \left|x_i\right|^{\alpha} + w_j \left|x_j\right|^{\alpha}\right)^{1/\alpha} \left(w_i \left|y_i\right|^{\beta} + w_j \left|y_j\right|^{\beta}\right)^{1/\beta} \right.\right.$$

$$\left.\left. - \left|w_i x_i y_i + w_j x_j y_j\right|\right] \left(w_i + w_j\right)^{1-\frac{1}{p}}\right\},$$

for any $p \geq 1$.

We also have for $n \geq 1$ that

$$\left[\left(\sum_{i=0}^{2n} w_i \left|x_i\right|^{\alpha}\right)^{1/\alpha} \left(\sum_{i=0}^{2n} w_i \left|y_i\right|^{\beta}\right)^{1/\beta} - \left|\sum_{i=0}^{2n} w_i x_i y_i\right|\right]^{q} \left(\sum_{i=0}^{2n} w_i\right)^{q\left(1-\frac{1}{p}\right)}$$

$$\geq \left[\left(\sum_{i=0}^{n} w_{2i} \left|x_{2i}\right|^{\alpha}\right)^{1/\alpha} \left(\sum_{i=0}^{n} w_{2i} \left|y_{2i}\right|^{\beta}\right)^{1/\beta} - \left|\sum_{i=0}^{n} w_{2i} x_{2i} y_{2i}\right|\right]^{q}$$

$$\times \left(\sum_{i=0}^{n} w_{2i}\right)^{q\left(1-\frac{1}{p}\right)}$$

$$+ \left[\left(\sum_{i=0}^{n-1} w_{2i+1} \left|x_{2i+1}\right|^{\alpha}\right)^{1/\alpha} \left(\sum_{i=0}^{n-1} w_{2i+1} \left|y_{2i+1}\right|^{\beta}\right)^{1/\beta} - \left|\sum_{i=0}^{n-1} w_{2i+1} x_{2i+1} y_{2i+1}\right|\right]^{q}$$

$$\times \left(\sum_{i=0}^{n-1} w_{2i+1}\right)^{q\left(1-\frac{1}{p}\right)}$$

(6.30)

for any $p, q \geq 1$.

## 6.5 Applications for Minkowski's inequality

Consider the $r$-Lebesgue space

$$L_w^r(\Omega, \nu) := \{f : \Omega \to \mathbb{C}, \ f \text{ is } \nu\text{-measurable and } \int_\Omega w \left|f\right|^r d\nu < \infty\},$$

for $r \geq 1$.

The following inequality is well known in the literature as *Minkowski's inequality*

$$\left(\int_\Omega w \left|f + g\right|^r d\nu\right)^{1/r} \leq \left(\int_\Omega w \left|f\right|^r d\nu\right)^{1/r} + \left(\int_\Omega w \left|g\right|^r d\nu\right)^{1/r} \tag{6.31}$$

for any $f, g \in L_w^r(\Omega, \nu)$.

Consider the functional $M_r(\cdot, w; f, g) : \mathcal{A}_\nu \to [0, \infty)$ defined by

$$M_r(A, w; f, g) := \left[\left(\int_A w \left|f\right|^r d\nu\right)^{1/r} + \left(\int_A w \left|g\right|^r d\nu\right)^{1/r}\right]^r - \int_A w \left|f + g\right|^r d\nu. \tag{6.32}$$

We use the following result, see Dragomir (2015):

**Lemma 6.5.1.** *Let* $f, g \in L_w^r(\Omega, \nu)$ *for* $r \geq 1$. *Then the functional* $M_r(\cdot, w; f, g) : \mathcal{A}_\nu \to [0, \infty)$ *defined by (6.32) is a supermeasure.*

For some Minkowski's inequality related functionals and their properties see Agarwal and Dragomir (1998 ), Dragomir (2009) and Everitt (1961).

For $p, q, r \geq 1$ consider the two parameter functional $M_{p,q,r}(\cdot, w; \Phi, f) : \mathcal{A}_\nu \to [0, \infty)$,

$$M_{p,q,r}(A, w; f, g) = M_r^q(A, w; f, g) \left(\int_A w d\nu\right)^{q\left(1 - \frac{1}{p}\right)}. \tag{6.33}$$

Observe that

$$M_{p,q,r}(A, w; f, g) \tag{6.34}$$

$$= \left(\left[\left(\frac{\int_\Omega w \left|f\right|^r d\nu}{\int_A w d\nu}\right)^{1/r} + \left(\frac{\int_\Omega w \left|g\right|^r d\nu}{\int_A w d\nu}\right)^{1/r}\right]^r - \frac{\int_\Omega w \left|f + g\right|^r d\nu}{\int_A w d\nu}\right)^q$$

$$\times \left(\int_A w d\nu\right)^{q\left(2 - \frac{1}{p}\right)}$$

In particular, for $q = p$ we have

$$M_{p,r}(A, w; f, g) \tag{6.35}$$

$$= \left( \left[ \left( \frac{\int_A w \, |f|^r \, d\nu}{\int_A w \, d\nu} \right)^{1/r} + \left( \frac{\int_A w \, |g|^r \, d\nu}{\int_A w \, d\nu} \right)^{1/r} \right]^r - \frac{\int_A w \, |f + g|^r \, d\nu}{\int_A w \, d\nu} \right)^p$$

$$\times \left( \int_A w \, d\nu \right)^{2p-1}.$$

**Theorem 6.5.2.** *Let $f, g \in L^r_w(\Omega, \nu)$ for $r \geq 1$ and $p, q \geq 1$. Then the functional $M_{p,q,r}(\cdot, w; f, g) : \mathcal{A}_\nu \to [0, \infty)$ defined by (6.33) is a supermeasure.*

The proof follows by Corollary 6.2.2 for $p, q \geq 1$.

Let $x = (x_i)_{i \in \mathbb{N}}$ and $x = (y_i)_{i \in \mathbb{N}}$ be sequences of complex numbers and $w = (w_i)_{i \in \mathbb{N}}$ a sequence of positive real numbers. Let $\Omega = \mathbb{N}$ and $\mathcal{P}_f(\mathbb{N})$ be the algebra of finite parts of natural numbers $\mathbb{N}$. By the monotonicity property of supermeasure on $\mathcal{P}_f(\mathbb{N})$ we have from the above results that the sequence

$$M_{n,p,r}(w; x, y) \tag{6.36}$$

$$:= \left( \left[ \left( \sum_{i=0}^n w_i \, |x_i|^r \right)^{1/r} + \left( \sum_{i=0}^n w_i \, |y_i|^r \right)^{1/r} \right]^r - \sum_{i=0}^n w_i \, |x_i + y_i|^r \right)$$

$$\times \left( \sum_{i=0}^n w_i \right)^{1 - \frac{1}{p}}$$

is monotonic nondecreasing and

$$M_{n,p,r}(w; x, y) \tag{6.37}$$

$$\geq \max_{0 \leq i \neq j \leq n} \left\{ \left( \left[ (w_i \, |x_i|^r + w_j \, |x_j|^r)^{1/r} + (w_i \, |y_i|^r + w_j \, |y_j|^r)^{1/r} \right]^r \right. \right.$$

$$\left. \left. - w_i \, |x_i + y_i|^r - w_j \, |x_j + y_j|^r \right) (w_i + w_j)^{1 - \frac{1}{p}} \right\}$$

for any $p \geq 1$.

We have the inequality

$$\left(\left[\left(\sum_{i=0}^{2n} w_i \left|x_i\right|^r\right)^{1/r} + \left(\sum_{i=0}^{2n} w_i \left|y_i\right|^r\right)^{1/r}\right]^r - \sum_{i=0}^{2n} w_i \left|x_i + y_i\right|^r\right)$$

$$\times \left(\sum_{i=0}^{2n} w_i\right)^{1-\frac{1}{p}}$$

$$\geq \left(\left[\left(\sum_{i=0}^{n} w_{2i} \left|x_{2i}\right|^r\right)^{1/r} + \left(\sum_{i=0}^{n} w_{2i} \left|y_{2i}\right|^r\right)^{1/r}\right]^r - \sum_{i=0}^{n} w_{2i} \left|x_{2i} + y_{2i}\right|^r\right)$$

$$\times \left(\sum_{i=0}^{n} w_{2i}\right)^{1-\frac{1}{p}}$$

$$+ \left(\left[\left(\sum_{i=0}^{n-1} w_{2i+1} \left|x_{2i+1}\right|^r\right)^{1/r} + \left(\sum_{i=0}^{n-1} w_{2i+1} \left|y_{2i+1}\right|^r\right)^{1/r}\right]^r\right.$$

$$\left. - \sum_{i=0}^{n-1} w_{2i+1} \left|x_{2i+1} + y_{2i+1}\right|^r\right) \times \left(\sum_{i=0}^{n-1} w_{2i+1}\right)^{1-\frac{1}{p}} \qquad (6.38)$$

for any $p \geq 1$.

---

## 6.6   Applications for Cauchy-Bunyakovsky-Schwarz's inequality

Consider the Hilbert space

$$L_w^2\left(\Omega, \nu\right) := \{f : \Omega \to \mathbb{C}, \ f \text{ is } \nu\text{-measurable and } \int_\Omega w \left|f\right|^2 d\nu < \infty\}.$$

The following inequality is well known in the literature as Cauchy-Bunyakovsky-Schwarz's (CBS) inequality

$$\left|\int_\Omega w f g d\nu\right| \leq \left(\int_\Omega w \left|f\right|^2 d\nu\right)^{1/2} \left(\int_\Omega w \left|g\right|^2 d\nu\right)^{1/2} \qquad (6.39)$$

where $f \in L_w^2\left(\Omega, \nu\right), \ g \in L_w^2\left(\Omega, \nu\right)$.

We consider the functional $H\left(\cdot, w; f, g\right) : \mathcal{A}_\nu \to [0, \infty)$ defined by

$$H\left(A, w; f, g\right) = \left(\int_A w \left|f\right|^2 d\nu\right)^{1/2} \left(\int_A w \left|g\right|^2 d\nu\right)^{1/2} - \left|\int_A w f g d\nu\right|. \qquad (6.40)$$

Taking into account that $H(A, w; f, g) = H_{\alpha,\beta}(A, w; f, g)$ for $\alpha = \beta = 2$, see (13.58), we have:

**Lemma 6.6.1.** *Let* $f \in L_w^2(\Omega, \nu)$, $g \in L_w^2(\Omega, \nu)$, *then the functional* $H(\cdot, w; f, g) : \mathcal{A}_\nu \to [0, \infty)$ *defined by (6.40) is a supermeasure.*

and

**Theorem 6.6.2.** *Let* $f \in L_w^2(\Omega, \nu)$, $g \in L_w^2(\Omega, \nu)$, *then the functional* $H_{p,q}(\cdot, w; f, g) : \mathcal{A}_\nu \to [0, \infty)$ *defined by*

$$H_{p,q}(A, w; f, g) \tag{6.41}$$

$$:= \left[ \left( \int_A w |f|^2 \, d\nu \right)^{1/2} \left( \int_A w |g|^2 \, d\nu \right)^{1/2} - \left| \int_A w f g \, d\nu \right| \right]^q$$

$$\times \left( \int_A w \, d\nu \right)^{q\left(1 - \frac{1}{p}\right)}$$

$$= \left[ \left( \frac{\int_A w |f|^2 \, d\nu}{\int_A w \, d\nu} \right)^{1/2} \left( \frac{\int_A w |g|^2 \, d\nu}{\int_A w \, d\nu} \right)^{1/2} - \left| \frac{\int_A w f g \, d\nu}{\int_A w \, d\nu} \right| \right]^q$$

$$\times \left( \int_A w \, d\nu \right)^{q\left(2 - \frac{1}{p}\right)},$$

*with* $p, q \geq 1$, *is a supermeasure.*

Now, consider the functional $L(\cdot, w; f, g) : \mathcal{A}_\nu \to [0, \infty)$ defined by

$$L(A, w; f, g) = \int_A w |f|^2 \, d\nu \int_A w |g|^2 \, d\nu - \left| \int_A w f g \, d\nu \right|^2. \tag{6.42}$$

We need the following result, see Dragomir (2015):

**Lemma 6.6.3.** *Let* $f \in L_w^2(\Omega, \nu)$, $g \in L_w^2(\Omega, \nu)$. *Then for any* $A, B \in \mathcal{A}_\nu$ *with* $A \cap B = \emptyset$ *we have*

$$L(A \cup B, w; f, g) \tag{6.43}$$
$$\geq L(A, w; f, g) + L(B, w; f, g)$$
$$+ \left( \det \begin{bmatrix} \left( \int_A w |f|^2 \, d\nu \right)^{1/2} & \left( \int_A w |g|^2 \, d\nu \right)^{1/2} \\ \left( \int_B w |f|^2 \, d\nu \right)^{1/2} & \left( \int_B w |g|^2 \, d\nu \right)^{1/2} \end{bmatrix} \right)^2.$$

**Corollary 6.6.4.** *Let* $f \in L_w^2(\Omega, \nu)$, $g \in L_w^2(\Omega, \nu)$. *The functional* $L(\cdot, w; f, g) : \mathcal{A}_\nu \to [0, \infty)$ *defined by (6.42) is a supermeasure.*

Consider the functional $L_{p,q}\left(\cdot, w; f, g\right) : \mathcal{A}_\nu \to [0, \infty)$ defined by

$$L_{p,q}\left(A, w; f, g\right) \qquad (6.44)$$

$$:= \left[\int_A w\left|f\right|^2 d\nu \int_A w\left|g\right|^2 d\nu - \left|\int_A wfgd\nu\right|^2\right]^q \left(\int_A wd\nu\right)^{q\left(1-\frac{1}{p}\right)}$$

$$= \left[\frac{\int_A w\left|f\right|^2 d\nu}{\int_A wd\nu} \frac{\int_A w\left|g\right|^2 d\nu}{\int_A wd\nu} - \left|\frac{\int_A wfgd\nu}{\int_A wd\nu}\right|^2\right]^q \left(\int_A wd\nu\right)^{q\left(3-\frac{1}{p}\right)},$$

where $p, q \geq 1$.

We have:

**Theorem 6.6.5.** *Let* $f \in L_w^2\left(\Omega, \nu\right)$, $g \in L_w^2\left(\Omega, \nu\right)$, *then the functional* $L_{p,q}\left(\cdot, w; f, g\right) : \mathcal{A}_\nu \to [0, \infty)$, *with* $p, q \geq 1$, *defined by (6.44) is a supermeasure.*

Let $f \in L_w^2\left(\Omega, \nu\right)$, $g \in L_w^2\left(\Omega, \nu\right)$. We can also consider the functional $Q\left(\cdot, w; f, g\right) : \mathcal{A}_\nu \to [0, \infty)$ defined by

$$Q\left(A, w; f, g\right) = \left[\int_A w\left|f\right|^2 d\nu \int_A w\left|g\right|^2 d\nu - \left|\int_A wfgd\nu\right|^2\right]^{1/2} \qquad (6.45)$$

$$= \sqrt{L\left(A, w; f, g\right)}.$$

We also need the following result, see Dragomir (2015):

**Lemma 6.6.6.** *Let* $f \in L_w^2\left(\Omega, \nu\right)$, $g \in L_w^2\left(\Omega, \nu\right)$, *then the functional* $H\left(\cdot, w; f, g\right) : \mathcal{A}_\nu \to [0, \infty)$ *defined by (6.45) is a supermeasure.*

Consider the functional $Q_{p,q}\left(\cdot, w; f, g\right) : \mathcal{A}_\nu \to [0, \infty)$ defined by

$$Q_{p,q}\left(A, w; f, g\right) \qquad (6.46)$$

$$:= \left[\int_A w\left|f\right|^2 d\nu \int_A w\left|g\right|^2 d\nu - \left|\int_A wfgd\nu\right|^2\right]^{\frac{q}{2}} \left(\int_A wd\nu\right)^{q\left(1-\frac{1}{p}\right)}$$

$$= \left[\frac{\int_A w\left|f\right|^2 d\nu}{\int_A wd\nu} \frac{\int_A w\left|g\right|^2 d\nu}{\int_A wd\nu} - \left|\frac{\int_A wfgd\nu}{\int_A wd\nu}\right|^2\right]^{\frac{q}{2}} \left(\int_A wd\nu\right)^{q\left(2-\frac{1}{p}\right)},$$

$p, q \geq 1$.

We have:

**Theorem 6.6.7.** *Let* $f \in L_w^2\left(\Omega, \nu\right)$, $g \in L_w^2\left(\Omega, \nu\right)$, *then the functional* $Q_{p,q}\left(\cdot, w; f, g\right) : \mathcal{A}_\nu \to [0, \infty)$, *with* $p, q \geq 1$, *defined by (6.46) is a supermeasure.*

For some CBS's inequality related functionals and their properties see Agarwal and Dragomir (1998 ), Dragomir (2009), Dragomir and Mond (1994/1996), Dragomir and Mond (1995) and Dragomir and Mond (1995).

Let $x = (x_i)_{i \in \mathbb{N}}$ and $x = (y_i)_{i \in \mathbb{N}}$ be sequences of complex numbers and $w = (w_i)_{i \in \mathbb{N}}$ a sequence of positive real numbers. Let $\Omega = \mathbb{N}$ and $\mathcal{P}_f(\mathbb{N})$ be the algebra of finite parts of natural numbers $\mathbb{N}$. By the monotonicity property of supermeasure on $\mathcal{P}_f(\mathbb{N})$ we have from the above results that the sequences

$$H_{n,p}(w; x, y) \tag{6.47}$$

$$:= \left[ \left( \sum_{i=0}^{n} w_i |x_i|^2 \right)^{1/2} \left( \sum_{i=0}^{n} w_i |y_i|^2 \right)^{1/2} - \left| \sum_{i=0}^{n} w_i x_i y_i \right| \right] \left( \sum_{i=0}^{n} w_i \right)^{1 - \frac{1}{p}}$$

and

$$L_{n,p}(w; x, y) := \left[ \sum_{i=0}^{n} w_i |x_i|^2 \sum_{i=0}^{n} w_i |y_i|^2 - \left| \sum_{i=0}^{n} w_i x_i y_i \right|^2 \right] \left( \sum_{i=0}^{n} w_i \right)^{1 - \frac{1}{p}}$$

$$\tag{6.48}$$

are monotonic nondecreasing and

$$H_{n,p}(w; x, y)$$

$$\geq \max_{0 \leq i \neq j \leq n} \left\{ \left[ \left( w_i |x_i|^2 + w_j |x_j|^2 \right)^{1/2} \left( w_i |y_i|^2 + w_j |y_j|^2 \right)^{1/2} - |w_i x_i y_i + w_j x_j y_j| \right] \right.$$

$$\left. \times (w_i + w_j)^{1 - \frac{1}{p}} \right\} \tag{6.49}$$

and

$$L_{n,p}(w; x, y)$$

$$\geq \max_{0 \leq i \neq j \leq n} \left\{ \left[ \left( w_i |x_i|^2 + w_j |x_j|^2 \right) \left( w_i |y_i|^2 + w_j |y_j|^2 \right) - |w_i x_i y_i + w_j x_j y_j|^2 \right] \right.$$

$$\left. \times (w_i + w_j)^{1 - \frac{1}{p}} \right\} \tag{6.50}$$

for any $p \geq 1$.

Finally, the sequence

$$Q_{n,p}(w; x, y) := \left[ \sum_{i=0}^{n} w_i |x_i|^2 \sum_{i=0}^{n} w_i |y_i|^2 - \left| \sum_{i=0}^{n} w_i x_i y_i \right|^2 \right]^{1/2} \tag{6.51}$$

$$\times \left( \sum_{i=0}^{n} w_i \right)^{1 - \frac{1}{p}}$$

is also monotonic nondecreasing and we have the bound

$$Q_{n,p}(w;x,y)$$
$$\geq \max_{0\leq i\neq j\leq n}\left\{\left[\left(w_i\left|x_i\right|^2+w_j\left|x_j\right|^2\right)\left(w_i\left|y_i\right|^2+w_j\left|y_j\right|^2\right)-\left|w_ix_iy_i+w_jx_jy_j\right|^2\right]^{1/2}\right.$$
$$\left.\times(w_i+w_j)^{1-\frac{1}{p}}\right\} \quad (6.52)$$

for any $p\geq 1$.

## 6.7   Applications for Čebyšev's inequality

We say that the pair of measurable functions $(f,g)$ are *synchronous* on $\Omega$ if

$$(f(x)-f(y))(g(x)-g(y))\geq 0 \quad (6.53)$$

for $\nu$-a.e. $x,y\in\Omega$. If the inequality reverses in (6.53), the functions are called *asynchronous* on $\Omega$.

If $(f,g)$ are synchronous on $\Omega$ and $f$, $g$, $fg\in L_w(\Omega,\nu)$ then the following inequality, that is known in the literature as *Čebyšev's inequality*, holds

$$\int_\Omega wd\nu\int_\Omega wfgd\nu\geq\int_\Omega wfd\nu\int_\Omega wgd\nu, \quad (6.54)$$

where $w(x)\geq 0$ for $\nu$ -a.e. (almost every) $x\in\Omega$.

We consider the *Čebyšev functional* $C(\cdot,w;f,g):\mathcal{A}_\nu\to[0,\infty)$ defined by

$$C(A,w;f,g):=\int_A wd\nu\int_A wfgd\nu-\int_A wfd\nu\int_A wgd\nu. \quad (6.55)$$

The following result is known in the literature as *Korkine's identity:*

$$C(A,w;f,g)=\frac{1}{2}\int_A\int_A w(x)w(y)(f(x)-f(y))(g(x)-g(y))d\nu(x)d\nu(y).$$
$$(6.56)$$

The proof is obvious by developing the right side of (6.56) and using Fubini's theorem.

We have the following result, see Dragomir (2015):

**Lemma 6.7.1.** *Let $(f,g)$ be synchronous on $\Omega$ and $f$, $g$, $fg\in L_w(\Omega,\nu)$. Then the Čebyšev functional defined by (6.55) is a supermeasure on $\mathcal{A}_\nu$.*

For $p,q\geq 1$ consider the two parameter functional $C_{p,q}(\cdot,w;\Phi,f):\mathcal{A}_\nu\to$

$[0, \infty)$,

$$C_{p,q}(A, w; f, g) \qquad (6.57)$$

$$:= C^q(A, w; f, g)\left(\int_A w d\nu\right)^{q\left(1-\frac{1}{p}\right)}$$

$$= \left(\frac{\int_A wfg d\nu}{\int_A w d\nu} - \frac{\int_A wf d\nu}{\int_A w d\nu} \cdot \frac{\int_A wg d\nu}{\int_A w d\nu}\right)^q \left(\int_A w d\nu\right)^{q\left(3-\frac{1}{p}\right)}.$$

**Theorem 6.7.2.** *Let $(f, g)$ be synchronous on $\Omega$ and $f, g, fg \in L_w(\Omega, \nu)$ and $p, q \geq 1$. Then the functional $C_{p,q}(\cdot, w; f, g) : \mathcal{A}_\nu \to [0, \infty)$ defined by (6.57) is a supermeasure.*

The proof follows by Corollary 6.2.2 for $p, q \geq 1$.

For some Čebyšev's inequality related functionals and their properties see Agarwal and Dragomir (1998) and Dragomir and Mond (1993/94).

Let $x = (x_i)_{i \in \mathbb{N}}$ and $x = (y_i)_{i \in \mathbb{N}}$ be synchronous sequences of real numbers and $w = (w_i)_{i \in \mathbb{N}}$ a sequence of positive real numbers. Let $\Omega = \mathbb{N}$ and $\mathcal{P}_f(\mathbb{N})$ be the algebra of finite parts of natural numbers $\mathbb{N}$. By the monotonicity property of supermeasure on $\mathcal{P}_f(\mathbb{N})$ we have from the above results that the sequence

$$C_{n,p}(w; x, y) := \left(\sum_{i=0}^n w_i \sum_{i=0}^n w_i x_i y_i - \sum_{i=0}^n w_i x_i \sum_{i=0}^n w_i y_i\right)\left(\sum_{i=0}^n w_i\right)^{1-\frac{1}{p}}$$

is monotonic nondecreasing and and we have the bound

$$C_{n,p}(w; x, y) \geq \frac{1}{2} \max_{0 \leq i \neq j \leq n}\left\{w_i w_j (x_i - x_j)(y_i - y_j)(w_i + w_j)^{1-\frac{1}{p}}\right\},$$

for any $p \geq 1$.

---

## 6.8 Applications for Hermite-Hadamard inequalities

Let $I$ be an interval consisting of more than one point and $f : I \to \mathbb{R}$ a convex function. If $a, b \in I$ with $a < b$, then we have the well-known *Hermite-Hadamard inequality*

$$f\left(\frac{a+b}{2}\right) \leq \frac{1}{b-a}\int_a^b f(t)\,dt \leq \frac{f(a) + f(b)}{2}. \qquad (6.58)$$

Suppose $f : I \to \mathbb{R}$ and for $f \in L[a, b]$ define the functionals

$$H([a, b]; f) := \int_a^b f(t)\,dt - (b-a)f\left(\frac{a+b}{2}\right)$$

and

$$L\left([a,b]\,;f\right) := \frac{f\left(a\right) + f\left(b\right)}{2}\left(b - a\right) - \int_{a}^{b} f\left(t\right)dt.$$

We have the following result concerning the properties of these mappings as functions of interval Dragomir and Pearce (2002):

**Lemma 6.8.1.** *Let* $f : I \to \mathbb{R}$ *be a convex function. Then*
*(i) For all* $a, b, c \in I$ *with* $a \le c \le b$, *we have*

$$0 \le H\left([a,c]\,;f\right) + H\left([c,b]\,;f\right) \le H\left([a,b]\,;f\right) \tag{6.59}$$

*and*

$$0 \le L\left([a,c]\,;f\right) + L\left([c,b]\,;f\right) \le L\left([a,b]\,;f\right), \tag{6.60}$$

*i.e. the functionals* $H\left(\cdot;f\right)$ *and* $L\left(\cdot;f\right)$ *are superadditive as functions of interval;*
*(ii) For all* $[c,d] \subseteq [a,b] \subseteq I$, *we have*

$$0 \le H\left([c,d]\,;f\right) \le H\left([a,b]\,;f\right) \tag{6.61}$$

*and*

$$0 \le L\left([c,d]\,;f\right) \le L\left([a,b]\,;f\right), \tag{6.62}$$

*i.e. the functionals* $H\left(\cdot;f\right)$ *and* $L\left(\cdot;f\right)$ *are monotonic nondecreasing as functions of interval.*

For $p, q \ge 1$ consider the two parameter functionals

$$H_{p,q}\left([a,b]\,;f\right) := (b-a)^{q\left(1-\frac{1}{p}\right)} H^{q}\left([a,b]\,;f\right) \tag{6.63}$$

$$= (b-a)^{q\left(1-\frac{1}{p}\right)} \left[\int_{a}^{b} f\left(t\right)dt - (b-a) f\left(\frac{a+b}{2}\right)\right]^{q}$$

and

$$L_{p,q}\left([a,b]\,;f\right) := (b-a)^{q\left(1-\frac{1}{p}\right)} L^{q}\left([a,b]\,;f\right) \tag{6.64}$$

$$= (b-a)^{q\left(1-\frac{1}{p}\right)} \left[\frac{f\left(a\right)+f\left(b\right)}{2}\left(b-a\right) - \int_{a}^{b} f\left(t\right)dt\right]^{q}.$$

**Theorem 6.8.2.** *Let* $f : I \to \mathbb{R}$ *be a convex function. Then for any* $p, q \ge 1$ *we have that the mappings* $H_{p,q}\left(\cdot;f\right)$ *and* $L_{p,q}\left(\cdot;f\right)$ *are superadditive and monotonic nondecreasing as functions of interval.*

The proof follows by Corollary 6.2.2 for $p, q \ge 1$.
For an arbitrary function $f : I \to \mathbb{R}$ we introduce the mapping

$$S\left([a,b]\,;f\right) := (b-a)\left[\frac{f\left(a\right)+f\left(b\right)}{2} - f\left(\frac{a+b}{2}\right)\right],$$

where $a, b \in I$ with $a < b$.
We have Dragomir and Pearce (2002):

**Lemma 6.8.3.** *Let $f : I \to \mathbb{R}$ a convex function. Then the mapping $S(\cdot; f)$ is a superadditive and monotonic nondecreasing function of interval.*

For $p, q \geq 1$ consider the two parameter functional

$$S_{p,q}([a,b]; f) := (b-a)^{q\left(2-\frac{1}{p}\right)} \left[\frac{f(a) + f(b)}{2} - f\left(\frac{a+b}{2}\right)\right]^q.$$

Using Corollary 6.2.2 for $p, q \geq 1$, we then have:

**Theorem 6.8.4.** *Let $f : I \to \mathbb{R}$ be a convex function. Then for any $p, q \geq 1$ we have that the mapping $S_{p,q}(\cdot; f)$ is superadditive and monotonic nondecreasing as function of interval.*

If we use the superadditivity of the functional $H_{p,q}([a,b]; f)$ with $p, q \geq 1$ on the intervals $\left[a, \frac{a+b}{2}\right]$ and $\left[\frac{a+b}{2}, b\right]$ then we get

$$(b-a)^{q\left(1-\frac{1}{p}\right)} \left[\int_a^b f(t)\, dt - (b-a) f\left(\frac{a+b}{2}\right)\right]^q$$

$$\geq \left(\frac{b-a}{2}\right)^{q\left(1-\frac{1}{n}\right)} \left[\int_a^{\frac{a+b}{2}} f(t)\, dt - \frac{b-a}{2} f\left(\frac{3a+b}{4}\right)\right]^q$$

$$+ \left(\frac{b-a}{2}\right)^{q\left(1-\frac{1}{p}\right)} \left[\int_{\frac{a+b}{2}}^b f(t)\, dt - \frac{b-a}{2} f\left(\frac{a+3b}{4}\right)\right]^q,$$

which is equivalent to

$$\left[\int_a^b f(t)\, dt - (b-a) f\left(\frac{a+b}{2}\right)\right]^q$$

$$\geq \frac{1}{2^{q\left(1-\frac{1}{p}\right)}} \left\{\left[\int_a^{\frac{a+b}{2}} f(t)\, dt - \frac{b-a}{2} f\left(\frac{3a+b}{4}\right)\right]^q \right.$$

$$\left. + \left[\int_{\frac{a+b}{2}}^b f(t)\, dt - \frac{b-a}{2} f\left(\frac{a+3b}{4}\right)\right]^q\right\} \quad (6.65)$$

for $p, q \geq 1$.

Similarly, we have

$$\left[\frac{f(a) + f(b)}{2}(b-a) - \int_a^b f(t)\, dt\right]^q$$

$$\geq \frac{1}{2^{q\left(1-\frac{1}{p}\right)}} \left\{\left[\frac{f(a) + f\left(\frac{a+b}{2}\right)}{4}(b-a) - \int_a^{\frac{a+b}{2}} f(t)\, dt\right]^q\right.$$

$$\left. + \left[\frac{f\left(\frac{a+b}{2}\right) + f(b)}{4}(b-a) - \int_{\frac{a+b}{2}}^b f(t)\, dt\right]^q\right\} \quad (6.66)$$

and

$$\left[\frac{f(a)+f(b)}{2}-f\left(\frac{a+b}{2}\right)\right]^q$$

$$\geq \frac{1}{2^{q\left(2-\frac{1}{p}\right)}}\left[\frac{f(a)+f\left(\frac{a+b}{2}\right)}{2}-f\left(\frac{3a+b}{4}\right)\right]^q$$

$$+\left[\frac{f\left(\frac{a+b}{2}\right)+f(b)}{2}-f\left(\frac{a+3b}{4}\right)\right]^q \quad (6.67)$$

for $p, q \geq 1$.

By the monotonicity property we also have the inequalities:

$$(b-a)^{1-\frac{1}{p}}\left[\int_a^b f(t)\,dt-(b-a)f\left(\frac{a+b}{2}\right)\right] \quad (6.68)$$

$$\geq (d-c)^{1-\frac{1}{p}}\left[\int_c^d f(t)\,dt-(d-c)f\left(\frac{c+d}{2}\right)\right]$$

$$(b-a)^{1-\frac{1}{p}}\left[\frac{f(a)+f(b)}{2}(b-a)-\int_a^b f(t)\,dt\right] \quad (6.69)$$

$$\geq (d-c)^{1-\frac{1}{p}}\left[\frac{f(c)+f(d)}{2}(d-c)-\int_c^d f(t)\,dt\right]$$

and

$$(b-a)^{2-\frac{1}{p}}\left[\frac{f(a)+f(b)}{2}-f\left(\frac{a+b}{2}\right)\right] \quad (6.70)$$

$$\geq (d-c)^{2-\frac{1}{p}}\left[\frac{f(c)+f(d)}{2}-f\left(\frac{c+d}{2}\right)\right]$$

for any $[c, d] \subset [a, b]$.

---

## 6.9   Applications for convex functions

Consider a convex function $f : I \subset \mathbb{R} \to \mathbb{R}$ defined on the interval $I$ of the real line $\mathbb{R}$ and two distinct elements $a, b \in I$ with $a < b$. We denote by $[a, b]$ the closed segment defined by $\{(1 - t)a + tb, t \in [0, 1]\}$. We also define the functional of interval

$$C_t([a, b]; f) := (1 - t)f(a) + tf(b) - f((1 - t)a + tb) \geq 0 \quad (6.71)$$

where $a, b \in I$ with $a < b$ and $t \in [0, 1]$ is fixed.

We have Dragomir (2015) and Dragomir (2012):

**Lemma 6.9.1.** *Let $f : I \to \mathbb{R}$ be a convex function and $t \in (0, 1)$. Then the mapping $C_t (\cdot; f)$ is a superadditive and monotonic nondecreasing function of interval.*

For $t = \frac{1}{2}$ we consider the functional

$$C([a, b] ; f) := C_{\frac{1}{2}}([a, b] ; f) = \frac{f(a) + f(b)}{2} - f\left(\frac{a + b}{2}\right).$$

**Corollary 6.9.2.** *Let $f : I \to \mathbb{R}$ be a convex function. Then the mapping $C(\cdot; f)$ is a superadditive and monotonic nondecreasing function of interval.*

We can also define the symmetric functional

$$
\begin{aligned}
D_t([a, b] ; f) \quad : \quad &= \frac{1}{2}\left[C_t([a, b] ; f) + C_{1-t}([a, b] ; f)\right] \\
&= \frac{f(a) + f(b)}{2} - \frac{1}{2}\left[f((1 - t) a + tb) + f((1 - t) b + ta)\right],
\end{aligned}
$$

where $t \in [0, 1]$ is fixed.

**Corollary 6.9.3.** *Let $f : I \to \mathbb{R}$ be a convex function and $t \in (0, 1)$. Then the mapping $D_t (\cdot; f)$ is a superadditive and monotonic nondecreasing function of interval.*

Perhaps the most interesting functional we can consider from the above is the following one:

$$
\begin{aligned}
\Theta([a, b] ; f) &:= \frac{f(a) + f(b)}{2} - \int_0^1 f((1 - t) a + tb) \, dt \qquad (6.72) \\
&= \frac{f(a) + f(b)}{2} - \frac{1}{b - a}\int_a^b f(s) \, ds \geq 0,
\end{aligned}
$$

which is related to the second Hermite-Hadamard inequality.

We observe that

$$\Theta([a, b] ; f) = \int_0^1 C_t([a, b] ; f) \, dt. \qquad (6.73)$$

**Corollary 6.0.4.** *Let $f . I \to \mathbb{R}$ be a convex function. Then the mapping $\Theta(\cdot; f)$ is a superadditive and monotonic nondecreasing function of interval.*

We can define the associated two parameter $p, q \geq 1$ functionals

$$
\begin{aligned}
C_{p,q,t}([a, b] ; f) &:= (b - a)^{q\left(1 - \frac{1}{p}\right)}\left[C_t([a, b] ; f)\right]^q \\
&= (b - a)^{q\left(1 - \frac{1}{p}\right)}\left[(1 - t) f(a) + t f(b) - f((1 - t) a + tb)\right]^q,
\end{aligned}
$$

$$C_{p,q}\left([a,b];f\right) := (b-a)^{q\left(1-\frac{1}{p}\right)} \left[C\left([a,b];f\right)\right]^q$$

$$= (b-a)^{q\left(1-\frac{1}{p}\right)} \left[\frac{f\left(a\right)+f\left(b\right)}{2} - f\left(\frac{a+b}{2}\right)\right]^q,$$

$$D_t\left([a,b];f\right) := \frac{1}{2^q}(b-a)^{q\left(1-\frac{1}{p}\right)}\left[C_t\left([a,b];f\right)+C_{1-t}\left([a,b];f\right)\right]^q$$

$$= (b-a)^{q\left(1-\frac{1}{p}\right)}$$

$$\times \left[\frac{f\left(a\right)+f\left(b\right)}{2} - \frac{1}{2}\left[f\left(\left(1-t\right)a+tb\right)+f\left(\left(1-t\right)b+ta\right)\right]\right]^q$$

and

$$\Theta_{p,q}\left([a,b];f\right) := (b-a)^{q\left(1-\frac{1}{p}\right)} \left[\Theta\left([a,b];f\right)\right]^q$$

$$= (b-a)^{q\left(1-\frac{1}{p}\right)} \left[\frac{f\left(a\right)+f\left(b\right)}{2} - \frac{1}{b-a}\int_a^b f\left(s\right)ds\right]^q.$$

Using Corollary 6.2.2 for $p,q \geq 1$, we then have:

**Theorem 6.9.5.** *Let* $f : I \to \mathbb{R}$ *be a convex function and* $t \in (0,1)$. *Then for any* $p,q \geq 1$ *we have that the mappings* $C_{p,q,t}\left(\cdot;f\right)$, $C_{p,q}\left(\cdot;f\right)$, $D_t\left(\cdot;f\right)$ *and* $\Theta_{p,q}\left(\cdot;f\right)$ *are superadditive and monotonic nondecreasing as functions of interval.*

Various inequalities similar with the ones from the previous section can be stated, however we do not present them here.

---

## Bibliography

Abramovich S. Convexity, subadditivity and generalized Jensen's inequality. *Ann. Funct. Anal.*, (4):183–194, 2013.

Agarwal R. P. and Dragomir S. S. The property of supermultiplicity for some classical inequalities and applications. *Comput. Math. Appl.*, (35):105–118, 1998.

Aldaz J. M. A measure-theoretic version of the Dragomir-Jensen inequality. *Proc. Amer. Math. Soc.*, (144):2391–2399, 2012.

Barić M., Matić J., and Pečarić J. On the bounds for the normalized Jensen functional and Jensen-Steffensen inequality. *Math. Inequal. Appl.*, (12):413–432, 2009.

Dragomir S. S., Pečarić J., and Persson L. E. Properties of some functionals related to Jensen's inequality. *Acta Math. Hungar.*, (70):129–143, 1996.

Dragomir S. S. Bounds for the normalised Jensen functional. *Bull. Aust. Math. Soc.*, (74):471–478, 2006.

Dragomir S. S. Some properties of quasilinearity and monotonicity for Hölder's and Minkowski's inequalities. *Tamkang J. Math.*, (26):21–24, 2009.

Dragomir S. S. Superadditivity of some functionals associated with Jensen's inequality for convex functions on linear spaces with applications. *Bull. Aust. Math. Soc.*, (82):44–61, 2010.

Dragomir S. S. Superadditivity and monotonicity of some functionals associated with the Hermite-Hadamard inequality for convex functions in linear spaces. *Rocky Mountain J. Math.*, (42):1447–1459, 2012.

Dragomir S. S. Supermeasures associated to some classical inequalities. *RGMIA Res. Rep. Coll.*, (18):1–21, 2015.

Dragomir S. S. and Milošević D. M. Two mappings in connection to Jensen's inequality. *Math. Balkanica*, (9):3–9, 1995.

Dragomir S. S. and Mond B. Some mappings associated with Cebysev's inequality for sequences of real numbers. *Bull. Allahabad Math. Soc.*, (8/9):37–55, 1993/94.

Dragomir S. S. and Mond B. On the superadditivity and monotonicity of Schwarz's inequality in inner product spaces. *Makedon. Akad. Nauk. Umet. Oddel. Mat.-Tehn. Nauk. Prilozi*, (15):5–22, 1994/1996.

Dragomir S. S. and Mond B. Some inequalities for Fourier coefficients in inner product spaces. *Period. Math. Hungar.*, (31):167–182, 1995.

Dragomir S. S. and Mond B. Some mappings associated with Cauchy-Buniakowski-Schwarz's inequality in inner product spaces. *Soochow J. Math.*, (21):413–426, 1995.

Dragomir S. S. and Pearce C. E. M. Quasilinearity and Hadamard's inequality. *Math. Inequal. Appl.*, (5):463–471, 2002.

Dragomir S. S., Cho Y. J., and Kim J. K. Subadditivity of some functionals associated to Jensen's inequality with applications. *Taiwanese J. Math.*, (15):1815–1828, 2011.

Everitt W. N. Properties of some functionals related to Jensen's inequality. *J. London Math. Soc.*, (36):145–158, 1961.

# Chapter 7

## Generalized Double Statistical Weighted Summability and Its Application to Korovkin Type Approximation Theorem

**Ugur Kadak**

*Department of Mathematics, Gazi University, Ankara, Turkey*

**Hemen Dutta**

*Department of Mathematics, Gauhati University, Guwahati, India*

## 7.1 Introduction

At the end of the nineteenth century, many researchers focused their attention on various alternative methods based on the theory of infinite series. These summation methods were combined under a single heading which is called *Summability Methods* and have been used efficiently in classical and modern mathematics, in particular, operator theory, probability theory, the theory of orthogonal series, the rate of convergence, analytic continuation, hypergeometric series, approximation theory, and fixed point theory. Further, it is worth pointing out that summability methods help us to generalize the limit of a sequence or series, and so provide us a technique to assign limits even to divergent sequences. With the rapid development of sequence spaces, the concept of statistical convergence, which was independently introduced by Fast Fast (1951) and Steinhaus Steinhaus (1951) in the same year 1951, attracts great attention of researchers. In 1935, this idea was introduced by Zygmund Zygmund (1959) under a different name as *almost convergence*.

During the last decades, the concepts of statistical convergence and statistical summability have been generalized, extended, and revisited in many directions. To focus on some extended versions, we refer the reader to the more recent results, e.g., $\lambda$-statistical convergence Mursaleen (2000), lacunary statistical convergence Et and Şengül (2014); Fridy and Orhan (1993), weighted statistical convergence Karakaya and Chishti (2009), $A$-statistical convergence Connor (1989); Mursaleen et al. (2012), weighted $A$-statistical convergence and weighted $A$-statistical summability Kolk (1993). The reader can also see Demirci and Kolay (2017); Kadak (2017b,c); Kadak et al. (2017a); Mohiuddine (2016), for more recent developments, such as the weighted statistical convergence based on generalized difference operator involving $(p, q)$-gamma function, statistical weighted $B$-summability and the relative weighted summability in modular function spaces.

Due to the rapid development of quantum calculus ($q$-calculus), many authors have focused on the concept of post-quantum calculus (or $(p, q)$-calculus) which has been used efficiently in many fields of science, in particular, approximation theory, differential equations, Lie groups, field theory, probability, oscillator algebra, hypergeometric series, and physical sciences (see, e.g., Burban and Klimyk (1994); Hounkonnou and Kyemba (2013); Sahai and Yadav (2007)). In recent years, the well-known linear positive operators have been transferred to the $(p, q)$-calculus and these analogues have been intensively investigated. It is caused both by the rapid development of applied analysis and by using such extended operators as efficient tools for the study of well-known positive linear positive operators. In the year 2015, Mursaleen et al. Mursaleen et al. (2015) introduced the $(p, q)$-analogue of Bernstein operator and studied the convergence behavior of the proposed operator. Very recently, Khan et al. Khan and Lobiyal (2017) have constructed Lupaş $(p, q)$-Bernstein operators and studied Bezier curves and surfaces involving $(p, q)$-integers. For more various approaches to $q$-calculus and $(p, q)$-calculus, we refer to Cai (2017); Hana et al. (2014); Kadak et al. (2017b); Kanat and Sofyalıoğlu (2018); Malik and Gupta (2017); Mursaleen et al. (2016).

In recent years there have been significant developments in the theory of infinite series as well as further developments in the sequence spaces. The concept of difference sequence space, which was firstly introduced by Kızmaz Kızmaz (1981), can be regarded as an extended version of the classical sequence spaces. Further this concept was examined and extended by several authors (see, for instance, Başar and Aydın (2004); Et and Nuray (2001); Kadak (2017a); Kadak and Baliarsingh (2015); Kirişci and Kadak (2017); Mursaleen and Alotaibi (2013)). Recently, Baliarsingh Baliarsingh (2013) has introduced certain difference sequence spaces based on fractional difference operator involving Euler gamma function. In the year 2016, Kadak Kadak (2016) introduced the weighted statistical convergence based on $(p, q)$-integers.

This chapter is devoted to extend the notions of weighted statistical summability and weighted statistical convergence which are largely used to obtain interesting applications to approximation theorems. First, we review some results on difference operators of double sequences, $(p, q)$-integers, $\alpha\beta$-statistical convergence, weighted statistical convergence and weighted statistical summability. We define a double difference operator with two integer orders via $(p, q)$-integers. We also introduce the concepts statistically weighted $\mathcal{N}$-summability, weighted $\Delta_{p,q}^{[n_1,n_2]}$-statistical convergence and weighted strongly $[\mathcal{N}]_r$-summability for $0 < r < \infty$ and, study some inclusion relations between newly proposed methods. Moreover, we establish a Korovkin type approximation theorem related to the statistically weighted $\mathcal{N}$-summability which will improve the existing methods. By using $(p, q)$-analogue of generalized bivariate Bleimann-Butzer-Hahn operators, we give an example which shows that proposed methods successfully work.

---

## 7.2   Notation and preliminaries

The aim of this section is to collect some auxiliary concepts and results which will be needed to gain a good understanding of the whole chapter.

### 7.2.1   Weighted statistical convergence

We begin by recalling some concepts concerning statistical convergence and its extended versions for single and double sequences.

By $\omega$, we denote the family of all real (or complex) valued sequences and each linear subspace of $\omega$ (with the included addition and scalar multiplication) is called a sequence space. It is well known that the spaces $\ell_\infty, c$ and $c_0$ are the subspaces of $\omega$ and normed by $\|x\|_\infty = \sup_{k \in \mathbb{N}} |x_k|$. With this supremum norm, it is proved that these spaces are all complete normed linear spaces.

Let $K \subset \mathbb{N}_0 := \mathbb{N} \cup \{0\}$ and

$$K_n = \{k : k \leqq n \quad \text{and} \quad k \in K\}.$$

Then, the number $\delta(K)$ defined by

$$\delta(K) = \lim_{n \to \infty} \frac{1}{n} |K_n|,$$

is called the *natural density* of $K$ (see Fast (1951)), provided that the limit exists. Here, and in what follows, $|K_n|$ denotes the cardinality of set $K_n$. A sequence $x = (x_k)$ is called statistically convergent to the number $L$, denoted by $st$-$\lim x_k = L$, if, for each $\epsilon > 0$, the set

$$K_\epsilon = \{k : k \in \mathbb{N} \quad \text{and} \quad |x_k - L| \geqq \epsilon\}$$

has natural density zero. Equivalently, we write

$$\lim_{n \to \infty} \frac{1}{n} \left| \{ k : k \leq n \quad \text{and} \quad |x_k - L| \geq \epsilon \} \right| = 0.$$

The concept of weighted statistical convergence has been investigated by many mathematicians, see Karakaya and Chishti (2009); Mursaleen et al. (2012), and they have several applications in different subjects of approximation theory, see Belen and Mohiuddine (2013); Kadak (2017b,c); Kadak et al. (2017a); Mohiuddine (2016); Savaş (2000). More recently, this notion was modified by Srivastava et al. Srivastava et al. (2012) and was further extended by Kadak Kadak (2016, 2017a).

Let $s = (s_k)_{k \in \mathbb{N}}$ be a sequence of nonnegative real numbers such that $s_0 > 0$ and $S_n = \sum_{k=0}^{n} s_k \to \infty$ as $n \to \infty$. We say that a sequence $x = (x_k)$ is weighted statistically convergent (or $S_N$-convergent) to the number $L$ if, for every $\epsilon > 0$,

$$\lim_{n \to \infty} \frac{1}{S_n} \left| \{ k \leq S_n : s_k |x_k - L| \geq \epsilon \} \right| = 0.$$

Very recently, Aktuğlu Aktuğlu (2014) has introduced the notion of $(\alpha, \beta)$-statistical convergence for single sequences with the help of two sequences $\{\alpha(n)\}_{n \in \mathbb{N}}$ and $\{\beta(n)\}_{n \in \mathbb{N}}$ of positive numbers which satisy the following conditions:

(i) $\alpha(n)$ and $\beta(n)$ are both non-decreasing sequences;

(ii) $\beta(n) \geq \alpha(n)$;

(iii) $\beta(n) - \alpha(n) \longrightarrow \infty$ as $n \longrightarrow \infty$.

In what follows, we denote by $\Lambda$ the set of pairs $(\alpha, \beta)$ satisfying the conditions (i) to (iii). A sequence $x = (x_k)$ is said to be $(\alpha, \beta)$-statistically convergent to $L$ if, for each $\varepsilon > 0$,

$$\lim_{n \to \infty} \frac{1}{\beta(n) - \alpha(n) + 1} \left| \left\{ k \in S_n^{(\alpha, \beta)} : |x_k - L| \geq \epsilon \right\} \right| = 0,$$

where $S_n^{(\alpha, \beta)} = [\alpha(n), \beta(n)]$ is a closed interval for $(\alpha, \beta) \in \Lambda$.

In 1900, Pringsheim Pringsheim (1900) introduced the notion of convergence for double sequences. A double sequence $(x_{m,n})_{m,n \in \mathbb{N}}$ is said to be convergent to the number $L$, denoted by $P - \lim x_{m,n} = L$, provided that given $\epsilon > 0$ there exists $n_0 \in \mathbb{N}$ such that $|x_{m,n} - L| < \epsilon$ whenever $m, n \geq n_0$. We shall describe such a convergent double sequence $(x_{m,n})$ more briefly as $P$-convergent. A double sequence $(x_{m,n})$ is called bounded if there exists a positive number $M$ such that $|x_{m,n}| \leq M$ for all $m, n \in \mathbb{N}$. We remark that in contrast to the case for single sequences, a convergent double sequence need not to be bounded. In 2003, the above-mentioned results were extended to statistical convergence by Mursaleen and Edely Mursaleen and Edely (2003).

Let $H^2 \subset \mathbb{N}^2 := \mathbb{N} \times \mathbb{N}$ be a two-dimensional set of positive integers, and let

$$H^2(m,n) = \{(j,k) \in \mathbb{N}^2 : j \leq m \text{ and } k \leq n\}.$$

We say that $H^2$ has a double natural density $\delta_2(H^2)$ defined by

$$\delta_2(H^2) = P - \lim_{m,n \to \infty} \frac{1}{mn} |H^2(m,n)|,$$

if it exists. A double sequence $x = (x_{m,n})$ of real numbers is said to be statistically convergent to $L$ in the Pringsheim sense, if for every $\epsilon > 0$,

$$\delta_2(H_\epsilon^2(m,n)) = 0$$

where

$$H_\epsilon^2(m,n) := \{(j,k), j \leq m \text{ and } k \leq n : |x_{j,k} - L| \geq \epsilon\}.$$

We denote it by $\text{st}_2\text{-}\lim x_{m,n} = L$. Note that every $P$-convergent double sequence is statistically convergent to the same limit point but not conversely. For further related results and generalizations, see, for instance Adams (1932), Altay and Başar (2005); Çakan and Altay (2006); Chen and Chang (2007); Hill (1940).

## 7.2.2 Difference operator of order $m$ based on $(p,q)$-integers

We begin by recalling the following notations and preliminary facts which are used throughout this chapter.

For any $p > 0$ and $q > 0$, the $(p,q)$-integers $[n]_{p,q}$ are defined by

$$
\begin{aligned}
[n]_{p,q} &:= p^{n-1} + p^{n-2}q + p^{n-3}q^2 + \cdots + pq^{n-2} + q^{n-1} \\
&= \begin{cases} \frac{p^n - q^n}{p-q} & (p \neq q \neq 1) \\ np^{n-1} & (p = q \neq 1) \\ [n]_q & (p = 1) \\ n & (p = q = 1) \end{cases},
\end{aligned}
$$

where $[n]_q$ denotes the $q$-integers and $n = 0, 1, 2, \ldots$. By simple calculations we can find that, $[n]_{p,q} = p^{n-1}[n]_{\frac{q}{p}}$.

The $(p,q)$-factorial is defined by

$$[0]_{p,q}! := 1 \quad \text{and} \quad [n]!_{p,q} = [1]_{p,q}[2]_{p,q} \cdots [n]_{p,q} \quad \text{if} \ n \geq 1$$

The $(p,q)$-binomial coefficient satisfies

$$\begin{bmatrix} n \\ r \end{bmatrix}_{p,q} = \frac{[n]_{p,q}!}{[r]_{p,q}! \, [n-r]_{p,q}!}, \qquad 0 \leq r \leq n.$$

Also, the formula for $(p,q)$-binomial expansion is as follows:

$$(ax + by)_{p,q}^n = \sum_{k=0}^{n} p^{\frac{(n-k)(n-k-1)}{2}} q^{\frac{k(k-1)}{2}} \begin{bmatrix} n \\ k \end{bmatrix}_{p,q} a^{n-k} b^k x^{n-k} y^k,$$

$$(x + y)_{p,q}^n = (x + y)(px + qy)(p^2 x + q^2 y) \cdots (p^{n-1}x + q^{n-1}y),$$

$$(x - y)_{p,q}^n = (x - y)(px - qy)(p^2 x - q^2 y) \cdots (p^{n-1}x - q^{n-1}y).$$

As pointed out in the preceding section, the concept of difference sequence space was introduced by Kızmaz Kı zmaz (1981) in 1981. Further, Et and Çolak Et and Çolak (1995) extended the difference sequence spaces by defining

$$\lambda(\Delta^m) = \{x = (x_k) : \Delta^m(x) \in \lambda\}$$

for $\lambda \in \{\ell_\infty, c, c_0\}$, where $m \in \mathbb{N}$, $\Delta^0 x = (x_k)$, $\Delta^m x = (\Delta^{m-1}x_k - \Delta^{m-1}x_{k+1})$ and

$$\Delta^m x_k = \sum_{i=0}^{m} (-1)^i \binom{m}{i} x_{k+i}.$$

Recently, Baliarsingh Baliarsingh (2013) has introduced certain difference sequence spaces based on the fractional difference operator $\Delta^{(r)}$ involving Euler gamma function.

For a positive proper fraction $r$, the fractional difference sequence $\Delta^{(r)}x$ of order $r$ was defined as

$$\Delta^{(r)}(x_k) = \sum_{i=0}^{\infty} (-1)^i \frac{\Gamma(r + 1)}{i!\Gamma(r - i + 1)} x_{k-i}, \quad (k \in \mathbb{N}_0).$$

This difference operator can be given by a triangle (see Baliarsingh (2013); Kirişci and Kadak (2017)):

$$\Delta_{nk}^{(r)} = \begin{cases} (-1)^{n-k} \frac{\Gamma(r+1)}{(n-k)!\Gamma(r-n+k+1)} & , \quad (0 \le k \le n), \\ 0 & , \quad (k > n). \end{cases}$$

In addition, Kirişci and Kadak Kirişci and Kadak (2017) have defined the sequence spaces $fdf$ and $fdf_0$ as the set of all sequences whose $\Delta^{(r)}$−transforms are in the spaces almost convergent sequence and almost null sequence spaces, respectively:

$$fdf = \left\{ x = (x_k) \in \omega : \exists \mathcal{L} \in \mathbb{C} \ni \lim_{m \to \infty} t_{mn}\left(\Delta^{(r)}x\right) = \mathcal{L} \quad \text{uniformly in } n \right\}$$

and

$$fdf_0 = \left\{ x = (x_k) \in \omega : \lim_{m \to \infty} t_{mn}\left(\Delta^{(r)}x\right) = 0 \quad \text{uniformly in } n \right\}$$

where

$$t_{mn}(\Delta^{(r)}x) = \frac{1}{m+1} \sum_{k=0}^{m} \sum_{j=0}^{n+k} \left[ \sum_{i=0}^{k-j} (-1)^i \frac{\Gamma(r+1)}{i!\Gamma(r+1-i)} \right] x_j,$$

for all $m, n \in \mathbb{N}$.

The following new generalizations of difference sequences of natural order $m$ with respect to $(p, q)$-integers were introduced by Kadak in Kadak (2016).

**Definition 7.2.1.** Kadak (2016) Let $0 < q < p \leq 1$ and $m \in \mathbb{N}$. Then, the operator $\Delta_{p,q}^{[m]} : \omega \to \omega$ is defined by

$$\Delta_{p,q}^{[0]}(x_k) = x_k, \quad \Delta_{p,q}^{[1]}(x_k) = x_k - x_{k-1}, \quad \Delta_{p,q}^{[2]}(x_k) = x_k - [2]_{p,q} x_{k-1} + x_{k-2}$$

and

$$\Delta_{p,q}^{[m]}(x_k) = \sum_{i=0}^{m} (-1)^i \begin{bmatrix} m \\ i \end{bmatrix}_{p,q} x_{k-i}.$$

**Definition 7.2.2.** Kadak (2016) Let $(\alpha, \beta) \in \Lambda$, $m \in \mathbb{N}$ and $0 < q_n < p_n \leq 1$ such that $\lim_n q_n = a$ and $\lim_n p_n = b$ where $a, b \in (0, 1]$.

(i) A sequence $x = (x_k)$ is said to be statistically $\Lambda_{p,q}^{[m]}$-summable to a number $L$, if

$$st - \lim_{n \to \infty} \Lambda_{p,q}^{[m]}(x_n) = L,$$

where

$$\Lambda_{p,q}^{[m]}(x_n) = \frac{1}{(\beta(n) - \alpha(n) + 1)^\gamma} \sum_{k \in P_n^{\alpha, \beta}} \Delta_{p,q}^{[m]}(x_k), \quad (0 < \gamma \leq 1).$$

(ii) A sequence $x = (x_k)$ is said to be $\Lambda_{p,q}^{[m]}$-statistically convergent of order $\gamma$ to a number $L$, if for every $\epsilon > 0$,

$$\lim_{n \to \infty} \frac{\left| \left\{ k \in P_n^{\alpha, \beta} : |\Delta_{p,q}^{[m]}(x_k) - L| \geq \epsilon \right\} \right|}{(\beta(n) - \alpha(n) + 1)^\gamma} = 0.$$

## 7.3 Some new definitions involving generalized double difference operator

Let $\omega^2$ be the space of all real double sequences of the form $x = (x_{mn})_{m,n \in \mathbb{N}_0}$. Let $x = (x_{mn}) \in \omega^2$ and $0 < q_1, q_2 < p_1, p_2 \leq 1$. For non-negative integers $n_1, n_2$, we define a generalized double difference sequence via the difference operator $\Delta_{p_1,q_1,p_2,q_2}^{[n_1,n_2]} : \omega^2 \to \omega^2$, defined by

$$\Delta_{p_1,q_1,p_2,q_2}^{[n_1,n_2]}(x_{k,l}) = \sum_{i_1=0}^{n_1} \sum_{i_2=0}^{n_2} (-1)^{i_1+i_2} \begin{bmatrix} n_1 \\ i_1 \end{bmatrix}_{p_1,q_1} \begin{bmatrix} n_2 \\ i_2 \end{bmatrix}_{p_2,q_2} x_{k-i_1, l-i_2}. \quad (7.1)$$

That is,

$$\Delta^{[n_1,n_2]}_{p_1,q_1,p_2,q_2}(x_{k,l})$$

$$= \sum_{i_1=0}^{n_1}(-1)^{i_1}\begin{bmatrix}n_1\\i_1\end{bmatrix}_{p_1,q_1}\left\{\sum_{i_2=0}^{n_2}(-1)^{i_2}\begin{bmatrix}n_2\\i_2\end{bmatrix}_{p_2,q_2}x_{k-i_1,l-i_2}\right\}$$

$$= x_{k,l} - [n_2]_{p_2,q_2}x_{k,l-1} + \frac{[n_2]_{p_2,q_2}[n_2-1]_{p_2,q_2}}{[2]_{p_2,q_2}!}x_{k,l-2} - [n_1]_{p_1,q_1}x_{k-1,l} + \cdots$$

$$+ [n_1]_{p_1,q_1}[n_2]_{p_2,q_2}x_{k-1,l-1} - \frac{[n_1]_{p_1,q_1}[n_2]_{p_2,q_2}[n_2-1]_{p_2,q_2}}{[2]_{p_2,q_2}!}x_{k-1,l-2} + \cdots$$

$$+ \frac{[n_1]_{p_1,q_1}[n_1-1]_{p_1,q_1}}{[2]_{p_1,q_1}!}x_{k-2,l} - \frac{[n_1]_{p_1,q_1}[n_1-1]_{p_1,q_1}[n_2]_{p_2,q_2}}{[2]_{p_1,q_1}!}x_{k-2,l-1} + \cdots$$

$$+ \frac{[n_1]_{p_1,q_1}[n_1-1]_{p_1,q_1}[n_2]_{p_2,q_2}[n_2-1]_{p_2,q_2}}{[2]_{p_1,q_1}![2]_{p_2,q_2}!}x_{k-2,l-2} + \cdots$$

$$= x_{k,l} - \frac{p_2^{n_2} - q_2^{n_2}}{p_2 - q_2}x_{k,l-1} + \frac{(p_2^{n_2} - q_2^{n_2})(p_2^{n_2-1} - q_2^{n_2-1})}{(p_2 + q_2)(p_2 - q_2)^2}x_{k,l-2}$$

$$- \frac{p_1^{n_1} - q_1^{n_1}}{p_1 - q_1}x_{k-1,l} + \cdots + \frac{(p_1^{n_1} - q_1^{n_1})(p_2^{n_2} - q_2^{n_2})}{(p_1 - q_1)(p_2 - q_2)}x_{k-1,l-1}$$

$$- \frac{(p_1^{n_1} - q_1^{n_1})(p_2^{n_2} - q_2^{n_2})(p_2^{n_2-1} - q_2^{n_2-1})}{(p_1 - q_1)(p_2 + q_2)(p_2 - q_2)^2}x_{k-1,l-2} + \cdots$$

$$+ \frac{(p_1^{n_1} - q_1^{n_1})(p_1^{n_1-1} - q_1^{n_1-1})}{(p_1 + q_1)(p_1 - q_1)^2}x_{k-2,l}$$

$$- \frac{(p_1^{n_1} - q_1^{n_1})(p_1^{n_1-1} - q_1^{n_1-1})(p_2^{n_2} - q_2^{n_2})}{(p_1 + q_1)(p_2 - q_2)(p_1 - q_1)^2}x_{k-2,l-1} + \cdots$$

$$+ \frac{(p_1^{n_1} - q_1^{n_1})(p_1^{n_1-1} - q_1^{n_1-1})(p_2^{n_2} - q_2^{n_2})(p_2^{n_2-1} - q_2^{n_2-1})}{(p_1 + q_1)(p_2 + q_2)(p_1 - q_1)^2(p_2 - q_2)^2}x_{k-2,l-2} + \cdots.$$

For example, if we take as $n_1 = n_2 = 1$, i.e. , $[n_i]_{p_i,q_i} = 1$ for $i = 1, 2$, then the difference operator $\Delta^{[1,1]}_{p_1,q_1,p_2,q_2}(x_{k,l})$ is reduced to the classical difference operator $\Delta^1(x_{k,l})$ of first order, defined by

$$\Delta^1(x_{k,l}) := x_{k,l} - x_{k,l-1} - x_{k-1,l} + x_{k-1,l-1}.$$

It is not hard to verify that, the double difference operator defined in (7.1) is linear over the real field $\mathbb{R}$ and also generalizes all the difference operators studied by Kızmaz Kızmaz (1981), Et and Çolak Et and Çolak (1995), Bektaş et al. Bektaş et al. (2004), Et and Nuray Et and Nuray (2001), Kadak Kadak (2016) and many others Başar and Aydın (2004); Kadak (2017a); Kadak and Baliarsingh (2015); Mursaleen and Alotaibi (2013). In order to show the effectiveness of the double difference operator in (7.1), we present an example to illustrate the relationship between proposed method and its ordinary form.

**Example 7.3.1.** Define the double sequence $(x_{k,l})$ by $x_{kl} = k + l$ for all $k, l \in \mathbb{N}$. Although the double sequence $(x_{k,l})$ is not convergent (in Pringsheim's sense), the classical double difference sequence $\Delta^2(x_{k,l})$ of second order is convergent to zero, that is, $P - \lim_{k,l \to \infty} \Delta^2(x_{k,l}) = 0$. On the other hand, since

$$\Delta_{p_1,q_1,p_2,q_2}^{[2,1]}(x_{k,l})$$
$$= x_{k,l} - x_{k,l-1} - [2]_{p_1,q_1} x_{k-1,l} + [2]_{p_1,q_1} x_{k-1,l-1} + x_{k-2,l} - x_{k-2,l-1}$$

then $(\Delta_{p_1,q_1,p_2,q_2}^{[2,1]}(x_{k,l}))$ is $P$-convergent to a number $L = 2 - (p_1 + q_1)$ for $0 < q_1, q_2 < p_1, p_2 \leq 1$. It is remarked that, depending on the choice of the parameters $p_1$ and $q_1$, $(\Delta_{p_1,q_1,p_2,q_2}^{[2,1]}(x_{k,l}))$ has infinitely many limit points. This problem, regarding proposed double difference operator, is based on the definition of $(p, q)$-integers. In order to reach to convergence results of the operator, we have to solve this problem. Currently are known two different ways to overcome such problems. The first way is choosing $p_i = q_i = 1$ for $i = 1, 2$ and hence we have an ordinary double difference operator $\Delta^2(x_{kl})$ of second order. The other way is to replace $p_1, p_2$ and $q_1, q_2$ by real sequences $(p_{n_1}), (p_{n_2})$ and $(q_{n_1}), (q_{n_2})$ respectively, satisfying

$$\lim_{n_1 \to \infty} p_{n_1} = L_1, \quad \lim_{n_1 \to \infty} q_{n_1} = L_2, \quad \lim_{n_2 \to \infty} p_{n_2} = \widetilde{L}_1, \quad \lim_{n_2 \to \infty} q_{n_2} = \widetilde{L}_2 \quad (7.2)$$

$$\lim_{n_1 \to \infty} (p_{n_1})^{n_1} = K_1, \quad \lim_{n_1 \to \infty} (q_{n_1})^{n_1} = K_2, \quad (7.3)$$

$$\lim_{n_2 \to \infty} (p_{n_2})^{n_2} = \widetilde{K}_1, \quad \lim_{n_2 \to \infty} (q_{n_2})^{n_2} = \widetilde{K}_2, \quad (7.4)$$

where $0 < q_{n_i} < p_{n_i} \leq 1$ and $L_i, K_i, \widetilde{L}_i, \widetilde{K}_i \in (0, 1]$ for $i = 1, 2$.

For instance, by taking $q_{n_1} = (\frac{n_1}{n_1+u}) < (\frac{n_1}{n_1+v}) = p_{n_1}$ for $0 < v < u$, we have

$$\lim_{n_1 \to \infty} p_{n_1} = 1, \ \lim_{n_1 \to \infty} q_{n_1} = 1, \quad \lim_{n_1 \to \infty} (q_{n_1})^{n_1} = e^{-u}, \quad \lim_{n_1 \to \infty} (p_{n_1})^{n_1} = e^{-v}.$$

Therefore,

$$\lim_{n_1,n_2 \to \infty} \Delta_{p_1,q_1,p_2,q_2}^{[2,1]}(x_{n_1,n_2}) = \lim_{n_1,n_2 \to \infty} \{(2 - (p_{n_1} + q_{n_1}))\} = 0,$$

which shows that $\Delta_{p_1,q_1,p_2,q_2}^{[2,1]}(x_{n_1,n_2}) \to 0$ as $n_1, n_2 \to \infty$.

*Remark* 7.3.2. Compared with the existing literature on this topic, our generalizations with respect to four parameters $p_1, q_1, p_2, q_2$ not only provide much more flexibility but also give a new perspective regarding the convergence (in any manner) of sequences.

Now, we present some new sets of double sequences as the set of all double sequences whose $\Delta_{p_{n_1},q_{n_1},p_{n_2},q_{n_2}}^{[n_1,n_2]}$-transforms are in $\ell_\infty^2, c^2$ and $c_0^2$, respectively.

**Definition 7.3.3.** Let $(p_{n_i})$ and $(q_{n_i})$ be sequences satisfying (7.2)-(7.4) for $i = 1, 2$ and $n_1, n_2 \in \mathbb{N}$. Then, we define

$$\ell^2_\infty(\Delta^{[n_1,n_2]}_{p,q}) := \left\{ (x_{n_1,n_2}) \in \omega^2 : \sup_{n_1,n_2 \in \mathbb{N}} \left| \Delta^{[n_1,n_2]}_{p_{n_1},q_{n_1},p_{n_2},q_{n_2}}(x_{n_1,n_2}) \right| < \infty \right\},$$

$$c^2(\Delta^{[n_1,n_2]}_{p,q}) := \left\{ (x_{n_1,n_2}) \in \omega^2 : \lim_{n_1,n_2 \to \infty} \left| \Delta^{[n_1,n_2]}_{p_{n_1},q_{n_1},p_{n_2},q_{n_2}}(x_{n_1,n_2}) - L \right| = 0 \right\},$$

$$c^2_0(\Delta^{[n_1,n_2]}_{p,q}) := \left\{ (x_{n_1,n_2}) \in \omega^2 : \lim_{n_1,n_2 \to \infty} \left| \Delta^{[n_1,n_2]}_{p_{n_1},q_{n_1},p_{n_2},q_{n_2}}(x_{n_1,n_2}) \right| = 0 \right\}$$

and

$$c^2_b(\Delta^{[n_1,n_2]}_{p,q}) := \ell^2_\infty(\Delta^{[n_1,n_2]}_{p,q}) \cap c^2(\Delta^{[n_1,n_2]}_{p,q}).$$

Here we note that these sequence spaces are complete linear spaces with the norm

$$\|x\|_{\Delta^{[n_1,n_2]}_{p,q}} = \sum_{i_1=0}^\infty \sum_{i_2=0}^\infty |x_{i_1,i_2}| + \sup_{n_1,n_2 \in \mathbb{N}} \left| \Delta^{[n_1,n_2]}_{p_{n_1},q_{n_1},p_{n_2},q_{n_2}}(x_{n_1,n_2}) \right|.$$

**Definition 7.3.4.** Kadak (2017b) Let $K^2 \subset \mathbb{N}^2$ be a two-dimensional subset of positive integers and $(\alpha, \beta) \in \Lambda$. Let $(s_{n_1})_{n_1 \in \mathbb{N}_0}$ and $(s_{n_2})_{n_2 \in \mathbb{N}_0}$ be two sequences of non-negative real numbers such that $s_0 > 0$,

$$S^{(\alpha,\beta)}_{n_1} = \sum_{i_1=\alpha(n_1)}^{\beta(n_1)} s_{i_1} \to \infty \quad \text{as} \quad n_1 \to \infty$$

and

$$S^{(\alpha,\beta)}_{n_2} = \sum_{i_2=\alpha(n_2)}^{\beta(n_2)} s_{i_2} \to \infty \quad \text{as} \quad n_2 \to \infty.$$

The lower and upper weighted double $\alpha\beta$-densities of the set $K^2$ are defined by

$$\underline{\delta}^{(\alpha,\beta)}_2(K^2) = P - \liminf_{n_1,n_2 \to \infty} \frac{1}{S^{(\alpha,\beta)}_{n_1} S^{(\alpha,\beta)}_{n_2}} \left| \left\{ k \leq S^{(\alpha,\beta)}_{n_1}, \ l \leq S^{(\alpha,\beta)}_{n_2} : (k,l) \in K^2 \right\} \right|$$

and

$$\overline{\delta}^{(\alpha,\beta)}_2(K^2) = P - \limsup_{n_1,n_2 \to \infty} \frac{1}{S^{(\alpha,\beta)}_{n_1} S^{(\alpha,\beta)}_{n_2}} \left| \left\{ k \leq S^{(\alpha,\beta)}_{n_1}, \ l \leq S^{(\alpha,\beta)}_{n_2} : (k,l) \in K^2 \right\} \right|,$$

respectively, provided that the $P$-limits exist. If $\underline{\delta}^{(\alpha,\beta)}_2(K^2) = \overline{\delta}^{(\alpha,\beta)}_2(K^2)$, we say that $K^2$ has weighted double $\alpha\beta$-density $\delta^{(\alpha,\beta)}_2(K^2)$ defined by

$$\delta^{(\alpha,\beta)}_2(K^2) = P - \lim_{n_1,n_2 \to \infty} \frac{1}{S^{(\alpha,\beta)}_{n_1} S^{(\alpha,\beta)}_{n_2}} \left| \left\{ k \leq S^{(\alpha,\beta)}_{n_1}, \ l \leq S^{(\alpha,\beta)}_{n_2} : (k,l) \in K^2 \right\} \right|$$

provided that the $P$-limit exists.

Throughout this chapter, we assume that $(p_{n_i})$ and $(q_{n_i})$ are sequences satisfying (7.2)-(7.4) for $i = 1, 2$ and $(\alpha, \beta) \in \Lambda$.

**Definition 7.3.5.** A double sequence $(x_{n_1, n_2})$ is said to be weighted $\widetilde{\mathcal{N}}$-summable to a number $L$, if

$$\widetilde{\mathcal{N}}\big(\Delta_{p,q}^{(\alpha,\beta)}(x_{n_1,n_2})\big)$$

$$= \frac{1}{S_{n_1}^{(\alpha,\beta)} S_{n_2}^{(\alpha,\beta)}} \sum_{i_1=\alpha(n_1)}^{\beta(n_1)} \sum_{i_2=\alpha(n_2)}^{\beta(n_2)} s_{i_1} s_{i_2} \, \Delta_{p_{n_1},q_{n_1},p_{n_2},q_{n_2}}^{[n_1,n_2]}(x_{i_1,i_2}) \to L,$$

as $n_1, n_2 \to \infty$, in Pringsheim's sense. By $[\widetilde{\mathcal{N}}(\Delta_{p,q}^{(\alpha,\beta)}); s_k, t_l]$ we denote the set of all weighted $\widetilde{\mathcal{N}}$-summable double sequences and we shall write $[\widetilde{\mathcal{N}}(\Delta_{p,q}^{(\alpha,\beta)}); s_{i_1}, s_{i_2}] - \lim_{n_1,n_2\to\infty} x_{n_1,n_2} = L$.

In particular, if we take $\alpha(r) = 1, \beta(r) = r$, $s_t = 1$ for each $t \in [1, r]$ $(t, r \in \mathbb{N})$, $n_1, n_2 = 1$ and $p_{n_1}, p_{n_2} \to 1$, $q_{n_1}, q_{n_2} \to 1$ as $n_1, n_2 \to \infty$, then weighted $\widetilde{\mathcal{N}}$-summability is reduced to ordinary summability of double difference sequences.

**Definition 7.3.6.** A double sequence $(x_{n_1, n_2})$ is statistically weighted $\widetilde{\mathcal{N}}$-summable to $L$ if, for every $\epsilon > 0$,

$$\delta_2(K^2(\epsilon)) = 0$$

where

$$K^2(\epsilon) := \big\{ (n_1, n_2) \in \mathbb{N}^2 : |\widetilde{\mathcal{N}}(\Delta_{p,q}^{(\alpha,\beta)}(x_{n_1,n_2})) - L| \geqq \epsilon \big\}.$$

That is,

$$\lim_{n_1,n_2\to\infty} \frac{1}{n_1 n_2} \Big| \big\{ (i_1, i_2); \, i_1 \leqq n_1, \, i_2 \leqq n_2 : |\widetilde{\mathcal{N}}(\Delta_{p,q}^{(\alpha,\beta)}(x_{i_1,i_2})) - L| \geqq \epsilon \big\} \Big| = 0.$$

In this case we write $st_2[\widetilde{\mathcal{N}}] - \lim x_{n_1,n_2} = L$.

**Definition 7.3.7.** A double sequence $(x_{n_1, n_2})$ is said to be weighted $\Delta_{p,q}^{[n_1,n_2]}$-statistically convergent to $L$, denoted by $[st_2(\Delta_{p,q}^{(\alpha,\beta)}); s_{i_1}, s_{i_2}] - \lim x_{n_1,n_2} = L$, if for every $\epsilon > 0$,

$$\delta_2^{(\alpha,\beta)}\big( (n_1, n_2) \in \mathbb{N}^2 : s_{n_1} s_{n_2} \, |\Delta_{p_{n_1},q_{n_1},p_{n_2},q_{n_2}}^{[n_1,n_2]}(x_{n_1,n_2}) - L| \geqq \epsilon \big) = 0,$$

or, equivalently, we may write

$$\lim_{n_1,n_2\to\infty} \frac{1}{S_{n_1}^{(\alpha,\beta)} S_{n_2}^{(\alpha,\beta)}}$$

$$\times \Big| \big\{ (i_1, i_2); i_1, i_2 \leq S_{n_1}^{(\alpha,\beta)}, S_{n_2}^{(\alpha,\beta)} : s_{i_1} s_{i_2} \, |\Delta_{p_{n_1},q_{n_1},p_{n_2},q_{n_2}}^{[n_1,n_2]}(x_{i_1,i_2}) \quad L| \geqq \epsilon \big\} \Big| = 0.$$

We denote by $[st_2(\Delta_{p,q}^{(\alpha,\beta)}); s_{i_1}, s_{i_2}]$, the set of all weighted $\Delta_{p,q}^{[n_1,n_2]}$-statistically convergent double sequences. Again, if we take $\alpha(r) = 1, \beta(r) = r$, $s_t = 1$ for each $t \in [1, r]$ $(t, r \in \mathbb{N})$, $n_1, n_2 = 1$ and $p_{n_1}, p_{n_2} \to 1$, $q_{n_1}, q_{n_2} \to 1$ as $n_1, n_2 \to \infty$, then weighted $\Delta_{p,q}^{[n_1,n_2]}$-statistical convergence reduces to ordinary statistical convergence of double difference sequences Mursaleen and Edely (2003) (see also Çakan and Altay (2006); Chen and Chang (2007)).

**Definition 7.3.8.** A double sequence $(x_{n_1,n_2})$ is said to be weighted strongly $[\widetilde{N}]_r$-summable $(0 < r < \infty)$ to a number $L$, denoted by $[\widetilde{N}_r(\Delta_{p,q}^{(\alpha,\beta)}); s_{i_1}, s_{i_2}]-\lim x_{n_1,n_2} = L$, if

$$\lim_{n_1,n_2 \to \infty} \frac{1}{S_{n_1}^{(\alpha,\beta)} S_{n_2}^{(\alpha,\beta)}} \sum_{i_1=\alpha(n_1)}^{\beta(n_1)} \sum_{i_2=\alpha(n_2)}^{\beta(n_2)} s_{i_1} s_{i_2} |\Delta_{p_{n_1},q_{n_1},p_{n_2},q_{n_2}}^{[n_1,n_2]} (x_{i_1,i_2}) - L|^r = 0.$$

To show the effectiveness of the proposed method, we describe now some important special cases.

(i) Let $\theta = (k_n)$ be an increasing sequence of positive integers such that $k_0 = 0$, $0 < k_n < k_{n+1}$ and $h_n = (k_n - k_{n-1}) \to \infty$ as $n \to \infty$. Under these conditions, $\theta$ is called a lacunary sequence. It is easily seen that statistically weighted $\widetilde{N}$-summability is reduced to the lacunary statistical summability for double sequences if we take $\alpha(r) = k_{r-1} + 1$, $\beta(r) = k_r$, $s_t = 1$ for each $t \in [k_{r-1} + 1, k_r]$ $(t, r \in \mathbb{N})$, $n_1, n_2 = 1$ and $p_{n_1}, p_{n_2} \to 1$, $q_{n_1}, q_{n_2} \to 1$ as $n_1, n_2 \to \infty$. Similarly, weighted $\Delta_{p,q}^{[n_1,n_2]}$-statistical convergence turns out to be lacunary statistical convergence under the same conditions (cf. Et and Şengül (2014); Fridy and Orhan (1993)).

(ii) Let $(\lambda_n)$ be a strictly increasing sequence of positive numbers such that $\lim_n \lambda_n = \infty$, $\lambda_{n+1} \leq \lambda_n + 1$ and $\lambda_1 = 1$. Let $\alpha(r) = r - \lambda_r + 1$ and $\beta(r) = r$, $s_t = 1$ for each $t \in [r - \lambda_r + 1, r]$ $(t, r \in \mathbb{N})$, $n_1, n_2 = 1$ and $p_{n_1}, p_{n_2} \to 1$, $q_{n_1}, q_{n_2} \to 1$ as $n_1, n_2 \to \infty$. Then, the statistically weighted $\widetilde{N}$-summability is reduced to the $\lambda$-statistically summability. Moreover, the weighted $\Delta_{p,q}^{[n_1,n_2]}$-statistical convergence is reduced to $\lambda$-statistical convergence under the same conditions (cf. Belen and Mohiuddine (2013); Mursaleen (2000)).

(iii) Suppose that $\alpha(r) = 1, \beta(r) = r$, $s_t = 1$ for each $t \in [1, r]$, $n_1, n_2 = 1$ and also $p_{n_1}, p_{n_2} \to 1$, $q_{n_1}, q_{n_2} \to 1$ as $n_1, n_2 \to \infty$. Then, the weighted strongly $[\widetilde{N}]_r$-summability $(0 < r < \infty)$ reduces to the strong summability of double sequences (cf. Connor (1989)).

## 7.4  Some inclusion relations

In this section we first prove the following relation between newly proposed methods of statistically weighted $\widetilde{\mathcal{N}}$-summability, weighted $\Delta_{p,q}^{[n_1,n_2]}$-statistical convergence and weighted strongly $[\widetilde{\mathcal{N}}]_r$-summability for $0 < r < \infty$.

**Theorem 7.4.1.** *Let $(p_{n_i})$ and $(q_{n_i})$ be sequences satisfying (7.2)-(7.4) for $i = 1, 2$ and $(\alpha, \beta) \in \Lambda$. Assume that*

$$s_{i_1} s_{i_2} \, |\Delta_{p_{n_1},q_{n_1},p_{n_2},q_{n_2}}^{[n_1,n_2]}(x_{i_1,i_2}) - L| \leq M$$

*for all $i_1, i_2 \in \mathbb{N}$. If a double sequence $(x_{n_1,n_2})$ is weighted $\Delta_{p,q}^{[n_1,n_2]}$-statistically convergent to the number $L$, then it is statistically weighted $\widetilde{\mathcal{N}}$-summable to the same limit $L$, but not conversely.*

*Proof.* Let $(p_{n_i})$ and $(q_{n_i})$ be sequences satisfying (7.2)-(7.4) for $i = 1, 2$ and $(\alpha, \beta) \in \Lambda$. Suppose that $s_{i_1} s_{i_2} |\Delta_{p_{n_1},q_{n_1},p_{n_2},q_{n_2}}^{[n_1,n_2]}(x_{i_1,i_2}) - L| \leq M$ holds for all $i_1, i_2 \in \mathbb{N}$. Let us set

$$E_\epsilon = \left\{ (i_1,i_2); \, i_1 \leq S_{n_1}^{(\alpha,\beta)}, \, i_2 \leq S_{n_2}^{(\alpha,\beta)} : s_{i_1}s_{i_2}|\Delta_{p_{n_1},q_{n_1},p_{n_2},q_{n_2}}^{[n_1,n_2]}(x_{i_1,i_2}) - L| \geq \epsilon \right\}$$

and

$$E_\epsilon^C = \left\{ (i_1,i_2); \, i_1 \leq S_{n_1}^{(\alpha,\beta)}, \, i_2 \leq S_{n_2}^{(\alpha,\beta)} : s_{i_1}s_{i_2}|\Delta_{p_{n_1},q_{n_1},p_{n_2},q_{n_2}}^{[n_1,n_2]}(x_{i_1,i_2}) - L| < \epsilon \right\}.$$

From the hypothesis, we immediately have $\lim_{n_1,n_2} \frac{|E_\epsilon|}{S_{n_1}^{(\alpha,\beta)} S_{n_2}^{(\alpha,\beta)}} = 0$. We thus find that

$$\left| \frac{1}{S_{n_1}^{(\alpha,\beta)} S_{n_2}^{(\alpha,\beta)}} \sum_{i_1 \in I_{n_1}^{(\alpha,\beta)}, i_2 \in I_{n_2}^{(\alpha,\beta)}} s_{i_1} s_{i_2} \, \Delta_{p_{n_1},q_{n_1},p_{n_2},q_{n_2}}^{[n_1,n_2]}(x_{i_1,i_2}) - L \right|$$

$$\leq \frac{1}{S_{n_1}^{(\alpha,\beta)} S_{n_2}^{(\alpha,\beta)}} \sum_{i_1 \in I_{n_1}^{(\alpha,\beta)}, i_2 \in I_{n_2}^{(\alpha,\beta)}} s_{i_1} s_{i_2} \, |\Delta_{p_{n_1},q_{n_1},p_{n_2},q_{n_2}}^{[n_1,n_2]}(x_{i_1,i_2}) - L|$$

$$+ |L| \left| \frac{1}{S_{n_1}^{(\alpha,\beta)} S_{n_2}^{(\alpha,\beta)}} \sum_{i_1 \in I_{n_1}^{(\alpha,\beta)}, i_2 \in I_{n_2}^{(\alpha,\beta)}} s_{i_1} s_{i_2} - 1 \right|$$

$$\leq \frac{1}{S_{n_1}^{(\alpha,\beta)} S_{n_2}^{(\alpha,\beta)}} \sum_{\substack{i_1 \in I_{n_1}^{(\alpha,\beta)}, i_2 \in I_{n_2}^{(\alpha,\beta)} \\ ((i_1,i_2) \subset D_\epsilon)}} s_{i_1} s_{i_2} \, |\Delta_{p_{n_1},q_{n_1},p_{n_2},q_{n_2}}^{[n_1,n_2]}(x_{i_1,i_2}) - L|$$

$$+ \frac{1}{S_{n_1}^{(\alpha,\beta)} S_{n_2}^{(\alpha,\beta)}} \sum_{\substack{i_1 \in I_{n_1}^{(\alpha,\beta)}, i_2 \in I_{n_2}^{(\alpha,\beta)} \\ ((i_1,i_2) \in E_\epsilon^C)}} s_{i_1} s_{i_2} \, |\Delta_{p_{n_1},q_{n_1},p_{n_2},q_{n_2}}^{[n_1,n_2]}(x_{i_1,i_2}) - L|$$

$$\leq \frac{M \, |E_\epsilon|}{S_{n_1}^{(\alpha,\beta)} S_{n_2}^{(\alpha,\beta)}} + \frac{\epsilon \, |F_\epsilon^C|}{S_{n_1}^{(\alpha,\beta)} S_{n_2}^{(\alpha,\beta)}} \to 0 + \epsilon \cdot 1 = \epsilon \quad (n_1, n_2 \to \infty),$$

where $I_r^{(\alpha,\beta)} = [\alpha(r), \beta(r)]$ is a closed interval for $(\alpha, \beta) \in \Lambda$ and $r \in \mathbb{N}$. Hence, $(x_{n_1,n_2})$ is weighted $\widetilde{\mathcal{N}}$-summable to $L$ which implies that $(x_{n_1,n_2})$ is statistically weighted $\widetilde{\mathcal{N}}$-summable to $L$.

Our next example will show that the converse is not true.      □

**Example 7.4.2.** *For each $n_1, n_2 \in \mathbb{N}$, define the double difference sequence $y_{n_1,n_2} = \Delta_{p_{n_1},q_{n_1},p_{n_2},q_{n_2}}^{[n_1,n_2]}(x_{n_1,n_2})$ by*

$$
y_{n_1,n_2} = \begin{cases}
1 & \begin{cases} n_1 = u^2 - u, u^2 - u + 1, \ldots, u^2 - 1; \ u = 2, 3, 4, \ldots \\ n_2 = v^2 - v, v^2 - v + 1, \ldots, v^2 - 1; \ v = 2, 3, 4, \ldots \end{cases}, \\
-uv & \begin{cases} n_1 = u^2, \ u = 2, 3, 4, \ldots \\ n_2 = v^2, \ v = 2, 3, 4, \ldots \end{cases}, \\
0 & \text{otherwise.}
\end{cases}
$$

*Further, let us take $\alpha(r) = 1, \beta(r) = r, s_t = 1$ for each $t \in [1, r]$ $(t, r \in \mathbb{N})$. Also, suppose that $p_{n_1}, p_{n_2} \to 1, q_{n_1}, q_{n_2} \to 1$ as $n_1, n_2 \to \infty$. It is clear that the double difference sequence $(y_{n_1,n_2})$ is divergent. On the other hand*

$$
\widetilde{\mathcal{N}}(\Delta_{p,q}^{(\alpha,\beta)}(x_{n_1,n_2}))
$$

$$
= \frac{1}{n_1 n_2} \sum_{i_1=1}^{n_1} \sum_{i_2=1}^{n_2} y_{i_1,i_2}
$$

$$
= \begin{cases}
\frac{(\eta+1)(\theta+1)}{n_1 n_2} & \begin{cases} n_1 = u^2 - u + \eta; \ u = 2, 3, 4, \ldots; \eta = 0, 1, 2, \ldots u - 1, \\ n_2 = v^2 - v + \theta; \ v = 2, 3, 4, \ldots; \theta = 0, 1, 2, \ldots v - 1, \end{cases} \\
0 & \text{otherwise.}
\end{cases}
$$

*Using the above choices and letting $n_1, n_2 \to \infty$ in the last equality, we find that the double difference sequence $(y_{n_1,n_2})$ is statistically weighted $\widetilde{\mathcal{N}}$-summable to zero. However, since*

$$
P - \lim_{n_1,n_2 \to \infty} \frac{1}{n_1 n_2} \left| \left\{ (i_1, i_2); \ i_1 \leq n_1, \ i_2 \leq n_2 : |y_{i_1,i_2}| \geqq \epsilon \right\} \right| \neq 0,
$$

*then $(y_{n_1,n_2})$ is not weighted $\Delta_{p,q}^{[n_1,n_2]}$-statistical convergent to zero.*

**Theorem 7.4.3.** *Let $(p_{n_i})$ and $(q_{n_i})$ be sequences satisfying (7.2)-(7.4) for $i = 1, 2$ and $(\alpha, \beta) \in \Lambda$.*

(a) *Suppose that $(x_{n_1,n_2})$ is weighted strongly $[\widetilde{\mathcal{N}}]_r$-summable $(0 < r < \infty)$ to $L$. If*

$$
0 < r < 1 \quad and \quad 0 \leq \left| \Delta_{p_{n_1},q_{n_1},p_{n_2},q_{n_2}}^{[n_1,n_2]}(x_{i_1,i_2}) - L \right| < 1
$$

*or*

$$
1 \leq r < \infty \quad and \quad 1 \leq \left| \Delta_{p_{n_1},q_{n_1},p_{n_2},q_{n_2}}^{[n_1,n_2]}(x_{i_1,i_2}) - L \right| < \infty
$$

*for all $i_1, i_2 \in \mathbb{N}$, then double sequence $(x_{n_1,n_2})$ is weighted $\Delta_{p,q}^{[n_1,n_2]}$-statistical convergent to $L$.*

(b) *Suppose that* $(x_{n_1,n_2})$ *is weighted* $\Delta_{p,q}^{[n_1,n_2]}$-*statistical convergent to L. Suppose also that*

$$s_{i_1} s_{i_2} \left| \Delta_{p_{n_1},q_{n_1},p_{n_2},q_{n_2}}^{[n_1,n_2]} (x_{i_1,i_2}) - L \right| \le M$$

*for all* $i_1, i_2 \in \mathbb{N}$. *If*

$$0 < r < 1 \quad and \quad 1 \le M < \infty \qquad or \qquad 1 \le r < \infty \quad and \quad 0 \le M < 1,$$

*then the double sequence* $(x_{n_1,n_2})$ *is weighted strongly* $[\widetilde{\mathcal{N}}]_r$-*summable to L.*

*Proof.* (a) Proof of this case is a routine verification, we choose to omit the details involved.

(b) By the hypothesis, one obtains, for each $\epsilon > 0$, that

$$P - \lim_{n_1,n_2 \to \infty} \frac{1}{S_{n_1}^{(\alpha,\beta)} S_{n_2}^{(\alpha,\beta)}} |E_\epsilon| = 0.$$

Since $s_{i_1} s_{i_2} \left| \Delta_{p_{n_1},q_{n_1},p_{n_2},q_{n_2}}^{[n_1,n_2]} (x_{i_1,i_2}) - L \right| \le M$ for all $i_1, i_2, n_1, n_2 \in \mathbb{N}$, then

$$\frac{1}{S_{n_1}^{(\alpha,\beta)} S_{n_2}^{(\alpha,\beta)}} \sum_{i_1 \in I_{n_1}^{(\alpha,\beta)}, i_2 \in I_{n_2}^{(\alpha,\beta)}} s_{i_1} s_{i_2} \left| \Delta_{p_{n_1},q_{n_1},p_{n_2},q_{n_2}}^{[n_1,n_2]} (x_{i_1,i_2}) - L \right|^r$$

$$= \frac{1}{S_{n_1}^{(\alpha,\beta)} S_{n_2}^{(\alpha,\beta)}} \sum_{\substack{i_1 \in I_{n_1}^{(\alpha,\beta)}, i_2 \in I_{n_2}^{(\alpha,\beta)} \\ ((i_1,i_2) \in E_\epsilon)}} s_{i_1} s_{i_2} \left| \Delta_{p_{n_1},q_{n_1},p_{n_2},q_{n_2}}^{[n_1,n_2]} (x_{i_1,i_2}) - L \right|^r$$

$$+ \frac{1}{S_{n_1}^{(\alpha,\beta)} S_{n_2}^{(\alpha,\beta)}} \sum_{\substack{i_1 \in I_{n_1}^{(\alpha,\beta)}, i_2 \in I_{n_2}^{(\alpha,\beta)} \\ ((i_1,i_2) \in E_\epsilon^C)}} s_{i_1} s_{i_2} \left| \Delta_{p_{n_1},q_{n_1},p_{n_2},q_{n_2}}^{[n_1,n_2]} (x_{i_1,i_2}) - L \right|^r$$

$$= \Gamma_1(n_1, n_2) + \Gamma_2(n_1, n_2)$$

where

$$\Gamma_1(n_1, n_2) = \frac{1}{S_{n_1}^{(\alpha,\beta)} S_{n_2}^{(\alpha,\beta)}} \sum_{\substack{i_1 \in I_{n_1}^{(\alpha,\beta)}, i_2 \in I_{n_2}^{(\alpha,\beta)} \\ ((i_1,i_2) \in E_\epsilon)}} s_{i_1} s_{i_2} \left| \Delta_{p_{n_1},q_{n_1},p_{n_2},q_{n_2}}^{[n_1,n_2]} (x_{i_1,i_2}) - L \right|^r$$

and

$$\Gamma_2(n_1, n_2) = \frac{1}{S_{n_1}^{(\alpha,\beta)} S_{n_2}^{(\alpha,\beta)}} \sum_{\substack{i_1 \in I_{n_1}^{(\alpha,\beta)}, i_2 \in I_{n_2}^{(\alpha,\beta)} \\ ((i_1,i_2) \in E_\epsilon^C)}} s_{i_1} s_{i_2} \left| \Delta_{p_{n_1},q_{n_1},p_{n_2},q_{n_2}}^{[n_1,n_2]} (x_{i_1,i_2}) - L \right|^r.$$

For the case $(i_1, i_2) \in E_\epsilon^C$, we get

$$
\begin{aligned}
\Gamma_2(n_1, n_2) &= \frac{1}{S_{n_1}^{(\alpha,\beta)} S_{n_2}^{(\alpha,\beta)}} \sum_{\substack{i_1 \in I_{n_1}^{(\alpha,\beta)}, i_2 \in I_{n_2}^{(\alpha,\beta)} \\ ((i_1, i_2) \in E_\epsilon^C)}} s_{i_1} s_{i_2} \left| \Delta_{p_{n_1}, q_{n_1}, p_{n_2}, q_{n_2}}^{[n_1, n_2]} (x_{i_1, i_2}) - L \right|^r \\
&\le \frac{1}{S_{n_1}^{(\alpha,\beta)} S_{n_2}^{(\alpha,\beta)}} \sum_{\substack{i_1 \in I_{n_1}^{(\alpha,\beta)}, i_2 \in I_{n_2}^{(\alpha,\beta)} \\ ((i_1, i_2) \in E_\epsilon^C)}} s_{i_1} s_{i_2} \left| \Delta_{p_{n_1}, q_{n_1}, p_{n_2}, q_{n_2}}^{[n_1, n_2]} (x_{i_1, i_2}) - L \right| \\
&= \frac{1}{S_{n_1}^{(\alpha,\beta)} S_{n_2}^{(\alpha,\beta)}} |E_\epsilon^C| \to \epsilon \quad (n_1, n_2 \to \infty).
\end{aligned}
$$

Also, for $(i_1, i_2) \in E_\epsilon$, we obtain

$$
\begin{aligned}
\Gamma_1(n_1, n_2) &= \frac{1}{S_{n_1}^{(\alpha,\beta)} S_{n_2}^{(\alpha,\beta)}} \sum_{\substack{i_1 \in I_{n_1}^{(\alpha,\beta)}, i_2 \in I_{n_2}^{(\alpha,\beta)} \\ ((i_1, i_2) \in E_\epsilon)}} s_{i_1} s_{i_2} \left| \Delta_{p_{n_1}, q_{n_1}, p_{n_2}, q_{n_2}}^{[n_1, n_2]} (x_{i_1, i_2}) - L \right|^r \\
&\le \frac{1}{S_{n_1}^{(\alpha,\beta)} S_{n_2}^{(\alpha,\beta)}} \sum_{\substack{i_1 \in I_{n_1}^{(\alpha,\beta)}, i_2 \in I_{n_2}^{(\alpha,\beta)} \\ ((i_1, i_2) \in E_\epsilon)}} s_{i_1} s_{i_2} \left| \Delta_{p_{n_1}, q_{n_1}, p_{n_2}, q_{n_2}}^{[n_1, n_2]} (x_{i_1, i_2}) - L \right| \\
&= \frac{M}{S_{n_1}^{(\alpha,\beta)} S_{n_2}^{(\alpha,\beta)}} |E_\epsilon| \to 0 \quad (n_1, n_2 \to \infty).
\end{aligned}
$$

Therefore, $(x_{n_1, n_2})$ is weighted strongly $[\widetilde{\mathcal{N}}]_r$-summable to $L$ and the proof of theorem is completed. $\qquad\square$

---

## 7.5 Korovkin type approximation theorem

Over the last three decades, a number of studies were carried out with the aim to obtain applications in approximation theory. It is known that the theory of approximation by positive linear operators is an important research field in mathematics such as functional analysis, measure theory, real analysis, harmonic analysis, probability theory, summability theory, and partial differential equations. The foundation of the theory of approximation by positive linear operators was introduced independently by Bohman Bohman (1952) and Korovkin Korovkin (1953). Furthermore, Korovkin established the necessary and sufficient conditions for the uniform convergence of a sequence of positive linear operators acting on $C[a, b]$ based on some test functions. In particular, the test functions $\{1, e^{-x}, e^{-2x}\}$ are used in logarithmic form and, $\{1, \sin x, \cos x\}$ are also used the trigonometric form of Korovkin type approximation theorem. Also, several kinds of test functions have been studied, in both one-and multi-dimensional cases. In recent years, by using the notion of statistical convergence with some of its extended versions, various approximation results have been obtained (see for examples Aktuğlu (2014); Kadak (2016, 2017b,c); Kadak et al. (2017a); Mohiuddine (2016); Mursaleen et al. (2012) and the references therein).

In this section, we shall obtain a Korovkin type approximation theorem for functions of two variables applying the notion of statistically weighted $\widetilde{\mathcal{N}}$-summability via the following two dimensional test functions

$$e_{ij} = \left(\frac{u}{1+u}\right)^i \left(\frac{v}{1+v}\right)^j \qquad i,j = 0,1,2. \tag{7.5}$$

The classical Korovkin first approximation theorem states as follows Korovkin (1953):

**Theorem 7.5.1.** *Let $(B_n)$ be a sequence of positive linear operators from $C[a,b]$ into itself. Then*

$$\lim_{n\to\infty} \|B_n(f;x) - f(x)\|_{C[a,b]} = 0, \text{ for all } f \in C[a,b],$$

*if and only if*

$$\lim_{n\to\infty} \|B_n(e_i;x) - e_i\|_{C[a,b]} = 0, \text{ for each } i = 0,1,2,$$

*where $e_0(x) = 1$, $e_1(x) = x$ and $e_2(x) = x^2$.*

The statistical version of Theorem 7.5.1 was given by Gadjiev and Orhan Gadjiev and Orhan (2002) as follows:

**Theorem 7.5.2.** *For any $f \in C[a,b]$,*

$$st - \lim_{n\to\infty} \|B_n(f;x) - f\|_\infty = 0$$

*if and only if*

$$st - \lim_{n\to\infty} \|B_n(e_i;x) - e_i\|_\infty = 0, \text{ for each } i = 0,1,2,$$

*where the test function $e_i(x) = x^i$.*

Let $\mathbb{R}_+^2 = [0,\infty) \times [0,\infty)$. By $C_B(\mathbb{R}_+^2)$, we denote the space of all bounded and continuous functions on $\mathbb{R}_+^2$ which is linear normed space with

$$\|f\|_{C_B(\mathbb{R}_+^2)} = \sup_{x,y\geq 0} |f(x,y)| \qquad \left(f \in C_B(\mathbb{R}_+^2)\right).$$

Then as usual, we say that $T$ is a positive linear operator provided that $f \geq 0$ implies $Tf \geq 0$. Also, we use the notation $T(f(u,v);x,y)$ for the value of $Tf$ at a point $(x,y) \in \mathbb{R}_+^2$.

**Theorem 7.5.3.** *Assume that* $p_1 = (p_{n_1}), p_2 = (p_{n_2})$ *and* $q_1 = (q_{n_1}), q_2 = (q_{n_2})$ *are sequences, satisfying*

$$st - \lim_{n_1, n_2 \to \infty} p_{n_1}, p_{n_2} = 1, \quad st - \lim_{n_1, n_2 \to \infty} q_{n_1}, q_{n_2} = 1 \tag{7.6}$$

$$st - \lim_{n_1 \to \infty} (p_{n_1})^{n_1} = \mu_1, \quad st - \lim_{n_2 \to \infty} (p_{n_2})^{n_2} = \mu_2 \tag{7.7}$$

$$st - \lim_{n_1 \to \infty} (q_{n_1})^{n_1} = \tilde{\mu}_1, \quad st - \lim_{n_2 \to \infty} (q_{n_2})^{n_2} = \tilde{\mu}_2 \tag{7.8}$$

*where* $q_{n_i} \in (0,1)$, $p_{n_i} \in (q_{n_i}, 1]$ *and* $\mu_i, \tilde{\mu}_i \in (0,1]$ *for* $i = 1, 2$. *Assume further that* $(T_{n_1, n_2})$ *is a double sequence of positive linear operators acting from* $C_B(\mathbb{R}_+^2)$ *into itself satisfying the following conditions:*

$$st_2[\tilde{\mathcal{N}}] - \lim_{n_1, n_2 \to \infty} \|T_{n_1, n_2}(e_{00}; x, y) - e_{00}\|_{C_B(\mathbb{R}_+^2)} = 0,$$

$$st_2[\tilde{\mathcal{N}}] - \lim_{n_1, n_2 \to \infty} \|T_{n_1, n_2}(e_{10}; x, y) - e_{10}\|_{C_B(\mathbb{R}_+^2)} = 0,$$

$$st_2[\tilde{\mathcal{N}}] - \lim_{n_1, n_2 \to \infty} \|T_{n_1, n_2}(e_{01}; x, y) - e_{01}\|_{C_B(\mathbb{R}_+^2)} = 0,$$

$$st_2[\tilde{\mathcal{N}}] - \lim_{n_1, n_2 \to \infty} \|T_{n_1, n_2}(e_{20}; x, y) - e_{20}\|_{C_B(\mathbb{R}_+^2)} = 0,$$

$$st_2[\tilde{\mathcal{N}}] - \lim_{n_1, n_2 \to \infty} \|T_{n_1, n_2}(e_{02}; x, y) - e_{02}\|_{C_B(\mathbb{R}_+^2)} = 0.$$

*Then, for all* $f \in C_B(\mathbb{R}_+^2)$,

$$st_2[\tilde{\mathcal{N}}] - \lim_{n_1, n_2 \to \infty} \|T_{n_1, n_2}(f; x, y) - f\|_{C_B(\mathbb{R}_+^2)} = 0$$

*holds where* $e_{ij} : \mathbb{R}_+^2 \to [0,1)$ *defined in (7.5) are two dimensional test functions.*

*Proof.* Let us take $f \in C_B(\mathbb{R}_+^2)$ and $(x, y) \in \mathbb{R}_+^2$ be fixed. Since $f$ is bounded on the whole plane there exists a positive constant $M$ such that $|f(x, y)| \leq M$ for all $(x, y) \in \mathbb{R}_+^2$, we have

$$|f(u, v) - f(x, y)| \leq 2M \qquad (u, v, x, y \geq 0). \tag{7.9}$$

Since $f$ is continuous on $\mathbb{R}_+^2$, for a given $\epsilon > 0$, there exists a number $\delta = \delta(\epsilon)$ such that

$$|f(u, v) - f(x, y)| < \varepsilon \tag{7.10}$$

whenever

$$\left| \frac{u}{1+u} - \frac{x}{1+x} \right| < \delta \quad \text{and} \quad \left| \frac{v}{1+v} - \frac{y}{1+y} \right| < \delta.$$

Also we obtain for all $u, v, x, y \geq 0$ satisfying

$$\sqrt{\left( \frac{u}{1+u} - \frac{x}{1+x} \right)^2 + \left( \frac{v}{1+v} - \frac{y}{1+y} \right)^2} \geq \delta$$

that

$$|f(u,v) - f(x,y)| \le \frac{2M}{\delta^2}(\psi(u;x) + \psi(v;y)) \tag{7.11}$$

where

$$\psi(u;x) = \left(\frac{u}{1+u} - \frac{x}{1+x}\right)^2 \quad \text{and} \quad \psi(v;y) = \left(\frac{v}{1+v} - \frac{y}{1+y}\right)^2.$$

Now, from the relations (7.9) to (7.11), we get for all $(x,y), (u,v) \in \mathbb{R}_+^2$ and $f \in C_B(\mathbb{R}_+^2)$ that

$$|f(u,v) - f(x,y)| < \frac{2M}{\delta^2}(\psi(u;x) + \psi(v;y)) \tag{7.12}$$

Then, in light of the linearity and positivity of $T_{n_1,n_2}$, we obtain from (7.12), that

$$\left| T_{n_1,n_2}(f(u,v); x, y) - f(x,y) \right|$$
$$= \left| T_{n_1,n_2}(f(u,v) - f(x,y); x, y) + f(x,y)[T_{n_1,n_2}(e_{00}; x, y) - e_{00}(x,y)] \right|$$
$$\le T_{n_1,n_2}(|f(u,v) - f(x,y)|; x, y) + M \, |T_{n_1,n_2}(e_{00}; x, y) - e_{00}(x,y)|$$
$$\le \left| T_{n_1,n_2}\left(\varepsilon + \frac{2M}{\delta^2}(\psi(u;x) + \psi(v;y)); x, y\right) \right| + M \, |T_{n_1,n_2}(e_{00}; x, y) - e_{00}|$$
$$\le \varepsilon + (\varepsilon + M) \, |T_{n_1,n_2}(e_{00}; x, y) - e_{00}| - \frac{4M}{\delta^2}\frac{x}{1+x}|T_{n_1,n_2}(e_{10}; x, y) - e_{10}|$$
$$- \frac{4M}{\delta^2}\frac{y}{1+y}|T_{n_1,n_2}(e_{01}) - e_{01}| + \frac{2M}{\delta^2}|T_{n_1,n_2}(e_{20} + e_{02}) - (e_{20} + e_{02})|$$
$$+ \frac{2M}{\delta^2}\left(\left(\frac{x}{1+x}\right)^2 + \left(\frac{y}{1+y}\right)^2\right)|T_{n_1,n_2}(e_{00}; x, y) - e_{00}(x,y)|$$
$$\le \varepsilon + \left(\varepsilon + M + \frac{4M}{\delta^2}\right)|T_{n_1,n_2}(e_{00}; x, y) - e_{00}| + \frac{4M}{\delta^2}|T_{n_1,n_2}(e_{10}) - e_{10}|$$
$$+ \frac{4M}{\delta^2}|T_{n_1,n_2}(e_{01}; x, y) - e_{01}| + \frac{2M}{\delta^2}|T_{n_1,n_2}(e_{20} + e_{02}) - (e_{20} + e_{02})|.$$

Taking the supremum over $(x,y) \in \mathbb{R}_+^2$ in the last inequality, we have

$$\|T_{n_1,n_2}(f; x, y) - f\|_{C_B(\mathbb{R}_+^2)}$$
$$\le \varepsilon + K \left[\|T_{n_1,n_2}(e_{00}; x, y) - e_{00}\|_{C_B(\mathbb{R}_+^2)} + \|T_{n_1,n_2}(e_{10}; x, y) - e_{10}\|_{C_B(\mathbb{R}_+^2)}\right.$$
$$\left. + \|T_{n_1,n_2}(e_{01}; x, y) - e_{01}\|_{C_B(\mathbb{R}_+^2)} + \|T_{n_1,n_2}(e_{20} + e_{02}) - (e_{20} + e_{02})\|_{C_B(\mathbb{R}_+^2)}\right] \tag{7.13}$$

where $K := \left\{\varepsilon + M + \frac{4M}{\delta^2}\right\}$. We now replace $T_{n_1,n_2}(\cdot; x, y)$ by

$$\tilde{N}_{n_1,n_2}(\cdot; x, y) - \frac{1}{S_{n_1}^{(\alpha,\beta)}S_{n_2}^{(\alpha,\beta)}} \sum_{i_1 \in I_{n_1}^{(\alpha,\beta)}, i_2 \in I_{n_2}^{(\alpha,\beta)}} s_{i_1}s_{i_2}\,\Delta_{p_{n_1},q_{n_1},p_{n_2},q_{n_2}}^{[n_1,n_2]}(T_{i_1,i_2}(\cdot; x, y))$$

on both sides of (7.13). For a given $\varepsilon' > 0$, we choose a number $\varepsilon > 0$ such that $\varepsilon < \varepsilon'$. Setting

$$\mathcal{A} := \left\{ (n_1, n_2) \in \mathbb{N}^2 : \| \widetilde{\mathcal{N}}_{n_1,n_2}(f; x, y) - f \|_{C_B(\mathbb{R}^2_+)} \geqq \varepsilon' \right\},$$

$$\mathcal{A}_0 := \left\{ (n_1, n_2) \in \mathbb{N}^2 : \| \widetilde{\mathcal{N}}_{n_1,n_2}(e_{00}; x, y) - e_{00} \|_{C_B(\mathbb{R}^2_+)} \geq \frac{\varepsilon' - \varepsilon}{4K} \right\},$$

$$\mathcal{A}_1 := \left\{ (n_1, n_2) \in \mathbb{N}^2 : \| \widetilde{\mathcal{N}}_{n_1,n_2}(e_{10}; x, y) - e_{10} \|_{C_B(\mathbb{R}^2_+)} \geq \frac{\varepsilon' - \varepsilon}{4K} \right\},$$

$$\mathcal{A}_2 := \left\{ (n_1, n_2) \in \mathbb{N}^2 : \| \widetilde{\mathcal{N}}_{n_1,n_2}(e_{01}; x, y) - e_{01} \|_{C_B(\mathbb{R}^2_+)} \geq \frac{\varepsilon' - \varepsilon}{4K} \right\}$$

and

$$\mathcal{A}_3 := \left\{ (n_1, n_2) \in \mathbb{N}^2 : \| \widetilde{\mathcal{N}}_{n_1,n_2}(e_{20} + e_{02}) - (e_{20} + e_{02}) \|_{C_B(\mathbb{R}^2_+)} \geq \frac{\varepsilon' - \varepsilon}{4K} \right\},$$

we see that the inclusion $\mathcal{A} \subset \cup_{j=0}^3 \mathcal{A}_i$ holds true and their double densities (in Pringsheim's sense) satisfy the following relation:

$$\delta_2(\mathcal{A}) \leq \delta_2(\mathcal{A}_0) + \delta_2(\mathcal{A}_1) + \delta_2(\mathcal{A}_2) + \delta_2(\mathcal{A}_3).$$

Letting $n_1, n_2 \to \infty$ and using the hypotheses, we obtain

$$st_2[\widetilde{\mathcal{N}}] - \lim_{n_1,n_2 \to \infty} \| T_{n_1,n_2}(f; x, y) - f \|_{C_B(\mathbb{R}^2_+)} = 0.$$

This completes the proof. $\qquad\square$

We now present an example associated with the $(p, q)$-analogue of generalized bivariate Bleimann-Butzer-Hahn (BBH) operators Mursaleen and Nasiruzzzaman (2017) satisfying the conditions of Theorem 7.5.3.

**Example 7.5.4.** *Let* $0 < q_{n_1}, q_{n_2} < p_{n_1}, p_{n_2} \leq 1$ *and* $f : \mathbb{R}^2_+ \to \mathbb{R}$. *The* $(p, q)$-*analogue of generalized bivariate Bleimann-Butzer-Hahn operators is defined by*

$$L_{n_1,n_2}^{(p_{n_1}, p_{n_2}),(q_{n_1}, q_{n_2})}(f; x, y) = \sum_{i_1, i_2 = 0}^{n_1, n_2} f_{i_1, i_2} \, [B_{n_1}^{i_1}(x)]_{p_{n_1}, q_{n_1}} \, [B_{n_2}^{i_2}(y)]_{p_{n_2}, q_{n_2}} \quad (7.14)$$

*where*

$$f_{i_1, i_2} = f \left( \frac{p_{n_1}^{n_1 - i_1 + 1} [i_1]_{p_{n_1}, q_{n_1}}}{[n_1 - i_1 + 1]_{p_{n_1}, q_{n_1}} q_{n_1}^{i_1}}, \frac{p_{n_2}^{n_2 - i_2 + 1} [i_2]_{p_{n_2}, q_{n_2}}}{[n_2 - i_2 + 1]_{p_{n_2}, q_{n_2}} q_{n_2}^{i_2}} \right),$$

$$[B_{n_1}^{i_1}(x)]_{p_{n_1}, q_{n_1}} = \begin{bmatrix} n_1 \\ i_1 \end{bmatrix}_{p_{n_1}, q_{n_1}} \frac{p_{n_1}^{\frac{(n_1 - i_1)(n_1 - i_1 - 1)}{2}} q_{n_1}^{\frac{i_1(i_1 - 1)}{2}}}{\prod_{j_1=0}^{n_1 - 1}(p_{n_1}^{j_1} + q_{n_1}^{j_1} x)} x^{i_1}, \qquad i_1 = 0, \dots, n_1$$

*and*

$$[B_{n_2}^{i_2}(y)]_{p_{n_2},q_{n_2}} = \begin{bmatrix} n_2 \\ i_2 \end{bmatrix}_{p_{n_2},q_{n_2}} \frac{p_{n_2}^{\frac{(n_2-i_2)(n_2-k_2-1)}{2}} q_{n_2}^{\frac{i_2(i_2-1)}{2}}}{\prod_{j_2=0}^{n_2-1}(p_{n_2}^{j_2}+q_{n_2}^{j_2}y)} y^{k_2}, \qquad i_2 = 0,\dots,n_2.$$

*It is easy to check that (7.14) is linear and positive. If we put $p_{n_1} = p_{n_2} = 1$, then we obtain q-bivariate BBH operators Ersan and Doğru (2009). Also, observe that*

$$L_{n_1,n_2}^{(p_{n_1},p_{n_2}),(q_{n_1},q_{n_2})}(e_{00};x,y) = 1;$$

$$L_{n_1,n_2}^{(p_{n_1},p_{n_2}),(q_{n_1},q_{n_2})}(e_{10};x,y) = \frac{[n_1]_{p_{n_1},q_{n_1}}}{[n_1+1]_{p_{n_1},q_{n_1}}}\frac{x}{1+x};$$

$$L_{n_1,n_2}^{(p_{n_1},p_{n_2}),(q_{n_1},q_{n_2})}(e_{01};x,y) = \frac{[n_2]_{p_{n_2},q_{n_2}}}{[n_2+1]_{p_{n_2},q_{n_2}}}\frac{y}{1+y};$$

$$L_{n_1,n_2}^{(p_{n_1},p_{n_2}),(q_{n_1},q_{n_2})}(e_{20};x,y) = \frac{x^2 p_{n_1} q_{n_1}^2 [n_1]_{p_{n_1},q_{n_1}}[n_1-1]_{p_{n_1},q_{n_1}}}{(1+x)(p_{n_1}+q_{n_1}x)[n_1+1]_{p_{n_1},q_{n_1}}^2}$$
$$+ \frac{p_{n_1}^{n_1+1}[n_1]_{p_{n_1},q_{n_1}}}{[n_1+1]_{p_{n_1},q_{n_1}}^2}\frac{x}{1+x};$$

$$L_{n_1,n_2}^{(p_{n_1},p_{n_2}),(q_{n_1},q_{n_2})}(e_{02};x,y) = \frac{y^2 p_{n_2} q_{n_2}^2 [n_2]_{p_{n_2},q_{n_2}}[n_2-1]_{p_{n_2},q_{n_2}}}{(1+y)(p_{n_2}+q_{n_2}y)[n_2+1]_{p_{n_2},q_{n_2}}^2}$$
$$+ \frac{p_{n_2}^{n_2+1}[n_2]_{p_{n_2},q_{n_2}}}{[n_2+1]_{p_{n_2},q_{n_2}}^2}\frac{y}{1+y}.$$

*Now we consider the following positive linear operators $A_{n_1,n_2} : C_B(\mathbb{R}_+^2) \to C_B(\mathbb{R}_+^2)$ such that*

$$A_{n_1,n_2}(f;x,y) = (1+y_{n_1,n_2})\, L_{n_1,n_2}^{(p_{n_1},p_{n_2}),(q_{n_1},q_{n_2})}(f;x,y) \qquad (7.15)$$

*where the double difference sequence $(y_{n_1,n_2})$ is defined as in Example 7.4.2. It is obvious that the double sequence $(A_{n_1,n_2})$ satisfies all conditions of Theorem 7.5.3. By taking into account the conditions (7.6)-(7.8) and using the similar technique in Theorem 3.3 in Mursaleen and Nasiruzzaman (2017), we obtain that*

$$st_2[\widetilde{\mathcal{N}}] - \lim_{n_1,n_2\to\infty} \|A_{n_1,n_2}(e_{00};x,y)-e_{00}\|_{C_B(\mathbb{R}_+^2)} = 0,$$

$$st_2[\widetilde{\mathcal{N}}] - \lim_{n_1,n_2\to\infty} \|A_{n_1,n_2}(e_{10};x,y)-e_{10}\|_{C_B(\mathbb{R}_+^2)} = 0,$$

$$st_2[\widetilde{\mathcal{N}}] - \lim_{n_1,n_2\to\infty} \|A_{n_1,n_2}(e_{01};x,y)-e_{01}\|_{C_B(\mathbb{R}_+^2)} = 0,$$

$$st_2[\widetilde{\mathcal{N}}] - \lim_{n_1,n_2\to\infty} \|A_{n_1,n_2}(e_{20};x,y)-e_{20}\|_{C_B(\mathbb{R}_+^2)} = 0,$$

$$st_2[\dot{\mathcal{N}}] - \lim_{n_1,n_2\to\infty} \|A_{n_1,n_2}(e_{02};x,y)-e_{02}\|_{C_B(\mathbb{R}_+^2)} = 0.$$

*Since $st_2[\widetilde{\mathcal{N}}] - \lim y_{n_1,n_2} = 0$, then for all $f \in C_B(\mathbb{R}_+^2)$*

$$st_2[\widetilde{\mathcal{N}}] - \lim_{n_1,n_2 \to \infty} \|A_{n_1,n_2}(f;x,y) - f\|_{C_B(\mathbb{R}_+^2)} = 0.$$

*This gives us the statistically weighted $\widetilde{\mathcal{N}}$-summability of the double opera-tors $A_{n_1,n_2}$ on $\mathbb{R}_+^2$. Thus, clearly, Theorem 7.5.3 does not work for the dou-ble sequence $(A_{n_1,n_2})$ of positive linear opeartors in classical, statistical and weighted statistical versions of Korovkin type approximation theorems.*

We note that, if we choose $\alpha(r) = 1$, $\beta(r) = r$, $s_t = 1$ for each $t \in [1,r]$ $(t, r \in \mathbb{N})$, $n_1, n_2 = 1$ and $p_{n_1}, p_{n_2} \to 1$, $q_{n_1}, q_{n_2} \to 1$ as $n_1, n_2 \to \infty$, then Theorem 7.5.3 immediately gives the following classical and statistical Ko-rovkin type results for a double sequence of positive linear operators defined on $C_B(\mathbb{R}_+^2)$.

**Corollary 7.5.5.** *Let $(T_{n_1,n_2})$ be a double sequence of positive linear operators acting from $C_B(\mathbb{R}_+^2)$ into itself satisfying the following conditions:*

$$st_2 - \lim_{n_1,n_2 \to \infty} \|\widetilde{\mathcal{D}}_{n_1,n_2}(e_{00};x,y) - e_{00}\|_{C_B(\mathbb{R}_+^2)} = 0,$$

$$st_2 - \lim_{n_1,n_2 \to \infty} \|\widetilde{\mathcal{D}}_{n_1,n_2}(e_{10};x,y) - e_{10}\|_{C_B(\mathbb{R}_+^2)} = 0,$$

$$st_2 - \lim_{n_1,n_2 \to \infty} \|\widetilde{\mathcal{D}}_{n_1,n_2}(e_{01};x,y) - e_{01}\|_{C_B(\mathbb{R}_+^2)} = 0,$$

$$st_2 - \lim_{n_1,n_2 \to \infty} \|\widetilde{\mathcal{D}}_{n_1,n_2}(e_{20};x,y) - e_{20}\|_{C_B(\mathbb{R}_+^2)} = 0,$$

$$st_2 - \lim_{n_1,n_2 \to \infty} \|\widetilde{\mathcal{D}}_{n_1,n_2}(e_{02};x,y) - e_{02}\|_{C_B(\mathbb{R}_+^2)} = 0.$$

*where $\widetilde{\mathcal{D}}_{n_1,n_2}(\cdot;x,y) = \frac{1}{n_1 n_2} \sum_{i_1=1}^{n_1} \sum_{i_2=1}^{n_2} \Delta_{p_{n_1},q_{n_1},p_{n_2},q_{n_2}}^{[1,1]}(T_{i_1,i_2}(\cdot;x,y))$ Then, for all $f \in C_B(\mathbb{R}_+^2)$,*

$$st_2 - \lim_{n_1,n_2 \to \infty} \|\widetilde{\mathcal{D}}_{n_1,n_2}(f;x,y) - f\|_{C_B(\mathbb{R}_+^2)} = 0.$$

**Corollary 7.5.6.** *Let $(T_{n_1,n_2})$ be a double sequence of positive linear operators acting from $C_B(\mathbb{R}_+^2)$ into itself satisfying the following conditions:*

$$P - \lim_{n_1,n_2 \to \infty} \|\widetilde{\mathcal{D}}_{n_1,n_2}(e_{00};x,y) - e_{00}\|_{C_B(\mathbb{R}_+^2)} = 0,$$

$$P - \lim_{n_1,n_2 \to \infty} \|\widetilde{\mathcal{D}}_{n_1,n_2}(e_{10};x,y) - e_{10}\|_{C_B(\mathbb{R}_+^2)} = 0,$$

$$P - \lim_{n_1,n_2 \to \infty} \|\widetilde{\mathcal{D}}_{n_1,n_2}(e_{01};x,y) - e_{01}\|_{C_B(\mathbb{R}_+^2)} = 0,$$

$$P - \lim_{n_1,n_2 \to \infty} \|\widetilde{\mathcal{D}}_{n_1,n_2}(e_{20};x,y) - e_{20}\|_{C_B(\mathbb{R}_+^2)} = 0,$$

$$P - \lim_{n_1,n_2 \to \infty} \|\widetilde{\mathcal{D}}_{n_1,n_2}(e_{02};x,y) - e_{02}\|_{C_B(\mathbb{R}_+^2)} = 0$$

*where $\widetilde{\mathcal{D}}_{n_1,n_2}(\cdot;x,y)$ is defined as above. Then, for all $f \in C_B(\mathbb{R}_+^2)$,*

$$P - \lim_{n_1,n_2 \to \infty} \|\widetilde{\mathcal{D}}_{n_1,n_2}(f;x,y) - f\|_{C_B(\mathbb{R}_+^2)} = 0.$$

# Bibliography

Adams C. R. On summability of double series. *Math. Ann.*, 2(34):215–230, 1932.

Aktuğlu H. Korovkin type approximation theorems proved via $\alpha\beta$-statistical convergence. *Journal of Computational and Applied Mathematics*, (259):174–181, 2014.

Altay B. and Başar F. Some new spaces of double sequences. *J. Math. Anal. Appl.*, 1(309):70–90, 2005.

Başar F. and Aydın C. Some new difference sequence spaces. *Appl. Math. Comput*, 3(157):677–693, 2004.

Baliarsingh P. Some new difference sequence spaces of fractional order and their dual spaces. *Appl. Math. Comput*, 18(219):9737–9742, 2013.

Bektaş C. A., Et M., and Çolak R. Generalized difference sequence spaces and their dual spaces. *J. Math. Anal. Appl.*, 2(292):423–432, 2004.

Belen C. and Mohiuddine S. A. Generalized statistical convergence and application. *Appl. Math. Comput.*, (219):9821–9826, 2013.

Bohman H. On approximation of continuous and of analytic functions. *Arkiv Math*, 4(2):43–56, 1952.

Burban I. M. and Klimyk A. U. $P, Q$ differentiation, $P, Q$ integration and $P, Q$ hypergeometric functions related to quantum groups. *Integr. Trans. Spec. Funct.*, (2):15–36, 1994.

Cai Q.-B. On ( p, q)-analogue of modified Bernstein-Schurer operators for functions of one and two variables. *Journal of Applied Mathematics and Computing*, 1-2(54):1–21, 2017.

Çakan C. and Altay B. Statistical boundedness and statistical core of double sequences. *J. Math. Anal. Appl.*, 2(317):690–697, 2006.

Chen C.-P. and Chang C.-T. Tauberian conditions under which the original convergence of double sequences follows from the statistical convergence of their weighted means. *J. Math. Anal. Appl.*, (332):1242–1248, 2007.

Connor J. On strong matrix summability with respect to a modulus and statistical convergence. *Canad. Math. Bull.*, (32):194–198, 1989.

Demirci K. and Kolay B. $A$-statistical relative modular convergence of positive linear operators. *Positivity*, 21(3):847–863, 2017.

Ersan S. and Doğru O. Statistical approximation properties of q-Bleimann, Butzer and Hahn operators. *Math. Comput. Modell.*, (49):1595–1606, 2009.

Et M. and Çolak R. On some generalized difference sequence spaces. *Soochow J. Math.*, 4(21):377–386, 1995.

Et M. and Şengül H. Some Cesàro-type summability spaces of order $\alpha$ and lacunary statistical convergence of order $\alpha$. *Filomat*, (28):1593–1602, 2014.

Et M. and Nuray F. $\Delta^m$-statistical convergence. *Indian J. Pure Appl. Math.*, 6(32):961–969, 2001.

Fast H. Sur la convergence statistique. *Colloq. Math.*, pages 241–244, 1951.

Fridy J. A. and Orhan C. Lacunary statistical convergence. *Pac. J. Math.*, (160):43–51, 1993.

Gadjiev A. D. and Orhan C. Some approximation theorems via statistical convergence. *Rocky Mountain J. Math.*, 4(32):129–138, 2002.

Hana L.-W., Chua Y., and Qiu Z.-Y. Generalized Bezier curves and surfaces based on Lupaş $q$-analogue of Bernstein functions in CAGD. *Journal of Computational and Applied Mathematics*, (261):352–363, 2014.

Hill J. D. On perfect summability of double sequences. *Bull. Am. Math. Soc.*, 2(46):327–331, 1940.

Hounkonnou M. N. and Kyemba J. D. B. $R(p,q)$-calculus: differentiation and integration. *SUT J. Math.*, 2(49):145–167, 2013.

Kadak U. On weighted statistical convergence based on $(p,q)$-integers and related approximation theorems for functions of two variables. *J. Math. Anal. Appl.*, 2(443):752–764, 2016.

Kadak U. Generalized weighted invariant mean based on fractional difference operator with applications to approximation theorems for functions of two variables. *Results in Mathematics*, 3(72):1181–1202, 2017a.

Kadak U. On relative weighted summability in modular function spaces and associated approximation theorems. *Positivity*, 21(4):1593–1614, 2017b.

Kadak U. Weighted statistical convergence based on generalized difference operator involving $(p,q)$-gamma function and its applications to approximation theorems. *J. Math. Anal. Appl.*, (448):1633–1650, 2017c.

Kadak U. and Baliarsingh P. On certain Euler difference sequence spaces of fractional order and related dual properties. *J. Nonlinear Sci. Appl.*, (8):997–1004, 2015.

Kadak U., Braha N. L., and Srivastava H. M. Statistical weighted B-summability and its applications to approximation theorems. *Appl. Math. Comput.*, (302):80–96, 2017a.

Kadak U., Mishra V. N., and Pandey S. Chlodowsky type generalization of (p,q)-Szasz operators involving Brenke type polynomials. *Revista de la Real Academia de Ciencias Exactas, Fsicas y Naturales. Serie A. Matemticas*, 2017b.

Kanat K. and Sofyalıoğlu M. Some approximation results for Stancu type Lupas Schurer operators based on (p, q)-integers. *Appl. Math. Comput.*, (317):129–142, 2018.

Karakaya V. and Chishti T. A. Weighted statistical convergence. *Iran. J. Sci. Technol. Trans. A. Sci.*, (33):219–223, 2009.

Khan K. and Lobiyal D. K. Bezier curves based on Lupaş $(p,q)$-analogue of Bernstein functions in CAGD. *Journal of Computational and Applied Mathematics*, (317):458–477, 2017.

Kı zmaz H. On certain sequence spaces. *Canad. Math. Bull.*, 2(24):169–176, 1981.

Kirişci M. and Kadak U. The method of almost convergence with operator of the form fractional order and applications. *J. Nonlinear Sci. Appl.*, (10):828–842, 2017.

Kolk E. Matrix summability of statistically convergent sequences. *Analysis*, (13):77–83, 1993.

Korovkin P. P. On convergence of linear positive operators in the spaces of continuous functions (Russian). *Doklady Akad. Nauk. S. S. S. R.*, 4(90):961–964, 1953.

Malik N. and Gupta V. Approximation by $(p,q)$-Baskakov-Beta operators. *Appl. Math. Comput.*, (293):49–56, 2017.

Mohiuddine S. A. Statistical weighted $A$ -summability with application to Korovkins type approximation theorem. *J. Inequal. Appl.*, 2016.

Mursaleen M. λ-statistical convergence. *Math. Slovaca*, (50):111–115, 2000.

Mursaleen M. and Alotaibi A. Generalized statistical convergence of difference sequences. *Adv. Difference Equ.*, 2013(212), 2013.

Mursaleen M., Ansari K. J., and Khan A. Erratum to "On $(p,q)$-analogue of Bernstein Operators". *Appl. Math. Comput.*, (266):874–882, 2015.

Mursaleen M. and Edely O. H. H. Statistical convergence of double sequences. *J. Math. Anal. Appl.*, 2(288):223–231, 2003.

Mursaleen M., Karakaya V., Ertürk M., and Gürsoy F. Weighted statistical convergence and its application to Korovkin type approximation theorem. *Appl. Math. Comput.*, (218):9132–9137, 2012.

Mursaleen M., Khan F., and Khan A. Approximation by $(p, q)$-Lorentz polynomials on a compact disk. *Complex Analysis and Operator Theory*, 8(10):1725–1740, 2016.

Mursaleen M. and Nasiruzzzaman M. Some approximation properties of bivariate Bleimann-Butzer-Hahn operators based on (p, q)-integers. *Boll. Unione Mat. Ital.*, (10):271–289, 2017.

Pringsheim A. Zur theorie der zweifach unendlichen Zahlenfolgen. *Math. Ann.*, (53):289–321., 1900.

Sahai V. and Yadav S. Representations of two parameter quantum algebras and $p, q$-special functions. *J. Math. Anal. Appl.*, (335):268–279, 2007.

Savaş E. Strong almost convergence and almost statistical convergence. *Hokkaido Math. J.*, 3(29):531–536, 2000.

Srivastava H. M., Mursaleen M., and Khan A. Generalized equi-statistical convergence of positive linear operators and associated approximation theorems. *Math. Comput. Model.*, (55):2041–2051, 2012.

Steinhaus H. Sur la convergence ordinaire et la convergence asymptotique. *Colloq. Math.*, (2):73–74, 1951.

Zygmund A. *Trigonometric Series*. 2nd ed., Cambridge University Press, Cambridge, UK., 1959.

# Chapter 8

## Birkhoff-James Orthogonality and Its Application in the Study of Geometry of Banach Space

**Kallol Paul**

*Department of Mathematics, Jadavpur University, Kolkata, West Bengal, India*

**Debmalya Sain**

*Department of Mathematics, Indian Institute of Science, Bengaluru, Karnataka, India*

## 8.1 Introduction

In spite of the less restrictive framework of Banach spaces, as compared to Hilbert spaces, geometry of Banach spaces is surprisingly rich and full of elegant results. Indeed, as we would endeavor to illustrate in this chapter, many important results of Hilbert space geometry has their precursors in the study of geometry of Banach spaces. The geometry of real finite-dimensional Banach spaces, popularly known as Minkowski Geometry, is referred to as a geometry

245

"next" to the Euclidean Geometry in the famous fourth Hilbert problem. As it stands today, geometry of Banach spaces is an important field of study in modern Mathematics. It is an active area of research that has close connections with many other fields of Mathematics, including Approximation Theory, Convex Analysis and Finsler Geometry. It is well known that many "natural" geometric properties may fail to hold in a general normed linear space unless the norm is derived from an inner product. One of the major striking differences between Euclidean Geometry and the geometry of Banach spaces is that there is no unique notion of orthogonality in the later case. In fact, there are several notions of orthogonality in a normed linear space, which are not equivalent in general. In this chapter, we deal solely with Birkhoff-James orthogonality Birkhoff (1935); James (47 a), arguably the most "natural" and important notion of orthogonality defined in a normed linear space. We would like to mention here, for the sake of completeness only, that there are several other notions of orthogonality in Banach space, including Roberts orthogonality, isosceles orthogonality, Pythagorean orthogonality, Carlsson orthogonality etc. We refer the readers to two excellent and detailed references Alonso et al. (2012); Amir (1986) in this context.

It is well known that given any two Banach spaces, not necessarily distinct, one can associate another Banach space, namely the Banach space of all bounded linear operators between the given Banach spaces, endowed with the usual operator norm. In this chapter, we would like to study various geometric properties of the operator space, using Birkhoff-James orthogonality techniques. Our scheme is to obtain some connections between orthogonality of bounded linear operators and orthogonality of norm attaining vectors of the ground space. We begin with a study of Birkhoff-James orthogonality in a general Banach space and then gradually proceed towards studying the operator space geometry. As it will be illustrated in this chapter, our exploration has several possible areas of application, including the study of the smoothness of bounded linear operators. Without further ado, let us now establish the relevant notations and terminologies, that would be used throughout the chapter.

### 8.1.1 Notations and terminologies

Let us reserve the symbols $\mathbb{X}, \mathbb{Y}$ for Banach spaces and $T, A$ for bounded linear operators acting between Banach spaces. Although completeness does not feature in our study, and the notion of orthogonality can be developed in a general normed linear space, we mostly consider Banach spaces only. We also reserve the symbol $\mathbb{H}$ for real Hilbert spaces, if not otherwise mentioned. Furthermore, throughout the chapter, we consider the Banach spaces to be real, i.e., defined over the scalar field $\mathbb{R}$. Let $B_\mathbb{X} = \{x \in \mathbb{X} : \|x\| \leq 1\}$ and $S_\mathbb{X} = \{x \in \mathbb{X} : \|x\| = 1\}$ be the unit ball and the unit sphere of the Banach space $\mathbb{X}$ respectively. Let $\mathbb{B}(\mathbb{X}, \mathbb{Y})(\mathbb{K}(\mathbb{X}, \mathbb{Y}))$ denote the set of all bounded (compact) linear operators from $\mathbb{X}$ to $\mathbb{Y}$. We write $\mathbb{B}(\mathbb{X}, \mathbb{Y}) = \mathbb{B}(\mathbb{X})$ and $\mathbb{K}(\mathbb{X}, \mathbb{Y}) = \mathbb{K}(\mathbb{X})$

if $X = Y$. It is well known that $\mathbb{B}(X, Y)$ and $\mathbb{K}(X, Y)$ are Banach algebras, endowed with the usual operator norm $\|T\| = \sup\{\|Tx\| : x \in S_X\}$. $T \in \mathbb{B}(X, Y)$ is said to attain norm at $x_0 \in S_X$ if $\|Tx_0\| = \|T\|$. Let $M_T$ denote the set of all vectors in $S_X$ at which $T$ attains norm, i.e.,

$$M_T = \{x \in S_X : \|Tx\| = \|T\|\}.$$

For any two elements $x, y \in X$, $x$ is said to be orthogonal to $y$ in the sense of Birkhoff-James, written as $x \perp_B y$, if

$$\|x\| \leq \|x + \lambda y\| \ \forall \lambda \in \mathbb{R}.$$

Similarly, for $T, A \in \mathbb{B}(X, Y)$, $T$ is said to be orthogonal to $A$, if

$$\|T\| \leq \|T + \lambda A\| \ \forall \lambda \in \mathbb{R}.$$

It is to be noted that if the norm is derived from an inner product then $x \perp_B y$ if and only if $x$ is orthogonal to $y$ in the usual sense of the inner product.
An element $x \in B_X$ is said to be an *extreme point* of $B_X$ if $x = (1 - t)y + tz$, with $y, z \in B_X$ and $t \in (0, 1)$, implies that $x = y = z$. $x$ is an *exposed point* of $B_X$ if there exists a hyperplane $H$ (subspace of codimension 1) such that $x + H$ supports $B_X$ at $x$ (i.e., $B_X$ lies on one side of $x + H$), with $(x + H) \cap S_X = \{x\}$. It is rather elementary to check that all extreme points are in $S_X$ and every exposed point is an extreme point but not conversely. A much deeper result is that the set of exposed points is dense in the set of extreme points (Klee Jr. (1958)). $X$ is *strictly convex* if every $x \in S_X$ is an extreme point of $B_X$, and $X$ is *smooth* if for every $x \in S_X$, there is a unique hyperplane supporting $B_X$ at $x$. Equivalently, as mentioned in James (47 b), a Banach space $X$ is smooth if and only if Birkhoff-James orthogonality is right additive in $X$, i.e., for any $x, y, z \in X$, $x \perp_B y$, $x \perp_B z$ implies that $x \perp_B (y + z)$.

If $X$ is an inner product space then $x \perp_B y$ implies $\|x\| < \|x + \lambda y\|$ for all scalars $\lambda \neq 0$. Motivated by this fact, the notion of strong orthogonality Paul et al. (2013); Sain et al. (2015b) in a Banach space is introduced in the following way:
**Strongly orthogonal in the sense of Birkhoff-James:** An element $x \in X$ is said to be strongly orthogonal to another element $y \in X$ in the sense of Birkhoff-James, written as $x \perp_{SB} y$, if

$$\|x\| < \|x + \lambda y\| \ \text{ for all scalars } \lambda \neq 0.$$

If $x \perp_{SB} y$ then $x \perp_B y$ but the converse is not true. In $l_\infty(\mathbb{R}^2)$ the element $(1, 0)$ is orthogonal to $(0, 1)$ in the sense of Birkhoff-James but not strongly orthogonal.

The next notion is crucial for obtaining a complete characterization of the Birkhoff-James orthogonality of bounded linear operators on finite-dimensional Banach spaces. It was introduced by Sain Sain (2017) in the following way:

For any two elements $x, y \in \mathbb{X}$, let us say that $y \in x^+$ if $\|x + \lambda y\| \geq \|x\|$ for all $\lambda \geq 0$. Accordingly, we say that $y \in x^-$ if $\|x + \lambda y\| \geq \|x\|$ for all $\lambda \leq 0$. Using this notion, Sain Sain (2017) completely characterized Birkhoff-James orthogonality of linear operators defined on finite-dimensional Banach spaces. In Sain (2017), Sain introduced the notion of left symmetric and right symmetric points in Banach spaces:

An element $x \in \mathbb{X}$ is said to be left symmetric (with respect to Birkhoff-James orthogonality) if $x \perp_B y$ implies $y \perp_B x$ for any $y \in \mathbb{X}$. Likewise, for an element $x \in \mathbb{X}$, $x$ is said to be right symmetric (with respect to Birkhoff-James orthogonality) if $y \perp_B x$ implies $x \perp_B y$ for any $y \in \mathbb{X}$. It is easy to observe that similar to the notion of Birkhoff-James orthogonality, the definitions of left symmetric and right symmetric points extend in an obvious way to bounded linear operators. In fact, this is the case we are more interested in. We deal with the problem both in the context of Hilbert spaces and Banach spaces and give some complete and some partial answers towards completely characterizing left symmetric and right symmetric bounded linear operators on a Banach space.

### 8.1.2  A brief motivation and history

Birkhoff-James orthogonality is intimately connected with various geometric properties of a Banach space, including strict convexity, uniform convexity and smoothness. Indeed, this rich connection strengthens the role of Birkhoff-James orthogonality in exploring Banach space geometry. We refer the readers to Amir (1986) for an excellent survey of some of the relevant works in this context. It was independently proved by Bhatia and Šemrl Bhatia and Šemrl (1999) and Paul Paul (1999) that for bounded linear operators $T, A$ acting on a Hilbert space $\mathbb{H}$, $T \perp_B A$ if and only if there exists $\{x_n\} \subset S_{\mathbb{H}}$ such that $\|Tx_n\| \longrightarrow \|T\|$ and $\langle Tx_n, Ax_n \rangle \longrightarrow 0$. In light of this beautiful result, it is evident that for a finite-dimensional Hilbert space $\mathbb{H}$, $T \perp_B A$ if and only if there exists $x \in S_{\mathbb{X}}$ such that $x \in M_T$ and $Tx \perp_B Ax$. Indeed, the standard compactness argument does the trick. This profound result has opened a new vista for studying the geometry of operator spaces, by exhibiting a tractable connection between Birkhoff-James orthogonality of bounded linear operators and Birkhoff-James orthogonality of some "special" elements of the ground space. Indeed, much of our emphasis in this chapter would be on obtaining various generalizations of this result and to explore possible areas of application.

Bhatia and Šemrl conjectured in their paper that if $\mathbb{X}$ is a finite-dimensional complex Banach space and $T, A \in \mathbb{B}(\mathbb{X})$ are such that $T \perp_B A$ then there exists $x \in S_{\mathbb{X}}$ such that $x \in M_T$ and $Tx \perp_B Ax$. Li and Schneider Li and Schneider

(2002) gave examples of finite-dimensional Banach spaces $\mathbb{X}$ in which there exist operators $T, A \in \mathbb{B}(\mathbb{X})$ such that $T \perp_B A$ but there exists no $x \in S_\mathbb{X}$ such that $x \in M_T$ and $Tx \perp_B Ax$, which showed that the conjecture by Bhatia and Šemrl is not true.

In spirit of the result obtained by Bhatia and Šemrl, let us say that $T$ *satisfies the Bhatia-Šemrl (BŠ) property* if for any $A \in \mathbb{B}(\mathbb{X})$, $T \perp_B A$ implies that there exists $x \in M_T$ such that $Tx \perp_B Ax$. Benítez et al. Benítez et al. (2007) proved that a finite-dimensional real Banach space $\mathbb{X}$ is an inner product space if and only if every $T \in \mathbb{B}(\mathbb{X})$ satisfies the BŠ property. In this chapter, along with other related questions, we would be considering a "localized version" of the above problem to characterize the set of bounded linear operators that satisfy the BŠ property. As it will be seen, the methods discussed here are completely ingenious and the arguments are general, in the sense that they are valid for any Banach space and not just for Hilbert spaces. Moreover, using the powerful parallelogram equality available to us in the setting of Hilbert spaces, we would illustrate that the theorem of Bhatia and Šemrl follow as a simple corollary to our result.

### 8.1.3 Schematics

- Relation between Birkhoff-James orthogonality in $\mathbb{B}(\mathbb{X})$ and $\mathbb{X}$.

- Characterization of inner product space (IPS) in terms of operator norm attainment.

- Operator orthogonality on finite-dimensional Banach spaces.

- Operator orthogonality on infinite-dimensional normed spaces.

- Operator orthogonality on Hilbert spaces.

- Operator orthogonality and smoothness of bounded linear operators.

- Characterization of operator orthogonality for finite-dimensional Banach spaces.

- Symmetry of operator orthogonality.

---

## 8.2 Relation between Birkhoff-James orthogonality in $\mathbb{B}(\mathbb{X})$ and $\mathbb{X}$.

We begin with considering the Birkhoff-James orthogonality of two linear operators $T, A$ in the setting of finite-dimensional Banach spaces. As it turns out, if $M_T$ is of a particularly nice form ($M_T = \pm D$, where $D$ is a compact connected subset of $S_{\mathbb{X}}$), then a tractable characterization of $T \perp_B A$ can be obtained. If $\mathbb{X}$ is an Euclidean space then the characterization of operator orthogonality obtained by Bhatia and Šemrl in Bhatia and Šemrl (1999) can actually be derived from the following general characterization Sain and Paul (2013).

**Theorem 8.2.1.** *Let $\mathbb{X}$ be a finite-dimensional Banach space. Let $T \in \mathbb{B}(\mathbb{X})$ be such that $T$ attains norm at only $\pm D$, where $D$ is a compact connected subset of $S_{\mathbb{X}}$. Then for $A \in \mathbb{B}(\mathbb{X})$ with $T \perp_B A$, there exists $x \in D$ such that $Tx \perp_B Ax$.*

*Proof.* If possible, suppose that there does not exist any $x \in D$ such that $Tx \perp_B Ax$. Let us now complete the proof in the following three steps, by the method of contradiction.

**Step 1:** In the first step let us show that $D = W_1 \cup W_2$, where,

$$W_1 = \{x \in D : \|Tx + \lambda Ax\| > \|T\| \ \forall \ \lambda > 0\},$$

$$W_2 = \{x \in D : \|Tx + \lambda Ax\| > \|T\| \ \forall \ \lambda < 0\}.$$

Let $x_0 \in D$ be arbitrary. Since $Tx_0$ is not orthogonal to $Ax_0$ in the sense of Birkhoff-James, there exists $\lambda_0 \in \mathbb{R}$ such that $\|Tx_0 + \lambda_0 Ax_0\| < \|Tx_0\| = \|T\|$. Now, either $\lambda_0 > 0$ or $\lambda_0 < 0$. Assume that $\lambda_0 < 0$.
Note that for $\lambda > 0$, $\exists \ t \in (0,1)$ such that

$$Tx_0 = t(Tx_0 + \lambda Ax_0) + (1-t)(Tx_0 + \lambda_0 Ax_0)$$
$$\Rightarrow \quad \|Tx_0\| < t\|(Tx_0 + \lambda Ax_0)\| + (1-t)\|Tx_0\|$$
$$\Rightarrow \quad \|Tx_0\| < \|(Tx_0 + \lambda Ax_0)\|$$

$$\text{Therefore,} \ \ \|Tx_0 + \lambda Ax_0\| > \|Tx_0\| = \|T\| \ \ \forall \ \lambda > 0.$$

If $\lambda_0 > 0$ then similarly it can be shown that

$$\|Tx_0 + \lambda Ax_0\| > \|Tx_0\| = \|T\| \ \ \forall \ \lambda < 0.$$

Thus, for $x \in D$, either $\|Tx + \lambda Ax\| > \|T\| \ \forall \lambda > 0$ or $\|Tx + \lambda Ax\| > \|T\| \ \forall \lambda < 0$ and so $D = W_1 \cup W_2$.

**Step 2:** Claim $W_1 \neq \phi$ and $W_2 \neq \phi$.

To show that $W_1 \neq \phi$, it is sufficient to prove that there exists $y_0 \in D$ such that
$$\|Ty_0 + \lambda Ay_0\| > \|Ty_0\| = \|T\| \ \forall \ \lambda > 0.$$

If possible, suppose that $W_1 = \phi$, i.e., for all $x \in D, \|Tx + \lambda Ax\| > \|Tx\| = \|T\| \ \forall \ \lambda < 0$. Since $Tx$ is not orthogonal to $Ax$ in the sense of Birkhoff-James, there exists $\lambda_0 > 0$ such that $\|Tx + \lambda_0 Ax\| < \|Tx\| = \|T\|$. By the convexity of the norm function, it now follows that

$$\|Tx + \lambda Ax\| < \|Tx\| = \|T\| \quad \forall \quad \lambda \in (0, \lambda_0).$$

Choose $\lambda_x$ such that $0 < \lambda_x < \min\{\lambda_0, 1\}$.
Consider the continuous function $g : S_{\mathbb{X}} \times [-1, 1] \longrightarrow \mathbb{R}$ defined by

$$g(x, \lambda) = \|Tx + \lambda Ax\|.$$

Then $g(x, \lambda_x) = \|Tx + \lambda_x Ax\| < \|T\|$ and so by the continuity of $g$, there exists $r_x, \delta_x > 0$ such that

$$g(y, \lambda) < \|T\| \quad \forall \quad y \in B(x, r_x) \cap S_{\mathbb{X}} \text{ and } \forall \, \lambda \in (\lambda_x - \delta_x, \lambda_x + \delta_x).$$

Let $y \in B(x, r_x) \cap S_{\mathbb{X}}$. Then for $\lambda \in (0, \lambda_x)$, there exists $t \in (0, 1)$ such that

$$
\begin{aligned}
& Ty + \lambda Ay = t(Ty) + (1 - t)(Ty + \lambda_x Ay) \\
\Rightarrow \quad & \|Ty + \lambda Ay\| \leq t\|Ty\| + (1 - t)\|Ty + \lambda_x Ay\| \\
\Rightarrow \quad & \|Ty + \lambda Ay\| < \|T\|
\end{aligned}
$$

Therefore, $g(y, \lambda) = \|Ty + \lambda Ay\| < \|T\| \quad \forall \quad y \in B(x, r_x) \cap S_{\mathbb{X}}$ and $\forall \lambda \in (0, \lambda_x)$. Since $g(x, \lambda) = g(-x, \lambda)$, it follows that $\|Ty + \lambda Ay\| < \|T\| \quad \forall \quad y \in B(-x, r_x) \cap S_{\mathbb{X}}$ and $\forall \lambda \in (0, \lambda_x)$.
Next, for $z \in S_{\mathbb{X}} \setminus (D \cap (-D))$, clearly $g(z, 0) = \|Tz\| < \|T\|$. So by the continuity of $g$, there exist open balls $B(z, r_z) \cap S_{\mathbb{X}}$ and $(-\delta_z, \delta_z)$ such that $g(y, \lambda) = \|Ty + \lambda Ay\| < \|T\| \,\forall\, y \in B(z, r_z) \cap S_{\mathbb{X}}$ and $\forall \lambda \in (-\delta_z, \delta_z)$.
Consider the open cover

$$\{B(x, r_x) \cap S_{\mathbb{X}}, \ B(-x, r_x) \cap S_{\mathbb{X}} : x \in D\} \cup \{B(z, r_z) \cap S_{\mathbb{X}} : z \in S_{\mathbb{X}}, z \notin D \cup (-D)\}$$

of $S_{\mathbb{X}}$. By the compactness of $S_{\mathbb{X}}$, this cover has a finite sub-cover of the form

$$S_{\mathbb{X}} \subset \cup_{i=1}^{n_1} B(x_i, r_{x_i}) \cup_{i=1}^{n_1} B(-x_i, r_{x_i}) \cup_{k=1}^{n_2} B(z_k, r_{z_k}) \cap S_{\mathbb{X}},$$

for some positive integers $n_1, n_2$.
Choose $\mu_0 \in (\cap_{i=1}^{n_1} (0, \lambda_{x_i})) \cap (\cap_{k=1}^{n_2} (-\delta_{z_k}, \delta_{z_k}))$.
Now, since $\mathbb{X}$ is finite-dimensional, $T + \mu_0 A$ attains norm at some $w_0 \in S_X$. However, it follows from the choice of $\mu_0$ that $\|T + \mu_0 A\| = \|(T + \mu_0 A)w_0\| < \|T\|$, which contradicts that $T \perp_B A$.
Thus, it is not possible that for all $x \in D, \|Tx + \lambda Ax\| > \|Tx\| = \|T\| \,\forall\, \lambda < 0$ and so $W_1 \neq \phi$.
Similar argument shows that $W_2 \neq \phi$.

**Step 3:** Finally it is shown that $W_1, W_2$ form a separation of $D$.

Clearly, $\overline{W_1} \cap W_2 = \phi$ and $W_1 \cap \overline{W_2} = \phi$. For, otherwise, one can find $x \in D$ such that $Tx \perp_B Ax$.

Therefore, we have, $D = W_1 \cup W_2$, $\overline{W_1} \cap W_2 = \phi$ and $W_1 \cap \overline{W_2} = \phi$. This establishes that $W_1, W_2$ form a separation of $D$. However, this clearly leads to a contradiction, since $D$ is connected.

Therefore, there exists some $x \in D$ such that $Tx \perp_B Ax$. This completes the proof of the theorem. $\qquad\square$

**Corollary 8.2.2.** *Let $\mathbb{X}$ be a finite-dimensional Banach space. Let $T \in \mathbb{B}(\mathbb{X})$ be such that $T$ attains norm at only $\pm x_0 \in S_{\mathbb{X}}$. Then for any $A \in \mathbb{B}(\mathbb{X})$, $T \perp_B A \Leftrightarrow Tx_0 \perp_B Ax_0$.*

**Corollary 8.2.3.** *Let $\mathbb{X}$ be a finite-dimensional Banach space. Let $T \in \mathbb{B}(\mathbb{X})$ be such that $T$ attains norm at all points of $S_{\mathbb{X}}$. Then for any $A \in \mathbb{B}(\mathbb{X})$, $T \perp_B A \Leftrightarrow$ There exists $x_0 \in S_{\mathbb{X}}$ such that $Tx_0 \perp_B Ax_0$.*

---

## 8.3    Characterization of IPS in terms of operator norm attainment

In this section, a complete characterization of the possible norm attainment set of a bounded linear operator, defined on a Hilbert space, is obtained. Thereafter, using the above Theorem 8.2.1 and Theorem 3.3 of Benitez et al. Benítez et al. (2007), the following characterization of finite-dimensional real inner product spaces is proved:

**Theorem 8.3.1.** *A finite-dimensional real Banach space $\mathbb{X}$ is an inner product space if and only if for any linear operator $T$ on $\mathbb{X}$, $T$ attains norm at $e_1, e_2 \in S_{\mathbb{X}}$ implies $T$ attains norm at $span\{e_1, e_2\} \cap S_{\mathbb{X}}$.*

*Proof.* Suppose that $\mathbb{X}$ is an inner product space and $T$ is a linear operator on $\mathbb{X}$. Claim that if $e_k \in S_{\mathbb{X}}$, $\|Te_k\| = \|T\|$, and $\lambda_k \in \mathbb{R}$, $k = 1, 2$, then $\|T(\lambda_1 e_1 + \lambda_2 e_2)\| = \|T\| \|\lambda_1 e_1 + \lambda_2 e_2\|$.

Applying the parallelogram equality,

$$
\begin{aligned}
2(\lambda_1^2 + \lambda_2^2)\|T\|^2 &= 2\|T(\lambda_1 e_1)\|^2 + 2\|T(\lambda_2 e_2)\|^2 \\
&= \|T(\lambda_1 e_1 + \lambda_2 e_2)\|^2 + \|T(\lambda_1 e_1 - \lambda_2 e_2)\|^2 \\
&\leq \|T\|^2(\|\lambda_1 e_1 + \lambda_2 e_2\|^2 + \|\lambda_1 e_1 - \lambda_2 e_2\|^2) \\
&= \|T\|^2(2\|\lambda_1 e_1\|^2 + 2\|\lambda_2 e_2\|^2) \\
&= 2(\lambda_1^2 + \lambda_2^2)\|T\|^2.
\end{aligned}
$$

So, the former inequality is actually an equality.

Since
$$\|T(\lambda_1 e_1 \pm \lambda_2 e_2)\| \leq \|T\| \, \|\lambda_1 e_1 \pm \lambda_2 e_2\|,$$
necessarily,
$$\|T(\lambda_1 e_1 \pm \lambda_2 e_2)\| = \|T\| \, \|\lambda_1 e_1 \pm \lambda_2 e_2\|.$$
This completes the proof of necessary part of the theorem.

Conversely, suppose $\mathbb{X}$ is a finite-dimensional Banach space such that any $T \in \mathbb{B}(\mathbb{X})$ attains norm at $e_1, e_2 \in S_{\mathbb{X}}$ implies that $T$ attains norm at $span\{e_1, e_2\} \cap S_{\mathbb{X}}$. Firstly, it is shown that any such operator $T$ attains norm only at $\pm D$, where $D$ is a connected subset of $S_{\mathbb{X}}$.

Note that, "$T \in \mathbb{B}(\mathbb{X})$ attains norm at $e_1, e_2 \in S_{\mathbb{X}}$ implies that $T$ attains norm at $span\{e_1, e_2\} \cap S_{\mathbb{X}}$" is equivalent to

$$\|Tx\| = \|T\| \|x\|, \ \|Ty\| = \|T\| \|y\| \Rightarrow \|T(\alpha x + \beta y)\| = \|T\| \|\alpha x + \beta y\|, \ \forall \alpha, \beta \in \mathbb{R}.$$

If $T$ attains norm only at $span\{e_1, e_2\} \cap S_{\mathbb{X}}$, then it is done. If not, then there exists some $e_3 \in S_{\mathbb{X}} - span\{e_1, e_2\}$ such that $T$ attains norm at $e_3$. Let us now show that $T$ attains norm at $span\{e_1, e_2, e_3\} \cap S_{\mathbb{X}}$.

Consider $z = \frac{1}{r}(\alpha e_1 + \beta e_2 + \gamma e_3) \in span\{e_1, e_2, e_3\} \cap S_{\mathbb{X}}$, where $\alpha, \beta, \gamma$ are scalars and $\|\alpha e_1 + \beta e_2 + \gamma e_3\| = r$.

Since $z$ can be written as linear combination of $\frac{\alpha e_1 + \beta e_2}{\|\alpha e_1 + \beta e_2\|}$ and $e_3$, by the hypothesis $T$ attains norm at $z$.

Continuing in this way, one can conclude that $T$ attains norm only at the unit sphere of some subspace of $\mathbb{X}$ and so, in particular, $T$ attains norm only at $\pm D$, where $D$ is a connected subset of $S_{\mathbb{X}}$.

So from Theorem 8.2.1, it follows that given any $T, A \in \mathbb{B}(\mathbb{X})$ with $T \perp_B A$, there exists $x \in S_{\mathbb{X}}$ such that $x \in M_T$ and $Tx \perp_B Ax$. Using the sufficient part of Theorem 3.3 of Benítez et al. Benítez et al. (2007), one can conclude that $\mathbb{X}$ is an inner product space.                                    $\square$

*Remark* 8.3.2. The necessary part of the theorem holds for any inner product space, real or complex, with any dimension, finite or infinite.

*Remark* 8.3.3. It is clear from the proof of the Theorem 8.3.1 that in case of a Hilbert space $M_T = D \cup (-D)$, where $D$ is a compact connected subset of the unit sphere and so from Theorem 8.2.1 it follows that $T$ always satisfies BŠ property if the space is a finite-dimensional Hilbert space.

---

## 8.4 Operator orthogonality on finite-dimensional Banach spaces

Continuing the study on Birkhoff-James orthogonality, the question (opposite direction of Theorem 8.2.1) that arises naturally is that if an operator

$T$ satisfies the BŠ property then whether $M_T = D \cup -D$. It was proved in Sain et al. (2015a) that if a linear operator $T$ on a two-dimensional Banach space $\mathbb{X}$ satisfies the BŠ property then $T$ attains norm only on $\pm D$, where $D$ is a non-empty compact connected subset of $S_{\mathbb{X}}$. Motivated by this result, it was conjectured in the same paper Sain et al. (2015a) that a linear operator $T$ on a finite-dimensional Banach space $\mathbb{X}$ satisfies the BŠ property if and only if $T$ attains norm only on $\pm D$, where $D$ is a non-empty compact connected subset of $S_{\mathbb{X}}$. Furthermore, it was also proved in Sain et al. (2015a) that if $\mathbb{X}$ is finite-dimensional strictly convex then the class of operators in $\mathbb{B}(\mathbb{X})$ that satisfies the BŠ property is dense in $\mathbb{B}(\mathbb{X})$.

An easily available class of bounded linear operators not satisfying the BŠ property consists of those operators $T$ for which $M_T$ can be partitioned into two non-empty sets which are contained in complementary subspaces of $\mathbb{X}$. This fact follows from the following proposition.

**Proposition 8.4.1.** *Let $T$ be a linear operator on a finite-dimensional Banach space $\mathbb{X}$. If $M_T$ can be partitioned into two non-empty sets which are contained in complementary subspaces of $\mathbb{X}$, then there is a linear operator $A$ on $\mathbb{X}$ such that $T \perp_B A$ but $Tx \not\perp_B Ax$ for any $x \in M_T$.*

*Proof.* Let $X_1$ and $X_2$ be subspaces of $\mathbb{X}$ such that $X_1 \cap X_2 = \{\theta\}$, $\mathbb{X} = X_1 \oplus X_2$, $M_T \subseteq X_1 \cup X_2$, and $M_T \cap X_1 \neq \emptyset \neq M_T \cap X_2$. Define $A$ to be the linear operator which equals $T$ on $X_1$ and $-T$ on $X_2$. Fix $x_1 \in M_T \cap X_1$ and $x_2 \in M_T \cap X_2$. If $\lambda > 0$, then $\|(T + \lambda A)x_1\| = (1 + \lambda)\|Tx_1\| = (1 + \lambda)\|T\|$, so $\|T + \lambda A\| > \|T\|$. Similarly, if $\lambda < 0$, then $\|(T + \lambda A)x_2\| = (1 - \lambda)\|Tx_2\| = (1 - \lambda)\|T\|$, so $\|T + \lambda A\| > \|T\|$. Thus, $T \perp_B A$. Finally, if $x \in M_T \subseteq X_1 \cup X_2$, then $Tx = Ax$ if $x \in X_1$ and $Tx = -Ax$ if $x \in X_2$, so $Tx \not\perp_B Ax$ in either case. $\square$

Next, it is shown that if $M_T$ is a countable set with more than two points in a smooth space, then $T$ does not satisfy the BŠ property. To prove this, the following lemma is needed.

**Lemma 8.4.2.** *Let $M$ be a countable subset of a Banach space $\mathbb{X}$ of dimension $n \geq 2$ such that any two distinct elements of $M$ are linearly independent. Then there exist $n-2$ elements $y_3, y_4, \ldots, y_n \in \mathbb{X}$ such that for every pair of distinct elements $v \neq w \in M$, the set $\{v, w, y_3, y_4, \ldots, y_n\}$ is a basis of $\mathbb{X}$.*

*Proof.* The lemma is proved using the fact that any proper linear subspace of $\mathbb{X}$ is nowhere dense, and hence by the Baire category theorem, no countable family of proper linear subspaces of $\mathbb{X}$ can exhaust $\mathbb{X}$.

Now, let $M$ be as in the statement of the lemma. The case $n = 2$ is trivially true, so assume $n > 2$. Consider the family of subspaces of $\mathbb{X}$ spanned by any two vectors chosen from $M$. This is a countable family of proper subspaces, so by the fact above, one can fix $y_3 \in \mathbb{X}$ which is not in any subspace spanned by any two elements of $M$. Then $\{v, w, y_3\}$ is a linearly independent set for all distinct $v \neq w \in M$. If $n = 3$, it is done. If $n > 3$, then apply the fact again to fix $y_4 \in \mathbb{X}$ which is not in any subspace spanned by any set of the form $\{v, w, y_3\}$, where $v, w$ are arbitrary members of $M$. Then once again, $\{v, w, y_3, y_4\}$ is a linearly independent set for all distinct $v \neq w \in M$. If $n = 4$, it is done. Otherwise, continue in this fashion until $n-2$ elements $y_3, y_4, \ldots, y_n$ are obtained such that $\{v, w, y_3, y_4, \ldots, y_n\}$ is a linearly independent set for all distinct $v \neq w \in M$. This establishes the lemma. $\qquad\square$

**Theorem 8.4.3.** *Let $T$ be a linear operator on a finite-dimensional smooth Banach space $\mathbb{X}$. If $M_T = \{x \in S_{\mathbb{X}} : \|Tx\| = \|T\|\}$ is a countable set with more than 2 points, then there is a linear operator $A$ on $\mathbb{X}$ such that $T \perp_B A$ but $Tx \not\perp_B Ax$ for any $x \in M_T$.*

*Proof.* Without loss of generality, assume that $\|T\| = 1$.

Suppose that $\dim \mathbb{X} = n$ and that the set $M_T \subseteq S_{\mathbb{X}}$ on which $T$ attains norm is countable with more than two points. Then one can write $M_T = \{\pm x : x \in M\}$, where $M \subseteq M_T$ is a set of pairwise linearly independent vectors. Since $M_T$ has more than two points, $M$ has at least two points. By Lemma 8.4.2, fix $n-2$ elements $y_3, \ldots, y_n \in \mathbb{X}$ such that $\{v, w, y_3, \ldots, y_n\}$ is a basis of $\mathbb{X}$ for every pair of distinct elements $v \neq w \subset M$. Fix $x_1, x_2 \in M$ with $x_1 \neq x_2$. Then $\{x_1, x_2, y_3, \ldots, y_n\}$ is a basis of $\mathbb{X}$, and so one can fix scalars $c_{x,j}$ ($x \in \mathbb{X}$, $1 \leq j \leq n$) such that

$$x = c_{x,1}x_1 + c_{x,2}x_2 + c_{x,3}y_3 + \cdots + c_{x,n}y_n, \qquad \text{for every } x \in \mathbb{X}.$$

Note that, if $x \in M$ and $x \neq x_1$, then the set $\{x_1, x, y_3, \ldots, y_n\}$ is a basis, and therefore linearly independent. It follows that if $x \in M \setminus \{x_1\}$ then $c_{x,2} \neq 0$.

For each $v \in \mathbb{X}$, let $A_v$ be the linear operator defined by setting $A_v x_1 = Tx_1$, $A_v x_2 = v$, and $A_v y_i = Ty_i$ for $i = 3, 4, \ldots, n$. Next, it is shown that there is $v$ such that $T \perp_B A_v$, but $Tx \not\perp_B A_v x$ for any $x \in M_T$.

For any $\lambda \geq 0$ and $v \in B(-Tx_2, 1)$,

$$\|T + \lambda A_v\| \geq \|(T + \lambda A_v)x_1\| = \|(1 + \lambda)Tx_1\| = (1 + \lambda) \geq \|T\|,$$

and

$$
\begin{aligned}
\|T - \lambda A_v\| &\geq \|(T - \lambda A_v)x_2\| = \|Tx_2 - \lambda v\| \\
&= \|(1 + \lambda)Tx_2 - \lambda(Tx_2 + v)\| \\
&\geq (1 + \lambda)\|Tx_2\| - \lambda\|Tx_2 + v\| \\
&\geq (1 + \lambda) - \lambda = 1 = \|T\|.
\end{aligned}
$$

Therefore, $T \perp_B A_v$ for each $v \in B(-Tx_2, 1)$.

Finally, it is shown that there is $v \in B(-Tx_2, 1)$ such that $Tx \not\perp_B A_v x$ for each $x \in M$.

First, $Tx_1 \not\perp_B A_v x_1$, since $A_v x_1 = Tx_1 \neq \theta$.

Next, let $x \in M \setminus \{x_1\}$. Define

$$H_x = \{y \in \mathbb{X} : Tx \perp_B y\}.$$

By smoothness of $\mathbb{X}$, $H_x$ is an $(n-1)$-dimensional linear subspace of $\mathbb{X}$, i.e., a hyperplane (it is the unique hyperplane such that $Tx + H_x$ supports $B_{\mathbb{X}}$ at $Tx$). Hence $H_x$ is a nowhere dense set in $\mathbb{X}$. Put

$$P_x = \{v \in \mathbb{X} \colon A_v x \in H_x\}.$$

As $A_v x = c_{x,1} T x_1 + c_{x,2} v + c_{x,3} T y_3 + \cdots + c_{x,n} T y_n$ and $c_{x,2} \neq 0$, so

$$P_x = \frac{1}{c_{x,2}}(H_x - (c_{x,1} T x_1 + c_{x,3} T y_3 + \cdots + c_{x,n} T y_n)).$$

Then for each $x \in M \smallsetminus \{x_1\}$, the set $P_x = \{v \colon Tx \perp_B A_v x\}$, being homeomorphic to $H_x$, is also nowhere dense. By the Baire category theorem, the non-empty open set $B(-Tx_2, 1)$ contains an element $v$ such that $v \notin P_x$ for all $x \in M \smallsetminus \{x_1\}$.

Thus, there exists $v \in B(-Tx_2, 1)$ such that $T \perp_B A_v$ but $Tx \not\perp_B A_v x$ for all $x \in M_T$. This completes the proof of the theorem. $\qquad\square$

*Remark* 8.4.4. The example given by Li and Schneider in Li and Schneider (2002) to negate the Bhatia-Šemrl conjecture is indeed a special case of Theorem 8.4.3.

*Remark* 8.4.5. Suppose that $M_T = \{\pm x_1, \pm x_2, \ldots, \pm x_k\}$, $k \geq 2$. If $x_1, x_2, \ldots, x_k$ are linearly independent, then the smoothness assumption in the Theorem 8.4.3 is not necessary, since one can apply Proposition 8.4.1 with $X_1 = \operatorname{span}\{x_1\}$ and $X_2 = \operatorname{span}\{x_2, x_3, \ldots, x_k\}$ as the pair of complementary subspaces.

The next theorem reveals that in a two-dimensional Banach space, it is possible to obtain a much better result. In order to prove the desired theorem, the following two lemmas are needed:

**Lemma 8.4.6.** *Let $\mathbb{X}$ be a two-dimensional Banach space and let $M$ be a closed subset of $S_{\mathbb{X}}$, with more than two connected components, satisfying $M = -M$. Then there exist linearly independent vectors $x_1, x_2 \in M$ such that $\frac{(1-t)x_1 + tx_2}{\|(1-t)x_1 + tx_2\|} \notin M$ for all $0 < t < 1$. Moreover, if $D_i$ denotes the connected component of $M$ containing $x_i$, $i = 1, 2$, then $D_2 \neq \pm - D_1$.*

*Proof.* Fix any component $A$ of $M$. Then $-A$ is also a component of $M$, since the mapping $x \mapsto -x$ is a homeomorphism of $M$. Since $M$ has more than two components, one can fix another component $B$ of $M$ which is disjoint from both $A$ and $-A$. Fix $y_0 \in A$, $y_1 \in B$. Then $y_0 \neq \pm y_1$, so $y_0, y_1$ are linearly independent.

Put $y_t = (1-t)y_0 + ty_1$, and $\hat{y}_t = y_t/\|y_t\|$. There is some $0 < t < 1$ with $\hat{y}_t \notin M$, since otherwise the continuous curve $\{\hat{y}_t \colon 0 \leq t \leq 1\} \subseteq M$ would meet the distinct components $A$ and $B$ of $M$. Hence by continuity of the mapping $t \mapsto \hat{y}_t$, the set $G = \{t \in [0,1] \colon \hat{y}_t \notin M\}$ is a non-empty open set contained in $(0,1)$. Fix a component open interval $(a, b)$ of the open set $G$, where $0 \leq a < b \leq 1$. Put $x_1 = \hat{y}_a$ and $x_2 = \hat{y}_b$. Then $x_1, x_2 \in M$, and $x_1, x_2$ are linearly independent. Let $v_t = (1 - t)x_1 + tx_2$ and $\hat{v}_t = v_t/\|v_t\|$.

Next, it is shown that $\hat{v}_t \notin M$ for $0 < t < 1$, as claimed in the lemma. Let $0 < t < 1$. Then $v_t = (1 - t)\hat{y}_a + t\hat{y}_b = py_a + qy_b$, where $p, q > 0$. Let $\alpha = p(1 - a) + q(1 - b)$, $\beta = pa + qb$, and $s = \beta/(\alpha + \beta) = (pa + qb)/(p + q)$. Then $v_t = \alpha y_0 + \beta y_1 = (\alpha + \beta)y_s$ and $a < s < b$, hence $\hat{v}_t = \hat{y}_s \notin M$.

Let $D_i$ be the component of $M$ containing $x_i$, $i = 1, 2$. Claim is that $D_2 \neq \pm D_1$.

First, $-D_1 \cap D_1 = \emptyset$, since otherwise $-D_1 = D_1$, and then for any $x \in S_{\mathbb{X}} \setminus D_1$, the two half-planes determined by the line spanned by $x$ would disconnect $D_1 = -D_1$ into two pieces. So $-D_1 \cap D_1 = \emptyset$.

Next, suppose that $D_2 = D_1$. Then $x_1, x_2 \in D_1$. For every $0 < t < 1$, since $\hat{v}_t \notin M \supseteq D_1$, one must have $-\hat{v}_t \in D_1$, since otherwise the two half spaces given by the line spanned by $\hat{v}_t$ would disconnect $D_1$ into two pieces, one containing $x_1$ and the other containing $x_2$. Since $x_1$ is a limit point of $\{\hat{v}_t : 0 < t < 1\}$, $-x_1$ is a limit point of $\{-\hat{v}_t : 0 < t < 1\} \subseteq D_1$. As $D_1$ is closed, one gets $-x_1 \in D_1$, which is impossible since $-D_1 \cap D_1 = \emptyset$. Thus, $D_2 \neq D_1$.

Finally, suppose that $D_2 = -D_1$. $M$ has more than two components, so one can fix $x \in M \setminus (-D_1 \cup D_1)$. Since $x_1$ and $x_2$ are linearly independent, $x = \alpha x_1 + \beta x_2$, with $\alpha \neq 0 \neq \beta$. Then $\alpha$ and $\beta$ cannot be of the same sign, since otherwise, one gets $x = (\alpha + \beta)v_t$, where $t = \beta/(\alpha + \beta) \in (0, 1)$, which implies (as $\|x\| = 1$) $\hat{v}_t = \pm x \in M$, contradicting $\hat{v}_t \notin M$. Without loss of generality, assume $\alpha > 0$, $\beta < 0$. Then $x = \alpha x_1 + (-\beta)(-x_2)$, so dividing by $\alpha + (-\beta)$, one can express $x$ as a positive scalar multiple of a convex combination of $x_1$ and $-x_2$. Since $x_1, -x_2 \in D_1$ while $x \notin \pm D_1$, the line of $x$ would then disconnect $D_1$, with $x_1$ in one piece and $-x_2$ in another piece, which is impossible since $D_1$ is connected. Thus $D_2 \neq -D_1$. $\qquad\square$

**Lemma 8.4.7.** *Let $\mathbb{X}$ be a two-dimensional Banach space and let $L$ be a non-empty open line segment with $L \subseteq S_{\mathbb{X}}$. Then for any $z_0 \in L$, there is a neighbourhood $V$ of $z_0$ in $\mathbb{X}$ such that $V \cap S_{\mathbb{X}} \subseteq L$.*

*Proof.* Fix $y_0 \neq \theta$ such that $z_0 \pm y_0 \in L$, and so $z_0 + \lambda y_0 \in L$ for $|\lambda| < 1$. Then $y_0$ and $z_0$ are linearly independent, and so the set

$$V = \{z_0 + \alpha y_0 + \beta z_0 : |\alpha| < \tfrac{1}{2}, |\beta| < \tfrac{1}{2}\}$$

is a neighbourhood of $z_0$ in $\mathbb{X}$, as $\mathbb{X}$ is two-dimensional.

Now let $z \in V \cap S_{\mathbb{X}}$. Then $z = z_0 + \alpha y_0 + \beta z_0$, for some $|\alpha| < \tfrac{1}{2}$, $|\beta| < \tfrac{1}{2}$. Since $|\frac{\alpha}{1+\beta}| < 1$, clearly $\frac{z}{1+\beta} = z_0 + \frac{\alpha}{1+\beta}y_0 \in L \subseteq S_{\mathbb{X}}$, which gives $\frac{\|z\|}{1+\beta} = 1$, and so $1 + \beta = \|z\| = 1$. Hence $\beta = 0$ and $z = z_0 + \alpha y_0 \in L$. $\qquad\square$

**Theorem 8.4.8.** *Let $T$ be a linear operator on a two-dimensional Banach space $\mathbb{X}$. If $M_T$ has more than two components then there is a linear operator $A$ on $\mathbb{X}$ such that $T \perp_B A$ but $Tx \not\perp_B Ax$ for any $x \in M_T$.*

*Proof.* Without loss of generality, assume $\|T\| = 1$. The theorem is proved in the following three steps.

**Step 1.** *Claim that there is an exposed point $u$ of $B_{\mathbb{X}}$ with $u \in S_{\mathbb{X}} \setminus T(M_T)$.*

The set $M_T$ satisfies the conditions of the set $M$ in Lemma 8.4.6, so applying the lemma, one can fix $x_1, x_2 \in M_T$ such that $\hat{v}_t \notin M_T$ for any $0 < t < 1$, where:

$$v_t = (1 - t)x_1 + tx_2 \quad \text{and} \quad \hat{v}_t = v_t/\|v_t\|.$$

Moreover, let $D_i$ be the component of $x_i$ for $i = 1, 2$. Then $D_2 \neq \pm D_1$.

Since $\|Tv_t\| \leq \|T\hat{v}_t\| < 1$ for all $0 < t < 1$, one gets $Tx_1 \neq Tx_2$ (since $Tx_1 = Tx_2$ implies $\|Tv_t\| = \|Tx_1\| = 1$ for all $t$). Then $Tx_1 \neq -Tx_2$, since otherwise $T$ would assume the constant value $Tx_1 = -Tx_2$ on the closed line segment joining $x_1$ and $-x_2$. In that case, there is a line segment contained within $M_T$, connecting the distinct components $D_1$ and $-D_2$, which is impossible. Thus, $Tx_1$ and $Tx_2$ are linearly independent. Hence $Tv_t \neq \theta$ for $0 \leq t \leq 1$, and one may define:

$$w_t = Tv_t/\|Tv_t\| = T\hat{v}_t/\|T\hat{v}_t\|.$$

It follows that if $t \neq t'$ are in $[0, 1]$, then $w_t$ and $w_{t'}$ are linearly independent.

Since the mapping $t \mapsto \|Tv_t\|$ is continuous, it assumes its minimum value $m = \inf_{0 \leq t \leq 1} \|Tv_t\|$ on a compact subset of $[0, 1]$ and so one may fix the smallest $t_0$ in $[0, 1]$ such that $\|Tv_{t_0}\| = m$. Then $0 < t_0 < 1$ and $\|Tv_{t_0}\| \leq \|Tv_t\|$ for all $0 \leq t \leq 1$, while $\|Tv_{t_0}\| < \|Tv_t\|$ for $0 \leq t < t_0$.

**Claim 1.** $w_{t_0} = Tv_{t_0}/\|Tv_{t_0}\|$ *is an extreme point of $B_{\mathbb{X}} = \{x: \|x\| \leq 1\}$.*

If not, fix an open line segment $L$ with $w_{t_0} \in L \subseteq S_{\mathbb{X}}$. By Lemma 8.4.7, fix a neighbourhood $V$ of $w_{t_0}$ in $\mathbb{X}$ such that $V \cap S_{\mathbb{X}} \subseteq L$. By continuity of the map $t \mapsto w_t$ at $t = t_0$, fix $0 < t_1 < t_0 < t_2 < 1$ such that if $t_1 \leq t \leq t_2$ then $w_t \in V$. Then for $t_1 < t < t_2$, $w_t \in V \cap S_{\mathbb{X}}$ and so $w_t \in L$. Thus, $w_{t_1}$, $w_{t_0}$, $w_{t_2}$ are all in $L$, and are pairwise independent. Since $0 < t_1 < t_0 < t_2 < 1$, it follows that:

$$Tv_{t_0} = (1 - \lambda)Tv_{t_1} + \lambda Tv_{t_2}, \quad \text{for some } 0 < \lambda < 1.$$

Divide the last equation by $\|Tv_{t_0}\|$ and put $\alpha = (1 - \lambda)(\|Tv_{t_1}\|/\|Tv_{t_0}\|) > 1 - \lambda$, $\beta = \lambda(\|Tv_{t_2}\|/\|Tv_{t_0}\|) \geq \lambda$, to express $w_{t_0}$ as a linear combination of $w_{t_1}$ and $w_{t_2}$:

$$w_{t_0} = \alpha w_{t_1} + \beta w_{t_2}, \quad \text{with } \alpha + \beta > 1.$$

Since $w_{t_0}$, $w_{t_1}$ and $w_{t_2}$ lie on a line (the segment $L$), it follows that:

$$w_{t_0} = \alpha' w_{t_1} + \beta' w_{t_2}, \quad \text{with } \alpha' + \beta' = 1.$$

But this contradicts the fact that $w_{t_0}$ can be expressed *uniquely* as a linear combination of $w_{t_1}$ and $w_{t_2}$ (due to linear independence of $w_{t_1}$ and $w_{t_2}$). This proves the claim, showing that $w_{t_0}$ is an extreme point of $B_{\mathbb{X}}$.

Next, it is shown that $w_{t_0} \notin T(M_T)$. Suppose that $w_{t_0} \in T(M_T)$. Fix $z \in M_T$ with $w_{t_0} = Tz$. Write $z = \alpha x_1 + \beta x_2 \in M_T$, using linear independence of $x_1, x_2$. Then

$$\alpha T x_1 + \beta T x_2 = w_{t_0} = \frac{T v_{t_0}}{\|T v_{t_0}\|} = \frac{1 - t_0}{\|T v_{t_0}\|} T x_1 + \frac{t_0}{\|T v_{t_0}\|} T x_2.$$

Therefore, $\alpha = \frac{1 - t_0}{\|T v_{t_0}\|}$ and $\beta = \frac{t_0}{\|T v_{t_0}\|}$, as $T x_1$ and $T x_2$ are linearly independent. Thus, we have,

$$\|z\| = \frac{\|(1 - t_0) x_1 + t_0 x_2\|}{\|T v_{t_0}\|} = \frac{\|v_{t_0}\|}{\|T v_{t_0}\|} = \frac{1}{\|T \hat{v}_{t_0}\|} > 1,$$

which contradicts that $z \in M_T$. Therefore, $w_{t_0} \notin T(M_T)$.

Now, $w_{t_0}$ is an element of the open set $U = \{\mu T x_1 + \nu T x_2 : \mu, \nu > 0\}$, comprising of positive linear combinations of the linearly independent vectors $T x_1$ and $T x_2$. Also, $T(M_T)$ is closed and $w_{t_0} \notin T(M_T)$. Therefore, $U \smallsetminus T(M_T)$ is an open set containing $w_{t_0}$. As the set of exposed points of $B_{\mathbb{X}}$ is dense in the set of extreme points, $U \smallsetminus T(M_T)$ contains an exposed point $u$ of $B_{\mathbb{X}}$ such that $u \in S_{\mathbb{X}} \smallsetminus T(M_T)$ (and moreover $u$ is a positive linear combination of $T x_1$ and $T x_2$). This completes the proof of Step 1.

**Step 2.** *Construct an $A \in \mathbb{B}(\mathbb{X})$ such that $T z \not\perp_B A z$ for any $z \in M_T$.*

Since $u \in S_{\mathbb{X}} \smallsetminus T(M_T)$ is an exposed point of $B_{\mathbb{X}}$, one can find a one-dimensional subspace $V_0$ of $\mathbb{X}$ such that $B_{\mathbb{X}}$ lies within one of the half-planes determined by the line $u + V_0$ and $(u + V_0) \cap S_{\mathbb{X}} = \{u\}$. Choose a unit vector $v \in V_0$. Then $\{u, v\}$ is a basis of $\mathbb{X}$ such that $\|u\| < \|u + \lambda v\|$ for all $\lambda \neq 0$.

Claim that if $w \in S_{\mathbb{X}}$ and $w \perp_B v$ then $w = \pm u$. Express $w$ as $w = au + bv$, where $a, b \in \mathbb{R}$. Since $w \perp_B v$, $a \neq 0$. If $b = 0$ then it is done. Assume $b \neq 0$. Then

$$1 = \|w\| = \|au + bv\| = |a| \|u + \tfrac{b}{a} v\| > |a|.$$

Again, for all $\lambda$,

$$1 \leq \|w + \lambda v\| = |a| \|u + \tfrac{\lambda + b}{a} v\|,$$

hence by choosing $\lambda = -b$, one gets $1 \leq |a|$, which is a contradiction. Thus, if $w \in S_{\mathbb{X}}$ and $w \perp_B v$ then $w = \pm u$.

Recall that $x_1 \in D_1$ and $-x_2 \in -D_2$, with $D_1$ and $-D_2$ being distinct components of $M_T$ and $x_1, -x_2$ linearly independent. Put $y_t = ((1 - t) x_1 - t x_2)/\|(1 - t) x_1 - t x_2\|$. Then $T$ cannot attain norm at every $y_t$, $0 \leq t \leq 1$ (since otherwise $\{y_t : 0 \leq t \leq 1\}$ would be a connected subset of $M_T$ meeting distinct components $D_1$ and $-D_2$ of $M_T$). Fix $0 < s < 1$ such that $y_s \in S_{\mathbb{X}} \smallsetminus M_T$, put $c_1 = s/\|(1 - s) x_1 - s x_2\|$, $c_2 = (1 - s)/\|(1 - s) x_1 - s x_2\|$, and $y = y_s$. Then $c_1, c_2 > 0$ and $y = c_2 x_1 - c_1 x_2 \in S_{\mathbb{X}} \smallsetminus M_T$. Define $A x_i = c_i v$, for $i = 1, 2$.

Claim that $A z \neq \theta$ for any $z \in M_T$. If not, let $z = d_1 x_1 + d_2 x_2 \in M_T$ be such that $A z = \theta$. Then $d_1 c_1 + d_2 c_2 = 0$ and so $c_2 z = d_1 y$. As both $y, z \in S_{\mathbb{X}}$, clearly $|c_2| = |d_1|$ and so $z = \pm y$. This is a contradiction, since $z \in M_T$, while $y \notin M_T$. Thus, $Az \neq 0$ for any $z \in M_T$.

Now, if $Tz \perp_B Az$ with $Tz \in S_\mathbb{X}$, then $Tz$ must be equal to $\pm u$, since $Az$ is a non-zero scalar multiple of $v$. It follows that $Tz \not\perp_B Az$ for any $z \in M_T$.

This completes the proof of Step 2.

**Step 3.** Claim that $T \perp_B A$.

Let $L = \{\alpha u \colon \alpha \in \mathbb{R}\}$. Then $Tx_1, Tx_2 \in \mathbb{X} \setminus L$, since $u \notin T(M_T)$ and $Tx_i \in L$ would imply $u = \pm Tx_i \in T(M_T)$. As $\mathbb{X}$ is two-dimensional, $\mathbb{X} \setminus L$ splits into two components (open half-planes). Let $L_1$ be the component containing $Tx_1$, and $L_2$ be the other component. Then $Tx_2 \in L_2$, since if $Tx_2 \in L_1$, then $u$ (which is a positive linear combination of $Tx_1$ and $Tx_2$) would belong to $L_1$. Without loss of generality, one may assume that $v \in L_1$.

Let
$$V = \{z \in S_\mathbb{X} \cap L_1 \colon \|z + \lambda v\| < \|z\| \text{ for some } \lambda < 0\},$$

and
$$W = \{z \in S_\mathbb{X} \cap L_1 \colon \|z + \lambda v\| < \|z\| \text{ for some } \lambda > 0\}.$$

Then both $V$ and $W$ are open in $S_\mathbb{X} \cap L_1$ and $V \neq \emptyset$ as $v \in V$. Also, $V \cap W = \emptyset$ and $V \cup W = S_\mathbb{X} \cap L_1$. Hence $W = \emptyset$, as $S_\mathbb{X} \cap L_1$ is connected, and so $Tx_1 \notin W$. Therefore, $\|Tx_1 + \lambda v\| \geq \|Tx_1\|$ for all $\lambda > 0$.

Similarly, considering $S_\mathbb{X} \cap L_2$, one gets $\|Tx_2 + \lambda v\| \geq \|Tx_2\|$ for all $\lambda < 0$.

Thus, for $\lambda > 0$,

$$\|T + \lambda A\| \geq \|Tx_1 + \lambda Ax_1\| \geq \|Tx_1\| = \|T\|,$$

and for $\lambda < 0$,

$$\|T + \lambda A\| \geq \|Tx_2 + \lambda Ax_2\| \geq \|Tx_2\| = \|T\|.$$

Hence $\|T + \lambda A\| \geq \|T\|$ for all $\lambda$, and so $T \perp_B A$.

This completes the proof of Step 3 and establishes the theorem. $\qquad \square$

The following example illustrates the above situation.

**Example 8.4.9.** *Consider* $(\mathbb{R}^2, \|.\|)$, *whose unit sphere is the regular hexagon with vertices at* $\pm(1, 0)$, $\pm(\frac{1}{2}, \frac{\sqrt{3}}{2})$, $\pm(-\frac{1}{2}, \frac{\sqrt{3}}{2})$. *Let*

$$T = \begin{pmatrix} \frac{3}{4} & -\frac{\sqrt{3}}{4} \\ \frac{\sqrt{3}}{4} & \frac{3}{4} \end{pmatrix}.$$

*Then* $\|T\| = 1$ *and* $T$ *attains norm only at the points* $\pm(1, 0)$, $\pm(\frac{1}{2}, \frac{\sqrt{3}}{2})$, $\pm(-\frac{1}{2}, \frac{\sqrt{3}}{2})$. *Define a linear operator* $A$ *by setting* $A(1, 0) = (1, 0)$ *and* $A(\frac{1}{2}, \frac{\sqrt{3}}{2}) = (-\frac{1}{2}, -\frac{\sqrt{3}}{2})$. *Then* $T \perp_B A$ *but there is no* $x \in S_\mathbb{X}$ *such that* $\|Tx\| = \|T\|$ *and* $Tx \perp_B Ax$.

The next theorem, which gives a complete characterization of operators on two-dimensional Banach spaces, that satisfy the BŠ property, follows from the Theorems 8.4.8 and 8.2.1. Notice that for a linear operator $T$ on a two-dimensional Banach space $\mathbb{X}$, $M_T$ has at most two components if and only if its projective identification $M_T/\{x \sim -x\}$ is connected in the projective space $S_{\mathbb{X}}/\{x \sim -x\} \equiv \mathbb{R}P^1$.

**Theorem 8.4.10.** *A linear operator $T$ on a two-dimensional Banach space $\mathbb{X}$ satisfies the BŠ property if and only if the set of unit vectors on which $T$ attains norm is connected in the corresponding projective space $\mathbb{R}P^1 \equiv S_{\mathbb{X}}/\{x \sim -x\}$.*

The obvious question that arises naturally, in connection with the above theorem, is regarding the possible extension of the above theorem to higher-dimensional Banach spaces. However, to the best of our knowledge, not much progress has been made in this direction and there remains ample scope for further research towards exploring the following conjecture:

**Conjecture 8.4.11.** *A linear operator $T$ on an $n$-dimensional Banach space $\mathbb{X}$ satisfies the BŠ property if and only if the set of unit vectors on which $T$ attains norm is connected in the corresponding projective space $\mathbb{R}P^{n-1} \equiv S_{\mathbb{X}}/\{x \sim -x\}$.*

The next theorem indicates that the set of operators satisfying the BŠ property is not meager in size. In fact, if the space $\mathbb{X}$ is finite-dimensional and strictly convex then the set of linear operators satisfying the BŠ property is dense in $\mathbb{B}(\mathbb{X})$. In order to prove this, the following lemma is required.

**Lemma 8.4.12.** *Let $\mathbb{X}$ be a reflexive strictly convex Banach space. Then the set of linear operators on $\mathbb{X}$ which attains norm at unique (up to multiplication by scalars) vector is dense in $\mathbb{B}(\mathbb{X})$.*

*Proof.* By the classic result of Lindenstrauss Lindenstrauss (1963), the set of norm attaining bounded linear operators on a reflexive space $\mathbb{X}$, is dense in $\mathbb{B}(\mathbb{X})$. Therefore, it suffices to prove that any linear operator, which attains norm, can be approximated by an operator which attains norm at a unique (up to multiplication by scalars) vector.

Let $S \in \mathbb{B}(\mathbb{X})$ attains norm and let $x \in M_S$. By strict convexity of $\mathbb{X}$, there exists a subspace $V$ of $\mathbb{X}$ such that $x + V$ is a supporting hyperplane to $B_{\mathbb{X}}$, that intersects $B_{\mathbb{X}}$ only at $x$. Hence if $\theta \neq y \in V$, then $\|\alpha x + y\| > \|\alpha x\|$ for all $\alpha \in \mathbb{R}$. Since $x \notin V$, any $z \in \mathbb{X}$ can be expressed in a unique way as $z = \alpha x + y$, $\alpha \in \mathbb{R}$, $y \in V$.

Let $\epsilon \in (0,1)$ be arbitrary. Define a linear operator $T$ on $\mathbb{X}$, by setting $Tx = Sx$ and $Ty = (1-\epsilon)Sy$ for $y \in V$. Then $\|T\| = \|S\|$. Indeed,

$$\begin{aligned}
\|T(\alpha x + y)\| &= \|\alpha Sx + (1-\epsilon)Sy\| \tag{8.1}\\
&= \|\epsilon \alpha Sx + (1-\epsilon)S(\alpha x + y)\|\\
&\leq \epsilon \|S\|\|\alpha x\| + (1-\epsilon)\|S\|\|(\alpha x + y)\|\\
&\leq \|S\|\|(\alpha x + y)\|.
\end{aligned}$$

This implies $\|T\| \leq \|S\|$. On the other hand, $\|T\| \geq \|Tx\| = \|Sx\| = \|S\|$.

From (8.1), it is clear that $T$ attains norm only at $x$ and its scalar multiples on $S_{\mathbb{X}}$, since the last inequality in (8.1) is strict, except for $y = \theta$.

One can also estimate $\|S - T\|$ easily:

$$
\begin{aligned}
\|(S - T)(\alpha x + y)\| &= \|\epsilon Sy\| \leq \epsilon \|S\| \|y\| \\
&= \epsilon \|S\| \|(\alpha x + y) + (-\alpha x)\| \\
&\leq \epsilon \|S\| (\|\alpha x + y\| + \|\alpha x\|) \\
&\leq 2\epsilon \|S\| \|\alpha x + y\|,
\end{aligned}
$$

and so $\|S - T\| \leq 2\epsilon \|S\|$. As $\epsilon \in (0, 1)$ is arbitrary, this completes the proof of the lemma. □

**Theorem 8.4.13.** *Suppose $\mathbb{X}$ is a finite-dimensional strictly convex Banach space. Then the set of linear operators satisfying the BŠ property is dense in $\mathbb{B}(\mathbb{X})$.*

*Proof.* From Theorem 8.2.1, it follows that the set of operators that attains norm at unique (up to multiplication by scalars) vector satisfy the BŠ property and so the proof of the theorem can be completed by applying Lemma 8.4.12. □

*Remark* 8.4.14. The example given below shows if the underlying Banach space is infinite-dimensional then there exists a bounded linear operator that attains norm at a unique (up to multiplication by scalars) vector but does not satisfy the BŠ property.

Consider $T \colon \ell_2 \to \ell_2$ defined by $Te_1 = -e_1$, and $Te_n = (1 - 1/n)e_n$ for $n \geq 2$, where $\{e_n \colon n \in \mathbb{N}\}$ is the usual orthonormal basis for the Hilbert space $\ell_2$. Then $T$ attains norm only at $\pm e_1$. Let $A = I$, the identity operator on $\ell_2$. It is easy to check that $T \perp_B A$. Indeed, $\|(T + \lambda A)e_1\| \geq \|T\|$ for all $\lambda \leq 0$ and $\|(T + \lambda A)e_n\| \geq \|T\|$, for all $\lambda \geq 1/n$. But $Te_1$ is not orthogonal to $Ae_1$ in the sense of Birkhoff-James.

So the method used in the proof of Theorem 8.4.13 cannot be followed if the space is infinite-dimensional.

*Remark* 8.4.15. Strict convexity in Theorem 8.4.13 is not necessary as it can be seen from the following example of a non-strictly convex space $\mathbb{X}$, in which the set of operators satisfying the BŠ property is dense in $\mathbb{B}(\mathbb{X})$.

Consider the real Banach space $(\mathbb{R}^n, \|.\|_\infty)$. Let $S \in \mathbb{B}(\mathbb{R}^n)$. Noting that every extreme point of the unit sphere is also an exposed point of the unit sphere for this space $(\mathbb{R}^n, \|.\|_\infty)$ and every linear operator on $\mathbb{R}^n$ attains norm on at least one extreme point of the unit sphere, it follows that $S$ attains norm on an exposed point of the unit sphere. As in the proof of the Lemma 8.4.12, one can construct a linear operator $T$ such that $T$ attains norm only on a unique (up to multiplication by scalars) vector such that $\|S - T\| < 2\epsilon \|S\|$ for any $\epsilon \in (0, 1)$. Then $T$ satisfies the BŠ property. Thus, the set of operators satisfying the BŠ property is dense in $\mathbb{B}(\mathbb{R}^n)$.

## 8.5 Operator orthogonality on infinite-dimensional normed spaces

Next, Birkhoff-James orthogonality of bounded linear operators is considered in the setting of infinite-dimensional Banach spaces. In Paul et al. (2016) a sufficient condition for a bounded linear operator to satisfy the BŠ property is obtained, when the space is infinite-dimensional. This exploration has deep connections with the study of smoothness of bounded linear operators. In fact, as an application of the concerned study, it is shown in Paul et al. (2016) that $T$ is a smooth point in $\mathbb{B}(\mathbb{X}, \mathbb{Y})$ if $T$ attains norm at unique (upto muliplication by scalars) vector $x \in S_{\mathbb{X}}$, $Tx$ is a smooth point of $\mathbb{Y}$ and $\sup_{y \in C} \|Ty\| < \|T\|$ for all closed subsets $C$ of $S_{\mathbb{X}}$, with $d(\pm x, C) > 0$.

In this direction, the first result Paul et al. (2016) is that if $T$ is a compact linear operator on a reflexive Banach space $\mathbb{X}$, with $M_T = D \cup (-D)$ ($D$ is a non-empty compact connected subset of $S_{\mathbb{X}}$), then for any $A \in \mathbb{K}(\mathbb{X}, \mathbb{Y})$, $T \perp_B A \Leftrightarrow Tx \perp_B Ax$ for some $x \in M_T$. To prove the result, the following lemma is needed.

**Lemma 8.5.1.** *Let* $T \in B(\mathbb{X}, \mathbb{Y})$ *and* $M_T = D \cup (-D)$ *($D$ is a non-empty compact connected subset of $S_{\mathbb{X}}$). Then for any* $A \in \mathbb{B}(\mathbb{X}, \mathbb{Y})$*, either there exists* $x \in M_T$ *such that* $Tx \perp_B Ax$ *or there exists* $\lambda_0 \neq 0$ *such that* $\|Tx + \lambda_0 Ax\| < \|Tx\| \; \forall \; x \in M_T$.

*Proof.* Suppose that there exists no $x \in M_T$ such that $Tx \perp_B Ax$. Let

$$W_1 = \{x \in D : \|Tx + \lambda_x Ax\| < \|T\| \text{ for some } \lambda_x > 0\}$$

$$\text{and } W_2 = \{x \in D : \|Tx + \lambda_x Ax\| < \|T\| \text{ for some } \lambda_x < 0\}.$$

Then it is easy to check that both $W_1, W_2$ are open sets in $D$ and $D = W_1 \cup W_2$. The connectedness of $D$ ensures that either $D = W_1$ or $D = W_2$.

Consider the case $D = W_1$. Then for each $x \in D$, there exists $\lambda_x \in (0, 1)$ such that $\|Tx + \lambda_x Ax\| < \|Tx\| = \|T\|$. By the convexity of the norm function, it now follows that

$$\|Tx + \lambda Ax\| < \|Tx\| = \|T\| \; \forall \; \lambda \in (0, \lambda_x).$$

Consider the continuous function $g : S_X \times [-1, 1] \longrightarrow \mathbb{R}$ defined by

$$g(x, \lambda) = \|Tx + \lambda Ax\|.$$

Now, $g(x, \lambda_x) = \|Tx + \lambda_x Ax\| < \|T\|$ and so by the continuity of $g$, there exists $r_x, \delta_x > 0$ such that $g(y, \lambda) < \|T\| \quad \forall \quad y \in B(x, r_x) \cap S_{\mathbb{X}}$ and $\forall \lambda \in$

$(\lambda_x - \delta_x, \lambda_x + \delta_x)$. Let $y \in B(x, r_x) \cap D$. Then for any $\lambda \in (0, \lambda_x)$,

$$Ty + \lambda Ay = \left(1 - \frac{\lambda}{\lambda_x}\right)Ty + \frac{\lambda}{\lambda_x}(Ty + \lambda_x Ay)$$

$$\Rightarrow \quad \|Ty + \lambda Ay\| < (1 - \frac{\lambda}{\lambda_x})\|T\| + \frac{\lambda}{\lambda_x}\|T\|$$

$$\Rightarrow \quad \|Ty + \lambda Ay\| < \|T\|$$

Therefore, $g(y, \lambda) < \|T\| \quad \forall \quad y \in B(x, r_x) \cap D$ and $\forall \lambda \in (0, \lambda_x)$.
Consider the open cover $\{B(x, r_x) \cap D : x \in D\}$ of $D$. By the compactness of $D$, this cover has a finite subcover $\{B(x_i, r_{x_i}) \cap D : i = 1, 2, \ldots, n\}$ so that

$$D \subset \cup_{i=1}^n B(x_i, r_{x_i}).$$

Choose $\lambda_0 \in \cap_{i=1}^n (0, \lambda_{x_i})$. Then for any $x \in M_T$, $\|Tx + \lambda_0 Ax\| < \|T\|$.
If $D = W_2$ then similarly one can show that there exists some $\lambda_0 < 0$ such that for any $x \in M_T$, $\|Tx + \lambda_0 Ax\| < \|T\|$.
This completes the proof of lemma. $\qquad \square$

**Theorem 8.5.2.** *Let $\mathbb{X}$ be a reflexive Banach space and $\mathbb{Y}$ be any normed space. Let $T \in \mathbb{K}(\mathbb{X}, \mathbb{Y})$ and $M_T = D \cup (-D)$ ($D$ is a non-empty compact connected subset of $S_{\mathbb{X}}$). Then for any $A \in \mathbb{K}(\mathbb{X}, \mathbb{Y}), T \perp_B A$ if and only if there exists $x \in M_T$ such that $Tx \perp_B Ax$.*

*Proof.* Suppose there exists no $x \in M_T$ such that $Tx \perp_B Ax$. Then by applying Lemma 8.5.1, there exists some $\lambda_0 \neq 0$ such that

$$\|Tx + \lambda_0 Ax\| < \|T\| \, \forall \, x \in M_T.$$

Without loss of generality, assume that $\lambda_0 > 0$.
For each $n \in \mathbb{N}$, the operator $(T + \frac{1}{n}A)$, being compact on a reflexive Banach space, attains norm. So, there exists $x_n \in S_{\mathbb{X}}$ such that $\|T + \frac{1}{n}A\| = \|(T + \frac{1}{n}A)x_n\|$.
Now, $\mathbb{X}$ is reflexive and so $B_{\mathbb{X}}$ is weakly compact. Hence there exists a subsequence $\{x_{n_k}\}$ of $\{x_n\}$ such that $x_{n_k} \rightharpoonup x_0$ (say) in $B_{\mathbb{X}}$ weakly. Without loss of generality, assume that $x_n \rightharpoonup x_0$ weakly. Then $T, A$ being compact, $Tx_n \longrightarrow Tx_0$ and $Ax_n \longrightarrow Ax_0$. As $T \perp_B A$, one gets $\|T + \frac{1}{n}A\| \geq \|T\| \, \forall \, n \in \mathbb{N}$ and so $\|Tx_n + \frac{1}{n}Ax_n\| \geq \|T\| \geq \|Tx_n\| \, \forall \, n \in \mathbb{N}$. Letting $n \longrightarrow \infty$ one gets $\|Tx_0\| \geq \|T\| \geq \|Tx_0\|$. Then $x_0 \in M_T$.

For any $\lambda > \frac{1}{n}$, one gets $\|Tx_n + \lambda Ax_n\| \geq \|Tx_n\|$, since otherwise,

$$Tx_n + \frac{1}{n}Ax_n = \left(1 - \frac{1}{n\lambda}\right)Tx_n + \left(\frac{1}{n\lambda}\right)(Tx_n + \lambda Ax_n)$$

$$\Rightarrow \|Tx_n + \frac{1}{n}Ax_n\| < \left(1 - \frac{1}{n\lambda}\right)\|Tx_n\| + \left(\frac{1}{n\lambda}\right)\|Tx_n\|$$

$$\Rightarrow \|Tx_n + \frac{1}{n}Ax_n\| < \|Tx_n\| \text{ , a contradiction.}$$

Choose $\lambda > 0$. Then there exists $n_0 \in \mathbb{N}$ such that $\lambda > \frac{1}{n_0}$ and so for all $n \geq n_0$ one gets

$$\|Tx_n + \lambda Ax_n\| \geq \|Tx_n\|.$$

Letting $n \longrightarrow \infty$,

$$\|Tx_0 + \lambda Ax_0\| \geq \|Tx_0\| \ldots \ldots \ldots (i)$$

Next claim is that $\|Tx_0 + \lambda Ax_0\| \geq \|Tx_0\|$ for each $\lambda < 0$. Suppose there exist some $\lambda_1 < 0$ such that $\|Tx_0 + \lambda_1 Ax_0\| < \|Tx_0\|$. Note that $\|Tx_0 + \lambda_0 Ax_0\| < \|Tx_0\|$. Then

$$Tx_0 = (1 - \frac{\lambda_0}{\lambda_0 - \lambda_1})(Tx_0 + \lambda_0 Ax_0) + (\frac{\lambda_0}{\lambda_0 - \lambda_1})(Tx_0 + \lambda_1 Ax_0)$$

$$\Rightarrow \|Tx_0\| < (1 - \frac{\lambda_0}{\lambda_0 - \lambda_1})\|Tx_0\| + (\frac{\lambda_0}{\lambda_0 - \lambda_1})\|Tx_0\|$$

$$\Rightarrow \|Tx_0\| < \|Tx_0\|, \text{ a contradiction.}$$

Thus, $\|Tx_0 + \lambda Ax_0\| \geq \|Tx_0\|$ for each $\lambda < 0$. This, along with (i), shows that $Tx_0 \perp_B Ax_0$. This completes the proof of the theorem $\qquad \square$

**Corollary 8.5.3.** *Let $T \in \mathbb{K}(\mathbb{X}, \mathbb{Y})$ and $M_{T^*} = D \cup (-D)$ ($D$ is a compact connected subset of $S_{\mathbb{Y}^*}$). Then for any $A \in \mathbb{K}(\mathbb{X}, \mathbb{Y})$, $T \perp_B A$ if and only if there exists $g \in M_{T^*}$ such that $T^*g \perp_B A^*g$.*

*Proof.* Noting that $T \perp_B A$ if and only if $T^* \perp_B A^*$ and $S_{\mathbb{Y}^*}$ is weak* compact, one can apply the above Theorem 8.5.2 to conclude that if $T \perp_B A$ then there exists $g \in M_{T^*}$ such that $T^*g \perp_B A^*g$. The other part is obvious. $\qquad \square$

*Remark* 8.5.4. Theorem 8.2.1 is a simple consequence of Theorem 8.5.2, since every finite-dimensional normed space is reflexive and every linear operator defined there is compact.

The following example shows that the above theorem can not be extended to bounded linear operators without any additional restriction on $T$.

**Example 8.5.5.** *Consider $T: \ell_2 \to \ell_2$ defined by $Te_1 = -e_1$ and $Te_n = (1 - 1/n)e_n$ for $n \geq 2$, where $\{e_n : n \in \mathbb{N}\}$ is the usual orthonormal basis for the Hilbert space $\ell_2$. Then $T$ attains norm only at $\pm e_1$. Let $A = I$, the identity operator on $\ell_2$. It is easy to check that $T \perp_B A$. Indeed, $\|(T + \lambda A)e_1\| \geq \|T\|$ for all $\lambda \leq 0$ and $\|(T + \lambda A)e_n\| \geq \|T\|$, for all $\lambda \geq 1/n$. But $Te_1$ is not orthogonal to $Ae_1$ in the sense of Birkhoff-James.*

An additional condition that is required for a bounded linear operator $T$ to satisfy the BŠ property is that $\sup\{\|Tx\| : x \in C\} < \|T\|$, for all closed subsets $C$ of $S_{\mathbb{X}}$ with $d(M_T, C) > 0$.

**Theorem 8.5.6.** *Let $\mathbb{X}$ be a Banach space, $T \in \mathbb{B}(\mathbb{X}, \mathbb{Y})$, $M_T = D \cup (-D)$ ($D$ is a non-empty compact connected subset of $S_{\mathbb{X}}$). If $\sup\{\|Tx\| \cdot x \in C\} < \|T\|$ for all closed subsets $C$ of $S_{\mathbb{X}}$ with $d(M_T, C) > 0$ then for any $A \in \mathbb{B}(\mathbb{X}, \mathbb{Y})$, $T \perp_B A$ if and only if there exists $z \in M_T$ such that $Tz \perp_B Az$.*

*Proof.* Assume that $M_T = D \cup (-D)$ ($D$ is a non-empty compact connected subset of $S_{\mathbb{X}}$) and $\sup\{\|Tx\| : x \in C\} < \|T\|$ for all closed subsets $C$ of $S_{\mathbb{X}}$ with $d(M_T, C) > 0$.

If $z \in M_T$ such that $Tz \perp_B Az$ then clearly $T \perp_B A$. For the other part, suppose that $T \perp_B A$ but there exists no $x \in M_T$ such that $Tx \perp_B Ax$. Then by applying Lemma 8.5.1, there exists some $\lambda_0 \neq 0$ such that

$$\|Tx + \lambda_0 Ax\| < \|T\| \; \forall \; x \in M_T.$$

Without loss of generality, assume that $\lambda_0 > 0$.

Now $x \longrightarrow \|Tx + \lambda_0 Ax\|$ is a real valued continuous function from $S_{\mathbb{X}}$ to $\mathbb{R}$. As $M_T$ is a compact subset of $S_{\mathbb{X}}$, this function attains its maximum on $M_T$. Then there exists $\epsilon_1 > 0$ such that

$$\|Tx + \lambda_0 Ax\| < \|T\| - \epsilon_1 \; \forall \; x \in M_T.$$

Choose $\epsilon_x = \|T\| - \epsilon_1 - \|Tx + \lambda_0 Ax\|$. For each $x \in M_T$, clearly $\epsilon_x > 0$ and so by continuity of the function $T + \lambda_0 A$ at the point $x$, one can find an open ball $B(x, r_x)$ such that

$$\|(T + \lambda_0 A)(z - x)\| < \epsilon_x \; \forall \; z \in B(x, r_x) \cap S_{\mathbb{X}}.$$

Then $\|(T + \lambda_0 A)z\| < \|T\| - \epsilon_1 \; \forall \; z \in B(x, r_x) \cap S_{\mathbb{X}}$.

Again, let $\lambda \in (0, \lambda_0)$. Then for all $z \in B(x, r_x) \cap S_{\mathbb{X}}$

$$
\begin{aligned}
Tz + \lambda Az &= (1 - \frac{\lambda}{\lambda_0})Tz + \frac{\lambda}{\lambda_0}(Tz + \lambda_0 Az) \\
\Rightarrow \|Tz + \lambda Az\| &\leq (1 - \frac{\lambda}{\lambda_0})\|Tz\| + \frac{\lambda}{\lambda_0}\|Tz + \lambda_0 Az\| \\
\Rightarrow \|Tz + \lambda Az\| &< (1 - \frac{\lambda}{\lambda_0})\|T\| + \frac{\lambda}{\lambda_0}(\|T\| - \epsilon_1) \\
&= \|T\| - \frac{\lambda}{\lambda_0}\epsilon_1.
\end{aligned}
$$

The compactness of $M_T$ ensures that the cover $\{B(x, r_x) \cap M_T : x \in M_T\}$ has a finite subcover $\{B(x_i, r_{x_i}) \cap M_T : i = 1, 2, \ldots n\}$ so that

$$M_T \subset \cup_{i=1}^n B(x_i, r_{x_i}).$$

So for $\lambda \in (0, \lambda_0)$ and $z \in \left( \cup_{i=1}^n B(x_i, r_{x_i}) \right) \cap S_{\mathbb{X}}$, one gets,

$$\|Tz + \lambda Az\| < \|T\| - \frac{\lambda}{\lambda_0}\epsilon_1.$$

Consider $C = \cap_{i=1}^n B(x_i, r_{x_i})^c$. Then $C$ is a closed subset of $S_{\mathbb{X}}$ with $C \cap M_T = \emptyset$. As $M_T$ is compact so $d(C, M_T) > 0$. By the hypothesis, $\sup\{\|Tz\| : z \in C\} < \|T\|$ and so there exists $\epsilon_2 > 0$ such that $\sup\{\|Tz\| : z \in C\} < \|T\| - \epsilon_2$.

Choose $0 < \widetilde{\lambda} < \min\{\lambda_0, \frac{\epsilon_2}{2\|A\|}\}$. Then for all $z \in C$,

$$
\begin{aligned}
\|Tz + \widetilde{\lambda}Az\| &\leq \|Tz\| + |\widetilde{\lambda}|\|Az\| \\
&< \|T\| - \epsilon_2 + |\widetilde{\lambda}|\|A\| \\
&< \|T\| - \frac{1}{2}\epsilon_2.
\end{aligned}
$$

Choose $\epsilon = \min\{\frac{1}{2}\epsilon_2, \frac{\widetilde{\lambda}}{\lambda_0}\epsilon_1\}$. Then for all $x \in S_{\mathbb{X}}$, one gets,

$$
\|Tx + \widetilde{\lambda}Ax\| < \|T\| - \epsilon.
$$

This shows that $\|T + \widetilde{\lambda}A\| < \|T\|$, which contradicts the fact that $T\perp_B A$. This completes the proof. $\qquad\square$

## 8.6 Operator orthogonality on Hilbert spaces.

As mentioned earlier, it follows from Bhatia and Šemrl (1999); Paul (1999) that if $T$ is a bounded linear operator on a finite-dimensional Hilbert space $\mathbb{H}$, then $T\perp_B A$ for $A \in \mathbb{B}(\mathbb{H})$ if and only if there exists $x \in M_T$ such that $\langle Tx, Ax \rangle = 0$. The result is not necessarily true if the space is infinite-dimensional. In the following theorem, the problem is settled for Hilbert spaces of any dimension.

**Theorem 8.6.1.** *Let $T \in \mathbb{B}(\mathbb{H})$. Then for any $A \in \mathbb{B}(\mathbb{H})$, $T\perp_B A \Leftrightarrow Tx_0 \perp Ax_0$ for some $x_0 \in M_T$ if and only if $M_T = S_{H_0}$, where $H_0$ is a finite-dimensional subspace of $\mathbb{H}$ and $\|T\|_{H_0^\perp} < \|T\|$.*

*Proof.* Without loss of generality, assume that $\|T\| = 1$.
**Necessary part.**
From Theorem 8.3.1, it follows that in case of a Hilbert space, the norm attaining set $M_T$ is always a unit sphere of some subspace of the space. First, it is shown that the subspace is finite-dimensional. Suppose $M_T$ is the unit sphere of an infinite-dimensional subspace $H_0$. Then there exists a set $\{e_n : n \in \mathbb{N}\}$ of orthonormal vectors in $H_0$. Extend the set to a complete orthonormal basis $\mathcal{B} = \{e_\alpha : \alpha \in \Lambda \supset \mathbb{N}\}$ of $\mathbb{H}$. For each $e_\alpha \in H_0 \cap \mathcal{B}$,

$$
\|T^*T\| = \|T\|^2 = \|Te_\alpha\|^2 = \langle T^*Te_\alpha, e_\alpha \rangle \leq \|T^*Te_\alpha\|\|e_\alpha\| \leq \|T^*T\|,
$$

so that by the equality condition of Schwarz's inequality, $T^*Te_\alpha = \lambda_\alpha e_\alpha$ for some scalar $\lambda_\alpha$. Thus, $\{Te_\alpha : e_\alpha \in H_0 \cap \mathcal{B}\}$ is a set of orthonormal vectors in

$\mathbb{H}$. Define $A : \mathcal{B} \longrightarrow \mathbb{H}$ in the following way:

$$A(e_n) = \frac{1}{n^2} T e_n, n \in \mathbb{N}$$

$$A(e_\alpha) = T e_\alpha, \ e_\alpha \in H_0 \cap \mathcal{B} - \{e_n : n \in \mathbb{N}\}$$

$$A(e_\alpha) = 0, \ e_\alpha \in \mathcal{B} - H_0 \cap \mathcal{B}$$

As $\{T e_\alpha : e_\alpha \in H_0 \cap \mathcal{B}\}$ is a set of orthonormal vectors in $\mathbb{H}$, it is easy to see that $A$ can be extended as a bounded linear operator on $\mathbb{H}$.

Now, for any scalar $\lambda$, $\|T + \lambda A\| \geq \|(T + \lambda A)e_n\| = \|(1 + \frac{\lambda}{n^2})T e_n\| = | 1 + \frac{\lambda}{n^2} | \|T\| \longrightarrow \|T\|$. Thus, $T \perp_B A$.

Next, it is shown that there exists no $x \in M_T$ such that $Tx \perp_B Ax$. Let $x = \sum_\alpha \langle x, e_\alpha \rangle e_\alpha \in M_T$. Then

$$\langle Tx, Ax \rangle = \sum_n \frac{1}{n^2} | \langle x, e_n \rangle |^2 \|T\|^2 + \sum_{\alpha \notin \mathbb{N}} | \langle x, e_\alpha \rangle |^2 \|T\|^2$$

and so $\langle Tx, Ax \rangle = 0$ if and only if $x = 0$. Thus, $T \perp_B A$ but there exists no $x \in M_T$ such that $Tx \perp_B Ax$. This is a contradiction and so $H_0$ must be finite-dimensional.

Next claim is that $\|T\|_{H_0^\perp} < \|T\|$. Suppose $\|T\|_{H_0^\perp} = \|T\|$. As $T$ does not attain norm on $H_0^\perp$ and $\|T\| = \sup \{\|Tx\| : x \in S_{H_0^\perp}\}$, there exists $\{e_n\}$ in $H_0^\perp$ such that $\|T e_n\| \longrightarrow \|T\|$. Clearly, $\mathbb{H} = H_0 \oplus H_0^\perp$.

Define $A : \mathbb{H} \longrightarrow \mathbb{H}$ in the following way:

$$Az = Tx, \ \text{where } z = x + y, x \in H_0, y \in H_0^\perp$$

Then it is easy to check $A$ is bounded on $\mathbb{H}$. Also, for any scalar $\lambda$, $\|T + \lambda A\| \geq \|(T + \lambda A)e_n\| = \|T e_n\|$ holds for each $n \in N$. Then $\|T + \lambda A\| \geq \|T\|$ for all $\lambda$, so that $T \perp_B A$. However, there exists no $x \in M_T$ such that $\langle Tx, Ax \rangle = 0$. This contradiction completes the necessary part of the theorem.

**Sufficient part.**

If $\langle Tx_0, Ax_0 \rangle = 0$ for some $x_0 \in M_T$, then $T \perp_B A$. Next, let $T \perp_B A$. Then by Paul Paul (1999), there exists $\{z_n\} \subset S_\mathbb{H}$ such that $\|T z_n\| \to \|T\|$ and $\langle T z_n, A z_n \rangle \longrightarrow 0$. For each $n \in \mathbb{N}$, $z_n = x_n + y_n$, where $x_n \in H_0, y_n \in H_0^\perp$. Then $\|z_n\|^2 = 1 = \|x_n\|^2 + \|y_n\|^2$ and so $\|x_n\| \leq 1 \ \forall \ n \in \mathbb{N}$. As $H_0$ is a finite-dimensional subspace, $\{x_n\}$ being bounded, $\{x_n\}$ has a convergent subsequence, converging to some element of $H_0$. Without loss of generality, assume that $x_n \longrightarrow x_0$ (say) in $H_0$ in norm. Now, for each non-zero element $x \in H_0$,

$$\|T^*T\| \|x\|^2 \leq \|T\|^2 \|x\|^2 = \|Tx\|^2 = \langle T^*Tx, x \rangle \leq \|T^*Tx\| \|x\| \leq \|T^*T\| \|x\|^2$$

and so $\langle T^*Tx, x \rangle = \|T^*Tx\| \|x\|$. By the equality condition of Schwarz's inequality, $T^*Tx = \lambda_x x$ for some $\lambda_x$.

Now, $\langle T^*Tx_n, y_n \rangle = \langle T^*Ty_n, x_n \rangle = 0$ and so

$$
\begin{aligned}
\langle T^*Tz_n, z_n \rangle &= \langle T^*Tx_n, x_n \rangle + \langle T^*Tx_n, y_n \rangle + \langle T^*Ty_n, x_n \rangle + \langle T^*Ty_n, y_n \rangle \\
\Rightarrow \lim \|Tz_n\|^2 &= \lim \|Tx_n\|^2 + \lim \|Ty_n\|^2 \\
\Rightarrow \|T\|^2 &= \|Tx_0\|^2 + \lim \|Ty_n\|^2 \\
\Rightarrow \lim \|Ty_n\|^2 &= \|T\|^2(1 - \|x_0\|^2) \\
\Rightarrow \lim \|Ty_n\|^2 &= \|T\|^2 \lim \|y_n\|^2 \dots\dots (1)
\end{aligned}
$$

By hypothesis, $sup\{\|Ty\| : y \in H_0{}^\perp, \|y\| = 1\} < \|T\|$ and so by (1), there does not exist any non-zero subsequence of $\{\|y_n\|\}$. So $y_n = 0 \,\forall\, n$ and $z_n = x_n \,\forall\, n$. Then $\langle Tz_n, Az_n \rangle \to 0 \Rightarrow \langle Tx_0, Ax_0 \rangle = 0$. This establishes the theorem.

$\square$

## 8.7 Operator orthogonality and smoothness of bounded linear operators

Smoothness of bounded linear operators have been studied in great detail by several mathematicians including Holub Holub (1973), Heinrich Heinrich (1975), Hennefeld Hennefeld (1979), Abatzoglou Abatzoglou (1979), Kittaneh and Younis Kittaneh and Younis (1990). Smoothness of bounded linear operators on some particular spaces like $\ell^p$ spaces, etc. have also been studied by Werner Werner (1992) and Deeb and Khalil Deeb and Khalil (1992). As an application of the results obtained in connection with operator orthogonality, a sufficient condition for smoothness of compact linear operators on a reflexive Banach space is given. Later on, a sufficient condition for smoothness of bounded linear operators on a reflexive Banach space is also obtained. Also, characterization of smoothness of compact operators on reflexive Banach spaces and compact as well as bounded linear operators on Hilbert spaces is provided.

**Theorem 8.7.1.** *Let $\mathbb{X}$ be a reflexive Banach space and $\mathbb{Y}$ be a normed space. Then $T \in \mathbb{K}(\mathbb{X}, \mathbb{Y})$ is smooth if $T$ attains norm at a unique (upto scalar multiplication) vector $x_0$ (say) of $S_{\mathbb{X}}$ and $Tx_0$ is a smooth point.*

*Proof.* Assume $T$ attains norm at a unique (upto scalar multiplication) vector $x_0$ (say) of $S_{\mathbb{X}}$ and $Tx_0$ is a smooth point. Claim that for any $P, Q \in K(\mathbb{X}, \mathbb{Y})$, if $T \perp_B P$ and $T \perp_B Q$ then $T \perp_B (P + Q)$. By Theorem 8.5.2, $Tx_0 \perp_B Px_0$ and $Tx_0 \perp_B Qx_0$. As $Tx_0$ is a smooth point so $Tx_0 \perp_B (Px_0 + Qx_0)$. Then $T \perp_B (P + Q)$.
This completes the proof. $\square$

*Remark* 8.7.2. This improves the result [Theorem 2.2 ] proved by Hennefeld Hennefeld (1979), in which the author assumed $\mathbb{X}$ to be a smooth reflexive Banach space with a Schauder basis.

Conversely, it is shown that the conditions are necessary.

**Theorem 8.7.3.** *Let $\mathbb{X}$ be a reflexive Banach space and $\mathbb{Y}$ be a normed space. If $T \in \mathbb{K}(\mathbb{X}, \mathbb{Y})$ is smooth then $T$ attains norm at a unique (upto scalar multiplication) vector $x_0$ (say) of $S_{\mathbb{X}}$ and $Tx_0$ is a smooth point.*

*Proof.* Since the space $\mathbb{X}$ is reflexive and $T$ is compact, there exists $x \in M_T$. It is shown that if $x_1, x_2 \in M_T$ then $x_1 = \pm x_2$. If possible, suppose that $x_1 \neq \pm x_2$. There exists a subspace $H_1$ of codimension 1 such that $x_1 \perp_B H_1$. There exists a scalar $a$ with $\mid a \mid \leq 1$ such that $ax_1 + x_2 \in H_1$. Again, there exists a subspace $H_2$ of $H_1$ with codimension 1 in $H_1$ such that $(ax_1 + x_2) \perp_B H_2$. Now, every element $z \in S_{\mathbb{X}}$ can be written uniquely as $z = \alpha x_1 + h_1$ for some scalar $\alpha$ and $h_1 \in H_1$. Also, $h_1$ can be written uniquely as $h_1 = \beta(ax_1 + x_2) + h_2$ for some scalar $\beta$ and $h_2 \in H_2$. Thus, $z = (\alpha + a\beta)x_1 + \beta x_2 + h_2$. Define operators $A_1, A_2 : \mathbb{X} \longrightarrow \mathbb{Y}$ in the following way:

$$A_1(z) = (\alpha + a\beta)Tx_1, \ \ A_2(z) = \beta Tx_2 + Th_2.$$

Clearly, both $A_1, A_2$ are compact linear operators. Then $T \perp_B A_1, T \perp_B A_2$ but $T = A_1 + A_2$, which shows that $T$ is not orthogonal to $A_1 + A_2$ in the sense of Birkhoff-James. This shows that $T$ is not smooth. Hence $T$ attains norm at unique( upto scalar multiplication ) vector $x_0 \in S_{\mathbb{X}}$.

Next, it is shown that $Tx_0$ is a smooth point in $\mathbb{Y}$. If possible, suppose that $Tx_0$ is not smooth. Then there exists $y, z \in \mathbb{Y}$ such that $Tx_0 \perp_B y, Tx_0 \perp_B z$ but $Tx_0$ is not orthogonal to $y + z$ in the sense of Birkhoff-James. There exists a hyperplane $H$ such that $x_0 \perp_B H$. Define two operators $A_1, A_2 : \mathbb{X} \longrightarrow \mathbb{Y}$ as follows:

$$A_1(ax_0 + h) = ay, \ \ A_2(ax_0 + h) = az.$$

Then it is easy to check that both $A_1, A_2$ are compact linear operators and $T \perp_B A_1, T \perp_B A_2$. But $T$ is not orthogonal to $A_1 + A_2$, otherwise since $M_T = \{\pm x_0\}$, by Theorem 8.5.2, $Tx_0 \perp_B (y + z)$, which is not possible. This contradiction shows that $Tx_0$ is a smooth point. this completes the proof of the theorem. □

**Corollary 8.7.4.** *$T \in \mathbb{K}(\mathbb{X}, \mathbb{Y})$ is a smooth point if and only if $T^*$ attains norm at a unique (upto scalar multiplication) vector $g$ (say) of $S_{\mathbb{Y}^*}$ and $T^*g$ is a smooth point.*

*Proof.* First note that $T$ is smooth if and only if $T^*$ is smooth. Then by using Corollary 8.5.3 and following the same method as above, one can show $T$ is a smooth point if and only if $T^*$ attains norm at a unique (upto scalar multiplication) vector $g$ (say) of $S_{\mathbb{Y}^*}$ and $T^*g$ is a smooth point. □

*Remark* 8.7.5. In Heinrich (1975) Heinrich proved necessary and sufficient conditions for smoothness of compact operators from a Banach space $\mathbb{X}$ to a Banach space $\mathbb{Y}$, using differentiability of the norm on a Banach space. In Theorem 8.7.1, Theorem 8.7.3 and Corollary 8.7.4, alternative proofs of the results, without assuming $\mathbb{Y}$ to be a Banach space, is given.

Next, a sufficient condition for smoothness of a bounded linear operator is given.

**Theorem 8.7.6.** *Let $\mathbb{X}$ be a Banach space and $\mathbb{Y}$ be a normed space. Then $T \in \mathbb{B}(\mathbb{X}, \mathbb{Y})$ is a smooth point if $T$ attains norm only at $\pm x_0$, $Tx_0$ is smooth and $\sup\{\|Tx\| : x \in C\} < \|T\|$ for all closed subsets $C$ of $S_\mathbb{X}$ with $d(\pm x_0, C) > 0$.*

*Proof.* Suppose $T$ attains norm only at $\pm x_0$, $Tx_0$ is smooth and $sup\{\|Tx\| : x \in C\} < \|T\|$ for all closed subsets $C$ of $S_\mathbb{X}$ with $d(\pm x_0, C) > 0$. Let $T \perp_B A_1$, $T \perp_B A_2$. Then by Theorem 8.5.6, $Tx_0 \perp_B A_1 x_0$, $Tx_0 \perp_B A_2 x_0$. As $Tx_0$ is a smooth point so $Tx_0 \perp_B (A_1 + A_2) x_0$. Therefore, we have, $T \perp_B (A_1 + A_2)$. Thus, $T$ is a smooth point.

$\square$

In the following theorem, a necessary condition for smoothness of bounded linear operator is given.

**Theorem 8.7.7.** *Let $\mathbb{X}$ be a Banach space and $\mathbb{Y}$ be a normed space. If $T \in \mathbb{B}(\mathbb{X}, \mathbb{Y})$ is a smooth point that attains norm only at $\pm x_0 \in S_\mathbb{X}$ then $\sup_{x \in H \cap S_\mathbb{X}} \|Tx\| < \|T\|$, where $H$ is a hyperplane such that $x_0 \perp_B H$.*

*Proof.* If possible, suppose that $sup_{x \in H \cap S_\mathbb{X}} \|Tx\| = \|T\|$. Then there exists $\{x_n\} \subset H \cap S_\mathbb{X}$ such that $\|Tx_n\| \longrightarrow \|T\|$. Every element $z \in S_\mathbb{X}$ can be written as $z = \alpha x_0 + h$ for some scalar $\alpha$ and $h \in H$. Define operators $A_1, A_2 : \mathbb{X} \longrightarrow \mathbb{Y}$ in the following way:

$$A_1(z) = \alpha Tx_0, \quad A_2(z) = Th.$$

It is easy to verify that both $A_1, A_2$ are bounded linear operators. Now, $\|T + \lambda A_1\| \geq \|(T + \lambda A_1) x_n\| = \|Tx_n\| \to \|T\|$, so that $T \perp_B A_1$. Similarly, $\|T + \lambda A_2\| \geq \|(T + \lambda A_2) x_0\| = \|Tx_0\| = \|T\|$, so that $T \perp_B A_2$. But $T = A_1 + A_2$, which shows that $T$ is not orthogonal to $A_1 + A_2$. This contradiction proves the result. $\square$

Another necessary condition for smoothness of a bounded linear operator is that $M_T = \{\pm x\}$, if $M_T \neq \emptyset$. It is easy to observe that the proof of this fact follows in the same way as in the proof of Theorem 8.7.3.

**Theorem 8.7.8.** *Let $\mathbb{X}$ be a Banach space, $\mathbb{Y}$ be a normed space and $T \in \mathbb{B}(\mathbb{X}, \mathbb{Y})$ be a smooth point. If $x_1, x_2 \in M_T$ then $x_1 = \pm x_2$.*

Abatzoglou Abatzoglou (1979) studied the smoothness of bounded linear operators on a Hilbert space. Here we explore the same topic, using an alternative approach. For this purpose, the following lemma is required.

**Lemma 8.7.9.** *Suppose $T$ is a bounded linear operator on a Hilbert space $\mathbb{H}$, that does not attain norm, i.e., $M_T = \emptyset$. Then there exists a sequence $\{e_n\}$ of orthonormal elements in $\mathbb{H}$ such that $\|Te_n\| \longrightarrow \|T\|$.*

*Proof.* Note that $T^*T$ is a positive operator, $\|T^*T\| = \|T\|^2$ and $T$ attains norm if and only if $T^*T$ attains norm. Also $\|T\|^2 \in \sigma(T^*T) = \sigma_{disc}(T^*T) \cup \sigma_{ess}(T^*T)$. By Theorem VII.10 of Reed and Simon Reed and Simon (1980), $\|T\|^2$ can't belong to $\sigma_{disc}(T^*T)$. So by Theorem VII.12(Weyl's criterion) of Reed and Simon Reed and Simon (1980), there exists $\{e_n\}$ of orthonormal vectors such that $\|(T^*T - \|T\|^2)e_n\| \longrightarrow 0$. Now, $|\langle (T^*T - \|T\|^2)e_n, e_n \rangle| \leq \|(T^*T - \|T\|^2)e_n\| \longrightarrow 0$. This implies $\langle (T^*T - \|T\|^2)e_n, e_n \rangle = \langle T^*Te_n, e_n \rangle - \langle \|T\|^2 e_n, e_n \rangle \longrightarrow 0$. Thus, $\|Te_n\| \longrightarrow \|T\|$. $\square$

**Theorem 8.7.10.** *Let $\mathbb{H}$ be a Hilbert space. Then $T \in \mathbb{B}(\mathbb{H})$ is a smooth point if and only if $T$ attains norm only at $\pm x_0$ and $\sup\{\|Ty\| : x_0 \perp y, y \in S_{\mathbb{H}}\} < \|T\|$.*

*Proof.* First, the necessary part is proved in the following three steps:
(i) T attains norm at some point of $S_{\mathbb{X}}$.
(ii) T attains norm at unique point $x_0 \in S_{\mathbb{X}}$.
(iii) $\sup\{\|Ty\| : x_0 \perp y, y \in S_{\mathbb{H}}\} < \|T\|$.

Claim (i): If $T$ does not attain norm then by Lemma 8.7.9, there exists a sequence $\{e_n\}$ of orthonormal elements such that $\|Te_n\| \longrightarrow \|T\|$. Clearly, there exists an orthonormal basis $\mathcal{B}$ containing $\{e_n : n = 1, 2, \ldots\}$. Define a linear operator $A_1$ on $H$ in the following way:
$A_1 e_{2n} = Te_{2n}$ and $A_1$ takes every other element of $\mathcal{B}$ to 0. Claim that $A_1$ is bounded. Every element $z$ in $H$ can be written as $z = x + y$, where $x = \sum \alpha_{2n} e_{2n}$ and $y \perp x$. Now, $\|A_1 z\| = \|Tx\| \leq \|T\|\|x\| \leq \|T\|$ for every $z$ with $\|z\| = 1$. Thus, $A_1$ is bounded. Consider another bounded operator $A_2 = T - A_1$. Next claim is that $T \perp_B A_1, T \perp_B A_2$. Now, $\|Te_{2n+1}\| \longrightarrow \|T\|$ and $\langle Te_{2n+1}, A_1 e_{2n+1} \rangle \longrightarrow 0$, and so by Lemma 2 of Paul (1999), $T \perp_B A_1$. Similarly, one can show that $T \perp_B A_2$. But $T$ is not orthogonal to $A_1 + A_2$ in the sense of Birkhoff-James. This contradicts the fact that $T$ is smooth and completes the proof of the claim.

Claim (ii): Suppose $x_1, x_2 \in M_T$ and $x_1 \neq \pm x_2$. By Theorem 8.3.1, the norm attaining set $M_T$ is a unit sphere of some subspace of $\mathbb{H}$. So, without loss of generality, assume that $x_1 \perp x_2$. Let $H_0 = \langle \{x_1, x_2\} \rangle$. Then $\mathbb{H} = H_0 \oplus H_0^\perp$. Define $A_1, A_2 : \mathbb{H} \longrightarrow \mathbb{H}$ in the following way:
$A_1(c_1 x_1 + c_2 x_2 + h) = c_1 Tx_1$, $A_2(c_1 x_1 + c_2 x_2 + h) = c_2 Tx_2 + Th$, where $h \in H_0^\perp$. Then, as before, it is easy to check that both $A_1, A_2$ are bounded linear operators and $T \perp_B A_1, T \perp_B A_2$. But $T = A_1 + A_2$ and so $T$ is not orthogonal to $A_1 + A_2$ in the sense of Birkhoff-James, which contradicts the fact that $T$ is smooth.

Claim (iii): Suppose $T$ attains norm only at $\pm x_0 \in S_{\mathbb{H}}$ and $sup\{\|Ty\| : x_0 \perp y, y \in S_{\mathbb{H}}\} = \|T\|$. Let $H_0 = \langle\{x_0\}\rangle$. Then $\mathbb{H} = H_0 \oplus H_0^{\perp}$. Define $A_1, A_2 : \mathbb{H} \longrightarrow \mathbb{H}$ in the following way:

Let $z = x + y \in \mathbb{H}$, where $x \in H_0, y \in H_0^{\perp}$. Then $A_1 z = Tx, A_2 z = Ty$. It is easy to check that both $A_1, A_2$ are bounded linear operators and $T \perp_B A_1, T \perp_B A_2$. But $T = A_1 + A_2$ and so $T$ is not orthogonal to $A_1 + A_2$ in the sense of Birkhoff-James, which contradicts the fact that $T$ is smooth.

For the sufficient part, assume $T$ attains norm only at $\pm x_0$ and $sup\{\|Ty\| : x_0 \perp y, y \in S_{\mathbb{H}}\} < \|T\|$. Let $T \perp_B A_i (i = 1, 2)$. Then by Theorem 8.6.1, $Tx_0 \perp A_1 x_0$ and $Tx_0 \perp A_2 x_0$. As $Tx_0$ is a smooth point of $\mathbb{H}$, we have, $Tx_0 \perp (A_1 + A_2) x_0$. Then $T \perp_B (A_1 + A_2)$. Thus, $T$ is smooth. $\square$

## 8.8 Characterization of operator orthogonality for finite-dimensional spaces

After all these discussions, it would be nice to have a complete characterization of the Birkhoff-James orthogonality of bounded linear operators defined on a Banach space. Indeed, that would be the culmination of an analogous Banach space theoretic study of operator orthogonality in the sense of Birkhoff-James, initiated by Bhatia and Šemrl in Bhatia and Šemrl (1999) and Paul et al., in Paul et al. (2008), in the context of Hilbert spaces.

In order to characterize Birkhoff-James orthogonality of linear operators defined on finite-dimensional Banach spaces, the notions $y \in x^+$ and $y \in x^-$, for any two elements $x, y$ in a normed space $\mathbb{X}$ were introduced in Sain (2017). First, some obvious but useful properties of this notion is mentioned, without giving explicit proofs.

**Proposition 8.8.1.** *Let $\mathbb{X}$ be a normed space and $x, y \in \mathbb{X}$. Then the following are true:*

*(i) Either $y \in x^+$ or $y \in x^-$.*
*(ii) $x \perp_B y$ if and only if $y \in x^+$ and $y \in x^-$.*
*(iii) $y \in x^+$ implies that $\eta y \in (\mu x)^+$ for all $\eta, \mu > 0$.*
*(iv) $y \in x^+$ implies that $-y \in x^-$ and $y \in (-x)^-$.*
*(v) $y \in x^-$ implies that $\eta y \in (\mu x)^-$ for all $\eta, \mu > 0$.*
*(vi) $y \in x^-$ implies that $y \in x^+$ and $y \in (-x)^+$.*

In the next theorem, this notion is used to give a characterization of Birkhoff-James orthogonality of linear operators defined on a finite-dimensional Banach space.

**Theorem 8.8.2.** *Let $\mathbb{X}$ be a finite-dimensional Banach space. Let $T, A \in$*

$\mathbb{B}(\mathbb{X})$. *Then $T \perp_B A$ if and only if there exists $x, y \in M_T$ such that $Ax \in Tx^+$ and $Ay \in Ty^-$.*

*Proof.* **Sufficient part.**
Suppose there exists $x, y \in M_T$ such that $Ax \in Tx^+$ and $Ay \in Ty^-$. For any $\lambda \geq 0, \|T + \lambda A\| \geq \|Tx + \lambda Ax\| \geq \|Tx\| = \|T\|$. Similarly, for any $\lambda \leq 0, \|T + \lambda A\| \geq \|Ty + \lambda Ay\| \geq \|Ty\| = \|T\|$. This proves that $T \perp_B A$.
**Necessary part.**
Let $T, A \in \mathbb{B}(\mathbb{X})$ be such that $T \perp_B A$. If possible, suppose that there does not exist $x, y \in M_T$ such that $Ax \in Tx^+$ and $Ay \in Ty^-$. Using $(i)$ of Proposition 8.8.1, it is easy to show that either of the following is true:
(i) $Ax \in Tx^+$ for each $x \in M_T$ and $Ax \notin Tx^-$ for any $x \in M_T$
(ii) $Ax \in Tx^-$ for each $x \in M_T$ and $Ax \notin Tx^+$ for any $x \in M_T$.
Without loss of generality, assume that $Ax \in Tx^+$ for each $x \in M_T$ and $Ax \notin Tx^-$ for any $x \in M_T$. Consider the function $g : S_{\mathbb{X}} \times [-1, 1] \longrightarrow \mathbb{R}$ defined by

$$g(x, \lambda) = \|Tx + \lambda Ax\|.$$

It is easy to check that $g$ is continuous. Given any $x \in M_T$, since $Ax \notin Tx^-$, there exists $\lambda_x < 0$ such that $g(x, \lambda_x) = \|Tx + \lambda_x Ax\| < \|Tx\| = \|T\|$. By continuity of $g$, there exists $r_x, \delta_x > 0$ such that

$$g(y, \lambda) < \|T\| \quad \text{for all} \quad y \in B(x, r_x) \cap S_{\mathbb{X}} \text{ and for all } \lambda \in (\lambda_x - \delta_x, \lambda_x + \delta_x).$$

Using the convexity property of the norm function, it is easy to show that $g(y, \lambda) = \|Ty + \lambda Ay\| < \|T\|$ for all $y \in B(x, r_x) \cap S_{\mathbb{X}}$ and for all $\lambda \in (\lambda_x, 0)$.

For any $z \in S_{\mathbb{X}} \setminus M_T$, clearly $g(z, 0) = \|Tz\| < \|T\|$. Thus, by continuity of $g$, there exist $r_z, \delta_z > 0$ such that $g(y, \lambda) = \|Ty + \lambda Ay\| < \|T\|$ for all $y \in B(z, r_z) \cap S_{\mathbb{X}}$ and for all $\lambda \in (-\delta_z, \delta_z)$.
Consider the open cover $\{B(x, r_x) \cap S_{\mathbb{X}} : x \in M_T\} \cup \{B(z, r_z) \cap S_{\mathbb{X}} : z \in S_{\mathbb{X}} \setminus M_T\}$ of $S_{\mathbb{X}}$. Since $\mathbb{X}$ is finite-dimensional, $S_{\mathbb{X}}$ is compact. This proves that the considered open cover has a finite subcover and so

$$S_{\mathbb{X}} \subset \cup_{i=1}^{n_1} B(x_i, r_{x_i}) \cup_{k=1}^{n_2} B(z_k, r_{z_k}) \cap S_{\mathbb{X}},$$

for some positive integers $n_1, n_2$, where each $x_i \in M_T$ and each $z_k \in S_{\mathbb{X}} \setminus M_T$.

Choose $\lambda_0 \in (\cap_{i=1}^{n_1}(\lambda_{x_i}, 0)) \bigcap (\cap_{k=1}^{n_2}(-\delta_{z_k}, \delta_{z_k}))$.
Since $\mathbb{X}$ is finite-dimensional, $T + \lambda_0 A$ attains its norm at some $w_0 \in S_{\mathbb{X}}$. Either $w_0 \in B(x_i, r_{x_i})$ for some $x_i \in M_T$ or $w_0 \in B(z_k, r_{z_k})$ for some $z_k \in S_{\mathbb{X}} \setminus M_T$. In either case, it follows from the choice of $\lambda_0$ that $\|T + \lambda_0 A\| = \|(T + \lambda_0 A)w_0\| < \|T\|$, which contradicts the primary assumption that $T \perp_B A$ and thereby completes the proof of the necessary part. $\square$

Theorem 8.2.1 can now be deduced as a corollary to the previous theorem.

**Corollary 8.8.3.** *Let $\mathbb{X}$ be a finite-dimensional Banach space. Let $T \in \mathbb{B}(\mathbb{X})$ be such that $M_T = \pm D$, where $D$ is a compact, connected subset of $S_{\mathbb{X}}$. Then for $A \in \mathbb{B}(\mathbb{X})$ with $T \perp_B A$, there exists $x \in D$ such that $Tx \perp_B Ax$.*

*Proof.* Since $T \perp_B A$, applying Theorem 8.8.2, it is evident that there exists $x, y \in M_T = \pm D$ such that $Ax \in Tx^+$ and $Ay \in Ty^-$. Moreover, it is easy to see that by applying $(iv)$ and $(vi)$ of Proposition 8.8.1, one may assume without loss of generality that $x, y \in D$. Then following the same line of arguments, as in the proof of Theorem 8.8.2, it can be proved that there exists $u_0 \in D$ such that $Au_0 \in Tu_0^+$ and $Au_0 \in Tu_0^-$, by using the connectedness of $D$. However, this is equivalent to $Tu_0 \perp_B Au_0$, completing the proof of Theorem 8.2.1. □

*Remark* 8.8.4. The previous theorem gives a complete characterization of Birkhoff-James orthogonality of linear operators defined on a finite-dimensional Banach space $\mathbb{X}$. Moreover, Theorem 8.8.2 is very useful computationally as well as from theoretical point of view. It should be noted that the main idea of the proof of Theorem 8.8.2 was already there in the proof of Theorem 8.2.1. However, complete characterization of Birkhoff-James orthogonality of linear operators on $\mathbb{X}$ could not be obtained in Theorem 8.2.1. This reveals the usefulness of the notions introduced in Sain (2017).

---

## 8.9 Symmetry of operator orthogonality

James James (47 a) proved that Birkhoff-James orthogonality is symmetric in a normed linear space $\mathbb{X}$, having dimension three or more, if and only if a compatible inner product can be defined on $\mathbb{X}$. Since $\mathbb{B}(\mathbb{X})$ is not an inner product space, it is interesting to study the symmetry of Birkhoff-James orthogonality of operators in $\mathbb{B}(\mathbb{X})$. It is very easy to observe that in $\mathbb{B}(\mathbb{X})$, $T \perp_B A$ may not imply $A \perp_B T$ or conversely. Consider $T = \begin{pmatrix} 1 & 0 & 0 \\ 0 & 1/2 & 0 \\ 0 & 0 & 1/2 \end{pmatrix}$

and $A = \begin{pmatrix} 0 & 0 & 0 \\ 0 & 1 & 0 \\ 0 & 0 & 0 \end{pmatrix}$ on $(\mathbb{R}^3, \|.\|_2)$. Then it can be shown using elementary arguments that $T \perp_B A$ but $A \not\perp_B T$. However, in Ghosh et al. (2016) it is proved that if $\mathbb{H}$ is a finite-dimensional Hilbert space and $T \in \mathbb{B}(\mathbb{H})$, then for all $A \in \mathbb{B}(\mathbb{H}), A \perp_B T \Rightarrow T \perp_B A$ if and only if $M_T = S_{\mathbb{H}}$. If $\mathbb{H}$ is an infinite-dimensional Hilbert space and $T \in \mathbb{K}(\mathbb{H})$, then for all $A \in \mathbb{B}(\mathbb{H}), A \perp_B T \Rightarrow T \perp_B A$ if and only if $T$ is the zero operator. It is also proved that $T \perp_B A \Rightarrow A \perp_B T$ for all $A \in \mathbb{B}(\mathbb{H})$ if and only if $T$ is the zero operator.

The left symmetry of Birkhoff-James orthogonality of linear operators defined on $\mathbb{X}$ is also explored. It is proved that $T \in \mathbb{B}(l_p^2)(p \geq 2, p \neq \infty)$ is left symmetric with respect to Birkhoff-James orthogonality if and only if $T$ is the zero operator.

We begin with a complete characterization of right symmetric linear operators on finite-dimensional Hilbert spaces.

**Theorem 8.9.1.** *Let $\mathbb{H}$ be a finite-dimensional Hilbert space and $T \in \mathbb{B}(\mathbb{H})$. Then for all $A \in \mathbb{B}(\mathbb{H}), A \perp_B T \Rightarrow T \perp_B A$ if and only if $M_T = S_{\mathbb{H}}$.*

*Proof.* Let $M_T = S_{\mathbb{H}}$ and $A \in \mathbb{B}(\mathbb{H})$ such that $A \perp_B T$. Then it follows from Theorem 3.3 of Benítez et al. (2007) that $\langle Ax_0, Tx_0 \rangle = 0$ for some $x_0 \in M_A$. Now, $M_T = S_{\mathbb{H}}$ and $\langle Tx_0, Ax_0 \rangle = 0$, so it is easy to see that that $T \perp_B A$. Conversely, let $M_T \neq S_{\mathbb{H}}$. Let dim $\mathbb{H} = n$. Without loss of generality, assume that $\|T\| = 1$. An operator $A \in \mathbb{B}(\mathbb{H})$ is exhibited such that $A \perp_B T$ but $T \not\perp_B A$.

By Theorem 8.3.1, it follows that in case of a Hilbert space $\mathbb{H}$, the norm attainment set $M_T$ is the unit sphere of some subspace of $\mathbb{H}$. Then $M_T = S_{\mathbb{H}_0}$ for some proper subspace $H_0$ of $\mathbb{H}$. Let $\{x_1, x_2, \ldots, x_m\}$, with m $< n$, be an orthonormal basis of $H_0$. Extend this basis to an orthonormal basis $\{x_1, x_2, \ldots, x_m, \ldots, x_n\}$ of $\mathbb{H}$. Clearly, $T(H_0) \neq \mathbb{H}$ and so there exists a unit vector $w_0$ such that $w_0 \perp T(H_0)$.

For the two vectors $w_0, Tx_{m+1}$, either $\|w_0 + \lambda Tx_{m+1}\| \geq 1$ for all $\lambda \geq 0$ or $\|w_0 + \lambda Tx_{m+1}\| \geq 1$ for all $\lambda \leq 0$.

Without any loss of generality, assume that $\|w_0 + \lambda Tx_{m+1}\| \geq 1$ for all $\lambda \geq 0$. Define a linear operator $A : \mathbb{H} \longrightarrow \mathbb{H}$ in the following way:

$$
\begin{aligned}
Ax_i &= -Tx_i, \quad i \in \{1, 2, \ldots, m\} \\
Ax_i &= w_0, \quad i = m+1 \\
Ax_i &= 0, \quad i \in \{m+2, \ldots, n\}
\end{aligned}
$$

Let $z = \sum_{i=1}^{n} c_i x_i \in S_{\mathbb{H}}$. Then $\|Az\|^2 = \|\sum_{i=1}^{m} c_i Tx_i\|^2 + c_{m+1}^2 = \sum_{i=1}^{n} c_i^2 + c_{m+1}^2 \leq 1$, which shows that $\|A\| = 1$.
For $\lambda > 0$,
$\|A + \lambda T\| \geq \|(A + \lambda T)x_{m+1}\| = \|w_0 + \lambda Tx_{m+1}\| > 1 = \|A\|$.
Also, for $\lambda < 0$,
$\|A + \lambda T\| \geq \|(A + \lambda T)x_m\| = \|(-1 + \lambda)Tx_m\| = |1 - \lambda|\|T\| > 1 = \|A\|$.
Therefore, $A \perp_B T$.
Finally, it is shown that $T \not\perp_B A$. If not, then by the Theorem 3.3 of Benítez et al. (2007), there exists $x \in S_{\mathbb{H}_0}$ such that $\langle Tx, Ax \rangle = 0$. However, this is not possible as $Ax = -Tx$ and $x \in M_T$. Thus, $T \not\perp_B A$.
This completes the proof. □

In the next theorem, a complete characterization of right symmetric bounded linear operators is obtained when the underlying space is an infinite-dimensional Hilbert space.

**Theorem 8.9.2.** *Let* $\mathbb{H}$ *be a infinite-dimensional Hilbert space and* $T \in \mathbb{K}(\mathbb{H})$. *Then for all* $A \in \mathbb{B}(\mathbb{H}), A \perp_B T \Rightarrow T \perp_B A$ *if and only if* $T$ *is the zero operator.*

*Proof.* If $T$ is the zero operator then there is nothing to prove. Conversely, if possible, let $T$ be a non-zero compact operator such that $A \perp_B T \Rightarrow T \perp_B A$ for all $A \in \mathbb{B}(\mathbb{H})$. Without any loss of generality, assume that $\|T\| = 1$. An operator $A \in \mathbb{B}(\mathbb{H})$ is exhibited such that $A \perp_B T$ but $T \not\perp_B A$. By Theorem 8.3.1, it follows that in case of a Hilbert space $\mathbb{H}$, the norm attainment set $M_T$ is the unit sphere of some subspace of $\mathbb{H}$. Since $T$ is a compact operator on infinite-dimensional Hilbert space $H$, $M_T \neq S_{\mathbb{H}}$. Hence $M_T$ is the unit sphere of some proper subspace $H_0$ of $\mathbb{H}$. Let $\{x_\alpha : \alpha \in \Lambda_1\}$ be an orthonormal basis of $H_0$. Extend this basis to an orthonormal basis $\{x_\alpha, y_\beta : \alpha \in \Lambda_1, \beta \in \Lambda_2\}$ of $\mathbb{H}$. Claim $T(H_0) \neq \mathbb{H}$. If not, then $T(B_{H_0}) = B_{\mathbb{H}}$, from which it follows that $B_{\mathbb{H}}$ is compact, which is clearly a contradiction. As the space $\mathbb{H}$ is a Hilbert space, corresponding to the subspace $T(H_0)$, by James James (47 a), there exists a unit vector $w_0$ such that $w_0 \perp T(H_0)$. Choose $\beta_0 \in \Lambda_2$. For the two vectors $w_0, Ty_{\beta_0}$, either $\|w_0 + \lambda Ty_{\beta_0}\| \geq 1$ for all $\lambda \geq 0$ or $\|w_0 + \lambda Ty_{\beta_0}\| \geq 1$ for all $\lambda \leq 0$. Without any loss of generality, assume that $\|w_0 + \lambda Ty_{\beta_0}\| \geq 1$ for all $\lambda \geq 0$.

Define a linear operator $A : \mathbb{H} \longrightarrow \mathbb{H}$ in the following way:

$$\begin{aligned} Ax_\alpha &= -Tx_\alpha, \quad \alpha \in \Lambda_1 \\ Ay_\beta &= w_0, \quad \beta = \beta_0 \\ Ay_\beta &= 0, \quad \beta \neq \beta_0 \end{aligned}$$

It is easy to see that $A$ is bounded.

Let $z = \sum c_{\alpha_i} x_{\alpha_i} + \sum d_{\beta_i} y_{\beta_i} \in S_{\mathbb{H}}$. Then

$$\|Az\|^2 = \|\sum c_{\alpha_i} Tx_{\alpha_i}\|^2 + d_{\beta_0}^2 = \sum c_{\alpha_i}^2 + d_{\beta_0}^2 \leq 1,$$

which shows that $\|A\| = 1$. For $\lambda > 0$, $\|A + \lambda T\| \geq \|(A + \lambda T)y_{\beta_0}\| = \|w_0 + \lambda Ty_{\beta_0}\| \geq 1 = \|A\|$ and for $\lambda < 0$, $\|A + \lambda T\| \geq \|(A + \lambda T)x_\alpha\| = \|(-1 + \lambda)Tx_\alpha\| = |1 - \lambda|\|T\| > 1 = \|A\|$. Therefore, $A \perp_B T$.

Next, it is shown that $T \not\perp_B A$. If not, then as before, there exists a sequence $\{x_n\} \subset S_{\mathbb{H}}$ such that $\|Tx_n\| \to \|T\|$ and $\langle Tx_n, Ax_n \rangle \to 0$. Since the unit ball of $\mathbb{H}$ is weakly compact, there exists a subsequence $\{x_{n_k}\}$ of $\{x_n\}$ such that $x_{n_k} \rightharpoonup x_0$ (say) in $B_{\mathbb{H}}$ weakly. Then $A$ being bounded, $Ax_{n_k} \rightharpoonup Ax_0$ and $T$ being compact, $Tx_{n_k} \to Tx_0$ and so $\|Tx_{n_k}\| \to \|Tx_0\|$. Therefore, $\langle Ax_{n_k}, Tx_{n_k} \rangle \to \langle Ax_0, Tx_0 \rangle$. As $\langle Tx_n, Ax_n \rangle \to 0$, we must have, $\langle Tx_0, Ax_0 \rangle = 0$. However, this is not possible as $Ax_0 = -Tx_0$ and $x_0 \subset M_T$. Thus, $T \not\perp_B A$. This completes the proof. $\square$

*Remark* 8.9.3. In case of a complex Hilbert space $\mathbb{H}$, Turnsek Turnsek (2005) characterized those $T \in \mathbb{B}(\mathbb{H})$ for which $T \perp_B A \Rightarrow A \perp_B T$ for all $A \in \mathbb{B}(\mathbb{H})$. It should be noted that the technique used here, both for finite-dimensional Hilbert spaces and for compact operators on infinite-dimensional Hilbert spaces, are completely different from the one used by Turnsek.

In the next theorem, left symmetric bounded linear operators on a Hilbert space of any dimension are characterized.

**Theorem 8.9.4.** *Let $\mathbb{H}$ be a Hilbert space and $T \in \mathbb{K}(\mathbb{H})$. Then for all $A \in \mathbb{B}(\mathbb{H})$, $T \perp_B A \Rightarrow A \perp_B T$ if and only if $T$ is the zero operator.*

*Proof.* One part of the proof is obvious, i.e., if $T$ is the zero operator then $\forall A \in \mathbb{B}(\mathbb{H})$, $T \perp_B A \Rightarrow A \perp_B T$. For the other part, suppose that $T$ is a non-zero compact operator. Without loss of generality, assume $\|T\| = 1$. An operator $A \in \mathbb{B}(\mathbb{H})$ is exhibited such that $T \perp_B A$ but $A \not\perp_B T$. Since $T$ is a compact operator on a Hilbert space, $T$ attains norm at $x \in S_\mathbb{H}$ (say). Then $\mathbb{H} = \langle \{x\} \rangle \oplus H_0$, where $H_0 = \langle \{x\} \rangle^\perp$. Let $\{x_\alpha : \alpha \in \Lambda\}$ be an orthonormal basis of $H_0$, then $\{x, x_\alpha : \alpha \in \Lambda\}$ is an orthonormal basis of $\mathbb{H}$.

**Case 1.** $T(H_0) = \{0\}$.

There exists $w \in S_\mathbb{H}$ such that $w \perp Tx$. Define a linear operator $A : \mathbb{H} \longrightarrow \mathbb{H}$ in the following way:

$$
\begin{aligned}
Ax &= w \\
Ax_\alpha &= \frac{w + Tx}{\sqrt{2}}, \quad \alpha = \alpha_0 \\
Ax_\alpha &= 0, \quad \alpha \neq \alpha_0
\end{aligned}
$$

It is easy to verify that $A$ is compact. Also, $T \perp_B A$, as $Tx \perp Ax$ and $x \in M_T$. Let $z = cx + \sum c_{\alpha_i} x_{\alpha_i} \in S_\mathbb{H}$. Then

$$
\|Az\|^2 = c^2 + c_{\alpha_0}^2 + \sqrt{2} c c_{\alpha_0} \leq 1 + \frac{1}{\sqrt{2}}
$$

and $\|Az\|^2 = 1 + \frac{1}{\sqrt{2}}$ if and only if $c = c_{\alpha_0} = \frac{1}{\sqrt{2}}$. Thus, $A$ attains norm only at $z = \pm \frac{1}{\sqrt{2}}(x + x_{\alpha_0})$.

Claim that $A \not\perp_B T$. If not, then there exists a sequence $\{x_n\} \subset S_\mathbb{H}$ such that $\|Ax_n\| \to \|A\|$ and $\langle Ax_n, Tx_n \rangle \to 0$. Since the unit ball of $\mathbb{H}$ is weakly compact, there exists a subsequence $\{x_{n_k}\}$ of $\{x_n\}$ such that $x_{n_k} \rightharpoonup x_0$ (say) in $B_\mathbb{H}$ weakly. Then $A, T$ being compact, $Ax_{n_k} \to Ax_0$, $\|Ax_{n_k}\| \to \|Ax_0\|$, $Tx_{n_k} \to Tx_0$ and $\langle Ax_{n_k}, Tx_{n_k} \rangle \to \langle Ax_0, Tx_0 \rangle$. So $x_0 = \pm \frac{1}{\sqrt{2}}(x + x_{\alpha_0})$ and $\langle Ax_0, Tx_0 \rangle = 0$. Then $\langle A(x + x_{\alpha_0}), T(x + x_{\alpha_0}) \rangle = 0$ implies that $\langle Tx, Tx \rangle = 0$, i.e., $\|T\| = 0$, which is a contradiction as $T \neq 0$. So $A \not\perp_B T$, as claimed.

**Case 2.** $T(H_0) \neq \{0\}$. Then there exists some $x_{\alpha_1}, \alpha_1 \in \Lambda$ such that $Tx_{\alpha_1} \neq 0$. Define a linear operator $A : \mathbb{H} \longrightarrow \mathbb{H}$ in the following way:

$$
\begin{aligned}
Ax &= 0 \\
Ax_\alpha &= Tx_\alpha, \quad \alpha = \alpha_1 \\
Ax_\alpha &= 0, \quad \alpha \neq \alpha_1
\end{aligned}
$$

Clearly, $A$ is compact as $A(\mathbb{H})$ is finite-dimensional and $T \perp_B A$, as $Tx \perp Ax$. Here $A$ attains norm only at $\pm x_{\alpha_1}$ and $\langle Ax_{\alpha_1}, Tx_{\alpha_1} \rangle \neq 0$. So $A \not\perp_B T$. This completes the proof. $\qquad \square$

**Corollary 8.9.5.** *Let* $\mathbb{H}$ *be a finite-dimensional inner product space and* $T \in$ $\mathbb{B}(\mathbb{H})$. *Then for all* $A \in \mathbb{B}(\mathbb{H}), T \perp_B A \Rightarrow A \perp_B T$ *if and only if* $T$ *is the zero operator.*

*Remark* 8.9.6. If $T$ is a bounded linear operator which attains norm, following the same method as Theorem 8.9.4, it can be shown that for all $A \in \mathbb{B}(\mathbb{H}), T \perp_B A \Rightarrow A \perp_B T$ if and only if $T$ is the zero operator.

*Remark* 8.9.7. Consider the normed linear spaces $(\mathbb{R}^2, \|.\|_p), (1 < p < \infty, p \neq 2)$. Then there are non-trivial elements $x = (\pm \frac{1}{2^{1/p}}, \pm \frac{1}{2^{1/p}})$, for which $x \perp_B y \Rightarrow y \perp_B x$ for all $y \in \mathbb{R}^2$.

*Remark* 8.9.8. Theorem 8.9.1, Theorem 8.9.2 and Theorem 8.9.4 are valid only when the underlying space is a Hilbert space. It is still unknown whether they continue to hold in the far more general setting of Banach spaces.

Next, consider the left symmetry of Birkhoff-James orthogonality of linear operators defined on a finite-dimensional Banach space $\mathbb{X}$. $T \in \mathbb{B}(\mathbb{X})$ is left symmetric if $T \perp_B A$ implies that $A \perp_B T$ for any $A \in \mathbb{B}(\mathbb{X})$. In the following theorem, a useful connection between left symmetry of an operator $T \in \mathbb{B}(\mathbb{X})$ and left symmetry of points in the corresponding norm attainment set $M_T$ is established.

**Theorem 8.9.9.** *Let* $\mathbb{X}$ *be a finite-dimensional strictly convex Banach space. If* $T \in \mathbb{B}(\mathbb{X})$ *is a left symmetric point then for each* $x \in M_T, Tx$ *is a left symmetric point.*

*Proof.* First observe that the theorem is trivially true if $T$ is the zero operator. Let $T$ be nonzero. Since $\mathbb{X}$ is finite-dimensional, $M_T$ is nonempty. If possible, suppose that there exists $x_1 \in M_T$ such that $Tx_1$ is not a left symmetric point. Since $T$ is nonzero, $Tx_1 \neq 0$. Then there exists $y_1 \in S_{\mathbb{X}}$ such that $Tx_1 \perp_B y_1$ but $y_1 \not\perp_B Tx_1$. Since $\mathbb{X}$ is strictly convex, $x_1$ is an exposed point of the unit ball $B_{\mathbb{X}}$. Let $H$ be the hyperplane of codimension 1 in $\mathbb{X}$ such that $x_1 \perp_B H$. Clearly, any element $x$ of $\mathbb{X}$ can be uniquely written in the form $x = \alpha_1 x_1 + h$, where $\alpha_1 \in \mathbb{R}$ and $h \in H$. Define a linear operator $A \in \mathbb{B}(\mathbb{X})$ in the following way:

$$Ax_1 = y_1, Ah = 0 \text{ for all } h \in H.$$

Since $x_1 \in M_T$ and $Tx_1 \perp_B Ax_1$, it follows that $T \perp_B A$. Since $T$ is left symmetric, it follows that $A \perp_B T$.
It is easy to check that $M_A = \{\pm x_1\}$, since $\mathbb{X}$ is strictly convex. Applying Theorem 8.2.1 to $A$, it follows from $A \perp_B T$ that $Ax_1 \perp_B Tx_1$, i.e., $y_1 \perp_B Tx_1$, contrary to the initial assumption that $y_1 \not\perp_B Tx_1$. This contradiction completes the proof of the theorem. $\square$

The proof of the following corollary is now obvious.

**Corollary 8.9.10.** *Let* $\mathbb{X}$ *be a finite-dimensional strictly convex Banach space such that the unit sphere* $S_{\mathbb{X}}$ *has no left symmetric point. Then* $\mathbb{B}(\mathbb{X})$ *can not have any nonzero left symmetric point.*

In the next theorem it is proved that if $\mathbb{X}$ is a finite-dimensional strictly convex and smooth Banach space, then a "large" class of operators can not be left symmetric in $\mathbb{B}(\mathbb{X})$.

**Theorem 8.9.11.** *Let $\mathbb{X}$ be a finite-dimensional strictly convex and smooth Banach space. Let $T \in \mathbb{B}(\mathbb{X})$ be such that there exists $x, y \in S_{\mathbb{X}}$ satisfying (i) $x \in M_T$, (ii) $y \perp_B x$, (iii) $Ty \neq 0$.*
*Then $T$ can not be left symmetric.*

*Proof.* There exists a hyperplane $H$ of codimension 1 in $\mathbb{X}$ such that $y \perp_B H$. Define a linear operator $A \in \mathbb{B}(\mathbb{X})$ in the following way:

$$Ay = Ty, \, A(H) = 0.$$

Since $y \perp_B x$ and $\mathbb{X}$ is smooth, it follows that $x \in H$, i.e., $Ax = 0$. Since $\mathbb{X}$ is strictly convex, as before it is easy to show that $M_A = \{\pm y\}$. Observe that $T \perp_B A$, since $x \in M_T$ and $Tx \perp_B Ax = 0$. However, $M_A = \{\pm y\}$, $Ay \not\perp_B Ty$ together implies that $A \not\perp_B T$. This completes the proof of the fact that $T$ can not be left symmetric. $\square$

**Corollary 8.9.12.** *Let $\mathbb{X}$ be a finite-dimensional strictly convex and smooth Banach space. Let $T \in \mathbb{B}(\mathbb{X})$ be invertible. Then $T$ can not be left symmetric.*

*Proof.* Since $\mathbb{X}$ is finite-dimensional, there exists $x \in S_{\mathbb{X}}$ such that $\|Tx\| = \|T\|$. From Theorem 2.3 of James James (47 b), it follows that there exists $y(\neq 0) \in \mathbb{X}$ such that $y \perp_B x$. Using the homogeneity property of Birkhoff-James orthogonality, assume without loss of generality that $\|y\| = 1$. Since $T$ is invertible, $Ty \neq 0$. Thus, all the conditions of the previous theorem are satisfied and hence $T$ can not be left symmetric. $\square$

Let $\mathbb{H}$ be a finite-dimensional Hilbert space. It was proved in 8.9.4 that $T \in \mathbb{B}(\mathbb{H})$ is a left symmetric point in $\mathbb{B}(\mathbb{H})$ if and only if $T$ is the zero operator. In the next example, it is shown that if $\mathbb{X}$ is a finite-dimensional Banach space, which is not a Hilbert space, then there may exist nonzero left symmetric operators in $\mathbb{B}(\mathbb{X})$.

**Example 8.9.13.** *Let $\mathbb{X}$ be the two-dimensional Banach space $l_1^2$. Let $T \in \mathbb{B}(\mathbb{X})$ be defined by $T(1,0) = (\frac{1}{2}, \frac{1}{2}), T(0,1) = (0,0)$. Claim that $T$ is left symmetric in $\mathbb{B}(\mathbb{X})$. Indeed, let $A \in \mathbb{B}(\mathbb{X})$ be such that $T \perp_B A$. Since $M_T = \{\pm(1,0)\}$, it follows that $(\frac{1}{2}, \frac{1}{2}) = T(1,0) \perp_B A(1,0)$. Since $(\frac{1}{2}, \frac{1}{2})$ is a left symmetric point in $\mathbb{X}$, it follows that $A(1,0) \perp_B T(1,0)$. Also, note that $\{\pm(1,0), \pm(0,1)\}$ are the only extreme points of $S_{\mathbb{X}}$. Since a linear operator defined on a finite-dimensional Banach space must attain norm at some extreme point of the unit sphere, either $(1,0) \in M_A$ or $(0,1) \in M_A$. In either case, there exists a unit vector $x$ such that $x \in M_A$ and $Ax \perp_B Tx$. However, this clearly implies that $A \perp_B T$, completing the proof of the fact that $T$ is a nonzero left symmetric point in $\mathbb{B}(\mathbb{X})$.*

Next, it is proved that in case of the strictly convex and smooth Banach spaces $l_p^2(p \geq 2, p \neq \infty), T \in \mathbb{B}(l_p^2)(p \geq 2, p \neq \infty)$ is left symmetric if and only if $T$ is the zero operator. Before proving the desired result, two easy propositions are stated. It may be noted that the proofs of both the propositions follow easily from ordinary calculus.

**Proposition 8.9.14.** *Let* $\mathbb{X}$ *be the real Banach space* $l_p^2(p \neq 1, \infty)$. $x \in S_{\mathbb{X}}$ *is a left symmetric point in* $\mathbb{X}$ *if and only if* $x \in \pm\{(1,0),(0,1),(\frac{1}{2^{1/p}},\frac{1}{2^{1/p}}),(\frac{1}{2^{1/p}},\frac{-1}{2^{1/p}})\}$.

**Proposition 8.9.15.** *Let* $\mathbb{X}$ *be the Banach space* $l_p^2(p \neq 1, \infty)$. *If* $x, y \in S_{\mathbb{X}}$ *are such that* $x \perp_B y$ *and* $y \perp_B x$ *then either of the following is true:*
*(i)* $x = \pm(1,0)$ *and* $y = \pm(0,1)$.
*(ii)* $x = \pm(0,1)$ *and* $y = \pm(1,0)$.
*(iii)* $x = \pm(\frac{1}{2^{1/p}},\frac{1}{2^{1/p}}), y = \pm(\frac{1}{2^{1/p}},\frac{-1}{2^{1/p}})$.
*(iv)* $x = \pm(\frac{1}{2^{1/p}},\frac{-1}{2^{1/p}}), y = \pm(\frac{1}{2^{1/p}},\frac{1}{2^{1/p}})$.

Next, these two propositions and some of the results already discussed are used to prove that $T \in \mathbb{B}(l_p^2)(p \geq 2, p \neq \infty)$ is left symmetric if and only if $T$ is the zero operator.

**Theorem 8.9.16.** *Let* $\mathbb{X}$ *be the two-dimensional Banach space* $l_p^2(p \geq 2, p \neq \infty)$. $T \in \mathbb{B}(\mathbb{X})$ *is left symmetric if and only if* $T$ *is the zero operator.*

*Proof.* If possible, suppose that $T \in \mathbb{B}(\mathbb{X})$ is a nonzero left symmetric point in $\mathbb{B}(\mathbb{X})$. Since Birkhoff-James orthogonality is homogeneous and $T$ is nonzero, assume without loss of generality that $\|T\| = 1$. Let $T$ attains norm at $x \in S_{\mathbb{X}}$. From Theorem 2.3 of James James (47 b), it follows that there exists $y \in S_{\mathbb{X}}$ such that $y \perp_B x$. Since $\mathbb{X}$ is strictly convex and smooth, applying Theorem 8.9.11, it is easy to see that $Ty = 0$. Theorem 8.9.9 ensures that $Tx$ must be a left symmetric point in $\mathbb{X}$. Thus, by applying Proposition 8.9.14,

$$Tx \in \pm\{(1,0),(0,1),(\frac{1}{2^{1/p}},\frac{1}{2^{1/p}}),(\frac{1}{2^{1/p}},\frac{-1}{2^{1/p}})\}.$$

Next claim that $x \perp_B y$.
From Theorem 2.3 of James James (47 b), it follows that there exists a real number $a$ such that $ay + x \perp_B y$. Since $y \perp_B x$ and $x, y \neq 0$, $\{x, y\}$ is linearly independent and hence $ay + x \neq 0$. Let $z = \frac{ay+x}{\|ay+x\|}$. Note that if $Tz = 0$ then $T$ is the zero operator. Let $Tz \neq 0$. Clearly, $\{y, z\}$ is a basis of $\mathbb{X}$.
Let $\|c_1z + c_2y\| = 1$, for some scalars $c_1, c_2$. Then $1 = \|c_1z + c_2y\| \geq |c_1|$. Since $\mathbb{X}$ is strictly convex, $1 > |c_1|$, if $c_2 \neq 0$. Also $\|T(c_1z+c_2y)\| = \|c_1Tz\| = |c_1| \|Tz\| \leq \|Tz\|$ and $\|T(c_1z+c_2y)\| = \|Tz\|$ if and only if $c_1 = \pm 1$ and $c_2 = 0$. This proves that $M_T = \{\pm z\}$. However, it is already assumed that $x \in M_T$.

Thus, $x = \pm z$. Since $z \perp_B y$, the claim is proved.

Thus, $x, y \in S_{\mathbb{X}}$ are such that $x \perp_B y$ and $y \perp_B x$. Therefore, by applying Proposition 8.9.15, the following information about $T$ can be obtained:

(i) $T$ attains norm at $x$, $x \perp_B y$, $y \perp_B x$, $Ty = 0$.

(ii) $x, y, Tx \in \pm\{(1,0), (0,1), (\frac{1}{2^{1/p}}, \frac{1}{2^{1/p}}), (\frac{1}{2^{1/p}}, \frac{-1}{2^{1/p}})\}$.

This effectively ensures that in order to prove that $T \in \mathbb{B}(\mathbb{X})$ is left symmetric if and only if $T$ is the zero operator, it is sufficient to consider 32 different operators that satisfy (i) and (ii) and show that none among them is left symmetric.

First consider one such typical linear operator and prove that it is not left symmetric.

Let $T \in \mathbb{B}(\mathbb{X})$ be defined by: $T(1,0) = (1,0), T(0,1) = (0,0)$. Define $A \in \mathbb{B}(\mathbb{X})$ by $A(1,0) = (0,1), A(0,1) = (1,1)$. Clearly, $T$ attains norm only at $\pm(1,0)$. Since $T(1,0) = (1,0) \perp_B (0,1) = A(1,0)$, it follows that $T \perp_B A$. Claim that $A \not\perp_B T$.

Now, $\|A(\frac{1}{2^{1/p}}, \frac{1}{2^{1/p}})\|^p = \frac{1}{2} + 2^{p-1} > 2 = \|A(0,1)\|^p$, since $p \geq 2$. This proves that $\pm(1,0), (0,1) \notin M_A$. It is also easy to observe that if $(\alpha, \beta) \in M_A$ then either $\alpha, \beta \geq 0$ or $\alpha, \beta \leq 0$.

For any $\alpha, \beta > 0$ and for sufficiently small negative $\lambda$,
$\|A(\alpha, \beta) + \lambda T(\alpha, \beta)\|^p = \|(\beta + \lambda\alpha, \alpha + \beta)\|^p = |\beta + \lambda\alpha|^p + |\alpha + \beta|^p < |\beta|^p + |\alpha + \beta|^p = \|A(\alpha, \beta)\|^p$.

Similarly, for any $\alpha, \beta < 0$ and for sufficiently small negative $\lambda$, $\|A(\alpha, \beta) + \lambda T(\alpha, \beta)\|^p < \|A(\alpha, \beta)\|^p$.

This proves that for any $w \in M_A, Tw \notin Aw^-$. Applying Theorem 8.8.2, it now follows that $A \not\perp_B T$. Thus, $T$ is not left symmetric in $\mathbb{B}(\mathbb{X})$, contradicting our initial assumption.

Next, a general method is described to prove that none among these 32 operators are left symmetric.

Let $T$ attains norm at $x$, $x \perp_B y, y \perp_B x, Ty = 0$ and $x, y, Tx \in \pm\{(1,0), (0,1), (\frac{1}{2^{1/p}}, \frac{1}{2^{1/p}}), (\frac{1}{2^{1/p}}, \frac{-1}{2^{1/p}})\}$. Define a linear operator $A \in \mathbb{B}(\mathbb{X})$ by $Ax = y, Ay = (1,0)$ or $(1,1)$ such that

(i) $A$ does not attain its norm at $\pm x, \pm y$.

(ii) $Tw \notin Aw^-$ for any $w \in M_A$.

Then, as before, it is easy to see that $T \perp_B A$ but $A \not\perp_B T$. Thus, $T$ is not left symmetric.

This completes the proof of the theorem. $\qquad\square$

*Remark* 8.9.17. It would be indeed interesting to extend the above theorem to higher-dimensional $l_p$ spaces and more generally, to finite-dimensional strictly convex and smooth Banach spaces, if possible.

As a final remark to the prospective readers of this book-chapter, we would like to emphasize that almost every result in this book-chapter has a strong undertone of geometric or intuitive insights. A reader should, in our opinion, pay attention to the corresponding geometric analogies, in order to appreciate the full strength of the analytic arguments presented here.

# Bibliography

Abatzoglou T. J. Norm derivatives on spaces of operators. *Math. Ann.*, 239:129–135, 1979.

Alonso J., Martini H., and Wu S. On Birkhoff orthogonality and isosceles orthogonality in normed linear spaces. *Aequat. Math.*, 83:153–189, 2012.

Amir D. *Characterizations of inner product spoaces*, volume 20. Birkhäuser, 1986.

Benítez C., Fernández M., and L. S. M. Orthogonality of matrices. *Linear Algebra and its Applications*, 422:155–163, 2007.

Bhatia R. and Šemrl P. Orthogonality of matrices and distance problem. *Linear Algebra and its Applications*, 287:77–85, 1999.

Birkhoff G. Orthogonality in linear metric spaces. *Dukc Math. J.*, 1:169–17?, 1935.

Deeb W. and Khalil R. Exposed and smooth points of some classes of operators in $L(l^p)$. *Journal of Functional Analysis*, 103(2):217–228, 1992.

Ghosh P., Sain D., and Paul K. Orthogonality of bounded linear operators. *Linear algebra and its applications*, 500:43–51, 2016.

Heinrich S. The differentiability of the norm in spaces of operators. *Functional Anal. Appl.*, 4(9):360–362, 1975.

Hennefeld J. Smooth, compact operators. *Proc. Amer. Math. Soc.*, 77(1):87–90, 1979.

Holub J. R. On the metric geometry of ideals of operators on Hilbert space. *Math. Ann.*, 201:157–163, 1973.

James R. C. Inner product in normed linear spaces. *Bull. Amer. Math. Soc.*, 53:559–566, 1947 a.

James R. C. Orthogonality and linear functionals in normed linear spaces. *Transactions of the American Mathematical Society*, 61:265–292, 1947 b.

Kittaneh F. and Younis R. Smooth points of certain operator spaces. *Int. Equations and Operator Theory*, 13:849–855, 1990.

Klee Jr. V. K. Extremal structure of convex sets.II. *Mathematische Zeitschrift*, 69:90–104, 1958.

Li C. K. and Schneider H. Orthogonality of matrices. *Linear Algebra and its Applications*, 47:115–122, 2002.

Lindenstrauss J. On operators which attain their norm. *Israel J. Math.*, 1:139–148, 1963.

Paul K. Translatable radii of an operator in the direction of another operator. *Scientiae Mathematicae*, 2:119–122, 1999.

Paul K., Hossein M., and Das K. C. Orthogonality on B(H,H) and Minimal-norm Operator. *Journal of Analysis and Applications*, 6:169–178, 2008.

Paul K., Sain D., and Ghosh P. Birkhoff-James orthogonality and smoothness of bounded linear operators. *Linear Algebra and its Applications*, 506:551–563, 2016.

Paul K., Sain D., and Jha K. On strong orthogonality and strictly convex normed linear spaces. *Journal of inequalities and applications*, 2013:242, 2013.

Reed M. and Simon B. *Methods of modern mathematical physics*, volume I. Academic Press, 1980.

Sain D. Birkhoff-James orthogonality of linear operators on finite dimensional Banach spaces. *Journal of Mathematical Analysis and Applications*, 447:860–866, 2017.

Sain D. and Paul K. Operator norm attainment and inner product spaces. *Linear Algebra and its Applications*, 439:2448–2452, 2013.

Sain D., Paul K., and Hait S. Operator norm attainment and Birkhoff-James orthogonality. *Linear Algebra and its Applications*, 476:85–97, 2015a.

Sain D., Paul K., and Jha K. Strictly convex space : Strong orthogonality and Conjugate diameters. *Journal of Convex Analysis*, 22:1215–1225, 2015b.

Turnsek A. On operators preserving James' orthogonality. *Linear Algebra and its Applications*, 407:189–195, 2005.

Werner W. Smooth points in some spaces of bounded operators. *Integral Equations Operator Theory*, 15:496–502, 1992.

# Chapter 9

## Fixed Point of Mappings on Metric Space Endowed with Graphic Structure

**Mujahid Abbas**

*Department of Mathematics, Government College University, Katchery Road, Lahore, Pakistan & Department of Mathematics and Applied Mathematics, University of Pretoria, Lynwood Road Pretoria-0002, South Africa*

**Talat Nazir**

*Department of Mathematics, COMSATS University Islamabad, Abbottabad Campus 22060, Pakistan*

## 9.1 Fixed point theory with graph structure

Extensions of Banach contraction principle have mainly been obtained either by generalizing the domain or by extending the contractive condition on the mappings or extending the range of the mappings. Nadler Nadler (2069) replaced the range of mapping with classes of non empty closed and bounded subsets and proved the Banach contraction principle for multivalued mapping in the setup of Hausdorff metric. Order oriented fixed point theory revolves around the idea of modifications of a domain of a mapping and is studied in the framework of partially ordered sets with appropriate mappings satisfying certain order conditions like order continuity, monotonicity or expansivity. The existence of fixed points of maps in the setup of partially ordered metric spaces has been studied by Ran and Reurings Ran and Reurings (2004). Later, many researchers have obtained fixed point results for single and multivalued mappings defined on partially ordered metric spaces (see for example Amini-Harandi and Emami (2010); Beg and Butt (2013b); Harjani and Sadarangani (2009); Nieto and Lopez (2005)).

The study of fixed point theory endowed with a graph occupies a prominent place in many aspects. Echenique Echenique (2005) studied fixed point theory by using the graphic structure. Espinola and Kirk Espinola and Kirk (2005) obtained some useful results on combining fixed point theory and graph theory. Jachymski Jachymski and Jozwik (2007) initiated a new approach of fixed point theory in metric space by replacing order structure with graph structure. In this way, the results proved in ordered metric spaces are generalized (see also Jachymski (2008)); in fact, in 2010, Gwodzdz-Lukawska and Jachymski Gwozdz-Lukawska and Jachymski (2009), established the Hutchinson Barnsley theory for finite family of mappings on metric space endowed with the directed graph. Bojor Bojor (2010) proved fixed point theorem of $\varphi-$ contraction mapping on a metric space endowed with a graph. Recently, Bojor Bojor (2012a) proved fixed point theorems for Reich type contractions on metric spaces with a graph. For more results in this direction, we refer to Abbas et al. (2014); Abbas and Nazir (2013); Aleomraninejad et al. (2012); Alfuraidan and Khamsi (2014); Bojor (2012b); Chifu and Petrusel (2012); Khan et al. (2014); Nicolae et al. (2011) and references mentioned therein.

We prove fixed point results of multivalued maps, defined on the family of closed and bounded subsets of a metric space endowed with a graph and satisfying graphic generalized $\phi-$ contractive conditions. We also study coincidence and common fixed point results of weakly commuting multivalued mappings. These results extend and strengthen various known results in Beg and Butt (2013a); Bojor (2010); Boyd and Wong (1969); Jachymski (2008); Jachymski and Jozwik (2007); Nadler (2069).

Consistent with Jachymski Jachymski (2008), let $(X, d)$ be a metric space and $\Delta$ denotes the diagonal of $X \times X$. Let $G$ be a directed graph, such that the set $V(G)$ of its vertices coincides with $X$ and $E(G)$ be the set of edges of the graph which contains all loops, that is, $\Delta \subseteq E(G)$. Also assume that the graph $G$ has no parallel edges and, thus, one can identify $G$ with the pair $(V(G), E(G))$.

**Definition 9.1.1.** An operator $f : X \to X$ is called a *Banach G-contraction* or simply *G*-contraction if

1. $f$ preserves edges of $G$, that is, for each $x, y \in X$ with $(x, y) \in E(G)$, we have $(f(x), f(y)) \in E(G)$,

2. $f$ decreases weights of edges of $G$, that is, there exists $h \in (0, 1)$ such that for all $x, y \in X$ with $(x, y) \in E(G)$, we have $d(f(x), f(y)) \leq hd(x, y)$.

If $x$ and $y$ are vertices of $G$, then a path in $G$ from $x$ to $y$ of length $k \in \mathbb{N}$ is a finite sequence $\{x_n\}$ ( $n \in \{0, 1, 2, ..., k\}$ ) of vertices such that $x_0 = x$, $x_k = y$ and $(x_{i-1}, x_i) \in E(G)$ for $i \in \{1, 2, ..., k\}$.

Notice that a graph $G$ is connected if there is a directed path between any two vertices and it is weakly connected if $\widetilde{G}$ is connected, where $\widetilde{G}$ denotes the undirected graph obtained from $G$ by ignoring the direction of edges. Denote by $G^{-1}$ the graph obtained from $G$ by reversing the direction of edges. Thus,

$$E\left(G^{-1}\right) = \{(x,y) \in X \times X : (y,x) \in E\left(G\right)\}.$$

It is more convenient to treat $\widetilde{G}$ as a directed graph for which the set of its edges is symmetric, under this convention; we have that

$$E(\widetilde{G}) = E(G) \cup E(G^{-1}).$$

If $G$ is such that $E(G)$ is symmetric, then for $x \in V(G)$, the symbol $[x]_G$ denotes the equivalence class of the relation $R$ defined on $V(G)$ by the rule:

$$uRv \text{ if there is a path in } G \text{ from } u \text{ to } v.$$

For an operator $f : X \to X$, by $F_f$ we denote the set of all fixed points of $f$. We set also

$$X_f := \{v \in X : (v, f(v)) \in E(G)\}.$$

Jachymski Jachymski and Jozwik (2007) used the following property:

(P) : for any sequence $\{x_n\}$ in $X$ such that $x_n \to x$ as $n \to \infty$ and $(x_n, x_{n+1}) \in E(G)$ implies that $(x_n, x) \in E(G)$.

**Theorem 9.1.2.** *Let $(X, d)$ be a complete metric space and let $G$ be a directed graph such that $V(G) = X$ and the triplet $(X, d, G)$ have property (P). If $f : X \to X$ is a G-contraction, then the following statements hold:*

*(1) $X_f \neq \emptyset$ if and only if $F_f \neq \emptyset$;*

*(2) if $G$ is weakly connected with $X_f \neq \emptyset$, then $f$ is a Picard operator, that is, $F_f = \{x^*\}$ and the sequence $\{f^n(x)\} \to x^*$ as $n \to \infty$, for all $x \in X$;*

*(3) for any $x \in X_f$, $f\mid_{[x]_{\widetilde{G}}}$ is a Picard operator;*

*(4) if $X_f \times X_f \subseteq E(G)$, then $f$ is a weakly Picard operator, that is, $F_f \neq \emptyset$ and, for each $x \in X$, the sequence $\{f^n(x)\} \to x^* \in F_f$ as $n \to \infty$.*

For detailed discussion on Picard operators, we refer to Berinde Berinde and Berinde (2007); Berinde (2003a,b, 2004, 2008, 2009).

Let $(X, d)$ be a metric space and $CB(X)$ be a class of all nonempty closed and bounded subsets of $X$. For $A, B \in CB(X)$, let

$$H(A, B) = \max\{\sup_{b \in B} d(b, A), \sup_{a \in A} d(a, B)\},$$

where $d(x, B) = \inf\{d(x, b) : b \in B\}$ is the distance from a point $x$ to the set $B$. The mapping $H$ is said to be the *Pompeiu-Hausdorff metric* induced by $d$.

Throughout this chapter, we assume that a directed graph $G$ has no parallel edge and $G$ is a weighted graph in the sense that each vertex $x$ is assigned the weight $d(x, x) = 0$ and each edge $(x, y)$ is assigned the weight $d(x, y)$.

Since $d$ is a metric on $X$, the weight assigned to each vertex $x$ to vertex $y$ need not be zero and, whenever a zero weight is assigned to some edge $(x, y)$, it reduces to a loop $(x, x)$ having weight 0. Further, in Pompeiu-Hausdorff metric induced by metric $d$, the Pompeiu-Hausdorff weight assigned to each $U, V \in CB(X)$ need not be zero that is, $H(U, V) > 0$ and, whenever a zero Pompeiu-Hausdorff weight is assigned to some $U, V \in CB(X)$, then it reduces to $U = V$.

**Definition 9.1.3.** Let $A$ and $B$ be two nonempty subsets of $X$. Then by:

(a) 'there is an edge between $A$ and $B$' means there is an edge between some $a \in A$ and $b \in B$ which we denote by $(A, B) \subset E(G)$.

(b) 'there is a path between $A$ and $B$' means there is a path between some $a \in A$ and $b \in B$.

In $CB(X)$, we define a relation $R$ in the following way:

For $A, B \in CB(X)$, we have $ARB$ if and only if, there is a path between $A$ and $B$.

We say that the relation $R$ on $CB(X)$ is transitive if there is a path between $A$ and $B$, and there is a path between $B$ and $C$, then there is a path between $A$ and $C$.

For $A \in CB(X)$, the equivalence class of $A$ induced by $R$ is denoted by

$$[A]_G = \{B \in CB(X) : ARB\}.$$

Now we consider the mapping $T : CB(X) \to CB(X)$ to study fixed points of graph contraction mappings. We define the following set:

$$X_T := \{U \in CB(X) : (U, T(U)) \subseteq E(G)\}.$$

**Definition 9.1.4.** Let $T : CB(X) \to CB(X)$ be a multivalued mapping. The mapping $T$ is said to be a graphic generalized $\phi$-contraction if it satisfies the following conditions:

(a) The edge between $A$ and $B$ implies there is an edge between $T(A)$ and $T(B)$ for all $A, B \in CB(X)$.

(b) The path between $A$ and $B$ implies there is a path between $T(A)$ and $T(B)$ for all $A, B \in CB(X)$.

(c) There exists an upper semi-continuous and nondecreasing function $\phi : \mathbb{R}_+ \to \mathbb{R}_+$ with $\phi(t) < t$ for each $t > 0$ with $\sum_{i=0}^{\infty} \phi^i(t)$ is convergent such that for all $A, B \in CB(X)$ with $(A, B) \subseteq E(G)$ implies

$$H(T(A), T(B)) \leq \phi(M_T(A, B)), \qquad (9.1)$$

where

$$
\begin{aligned}
M_T(A, B) \;=\; & \max\{H(A, B), H(A, T(A)), H(B, T(B)), \\
& \frac{H(A, T(B)) + H(B, T(A))}{2}, H(T^2(A), T(A)), \\
& H(T^2(A), B), H(T^2(A), T(B))\}.
\end{aligned}
$$

**Example 9.1.5.** *1. Any constant mapping $T : CB(X) \to CB(X)$ is a graphic generalized $\phi$-contraction for $\Delta \subset E(G)$.*

*2. Any graphic generalized $\phi$-contraction map for a graph $G$ is also a graphic generalzied $\phi$-contraction for graph $G_0$, where the graph $G_0$ is defined by $E(G_0) = X \times X$.*

It is obvious that if $T : CB(X) \to CB(X)$ is a graphic generalzied $\phi$-contraction for graph $G$, then $T$ is also graphic generalzied $\phi$-contraction for the graphs $G^{-1}$ and $\widetilde{G}$.

A graph $G$ is said to have property:

(P\*) : if for any sequence $\{X_n\}$ in $CB(X)$ with $X_n \to X$ as $n \to \infty$, there exists an edge between $X_n$ and $X_{n+1}$ for $n \in \mathbb{N}$, implies that there is a subsequence $\{X_{n_k}\}$ of $\{X_n\}$ with an edge between $X_{n_k}$ and $X$ for $n \in \mathbb{N}$.

**Definition 9.1.6.** Let $T : CB(X) \to CB(X)$. The set $A \in CB(X)$ is said to be a fixed point of $T$ if $T(A) = A$. The set of all fixed points of $T$ is denoted by $Fix(T)$.

A subset $\Upsilon$ of $CB(X)$ is said to be complete if for $X, Y \in \Upsilon$, there is an edge between $X$ and $Y$.

**Definition 9.1.7.** Beg and Butt (2013a) A metric space $(X, d)$ is called an $\varepsilon$−chainable metric space for some $\varepsilon > 0$ if for given $x, y \in X$, there is $n \in \mathbb{N}$ and a sequence $\{x_n\}$ such that

$$x_0 = x, \ x_n = y \text{ and } d(x_{i-1}, x_i) < \varepsilon \text{ for } i = 1, ..., n.$$

**Theorem 9.1.8.** *Assad and Kirk (1972); Nadler (2069) If $U, V \in CB(X)$ with $H(U, V) < \varepsilon$, then for each $u \in U$ there exists an element $v \in V$ such that $d(u, v) < \varepsilon$.*

---

## 9.2 Fixed points of graphic contraction mappings

In this section, we obtain fixed point results for maps $T : CB(X) \to CB(X)$ satisfying certain graph contraction conditions. First, we present the following result.

**Theorem 9.2.1.** *Let $(X, d)$ be a complete metric space endowed with a directed graph $G$ such that $V(G) = X$ and $E(G) \supseteq \Delta$. If $T : CB(X) \to CB(X)$ is a graph $\phi$ contraction mapping such that the relation $R$ on $CB(X)$ is transitive, then following statements hold:*

*(i) if $Fix(T)$ is complete, then the Pompeiu-Hausdorff weight assigned to the $U, V \in Fix(T)$ is 0.*

*(ii) $X_T \neq \emptyset$ provided that $Fix(T) \neq \emptyset$.*

*(iii) If $X_T \neq \emptyset$ and the weakly connected graph $G$ satisfies the property $(P^*)$, then $T$ has a fixed point.*

*(iv) $Fix(T)$ is complete if and only if $Fix(T)$ is a singleton.*

*Proof.* To prove (i), let $X_1, X_2 \in Fix(T)$. Suppose that the Pompeiu-Hausdorff weight assigned to the $X_1$ and $X_2$ is not zero. Since $T$ is a graphic generalzied $\phi$-contraction, we have

$$
\begin{aligned}
H(X_1, X_2) &= H(T(X_1), T(X_2)) \\
&\leq \phi(M_T(X_1, X_2)), \\
&< H(X_1, X_2),
\end{aligned}
$$

where

$$
\begin{aligned}
M_T(X_1, X_2) &= \max\{H(X_1, X_2), H(X_1, T(X_1)), H(X_2, T(X_2)), \\
&\quad \frac{H(X_1, T(X_2)) + H(X_2, T(X_1))}{2}, H(T^2(X_1), T(X_1)), \\
&\quad H(T^2(X_1), X_2), H(T^2(X_1), T(X_2))\} \\
&= \max\{H(X_1, X_2), 0, 0, H(X_1, X_2), 0, H(X_1, X_2), H(X_1, X_2)\} \\
&= H(X_1, X_2),
\end{aligned}
$$

that is

$$
\begin{aligned}
H(X_1, X_2) &\leq \phi(H(X_1, X_2)), \\
&< H(X_1, X_2),
\end{aligned}
$$

a contradiction. Hence (i) is proved.

To prove (ii), let $Fix(T) \neq \emptyset$. Then there exists $X^* \in CB(X)$ such that $T(X^*) = X^*$. Since $\Delta \subseteq E(G)$ and $X^*$ is nonempty, we conclude that $X_T \neq \emptyset$.

To prove (iii), Let $U \in X_T$. As $T$ is a graph $\phi$-contraction and $A, B \in CB(X)$, it follows by the hypothesis, $CB(X) \subseteq [A]_{\widetilde{G}} = P(X)$, where $P(X)$ denotes the power set of $X$ and so, $T(A) \in [A]_{\widetilde{G}}$. Now for $A \in CB(X)$ and $B \in [A]_{\widetilde{G}}$, there exists a path $\{x_i\}_{i=0}^n$ from some $x \in A = A_0$ and to $y \in T(A) = A_1$, that is, $x_0 = x$ and $x_n = y$ and $(x_{i-1}, x_i) \in E(\widetilde{G})$ for $i = 1, 2, ..., n$ such that $x_0 \in A_0, x_1 \in A_1, ..., x_n \in A_n = T(A_{n-1})$, where each $A_i \in CB(X)$. Since $T$ is graphic generalized $\phi$-contraction for graph $\widetilde{G}$, therefore for $m \in \{0, 1, 2, ...\}$, we have

$$
\begin{aligned}
H(A_{m+1}, A_{m+2}) &= H(T(A_m), T(A_{m+1})) \\
&\leq \phi(M_T(A_m, A_{m+1})),
\end{aligned}
$$

where

$$M_T(A_m, A_{m+1}) = \max\{H(A_m, A_{m+1}), H(A_m, T(A_m)), H(A_{m+1}, T(A_{m+1})),$$
$$\frac{H(A_m, T(A_{m+1})) + H(A_{m+1}, T(A_m))}{2},$$
$$H(T^2(A_m), T(A_m)), H(T^2(A_m), A_{m+1}),$$
$$H(T^2(A_m), T(A_{m+1}))\}$$

$$= \max\{H(A_m, A_{m+1}), H(A_m, A_{m+1}), H(A_{m+1}, A_{m+2}),$$
$$\frac{H(A_m, A_{m+2}) + H(A_{m+1}, A_{m+1})}{2},$$
$$H(A_{m+2}, A_{m+1}), H(A_{m+2}, A_{m+1}), H(A_{m+2}, A_{m+2})\}$$
$$\leq \max\{H(A_m, A_{m+1}), H(A_{m+1}, A_{m+2}),$$
$$\frac{H(A_m, A_{m+1}) + H(A_{m+1}, A_{m+2})}{2}\}$$
$$= \max\{H(A_m, A_{m+1}), H(A_{m+1}, A_{m+2})\}.$$

Thus, we have

$$H(A_{m+1}, A_{m+2}) \leq \phi(\max\{H(A_m, A_{m+1}), H(A_{m+1}, A_{m+2})\})$$
$$= \phi(H(A_m, A_{m+1}))$$

for all $m \in \mathbb{N} \cup \{0\}$. Thus

$$H(T(A_0), T(A_1)) = H(A_1, A_2) \leq \phi(H(A_0, A_1)),$$
$$H(T(A_1), T(A_2)) = H(A_2, A_1) \leq \phi(H(A_1, A_2)),$$
$$\cdots,$$
$$H(T(A_n), T(A_{n+1})) = H(A_{n+1}, A_{n+2}) \leq \phi(H(A_n, A_{n+1})),$$

and so we obtain

$$H(T^n(A), T^{n+1}(A)) \leq \phi(H(T^{n-1}(A), T^n(A)))$$
$$\leq \cdots$$
$$\leq \phi^n(H(A, T(A)))$$

for all $n \in \mathbb{N} \cup \{0\}$. Now for $m, n \in \mathbb{N}$ with $m > n$,

$$H(T^n(A), T^m(A)) \leq H(T^n(A), T^{n+1}(A)) + H(T^{n+1}(A), T^{n+2}(A))$$
$$+ ... + H(T^{m-1}(A), T^m(A))$$
$$\leq \phi^n(H(A, T(A))) + \phi^{n+1}(H(A, T(A)))$$
$$+ ... + \phi^{m-1}(H(A, T(A))).$$

On taking upper limit as $n, m \to \infty$, we get $H(T^n(A), T^m(A))$ converges to 0. Since $(X, d)$ is complete, we have $T^n(A) \to U^*$ as $n \to \infty$ for some $U^* \in CB(X)$. There exists an edge between $U$ and $T(U)$, the fact that $T$ is a graphic generalized $\phi$-contraction implies there is an edge between $T^n(U)$ and $T^{n+1}(U)$ for all $n \in \mathbb{N}$. By property (P*), there exists a subsequence $\{T^{n_k}(U)\}$ such that there is an edge between $T^{n_k}(U)$ and $U^*$ for every $n \in \mathbb{N}$. By transitivity of the relation $R$, there is a path in $G$ (and hence also in $\widetilde{G}$) between $U$ and $U^*$. Thus $U \in [U]_{\widetilde{G}}$. Now

$$H(T^{n_k+1}(U), T(U^*)) \leq \phi(M_T(T^{n_k}(U), U^*)),$$

where

$$M_T(A, B) = \max\{H(T^{n_k}(U), U^*), H(T^{n_k}(U), T^{n_k+1}(U)), H(U^*, T(U^*)),$$
$$\frac{H(T^{n_k}(U), T(U^*)) + H(U^*, T^{n_k+1}(U))}{2},$$
$$H(T^{n_k+2}(U), T^{n_k+1}(U)), H(T^{n_k+2}(U), U^*),$$
$$H(T^{n_k+2}(U), T(U^*))\}.$$

Now $T^{n_k}(U) \to U^*$ as $n \to \infty$ implies on taking upper limit as $n \to \infty$, $T^{n_k+1}(U) \to T(U^*)$ as $n \to \infty$. Thus we obtain that $U^* = T(U^*)$.

Finally to prove (iv), suppose the set $Fix(T)$ is complete. We are to show that $Fix(T)$ is singleton. Assume on contrary that there exist $U, V \in CB(X)$ such that $U, V \in Fix(T)$ and $U \neq V$. By completeness of $Fix(T)$, there exists an edge between $U$ and $V$. As $T$ is a graphic generalized $\phi$-contraction, so we have

$$\begin{aligned} 0 \; &< \; H(U, V) \\ &= \; H(T(U), T(V)) \leq \phi(M_T(U, V)), \end{aligned}$$

where
$$\begin{aligned} M_T(U, V) \; &= \; \max\{H(U, V), H(U, T(U)), H(V, T(V)), \\ &\quad \frac{H(U, T(V)) + H(V, T(U))}{2}, H(T^2(U), T(U)), \\ &\quad H(T^2(U), V), H(T^2(U), T(V))\} \\ &= \; \max\{H(U, V), 0, 0, H(U, V), 0, H(U, V), H(U, V)\} \\ &= \; H(U, V), \end{aligned}$$

that is,

$$\begin{aligned} 0 \; &< \; H(U, V) \\ &\leq \; \phi(M_T(U, V)) \\ &= \; \phi(H(U, V)) \\ &< \; H(U, V), \end{aligned}$$

a contradiction. Hence $U = V$.

Conversely, if $Fix(T)$ is singleton, then obviously $Fix(T)$ is complete. $\quad\square$

**Example 9.2.2.** *Let* $X = V(G) = \{1, 2, ..., n\} = V(G)$ *for some* $n > 2$ *and*

$$
\begin{aligned}
E(G) \quad = \quad & \{((1,1),(2,2),...,(n,n), \\
& (1,2),(1,3),...,(1,n), \\
& (2,2),(2,3),...,(2,n), \\
& \quad \cdot \quad \cdot \quad \cdot, \\
& (n-1,n)\}.
\end{aligned}
$$

*Consider a metric* $d : X \times X \to \mathbb{R}_+$ *defined by*

$$
d(x,y) = \begin{cases}
0 & \text{if } x = y, \\
\frac{1}{n+1} & \text{if } x, y \in \{1, 2\} \ x \neq y, \\
\frac{n+1}{n+2} & \text{otherwise.}
\end{cases}
$$

*Furthermore,*

$$
H(A,B) = \begin{cases}
0 & \text{if } A = B. \\
\dfrac{n+1}{n+2} & \text{if } A \text{ or } B \text{ (or both)} \nsubseteq \{0,1\} \text{ with } A \neq B, \\
\dfrac{1}{n+1} & \text{if } A, B \subseteq \{1,2\} \text{ with } A \neq B,
\end{cases}
$$

*The graph* $G$ *for* $n = 5$ *is shown in Figure 9.1. The mapping* $T : CB(X) \to CB(X)$ *is defined as*

$$
T(U) = \begin{cases}
\{1\}, & \text{if } U \subseteq \{1,2\}, \\
\{1,2\}, & \text{if } U \nsubseteq \{1,2\}.
\end{cases}
$$

*Note that, for all* $A, B \in CB(X)$ *with* $(A, B) \subseteq E(G)$ *implies* $(T(A), T(B)) \subseteq E(G)$. *Also there is a path between* $A$ *and* $B$ *implies that there is a path between* $T(A)$ *and* $T(B)$.

*Consider* $\phi : \mathbb{R}_+ \to \mathbb{R}_+$ *defined by*

$$
\phi(t) = \begin{cases}
\dfrac{3t}{4}, & \text{if } t \in [0, \dfrac{3}{2}), \\
\dfrac{2^{n+1}t - 3}{2^{n+1}}, & \text{if } t \in [\dfrac{2^n + 1}{2^n}, \dfrac{2^{n+1} + 1}{2^{n+1}}], \ n \in \mathbb{N}.
\end{cases}
$$

*Note that* $\phi$ *is upper semi-continuous and nondecreasing on* $\mathbb{R}_+$ *and* $\phi(t) < t$ *for all* $t > 0$.

*Now for all* $A, B \in CB(X)$, *we consider the following cases:*

*(i) for* $A, B \subseteq \{1, 2\}$, *we have* $H(T(A), T(B)) = 0$.

*(ii) If* $A \subseteq \{\{1\}, \{2\}, \{1, 2\}\}$ *and* $B \nsubseteq \{\{1\}, \{2\}, \{1, 2\}\}$, *then we have*

$$
\begin{aligned}
H(T(A), T(B)) \quad &= \quad H(\{1\}, \{1,2\}) = \frac{1}{n+1} \\
&< \quad \frac{3(n+1)}{4(n+2)} - \phi\left(\frac{n+1}{n+2}\right) \\
&= \quad \phi(H(A,B)) \leq \phi(M_T(A,B))
\end{aligned}
$$

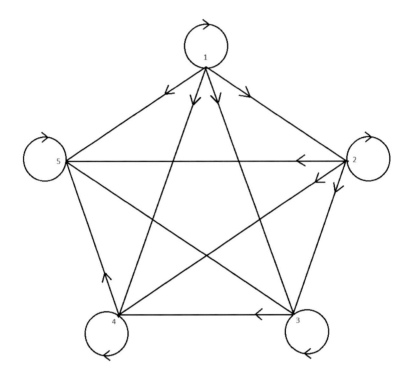

**FIGURE 9.1**: A complete graph with $n = 5$.

*(iii)* *In case* $A, B \subsetneqq \{\{1\}, \{2\}, \{1, 2\}\}$, *we have*

$$H(T(A), T(B)) = H(\{1, 2\}, \{1, 2\}) = 0.$$

*Thus, for all* $A, B \in CB(X)$ *having an edge between* $A$ *and* $B$, *(9.1) is satisfied and so* $T$ *is a graphic generalized* $\phi$-*contraction. All the conditions of Theorem 9.2.1 are satisfied. Hence* $\{1\}$ *is the fixed point of* $T$ *and* $Fix(T)$ *is complete.*

**Theorem 9.2.3.** *Let* $(X, d)$ *be a complete metric space endowed with a directed graph* $G$ *such that* $V(G) = X$ *and* $E(G) \supseteq \Delta$. *If we replace (9.1) by the statement that there there exists* $\lambda \in [0, 1)$ *such that for all* $A, B \in CB(X)$ *with there is an edge between* $A$ *and* $B$ *implies*

$$H(T(A), T(B)) \le \lambda M_T(A, B),$$

$$\text{where } M_T(A,B) = \max\{H(A,B), H(A,T(A)), H(B,T(B)),$$
$$\frac{H(A,T(B)) + H(B,T(A))}{2}, H(T^2(A), T(A)),$$
$$H(T^2(A), B), H(T^2(A), T(B))\},$$

*then the conclusions obtained in Theorem 9.2.1 remain true.*

The following theorem is a direct consequence of Theorem 9.2.1.

**Theorem 9.2.4.** *Let $(X,d)$ be a complete metric space endowed with a directed graph $G$ such that $V(G) = X$ and $E(G) \supseteq \Delta$. If $G$ is weakly connected and $T : CB(X) \to CB(X)$ is generalized $\phi$−contraction mapping with $(A_0, A_1) \subset E(G)$ for some $A_1 \in T(A_0)$, then $T$ has a fixed point.*

**Theorem 9.2.5.** *Let $(X,d)$ be a $\varepsilon$−chainable complete metric space for some $\varepsilon > 0$, $T : CB(X) \to CB(X)$ and $\phi : \mathbb{R}^+ \to \mathbb{R}^+$ be an upper semi-continuous and nondecreasing function with $\phi(t) < t$ for each $t > 0$ with*

$$0 < H(A,B) < \varepsilon.$$

*If*

$$H(T(A), T(B)) \le \phi(M_T(A,B)) \text{ for all } A, B \in CB(X),$$

$$\text{where } M_T(A,B) = \max\{H(A,B), H(A,T(A)), H(B,T(B)),$$
$$\frac{H(A,T(B)) + H(B,T(A))}{2}, H(T^2(A), T(A)),$$
$$H(T^2(A), B), H(T^2(A), T(B))\}.$$

*Then $T$ has a fixed point.*

*Proof.* By Theorem 9.1.8, from $H(A,B) < \varepsilon$, we have for each $a \in A$, there is $b \in B$ such that $d(a,b) < \varepsilon$. Consider the graph $G$ as $V(G) = X$ and

$$E(G) = \{(x,y) \in X \times X : 0 < d(x,y) < \varepsilon\}.$$

Then the $\varepsilon$−chainability of $(X,d)$ implies that $G$ is connected. For $(A,B) \subset E(G)$, we have from the hypothesis

$$H(T(A), T(B)) < \phi(M_T(A,B)),$$

$$\text{where } M_T(A,B) = \max\{H(A,B), H(A,T(A)), H(B,T(B)),$$
$$\frac{H(A,T(B)) + H(B,T(A))}{2}, H(T^2(A), T(A)),$$
$$H(T^2(A), B), H(T^2(A), T(B))\}.$$

Thus $T$ is a graphic generalized $\phi$−contraction mapping. Further, the $G$ has property (P*). Indeed, if $\{X_n\}$ in $CB(X)$ with $X_n \to X$ as $n \to \infty$ and $(X_n, X_{n+1}) \subset E(G)$ for $n \in \mathbb{N}$, implies there is a subsequence $\{X_{n_k}\}$ of $\{X_n\}$ such that $(X_{n_k}, X) \subset E(G)$ for $n \in \mathbb{N}$. Thus by Theorem 9.2.1 (iii), $T$ has a fixed point. $\square$

**Remark**

1. Theorem 12.5 generalizes Theorem 2.1 in Abbas et al. (2015), and extend and improves Theorem 9.2.1 in Boyd and Wong (1969) from single valued to multivalued mappings.

2. If $E(G) := X \times X$, then graph $G$ is connected and Theorem 9.2.1 improves and generalizes Theorem 2.5 in Beg and Butt (2013a), Theorems 2.1-2.3 in Bojor (2010) and Theorem 3.1 in Jachymski and Jozwik (2007).

3. If $E(G) := X \times X$, then graph $G$ is connected and Theorem 9.2.3 improves, generalizes and extends Theorem 2.5 in Beg and Butt (2013a), Theorem 3.2 in Nadler (2069) and Theorem 3.1 in Jachymski and Jozwik (2007).

4. If $E(G) := X \times X$, then graph $G$ is connected and Theorem 9.2.4 improves and generalizes Theorem 3.2 in Nadler (2069) and Theorem 3.1 in Jachymski and Jozwik (2007).

5. Theorem 9.2.5 improves and extends Theorem 5.1 in Edelstein (1961) and Banach contraction theorem.

---

## 9.3 Weakly commuting multivalued mappings endowed with directed graph

In this section, we give definitions and results for coincidence and common fixed points of multivalued mappings that satisfying the weakly commuting condition. Wardowski Wardowski (2012) established a new type of contraction called $F$-contraction and proved a fixed point result as a generalization of the Banach contraction principle. Beg and Butt Beg and Butt (2013a) proved the existence of fixed points of multivalued mapping in metric spaces endowed with a graph $G$. Abbas et al. Abbas et al. (2016) established the existence of common fixed points of multivalued graphic $F$-contraction mappings on a metric space.

We prove some coincidence and common fixed point results for discontinuous pairs of multivalued graphic generalized $\phi$-contraction mappings defined on the family of closed and bounded subsets of a metric space endowed with a graph $G$. It is worth mentioning that these results are obtained without appealing to any form of continuity of mappings involved herein. These results extend and strengthen various comparable results in the existing literature Abbas et al. (2015); Beg and Butt (2013a); Bojor (2010); Jachymski (2008); Jachymski and Jozwik (2007); Nadler (2069).

**Definition 9.3.1.** Let $T_1, T_2 : CB(X) \to CB(X)$ be two multivalued mappings. The set $Y \in CB(X)$ is said to be a coincidence point of $T_1$ and $T_2$, if $T_1(Y) = T_2(Y)$. The set of all coincidence points of $T_1$ and $T_2$ is denoted by $CP(T_1, T_2)$.

**Definition 9.3.2.** Two maps $T_1, T_2 : CB(X) \to CB(X)$ are said to be weakly compatible if the commute at their coincidence point.

For more details to the weakly compatible maps, we refer the reader to Abbas and Jungck (2008); Jungck (1988, 1996).

Now we give the following definition:

**Definition 9.3.3.** Let $T_1, T_2 : CB(X) \to CB(X)$. The pair $(T_1, T_2)$ of maps is said to be a graphic generalized $\phi$-contraction pair if the following conditions hold:

(i) For every $U$ in $CB(X)$, $(T_1(U), U) \subseteq E(G)$ and $(U, T_2(U)) \subseteq E(G)$.

(ii) There exists an upper semi-continuous and nondecreasing function $\phi :$ $\mathbb{R}_+ \to \mathbb{R}_+$ with $\phi(t) < t$ for each $t > 0$ with $\sum_{i=0}^{\infty} \phi^i(t)$ is convergent such that for all $A, B \in CB(X)$ with $(A, B) \subseteq E(G)$ implies

$$H(T_1(A), T_1(B)) \leq \phi(M_{T_1, T_2}(A, B)) \tag{9.2}$$

holds, where

$$
\begin{aligned}
&M_{T_1, T_2}(A, B) \\
&= \max\{H(T_2(A), T_2(B)), H(T_1(A), T_2(A)), H(T_1(B), T_2(B)), \\
&\quad \frac{H(T_1(A), T_2(B)) + H(T_1(B), T_2(A))}{2}\}.
\end{aligned}
$$

It is obvious that if a pair $(T_1, T_2)$ of multivalued mappings on $CB(X)$ is a generalized graphic $\phi$-contraction for a graph $G$, then $(T_1, T_2)$ is also generalized graphic $\phi$-contraction for the graphs $G^{-1}$ and $\widetilde{G}$.

---

## 9.4 Coincidence and common fixed points

In this section, we obtain coincidence and common fixed point results for pair of multivalued selfmaps on $CB(X)$ satisfying generalized graphic $\phi$-contraction conditions on $CB(X)$ endowed with a directed graph.

**Theorem 9.4.1.** *Let $(X, d)$ be a metric space endowed with a directed graph $G$ with $V(G) = X$, $E(G) \supseteq \Delta$ and $T_1, T_2 : CB(X) \to CB(X)$ a generalized graphic $\phi$-contraction pair such that the range of $T_2$ contains the range of $T_1$. Then the following statements hold:*

(i) $CP(T_1, T_2) \neq \emptyset$ provided that $G$ is weakly connected which satisfies the property $(P^*)$ and $T_2(X)$ is complete subspace of $CB(X)$.

(ii) If $CP(T_1, T_2)$ is complete, then the Pompeiu-Hausdorff weight assigned to the $T_1(U)$ and $T_1(V)$ is $0$ for all $U, V \in CP(T_1, T_2)$.

(iii) If $CP(T_1, T_2)$ is complete and $T_1$ and $T_2$ are weakly compatible, then $Fix(T_1) \cap Fix(T_2)$ is a singleton.

(iv) $Fix(T_1) \cap Fix(T_2)$ is complete if and only if $Fix(T_1) \cap Fix(T_2)$ is a singleton.

*Proof.* To prove (i), let $A_0 \in CB(X)$ be an arbitrary set. Since range of $T_2$ contains the range of $T_1$, pick $A_1$ in $CB(X)$ such that $T_1(A_0) = T_2(A_1)$. Continuing this way, having chosen $A_n$ in $CB(X)$, we obtain $A_{n+1}$ in $CB(X)$ such that $T_1(x_n) = T_2(x_{n+1})$ for $n \in \mathbb{N}$. The inclusions $(A_{n+1}, T_2(A_{n+1})) \subseteq E(G)$ and $(T_2(A_{n+1}), A_n) = (T_1(A_n), A_n) \subseteq E(G)$ imply that $(A_{n+1}, A_n) \subseteq E(G)$.

We assume that $T_1(A_n) \neq T_1(A_{n+1})$ for all $n \in \mathbb{N}$. If not, then $T_1(A_{2k}) = T_1(A_{2k+1})$ holds for some $k$ which implies that $T_2(A_{2k+1}) = T_1(A_{2k+1})$ and hence $A_{2k+1} \in CP(T_1, T_2)$. Since $(A_{n+1}, A_n) \subseteq E(G)$ for all $n \in \mathbb{N}$. By using (9.2), we have

$$
\begin{aligned}
& H(T_2(A_{n+1}), T_2(A_{n+2})) \\
= \; & H(T_1(A_n), T_1(A_{n+1})) \\
\leq \; & \phi(M_{T_1,T_2}(A_n, A_{n+1})),
\end{aligned}
\tag{9.3}
$$

where

$$
\begin{aligned}
& M_{T_1,T_2}(A_n, A_{n+1}) \\
= \; & \max\{H(T_2(A_n), T_2(A_{n+1})), \\
& H(T_1(A_n), T_2(A_n)), H(T_1(A_{n+1}), T_2(A_{n+1})), \\
& \frac{H(T_1(A_n), T_2(A_{n+1})) + H(T_1(A_{n+1}), T_2(A_n))}{2}\} \\
= \; & \max\{H(T_2(A_n), T_2(A_{n+1})), \\
& H(T_2(A_{n+1}), T_2(A_n)), H(T_2(A_{n+2}), T_2(A_{n+1})), \\
& \frac{H(T_2(A_{n+1}), T_2(A_{n+1})) + H(T_2(A_{n+2}), T_2(A_n))}{2}\} \\
\leq \; & \max\{H(T_2(A_n), T_2(A_{n+1})), H(T_2(A_{n+1}), T_2(A_{n+2})), \\
& \frac{H(T_2(A_{n+2}), T_2(A_{n+1})) + H(T_2(A_{n+1}), T_2(A_n))}{2}\} \\
= \; & \max\{H(T_2(A_n), T_2(A_{n+1})), H(T_2(A_{n+1}), T_2(A_{n+2}))\}.
\end{aligned}
$$

Thus, we have

$$
\begin{aligned}
&H(T_2\left(A_{n+1}\right), T_2\left(A_{n+2}\right)) \\
\leq\ & \phi\left(\max\{H\left(T_2\left(A_n\right), T_2\left(A_{n+1}\right)\right), H\left(T_2\left(A_{n+1}\right), T_2\left(A_{n+2}\right)\right)\}\right) \\
\leq\ & \phi(H\left(T_2\left(A_n\right), T_2\left(A_{n+1}\right)\right)),
\end{aligned}
$$

that is,

$$
H(T_2\left(A_{n+1}\right), T_2\left(A_{n+2}\right)) \leq \phi(H\left(T_2\left(A_n\right), T_2\left(A_{n+1}\right)\right))
$$

for all $n \in \mathbb{N} \cup \{0\}$. Thus

$$
\begin{aligned}
H(T_2\left(A_{n+1}\right), T_2\left(A_{n+2}\right)) &\leq \phi\left(H(T_2\left(A_n\right), T_2\left(A_{n+1}\right))\right) \\
&\leq \cdots \\
&\leq \phi^{n+1}(H(T_2\left(A_0\right), T_2\left(A_1\right)))
\end{aligned}
$$

for all $n \in \mathbb{N} \cup \{0\}$. Now for $m, n \in \mathbb{N}$ with $m > n$,

$$
\begin{aligned}
&H\left(T_2\left(A_n\right), T_2\left(A_m\right)\right) \\
\leq\ & H\left(T_2\left(A_n\right), T_2\left(A_{n+1}\right)\right) + H\left(T_2\left(A_{n+1}\right), T_2\left(A_{n+2}\right)\right) \\
& +\ldots+ H``\left(T_2\left(A_{m-1}\right), T_2\left(A_m\right)\right) \\
\leq\ & \phi^n(H\left(T_2\left(A_0\right), T_2\left(A_1\right)\right)) + \phi^{n+1}(H\left(T_2\left(A_0\right), T_2\left(A_1\right)\right)) \\
& +\ldots+ \phi^{m-1}(H\left(T_2\left(A_0\right), T_2\left(A_1\right)\right)).
\end{aligned}
$$

On taking upper limit as $n, m \to \infty$, we get $H\left(T_2\left(A_n\right), T_2\left(A_m\right)\right) \to 0$. Therefore $\{T_2\left(A_n\right)\}$ is a Cauchy sequence in $T_2\left(X\right)$. By the completeness of $\left(T_2\left(X\right), d\right)$ in $CB\left(X\right)$, we have $T_2\left(A_n\right) \to V$ as $n \to \infty$ for some $V \in CB\left(X\right)$. Also, we can find $U$ in $CB\left(X\right)$ such that $T_2(U) = V$.

We assume that $T_1(U) = T_2(U)$. If not, then by $(T_2\left(A_{n+1}\right), T_2\left(A_n\right)) \subseteq E\left(G\right)$ and the property (P*), there exists a subsequence $\{T_2\left(A_{n_k+1}\right)\}$ of $\{T_2\left(A_{n+1}\right)\}$ such that $(T_2\left(U\right), T_2\left(A_{n_k+1}\right)) \subseteq E\left(G\right)$ for every $n \in_+$ and Pompeiu-Hausdorff metric $H : CB\left(X\right) \to \mathbb{R}$.

As $(U, T_2\left(U\right)) \subseteq E\left(G\right)$ and $(T_2\left(A_{n_k+1}\right), A_{n_k}) = (S\left(A_{n_k}\right), A_{n_k}) \subseteq E\left(G\right)$, we have $(U, A_{n_k}) \subseteq E\left(G\right)$. By (9.2), we obtain

$$
\begin{aligned}
H(T_1\left(U\right), T_2\left(A_{n_k+1}\right)) &= H(T_1\left(U\right), T_1\left(A_{n_k}\right)) \\
&\leq \phi\left(M_{T_1, T_2}\left(U, A_{n_k}\right)\right), \tag{9.4}
\end{aligned}
$$

where

$$
\begin{aligned}
M_{T_1, T_2}\left(U, A_{n_k}\right) =\ & \max\{H(T_2\left(U\right), T_2\left(A_{n_k}\right)), \\
& H(T_1\left(U\right), T_2\left(U\right)), H(T_1\left(A_{n_k}\right), T_2\left(A_{n_k}\right)), \\
& \frac{H(T_1\left(U\right), T_2\left(A_{n_k}\right)) + H(T_1\left(A_{n_k}\right), T_2\left(U\right))}{2}\} \\
=\ & \max\{H(U, T_2\left(A_{n_k}\right)), \\
& H(T_1\left(U\right), T_2\left(U\right)), H(T_2\left(A_{n_k+1}\right), T_2\left(A_{n_k}\right)), \\
& \frac{H(T_1\left(U\right), T_2\left(A_{n_k}\right)) + H(T_2\left(A_{n_k+1}\right), T_1\left(U\right))}{2}\}.
\end{aligned}
$$

Now the following cases arise:

If $M_{T_1,T_2}(U, A_{n_k}) = H(T_2(U), T_2(A_{n_k}))$, then (9.4) implies that

$$H(T_1(U), T_1(A_{n_k+1})) \leq \phi(H(T_2(U), T_2(A_{n_k}))).$$

Taking upper limit as $k \to \infty$ implies that

$$H(T_1(U), T_2(U)) \leq \phi(H(T_2(U), T_2(U))),$$

a contradiction.

In case $M_{T_1,T_2}(U, A_{n_k}) = H(T_1(U), T_2(U))$, then

$$H(T_1(U), T_2(U)) \leq \phi(H(T_1(U), T_2(U))) < H(T_1(U), T_2(U)),$$

a contradiction.

When $M_{T_1,T_2}(U, A_{n_k}) = H(T_2(A_{n_k+1}), T_2(A_{n_k}))$, then (9.4) gives that

$$H(T_1(U), T_2(A_{n_k+1})) \leq \phi(H(T_2(A_{n_k+1}), T_2(A_{n_k}))).$$

Taking upper limit as $k \to \infty$, implies that

$$H(T_1(U), T_2(U)) \leq \phi(H(T_2(U), T_2(U))),$$

a contradiction.

Finally, if $M_{T_1,T_2}(U, A_{n_k}) = \dfrac{H(T_1(U), T_2(A_{n_k})) + H(T_2(A_{n_k+1}), T_2(U))}{2}$, then by (9.4) we have

$$H(T_1(U), T_2(A_{n_k+1})) \leq \phi(\frac{H(T_1(U), T_2(A_{n_k})) + H(T_2(A_{n_k+1}), T_2(U))}{2}).$$

By taking upper limit as $k \to \infty$ gives

$$\begin{aligned}
H(T_1(U), T_2(U)) &\leq \phi(\frac{H(T_1(U), T_2(U)) + H(T_2(U), T_2(U))}{2}) \\
&\leq \phi(\frac{H(T_1(U), T_2(U))}{2}) \\
&< \frac{H(T_1(U), T_2(U))}{2},
\end{aligned}$$

a contradiction.

Hence $T_1(U) = T_2(U)$, that is, $U \in CP(T_1, T_2)$.

To prove (ii) Let $U, V \in CP(T_1, T_2)$. Suppose that the Pompeiu-Hausdorff weight assigned to the $T_1(U)$ and $T_1(V)$ is not zero. Since pair $(T_1, T_2)$ is a generalized graphic $F$-contraction, we have

$$H(T_2(U), T_1(V)) \leq \phi(M_{T_1,T_2}(U, V)), \tag{9.5}$$

where

$$
\begin{aligned}
M_{T_1,T_2}(U,V) &= \max\{H(T_2(U),T_2(V)), \\
&\quad H(T_1(U),T_2(U)),H(T_1(V),T_2(V)), \\
&\quad \frac{H(T_1(U),T_2(V))+H(T_1(V),T_2(U))}{2}\} \\
&= \max\{H(T_1(U),T_1(V)), \\
&\quad H(T_1(U),T_1(U)),H(T_1(V),T_2(V)), \\
&\quad \frac{H(T_1(U),T_1(V))+H(T_1(V),T_1(U))}{2}\} \\
&= H(T_1(U),T_1(V)),
\end{aligned}
$$

that is,

$$
\begin{aligned}
H(T_1(U),T_2(V)) &\leq \phi(H(T_1(U),T_1(V))) \\
&< H(T_1(U),T_2(V)),
\end{aligned}
$$

a contradiction and hence (ii) is proved.

To prove (iii) First we show that $Fix(T_1)\cap Fix(T_2)$ is nonempty. If $W = T_1(U) = T_2(U)$, then we have $T_2(W) = T_2T_1(U) = T_1T_2(U) = T_1(W)$ which shows that $W \in CP(T_1,T_2)$. Thus by (ii), the Pompeiu-Hausdorff weight assign to $T_1(U)$ and $T_1(W)$ is zero . Hence $W = T_1(W) = T_2(W)$, that is, $W \in Fix(T_1)\cap Fix(T_2)$. As $CP(T_1,T_2)$ is singleton set, $Fix(T_1)\cap Fix(T_2)$ is also singleton.

Now to prove (iv). Suppose the set $Fix(T_1)\cap Fix(T_2)$ is complete. We now show that $Fix(T_2)\cap Fix(T_1)$ is singleton. Assume on contrary that there exist $U$ and $V$ in $CB(X)$ such that $U,V \in Fix(T_1)\cap Fix(T_2)$ with $U \neq V$. By the completeness property of $Fix(T_1)\cap Fix(T_2)$, there exists an edge between $U$ and $V$. Also the pair $(T_1,T_2)$ is a graph $F$-contraction, so we have

$$
\begin{aligned}
&H(U,V) \\
&= \phi(H(T_1(U),T_1(V))) \\
&\leq \phi(M_{T_1,T_2}(U,V)), \qquad\qquad (9.6)
\end{aligned}
$$

where

$$
\begin{aligned}
M_{T_1,T_2}(U,V) &= H(T_2(U),T_2(V)),H(T_1(U),T_2(U)),H(T_1(V),T_2(V)), \\
&\quad \frac{H(T_1(U),T_2(V))+H(T_1(V),T_2(U))}{2}\} \\
&= F(\max\{H(U,V),H(U,U),H(V,V), \\
&\quad \frac{H(U,V)+H(V,U)}{2}\} \\
&= H(U,V).
\end{aligned}
$$

Hence

$$
H(U,V) \leq \phi(H(U,V)),
$$

which is a contradiction. Thus $U = V$. Conversely, if $Fix(T_1) \cap Fix(T_2)$ is singleton, then since $E(G) \supseteq \Delta$, $F(T_1) \cap F(T_2)$ is trivially a complete set. $\square$

**Example 9.4.2.** *Let $X = \{1, 2, ..., n\} = V(G)$ for $n > 2$ and*

$$E(G) = \{(1,1), (2,2), ..., (n,n),$$
$$(1,2), ..., (1,n)\}.$$

*On $V(G)$, the metric $d : X \times X \to \mathbb{R}_+$ and Pompeiu-Hausdorff metric $H : CB(X) \to \mathbb{R}_+$ are defined by*

$$d(x,y) = \begin{cases} 0 & \text{if } x = y, \\ \dfrac{1}{n} & \text{if } x \in \{1,2\} \text{ with } x \neq y, \\ \dfrac{n}{n+1} & \text{otherwise} \end{cases}$$

*and*

$$H(A,B) = \begin{cases} \dfrac{1}{n} & \text{if } A, B \subseteq \{1,2\} \text{ with } A \neq B, \\ \dfrac{n}{n+1} & \text{if } A \text{ or } B \text{ (or both)} \nsubseteq \{1,2\} \text{ with } A \neq B, \\ 0 & \text{if } A = B. \end{cases}$$

*Define $T_1, T_2 : CB(X) \to CB(X)$ as follows:*

$$T_1(U) = \begin{cases} \{1\}, & \text{if } U = \{1\}, \\ \{1,2\}, & \text{if } U \neq \{1\} \end{cases}$$
$$T_2(U) = \begin{cases} \{1\}, & \text{if } U = \{1\}, \\ \{1,...,n\}, & \text{if } U \neq \{1\}. \end{cases}$$

*Note that, $(T_1(A), A) \subseteq E(G)$ and $(A, T_2(A)) \subseteq E(G)$ for all $A \in CB(X)$. Take $\phi : \mathbb{R}_+ \to \mathbb{R}_+$ defined as*

$$\phi(t) = \begin{cases} \dfrac{2t}{3}, & \text{if } t \in [0, \dfrac{5}{2}), \\ \dfrac{2^{n+1}t}{3^{n+1}}, & \text{if } t \in [\dfrac{2^{2n}+1}{2^n}, \dfrac{2^{2n+1}+1}{2^{n+1}}], \ n \in \mathbb{N}. \end{cases}$$

*Clearly $\phi$ is upper semi-continuous and nondecreasing on $\mathbb{R}_+$ and $\phi(t) < t$ for all $t > 0$.*

*For all $A, B \in CB(X)$ with $T_1(A) \neq T_1(B)$, we consider the following cases:*

*(I) If $A = \{1\}$ and $B \neq \{1\}$, then we have*

$$
\begin{aligned}
H(T_1(A), T_1(B)) &= \frac{1}{n} \\
&< \frac{2n}{3n+3} \\
&= \phi\left(\frac{n}{n+1}\right) \\
&= \phi\left(H(T_1(B), T_2(B))\right) \\
&\leq \phi\left(M_{T_1, T_2}(A, B)\right).
\end{aligned}
$$

*(II) If $A \neq \{1\}$ and $B = \{1\}$, then we have*

$$
\begin{aligned}
H(T_1(A), T_1(B)) &= \frac{1}{n} \\
&< \frac{2n}{3n+3} \\
&= \phi\left(\frac{n}{n+1}\right) \\
&= \phi\left(H(T_1(A), T_2(A))\right) \\
&\leq \phi(M_{T_1, T_2}(A, B)).
\end{aligned}
$$

*Hence for all $A, B \in CB(X)$ having an edge between $A$ and $B$, (9.2) is satisfied. Thus all the conditions of Theorem 9.4.1 are satisfied. Moreover, $T_1$ and $T_2$ have a common fixed point and $Fix(T_1) \cap Fix(T)$ is complete in $CB(X)$.*

**Theorem 9.4.3.** *Let $(X, d)$ be a $\varepsilon$-chainable complete metric space for some $\varepsilon > 0$ and $T_1, T_2 : CB(X) \to CB(X)$ multivalued mappings. Suppose that for all $A, B \in CB(X)$ with*

$$
0 < H(T_1(A), T_1(B)) < \varepsilon
$$

*there exists an upper semi-continuous and nondecreasing function $\phi : \mathbb{R}_+ \to \mathbb{R}_+$ with $\phi(t) < t$ for each $t > 0$ with $\sum_{i=0}^{\infty} \phi^i(t)$ is convergent such that*

$$
H(T_1(A), T_1(B)) \leq \phi(M^*(A, B))
$$

*holds, where*

$$
\begin{aligned}
M^*(A, B) = \ &\max\{H(T_2(A), T_2(B)), H(T_1(A), T_2(A)), H(T_1(B), T_2(B)), \\
&\frac{H(T_1(A), T_2(B)) + H(T_1(B), T_2(A))}{2}\}.
\end{aligned}
$$

*Then $T_1$ and $T_2$ have a common fixed point provided that $T_1$ and $T_2$ are weakly compatible.*

*Proof.* As $H(A, B) < \varepsilon$, for each $a \in A$ take an element $b \in B$ such that $d(a, b) < \varepsilon$. Consider the graph $G$ with $V(G) = X$ and

$$E(G) = \{(u, v) \in X \times X : 0 < d(u, v) < \varepsilon\}.$$

Then the $\varepsilon$−chainability of $(X, d)$ implies that $G$ is connected. For $(A, B) \subset E(G)$, we have

$$H(T_1(A), T_2(B)) \leq \phi(M^*(A, B)),$$

where

$$M^*(A, B) = \max\{H(T_1(A), T_2(B)), H(T_1(A), T(A)), H(T_1(B), T_2(B)),$$
$$\frac{H(T_1(A), T_2(B)) + H(T_1(B), T_2(A))}{2}\}$$

which implies that pair $(T_1, T_2)$ is graphic generalized $\phi$−contraction.

Also, $G$ has property (P*). Indeed, if $\{X_n\}$ in $CB(X)$ with $X_n \to X$ as $n \to \infty$ and $(X_n, X_{n+1}) \subset E(G)$ for $n \in \mathbb{N}$ implies that there is a subsequence $\{X_{n_k}\}$ of $\{X_n\}$ such that $(X_{n_k}, X) \subset E(G)$ for $n \in \mathbb{N}$. So by Theorem 9.4.1 (iii), $T_1$ and $T_2$ have a common fixed point. □

**Theorem 9.4.4.** *Let $(X, d)$ be a complete metric space endowed with directed graph $G$ such that $V(G) = X$ and $E(G) \supseteq \Delta$. Suppose that the mapping $T : CB(X) \to CB(X)$ satisfies the following:*

(a) *For every $V$ in $CB(X)$, $(V, T(V)) \subset E(G)$.*

(b) *There exists an upper semi-continuous and nondecreasing function $\phi :$ $\mathbb{R}_+ \to \mathbb{R}_+$ with $\phi(t) < t$ for each $t > 0$ with $\sum_{i=0}^{\infty} \phi^i(t)$ is convergent such that for all $A, B \in CB(X)$ with $(A, B) \subseteq E(G)$ implies*

$$H(T(A), T(B)) \leq \phi(M_T(A, B)),$$

*holds, where*

$$M_T(A, B) = \max\{H(A, B), H(A, T(A)), H(B, T(B)),$$
$$\frac{H(A, T(B)) + H(B, T(A))}{2}\}.$$

*Then the following statements hold:*

(i) *If $Fix(T)$ is complete, then the Pompeiu-Hausdorff weight assigned to the $U, V \in Fix(T)$ is 0.*

(ii) *If the weakly connected graph $G$ satisfies the property (P*), then $T$ has a fixed point.*

(iii) *$Fix(T)$ is singleton if and only if $Fix(T)$ is a complete.*

*Proof.* Take $T_2 = I$ (identity map) and $T = T_1$ in (9.2), then Theorem 9.4.4 follows from Theorem 9.4.1. □

## Remark

1. Theorem 9.4.1 with the graph $G$ improves, extends and generalizes Theorem 2.1 in Abbas et al. (2015).

2. If $E(G) := X \times X$, then the graph $G$ is connected and Theorem 9.4.1 generlizes and improves Theorem 2.1 in Abbas et al. (2015), Theorem 2.1 in Beg and Butt (2013a), Theorem 3.1 in Klim and Wardowski (2015), Theorem 3.1 in Jachymski and Jozwik (2007) and Theorem 3.1 in Sgroi and Vetro (2013).

3. If $E(G) := X \times X$, the graph $G$ is connected and Theorem 9.4.3 extends and generalizes Theorem 2.5 in Beg and Butt (2013a), Theorem 3.2 in Nadler (2069), Theorem 5.1 in Edelstein (1961) and Theorem 3.1 in Jachymski and Jozwik (2007).

4. If $E(G) := X \times X$, then the $G$ becomes connected and Theorem 9.4.4 improves and generalizes Theorem 2.1 in Beg and Butt (2013a), Theorem 3.2 in Nadler (2069) and Theorem 3.1 in Jachymski and Jozwik (2007).

5. If we take $E(G) := X \times X$, then the results of Abbas et al. (2015); Beg and Butt (2013a); Edelstein (1961); Jachymski and Jozwik (2007); Nadler (2069); Sgroi and Vetro (2013) are generalized for $\phi$-contraction mapping.

---

# Bibliography

Abbas M., Alfuraidan M.R., Khan A.R., and Nazir T. Fixed point results for set-contractions on metric spaces with a directed graph. *Fixed Point Theory Appl.*, 14:1–9, 2015.

Abbas M., Alfuraidan M.R., and Nazir T. Common fixed points of multivalued $F$-contractions on metric spaces with a directed graph. *Carpathian J. Math.*, 32(1):1–12, 2016

Abbas M., Ali B., and Petrusel G. Fixed points of set-valued contractions in partial metric spaces endowed with a graph. *Carpathian J. Math.*, 30(2):129–137, 2014.

Abbas M. and Jungck G. Common fixed point results for noncommuting mappings without continuity in cone metric spaces. *J. Math. Anal. Appl.*, 341:416–420, 2008.

Abbas M. and Nazir T. Common fixed point of a power graphic contraction pair in partial metric spaces endowed with a graph. *Fixed Point Theory Appl.*, 20:1–8, 2013.

Aleomraninejad S.M.A., Rezapoura Sh., and Shahzad N. Some fixed point results on a metric space with a graph. *Topology Appl.*, 159:416–420, 2012.

Alfuraidan M. R. and Khamsi M.A. Caristi fixed point theorem in metric spaces with a graph. *Abstract Appl. Anal.*, 303484:1–15, 2014.

Amini-Harandi A. and Emami H. A fixed point theorem for contraction type maps in partially ordered metric spaces and application to ordinary differential equations. *Nonlinear Anal.*, 72:2238–2242, 2010.

Assad N.A. and Kirk W.A. Fixed point theorems for set-valued mappings of contractive type. *Pacific. J. Math.*, 43:533–562, 1972.

Beg I. and Butt A.R. Fixed point of set-valued graph contractive mappings. *J. Ineq. Appl.*, 52:1–7, 2013a.

Beg I. and Butt A.R. Fixed point theorems for set valued mappings in partially ordered metric spaces. *Inter. J. Math. Sci.*, 7(2):66–68, 2013b.

Berinde M. and Berinde V. On a general class of multivalued weakly Picard mappings. *J. Math. Anal. Appl.*, 326:772–782, 2007.

Berinde V. Approximating fixed points of weak $\phi$-contractions using the Picard iteration. *Fixed Point Theory*, 2003:131–147, 2003a.

Berinde V. On the approximation of fixed points of weak contractive mappings. *Carpathian J. Math.*, 19(1):7–22, 2003b.

Berinde V. Approximating fixed points of weak contractions using the Picard iteration. *Nonlinear Anal. Forum*, 9(1):43–53, 2004.

Berinde V. General constructive fixed point theorems for Ciric-type almost contractions in metric spaces. *Carpathian J. Math.*, 24(2):10–19, 2008.

Berinde V. Iterative approximation of fixed points. *Springer-Verlag, Berlin–Heidelberg*, 2009.

Bojor F. Fixed point of $\varphi$-contraction in metric spaces endowed with a graph. *Annals Uni. Craiova, Math. Comp. Sci. Series*, 37(4):85–92, 2010.

Bojor F. Fixed point theorems for Reich type contractions on metric spaces with a graph. *Nonlinear Anal.*, 75:3895–3901, 2012a.

Bojor F. On Jachymski's theorem. *Annals Uni. Craiova, Math. Comp. Sci. Series*, 40(1):23–28, 2012b.

Boyd D.W. and Wong J.S.W. On nonlinear contractions. *Proc. Amer. Math. Soc.*, 20:458–464, 1969.

Chifu C. I. and Petrusel G. R. General constructive fixed point theorems for Ciric-type almost contractions in metric spaces. *Fixed Point Theory Appl.*, 161:1–10, 2012.

Echenique F. A short and constructive proof of Tarski's fixed point theorem. *Internat. J. Game Theory*, 33:215–218, 2005.

Edelstein M. An extension of Banach's contraction principle. *Proc. Amer. Math. Soc.*, 12:7–10, 1961.

Espinola R. and Kirk W.A. Fixed point theorems in $R$-trees with applications to graph theory. *Topology Appl.*, 33:1046–1055, 2005.

Gwozdz-Lukawska G. and Jachymski J. IFS on a metric space with a graph structure and extensions of the Kelisky-Rivlin theorem. *J. Math. Anal. Appl.*, 356:453–463, 2009.

Harjani J. and Sadarangani K. Fixed point theorems for weakly contractive mappings in partially ordered sets. *Nonlinear Anal.*, 71:3403–3410, 2009.

Jachymski J. The contraction principle for mappings on a metric space with a graph. *Proc. Amer. Math. Soc.*, 136:1359–1373, 2008.

Jachymski J. and Jozwik I. Nonlinear contractive conditions: a comparison and related problems. *Banach Center Publ.*, 77:123–146, 2007.

Jungck G. Common fixed points for commuting and compatible maps on compacta. *Proc. Amer. Math. Soc.*, 103:1046–1055, 1988.

Jungck G. Common fixed points for noncontinuous nonself maps on nonmetric spaces. *Far East J. Math. Sci.*, 4:199–215, 1996.

Khan A.R., Abbas M., Nazir T., and Ionescu C. Fixed points of multivalued mappings in partial metric spaces. *Abstract Appl. Anal.*, 2014(230708):1–9, 2014.

Klim D. and Wardowski D. Fixed points of dynamic processes of set-valued $F$-contractions and application to functional equations. *Fixed Point Theory Appl.*, 22:1–9, 2015.

Nadler S.B. Multivalued contraction mappings. *Pacific J. Math.*, 30:475–488, 2069.

Nicolae A., O'Regan D., and A. Petrusel A. Fixed point theorems for single-valued and multivalued generalized contractions in metric spaces endowed with a graph. *J. Georgian Math. Soc.*, 18:307–327, 2011.

Nieto J.J. and Lopez R.R. Contractive mapping theorems in partially ordered sets and applications to ordinary differential equations. *Order*, 22:223–239, 2005.

Ran A.C.M. and Reurings M.C.B. A fixed point theorem in partially ordered sets and some application to matrix equations. *Proc. Amer. Math. Soc.*, 132:1435–1443, 2004.

Sgroi M. and Vetro C. Multi-valued $F-$Contractions and the Solution of certain functional and integral equations. *FILOMAT*, 27(7):1259–1268, 2013.

Wardowski D. Fixed points of new type of contractive mappings in complete metric spaces. *Fixed Point Theory Appl.*, 94:1–14, 2012.

# Chapter 10

## Some Subclasses of Analytic Functions and Their Properties

**Nizami Mustafa**

*Kafkas University, Faculty of Science and Letters, Department of Mathematics, Kars, Turkey*

**Veysel Nezir**

*Kafkas University, Faculty of Science and Letters, Department of Mathematics, Kars, Turkey*

**Hemen Dutta**

*Gauhati University, Department of Mathematics, Guwahati, India*

## 10.1  Introduction

Let $A$ be the class of analytic functions $f(z)$ in the open unit disk $U = \{z \in \mathbb{C} : |z| < 1\}$, normalized by $f(0) = 0 = f'(0) - 1$, in the form

$$f(z) = z + a_2 z^2 + a_3 z^3 + \cdots + a_n z^n + \cdots = z + \sum_{n=2}^{\infty} a_n z^n, \ a_n \in \mathbb{C}. \quad (10.1)$$

It is well-known that a function $f : \mathbb{C} \to \mathbb{C}$ is said to be univalent if the following condition is satisfied: $z_1 = z_2$ if $f(z_1) = f(z_2)$ or $f(z_1) \neq f(z_2)$ if $z_1 \neq z_2$.

We will denote the family of all functions in $A$ by $S$ which are univalent in $U$.

Let $T$ denote the subclass of all functions $f(z)$ in $A$ of the form

$$f(z) = z - a_2 z^2 - a_3 z^3 - \cdots - a_n z^n - \cdots = z - \sum_{n=2}^{\infty} a_n z^n, \ a_n \geq 0. \quad (10.2)$$

Some of the important and well-investigated subclasses of the univalent functions class $S$ include the classes $S^*(\alpha)$ and $C(\alpha)$, respectively, starlike and convex of order $\alpha$ ($\alpha \in [0, 1)$) in the open unit disk $U$.

By definition, we have (see for details, Duren (1983), Goodman (1983), Owa (1985) also Srivastava and Owa (1992))

$$S^*(\alpha) = \left\{ f \in A : \Re \left( \frac{z f'(z)}{f(z)} \right) > \alpha, \ z \in U \right\}, \ \alpha \in [0, 1),$$

and

$$C(\alpha) = \left\{ f \in A : \Re \left( 1 + \frac{z f''(z)}{f'(z)} \right) > \alpha, \ z \in U \right\}, \ \alpha \in [0, 1).$$

Note that we will use $TS^*(\alpha) = S^*(\alpha) \cap T$ and $TC(\alpha) = C(\alpha) \cap T$.

Interesting generalization of the functions classes $S^*(\alpha)$ and $C(\alpha)$, are classes $S^*(\alpha, \beta)$ and $C(\alpha, \beta)$, which defined by

$$S^*(\alpha, \beta) = \left\{ f \in A : \Re \left( \frac{z f'(z)}{\beta z f'(z) + (1 - \beta) f(z)} \right) > \alpha, \ z \in U \right\}, \ \alpha, \beta \in [0, 1)$$

and

$$C(\alpha, \beta) = \left\{ f \in A : \Re \left( \frac{f'(z) + z f''(z)}{f'(z) + \beta z f''(z)} \right) > \alpha, \ z \in U \right\}, \alpha, \beta \in [0, 1),$$

respectively.

We will denote $TS^*(\alpha, \beta) = S^*(\alpha, \beta) \cap T$ and $TC(\alpha, \beta) = C(\alpha, \beta) \cap T$.

These classes $TS^*(\alpha, \beta)$ and $TC(\alpha, \beta)$ were extensively studied by Altıntaş and Owa Altıntaş and Owa (1988), Porwal Porwal (2014), and certain conditions for hypergeometric functions and generalized Bessel functions for these classes were studied by Moustafa Moustafa (2009) and Porwal and Dixit Porwal and Dixit (2013).

The coefficient problems for the subclasses $TS^*(\alpha, \beta)$ and $TC(\alpha, \beta)$ were investigated by Altıntaş and Owa in Altıntaş and Owa (1988). They, also investigated properties like starlikeness and convexity of these classes.

Also, the coefficient problems, representation formula and distortion theorems for these subclasses $S^*(\alpha, \beta, \mu)$ and $C^*(\alpha, \beta, \mu)$ of the analytic functions were given by Owa and Aouf in Owa and Aouf (1988). In Kadioğlu (2003), results of Silverman were extended by Kadioğlu. Researches on subclasses of analytic functions have been of interest for many mathematicians. In this field, recently different type of subclasses have been introduced and properties of analytic functions with some conditions have been studied. The editor of this book, Hemen Dutta is also involved in many works in the field of special functions and like his recent joint studies ("S. O. Olatunji and H. Dutta. *Fekete-Szegő problem for certain analytic functions defined by q-derivative operator with respect to other points*, submitted" and "S. O. Olatunji, H. Dutta, and M. A. Adeniyi. *Subclasses of bi-Sakaguchi function associated with q-difference operator*. Thai J. Math., in editing."), there is a crucial interest in the field for the others.

As many researchers working on the area of special functions, the studies mentioned above inspire us to construct some new subclasses of analytic functions and in this chapter we investigate their properties. Hence, inspired by the studies mentioned above, we define a unification of the functions classes $S^*(\alpha, \beta)$ and $C(\alpha, \beta)$ as follows.

**Definition 10.1.1.** A function $f \in A$ given by (10.1) is said to be in the class $S^*C(\alpha, \beta; \gamma)$, $\alpha, \beta \in [0, 1)$, $\gamma \in [0, 1]$ if the following condition is satisfied

$$\Re \left\{ \frac{zf'(z) + \gamma z^2 f''(z)}{\gamma z \left(f'(z) + \beta z f''(z)\right) + (1 - \gamma)\left(\beta z f'(z) + (1 - \beta)f(z)\right)} \right\} > \alpha, z \in U.$$

Also, we will denote $TS^*C(\alpha, \beta; \gamma) = S^*C(\alpha, \beta; \gamma) \cap T$.

In special case, we have:
$S^*C(\alpha, \beta; 0) = S^*(\alpha, \beta)$; $S^*C(\alpha, \beta; 1) = C(\alpha, \beta)$; $S^*C(\alpha, 0; 0) = S^*(\alpha)$; $S^*C(\alpha, 0; 1) = C(\alpha)$; $TS^*C(\alpha, \beta; 0) = TS^*(\alpha, \beta)$; $TS^*C(\alpha, \beta; 1) = TC(\alpha, \beta)$; $TS^*C(\alpha, 0; 0) = TS^*(\alpha)$; $TS^*C(\alpha, 0; 1) = TC(\alpha)$.

Suitably specializing the parameters we note that

1) $S^*C(\alpha, 0; 0) = S^*(\alpha)$ Speicher (1998)

2) $S^*C(\alpha, 0; 1) = C(\alpha)$ Speicher (1998)

3) $TS^*C(\alpha, \beta; 0) = TS^*(\alpha, \beta)$ Kilbas et al. (2006), Blutzer and Torvik (1996), Altıntaş et al. (2004) and Porwal (2014)

4) $TS^*C(\alpha,0;0) = TS^*(\alpha)$ Speicher (1998)

5) $TS^*C(\alpha,\beta;1) = TC(\alpha,\beta)$ Altıntaş and Owa (1988) and Porwal (2014)

6) $TS^*C(\alpha,0;1) = TC(\alpha)$ Speicher (1998)

In this chapter, we investigate characteristic properties of the classes $S^*C(\alpha,\beta;\gamma)$ and $TS^*C(\alpha,\beta;\gamma)$, $\alpha,\beta \in [0,1)$, $\gamma \in [0,1]$ and also introduce some more subclasses of analytic functions in the open unit disk in the complex plane. We find upper bound estimates for some functions belonging to the subclasses mentioned in their sections below. Also, for some other functions in these classes, several coefficient inequalities are given. Furthermore, in the final section, two recent general $p-$valent integral operators in the unit disc $\mathbb{U}$ are introduced and the properties of $p-$valent starlikeness and $p-$valent convexity of these integral operators of $p-$valent functions on some classes of $\beta-$uniformly $p-$valent starlike and $\beta-$uniformly $p-$valent convex functions of complex order and type $\alpha$ $(0 \leq \alpha < p)$ including theirs in some special cases are obtained. Also, as special cases, the properties of $p-$valent starlikeness and $p-$valent convexity of the operators $\int_0^z pt^{p-1} \left(\frac{f(t)}{t^p}\right)^\delta dt$ and $\int_0^z pt^{p-1} \left(\frac{g't)}{pt^{p-1}}\right)^\delta dt$ are given.

We need to note that this chapter has been formed by the works Deniz et al. (2016); Mustafa (2017a,b); Topkaya and Mustafa (2017).

---

## 10.2    Conditions for subclasses $S^*C(\alpha,\beta;\gamma)$ and $TS^*C(\alpha,\beta;\gamma)$

In this section, we will examine some characteristic properties of the subclasses $S^*C(\alpha,\beta;\gamma)$ and $TS^*C(\alpha,\beta;\gamma)$ of analytic functions in the open unit disk.

A sufficient condition for the functions in the subclass $S^*C(\alpha,\beta;\gamma)$ is given by the following theorem.

**Theorem 10.2.1.** *Let $f \in A$. Then, the function $f(z)$ belongs to the class $S^*C(\alpha,\beta;\gamma)$, $\alpha,\beta \in [0,1)$, $\gamma \in [0,1]$ if the following condition is satisfied*

$$\sum_{n=2}^{\infty} (1 + (n-1)\gamma)(n - \alpha - (n-1)\alpha\beta)|a_n| \leq 1 - \alpha. \qquad (10.3)$$

*The result is sharp for the functions*

$$f_n(z) = z + \frac{1-\alpha}{(1+(n-1)\gamma)(n-\alpha-(n-1)\alpha\beta)}z^n, \ z \in U, \ n = 2,3,\dots .$$
$$(10.4)$$

*Proof.* From the Definition 10.1.1, a function $f \in S^*C(\alpha, \beta; \gamma)$, $\alpha, \beta \in [0, 1), \gamma \in [0, 1]$ if and only if

$$\Re \left\{ \frac{zf'(z) + \gamma z^2 f''(z)}{\gamma z \left( f'(z) + \beta z f''(z) \right) + (1 - \gamma) \left( \beta z f'(z) + (1 - \beta) f(z) \right)} \right\} > \alpha. \quad (10.5)$$

We can easily show that condition (10.5) holds true if

$$\left| \frac{zf'(z) + \gamma z^2 f''(z)}{\gamma z \left( f'(z) + \beta z f''(z) \right) + (1 - \gamma) \left( \beta z f'(z) + (1 - \beta) f(z) \right)} - 1 \right| \leq 1 - \alpha.$$

Now, let us show that this condition is satisfied under the hypothesis (10.3) of the theorem. By simple computation, we write

$$\left| \frac{zf'(z) + \gamma z^2 f''(z)}{\gamma z \left( f'(z) + \beta z f''(z) \right) + (1 - \gamma) \left( \beta z f'(z) + (1 - \beta) f(z) \right)} - 1 \right|$$

$$= \left| \frac{\sum\limits_{n=2}^{\infty} (1 + (n-1)\gamma)(n-1)(1-\beta) a_n z^n}{z + \sum\limits_{n=2}^{\infty} (1 + (n-1)\gamma)(1 + (n-1)\beta) a_n z^n} \right|$$

$$\leq \frac{\sum\limits_{n=2}^{\infty} (1 + (n-1)\gamma)(n-1)(1-\beta) |a_n|}{1 - \sum\limits_{n=2}^{\infty} (1 + (n-1)\gamma)(1 + (n-1)\beta) |a_n|}.$$

Last expression of the above inequality is bounded by $1 - \alpha$ if

$$\sum_{n=2}^{\infty} (1 + (n-1)\gamma)(n-1)(1-\beta) |a_n|$$

$$\leq (1 - \alpha) \left\{ 1 - \sum_{n=2}^{\infty} (1 + (n-1)\gamma)(1 + (n-1)\beta) |a_n| \right\},$$

which is equivalent to

$$\sum_{n=2}^{\infty} (1 + (n-1)\gamma)(n - \alpha - (n-1)\alpha\beta) |a_n| \leq 1 - \alpha.$$

Also, we can easily see that the equality in (10.3) is satisfied by the functions given by (10.4).

Thus, the proof of Theorem 10.2.1 is completed. $\qquad\square$

Setting $\gamma = 0$ and $\gamma = 1$ in Theorem 10.2.1, we can readily deduce the following results, respectively.

**Corollary 10.2.2.** *The function $f(z)$ definition by (10.1) belongs to the class $S^*(\alpha, \beta)$, $\alpha, \beta \in [0, 1)$ if the following condition is satisfied*

$$\sum_{n=2}^{\infty} (n - \alpha - (n-1)\alpha\beta) |a_n| \le 1 - \alpha.$$

*The result is sharp for the functions*

$$f_n(z) = z + \frac{1-\alpha}{n - \alpha - (n-1)\alpha\beta} z^n, \ z \in U, \ n = 2, 3, \dots .$$

**Corollary 10.2.3.** *The function $f(z)$ definition by (10.1) belongs to the class $C(\alpha, \beta)$, $\alpha, \beta \in [0, 1)$ if the following condition is satisfied*

$$\sum_{n=2}^{\infty} n\,(n - \alpha - (n-1)\alpha\beta) |a_n| \le 1 - \alpha.$$

*The result is sharp for the functions*

$$f_n(z) = z + \frac{1-\alpha}{n\,(n - \alpha - (n-1)\alpha\beta)} z^n, \ z \in U, \ n = 2, 3, \dots .$$

**Corollary 10.2.4.** *The function $f(z)$ definition by (10.1) belongs to the class $S^*(\alpha)$, $\alpha \in [0, 1)$ if the following condition is satisfied*

$$\sum_{n=2}^{\infty} (n - \alpha) |a_n| \le 1 - \alpha.$$

*The result is sharp for the functions*

$$f_n(z) = z + \frac{1-\alpha}{n - \alpha} z^n, \ z \in U, \ n = 2, 3, \dots .$$

**Corollary 10.2.5.** *The function $f(z)$ definition by (10.1) belongs to the class $C(\alpha)$, $\alpha \in [0, 1)$ if the following condition is satisfied*

$$\sum_{n=2}^{\infty} n\,(n - \alpha) |a_n| \le 1 - \alpha.$$

*The result is sharp for the functions*

$$f_n(z) = z + \frac{1-\alpha}{n\,(n - \alpha)} z^n, \ z \in U, \ n = 2, 3, \dots .$$

*Remark* 10.2.6. Further consequences of the properties given by Corollary 10.2.4 and Corollary 10.2.5 can be obtained for each of the classes studied by earlier researches, by specializing the various parameters involved. Many of these consequences were proved by earlier researches on the subject (cf., e.g., Speicher (1998)).

For the function in the class $TS^*C(\alpha, \beta; \gamma)$, the converse of Theorem 10.2.1 is also true.

**Theorem 10.2.7.** *Let $f \in T$. Then, the function $f(z)$ belongs to the class $TS^*C(\alpha, \beta; \gamma)$, $\alpha, \beta \in [0, 1)$, $\gamma \in [0, 1]$ if and only if*

$$\sum_{n=2}^{\infty} (1 + (n-1)\gamma)(n - \alpha - (n-1)\alpha\beta)a_n \leq 1 - \alpha. \qquad (10.6)$$

*The result is sharp for the functions*

$$f_n(z) = z - \frac{1 - \alpha}{(1 + (n-1)\gamma)(n - \alpha - (n-1)\alpha\beta)} z^n, \ z \in U, \ n = 2, 3, \dots .$$
$$(10.7)$$

*Proof.* The proof of the sufficiency of the theorem can be proved similarly to the proof of Theorem 10.2.1.

We will prove only the necessity of the theorem.

Assume that $f \in TS^*C(\alpha, \beta; \gamma)$, $\alpha, \beta \in [0, 1)$, $\gamma \in [0, 1]$; that is,

$$\Re \left\{ \frac{zf'(z) + \gamma z^2 f''(z)}{\gamma z (f'(z) + \beta z f''(z)) + (1 - \gamma)(\beta z f'(z) + (1 - \beta)f(z))} \right\} > \alpha, \ z \in U.$$

By simple computation, we write

$$\Re \left\{ \frac{zf'(z) + \gamma z^2 f''(z)}{\gamma z (f'(z) + \beta z f''(z)) + (1 - \gamma)(\beta z f'(z) + (1 - \beta)f(z))} \right\}$$

$$= \Re \left\{ \frac{z - \sum_{n=2}^{\infty} n(1 + (n-1)\gamma) a_n z^n}{z - \sum_{n=2}^{\infty} (1 + (n-1)\gamma)(1 + (n-1)\beta) a_n z^n} \right\} > \alpha.$$

The last expression in the brackets of the above inequality is real if $z$ is chosen as real. Hence, from the previous inequality letting $z \to 1$ through real values, we obtain

$$1 - \sum_{n=2}^{\infty} n(1 + (n-1)\gamma) a_n \geq \alpha \left\{ 1 - \sum_{n=2}^{\infty} (1 + (n-1)\gamma)(1 + (n-1)\beta) a_n \right\}.$$

This follows

$$\sum_{n=2}^{\infty} (1 + (n-1)\gamma)(n - \alpha - (n-1)\alpha\beta)a_n \leq 1 - \alpha,$$

which is the same as the condition (10.6).

Also, it is clear that the equality in (10.6) is satisfied by the functions given by (10.7)

Thus, the proof of Theorem 10.2.7 is completed. □

Taking $\gamma = 0$ and $\gamma = 1$ in Theorem 10.2.7, we can readily deduce the following results, respectively.

**Corollary 10.2.8.** *The function $f(z)$ definition by (10.2) belongs to the class $TS^*(\alpha, \beta)$, $\alpha, \beta \in [0, 1)$ if and only if*

$$\sum_{n=2}^{\infty} (n - \alpha - (n - 1)\alpha\beta)a_n \leq 1 - \alpha.$$

*The result is sharp for the functions*

$$f_n(z) = z - \frac{1 - \alpha}{n - \alpha - (n - 1)\alpha\beta} z^n, \ z \in U, \ n = 2, 3, \dots .$$

**Corollary 10.2.9.** *The function definition by (10.2) belongs to the class $TC(\alpha, \beta)$, $\alpha, \beta \in [0, 1)$ if and only if*

$$\sum_{n=2}^{\infty} n \left(n - \alpha - (n - 1)\alpha\beta\right)a_n \leq 1 - \alpha.$$

*The result is sharp for the functions*

$$f_n(z) = z - \frac{1 - \alpha}{n \left(n - \alpha - (n - 1)\alpha\beta\right)} z^n, \ z \in U, \ n = 2, 3, \dots .$$

Setting $\beta = 0$ in Corollary 10.2.8 and 10.2.9, we can readily deduce the following results, respectively.

**Corollary 10.2.10.** *The function $f(z)$ definition by (10.2) belongs to the class $TS^*(\alpha)$, $\alpha \in [0, 1)$ if and only if*

$$\sum_{n=2}^{\infty} (n - \alpha)a_n \leq 1 - \alpha.$$

*The result is sharp for the functions*

$$f_n(z) = z - \frac{1 - \alpha}{n - \alpha} z^n, \ z \in U, \ n = 2, 3, \dots .$$

**Corollary 10.2.11.** *The function $f(z)$ definition by (10.2) belongs to the class $TC(\alpha)$, $\alpha \in [0, 1)$ if and only if*

$$\sum_{n=2}^{\infty} n \left(n - \alpha\right)a_n \leq 1 - \alpha.$$

*The result is sharp for the functions*

$$f_n(z) = z - \frac{1 - \alpha}{n \left(n - \alpha\right)} z^n, \ z \in U, \ n = 2, 3, \dots .$$

*Remark* 10.2.12. The results obtained by Corollary 10.2.10 and Corollary 10.2.11 would reduce to known results in Altıntaş and Owa (1988).

*Remark* 10.2.13. Further consequences of the properties given by Corollary 10.2.10 and Corollary 10.2.11 can be obtained for each of the classes studied by earlier researches, by specializing the various parameters involved. Many of these consequences were proved by earlier researches on the subject (cf., e.g., Speicher (1998)).

From Theorem 10.2.7, we obtain the following theorem on the coefficient bound estimates.

**Theorem 10.2.14.** *Let the function definition by (10.2)* $f \in TS^*C(\alpha, \beta; \gamma)$, $\alpha, \beta \in [0, 1)$, $\gamma \in [0, 1]$. *Then*

$$a_n \leq \frac{1 - \alpha}{(1 + (n-1)\gamma)(n - \alpha - (n-1)\alpha\beta)}, n = 2, 3, \dots .$$

**Corollary 10.2.15.** *Let the function definition by (10.2)* $f \in TS^*(\alpha, \beta)$, $\alpha, \beta \in [0, 1)$. *Then*

$$a_n \leq \frac{1 - \alpha}{n - \alpha - (n-1)\alpha\beta}, n = 2, 3, \dots .$$

**Corollary 10.2.16.** *Let the function definition by (10.2)* $f \in TC(\alpha, \beta)$, $\alpha, \beta \in [0, 1)$. *Then*

$$a_n \leq \frac{1 - \alpha}{n(n - \alpha - (n-1)\alpha\beta)}, n = 2, 3, \dots .$$

Setting $\beta = 0$ in Corollary 10.2.15 and 10.2.16, we can readily deduce the following results, respectively.

**Corollary 10.2.17.** *Let the function definition by (10.2)* $f \in TS^*(\alpha)$, $\alpha \in [0, 1)$. *Then*

$$a_n \leq \frac{1 - \alpha}{n - \alpha}, n = 2, 3, \dots .$$

**Corollary 10.2.18.** *Let the function definition by (10.2)* $f \in TC(\alpha)$, $\alpha \in [0, 1)$. *Then*

$$a_n \leq \frac{1 - \alpha}{n(n - \alpha)}, n = 2, 3, \dots .$$

*Remark* 10.2.19. Further consequences on the coefficient bound estimates given by Corollary 10.2.17 and Corollary 10.2.18 can be obtained for each of the classes studied by earlier researches, by specializing the various parameters involved.

## 10.3 Conditions for analytic functions involving gamma function

In this section, we will examine geometric properties of analytic functions involving gamma function. For these functions, we give conditions to be in these classes $S^*C(\alpha, \beta; \gamma)$ and $TS^*C(\alpha, \beta; \gamma)$. Let us define the function $F_{\lambda,\mu} : \mathbb{C} \to \mathbb{C}$ by

$$F_{\lambda,\mu}(z) = z + \sum_{n=2}^{\infty} \frac{\Gamma(\mu)}{\Gamma(\lambda(n-1) + \mu)} \frac{e^{-1/\mu}}{(n-1)!} z^n = z + (W_{\lambda,\mu}(z) - z) e^{-1/\mu},$$

(10.8)

$$z \in U, \lambda > -1, \mu > 0,$$

where $\Gamma(\mu)$ is Euler gamma function and $W_{\lambda,\mu}(z)$ is normalized Wright function (see, for details Wright (1933)).

We define also the function

$$G_{\lambda,\mu}(z) = 2z - F_{\lambda,\mu}(z) = z - \sum_{n=2}^{\infty} \frac{\Gamma(\mu)}{\Gamma(\lambda(n-1) + \mu)} \frac{e^{-1/\mu}}{(n-1)!} z^n, \quad z \in U. \quad (10.9)$$

It is clear that $F_{\lambda,\mu} \in A$ and $G_{\lambda,\mu} \in T$, respectively.

We will give sufficient condition for the function $F_{\lambda,\mu}(z)$ defined by (10.8), belonging to the class $S^*C(\alpha, \beta; \gamma)$, and necessary and sufficient condition for the function $G_{\lambda,\mu}(z)$ defined by (10.9), belonging to the class $TS^*C(\alpha, \beta; \gamma)$, respectively.

**Theorem 10.3.1.** *Let $\lambda \geq 1, \mu > 0.462$ and the following condition is satisfied*

$$\{(1 - \alpha\beta)\gamma + [1 - \alpha\beta + (2 - (1 + \beta)\alpha)\gamma]\mu\}\mu^{-2}e^{1/\mu} \leq 1 - \alpha. \quad (10.10)$$

*Then, the function $F_{\lambda,\mu}(z)$ defined by (10.8) is in the class $S^*C(\alpha, \beta; \gamma)$, $\alpha, \beta \in [0, 1), \gamma \in [0, 1]$.*

*Proof.* Since $F_{\lambda,\mu} \in A$ and

$$F_{\lambda,\mu}(z) = z + \sum_{n=2}^{\infty} \frac{\Gamma(\mu)}{\Gamma(\lambda(n-1) + \mu)} \frac{e^{-1/\mu}}{(n-1)!} z^n,$$

in view of Theorem 10.2.1, it suffices show that

$$\sum_{n=2}^{\infty} (1 + (n-1)\gamma)(n - \alpha - (n-1)\alpha\beta) \frac{\Gamma(\mu)}{\Gamma(\lambda(n-1) + \mu)} \frac{e^{-1/\mu}}{(n-1)!} \leq 1 - \alpha.$$

(10.11)

Let

$$L_1(p; \alpha, \beta; \gamma) = \sum_{n=2}^{\infty} (1 + (n-1)\gamma)(n - \alpha - (n-1)\alpha\beta) \frac{\Gamma(\mu)}{\Gamma(\lambda(n-1) - \mu)} \frac{e^{-1/\mu}}{(n-1)!}.$$

Under hypothesis $\lambda \geq 1$, the inequality $\Gamma(n - 1 + \mu) \leq \Gamma(\lambda(n-1) + \mu)$, $n \in \mathbb{N}$ holds for $\mu > 0.462$, which is equivalent to

$$\frac{\Gamma(\mu)}{\Gamma(\lambda(n-1) + \mu)} \leq \frac{1}{(\mu)_{n-1}}, \quad n \in \mathbb{N} \tag{10.12}$$

where $(\mu)_n = \Gamma(n+\mu)/\Gamma(\mu) = \mu(\mu+1)\cdots(\mu+n-1)$, $(\mu)_0 = 1$ is Pochhammer (or Appell) symbol, defined in terms of Euler gamma function.

Also, the inequality

$$(\mu)_{n-1} = \mu(\mu+1)\cdots(\mu+n-2) \geq \mu^{n-1}, \quad n \in \mathbb{N} \tag{10.13}$$

is true, which is equivalent to $1/(\mu)_{n-1} \leq 1/\mu^{n-1}$, $n \in \mathbb{N}$.
Setting

$$(1 + (n-1)\gamma)(n - \alpha - (n-1)\alpha\beta)$$
$$= (n-2)(n-1)(1 - \alpha\beta)\gamma + (n-1)(1 - \alpha\beta + (2 - (1+\beta)\alpha)\gamma) + 1 - \alpha$$

and using (10.12), (10.13), we can easily write that

$$L_1(p; \alpha, \beta; \gamma) \leq$$

$$\sum_{n=2}^{\infty} \frac{\{(n-2)(n-1)(1 - \alpha\beta)\gamma + (n-1)(1 - \alpha\beta + (2 - (1+\beta)\alpha)\gamma) + 1 - \alpha\}e^{-1/\mu}}{\mu^{n-1}(n-1)!}$$

$$= \sum_{n=3}^{\infty} \frac{(1 - \alpha\beta)\gamma}{(n-3)!} \frac{e^{-1/\mu}}{\mu^{n-1}} + \sum_{n=2}^{\infty} \frac{1 - \alpha\beta + (2 - (1+\beta)\alpha)\gamma}{(n-2)!} \frac{e^{-1/\mu}}{\mu^{n-1}} + \sum_{n=2}^{\infty} \frac{(1 - \alpha)e^{-1/\mu}}{\mu^{n-1}(n-1)!}$$

Thus,

$$L_1(p; \alpha, \beta; \gamma) \leq \frac{(1 - \alpha\beta)\gamma}{\mu^2} + \frac{1 - \alpha\beta + (2 - (1+\beta)\alpha)\gamma}{\mu} + (1-\alpha)\left(1 - e^{-1/\mu}\right).$$

From the last inequality we easily see that the inequality (10.11) is true if last expression is bounded by $1 - \alpha$, which is equivalent to (10.10).

Thus, the proof of Theorem 10.3.1 is completed. □

Taking $\gamma = 0$ and $\gamma = 1$ in Theorem 10.3.1, we arrive at the following results.

**Corollary 10.3.2.** *Let $\lambda \geq 1, \mu > 0.462$ and the following condition is satisfied*

$$(1 - \alpha\beta)\mu^{-1}e^{1/\mu} \leq 1 - \alpha.$$

*Then, the function $F_{\lambda,\mu}(z)$ defined by (10.8) is in the class $S^*(\alpha, \beta)$, $\alpha, \beta \in [0,1)$.*

**Corollary 10.3.3.** *Let* $\lambda \geq 1, \mu > 0.462$ *and the following condition is satisfied*

$$\left\{ (1 - \alpha\beta)\,\mu^{-2} + (3 - 2\alpha\beta - \alpha)\,\mu^{-1} \right\} e^{1/\mu} \leq 1 - \alpha.$$

*Then, the function* $F_{\lambda,\mu}(z)$ *defined by (10.8) is in the class* $C(\alpha,\beta), \alpha, \beta \in [0,1)$.

*Remark* 10.3.4. Further consequences of the results given by Corollary 10.3.2 and Corollary 10.3.3 can be obtained for each of the classes, by specializing the various parameters involved.

**Theorem 10.3.5.** *Let* $\lambda \geq 1, \mu > 0.462$, *then the function* $G_{\lambda,\mu}(z)$ *defined by (10.9) belongs to the class* $TS^*C(\alpha,\beta;\gamma),\ \alpha,\beta \in [0,1)\,, \gamma \in [0,1]$ *if*

$$\left\{ (1 - \alpha\beta)\,\gamma + [1 - \alpha\beta + (2 - (1+\beta)\alpha)\,\gamma]\,\mu \right\} \mu^{-2} e^{1/\mu} \leq 1 - \alpha. \tag{10.14}$$

*Proof.* The proof of Theorem 10.3.5 is the same as the proof of Theorem 10.3.1. Therefore, the details of the proof of Theorem 10.3.5 may be omitted. □

*Remark* 10.3.6. Further consequences of the results given by Theorem 10.3.5 can be obtained for each of the classes, by specializing the various parameters involved.

---

## 10.4 Integral operators of functions $F_{\lambda,\mu}(z)$ and $G_{\lambda,\mu}(z)$

In this section, we will examine some inclusion properties of integral operators associated with the functions $F_{\lambda,\mu}(z)$ and $G_{\lambda,\mu}(z)$ as follows:

$$\hat{F}_{\lambda\mu}(z) = \int_0^z \frac{F_{\lambda\mu}(t)}{t}\,dt \text{ and } \hat{G}_{\lambda,\mu}(z) = \int_0^z \frac{G_{\lambda,\mu}(t)}{t}\,dt \tag{10.15}$$

**Theorem 10.4.1.** *Let* $\lambda \geq 1, \mu > 0.462$ *and the following condition is satisfied*

$$\left\{ (1 - \alpha\beta)\,\gamma\mu^{-1} + (1 - \beta)\,(1 - \gamma)\,\alpha \left( 1 - e^{-1/\mu} \right) \right\} e^{1/\mu} \leq 1 - \alpha. \tag{10.16}$$

*Then, the function* $\hat{F}_{\lambda,\mu}(z)$ *defined by (10.15) is in the class* $S^*C(\alpha,\beta;\gamma),\ \alpha,\beta \in [0,1)\,, \gamma \in [0,1]$.

*Proof.* Since

$$\hat{F}_{\lambda,\mu}(z) = z + \sum_{n=2}^{\infty} \frac{\Gamma(\mu)}{\Gamma(\lambda(n-1)+\mu)} \frac{e^{-1/\mu}}{n!} z^n, z \in U$$

according to Theorem 10.2.1, the function $\hat{F}_{\lambda,\mu}(z)$ belongs to the class $S^*C(\alpha, \beta; \gamma)$ if the following condition is satisfied

$$\sum_{n=2}^{\infty} (1 + (n-1)\gamma)(n - \alpha - (n-1)\alpha\beta) \frac{\Gamma(\mu)}{\Gamma(\lambda(n-1)+\mu)} \frac{e^{-1/\mu}}{n!} \leq 1 - \alpha.$$

(10.17)

Let

$$L_2(p; \alpha, \beta; \gamma) = \sum_{n=2}^{\infty} (1 + (n-1)\gamma)(n - \alpha - (n-1)\alpha\beta) \frac{\Gamma(\mu)}{\Gamma(\lambda(n-1)+\mu)} \frac{e^{-1/\mu}}{n!}.$$

Setting

$$(1 + (n-1)\gamma)(n - \alpha - (n-1)\alpha\beta)$$
$$= (n-1)n(1-\alpha\beta)\gamma + n((1-\alpha\beta)(1-\gamma) + (1-\alpha)\gamma) - (1-\beta)(1-\gamma)\alpha$$

and by simple computation, we obtain

$$L_2(p; \alpha, \beta; \gamma) =$$

$$\sum_{n=2}^{\infty} \{(n-1)n(1-\alpha\beta)\gamma + n(1-\beta)(1-\gamma)\alpha$$

$$-(1-\beta)(1-\gamma)\alpha + n(1-\alpha)\} \frac{\Gamma(\mu)}{\Gamma(\lambda(n-1)+\mu)} \frac{e^{-1/\mu}}{n!}$$

$$= \sum_{n=2}^{\infty} \frac{(1-\alpha\beta)\gamma}{(n-2)!} \frac{\Gamma(\mu)e^{-1/\mu}}{\Gamma(\lambda(n-1)+\mu)} + \sum_{n=2}^{\infty} \frac{(1-\beta)(1-\gamma)\alpha}{(n-1)!} \frac{\Gamma(\mu)e^{-1/\mu}}{\Gamma(\lambda(n-1)+\mu)} -$$

$$\sum_{n=2}^{\infty} \frac{(1-\beta)(1-\gamma)\alpha}{n!} \frac{\Gamma(\mu)e^{-1/\mu}}{\Gamma(\lambda(n-1)+\mu)} + \sum_{n=2}^{\infty} \frac{1-\alpha}{(n-1)!} \frac{\Gamma(\mu)e^{-1/\mu}}{\Gamma(\lambda(n-1)+\mu)}.$$

Thus,

$$L_2(p; \alpha, \beta; \gamma) \leq \sum_{n=2}^{\infty} \frac{(1-\alpha\beta)\gamma}{(n-2)!} \frac{\Gamma(\mu)e^{-1/\mu}}{\Gamma(\lambda(n-1)+\mu)}$$

$$+ \sum_{n=2}^{\infty} \frac{(1-\beta)(1-\gamma)\alpha}{(n-1)!} \frac{\Gamma(\mu)e^{-1/\mu}}{\Gamma(\lambda(n-1)+\mu)}$$

$$+ \sum_{n=2}^{\infty} \frac{1-\alpha}{(n-1)!} \frac{\Gamma(\mu)e^{-1/\mu}}{\Gamma(\lambda(n-1)+\mu)}.$$

From (10.12) and (10.13), w have

$$L_2(p; \alpha, \beta; \gamma) \leq \sum_{n=2}^{\infty} \frac{(1 - \alpha\beta) \gamma}{(n-2)!} \frac{e^{-1/\mu}}{\mu^{n-1}} + \sum_{n=2}^{\infty} \frac{(1 - \beta)(1 - \gamma)\alpha}{(n-1)!} \frac{e^{-1/\mu}}{\mu^{n-1}}$$

$$+ \sum_{n=2}^{\infty} \frac{1 - \alpha}{(n-1)!} \frac{e^{-1/\mu}}{\mu^{n-1}}$$

$$= \frac{(1 - \alpha\beta) \gamma}{\mu} + (1 - \beta)(1 - \gamma)\alpha \left(1 - e^{-1/\mu}\right)$$

$$+ (1 - \alpha) \left(1 - e^{-1/\mu}\right).$$

Therefore, inequality (10.17) holds true if

$$\frac{(1 - \alpha\beta) \gamma}{\mu} + (1 - \beta)(1 - \gamma)\alpha \left(1 - e^{-1/\mu}\right) + (1 - \alpha) \left(1 - e^{-1/\mu}\right) \leq 1 - \alpha,$$

which is equivalent to (10.16).

Thus, the proof of Theorem 10.4.1 is completed. $\qquad\square$

Taking $\gamma = 0$ and $\gamma = 1$ in Theorem 10.4.1, we arrive at the following results.

**Corollary 10.4.2.** *Let $\lambda \geq 1, \mu > 0.462$ and the following condition is satisfied*

$$(1 - \beta) \alpha \left(e^{1/\mu} - 1\right) \leq 1 - \alpha.$$

*Then, the function $\hat{F}_{\lambda,\mu}(z)$ defined by (10.15) is in the class $S^*(\alpha, \beta), \alpha, \beta \in [0, 1)$.*

**Corollary 10.4.3.** *Let $\lambda \geq 1, \mu > 0.462$ and the following condition is satisfied*

$$(1 - \alpha\beta) \mu^{-1} e^{1/\mu} \leq 1 - \alpha.$$

*Then, the function $\hat{F}_{\lambda,\mu}(z)$ defined by (10.15) is in the class $C(\alpha, \beta), \alpha, \beta \in [0, 1)$.*

*Remark* 10.4.4. Further consequences of the results given by Corollary 10.4.2 and Corollary 10.4.3 can be obtained for each of the classes, by specializing the various parameters involved.

**Theorem 10.4.5.** *Let $\lambda \geq 1, \mu > 0.462$, then the function $\hat{G}_{\lambda,\mu}(z)$ defined by (10.15) belongs to the class $TS^*C(\alpha, \beta; \gamma)$, $\alpha, \beta \in [0, 1), \gamma \in [0, 1]$ if*

$$\left\{(1 - \alpha\beta) \gamma\mu^{-1} + (1 - \beta)(1 - \gamma)\alpha \left(1 - e^{-1/\mu}\right)\right\} e^{1/\mu} \leq 1 - \alpha.$$

*Proof.* The proof of Theorem 10.4.5 is the same as the proof of Theorem 10.4.1. Therefore, the details of the proof of Theorem 10.4.5 may be omitted. $\qquad\square$

Taking $\gamma = 0$ and $\gamma = 1$ in Theorem 10.4.5, we can readily deduce the following results, respectively.

**Corollary 10.4.6.** *Let* $\lambda \geq 1, \mu > 0.462$, *then the function* $\hat{G}_{\lambda,\mu}(z)$ *defined by (10.15) belongs to the class* $TS^*(\alpha, \beta)$, $\alpha, \beta \in [0, 1)$ *if*

$$(1 - \beta)\,\alpha \left(e^{1/\mu} - 1\right) \leq 1 - \alpha.$$

**Corollary 10.4.7.** *Let* $\lambda \geq 1, \mu > 0.462$, *then the function* $\hat{G}_{\lambda,\mu}(z)$ *defined by (4.1) belongs to the class* $TC(\alpha, \beta)$, $\alpha, \beta \in [0, 1)$ *if*

$$(1 - \alpha\beta)\,\mu^{-1}e^{1/\mu} \leq 1 - \alpha.$$

*Remark* 10.4.8. Further consequences of the results given by Corollary 10.4.6 and Corollary 10.4.7 can be obtained for each of the classes, by specializing the various parameters involved.

---

## 10.5 Various properties of certain subclasses of analytic functions of complex order

This subsection considers two new classes of analytic functions in the open unit disk in the complex plane. Several interesting properties of the functions belonging to these classes are examined. Here, sufficient, and necessary and sufficient conditions for the functions belonging to these classes are also given. Moreover, coefficient bound estimates for the functions belonging to these classes are obtained.

Let $A$ be the class of analytic functions $f(z)$ in the open unit disk $U = \{z \in \mathbb{C} : |z| < 1\}$, normalized by $f(0) = 0 = f'(0) - 1$ of the form

$$f(z) = z + a_2 z^2 + a_3 z^3 + \cdots + a_n z^n + \cdots = z + \sum_{n=2}^{\infty} a_n z^n, a_n \in \mathbb{C}, \quad (10.18)$$

and $S$ denote the class of all functions in $A$ which are univalent in $U$.

Also, let us define by $T$ the subclass of all functions $f(z)$ in $A$ of the form

$$f(z) = z - a_2 z^2 - a_3 z^3 - \cdots - a_n z^n - \cdots = z - \sum_{n=2}^{\infty} a_n z^n, a_n \geq 0. \quad (10.19)$$

Furthermore, we will denote by $S^*(\alpha)$ and $C(\alpha)$ the subclasses of $S$ that are, respectively, starlike and convex functions of order $\alpha$ ($\alpha \in [0, 1)$). By definition (see for details, Duren (1983), Goodman (1983), also Srivastava and Owa (1992))

$$S^*(\alpha) = \left\{ f \in S : \Re\left(\frac{zf'(z)}{f(z)}\right) > \alpha, \; z \in U \right\}, \; \alpha \in [0, 1), \quad (10.20)$$

and

$$C(\alpha) = \left\{ f \in S : \Re \left( 1 + \frac{zf''(z)}{f'(z)} \right) > \alpha, \; z \in U \right\}, \; \alpha \in [0,1). \qquad (10.21)$$

For convenience, $S^* = S^*(0)$ and $C = C(0)$ are, respectively, starlike and convex functions in $U$. It is easy to verify that $C \subset S^* \subset S$. For details on these classes, one could refer to the monograph by Goodman Goodman (1983).

Note that, we will use $TS^*(\alpha) = T \cap S^*(\alpha)$, $TC(\alpha) = T \cap C(\alpha)$ and in the special case, we have $TS^* = T \cap S^*$, $TC = T \cap C$ for $\alpha = 0$.

An interesting unification of the functions classes $S^*(\alpha)$ and $C(\alpha)$ is provided by the class $S^*C(\alpha, \beta)$ of functions $f \in S$, which also satisfies the following condition:

$$\Re \left\{ \frac{zf'(z) + \beta z^2 f''(z)}{\beta z f'(z) + (1 - \beta)f(z)} \right\} > \alpha, \alpha \in [0, 1), \beta \in [0, 1], \; z \in U.$$

In special case, for $\beta = 0$ and $\beta = 1$, respectively, we have $S^*C(\alpha, 0) = S^*(\alpha)$ and $S^*C(\alpha, 1) = C(\alpha)$, in terms of the simpler classes $S^*(\alpha)$ and $C(\alpha)$, defined by 10.20 and 10.21, respectively. Also, we will use $TS^*C(\alpha, \beta) = T \cap S^*C(\alpha, \beta)$.

The class $TS^*C(\alpha, \beta)$ was investigated by Altintaş *et al.* Blutzer and Torvik (1996) and Altıntaş et al. (2004) (in a more general way $T_n S^* C(p, \alpha, \beta)$) and (subsequently) by Irmak *et al.* Irmak et al. (1997). In particular, the class $T_n S^* C(1, \alpha, \beta)$ was considered earlier by Altıntaş Kilbas et al. (2006).

Inspired by the aforementioned works, we define a subclass of analytic functions as follows.

**Definition 10.5.1.** A function $f \in S$ given by 10.18 is said to be in the class $(S^*C)^\tau(\alpha, \beta)$, $\alpha \in [0, 1)$, $\beta \in [0, 1]$, $\tau \in \mathbb{C}^* = \mathbb{C} - \{0\}$ if the following condition is satisfied

$$\Re \left\{ 1 + \frac{1}{\tau} \left[ \frac{zf'(z) + \beta z^2 f''(z)}{\beta z f'(z) + (1 - \beta)f(z)} - 1 \right] \right\} > \alpha, z \in U, \alpha \in [0, 1), \beta \in [0, 1], \tau \in \mathbb{C}^*.$$

In special case, we have $(S^*C)^1(\alpha, \beta) = S^*C(\alpha, \beta)$ for $\tau = 1$. Note that, we will use $T(S^*C)^\tau(\alpha, \beta) = T \cap (S^*C)^\tau(\alpha, \beta)$. Also, we have $T(S^*C)^1(\alpha, \beta) = TS^*C(\alpha, \beta)$ and $T(S^*C)^\tau(\alpha, 0) = T \cap S^*(\alpha; \tau) = TS^*(\alpha; \tau)$, $T(S^*C)^\tau(\alpha, 1) = T \cap C(\alpha; \tau) = TC(\alpha; \tau)$.

In this section, two new subclasses $(S^*C)^\tau(\alpha, \beta)$ and $T(S^*C)^\tau(\alpha, \beta)$ of the analytic functions in the open unit disk are introduced. Various characteristic properties of the functions belonging to these classes are examined. Sufficient conditions for the analytic functions belonging to the class

$(S^*C)^\tau(\alpha, \beta)$, and necessary and sufficient conditions for those belonging to the class $T(S^*C)^\tau(\alpha, \beta)$ are also given. Furthermore, the various properties like order of starlikeness and radius of convexity of the subclasses $TC(\alpha; \tau)$ and $TS^*(\alpha; \tau)$, respectively, and radii of starlikeness and convexity of the subclasses $(S^*C)^\tau(\alpha, \beta)$ and $T(S^*C)^\tau(\alpha, \beta)$ are examined.

In this section, we will examine some inclusion results of the subclasses $(S^*C)^\tau(\alpha, \beta)$ and $T(S^*C)^\tau(\alpha, \beta)$ of analytic functions in the open unit disk. Also, we give coefficient bound estimates for the functions belonging to these classes.

A sufficient condition for the functions in class $(S^*C)^\tau(\alpha, \beta)$ is given by the following theorem.

**Theorem 10.5.2.** *Let $f \in A$. Then, the function $f(z)$ belongs to the class $(S^*C)^\tau(\alpha, \beta)$, $\alpha \in [0, 1)$, $\beta \in [0, 1]$, $\tau \in \mathbb{C}^* = \mathbb{C} - \{0\}$  if the following condition is satisfied*

$$\sum_{n=2}^\infty [n + (1 - \alpha)|\tau| - 1][1 + (n - 1)\beta]|a_n| \leq (1 - \alpha)|\tau|.$$

*The result is sharp for the functions*

$$f_n(z) = z + \frac{(1 - \alpha)|\tau|}{[n + (1 - \alpha)|\tau| - 1][1 + (n - 1)\beta]} z^n, \ z \in U, \ n = 2, 3, \ldots.$$

*Proof.* According to Definition 10.5.1, a function $f(z)$ is in the class $(S^*C)^\tau(\alpha, \beta)$, $\alpha \in [0, 1)$, $\beta \in [0, 1]$, $\tau \in \mathbb{C}^* = \mathbb{C} - \{0\}$ if and only if

$$\Re\left\{1 + \frac{1}{\tau}\left[\frac{zf'(z) + \beta z^2 f''(z)}{\beta z f'(z) + (1 - \beta)f(z)} - 1\right]\right\} > \alpha.$$

It suffices to show that

$$\left|\frac{1}{\tau}\left[\frac{zf'(z) + \beta z^2 f''(z)}{\beta z f'(z) + (1 - \beta)f(z)} - 1\right]\right| < 1 - \alpha. \tag{10.22}$$

Considering (10.18), by simple computation, we write

$$\left|\frac{1}{\tau}\left[\frac{zf'(z) + \beta z^2 f''(z)}{\beta z f'(z) + (1 - \beta)f(z)} - 1\right]\right| = \left|\frac{1}{\tau}\frac{\sum_{n=2}^\infty(n - 1)[1 + \beta(n - 1)]a_n z^n}{z + \sum_{n=2}^\infty[1 + \beta(n - 1)]a_n z^n}\right|$$

$$\leq \frac{\sum_{n=2}^\infty(n - 1)[1 + \beta(n - 1)]|a_n|}{|\tau|\{1 - \sum_{n=2}^\infty[1 + \beta(n - 1)]|a_n|\}}.$$

Last expression is bounded above by $1 - \alpha$ if

$$\sum_{n=2}^\infty(n - 1)[1 + \beta(n - 1)]|a_n| \leq |\tau|(1 - \alpha)\left\{1 - \sum_{n=2}^\infty[1 + \beta(n - 1)]|a_n|\right\},$$

which is equivalent to

$$\sum_{n=2}^{\infty} [n + (1 - \alpha) |\tau| - 1] [1 + (n - 1)\beta] |a_n| \le (1 - \alpha) |\tau|. \qquad (10.23)$$

Hence, the inequality (10.22) is true if the condition (10.23) is satisfied.

Thus, the proof of Theorem 10.5.2 is completed. □

Setting $\tau = 1$ in Theorem 10.5.2, we arrive at the following corollary.

**Corollary 10.5.3.** *The function $f(z)$ definition by (10.18) belongs to the class $S^*C(\alpha, \beta)$, $\alpha \in [0, 1)$, $\beta \in [0, 1]$ if the following condition is satisfied*

$$\sum_{n=2}^{\infty} (n - \alpha) [1 + \beta(n - 1)] |a_n| \le 1 - \alpha.$$

*The result is sharp for the functions*

$$f_n(z) = z + \frac{1 - \alpha}{(n - \alpha) [1 + \beta(n - 1)]} z^n, \ z \in U, \ n = 2, 3, \ldots .$$

*Remark* 10.5.4. The result obtained in Corollary 10.5.3 verifies Corollary 2.2 in Topkaya and Mustafa (2017).

Choose $\beta = 0$ in Corollary 10.5.3, we have the following result.

**Corollary 10.5.5.** *(see (Speicher, 1998, p. 110, Theorem 1)) The function $f(z)$ definition by (10.18) belongs to the class $S^*(\alpha)$, $\alpha \in [0, 1)$ if the following condition is satisfied*

$$\sum_{n=2}^{\infty} (n - \alpha) |a_n| \le 1 - \alpha.$$

*The result is sharp for the functions*

$$f_n(z) = z + \frac{1 - \alpha}{n - \alpha} z^n, \ z \in U, \ n = 2, 3, \ldots .$$

Taking $\beta = 1$ in Corollary 10.5.3, we arrive at the following result.

**Corollary 10.5.6.** *(see (Speicher, 1998, p. 110, Corollary of Theorem 1)) The function $f(z)$ definition by (10.18) belongs to the class $C(\alpha)$, $\alpha \in [0, 1)$ if the following condition is satisfied*

$$\sum_{n=2}^{\infty} n(n - \alpha) |a_n| \le 1 - \alpha.$$

*The result is sharp for the functions*

$$f_n(z) = z + \frac{1 - \alpha}{n(n - \alpha)} z^n, \ z \in U, \ n = 2, 3, \ldots .$$

*Remark* 10.5.7. The results obtained in Corollary 10.5.5 and 10.5.6 verify Corollaries 2.3 and 2.4 in Topkaya and Mustafa (2017), respectively.

From the following theorem, we see that for the functions in the class $T(S^*C)^\tau(\alpha, \beta)$, $\alpha \in [0, 1)$, $\beta \in [0, 1]$, $\tau \in \mathbb{R}^* = \mathbb{R} - \{0\}$ the converse of Theorem 10.5.2 is also true.

**Theorem 10.5.8.** *Let $f \in T$. Then, the function $f(z)$ belongs to the class $T(S^*C)^\tau(\alpha, \beta)$, $\alpha \in [0, 1)$, $\beta \in [0, 1]$, $\tau \in \mathbb{R}^* = \mathbb{R} - \{0\}$ if and only if*

$$\sum_{n=2}^{\infty} [n + (1 - \alpha) |\tau| - 1] [1 + (n - 1)\beta] a_n \leq (1 - \alpha) |\tau|.$$

*The result is sharp for the functions*

$$f_n(z) = z - \frac{(1 - \alpha) |\tau|}{[n + (1 - \alpha) |\tau| - 1] [1 + (n - 1)\beta]} z^n, \ z \in U, \ n = 2, 3, \dots \, ..$$

*Proof.* The proof of the sufficiency of the theorem can be proved similarly to the proof of Theorem 10.5.2. Therefore, we will prove only the necessity of the theorem.

Assume that $f \in T(S^*C)^\tau(\alpha, \beta)$, $\alpha \in [0, 1)$, $\beta \in [0, 1]$, $\tau \in \mathbb{R}^* = \mathbb{R} - \{0\}$; that is,

$$\Re \left\{ 1 + \frac{1}{\tau} \left[ \frac{zf'(z) + \beta z^2 f''(z)}{\beta z f'(z) + (1 - \beta)f(z)} - 1 \right] \right\} > \alpha, \ z \in U. \tag{10.24}$$

Using (10.19) and (10.24), we can easily show that

$$\Re \left\{ \frac{-\sum_{n=2}^{\infty} (n - 1) [1 + \beta(n - 1)] a_n z^n}{\tau \{z - \sum_{n=2}^{\infty} [1 + \beta(n - 1)] a_n z^n\}} \right\} > \alpha - 1.$$

The expression

$$\frac{-\sum_{n=2}^{\infty} (n - 1) [1 + \beta(n - 1)] a_n z^n}{\tau \{z - \sum_{n=2}^{\infty} [1 + \beta(n - 1)] a_n z^n\}}$$

is real if $z$ is chosen to be real.

Thus, from the previous inequality letting $z \to 1$ through real values, we have

$$\frac{-\sum_{n=2}^{\infty} (n - 1) [1 + \beta(n - 1)] a_n}{\tau \{1 - \sum_{n=2}^{\infty} [1 + \beta(n - 1)] a_n\}} \geq \alpha - 1. \tag{10.25}$$

We will examine of the last inequality depending on the different cases of the sing of the parameter $\tau$ as follows.

Let us $\tau > 0$. Then, from (10.25), we have

$$-\sum_{n=2}^{\infty} (n - 1) [1 + \beta(n - 1)] a_n \geq (\alpha - 1)\tau \left\{ 1 - \sum_{n=2}^{\infty} [1 + \beta(n - 1)] a_n \right\},$$

which is equivalent to

$$\sum_{n=2}^{\infty} [n + (1 - \alpha)\tau - 1] [1 + \beta(n - 1)] a_n \leq (1 - \alpha)\tau. \qquad (10.26)$$

Now, let us $\tau < 0$. Then, since $\tau = -|\tau|$, from (10.25), we get

$$\frac{\sum_{n=2}^{\infty} (n - 1) [1 + \beta(n - 1)] a_n}{|\tau| \{1 - \sum_{n=2}^{\infty} [1 + \beta(n - 1)] a_n\}} \geq \alpha - 1;$$

that is,

$$\sum_{n=2}^{\infty} (n - 1) [1 + \beta(n - 1)] a_n \geq (\alpha - 1) |\tau| \left\{ 1 - \sum_{n=2}^{\infty} [1 + \beta(n - 1)] a_n \right\}.$$

Therefore,

$$\sum_{n=2}^{\infty} [n + (\alpha - 1) |\tau| - 1] [1 + \beta(n - 1)] a_n \geq -(1 - \alpha) |\tau|.$$

Since $\alpha < 1$ (or $1 - \alpha > \alpha - 1$), from the last inequality, we have

$$\sum_{n=2}^{\infty} [n + (1 - \alpha) |\tau| - 1] [1 + \beta(n - 1)] a_n \geq -(1 - \alpha) |\tau|. \qquad (10.27)$$

Thus, from (10.27) and (10.28), the proof of the necessity of theorem; that is, the proof of theorem is completed. □

Special case of Theorem 10.5.8 has been proved by Altintaş et al Blutzer and Torvik (1996), $\tau = 1$ (there $p = n = 1$).

Setting $\tau = 1$ in Theorem 2.2, we arrive at the following corollary.

**Corollary 10.5.9.** *The function $f(z)$ definition by (10.19) belongs to the class $TS^*C(\alpha, \beta)$, $\alpha \in [0, 1)$, $\beta \in [0, 1]$ if and only if*

$$\sum_{n=2}^{\infty} (n - \alpha) [1 + \beta(n - 1)] a_n \leq 1 - \alpha.$$

*Remark* 10.5.10. The result obtained in Corollary 10.5.9 verifies Theorem 1 in Blutzer and Torvik (1996).

Taking $\beta = 0$ in Corollary 10.5.9, we have the following result.

**Corollary 10.5.11.** *(see (Speicher, 1998, p. 110, Theorem 2)) The function $f(z)$ definition by (10.19) belongs to the class $TS^*(\alpha)$, $\alpha \in [0, 1)$ if and only if*

$$\sum_{n=2}^{\infty} (n - \alpha) a_n \leq 1 - \alpha.$$

Choose $\beta = 1$ in Corollary 10.5.9, we arrive at the following result.

**Corollary 10.5.12.** *(see (Speicher, 1998, 9, p. 111, Corollary 2)) The function $f(z)$ definition by (10.19) belongs to the class $TC(\alpha)$, $\alpha \in [0,1)$ if and only if*

$$\sum_{n=2}^{\infty} n(n-\alpha)a_n \leq 1-\alpha.$$

*Remark* 10.5.13. The results obtained in Corollaries 10.5.11 and 10.5.12 verify Corollaries 2.7 and 2.8 in Topkaya and Mustafa (2017), respectively.

**Corollary 10.5.14.** *The function $f(z)$ definition by (10.2) belongs to the class $TS^*(\alpha;\tau)$, $\alpha \in [0,1)$, $\tau \in \mathbb{R}^*$ if and only if*

$$\sum_{n=2}^{\infty} [n+(1-\alpha)|\tau|-1]a_n \leq (1-\alpha)|\tau|.$$

**Corollary 10.5.15.** *The function $f(z)$ definition by (10.19) belongs to the class $TC(\alpha;\tau)$, $\alpha \in [0,1)$, $\tau \in \mathbb{R}^*$ if and only if*

$$\sum_{n=2}^{\infty} n[n+(1-\alpha)|\tau|-1]a_n \leq (1-\alpha)|\tau|.$$

On the coefficient bound estimates of the functions belonging in the class $T(S^*C)^\tau(\alpha,\beta)$, we give the following result.

**Theorem 10.5.16.** *Let the function $f(z)$ definition by (10.19) belong to the class $T(S^*C)^\tau(\alpha,\beta)$, $\alpha \in [0,1)$, $\beta \in [0,1]$, $\tau \in \mathbb{R}^* = \mathbb{R} - \{0\}$. Then,*

$$\sum_{n=2}^{\infty} |a_n| \leq \frac{(1-\alpha)|\tau|}{(1+\beta)[1+(1-\alpha)|\tau|]} \tag{10.28}$$

*and*

$$\sum_{n=2}^{\infty} n|a_n| \leq \frac{2(1-\alpha)|\tau|}{(1+\beta)[1+(1-\alpha)|\tau|]}. \tag{10.29}$$

*Proof.* Using Theorem 10.5.8, we obtain

$$[1+(1-\alpha)|\tau|](1+\beta)\sum_{n=2}^{\infty}|a_n|$$

$$\leq \sum_{n=2}^{\infty}[n+(1-\alpha)|\tau|-1][1+(n-1)\beta]|a_n| \leq (1-\alpha)|\tau|.$$

From here, we can easily show that (10.28) is true.

Similarly, we write

$$(1 + \beta) \sum_{n=2}^{\infty} [n + (1 - \alpha) |\tau| - 1] |a_n|$$

$$\leq \sum_{n=2}^{\infty} [n + (1 - \alpha) |\tau| - 1] [1 + (n - 1)\beta] |a_n| \leq (1 - \alpha) |\tau| \, ;$$

that is,

$$(1 + \beta) \sum_{n=2}^{\infty} n |a_n| \leq (1 - \alpha) |\tau| + [1 - (1 - \alpha) |\tau|] (1 + \beta) \sum_{n=2}^{\infty} |a_n|$$

Using (10.28) in the last inequality, we obtain

$$(1 + \beta) \sum_{n=2}^{\infty} n |a_n| \leq \frac{2(1 - \alpha) |\tau|}{1 + (1 - \alpha) |\tau|},$$

which immediately yields the inequality (10.29).

Thus, the proof of Theorem 10.5.16 is completed.          □

Setting $\tau = 1$ in Theorem 10.5.16, we obtain the following corollary.

**Corollary 10.5.17.** *Let the function $f(z)$ definition by (10.19) belong to the class $TS^*C(\alpha, \beta)$, $\alpha \in [0, 1)$, $\beta \in [0, 1]$. Then,*

$$\sum_{n=2}^{\infty} |a_n| \leq \frac{1 - \alpha}{(2 - \alpha)(1 + \beta)}$$

*and*

$$\sum_{n=2}^{\infty} n |a_n| \leq \frac{2(1 - \alpha)}{(2 - \alpha)(1 + \beta)}.$$

*Remark* 10.5.18. The result obtained in the Corollary 10.5.14 verifies Lemma 2 (with $n = p = 1$) of Blutzer and Torvik (1996).

From Theorem 10.5.8, for the coefficient bound estimates, we arrive at the following result.

**Corollary 10.5.19.** *Let $f \in T(S^*C)^{\tau}(\alpha, \beta), \alpha \in [0, 1), \beta \in [0, 1], \tau \in \mathbb{R}^* = \mathbb{R} - \{0\}$. Then,*

$$|a_n| \leq \frac{(1 - \alpha) |\tau|}{[n + (1 - \alpha) |\tau| - 1] [1 + (n - 1)\beta]}, \quad n = 2, 3, \dots .$$

*Remark* 10.5.20. Numerous consequences of Corollary 10.5.15 can indeed be deduced by specializing the various parameters involved. Many of these consequences were proved by earlier workers on the subject (see, for example, Speicher (1998); Ahmad (et al.); Kilbas et al. (2006)).

### 10.5.1 Order of starlikeness and radius of convexity for classes $TC(\alpha; \tau)$ and $TS^*(\alpha; \tau)$

In this section, we will examine some properties like order of starlikeness and radius of convexity of the subclasses $TC(\alpha; \tau)$ and $TS^*(\alpha; \tau)$.

On this, we can give the following theorem.

**Theorem 10.5.21.** *If $f \in TC(\alpha; \tau)$, $\alpha \in [0, 1)$, $\tau \in \mathbb{R}^*$, then the function $f(z)$ belongs to the class $TS^* \left( \frac{2+(1-\alpha)(|\tau|-1)}{2+(1-\alpha)|\tau|}; \tau \right)$; that is*

$$f \in TS^* \left( \frac{2 + (1 - \alpha)(|\tau| - 1)}{2 + (1 - \alpha)|\tau|}; \tau \right).$$

*The result is sharp for the functions*

$$f_n(z) = z - \frac{(1 - \alpha)|\tau|}{n(n + (1 - \alpha)|\tau| - 1)} z^n, \ z \in U, \ n = 2, 3, \dots \, .$$

*Proof.* In view of Corollary 10.5.14 and Corollary 10.5.15, we must prove that

$$\sum_{n=2}^{\infty} \frac{n - 1 + \left[1 - \frac{2+(1-\alpha)(|\tau|-1)}{2+(1-\alpha)|\tau|}\right]|\tau|}{\left[1 - \frac{2+(1-\alpha)(|\tau|-1)}{2+(1-\alpha)|\tau|}\right]|\tau|} a_n \leq 1 \tag{10.30}$$

if

$$\sum_{n=2}^{\infty} \frac{n[n - 1 + (1 - \alpha)|\tau|]}{(1 - \alpha)|\tau|} a_n \leq 1.$$

It suffices to show that

$$\frac{n[n - 1 + (1 - \alpha)|\tau|]}{(1 - \alpha)|\tau|} \geq \frac{n - 1 + \left[1 - \frac{2+(1-\alpha)(|\tau|-1)}{2+(1-\alpha)|\tau|}\right]|\tau|}{\left[1 - \frac{2+(1-\alpha)(|\tau|-1)}{2+(1-\alpha)|\tau|}\right]|\tau|} \tag{10.31}$$

for all $n = 2, 3, \dots$ .

Last inequality is equivalent to

$$\frac{n[n - 1 + (1 - \alpha)|\tau|] - (1 - \alpha)|\tau|}{1 - \alpha} \geq (n - 1)[2 + (1 - \alpha)|\tau|].$$

Taking into account that $\alpha \geq 0$ $\left(\text{or } \frac{1}{1-\alpha} \geq 1\right)$, it suffices to show that

$$n[n - 1 + (1 - \alpha)|\tau|] - (1 - \alpha)|\tau| \geq (n - 1)[2 + (1 - \alpha)|\tau|]$$

which is equivalent to $n(n - 1) \geq 2(n - 1)$; that is, $n > 2$. Thus, the inequality (10.31) is provided for all $n = 2, 3, \dots$ .

With this the proof of Theorem 10.5.21 is completed. $\qquad \square$

**Note 1.** *There is no converse to Theorem 10.5.21. That is, a function in* $TS^*(\alpha; \tau)$ *need not be convex. To show this, we need only find coefficients* $a_n$, $n = 2, 3, \ldots$ *for which*

$$\sum_{n=2}^{\infty} \frac{n - 1 + (1 - \alpha)|\tau|}{(1 - \alpha)|\tau|} a_n \leq 1 \text{ and } \sum_{n=2}^{\infty} \frac{n(n - 1 + |\tau|)}{|\tau|} a_n \geq 1 \qquad (10.32)$$

*Note that the functions* $f_n(z) = z - \frac{(1-\alpha)|\tau|}{n-1+(1-\alpha)|\tau|} z^n$, *for* $n \geq \left[\!\!\left[\frac{1}{(1-\alpha)|\tau|}\right]\!\!\right] + 1$ *all satisfy both inequalities of (10.32), where* $[\![x]\!]$ *is the exact value of number* $x$.

By considering of the above note, we now determine the radius of convexity for functions in $TS^*(\alpha; \tau)$. The following theorem is about this.

**Theorem 10.5.22.** *If* $f \in TS^*(\alpha; \tau)$, $\alpha \in [0, 1)$, $\tau \in \mathbb{R}^*$, *then the radius of convexity of the function* $f(z)$ *is* $r(\alpha; \tau) = \inf\left\{\left[\frac{n-1+(1-\alpha)|\tau|}{(1-\alpha)|\tau|}\right]^{1/(n-1)} : n = 2, 3, \ldots\right\}$; *that is, the function* $f(z)$ *is convex in the disk* $U_{r(\alpha;\tau)} = \{z : |z| < r(\alpha; \tau)\}$.

*Proof.* Let $f \in TS^*(\alpha; \tau)$, $\alpha \in [0, 1)$, $\tau \in \mathbb{R}^*$. It suffices that $|zf''(z)/f'(z)| \leq 1$ for $|z| < r(\alpha; \tau)$.

By simple computation, we have

$$\left|\frac{zf''(z)}{f'(z)}\right| = \left|\frac{-\sum_{n=2}^{\infty} n(n-1)a_n z^{n-1}}{1 - \sum_{n=2}^{\infty} na_n z^{n-1}}\right| \leq \frac{\sum_{n=2}^{\infty} n(n-1)a_n |z|^{n-1}}{1 - \sum_{n=2}^{\infty} na_n |z|^{n-1}}.$$

The last expression is bounded above by 1 if

$$\sum_{n=2}^{\infty} n(n-1)a_n |z|^{n-1} \leq 1 - \sum_{n=2}^{\infty} na_n |z|^{n-1},$$

which is equivalent to

$$\sum_{n=2}^{\infty} n^2 a_n |z|^{n-1} \leq 1. \qquad (10.33)$$

According to Corollary 10.5.15,

$$\sum_{n=2}^{\infty} \frac{n - 1 + (1 - \alpha)|\tau|}{(1 - \alpha)|\tau|} a_n \leq 1.$$

Hence, inequality (10.33) will be true if

$$n^2 |z|^{n-1} \leq \frac{n - 1 + (1 - \alpha)|\tau|}{(1 - \alpha)|\tau|}, \quad n = 2, 3, \ldots.$$

Solving the last inequality for $|z|$, we obtain

$$|z| \leq \left[ \frac{n - 1 + (1 - \alpha) |\tau|}{n^2 (1 - \alpha) |\tau|} \right]^{1/(n-1)} , \quad n = 2, 3, \dots .$$

From here, we obtain the desired result. Thus, the proof of Theorem 10.5.22 is completed. $\quad\square$

## 10.6 Upper bound for second Hankel determinant of certain subclass of analytic and bi-univalent functions

In this section, for $f \in S$, we consider the following class of analytic functions

$$Q(\alpha, \beta) = \left\{ f \in S : \Re \left[ (1 - \beta) \frac{f(z)}{z} + \beta f'(z) \right] > \alpha, z \in U \right\},$$

$$\alpha \in [0, 1), \beta \in [0, 1].$$

In 1995, Ding et al. Ding et al. (1995) have introduced and investigated the class $Q(\alpha, \beta)$. It is well-known that (see Duren (1983)) every function $f \in S$ has an inverse $f^{-1}$, defined by

$$f^{-1}(f(z)) = z, \ z \in U$$

and

$$f(f^{-1}(w)) = w, \ w \in D_{r_0} = \{ w \in \mathbb{C} : \ |w| < r_0(f) \}, \ r_0(f) \geq 1/4,$$

where

$$f^{-1}(w) = w - a_2 w^2 + \left( 2a_2^2 - a_3 \right) w^3 - \left( 5a_2^3 - 5a_2 a_3 + a_4 \right) w^4 + \cdots, \ w \in D_{r_0}.$$
$$(10.34)$$

A function $f \in A$ is called bi-univalent in $U$ if both $f$ and $f^{-1}$ are univalent in the definition sets. Let $\Sigma$ denote the class of bi-univalent functions in $U$ given by (10.1). For a short history and examples of functions in the class $\Sigma$, see Srivastava et al. (2010).

Firstly, Lewin Lewin (1967) introduced the class of bi-univalent functions, obtaining the estimate $|a_2| \leq 1.51$. Subsequently, Brannan and Clunie Brannan and Clunie (1980) developed the result of Lewin to $|a_2| \leq \sqrt{2}$ for $f \in \Sigma$. Accordingly, Netanyahu Netanyahu (1969) showed that $|a_2| \leq \frac{4}{3}$. Earlier, Brannan and Taha Keogh and Merkes (1969) introduced certain subclasses of bi-univalent function class $\Sigma$, namely bi-starlike function of order $\alpha$ denoted $S_\Sigma^*(\alpha)$ and bi-convex function of order $\alpha$ denoted $C_\Sigma(\alpha)$ corresponding to the function classes $S^*(\alpha)$ and $C(\alpha)$, respectively. For each of the function classes

$S_\Sigma^*(\alpha)$ and $C_\Sigma(\alpha)$, non-sharp estimates on the first two Taylor-Maclaurin coefficients were found in Brannan and Taha (1988); Taha (1980). Many researchers (see Srivastava et al. (2010); Xu et al. (2012a,b)) have introduced and investigated several interesting subclasses of bi-univalent function class $\Sigma$ and they have found non-sharp estimates on the first two Taylor-Maclaurin coefficients. However, the coefficient problem for each of the Taylor-Maclaurin coefficients $|a_n|$, $n = 3, 4, \ldots$ is still an open problem (see, for example, Lewin (1967); Netanyahu (1969)).

An analytic function $f$ is bi-starlike of Ma-Minda type or bi-convex of Ma-Minda type if both $f$ and $f^{-1}$ are, respectively, Ma-Minda starlike and convex. These classes are denoted, respectively, by $S_\Sigma^*(\phi)$ and $C_\Sigma(\phi)$. In the sequel, it is assumed that the function $\phi$ is an analytic function with positive real part in $U$, satisfying $\phi(0) = 1$, $\phi'(0) > 0$ and $\phi(U)$ is symmetric with respect to the real axis. Such a function has a series expansion of the following form:

$$\phi(z) = 1 + b_1 z + b_2 z^2 + b_3 z^3 + \cdots = 1 + \sum_{n=1}^{\infty} b_n z^n, \ b_1 > 0.$$

An analytic function $f$ is subordinate to an analytic function $\phi$, written $f(z) \prec \phi(z)$, provided that there is an analytic function (that is, Schwarz function) $\omega$ defined on $U$ with $\omega(0) = 0$ and $|\omega(z)| < 1$ satisfying $f(z) = \phi(\omega(z))$. Ma and Minda Ma and Minda (1992) unified various subclasses of starlike and convex functions for which either of the quantity $\frac{zf'(z)}{f(z)}$ or $1 + \frac{zf''(z)}{f'(z)}$ is subordinate to a more general function. For this purpose, they considered an analytic function $\phi$ with positive real part in $U, \phi(0) = 1$, $\phi'(0) > 0$ and $\phi$ maps $U$ onto a region starlike with respect to 1 and symmetric with respect to the real axis. The class of Ma-Minda starlike and Ma-Minda convex functions consists of functions $f \in A$ satisfying the subordination $\frac{zf'(z)}{f(z)} \prec \phi(z)$ and $1 + \frac{zf''(z)}{f'(z)} \prec \phi(z)$, respectively.

In 1976, Noonan and Thomas Noonan and Thomas (1976) defined the $q^{\text{th}}$ Hankel determinant of $f$ for $q \in \mathbb{N}$ by

$$H_q(n) = \begin{vmatrix} a_n & \cdots & a_{n+q-1} \\ \cdot & \cdots & \cdot \\ a_{n+q-1} & \cdots & a_{n+2q-2} \end{vmatrix}.$$

For $q = 2$ and $n = 1$, Fekete and Szegö Fekete and Szegö (1933) considered the Hankel determinant of $f$ as $H_2(1) = \begin{vmatrix} a_1 & a_2 \\ a_2 & a_3 \end{vmatrix} = a_1 a_3 - a_2^2$. They made an earlier study for the estimates of $|a_3 - \mu a_2^2|$ when $a_1 = 1$ with real $\mu \in \mathbb{R}$. The well-known result due to them states that if $f \in A$, then

$$|a_3 - \mu a_2^2| \leq \begin{cases} 3 - 4\mu & \text{if } \mu \in (-\infty, 0], \\ 1 + 2\exp\left(\frac{-2\mu}{1-\mu}\right) & \text{if } \mu \in [0, 1), \\ 4\mu - 3 & \text{if } \mu \in [1, +\infty). \end{cases}$$

Furthermore, Hummel Hummel (1957, 1960) obtained sharp estimates for $\left|a_3 - \mu a_2^2\right|$ when $f$ is a convex function and also Keogh and Merkes Keogh and Merkes (1969) obtained sharp estimates for $\left|a_3 - \mu a_2^2\right|$ when $f$ is a close-to-convex function, starlike and convex function in $U$.

Recently, the upper bounds of $|H_2(2)| = \left|a_2 a_4 - a_3^2\right|$ for the classes $S_\Sigma^*(\alpha)$ and $C_\Sigma(\alpha)$ were obtained by Deniz et al. Deniz et al. (2015a). Very soon, Orhan et al. Orhan et al. (2016) reviewed the study of bounds for the second Hankel determinant of the subclass $M_\Sigma^\alpha(\beta)$ of bi-univalent functions.

Chebyshev polynomials, which are used by us in this paper, play a considerable act in numerical analysis and mathematical physics. It is well-known that the Chebyshev polinomials are four kinds. The most of research articles related to specific orthogonal polynomials of Chebyshev family, contain essentially results of Chebyshev polynomials of first and second kinds $T_n(x)$ and $U_n(x)$, and their numerous uses in different applications (see Doha (1994); Mason (1967)).

The well-known kinds of the Chebyshev polynomials are the first and second kinds. In the case of real variable $x$ on $(-1, 1)$, the first and second kinds of the Chebyshev polynomials are defined by

$$T_n(x) = \cos(n \arccos x),$$

$$U_n(x) = \frac{\sin\left[(n+1)\arccos x\right]}{\sin(\arccos x)} = \frac{\sin\left[(n+1)\arccos x\right]}{\sqrt{1 - x^2}}.$$

We consider the function

$$G(t, z) = \frac{1}{1 - 2tz + z^2}, \quad t \in \left(\frac{1}{2}, 1\right), z \in U.$$

It is well-known that if $t = \cos\alpha$, $\alpha \in \left(0, \frac{\pi}{3}\right)$, then

$$G(t, z) = 1 + \sum_{n=1}^\infty \frac{\sin[(n+1)\alpha]}{\sin\alpha} z^n = 1 + 2\cos\alpha z + \left(3\cos^2\alpha - \sin^2\alpha\right) z^2$$
$$+ \left(8\cos^3\alpha - 4\cos\alpha\right) z^3 + \cdots, \quad z \in U.$$

That is,

$$G(t, z) = 1 + U_1(t)z + U_2(t)z^2 + U_3(t)z^3 + \cdots, \quad t \in \left(\frac{1}{2}, 1\right), z \in U, \quad (10.35)$$

where $U_n(t) = \frac{\sin[(n+1)\arccos t]}{\sqrt{1-t^2}}$, $n \in N$ are the second kind Chebyshev polynomials. From the definition of the second kind Chebyshev polynomials, we easily obtain that $U_1(t) = 2t$ .

Also, it is well-known that

$$U_{n+1}(t) = 2tU_n(t) - U_{n-2}(t)$$

for all $n \in N$. From here, we can easily obtain
$U_2(t) = 4t^2 - 1$, $U_3(t) = 8t^3 - 4t, \ldots$ .

Inspired by the aforementioned works, making use of the Chebyshev polynomials, we define a subclass of bi-univalent functions $\Sigma$ as follows.

**Definition 10.6.1.** A function $f \in \Sigma$ given by (10.1) is said to be in the class $Q_\Sigma(G, \beta, t)$, $\beta \geq 0, t \in \left(\frac{1}{2}, 1\right)$, where $G$ is an analytic function given by (10.35) if the following conditions are satisfied

$$(1 - \beta) \frac{f(z)}{z} + \beta f'(z) \prec G(t, z), \ z \in U$$

and

$$(1 - \beta) \frac{g(w)}{w} + \beta g'(w) \prec G(t, w), \ w \in D_{r_0}$$

where $g = f^{-1}$.

*Remark* 10.6.2. Taking $\beta = 1$, we have $Q_\Sigma(G, 1, t) = \Re_\Sigma(G, t)$, $t \in \left(\frac{1}{2}, 1\right)$; that is,
$f \in \Re_\Sigma(G, t) \Leftrightarrow f'(z) \prec G(t, z), \ z \in U$ and $g'(w) \prec G(t, w), \ w \in D_{r_0}$,
where $g = f^{-1}$.

*Remark* 10.6.3. Taking $\beta = 0$, we have $Q_\Sigma(G, 0, t) = N_\Sigma(G, t)$, $t \in \left(\frac{1}{2}, 1\right)$; that is,
$f \in N_\Sigma(G, t) \Leftrightarrow \frac{f(z)}{z} \prec G(t, z), \ z \in U$ and $\frac{g(w)}{w} \prec G(t, w), \ w \in D_{r_0}$,
where $g = f^{-1}$.

The object of this section is to determine the second Hankel determinant for the function class $Q_\Sigma(G, \beta, t)$ and its special classes, and is to give upper bound estimate for $|H_2(2)|$.

In order to prove our main results, we shall need the following lemma.

**Lemma 10.6.4.** *(Duren (1983)) Let* P *be the class of all analytic functions* $p(z)$ *of the form*

$$p(z) = 1 + p_1 z + p_2 z^2 + \cdots = 1 + \sum_{n=1}^{\infty} p_n z^n$$

*satisfying* $\Re(p(z)) > 0$, $z \in U$ *and* $p(0) = 1$. *Then,* $|p_n| \leq 2$, *for every* $n = 1, 2, 3, \ldots$ . *This inequality is sharp for each* $n$.
*Moreover,*

$$2p_2 = p_1^2 + \left(4 - p_1^2\right) x,$$

$$4p_3 = p_1^3 + 2\left(4 - p_1^2\right) p_1 x - \left(4 - p_1^2\right) p_1 x^2 + 2\left(4 - p_1^2\right)\left(1 - |x|^2\right) z,$$

*for some* $x$, $z$ *with* $|x| \leq 1$, $|z| \leq 1$.

### 10.6.1   Upper bound estimate for second Hankel determinant of class $Q_\Sigma(G, \beta, t)$

In this section, we prove the following theorem on upper bound of the second Hankel determinant of the function class $Q_\Sigma(G, \beta, t)$.

**Theorem 10.6.5.** *Let the function $f(z)$ given by (10.1) be in the class $Q_\Sigma(G, \beta, t)$, $\beta \in [0, 1]$, $t \in \left(\frac{1}{2}, 1\right)$, where the function $G$ is an analytic function given by (10.35). Then,*

$$
\left|a_2 a_4 - a_3^2\right| \leq
\begin{cases}
\max\left\{\frac{4t^2}{(1+2\beta)^2}, H(t, 2-)\right\}, & \begin{cases} \text{if } \Delta(\beta, t) \geq 0 \text{ and } c(\beta, t) \geq 0, \\ \text{if } \Delta(\beta, t) > 0 \text{ and } c(\beta, t) < 0, \\ \text{if } \Delta(\beta, t) \leq 0 \text{ and } c(\beta, t) \leq 0, \end{cases} \\[4pt]
\max\left\{H(t, \beta), H(t, 2-)\right\}, & \text{if } \Delta(\beta, t) < 0 \text{ and } c(\beta, t) > 0,
\end{cases}
$$

*where*

$$
H(t, 2-) = \frac{8t^2 \left|\left(2t^2 - 1\right)(1 + \beta)^3 - 2t^2(1 + 3\beta)\right|}{(1 + \beta)^4 (1 + 3\beta)},
$$

$$
H(t, \beta) = \frac{4t^2}{(1 + 2\beta)^2} - \frac{c^2(\beta, t)}{8(1 + \beta)^4 (1 + 2\beta)^2 (1 + 3\beta) \Delta(\beta, t)},
$$

$$
\Delta(\beta, t) = 16t^2 \left|\left(2t^2 - 1\right)(1 + \beta)^3 - 2t^2(1 + 3\beta)\right| (1 + 2\beta)^2 -
$$
$$
8t \left[(1 + 3\beta) t^2 + (1 + \beta)(1 + 2\beta)\left(4t^2 - 1\right)\right] (1 + \beta)^2 (1 + 2\beta) - 8t^2 \beta^2 (1 + \beta)^3,
$$

$$
c(\beta, t) = 8t \left[\left(5t^2 - 1\right)(1 + 3\beta) + 2\left(4t^2 - 1\right)\beta^2\right] (1 + \beta)^2 (1 + 2\beta) -
$$
$$
8t^2 \left[(1 + \beta)^2 + (2 + \beta)\beta\right] (1 + \beta)^3.
$$

*Proof.* Let $f \in Q_\Sigma(G, \beta, t)$, $\beta \in [0, 1]$, $t \in \left(\frac{1}{2}, 1\right)$ and $g = f^{-1}$. Then, according to Definition 10.6.1, there are analytic functions $\omega : U \to U$, $\varpi : D_{r_0} \to D_{r_0}$ with $\omega(0) = 0 = \varpi(0)$, $|\omega(z)| < 1$, $|\varpi(w)| < 1$ satisfying the following conditions

$$
(1 - \beta) \frac{f(z)}{z} + \beta f'(z) = G(t, \omega(z)), \quad z \in U \tag{10.36}
$$

and

$$
(1 - \beta) \frac{g(w)}{w} + \beta g'(w) = G(t, \varpi(w)), \quad w \in D_{r_0}. \tag{10.37}
$$

Let also the functions $p, q \in P$ be defined as follows;

$$
p(z) := \frac{1 + \omega(z)}{1 - \omega(z)} = 1 + p_1 z + p_2 z^2 + \cdots = 1 + \sum_{n=1}^{\infty} p_n z^n
$$

and

$$
q(w) := \frac{1 + \varpi(w)}{1 - \varpi(w)} = 1 + q_1 w + q_2 w^2 + \cdots = 1 + \sum_{n=1}^{\infty} q_n w^n.
$$

It follows that

$$\omega(z) := \frac{p(z)-1}{p(z)+1} = \frac{1}{2}\left[p_1 z + \left(p_2 - \frac{p_1^2}{2}\right)z^2 + \left(p_3 - p_1 p_2 + \frac{p_1^3}{4}\right)z^3 + \cdots\right]$$
(10.38)

and

$$\varpi(w) := \frac{q(w)-1}{q(w)+1} = \frac{1}{2}\left[q_1 w + \left(q_2 - \frac{q_1^2}{2}\right)w^2 + \left(q_3 - q_1 q_2 + \frac{q_1^3}{4}\right)w^3 + \cdots\right].$$
(10.39)

From (10.38) and (10.39), considering (10.35), we can easily show that

$$G(t, \omega(z)) = 1 + \frac{U_1(t)}{2}p_1 z + \left[\frac{U_1(t)}{2}\left(p_2 - \frac{p_1^2}{2}\right) + \frac{U_2(t)}{4}p_1^2\right]z^2 +$$
$$\left[\frac{U_1(t)}{2}\left(p_3 - p_1 p_2 + \frac{p_1^3}{4}\right) + \frac{U_2(t)}{2}p_1\left(p_2 - \frac{p_1^2}{2}\right) + \frac{U_3(t)}{8}p_1^3\right]z^3 + \cdots$$
(10.40)

and

$$G(t, \varpi(w)) = 1 + \frac{U_1(t)}{2}q_1 w + \left[\frac{U_1(t)}{2}\left(q_2 - \frac{q_1^2}{2}\right) + \frac{U_2(t)}{4}q_1^2\right]w^2 +$$
$$\left[\frac{U_1(t)}{2}\left(q_3 - q_1 q_2 + \frac{q_1^3}{4}\right) + \frac{U_2(t)}{2}q_1\left(q_2 - \frac{q_1^2}{2}\right) + \frac{U_3(t)}{8}q_1^3\right]w^3 + \cdots.$$
(10.41)

From (10.36), (10.40) and (10.37), (10.41), we can easily obtain that

$$(1+\beta)a_2 = \frac{U_1(t)}{2}p_1,$$
(10.42)

$$(1+2\beta)a_3 = \frac{U_1(t)}{2}\left(p_2 - \frac{p_1^2}{2}\right) + \frac{U_2(t)}{4}p_1^2,$$
(10.43)

$$(1+3\beta)a_4 = \frac{U_1(t)}{2}\left(p_3 - p_1 p_2 + \frac{p_1^3}{4}\right) + \frac{U_2(t)}{2}p_1\left(p_2 - \frac{p_1^2}{2}\right) + \frac{U_3(t)}{8}p_1^3$$
(10.44)

and

$$-(1+\beta)a_2 = \frac{U_1(t)}{2}q_1,$$
(10.45)

$$(1+2\beta)\left(2a_2^2 - a_3\right) = \frac{U_1(t)}{2}\left(q_2 - \frac{q_1^2}{2}\right) + \frac{U_2(t)}{4}q_1^2,$$
(10.46)

$$-(1+3\beta)\left(5a_2^3 - 5a_2 a_3 + a_4\right) = \frac{U_1(t)}{2}\left(q_3 - q_1 q_2 + \frac{q_1^3}{4}\right)$$
$$+\frac{U_2(t)}{2}q_1\left(q_2 - \frac{q_1^2}{2}\right) + \frac{U_3(t)}{8}q_1^3. \quad (10.47)$$

From (10.42) and (10.45), we obtain that

$$\frac{U_1(t)}{2\left(1+\beta\right)}p_1 = a_2 = -\frac{U_1(t)}{2\left(1+\beta\right)}q_1. \tag{10.48}$$

Subtracting (10.46) from (10.43) and considering (10.48), we can easily obtain that

$$a_3 = a_2^2 + \frac{U_1(t)}{4\left(1+2\beta\right)}\left(p_2 - q_2\right) = \frac{U_1^2(t)}{4\left(1+\beta\right)^2}p_1^2 + \frac{U_1(t)}{4\left(1+2\beta\right)}\left(p_2 - q_2\right). \tag{10.49}$$

On the other hand, subtracting (10.47) from (10.44) and considering (10.48) and (10.49), we get

$$\begin{aligned}a_4 = &\frac{5U_1^2(t)p_1\left(p_2 - q_2\right)}{16\left(1+\beta\right)\left(1+2\beta\right)} + \frac{U_1(t)\left(p_3 - q_3\right)}{4\left(1+3\beta\right)}\\ &+ \frac{U_2(t) - U_1(t)}{4\left(1+3\beta\right)}p_1\left(p_2 + q_2\right) + \frac{U_1(t) - 2U_2(t) + U_3(t)}{8\left(1+3\beta\right)}p_1^3.\end{aligned} \tag{10.50}$$

Thus, from (10.48), (10.49) and (10.50), we can easily establish that

$$\begin{aligned}a_2a_4 - a_3^2 = &\frac{U_1^3(t)p_1^2\left(p_2 - q_2\right)}{32\left(1+\beta\right)^2\left(1+2\beta\right)} + \frac{U_1^2(t)p_1\left(p_3 - q_3\right)}{8\left(1+\beta\right)\left(1+3\beta\right)}\\ &+ \frac{\left[U_2(t) - U_1(t)\right]U_1(t)}{8\left(1+\beta\right)\left(1+3\beta\right)}p_1^2\left(p_2 + q_2\right) - \frac{U_1^2(t)\left(p_2 - q_2\right)^2}{16\left(1+2\beta\right)^2}\\ &+ \frac{U_1(t)p_1^4}{16\left(1+\beta\right)^4\left(1+3\beta\right)}\left\{\left[U_1(t) - 2U_2(t) + U_3(t)\right]\left(1+\beta\right)^3 - U_1^3(t)\left(1+3\beta\right)\right\}.\end{aligned} \tag{10.51}$$

According to Lemma 10.6.4, we have

$$2p_2 = p_1^2 + \left(4 - p_1^2\right)x \text{ and } 2q_2 = q_1^2 + \left(4 - q_1^2\right)y \tag{10.52}$$

and

$$\begin{aligned}4p_3 &= p_1^3 + 2\left(4 - p_1^2\right)p_1x - \left(4 - p_1^2\right)p_1x^2 + 2\left(4 - p_1^2\right)\left(1 - |x|^2\right)z\\ 4q_3 &= q_1^3 + 2\left(4 - q_1^2\right)q_1y - \left(4 - q_1^2\right)q_1y^2 + 2\left(4 - q_1^2\right)\left(1 - |y|^2\right)w,\end{aligned} \tag{10.53}$$

for some $x, y, z, w$ with $|x| \leq 1$, $|y| \leq 1$, $|z| \leq 1$, $|w| \leq 1$.

Since (see ((10.48))) $p_1 = -q_1$, from (10.52) and (10.53), we get

$$p_2 - q_2 = \frac{4 - p_1^2}{2}\left(x - y\right), \ p_2 + q_2 = p_1^2 + \frac{4 - p_1^2}{2}\left(x + y\right) \tag{10.54}$$

and

$$\begin{aligned}p_3 - q_3 = &\frac{p_1^3}{2} + \frac{\left(4 - p_1^2\right)p_1}{2}\left(x + y\right) - \frac{\left(4 - p_1^2\right)p_1}{4}\left(x^2 + y^2\right) +\\ &\frac{4 - p_1^2}{2}\left[\left(1 - |x|^2\right)z - \left(1 - |y|^2\right)w\right].\end{aligned} \tag{10.55}$$

According to Lemma 10.6.4, we may assume without any restriction that $\tau \in [0, 2]$, where $\tau = |p_1|$.

Thus, substituting the expressions (10.54) and (10.55) in (10.51) and using triangle inequality, letting $|x| = \xi$, $|y| = \eta$, we can easily obtain that

$$\left| a_2 a_4 - a_3^2 \right| \le c_1(t, \tau) (\xi + \eta)^2 + c_2(t, \tau) (\xi^2 + \eta^2) + c_3(t, \tau) (\xi + \eta) + c_4(t, \tau)$$
$$:= F(\xi, \eta), \tag{10.56}$$

where

$$c_1(t, \tau) = \frac{U_1^2(t) (4 - \tau^2)^2}{64 (1 + 2\beta)^2} \ge 0, \quad c_2(t, \tau) = \frac{U_1^2(t) \tau (\tau - 2) (4 - \tau^2)}{32 (1 + \beta) (1 + 3\beta)} \le 0,$$

$$c_3(t, \tau) = \frac{U_1^3(t) \tau^2 (4 - \tau^2)}{64 (1 + \beta)^2 (1 + 2\beta)} + \frac{U_1(t) U_2(t) \tau^2 (4 - \tau^2)}{16 (1 + \beta) (1 + 3\beta)} \ge 0,$$

$$c_4(t, \tau) = \frac{U_1(t) \left| (1 + \beta)^3 U_3(t) - (1 + 3\beta) U_1^3(t) \right|}{16 (1 + \beta)^4 (1 + 3\beta)} \tau^4 + \frac{U_1^2(t) \tau (4 - \tau^2)}{8 (1 + \beta) (1 + 3\beta)} \ge 0,$$

$t \in \left( \frac{1}{2}, 1 \right), \tau \in [0, 2].$

Now, we need to maximize the function $F(\xi, \eta)$ on the closed square $\Omega = \{(\xi, \eta) : \xi, \eta \in [0, 1]\}$ for $\tau \in [0, 2]$. Since the coefficients of the function $F(\xi, \eta)$ are dependent on variable $\tau$ for fixed value of $t$, we must investigate the maximum of $F(\xi, \eta)$ respect to $\tau$ taking into account these cases $\tau = 0$, $\tau = 2$ and $\tau \in (0, 2)$.

Let $\tau = 0$. Then, we write

$$F(\xi, \eta) = c_1(t, 0) = \frac{U_1^2(t)}{4 (1 + 2\beta)^2} (\xi + \eta)^2.$$

It is clear that the maximum of the function $F(\xi, \eta)$ occurs at $(\xi, \tau) = (1, 1)$, and

$$\max \{F(\xi, \eta) : \xi, \eta \in [0, 1]\} = F(1, 1) = \frac{U_1^2(t)}{(1 + 2\beta)^2}. \tag{10.57}$$

Now, let $\tau = 2$. In this case, $F(\xi, \eta)$ is a constant function (respect to $\tau$) as follows:

$$F(\xi, \eta) = c_4(t, 2) = \frac{U_1(t) \left| (1 + \beta)^3 U_3(t) - (1 + 3\beta) U_1^3(t) \right|}{(1 + \beta)^4 (1 + 3\beta)}. \tag{10.58}$$

In the case $\tau \in (0, 2)$, we will examine the maximum of the function $F(\xi, \eta)$ taking into account the sing of $\Lambda(\xi, \eta) = F_{\xi\xi}(\xi, \eta) F_{\eta\eta}(\xi, \eta) - [F_{\xi\eta}(\xi, \eta)]^2$.

By simple computation, we can easily see that

$$\Lambda(\xi, \eta) = 4 c_2(t, \tau) [2 c_1(t, \tau) + c_2(t, \tau)].$$

Since $c_2(t,\tau) < 0$ for all $t \in \left(\frac{1}{2}, 1\right), \tau \in (0,2)$ and

$$2c_1(t,\tau) + c_2(t,\tau) = \frac{U_1^2(t)(4 - \tau^2)(2 - \tau)}{32(1 + \beta)(1 + 2\beta)^2(1 + 3\beta)} \varphi(\tau),$$

where $\varphi(\tau) = 2(1 + \beta)(1 + 3\beta) - \beta^2\tau$ since $2 - \beta^2\tau > 0$, $\varphi(\tau) > 6\beta^2 + 8\beta > 0$; that is $2c_1(t,\tau) + c_2(t,\tau) > 0$ for all $\tau \in (0,2)$ and $\beta \in [0,1]$, we conclude that $\Lambda(\xi,\eta) < 0$ for all $(\xi,\eta) \in \Omega$. Consequently, the function $F(\xi,\eta)$ cannot have a local maximum in $\Omega$. Therefore, we must investigate the maximum of the function $F(\xi,\eta)$ on the boundary of the square $\Omega$.

Let

$$\partial\Omega = \{(0,\eta) : \eta \in [0,1]\} \cup \{(\xi,0) : \xi \in [0,1]\}$$
$$\cup \{(1,\eta) : \eta \in [0,1]\} \cup \{(\xi,1) : \xi \in [0,1]\}.$$

We can easily show that the maximum of the function $F(\xi,\eta)$ on the boundary $\partial\Omega$ of the square $\Omega$ occurs at $(\xi,\eta) = (1,1)$, and

$$\max\{F(\xi,\eta) : (\xi,\eta) \in \partial\Omega\} - F(1,1) -$$
$$4c_1(t,\tau) + 2[c_2(t,\tau) + c_3(t,\tau)] + c_4(t,\tau), t \in \left(\frac{1}{2}, 1\right), \tau \in (0,2).$$
$$(10.59)$$

Now, let us define the function $H : (0,2) \to \mathbb{R}$ as follows:

$$H(t,\tau) = 4c_1(t,\tau) + 2[c_2(t,\tau) + c_3(t,\tau)] + c_4(t,\tau) \qquad (10.60)$$

for fixed value of $t$.

Substituting the value $c_j(t,\tau), j = 1,2,3,4$ in the (10.60), we obtain

$$H(t,\tau) = \frac{U_1^2(t)}{(1 + 2\beta)^2} + \frac{\Delta(\beta,t)\tau^4 + 4c(\beta,t)\tau^2}{32(1 + \beta)^4(1 + 2\beta)^2(1 + 3\beta)},$$

where

$$\Delta(\beta,t) = 2U_1(t)\left|(1 + \beta)^3 U_3(t) - (1 + 3\beta)U_1^3(t)\right|(1 + 2\beta)^2 -$$
$$U_1(t)\left[(1 + 3\beta)U_1^2(t) + 4(1 + \beta)(1 + 2\beta)U_2(t)\right](1 + \beta)^2(1 + 2\beta)$$
$$- 2U_1^2(t)\beta^2(1 + \beta)^3,$$

$$c(\beta,t) = U_1(t)\left[U_1^2(t)(1 + 3\beta) + 4U_2(t)(1 + \beta)(1 + 2\beta)\right](1 + \beta)^2(1 + 2\beta) -$$
$$2U_1^2(t)\left[(1 + \beta)^2 + (2 + \beta)\beta\right](1 + \beta)^3.$$

Now, we must investigate the maximum (respect to $\tau$) of the function $H(t,\tau)$ in the interval $(0,2)$ for fixed value of $t$.

By simple computation, we can easily show that

$$H'(t,\tau) - \frac{\Delta(\beta,t)\tau^2 + 2c(\beta,t)}{8(1 + \beta)^4(1 + 2\beta)^2(1 + 3\beta)}\tau.$$

We will examine the sign of the function $H'(t, \tau)$ depending on the different cases of the signs of $\Delta(t, \tau)$ and $c(t, \tau)$ as follows.

($\imath$) Let $\Delta(\beta, t) \geq 0$ and $c(\beta, t) \geq 0$, then $H'(t, \tau) \geq 0$, so $H(t, \tau)$ is an increasing function. Therefore,

$$\max \{H(t, \tau) : \tau \in (0, 2)\} = H(t, 2-) = \frac{U_1(t) \left| (1 + \beta)^3 U_3(t) - (1 + 3\beta) U_1^3(t) \right|}{(1 + \beta)^4 (1 + 3\beta)}. \tag{10.61}$$

That is,

$$\max \{\max \{F(\xi, \eta) : \xi, \eta \in [0, 1]\} : \tau \in (0, 2)\} = H(t, 2-).$$

($\imath\imath$) Let $\Delta(\beta, t) > 0$ and $c(\beta, t) < 0$, then $\tau_0 = \sqrt{\frac{-2c(\beta,t)}{\Delta(\beta,t)}}$ is a critical point of the function $H(t, \tau)$. We assume that $\tau_0 \in (0, 2)$. Since $H''(t, \tau_0) > 0$, $\tau_0$ is a local minimum point of the function $H(t, \tau)$. That is, the function $H(t, \tau)$ cannot have a local maximum.

($\imath\imath\imath$) Let $\Delta(\beta, t) \leq 0$ and $c(\beta, t) \leq 0$, then $H'(t, \tau) \leq 0$. Thus, $H(t, \tau)$ is an decreasing function on the interval $(0, 2)$.

Therefore,

$$\max \{H(t, \tau) : \tau \in (0, 2)\} = H(t, 0+) = 4c_1(t, 0) = \frac{U_1^2(t)}{(1 + 2\beta)^2}. \tag{10.62}$$

($\imath v$) Let $\Delta(\beta, t) < 0$ and $c(\beta, t) > 0$, then $\tau_0$ is a critical point of the function $H(t, \tau)$. We assume that $\tau_0 \in (0, 2)$. Since $H''(t, \tau_0) < 0$, $\tau_0$ is a local maximum point of the function $H(t, \tau)$ and maximum value occurs at $\tau = \tau_0$.

Therefore,

$$\max \{H(t, \tau) : \tau \in (0, 2)\} = H(t, \tau_0) \equiv H(t, \beta), \tag{10.63}$$

where

$$H(t, \beta) = \frac{4t^2}{(1 + 2\beta)^2} - \frac{c^2(\beta, t)}{8(1 + \beta)^4 (1 + 2\beta)^2 (1 + 3\beta) \Delta(\beta, t)}.$$

Thus, from (10.57)-(10.63), the proof of Theorem 10.6.5 is completed. $\square$

In the special cases from Theorem 10.6.5, we arrive at the following results.

**Corollary 10.6.6.** *Let the function $f(z)$ given by (10.1) be in the class $Q_\Sigma(G, 0, t) = N_\Sigma(G, t)$, $t \in \left(\frac{1}{2}, 1\right)$, where the function $G$ is an analytic function given by (10.35). Then,*

$$\left| a_2 a_4 - a_3^2 \right| \leq 8t^2.$$

**Corollary 10.6.7.** *Let the function $f(z)$ given by (10.1) be in the class $Q_\Sigma(G,1,t) = \Re_\Sigma(G,t)$, $t \in \left(\frac{1}{2},1\right)$, where the function $G$ is an analytic function given by (10.35). Then,*

$$
\left|a_2a_4 - a_3^2\right| \leq \begin{cases} t^2\left(1-t^2\right), & \text{if } t \in \left(\frac{1}{2}, t_0\right], \\ \max\left\{H(t), t^2\left(1-t^2\right)\right\}, & \text{if } t \in (t_0, t_1], \\ t^2\left(1-t^2\right), & \text{if } t \in \left(t_1, \frac{\sqrt{5}}{3}\right], \\ \frac{4}{9}t^2, & \text{if } t \in \left(\frac{\sqrt{5}}{3}, 1\right), \end{cases}
$$

*where*

$$
H(t) = \frac{4}{9}t^2 + \frac{\phi(t)}{\sigma(t)}, \phi(t) = 64t\left(42t^2 - 7t - 9\right)^2, \sigma(t) = 18t^3 + 42t^2 + t - 27
$$

*and $t_0 \cong 0.55368$, $t_1 \cong 0.69471$ are the positive numerical roots of the equations $42t^2 - 7t - 9 = 0, 18t^3 + 42t^2 + t - 27 = 0$, respectively.*

---

## 10.7 Fekete-Szegö problem for certain subclass of analytic and bi-univalent functions

Let us $\Sigma$ denote the class of bi-univalent functions in $U$ given 10.1 as in our introduction section 15.1.

Examples of functions in the class $\Sigma$ are

$$
\frac{z}{1-z}, \quad \ln\frac{1}{1-z}, \quad \ln\sqrt{\frac{1+z}{1-z}}.
$$

However, the familiar Koebe function is not a member of $\Sigma$. Other common examples of functions in $A$ such as

$$
\frac{2z - z^2}{2} \quad \text{and} \quad \frac{z}{1-z^2}
$$

are also not members of $\Sigma$.

Earlier, Brannan and Taha Brannan and Taha (1988) introduced certain subclasses of bi-univalent function class $\Sigma$, namely bi-starlike function of order $\alpha$ denoted $S_\Sigma^*(\alpha)$ and bi-convex function of order $\alpha$ denoted $C_\Sigma(\alpha)$ corresponding to the function classes $S^*(\alpha)$ and $C(\alpha)$, respectively. Thus, following Brannan and Taha Brannan and Taha (1988), a function $f \in \Sigma$ is in the classes $S_\Sigma^*(\alpha)$ and $C_\Sigma(\alpha)$, respectively, if each of the following conditions are satisfied:

$$
\Re\left(\frac{zf'(z)}{f(z)}\right) > \alpha, z \in U, \Re\left(\frac{zg'(w)}{g(w)}\right) > \alpha, w \in D
$$

and

$$\Re\left(1 + \frac{zf'(z)}{f(z)}\right) > \alpha, z \in U, \ \Re\left(1 + \frac{zg'(w)}{g(w)}\right) > \alpha, w \in D.$$

For a brief history and interesting examples of functions which are in the class $\Sigma$, together with various other properties of this bi-univalent function class, one can refer the work of Srivastava et al. Srivastava et al. (2010) and references therein. In Srivastava et al. (2010), Srivastava et al. reviewed the study of coefficient problems for bi-univalent functions. Also, various subclasses of bi-univalent function class were introduced and non-sharp estimates on the first two coefficients in the Taylor-Maclaurin series expansion 10.1 were found in several recent investigations (see, for example, Ali et al. (2012); Bulut (2013); Çağlar and Aslan (2016); Çağlar et al. (2013); Frasin and Aouf (2011); Goyal and Goswami (2012); Hayami and Owa (2012); Kumar et al. (2012); Magesh and J. Yamini (2013); Murugusundaramoorthy et al. (2013); Prema and Keerthi (2013); Srivastava (2012); Srivastava et al. (2013a,b); Xu et al. (2012a,b).

An analytic function $f$ is subordinate to an analytic function $\phi$, written $f(z) \prec \phi(z)$, provided there is an analytic function $u : U \to U$ with $u(0) = 0$ and $|u(z)| < 1$ satisfying $f(z) = \phi(u(z))$ (see, for example, Kim and Srivastava (2008)).

Ma and Minda Ma and Minda (1994) unified various subclasses of starlike and convex functions for which either of the quantity $\frac{zf'(z)}{f(z)}$ or $1 + \frac{zf''(z)}{f'(z)}$ is subordinate to a more superordinate function. For this purpose, they considered an analytic function $\phi$ with positive real part in $U$, with $\phi(0) = 1$, $\phi'(0) > 0$ and $\phi$ maps $U$ onto a region starlike with respect to 1 and symmetric with respect to the real axis. The class of Ma-Minda starlike and Ma-Minda convex functions consists of functions $f \in A$ satisfying the subordination $\frac{zf'(z)}{f(z)} \prec \phi(z)$ and $1 + \frac{zf''(z)}{f'(z)} \prec \phi(z)$, respectively. These classes denoted, respectively, by $S^*(\phi)$ and $C(\phi)$.

An analytic function $f \in S$ is said to be bi- starlike of Ma-Minda type or bi- convex of Ma-Minda type if both $f$ and $f^{-1}$ are, respectively, Ma-Minda starlike or Ma-Minda convex functions. These classes are denoted, respectively, by $S^*_\Sigma(\phi)$ and $C_\Sigma(\phi)$. In the sequel, it is assumed that $\phi$ is an analytic function with positive real part in $U$, satisfying $\phi(0) = 1$, $\phi'(0) > 0$ and $\phi(U)$ is starlike with respect to 1 and symmetric with respect to the real axis. Such a function has a series expansion of the following form:

$$\phi(z) = 1 + b_1 z + b_2 z^2 + b_3 z^3 + \cdots, \ b_1 > 0. \tag{10.64}$$

One of the important tools in the theory of analytic functions is the functional $H_2(1) = a_3 - a_2^2$ which is known as the Fekete-Szegö functional and one usually considers the further generalized functional $a_3 - \mu a_2^2$, where $\mu$ is some real number (see Fekete and Szegö (1933)). Estimating for the upper bound of $|a_3 - \mu a_2^2|$ is known as the Fekete-Szegö problem. In 1969, Keogh and Merkes Keogh and Merkes (1969) solved the Fekete-Szegö problem for the class starlike and convex functions. Someone can see the Fekete-Szegö problem for the classes of starlike functions of order $\alpha$ and convex functions of order $\alpha$ at special cases in the paper of Orhan et al. Orhan et al. (2010). On the other hand, recently, Çağlar and Aslan (see Çağlar and Aslan (2016)) have obtained Fekete-Szegö inequality for a subclass of bi-univalent functions. Also, Zaprawa (see Zaprawa (2014a,b)) studied on Fekete-Szegö problem for some subclasses of bi-univalent functions. In special cases, he studied the Fekete-Szegö problem for the subclasses bi-starlike functions of order $\alpha$ and bi-convex functions of order $\alpha$.

Motivated by the aforementioned works, we define a new subclass of bi-univalent functions $\Sigma$ as follows.

**Definition 10.7.1.** A function $f \in \Sigma$ given by 10.1 is said to be in the class $M_\Sigma(\phi, \beta)$, $\beta \geq 0$, where $\phi$ is an analytic function given by (10.64), if the following conditions are satisfied:

$$\left(\frac{zf'(z)}{f(z)}\right)^\beta \left(1 + \frac{zf''(z)}{f'(z)}\right)^{1-\beta} \prec \phi(z), \ z \in U,$$

$$\left(\frac{zy'(w)}{g(w)}\right)^\beta \left(1 + \frac{zg''(w)}{g'(w)}\right)^{1-\rho} \prec \phi(w), \ w \in D,$$

where $g = f^{-1}$.

*Remark* 10.7.2. Choose $\beta = 1$ in Definition 10.7.1 we have $M_\Sigma(\phi, 1) = S_\Sigma^*(\phi)$; that is,

$\frac{zf'(z)}{f(z)} \prec \phi(z), z \in U$ and $\frac{zg'(w)}{g(w)} \prec \phi(w), w \in D$

if and only if $f \in S_\Sigma^*(\phi)$, where $g = f^{-1}$.

*Remark* 10.7.3. Choose $\beta = 0$ in Definition 10.7.1, we have $M_\Sigma(\phi, 0) = C_\Sigma(\phi)$; that is,

$1 + \frac{zf'(z)}{f(z)} \prec \phi(z), z \in U$ and $1 + \frac{zg'(w)}{g(w)} \prec \phi(w), w \in D$

if and only if $f \in K_\Sigma(\phi)$, where $g = f^{-1}$.

*Remark* 10.7.4. These classes $S_\Sigma^*(\phi)$ and $C_\Sigma(\phi)$ was investigated by Ma and Minda Ma and Minda (1994).

To prove our main results, we require the following lemmas.

**Lemma 10.7.5.** *(See, for example, Pommerenke (1975)) If $p \in P$, then the estimates $|p_n| \leq 2, n = 1, 2, 3, \ldots$ are sharp, where P is the family of all functions $p$, analytic in $U$ for which $p(0) = 1$ and $\Re(p(z)) > 0$ ($z \in U$), and*

$$p(z) = 1 + p_1 z + p_2 z^2 + \cdots, \ z \subset U. \tag{10.65}$$

**Lemma 10.7.6.** *(See, for example, Grenander and Szegö (1958)) If the function $p \in P$ is given by the series (10.64), then*

$$2p_2 = p_1^2 + \left(4 - p_1^2\right) x,$$

$$4p_3 - p_1^3 + 2\left(4 - p_1^2\right) p_1 x - \left(4 - p_1^2\right) p_1 x^2 + 2\left(4 - p_1^2\right)\left(1 - |x|^2\right) z$$

*for some $x$ and $z$ with $|x| \leq 1$ and $|z| \leq 1$.*

The object of this section is to find the upper bound estimate for the Fekete-Szegö functional $\left| a_3 - \mu a_2^2 \right|$ for the class $M_\Sigma(\phi, \beta)$, $\beta \in [0, 1]$ and upper bound estimates for the initial coefficients $|a_2|$ and $|a_3|$ of the functions belonging to this class .

## 10.7.1   Fekete-Szegö problem for class $M_\Sigma(\phi, \beta)$

In this section, we prove the following theorem on upper bound of the Fekete-Szegö functional for the functions belonging to the class $M_\Sigma(\phi, \beta)$.

**Theorem 10.7.7.** *Let the function $f(z)$ given by 10.1 be in the class $M_\Sigma(\phi, \beta)$, $\beta \in [0, 1]$, where $\phi$ is an analytic function given by (10.64) and $\mu \in \mathbb{C}$ . Then,*

$$\left| a_3 - \mu a_2^2 \right| \leq \begin{cases} \frac{b_1}{2(3 - 2\beta)}, & \text{if } |1 - \mu| \in [0, \mu_0), \\ \frac{b_1^2}{(2-\beta)^2} |1 - \mu|, & \text{if } |1 - \mu| \in [\mu_0, +\infty), \end{cases}$$

*where $\mu_0 = \frac{(2-\beta)^2}{4b_1(3-2\beta)}$.*

*Proof.* Let $f \in M_\Sigma(\phi, \beta)$, $\beta \in [0, 1]$ and $g = f^{-1}$, $\phi$ be an analytic function given by (10.64). Then, in view of Definition 10.7.1, there are analytic functions $u : U \to U$, $v : D \to D$ with $u(0) = 0 = v(0)$, $|u(z)| < 1$, $|v(w)| < 1$ and satisfying

$$\left( \frac{z f'(z)}{f(z)} \right)^\beta \left( 1 + \frac{z f''(z)}{f'(z)} \right)^{1-\beta} = \phi(u(z)) \text{ and}$$

$$\left( \frac{w g'(w)}{g(w)} \right)^\beta \left( 1 + \frac{w g''(w)}{g'(w)} \right)^{1-\beta} = \phi(v(w)). \tag{10.66}$$

Let us define the functions $p(z)$ and $q(w)$ by
$p(z) := \frac{1+u(z)}{1-u(z)} = 1 + \sum_{n=1}^\infty p_n z^n, z \in U$ and $q(w) := \frac{1+v(w)}{1-v(w)} = 1 + \sum_{n=1}^\infty q_n w^n, w \in D.$

It follows that

$$u(z) = \frac{p(z) - 1}{p(z) + 1} = \frac{1}{2} \left\{ p_1 z + \left[ p_2 - \frac{p_1^2}{2} \right] z^2 + \left[ p_3 - p_1 p_2 + \frac{p_1^3}{4} \right] z^3 + \cdots \right\} \tag{10.67}$$

and

$$v(w) = \frac{q(w) - 1}{q(w) + 1} = \frac{1}{2} \left\{ q_1 w + \left[ q_2 - \frac{q_1^2}{2} \right] w^2 + \left[ q_3 - q_1 q_2 + \frac{q_1^3}{4} \right] w^3 + \cdots \right\}. \tag{10.68}$$

Using (10.67) and (10.68) in (10.64), we can easily write

$$\phi(u(z)) = 1 + \frac{b_1 p_1}{2} z + \left[ \frac{b_1}{2} \left( p_2 - \frac{p_1^2}{2} \right) + \frac{1}{4} b_2 p_1^2 \right] z^2$$
$$+ \left[ \frac{b_1}{2} \left( p_3 - p_1 p_2 + \frac{p_1^3}{4} \right) + \frac{b_2 p_1}{2} \left( p_2 - \frac{p_1^2}{2} \right) + \frac{b_3 p_1^3}{8} \right] z^3 + \cdots \tag{10.69}$$

and

$$\phi(v(w)) = 1 + \tfrac{b_1 q_1}{2} w + \left[ \tfrac{b_1}{2}\left(q_2 - \tfrac{q_1^2}{2}\right) + \tfrac{1}{4} b_2 q_1^2 \right] w^2$$
$$+ \left[ \tfrac{b_1}{2}\left(q_3 - q_1 q_2 + \tfrac{q_1^3}{4}\right) + \tfrac{b_2 q_1}{2}\left(q_2 - \tfrac{q_1^2}{2}\right) + \tfrac{b_3 q_1^3}{8} \right] w^3 + \cdots. \tag{10.70}$$

Also, using (10.69) and (10.70) in (10.66) and equating the coefficients, we get

$$(2 - \beta) a_2 = \frac{b_1 p_1}{2}, \tag{10.71}$$

$$2(3 - 2\beta) a_3 + \frac{1}{2}\left[(\beta - 2)^2 - 3(4 - 3\beta)\right] a_2^2 = \frac{b_1}{2}\left(p_2 - \frac{p_1^2}{2}\right) + \frac{1}{4} b_2 p_1^2, \tag{10.72}$$

and

$$-(2 - \beta) a_2 = \frac{b_1 q_1}{2}, \tag{10.73}$$

$$-2(3 - 2\beta) a_3 + \frac{1}{2}\left(\beta^2 - 11\beta + 16\right) a_2^2 = \frac{b_1}{2}\left(q_2 - \frac{q_1^2}{2}\right) + \frac{1}{4} b_2 q_1^2, \tag{10.74}$$

From (10.71) and (10.73), we have

$$a_2 = \frac{b_1 p_1}{2(2 - \beta)} = \frac{-b_1 q_1}{2(2 - \beta)}, \tag{10.75}$$

it follows that

$$p_1 = -q_1. \tag{10.76}$$

By subtracting from (10.72) to (10.74) and next considering (10.75) and (10.76), we can easily obtain

$$a_3 = a_2^2 + \frac{b_1(p_2 - q_2)}{8(3 - 2\beta)} = \frac{b_1^2 p_1^2}{4(2 - \beta)^2} + \frac{b_1(p_2 - q_2)}{8(3 - 2\beta)}. \tag{10.77}$$

From (10.77) and (10.75), we find that

$$a_3 - \mu a_2^2 = (1 - \mu) a_2^2 + \frac{b_1(p_2 - q_2)}{8(3 - 2\beta)}. \tag{10.78}$$

Since $p_1 = -q_1$, according to Lemma 10.7.5,

$$p_2 - q_2 = \frac{4 - p_1^2}{2}(x - y) \tag{10.79}$$

for some $x, y$ with $|x| \leq 1$, $|y| \leq 1$ and $p_1 \in [0, 2]$.

In this case, since $p_1 \in [0, 2]$, we may assume without any restriction that $t \in [0, 2]$, where $t = |p_1|$.

Substituting the expression (10.79) in (10.78) and using triangle inequality, taking $|x| = \xi$, $|y| = \eta$, we can easily obtain that

$$\left|a_3 - \mu a_2^2\right| \leq d_1(t) + d_2(t)(\xi + \eta) = F(\xi, \eta), \tag{10.80}$$

where $d_1(t) = |1 - \mu| \frac{b_1^2 t^2}{4(2-\beta)^2} \geq 0$ and $d_2(t) = \frac{b_1(4-t^2)}{16(3-2\beta)} \geq 0$.

From (10.80), we can write

$$\left| a_3 - \mu a_2^2 \right| \leq \max \left\{ \max \left\{ F(\xi, \eta) : (\xi, \eta) \in \Omega \right\} : t \in [0, 2] \right\}, \tag{10.81}$$

where $\Omega = \left\{ (\xi, \eta) \in \mathbb{R}^2 : \xi, \eta \in [0, 1] \right\}$.

It is clear that

$$\max \left\{ F(\xi, \eta) : (\xi, \eta) \in \Omega \right\} = F(1, 1) = d_1(t) + 2d_2(t) = c_1(\phi, \beta, \mu) t^2 + c_2(\phi, \beta),$$

where

$$c_1(\phi, \beta, \mu) = \frac{b_1^2}{4(2-\beta)^2} \left[ |1 - \mu| - \frac{(2-\beta)^2}{4b_1(3-2\beta)} \right], c_2(\phi, \beta) = \frac{b_1}{4(3-2\beta)}.$$

Let us define the function $G : \mathbb{R} \to \mathbb{R}$ as follows

$$G(t) = c_1(\phi, \beta, \mu) t^2 + c_2(\phi, \beta), \; t \in [0, 2]. \tag{10.82}$$

Differentiating both sides of (10.82), we have

$$G'(t) = 2c_1(\phi, \beta, \mu) t.$$

It is clear that $G'(t) < 0$ if $|1 - \mu| \in \left[ 0, \frac{(2-\beta)^2}{4b_1(3-2\beta)} \right)$. Thus, the function $G(t)$ is a strictly decreasing function if $|1 - \mu| \in [0, \mu_0)$, where $\mu_0 = \frac{(2-\beta)^2}{4b_1(3-2\beta)}$.

Therefore,

$$\max \left\{ G(t) : t \in [0, 2] \right\} = G(0) = 2d_2(0) = \frac{b_1}{2(3-2\beta)}. \tag{10.83}$$

Also, $G'(t) \geq 0$; that is, the function $G(t)$ is an increasing function for $|1 - \mu| \geq \mu_0$.

Therefore,

$$\max \left\{ G(t) : t \in [0, 2] \right\} = G(2) = d_1(2) = \frac{|1 - \mu| b_1^2}{(2-\beta)^2}. \tag{10.84}$$

Thus, from (10.80)-(10.84), we obtain that

$$\left| a_3 - \mu a_2^2 \right| \leq \begin{cases} \frac{b_1}{2(3-2\beta)}, & \text{if } |1 - \mu| \in [0, \mu_0), \\ \frac{b_1^2}{(2-\beta)^2} |1 - \mu|, & \text{if } |1 - \mu| \in [\mu_0, +\infty), \end{cases}$$

where $\mu_0 = \frac{(2-\beta)^2}{4b_1(3-2\beta)}$.

Thus, the proof of Theorem 10.7.7 is completed. □

In the special cases from Theorem 10.7.7, we arrive at the following results.

**Corollary 10.7.8.** *Let the function $f(z)$ given by 10.1 be in the class $S_\Sigma^*(\phi)$, where $\phi$ is an analytic function given by (10.64) and $\mu \in \mathbb{C}$ . Then,*

$$\left| a_3 - \mu a_2^2 \right| \leq \begin{cases} \frac{b_1}{2}, & \text{if } |1 - \mu| \in \left[0, \frac{1}{4b_1}\right), \\ |1 - \mu| b_1^2, & \text{if } |1 - \mu| \in \left[\frac{1}{4b_1}, +\infty\right). \end{cases}$$

**Corollary 10.7.9.** *Let the function $f(z)$ given by 10.1 be in the class $C_\Sigma(\phi)$, where $\phi$ is an analytic function given by (10.64) and $\mu \in \mathbb{C}$ . Then,*

$$\left| a_3 - \mu a_2^2 \right| \leq \begin{cases} \frac{b_1}{6}, & \text{if } |1 - \mu| \in \left[0, \frac{1}{3b_1}\right), \\ \frac{|1-\mu| b_1^2}{4}, & \text{if } |1 - \mu| \in \left[\frac{1}{3b_1}, +\infty\right). \end{cases}$$

The following theorem is direct result of Theorem 10.7.7.

**Theorem 10.7.10.** *Let the function $f(z)$ given by 10.1 be in the class $M_\Sigma(\phi, \beta)$, $\beta \in [0, 1]$. Then,*

$$|a_2| \leq \frac{b_1}{2 - \beta} \text{ and } |a_3| \leq \begin{cases} \frac{b_1}{2(3 - 2\beta)}, & \text{if } b_1 < \frac{(2-\beta)^2}{4(3-2\beta)}, \\ \frac{b_1^2}{(2-\beta)^2}, & \text{if } b_1 \geq \frac{(2-\beta)^2}{4(3-2\beta)}. \end{cases}$$

*Moreover,*

$$\left| a_3 - a_2^2 \right| \leq \frac{b_1}{2(3 - 2\beta)}.$$

*Proof.* Let us $f \in M_\Sigma(\phi, \beta)$, $\beta \in [0, 1]$ and $g = f^{-1}$, $\phi$ is an analytic function given by (10.64). Then, from (10.75) the first estimate of theorem is clear. The second and third estimates are direct results of Theorem 10.7.7 for $\mu = 0$ and $\mu = 1$, respectively . $\qquad \square$

In the special cases from Theorem 10.7.10 , we arrive at the following results.

**Corollary 10.7.11.** *Let the function $f(z)$ given by Equation 10.1 be in the class $S_\Sigma^*(\phi)$, where $\phi$ is an analytic function given by (10.64). Then,*

$$|a_2| \leq b_1 \text{ and } |a_3| \leq \begin{cases} \frac{b_1}{2}, & \text{if } b_1 < \frac{1}{4}, \\ b_1^2, & \text{if } b_1 \geq \frac{1}{4}. \end{cases}$$

*Moreover,*

$$\left| a_3 - a_2^2 \right| \leq \frac{b_1}{2}.$$

**Corollary 10.7.12.** *Let the function $f(z)$ given by Equation 10.1 be in the class $C_\Sigma(\phi)$, where $\phi$ is an analytic function given by (10.64). Then,*

$$|a_2| \leq \frac{b_1}{2} \text{ and } |a_3| \leq \begin{cases} \frac{b_1}{6}, & \text{if } b_1 < \frac{1}{3}, \\ \frac{b_1^2}{4}, & \text{if } b_1 \geq \frac{1}{3}. \end{cases}$$

## 10.7.2   Concluding remarks

If the function $\phi(z)$, aforementioned in study, is given by

$$\phi(z) := \frac{1+az}{1+bz} = 1+(a-b)z-b(a-b)z^2+b^2(a-b)z^3+\cdots \quad (-1\le b < a \le 1),$$
(10.85)

then $b_1 = (a-b)$, $b_2 = -b(a-b)$ and $b_3 = b^2(a-b)$.

Taking $a = 1-2\alpha$, $b = -1$ in (10.85), we have

$$\phi(z) = \frac{1+(1-2\alpha)z}{1-z} = 1+2(1-\alpha)z+2(1-\alpha)z^2+2(1-\alpha)z^3+\cdots \quad (0 \le \alpha < 1).$$
(10.86)

Hence, $b_1 = b_2 = b_3 = 2(1-\alpha)$.

Choosing $\phi(z)$ of the form (10.85) and (10.86) in Theorem 2.1, we can readily deduce the following results, respectively.

**Corollary 10.7.13.** *Let the function $f(z)$ given by Equation 10.1 be in the class $M_\Sigma\left(\frac{1+az}{1+bz}, \beta\right)$, $-1 \le b < a \le 1, 0 \le \beta \le 1$ and $\mu \in \mathbb{C}$. Then,*

$$\left|a_3 - \mu a_2^2\right| \le \begin{cases} \frac{a-b}{2(3-2\beta)}, & \text{if } |1-\mu| \in [0, \mu_0), \\ \frac{|1-\mu|(a-b)^2}{(2-\beta)^2}, & \text{if } |1-\mu| \in [\mu_0, +\infty), \end{cases}$$

*where $\mu_0 = \frac{(2-\beta)^2}{4(a-b)(3-2\beta)}$.*

**Corollary 10.7.14.** *Let the function $f(z)$ given by Equation 10.1 be in the class $M_\Sigma\left(\frac{1+(1-2\alpha)z}{1-z}, \beta\right) = M_\Sigma(\alpha, \beta), 0 \le \alpha < 1, 0 \le \beta \le 1$ and $\mu \in \mathbb{C}$. Then,*

$$\left|a_3 - \mu a_2^2\right| \le \begin{cases} \frac{1-\alpha}{3-2\beta}, & \text{if } |1-\mu| \in [0, \mu_0), \\ \frac{4|1-\mu|(1-\alpha)^2}{(2-\beta)^2}, & \text{if } |1-\mu| \in [\mu_0, +\infty), \end{cases}$$

*where $\mu_0 = \frac{(2-\beta)^2}{8(1-\alpha)(3-2\beta)}$*

*Also, taking $\alpha = 0$ in (10.86), we get*

$$\phi(z) = \frac{1+z}{1-z} = 1 + 2z + 2z^2 + 2z^3 + \cdots .$$
(10.87)

*Hence, $b_1 = b_2 = b_3 = 2$.*

Choosing $\phi(z)$ of the form (10.87) in Theorem 10.7.7, we arrive at the following corollary.

**Corollary 10.7.15.** *Let the function $f(z)$ given by Equation 10.1 be in the class $M_\Sigma\left(\frac{1+z}{1-z}, \beta\right), 0 \le \beta \le 1$ and $\mu \in \mathbb{C}$. Then,*

$$\left|a_3 - \mu a_2^2\right| \le \begin{cases} \frac{1}{3-2\beta}, & \text{if } |1-\mu| \in [0, \mu_0), \\ \frac{4|1-\mu|}{(2-\beta)^2}, & \text{if } |1-\mu| \in [\mu_0, +\infty), \end{cases}$$

*where $\mu_0 = \frac{(2-\beta)^2}{8(3-2\beta)}$.*

Choosing $\phi(z)$ of the form (10.85) and (10.86) in Theorem 10.7.7, we can readily deduce the following results, respectively.

**Corollary 10.7.16.** *Let the function* $f(z)$ *given by Equation 10.1 be in the class* $S_\Sigma^* \left( \frac{1+az}{1+bz} \right)$, $-1 \le b < a \le 1$. *Then,*

$$|a_2| \le \frac{a-b}{2-\beta} \text{ and } |a_3| \le \begin{cases} \frac{a-b}{2(3-2\beta)}, & \text{if } a < b + b_0, \\ \frac{(a-b)^2}{(2-\beta)^2}, & \text{if } a \ge b + b_0, \end{cases}$$

*where* $b_0 = \frac{(2-\beta)^2}{4(3-2\beta)}$.
*Moreover,*

$$\left| a_3 - a_2^2 \right| \le \frac{a-b}{2(3-2\beta)}.$$

**Corollary 10.7.17.** *Let the function* $f(z)$ *given by Equation 10.1 be in the class* $S_\Sigma^* \left( \frac{1+(1-2\alpha)z}{1-z} \right) = S_\Sigma^*(\alpha)$, $0 \le \alpha < 1$. *Then,*

$$|a_2| \le \frac{2(1-\alpha)}{2-\beta} \text{ and } |a_3| \le \begin{cases} \frac{1-\alpha}{3-2\beta}, & \text{if } \alpha \le 1 - \alpha_0, \\ \frac{4(1-\alpha)^2}{(2-\beta)^2}, & \text{if } \alpha > 1 - \alpha_0, \end{cases}$$

*where* $\alpha_0 = \frac{(2-\beta)^2}{8(3-2\beta)}$.
*Moreover,*

$$\left| a_3 - a_2^2 \right| \le \frac{1-\alpha}{3-2\beta}.$$

## 10.8 Some starlikeness and convexity properties for two new $p$-valent integral operators

Let $\mathcal{A}_p$ denote the class of the form

$$f(z) = z^p + \sum_{k=p+1}^{\infty} a_k z^k, \quad (p \in \mathbb{N} = \{1, 2, ..., \}), \tag{10.88}$$

which are analytic in the open disc $\mathbb{U} = \{z \in \mathbb{C} : |z| < 1\}$.

A function $f \in S_p^*(\gamma, \alpha)$ is $p$-valently starlike of complex order $\gamma$ $(\gamma \in \mathbb{C} - \{0\})$ and type $\alpha$ $(0 \le \alpha < p)$, that is, $f \in S_p^*(\gamma, \alpha)$, if it satisfies the following condition;

$$\Re \left\{ p + \frac{1}{\gamma} \left( \frac{z f'(z)}{f(z)} - p \right) \right\} > \alpha, \quad (z \in \mathbb{U}). \tag{10.89}$$

Furthermore, a function $f \in C_p(\gamma, \alpha)$ is $p$-valently convex of complex order $\gamma$ $(\gamma \in \mathbb{C} - \{0\})$ and type $\alpha$ $(0 \le \alpha < p)$, that is, $f \in C_p(\gamma, \alpha)$, if it satisfies the following condition;

$$\Re \left\{ p + \frac{1}{\gamma} \left( 1 + \frac{z f''(z)}{f'(z)} - p \right) \right\} > \alpha, \quad (z \in \mathbb{U}). \tag{10.90}$$

In particular cases, for $p = 1$ in the classes $\mathcal{S}_p^*(\gamma, \alpha)$ and $\mathcal{C}_p(\gamma, \alpha)$, we obtain the classes $\mathcal{S}^*(\gamma, \alpha)$ and $\mathcal{C}(\gamma, \alpha)$ of starlike functions of complex order $\gamma \, (\gamma \in \mathbb{C} - \{0\})$ and type $\alpha \, (0 \leq \alpha < p)$ and convex functions of complex order $\gamma \, (\gamma \in \mathbb{C} - \{0\})$ and type $\alpha \, (0 \leq \alpha < 1)$, respectively, which were introduced and studied by Frasin Frasin (2006). Also, for $\alpha = 0$ in the classes $\mathcal{S}_p^*(\gamma, \alpha)$ and $\mathcal{C}_p(\gamma, \alpha)$, we obtain the classes $\mathcal{S}_p^*(\gamma)$ and $\mathcal{C}_p(\gamma)$, which are called $p-$valently starlike of complex order $\gamma \, (\gamma \in \mathbb{C} - \{0\})$ and $p-$valently convex of complex order $\gamma \, (\gamma \in \mathbb{C} - \{0\})$, respectively. Setting $p = 1$ and $\alpha = 0$, we obtain the classess $\mathcal{S}^*(\gamma)$ and $\mathcal{C}(\gamma)$. The class $\mathcal{S}^*(\gamma)$ of starlike functions of complex order $\gamma \, (\gamma \in \mathbb{C} - \{0\})$ was defined by Nasr and Aouf (see Nasr and Aouf (1985)) while the class $\mathcal{C}(\gamma)$ of convex functions of complex order $\gamma \, (\gamma \in \mathbb{C} - \{0\})$ was considered earlier by Wiatrowski (see Wiatrowski (1971)). Note that $\mathcal{S}_p^*(1, \alpha) = \mathcal{S}_p^*(\alpha)$ and $\mathcal{C}_p(1, \alpha) = \mathcal{C}_p(\alpha)$ are, respectively, the classes of $p-$valently starlike and $p-$valently convex functions of order $\alpha \, (0 \leq \alpha < p)$ in $\mathbb{U}$. In special cases, $\mathcal{S}_p^*(0) = \mathcal{S}_p^*$ and $\mathcal{C}_p(0) = \mathcal{C}_p$ are, respectively, the familiar classes of $p-$valently starlike and $p-$valently convex functions in $\mathbb{U}$. Also, we note that $\mathcal{S}_1^*(\alpha) = \mathcal{S}^*(\alpha)$ and $\mathcal{C}_1(\alpha) = \mathcal{C}(\alpha)$ are, respectively, the usual classes of starlike and convex functions of order $\alpha \, (0 \leq \alpha < 1)$ in $\mathbb{U}$. In special cases, $\mathcal{S}_1^*(0) = \mathcal{S}^*$ and $\mathcal{C}_1 = \mathcal{C}$ are, respectively, the familiar classes of starlike and convex functions in $\mathbb{U}$.

A function $f \in \beta - \mathcal{US}_p(\alpha)$ is $\beta-$uniformly $p-$valently starlike of order $\alpha$ $(0 \leq \alpha < p)$, that is, $f \in \beta - \mathcal{US}_p(\alpha)$, if it satisfies the following condition;

$$\Re \left\{ \frac{zf'(z)}{f(z)} \right\} > \beta \left| \frac{zf'(z)}{f(z)} - p \right| + \alpha, \quad (\beta \geq 0, \ z \in \mathbb{U}). \tag{10.91}$$

Furthermore, a function $f \in \beta - \mathcal{UC}_p(\alpha)$ is $\beta-$uniformly $p-$valently convex of order $\alpha$ $(0 \leq \alpha < p)$, that is, $f \in \beta - \mathcal{UC}_p(\alpha)$, if it satisfies the following condition;

$$\Re \left\{ 1 + \frac{zf''(z)}{f'(z)} \right\} > \beta \left| 1 + \frac{zf''(z)}{f'(z)} - p \right| + \alpha, \quad (\beta \geq 0, \ z \in \mathbb{U}). \tag{10.92}$$

These classes generalize various other classes which are worthy to mention here. For example $p = 1$, the classes $\beta - \mathcal{US}(\alpha)$ and $\beta - \mathcal{UC}(\alpha)$ introduced by Bharti, Parvatham and Swaminathan (see Bharati et al. (1997)). Also, the class $\beta - \mathcal{UC}_1(0) = \beta - \mathcal{UCV}$ is the known class of $\beta-$*uniformly convex* functions Kanas and Wisniowska (1999). Using the Alexander type relation, we can obtain the class $\beta - \mathcal{US}_p(\alpha)$ in the following way:

$$f \in \beta - \mathcal{UC}_p(\alpha) \Leftrightarrow \frac{zf'}{p} \in \beta - \mathcal{US}_p(\alpha).$$

The class $1 - \mathcal{UC}_1(0) = \mathcal{UCV}$ of uniformly convex functions was defined by Goodman Goodman (1991) while the class $1 - \mathcal{US}_1(0) = \mathcal{SP}$ was considered by Rønning Rønning (1991).

For $f \in \mathcal{A}_p$ given by (10.88) and $g(z)$ given by

$$g(z) = z^p + \sum_{k=p+1}^{\infty} b_k z^k \qquad (10.93)$$

their convolution (or Hadamard product), denoted by $(f * g)$, is defined as

$$(f * g)(z) = z^p + \sum_{k=p+1}^{\infty} a_k b_k z^k = (g * f)(z), \qquad (z \in \mathbb{U}).$$

For a function $f$ in $\mathcal{A}_p$, in Deniz and Orhan (2011), the authors defined the *multiplier transformations* $\mathcal{D}_{p,\lambda,\mu}^m$ as follows:

**Definition 10.8.1.** Let $f \in \mathcal{A}_p$. For the parameters $\lambda, \mu \in \mathbb{R}$; $0 \le \mu \le \lambda$ and $m \in \mathbb{N}_0 = \mathbb{N} \cup \{0\}$, define the multiplier transformations $\mathcal{D}_{p,\lambda,\mu}^m$ on $\mathcal{A}_p$ by the following:

$$\mathcal{D}_{p,\lambda,\mu}^0 f(z) = f(z)$$
$$\mathcal{D}_{p,\lambda,\mu}^1 f(z) = \mathcal{D}_{p,\lambda,\mu} f(z)$$
$$= \frac{1}{p} \left[ \lambda \mu z^2 f''(z) + (\lambda - \mu + (1-p)\lambda\mu) \, z f'(z) + p(1 - \lambda + \mu) f(z) \right]$$
$$\vdots$$
$$\mathcal{D}_{p,\lambda,\mu}^m f(z) = \mathcal{D}_{p,\lambda,\mu} \left( \mathcal{D}_{p,\lambda,\mu}^{m-1} \right)$$

for $z \in \mathbb{U}$ and $p \in \mathbb{N} := \{1, 2, ...\}$.

If $f(z)$ is given by (10.88), then from the definition of the multiplier transformations $\mathcal{D}_{p,\lambda,\mu}^m f(z)$, we can easily see that

$$\mathcal{D}_{p,\lambda,\mu}^m f(z) = z^p + \sum_{k=p+1}^{\infty} \Phi_p^k(m, \lambda, \mu) a_k z^k$$

where

$$\Phi_p^k(m, \lambda, \mu) = \left[ \frac{(k-p)(\lambda\mu k + \lambda - \mu) + p}{p} \right]^m.$$

By using the operator $\mathcal{D}_{p,\lambda,\mu}^m f(z)$ $(m \in \mathbb{N}_0)$, we introduce the new classes $\beta - \mathcal{US}_p(m, \lambda, \mu, \gamma, \alpha)$ and $\beta - \mathcal{UC}_p(m, \lambda, \mu, \gamma, \alpha)$ as follows:

$$\beta - \mathcal{US}_p(m, \lambda, \mu, \gamma, \alpha) = \left\{ f \in \mathcal{A}_p : \, \mathcal{D}_{p,\lambda,\mu}^m f(z) \in \beta - \mathcal{US}_p(\alpha) \right\}$$

and

$$\beta - \mathcal{UC}_p(m, \lambda, \mu, \gamma, \alpha) = \left\{ f \in \mathcal{A}_p : \, \mathcal{D}_{p,\lambda,\mu}^m f(z) \in \beta - \mathcal{UC}_p(\alpha) \right\}$$

where $f \in \mathcal{A}_p$, $0 \leq \alpha < p$, $\beta \geq 0$ and $\gamma \in \mathbb{C} - \{0\}$.

We note that by specializing the parameters $m$, $p$, $\gamma$, $\beta$ and $\alpha$ in the classes $\beta - \mathcal{US}_p(m, \lambda, \mu, \gamma, \alpha)$ and $\beta - \mathcal{UC}_p(m, \lambda, \mu, \gamma, \alpha)$, these classes are reduced to several well-known subclasses of analytic functions. For example, for $m = 0$ the classes $\beta - \mathcal{US}_p(m, \lambda, \mu, \gamma, \alpha)$ and $\beta - \mathcal{UC}_p(m, \lambda, \mu, \gamma, \alpha)$ are reduced to the classes $\beta - \mathcal{US}_p(\gamma, \alpha)$ and $\beta - \mathcal{UC}_p(\gamma, \alpha)$, respectively. Someone can find more information about these classes in Deniz, Orhan and Sokol Deniz et al. (2015b), Deniz, Çağlar and Orhan Deniz et al. (2011) and Orhan, Deniz and Raducanu Orhan et al. (2010).

**Definition 10.8.2.** Let $l = (l_1, l_2, ..., l_n) \in \mathbb{N}_0^n$, $\delta = (\delta_1, \delta_2, ..., \delta_n) \in \mathbb{R}_+^n$ for all $i = \overline{1, n}$, $n \in \mathbb{N}$. We define the following general integral operators

$$\mathcal{I}_{n,p,l}^{\delta,\lambda,\mu}(f_1, f_2, ..., f_n) : \mathcal{A}_p^n \to \mathcal{A}_p$$

$$\mathcal{I}_{n,p,l}^{\delta,\lambda,\mu}(f_1, f_2, ..., f_n) = \mathcal{F}_{n,p,l}^{\delta,\lambda,\mu}(z),$$

$$\mathcal{F}_{n,p,l}^{\delta,\lambda,\mu}(z) = \int_0^z pt^{p-1} \prod_{i=1}^n \left( \frac{\mathcal{D}_{p,\lambda,\mu}^{l_i} f_i(t)}{t^p} \right)^{\delta_i} dt \qquad (10.94)$$

and

$$\mathcal{J}_{n,p,l}^{\delta,\lambda,\mu}(g_1, g_2, ..., g_n) : \mathcal{A}_p^n \to \mathcal{A}_p$$

$$\mathcal{J}_{n,p,l}^{\delta,\lambda,\mu}(g_1, g_2, ..., g_n) = \mathcal{G}_{n,p,l}^{\delta,\lambda,\mu}(z),$$

$$\mathcal{G}_{n,p,l}^{\delta,\lambda,\mu}(z) = \int_0^z pt^{p-1} \prod_{i=1}^n \left( \frac{\left( \mathcal{D}_{p,\lambda,\mu}^{l_i} g_i(t) \right)'}{pt^{p-1}} \right)^{\delta_i} dt \qquad (10.95)$$

where $f_i, g_i \in \mathcal{A}_p$ for all $i = \overline{1, n}$ and $\mathcal{D}_{p,\lambda,\mu}^l$ is defined in Definition 10.8.1.

*Remark* 10.8.3. We note that if $l_1 = l_2 = ... = l_n = 0$, then the integral operator $\mathcal{F}_{n,p,l}^{\delta,\lambda,\mu}(z)$ is reduced to the operator $F_p(z)$ which was studied by Frasin (see Frasin (2010)). Upon setting $p = 1$ in the operator (10.94), we can obtain the integral operator $\mathbb{F}_n(z)$ which was studied by Oros G.I. and Oros G.A. (see Oros and Oros (2010)). For $p = 1$ and $l_1 = l_2 = ... = l_n = 0$ in (10.94), the integral operator $\mathcal{F}_{n,p,l}^{\delta,\lambda,\mu}(z)$ is reduced to the operator $F_m(z)$ which was studied by Breaz D. and Breaz N. (see Breaz and Breaz (2002)). Observe that when $p = n = 1$, $l_1 = 0$ and $\delta_1 = \delta$, we obtain the integral operator $I_\delta(f)(z)$ which was studied by Pescar and Owa (see Pescar and Owa (2000)), for $\delta_1 = \delta \in [0, 1]$ special case of the operator $I_\delta(f)(z)$ was studied by Miller, Mocanu and Reade (see Miller et al. (1978)). For $p = n = 1$, $l_1 = 0$ and $\delta_1 = 1$ in (10.94), we have Alexander integral operator $I(f)(z)$ in Alexander (1915).

*Remark* 10.8.4. For $l_1 = l_2 = ... = l_n = 0$ in (10.95) the integral operator $\mathcal{G}_{n,p,l}^{\delta,\lambda,\mu}(z)$ is reduced to the operator $G_p(z)$ which was studied by Frasin (see Frasin (2010)). For $p = 1$ and $l_1 = l_2 = ... = l_n = 0$ in (10.95), the integral operator $\mathcal{G}_{n,p,l}^{\delta,\lambda,\mu}(z)$ is reduced to the operator $G_{\delta_1,\delta_2,...,\delta_m}(z)$ which was studied by Breaz D., Owa and Breaz N. (see Breaz et al. (2008)). If $p = n = 1$, $l_1 = 0$ and $\delta_1 = \delta$, we obtain the integral operator $G(z)$ which was introduced and studied by Pfaltzgraff (see Pfaltzgraff (1975)) and Kim and Merkes (see Kim and Merkes (1972)).

In this section, we consider the integral operators $\mathcal{F}_{n,p,l}^{\delta,\lambda,\mu}(z)$ and $\mathcal{G}_{n,p,l}^{\delta,\lambda,\mu}(z)$ defined by (10.94) and (10.95), respectively, and study their properties on the classes $\beta - \mathcal{US}_p(m, \lambda, \mu, \gamma, \alpha)$ and $\beta - \mathcal{UC}_p(m, \lambda, \mu, \gamma, \alpha)$. As special cases, the order of $p-$valently convexity and $p-$valently starlikeness of the operators $\int_0^z pt^{p-1}\left(\frac{f(t)}{t^p}\right)^\delta dt$ and $\int_0^z pt^{p-1}\left(\frac{g't)}{pt^{p-1}}\right)^\delta dt$ are given.

---

## 10.9 Convexity of integral operators $\mathcal{F}_{n,p,l}^{\delta,\lambda,\mu}(z)$ and $\mathcal{G}_{n,p,l}^{\delta,\lambda,\mu}(z)$

First, in this section we prove a sufficient condition for the integral operator $\mathcal{F}_{p,m,l,\mu}(z)$ to be $p-$valently convex of complex order.

**Theorem 10.9.1.** *Let* $l = (l_1, l_2, ...l_n) \in \mathbb{N}_0^n$, $\delta = (\delta_1, \delta_2, ..., \delta_n) \in \mathbb{R}_+^n$, $0 \leq \alpha_i < p$, $\gamma \in \mathbb{C} - \{0\}$ *such that* $0 < \sum_{i=1}^n \delta_i (p - \alpha_i) \leq p$, $\beta_i \geq 0$ *and* $f_i \in \beta_i - \mathcal{US}_p(l_i, \lambda, \mu, \gamma, \alpha_i)$ *for all* $i = \overline{1, n}$. *Then, the integral operator* $\mathcal{F}_{n,p,l}^{\delta,\lambda,\mu}$ *defined by (10.94) is* $p-$*valently convex of complex order* $\gamma$ ($\gamma \in \mathbb{C} - \{0\}$) *and type* $p - \sum_{i=1}^n \delta_i (p - \alpha_i)$, *that is,* $\mathcal{F}_{n,p,l}^{\delta,\lambda,\mu} \in \mathcal{C}_p(\gamma, p - \sum_{i=1}^n \delta_i (p - \alpha_i))$.

*Proof.* From the definition (10.94), we observe that $\mathcal{F}_{n,p,l}^{\delta,\lambda,\mu}(z) \in \mathcal{A}_p$. On the other hand, it is easy to see that

$$\left[\mathcal{F}_{n,p,l}^{\delta,\lambda,\mu}(z)\right]' = pz^{p-1} \prod_{i=1}^n \left(\frac{\mathcal{D}_{p,\lambda,\mu}^{l_i} f_i(z)}{z^p}\right)^{\delta_i}. \tag{10.96}$$

Now we differentiate (10.96) logarithmically and we easily obtain

$$p + \frac{1}{\gamma}\left(\frac{z\left[\mathcal{F}_{n,p,l}^{\delta,\lambda,\mu}(z)\right]''}{\left[\mathcal{F}_{n,p,l}^{\delta,\lambda,\mu}(z)\right]'} + 1 - p\right) = p + \sum_{i=1}^n \delta_i \left(p + \frac{1}{\gamma}\left(\frac{z\left(\mathcal{D}_{p,\lambda,\mu}^{l_i} f_i\right)'(z)}{\left(\mathcal{D}_{p,\lambda,\mu}^{l_i} f_i\right)(z)} - p\right)\right)$$
$$- p\sum_{i=1}^n \delta_i. \tag{10.97}$$

Then, we calculate the real part of both sides of (10.97) and obtain

$$
\Re\left\{p+\frac{1}{\gamma}\left(\frac{z\left[\mathcal{F}_{n,p,l}^{\delta,\lambda,\mu}(z)\right]''}{\left[\mathcal{F}_{n,p,l}^{\delta,\lambda,\mu}(z)\right]'}+1-p\right)\right\}
\tag{10.98}
$$
$$
=\sum_{i=1}^{n}\delta_{i}\Re\left\{p+\frac{1}{\gamma}\left(\frac{z\left(\mathcal{D}_{p,\lambda,\mu}^{l_i}f_i\right)'(z)}{\left(\mathcal{D}_{p,\lambda,\mu}^{l_i}f_i\right)(z)}-p\right)\right\}-p\sum_{i=1}^{n}\delta_{i}+p.
$$

Since $f_i \in \beta_i - \mathcal{US}_p(l_i,\lambda,\mu,\gamma,\alpha_i))$ for all $i=\overline{1,n}$ from (10.98), we have

$$
\Re\left\{p+\frac{1}{\gamma}\left(\frac{z\left[\mathcal{F}_{n,p,l}^{\delta,\lambda,\mu}(z)\right]''}{\left[\mathcal{F}_{n,p,l}^{\delta,\lambda,\mu}(z)\right]'}+1-p\right)\right\}
\tag{10.99}
$$
$$
>\sum_{i=1}^{n}\frac{\delta_i\beta_i}{|\gamma|}\left|\frac{z\left(\mathcal{D}_{p,\lambda,\mu}^{l_i}f_i\right)'(z)}{\left(\mathcal{D}_{p,\lambda,\mu}^{l_i}f_i\right)(z)}-p\right|+p-\sum_{i=1}^{n}\delta_i(p-\alpha_i).
$$

Because $\sum_{i=1}^{n}\frac{\delta_i\beta_i}{|\gamma|}\left|\frac{z\left(\mathcal{D}_{p,\lambda,\mu}^{l_i}f_i\right)'(z)}{\left(\mathcal{D}_{p,\lambda,\mu}^{l_i}f_i\right)(z)}-p\right|>0$, from (10.99), we obtain

$$
\Re\left\{p+\frac{1}{\gamma}\left(\frac{z\left[\mathcal{F}_{n,p,l}^{\delta,\lambda,\mu}(z)\right]''}{\left[\mathcal{F}_{n,p,l}^{\delta,\lambda,\mu}(z)\right]'}+1-p\right)\right\}>p-\sum_{i=1}^{n}\delta_i(p-\alpha_i).
$$

Therefore, the operator $\mathcal{F}_{n,p,l}^{\delta,\lambda,\mu}(z)$ is $p$−valently convex of complex order $\gamma\,(\gamma\in\mathbb{C}-\{0\})$ and type $p-\sum_{i=1}^{n}\delta_i(p-\alpha_i)$. The proof of Theorem 10.9.1 is completed. $\square$

*Remark* 10.9.2.

1. Letting $\gamma=1$ and $l_i=0$ for all $i=\overline{1,n}$ in Theorem 10.9.1, we obtain Theorem 2.1 in Frasin (2010).

2. Letting $p=1$, $\gamma=1$ and $l_i=0$ for all $i=\overline{1,n}$ in Theorem 10.9.1, we obtain Theorem 1 in Browder (1971).

3. Letting $p=1$, $\gamma=1$ and $\alpha_i=l_i=0$ for all $i=\overline{1,n}$ in Theorem 10.9.1, we obtain Theorem 2.5 in Breaz and Breaz (2006).

4. Letting $p=1$, $\beta=0$ and $l_i=0$ for all $i=\overline{1,n}$ in Theorem 10.9.1, we obtain Theorem 1 in Bulut (2008).

5. Letting $p=1$, $\beta=0$, $\alpha_i=\alpha$ and $l_i=0$ for all $i=\overline{1,n}$ in Theorem 10.9.1, we obtain Theorem 1 in Breaz and Güney (2008).

6. Letting $p = 1$, $\beta = 0$, $\alpha_i = 0$ and $l_i = 0$ for all $i = \overline{1,n}$ in Theorem 10.9.1, we obtain Theorem 1 in Breaz et al. (2009).

Putting $n = 1$, $l_1 = 0$, $\delta_1 = \delta$, $\alpha_1 = \alpha$, $\beta_1 = \beta$ and $f_1 = f$ in Theorem 10.9.1, we have

**Corollary 10.9.3.** *Let* $\delta > 0$, $0 \leq \alpha < p$, $\beta \geq 0$, $\gamma \in \mathbb{C} - \{0\}$ *and* $f \in \beta - \mathcal{US}_p(\gamma, \alpha)$. *If* $\delta \in (0, p / (p - \alpha)]$, *then* $\int_0^z pt^{p-1} \left( \frac{f(t)}{t^p} \right)^{\delta} dt$ *is convex of complex order* $\gamma$ $(\gamma \in \mathbb{C} - \{0\})$ *and type* $p - \delta (p - \alpha)$ *in* $\mathbb{U}$.

**Theorem 10.9.4.** *Let* $l = (l_1, l_2, ..., l_n) \in \mathbb{N}_0^n$, $\delta = (\delta_1, \delta_2, ..., \delta_n) \in \mathbb{R}_+^n$, $0 \leq \alpha_i < p$, $\beta_i \geq 0$, $\gamma \in \mathbb{C} - \{0\}$ *and* $f_i \in \beta_i - \mathcal{US}_p(l_i, \lambda, \mu, \gamma, \alpha_i)$ *for all* $i = \overline{1,n}$. *If*

$$\left| \frac{z \left( \mathcal{D}_{p,\lambda,\mu}^{l_i} f_i \right)'(z)}{\left( \mathcal{D}_{p,\lambda,\mu}^{l_i} f_i \right)(z)} - p \right| > -\frac{p + \sum_{i=1}^n \delta_i (\alpha_i - p)}{\sum_{i=1}^n \frac{\delta_i \beta_i}{|\gamma|}} \qquad (10.100)$$

*for all* $i = \overline{1,n}$, *then the integral operator* $\mathcal{F}_{n,p,l}^{\delta,\lambda,\mu}(z)$ *defined by (10.94) is* $p-$*valently convex of complex order* $\gamma$ $(\gamma \in \mathbb{C} - \{0\})$.

*Proof.* From (10.99) and (10.100), we easily get $\mathcal{F}_{n,p,l}^{\delta,\lambda,\mu}(z)$ is $p-$valently convex of complex order $\gamma$. $\qquad \square$

From Theorem 10.9.4, we easily get:

**Corollary 10.9.5.** *Let* $l = (l_1, l_2, ..., l_n) \in \mathbb{N}_0^n$, $\delta = (\delta_1, \delta_2, ..., \delta_n) \in \mathbb{R}_+^n$, $0 \leq \alpha_i < p$, $\beta_i \geq 0$, $\gamma \in \mathbb{C} - \{0\}$ *and* $f_i \in \beta_i - \mathcal{US}_p(l_i, \lambda, \mu, \gamma, \alpha_i)$ *for all* $i = \overline{1,n}$. *If* $\mathcal{D}_{p,\lambda,\mu}^{l_i} f_i \in \mathcal{S}_p^*(\sigma)$, *where* $\sigma = p - (p - \sum_{i=1}^n \delta_i (p - \alpha_i)) / \sum_{i=1}^n \frac{\delta_i \beta_i}{|\gamma|}$; $0 \leq \sigma < p$ *for all* $i = \overline{1,n}$, *then the integral operator* $\mathcal{F}_{n,p,l}^{\delta,\lambda,\mu}(z)$ *is* $p-$*valently convex of complex order* $\gamma$ $(\gamma \in \mathbb{C} - \{0\})$.

Putting $n = 1$, $l_1 = 0$, $\delta_1 = \delta$, $\alpha_1 = \alpha$, $\beta_1 = \beta$ and $f_1 = f$ in Corollary 10.9.5, we have:

**Corollary 10.9.6.** *Let* $\delta > 0$, $0 \leq \alpha < p$, $\beta > 0$, $\gamma \in \mathbb{C} - \{0\}$ *and* $f \in \mathcal{S}_p^*(\rho)$ *where* $\rho = [\delta(p\beta + (p - \alpha)|\gamma|) - p|\gamma|] / \delta\beta$; $0 \leq \rho < p$, *then the integral operator* $\int_0^z pt^{p-1} \left( \frac{f(t)}{t^p} \right)^{\delta} dt$ *is* $p-$*valently convex of complex order* $\gamma$ $(\gamma \in \mathbb{C} - \{0\})$ *in* $\mathbb{U}$.

Next, we give a sufficient condition for the integral operator $\mathcal{G}_{p,m,l,\mu}(z)$ to be $p-$valently convex of complex order.

**Theorem 10.9.7.** *Let* $l = (l_1, l_2, ...l_n) \in \mathbb{N}_0^n$, $\delta = (\delta_1, \delta_2, ..., \delta_n) \in \mathbb{R}_+^n$, $0 \leq \alpha_i < p$, $\gamma \in \mathbb{C} - \{0\}$ *such that* $0 < \sum_{i=1}^n \delta_i (p - \alpha_i) \leq p$, $\beta_i \geq 0$ *and* $f_i \in \beta_i - \mathcal{UC}_p(l_i, \lambda, \mu, \gamma, \alpha_i)$ *for all* $i = \overline{1,n}$. *Then the integral operator* $\mathcal{G}_{n,p,l}^{\delta,\lambda,\mu}$ *defined by (10.95) is* $p-$*valently convex of complex order* $\gamma$ $(\gamma \in \mathbb{C} - \{0\})$ *and type* $p - \sum_{i=1}^n \delta_i (p - \alpha_i)$, *that is,* $\mathcal{F}_{n,p,l}^{\delta,\lambda,\mu} \in \mathcal{C}_p(\gamma, p - \sum_{i=1}^n \delta_i (p - \alpha_i))$.

*Proof.* From the definition (10.95), we observe that $\mathcal{G}_{n,p,l}^{\delta,\lambda,\mu}(z) \in \mathcal{A}_p$. On the other hand, it is easy to see that

$$\left[\mathcal{G}_{n,p,l}^{\delta,\lambda,\mu}(z)\right]' = pz^{p-1} \prod_{i=1}^{n} \left( \frac{\left(\mathcal{D}_{p,\lambda,\mu}^{l_i} g_i(z)\right)'}{pz^{p-1}} \right)^{\delta_i}. \tag{10.101}$$

Now, we differentiate (10.101) logarithmically and then do some simple calculations, we have

$$\Re\left\{ p + \frac{1}{\gamma} \left( \frac{z\left[\mathcal{G}_{n,p,l}^{\delta,\lambda,\mu}(z)\right]''}{\left[\mathcal{G}_{n,p,l}^{\delta,\lambda,\mu}(z)\right]'} + 1 - p \right) \right\} \tag{10.102}$$

$$= \sum_{i=1}^{n} \delta_i \Re\left\{ p + \frac{1}{\gamma} \left( 1 + \frac{z\left(\mathcal{D}_{p,\lambda,\mu}^{l_i} g_i\right)''(z)}{\left(\mathcal{D}_{p,\lambda,\mu}^{l_i} g_i\right)'(z)} - p \right) \right\} - p\sum_{i=1}^{n} \delta_i + p.$$

Since $g_i \in \beta_i - \mathcal{UC}_p(l_i, \lambda, \mu, \gamma, \alpha_i)$ for all $i = \overline{1, n}$ from (10.102), we have

$$\Re\left\{ p + \frac{1}{\gamma} \left( \frac{z\left[\mathcal{G}_{n,p,l}^{\delta,\lambda,\mu}(z)\right]''}{\left[\mathcal{G}_{n,p,l}^{\delta,\lambda,\mu}(z)\right]'} + 1 - p \right) \right\} \tag{10.103}$$

$$> p - p\sum_{i=1}^{n} \delta_i + \sum_{i=1}^{n} \delta_i \left\{ \beta_i \left| \frac{1}{\gamma} \left( \frac{z\left(\mathcal{D}_{p,\lambda,\mu}^{l_i} g_i\right)''(z)}{\left(\mathcal{D}_{p,\lambda,\mu}^{l_i} g_i\right)'(z)} + 1 - p \right) \right| + \alpha_i \right\}$$

$$= p - \sum_{i=1}^{n} \delta_i(p - \alpha_i) + \sum_{i=1}^{n} \frac{\delta_i \beta_i}{|\gamma|} \left| \frac{z\left(\mathcal{D}_{p,\lambda,\mu}^{l_i} g_i\right)''(z)}{\left(\mathcal{D}_{p,\lambda,\mu}^{l_i} g_i\right)'(z)} + 1 - p \right|$$

$$> p - \sum_{i=1}^{n} \delta_i(p - \alpha_i).$$

Therefore, the operator $\mathcal{G}_{n,p,l}^{\delta,\lambda,\mu}(z)$ is $p$−valently convex of complex order $\gamma \, (\gamma \in \mathbb{C} - \{0\})$ and type $p - \sum_{i=1}^{n} \delta_i(p - \alpha_i)$. This evidently completes the proof of Theorem 10.9.7. □

*Remark* 10.9.8.

1. Letting $\gamma = 1$ and $l_i = 0$ for all $i = \overline{1, m}$ in Theorem 10.9.7, we obtain Theorem 3.1 in Frasin (2010).

2. Letting $p = 1$, $\beta = 0$ and $l_i = 0$ for all $i = \overline{1, m}$ in Theorem 10.9.7, we obtain Theorem 3 in Bulut (2008).

3. Letting $p = 1$, $\beta = 0$, $\alpha_i = \mu$ and $l_i = 0$ for all $i = \overline{1, m}$ in Theorem 10.9.7, we obtain Theorem 3 in Breaz and Güney (2008).

4. Letting $p = 1$, $\beta = 0$, $\alpha_i = 0$ and $l_i = 0$ for all $i = \overline{1, m}$ in Theorem 10.9.7, we obtain Theorem 2 in Breaz et al. (2009).

Putting $n = 1$, $l_1 = 0$, $\delta_1 = \delta$, $\alpha_1 = \alpha$, $\beta_1 = \beta$ and $g_1 = g$ in Theorem 10.9.7, we have

**Corollary 10.9.9.** *Let $\delta > 0$, $0 \leq \alpha < p$, $\beta \geq 0$, $\gamma \in \mathbb{C} - \{0\}$ and $g \in \beta - \mathcal{UC}_p(\gamma, \alpha)$. If $\delta \in (0, p / (p - \alpha)]$, then $\int_0^z pt^{p-1} \left( \frac{g't)}{pt^{p-1}} \right)^\delta dt$ is $p-$valently convex of complex order $\gamma$ ($\gamma \in \mathbb{C} - \{0\}$) and type $p - \delta (p - \alpha)$ in $\mathbb{U}$.*

**Theorem 10.9.10.** *Let $l = (l_1, l_2, ..., l_n) \in \mathbb{N}_0^n$, $\delta = (\delta_1, \delta_2, ..., \delta_n) \in \mathbb{R}_+^n$, $0 \leq \alpha_i < p$, $\beta_i \geq 0$, $\gamma \in \mathbb{C} - \{0\}$ and $g_i \in \beta_i - \mathcal{UC}_p(l_i, \lambda, \mu, \gamma, \alpha_i)$ for all $i = \overline{1, n}$. If*

$$\left| \frac{z \left( \mathcal{D}_{p,\lambda,\mu}^{l_i} g_i \right)'' (z)}{\left( \mathcal{D}_{p,\lambda,\mu}^{l_i} g_i \right)' (z)} + 1 - p \right| > - \frac{p + \sum_{i=1}^n \delta_i (\alpha_i - p)}{\sum_{i=1}^n \frac{\delta_i \beta_i}{|\gamma|}} \qquad (10.104)$$

*for all $i = \overline{1, n}$, then the integral operator $\mathcal{G}_{n,p,l}^{\delta, \lambda, \mu}(z)$ defined by (10.95) is $p-$valently convex of complex order $\gamma$ ($\gamma \in \mathbb{C} - \{0\}$).*

*Proof.* From (10.103) and (10.104), we easily get $\mathcal{G}_{n,p,l}^{\delta, \lambda, \mu}(z)$ is $p-$valently convex of complex order $\gamma$. □

From Theorem 10.9.10, we easily get:

**Corollary 10.9.11.** *Let $l = (l_1, l_2, ..., l_n) \in \mathbb{N}_0^n$, $\delta = (\delta_1, \delta_2, ..., \delta_n) \in \mathbb{R}_+^n$, $0 \leq \alpha_i < p$, $\beta_i \geq 0$, $\gamma \in \mathbb{C} - \{0\}$ and $g_i \in \beta_i - \mathcal{UC}_p(l_i, \lambda, \mu, \gamma, \alpha_i)$ for all $i = \overline{1, n}$. If $\mathcal{D}_{p,\lambda,\mu}^{l_i} g_i \in \mathcal{C}_p(\sigma)$, where $\sigma = p - (p - \sum_{i=1}^n \delta_i (p - \alpha_i)) / \sum_{i=1}^n \frac{\delta_i \beta_i}{|\gamma|}$; $0 \leq \sigma < p$ for all $i = \overline{1, n}$, then the integral operator $\mathcal{G}_{n,p,l}^{\delta, \lambda, \mu}(z)$ is $p-$valently convex of complex order $\gamma$ ($\gamma \in \mathbb{C} - \{0\}$).*

Putting $n = 1$, $l_1 = 0$, $\delta_1 = \delta$, $\alpha_1 = \alpha$, $\beta_1 = \beta$ and $g_1 = g$ in Corollary 10.9.11, we have:

**Corollary 10.9.12.** *Let $\delta > 0$, $0 \leq \alpha < p$, $\beta > 0$, $\gamma \in \mathbb{C} - \{0\}$ and $g \in \mathcal{C}(\rho)$ where $\rho = [\delta(p\beta + (p - \alpha) |\gamma|) - p |\gamma|] / \delta\beta$; $0 \leq \rho < p$, then the integral operator $\int_0^z pt^{p-1} \left( \frac{g't)}{pt^{p-1}} \right)^\delta dt$ is convex of complex order $\gamma$ ($\gamma \in \mathbb{C} - \{0\}$) in $\mathbb{U}$.*

## 10.10    Starlikeness of integral operators $\mathcal{F}_{n,p,l}^{\delta,\lambda,\mu}(z)$ and $\mathcal{G}_{n,p,l}^{\delta,\lambda,\mu}(z)$

In this section, we will give the sufficient conditions for the integral operators $\mathcal{F}_{n,p,l}^{\delta,\lambda,\mu}$ and $\mathcal{G}_{n,p,l}^{\delta,\lambda,\mu}(z)$ to be $p-$valently convex of complex order.
Let

$$H(\mathbb{U}) = \{f : \mathbb{U} \to \mathbb{C} : \ f \text{ analytic}\}$$

$$H[a,n] = \{f \in H(\mathbb{U}) : \ f(z) = a + a_n z^n + a_{n+1} z^{n+1} + ....,$$

$$z \in \mathbb{U}, \ a \in \mathbb{C}, \ n \in \mathbb{N}_0\}.$$

In order to prove our main results, we shall need the following lemma due to S. S. Miller and P. T. Mocanu Miller and Mocanu (1978).

**Lemma 10.10.1.** *Let the function $\psi : \mathbb{C}^2 \times \mathbb{U} \to \mathbb{U}$ satisfy*

$$\Re\,\psi(i\rho, \sigma; z) \leq 0$$

*for all $\rho, \sigma \in \mathbb{R}$, $n \geq 1$ with $\sigma \leq -\frac{n}{2}(1 + \rho^2)$. If $P \in H[1,n]$ and $\Re\,\psi(P(z), zP'(z); z) > 0$ for every $z \in \mathbb{U}$, then*

$$\Re\,P(z) > 0.$$

**Lemma 10.10.2.** *Let $n \in \mathbb{N}$, $\kappa \in \mathbb{R}$, $u, v \in \mathbb{C}$ such that $\Im v \leq 0$, $\Re\,(u - \kappa v) \geq 0$. Assume the following condition*

$$\Re \left\{ P(z) + \frac{zP'(z)}{u - vP(z)} \right\} > \kappa, \quad (z \in \mathbb{U})$$

*is satisfied such that $P \in H\,[P(0), n]$, $P(0) \in \mathbb{R}$ and $P(0) > \kappa$. Then,*

$$\Re\,P(z) > \kappa, \quad (z \in \mathbb{U}).$$

*Proof.* Firstly, we consider the function $R : \mathbb{U} \to \mathbb{C}$,

$$R(z) = \frac{P(z) - \kappa}{P(0) - \kappa}.$$

Then, $R(z) \in H\,[1, n]$. Furthermore, since $P(0) - \kappa > 0$ and

$$\Re \left\{ P(z) + \frac{zP'(z)}{u - vP(z)} \right\} > \kappa, \quad (z \in \mathbb{U}),$$

we have

$$\Re \left\{ R(z) + \frac{zR'(z)}{u - v\kappa - v\,(P(0) - \kappa)\,R(z)} \right\} > 0, \quad (z \in \mathbb{U}).$$

Now, we define the function $\psi$ as follows

$$\psi(R(z), zR'(z); z) = R(z) + \frac{zR'(z)}{u - v\kappa - v\left(P(0) - \kappa\right)R(z)}.$$

Thus,

$$\Re\psi(R(z), zR'(z); z) > 0.$$

We can now use Lemma 10.10.1 and must show that the following condition

$$\Re\psi(i\rho, \sigma; z) \leq 0$$

is satisfied for $\rho \leq 0$, $\sigma \leq -\frac{1+\rho^2}{2}$ and $z \in \mathbb{U}$. Indeed, from hypotheses, we obtain

$$
\begin{aligned}
\Re\psi(i\rho, \sigma; z) &= \Re\frac{\sigma}{u - v\kappa - v\left(P(0) - \kappa\right)\rho i} \\
&= \Re\frac{\sigma}{u_1 + iu_2 - (v_1 + iv_2)\kappa - (v_1 + iv_2)\left(P(0) - \kappa\right)\rho i} \\
&= \frac{\sigma\left[u_1 - v_1\kappa + v_2\rho\left(P(0) - \kappa\right)\right]}{\left[u_1 - v_1\kappa + v_2\rho\left(P(0) - \kappa\right)\right]^2 + \left[u_2 - v_2\kappa + v_1\rho\left(P(0) - \kappa\right)\right]^2} \leq 0.
\end{aligned}
$$

Hence, from Lemma 10.10.1, we get $\Re R(z) > 0$. Moreover, from the definition of $R(z)$, we obtain

$$\Re P(z) > \kappa, \quad (z \in \mathbb{U}).$$

$\square$

Now, we prove the following theorem using Lemma 10.10.2

**Theorem 10.10.3.** *Let* $l = (l_1, l_2, ...l_n) \in \mathbb{N}_0^n$, $\delta = (\delta_1, \delta_2, ..., \delta_n) \in \mathbb{R}_+^n$, $0 \leq \alpha_i < p$, $\gamma \in \mathbb{C} - \{0\}$ *such that* $0 < \sum_{i=1}^n \delta_i(p - \alpha_i) \leq p$, $\Im\gamma \geq 0$, $\Re\gamma \leq \frac{p}{\sum_{i=1}^n \delta_i(p-\alpha_i)}$, $\beta_i \geq 0$ *and* $f_i \in \mathcal{B}_i - \mathcal{US}_p(l_i, \lambda, \mu, \gamma, \alpha_i)$ *for all* $i = \overline{1, n}$. *Then the integral operator* $\mathcal{F}_{n,p,l}^{\delta,\lambda,\mu}$ *defined by (10.94) is* $p$-*valently starlike of complex order* $\gamma$ $(\gamma \in \mathbb{C} - \{0\})$ *and type* $p - \sum_{i=1}^n \delta_i(p - \alpha_i)$, *that is,* $\mathcal{F}_{n,p,l}^{\delta,\lambda,\mu} \in \mathcal{S}_p^*(\gamma, p - \sum_{i=1}^n \delta_i(p - \alpha_i))$.

*Proof.* We define the analytic function $q : \mathbb{U} \to \mathbb{C}$, $q(0) = p$ as follows

$$q(z) = p + \frac{1}{\gamma}\left(\frac{z\left[\mathcal{F}_{n,p,l}^{\delta,\lambda,\mu}(z)\right]'}{\left[\mathcal{F}_{n,p,l}^{\delta,\lambda,\mu}(z)\right]} - p\right).$$

Thus, we obtain

$$p + \gamma(q(z) - p) = \frac{z\left[\mathcal{F}_{n,p,l}^{\delta,\lambda,\mu}(z)\right]'}{\left[\mathcal{F}_{n,p,l}^{\delta,\lambda,\mu}(z)\right]}$$

$$\Rightarrow \quad \frac{\gamma z q'(z)}{p(1-\gamma)+\gamma q(z)} = 1 + \frac{z\left[\mathcal{F}_{n,p,l}^{\delta,\lambda,\mu}(z)\right]''}{\left[\mathcal{F}_{n,p,l}^{\delta,\lambda,\mu}(z)\right]'} - \frac{z\left[\mathcal{F}_{n,p,l}^{\delta,\lambda,\mu}(z)\right]'}{\left[\mathcal{F}_{n,p,l}^{\delta,\lambda,\mu}(z)\right]}$$

$$\Rightarrow \quad p + \gamma\left(q(z)-p\right) + \frac{\gamma z q'(z)}{p(1-\gamma)+\gamma q(z)} = 1 + \frac{z\left[\mathcal{F}_{n,p,l}^{\delta,\lambda,\mu}(z)\right]''}{\left[\mathcal{F}_{n,p,l}^{\delta,\lambda,\mu}(z)\right]'}$$

$$\Rightarrow \quad q(z) + \frac{z q'(z)}{p(1-b)+bq(z)} = p + \frac{1}{\gamma}\left[1-p+\frac{z\left[\mathcal{F}_{n,p,l}^{\delta,\lambda,\mu}(z)\right]''}{\left[\mathcal{F}_{n,p,l}^{\delta,\lambda,\mu}(z)\right]'}\right]$$

When we consider this last equality and the inequality (10.97), we can write

$$q(z) + \frac{z q'(z)}{p(1-\gamma)+\gamma q(z)} = p + \sum_{i=1}^{n}\delta_i\left(p+\frac{1}{\gamma}\left(\frac{z\left(D_{p,\lambda,\mu}^{l_i}f_i\right)'(z)}{D_{p,\lambda,\mu}^{l_i}f_i(z)}-p\right)\right) - p\sum_{i=1}^{n}\delta_i.$$

Similarly to the proof of the Theorem 10.9.1, it can be easly seen that

$$\Re\left\{q(z)+\frac{z q'(z)}{p(1-\gamma)+\gamma q(z)}\right\} > p - \sum_{i=1}^{n}\delta_i\left(p-\alpha_i\right).$$

Here, $q(0) = p > p - \sum_{i=1}^{n}\delta_i\left(p-\alpha_i\right)$ and the function $q$ is analytic on $\mathbb{U}$. Also, when we write $\kappa = p - \sum_{i=1}^{n}\delta_i\left(p-\alpha_i\right)$, $u = p(1-\gamma)$ and $v = -\gamma$, we find $\Im v \le 0$ and $\Re\left(u-\kappa v\right) \ge 0$. Hence, all the conditions of Lemma 10.10.1 are satisfied and so

$$\Re q(z) = \Re\left\{p+\frac{1}{\gamma}\left(\frac{z\left[\mathcal{F}_{n,p,l}^{\delta,\lambda,\mu}(z)\right]'}{\left[\mathcal{F}_{n,p,l}^{\delta,\lambda,\mu}(z)\right]}-p\right)\right\} > p - \sum_{i=1}^{n}\delta_i\left(p-\alpha_i\right).$$

Thus, the proof of the theorem is completed.                   $\square$

Putting $n=1$, $l_1 = 0$, $\delta_1 = \delta$, $\alpha_1 = \alpha$, $\beta_1 = \beta$ and $f_1 = f$ in Theorem 10.10.3, we have

**Corollary 10.10.4.** *Let* $\delta > 0$, $0 \le \alpha < p$, $\beta \ge 0$, $\gamma \in \mathbb{C} - \{0\}$, $\Im\gamma \ge 0$, $\Re\gamma \le \frac{p}{\delta(p-\alpha)}$ *and* $f \in \beta - \mathcal{US}_p(\gamma,\alpha)$. *If* $\delta \in \left(0, \frac{p}{p-\alpha}\right]$ *then* $\int_0^z p t^{p-1}\left(\frac{f(t)}{t^p}\right)^{\delta}dt \in \mathcal{S}_p^*(\gamma, p - \delta\left(p-\alpha\right))$.

From Theorem 10.10.3, we obtain the following result.

**Theorem 10.10.5.** *Let* $l = (l_1, l_2, ...l_n) \in \mathbb{N}_0^n$, $\delta = (\delta_1, \delta_2, ..., \delta_n) \in \mathbb{R}_+^n$, $0 \le \alpha_i < p$, $\gamma \in \mathbb{C} - \{0\}$ *such that* $0 < \sum_{i=1}^{n}\delta_i\left(p-\alpha_i\right) \le p$, $\Im\gamma \ge 0$, $\Re\gamma \le \frac{p}{\sum_{i=1}^{n}\delta_i(p-\alpha_i)}$, $\beta_i \ge 0$ *and* $f_i \in \beta_i - \mathcal{US}_p(l_i, \lambda, \mu, \gamma, \alpha_i)$ *for all* $i =$

$\overline{1,n}$. If the inequality (10.100) is satisfied for all $i = \overline{1,n}$, then the integral operator $\mathcal{F}_{n,p,l}^{\delta,\lambda,\mu}(z)$ defined by (10.94) is $p$−valently starlike of complex order $\gamma\,(\gamma \in \mathbb{C} - \{0\})$.

From Theorem 10.10.5, we get the following result.

**Corollary 10.10.6.** *Let* $l = (l_1, l_2, ...l_n) \in \mathbb{N}_0^n$, $\delta = (\delta_1, \delta_2, ..., \delta_n) \in \mathbb{R}_+^n$, $0 \leq \alpha_i < p$, $\gamma \in \mathbb{C} - \{0\}$ *such that* $0 < \sum_{i=1}^n \delta_i\,(p - \alpha_i) \leq p$, $\Im\gamma \geq 0$, $\Re\gamma \leq \frac{p}{\sum_{i=1}^n \delta_i(p-\alpha_i)}$, $\beta_i \geq 0$ *and* $f_i \in \mathcal{B}_i - \mathcal{US}_p(l_i, \lambda, \mu, \gamma, \alpha_i)$ *for all* $i = \overline{1,n}$. *If* $\mathcal{D}_{p,\lambda,\mu}^{l_i} f_i \in \mathcal{S}_p^*(\sigma)$, *where* $\sigma = p - (p - \sum_{i=1}^n \delta_i\,(p - \alpha_i))\,/\sum_{i=1}^n \frac{\delta_i\beta_i}{|\gamma|}$; $0 \leq \sigma < p$ *for all* $i = \overline{1,n}$, *then the integral operator* $\mathcal{F}_{n,p,l}^{\delta,\lambda,\mu}(z)$ *is* $p$−*valently starlike of complex order* $\gamma\,(\gamma \in \mathbb{C} - \{0\})$.

Next, we give a sufficient condition for the integral operator $\mathcal{G}_{p,m,l,\mu}(z)$ to be $p$−valently starlike of complex order.

**Theorem 10.10.7.** *Let* $l = (l_1, l_2, ...l_n) \in \mathbb{N}_0^n$, $\delta = (\delta_1, \delta_2, ..., \delta_n) \in \mathbb{R}_+^n$, $0 \leq \alpha_i < p$, $\gamma \in \mathbb{C} - \{0\}$ *such that* $0 < \sum_{i=1}^n \delta_i\,(p - \alpha_i) \leq p$, $\Im\gamma \geq 0$, $\Re\gamma \leq \frac{p}{\sum_{i=1}^n \delta_i(p-\alpha_i)}$, $\beta_i \geq 0$ *and* $f_i \in \mathcal{B}_i - \mathcal{UC}_p(l_i, \lambda, \mu, \gamma, \alpha_i)$ *for all* $i = \overline{1,n}$. *Then the integral operator* $\mathcal{G}_{n,p,l}^{\delta,\lambda,\mu}$ *defined by (10.95) is* $p$−*valently starlike of complex order* $\gamma\,(\gamma \in \mathbb{C} - \{0\})$ *and type* $p - \sum_{i=1}^n \delta_i\,(p - \alpha_i)$, *that is,* $\mathcal{G}_{n,p,l}^{\delta,\lambda,\mu} \in \mathcal{S}_p^*(\gamma, p - \sum_{i=1}^n \delta_i\,(p - \alpha_i))$.

*Proof.* Let us define the analytic function $q : \mathbb{U} \to \mathbb{C}$ given by

$$q(z) = p + \frac{1}{\gamma}\left(\frac{z\left(\mathcal{G}_{n,p,l}^{\delta,\lambda,\mu}(z)\right)'}{\left(\mathcal{G}_{n,p,l}^{\delta,\lambda,\mu}(z)\right)} - p\right).$$

Then, we follow the same steps as in the proof of Theorem 10.10.3, so we omit the details involved in this case. □

Putting $n = 1$, $l_1 = 0$, $\delta_1 = \delta$, $\alpha_1 = \alpha$, $\beta_1 = \beta$ and $g_1 = g$ in Theorem 10.10.7, we have

**Corollary 10.10.8.** *Let* $\delta > 0$, $0 \leq \alpha < p$, $\beta \geq 0$, $\gamma \in \mathbb{C} - \{0\}$, $\Im\gamma \geq 0$, $\Re\gamma \leq \frac{p}{\delta(p-\alpha)}$ *and* $f \in \mathcal{B} - \mathcal{UC}_p(\gamma, \alpha)$. *If* $\delta \in \left(0, \frac{p}{p-\alpha}\right]$, *then* $\int_0^z p t^{p-1}\left(\frac{g'(t)}{pt^{p-1}}\right)^\delta dt \in \mathcal{S}_p^*(\gamma, p - \delta\,(p - \alpha))$.

From Theorem 10.10.7, we obtain the following result.

**Theorem 10.10.9.** *Let* $l = (l_1, l_2, ...l_n) \in \mathbb{N}_0^n$, $\delta = (\delta_1, \delta_2, ..., \delta_n) \in \mathbb{R}_+^n$, $0 \leq \alpha_i < p$, $\gamma \in \mathbb{C} - \{0\}$ *such that* $0 < \sum_{i=1}^n \delta_i\,(p - \alpha_i) \leq p$, $\Im\gamma \geq 0$, $\Re\gamma \leq \frac{p}{\sum_{i=1}^n \delta_i(p-\alpha_i)}$, $\beta_i \geq 0$ *and* $f_i \in \mathcal{B}_i - \mathcal{US}_p(l_i, \lambda, \mu, \gamma, \alpha_i)$ *for all* $i = \overline{1,n}$. *If the inequality (10.104) is satisfied for all* $i = \overline{1,n}$, *then the integral operator* $\mathcal{G}_{n,p,l}^{\delta,\lambda,\mu}(z)$ *defined by (10.95) is* $p$−*valently starlike of complex order* $\gamma\,(\gamma \in \mathbb{C} - \{0\})$.

We obtain the following corollary using Theorem 10.10.9.

**Corollary 10.10.10.** *Let* $l = (l_1, l_2, \dots l_n) \in \mathbb{N}_0^n$, $\delta = (\delta_1, \delta_2, \dots, \delta_n) \in \mathbb{R}_+^n$, $0 \le \alpha_i < p$, $\gamma \in \mathbb{C} - \{0\}$ *such that* $0 < \sum_{i=1}^n \delta_i (p - \alpha_i) \le p$, $\Im \gamma \ge 0$, $\Re \gamma \le \frac{p}{\sum_{i=1}^n \delta_i (p - \alpha_i)}$, $\beta_i \ge 0$ *and* $f_i \in \beta_i - \mathcal{US}_p(l_i, \lambda, \mu, \gamma, \alpha_i)$ *for all* $i = \overline{1, n}$. *If* $\mathcal{D}_{p,\lambda,\mu}^{l_i} f_i \in \mathcal{C}_p(\sigma)$, *where* $\sigma = p - \left(p - \sum_{i=1}^n \delta_i (p - \alpha_i)\right) \big/ \sum_{i=1}^n \frac{\delta_i \beta_i}{|\gamma|}$; $0 \le \sigma < p$ *for all* $i = \overline{1, n}$, *then the integral operator* $\mathcal{G}_{n,p,l}^{\delta,\lambda,\mu}(z)$ *is* $p-$*valently starlike of complex order* $\gamma \left(\gamma \in \mathbb{C} - \{0\}\right)$.

---

# Bibliography

Alexander J. W. Functions which map the interior of the unit circle upon simple regions. *The Annals of Mathematics*, 17(1):12–22, 1915.

Ali R. M., Lee S. K., Ravichandran V. and Supramanian S. Coefficient estimates for bi-univalent Ma-Minda starlike and convex functions. *Applied Mathematics Letters*, 25(3):344–351, 2012.

Altıntaş O. On a subclass of certain starlike functions with negative coefficients. *Math. Japon.*, 36:489–495, 1991.

Altıntaş O., Irmak H. and Srivastava H. M. Fractional calculus and certain starlike functions with negative coefficients. *Computers & Mathematics with Applications*, 30(2):9–15, 1995.

Altıntaş O. and Owa S. On subclasses of univalent functions with negative coefficients. *East Asian Mathematical Journal*, 4:41–56, 1988.

Altıntaş O., Özkan Ö. and Srivastava H. M. Neighborhoods of a certain family of multivalent functions with negative coefficients. *Computers & Mathematics with Applications*, 47(10-11):1667–1672, 2004.

Bharati R., Parvatham R. and Swaminathan A. On subclasses of uniformly convex functions and corresponding class of starlike functions. *Tamkang Journal of Mathematics*, 28:17–32, 1997.

Brannan D. A. and Clunie J. *Aspects of Contemporary Complex Analysis*. Academic Press, 1980.

Brannan D. A. and Taha T. S. On some classes of bi-univalent functions. In *Mathematical Analysis and Its Applications*, pages 53–60. Elsevier, 1988.

Breaz D. A convexity properties for an integral operator on the classes $s_p(\alpha)$. *Gen. Math*, 15(2-3):177–183, 2007.

Breaz D., Aouf M. K. and Breaz N. Some properties for integral operators on some analytic functions with complex order. *Acta. Math. Acad. Paedagog. Nyhazi*, 25:39–43, 2009.

Breaz D. and Breaz N. Two integral operators. *Studia Universitatis Babes-Bolyai, Mathematica*, 47(3):13–19, 2002.

Breaz D. and Breaz N. Some convexity properties for a general integral operator. *Journal of Inequalities in Pure and Applied Mathematics*, 7(5), 2006.

Breaz D. and Güney H. Ö. The integral operator on the classes $s_\alpha^*(b)$ and $c_\alpha(b)$. *J. Math. Inequal*, 2:97–100, 2008.

Breaz D., Owa S. and Breaz N. A new integral univalent operator. *Acta Univ. Apulensis Math. Inform*, 16:11–16, 2008.

Bulut S. A note on the paper of Breaz and Güney. *J. Math. Ineq*, 2(4):549–553, 2008.

Bulut S. Coefficient estimates for a class of analytic and bi-univalent functions. *Novi Sad J. Math*, 43(2):59–65, 2013.

Çağlar M. and Aslan S. Fekete-Szegö inequalities for subclasses of bi-univalent functions satisfying subordinate conditions. In *AIP Conference Proceedings*, volume 1726, page 020078. AIP Publishing, 2016.

Çağlar M., Orhan H. and Yağmur N. Coefficient bounds for new subclasses of bi-univalent functions. *Filomat*, 27(7):1165–1171, 2013.

Deniz E. Çağlar M. and Orhan H. Some convexity properties for two new p-valent integral operators. *Hacettepe Journal of Mathematics and Statistics*, 40(6), 2011.

Deniz E. Çağlar M. and Orhan H. Second Hankel determinant for bi-starlike and bi-convex functions of order $\beta$. *Applied Mathematics and Computation*, 271:301–307, 2015.

Deniz E., Deniz E. and Mustafa N. Some starlikeness and convexity properties for two new p-valent integral operators. *Hacettepe Journal of Mathematics and Statistics*, 45(5):1367–1378, 2016.

Deniz E. and Orhan H. Certain subclasses of multivalent functions defined by new multiplier transformations. *Arabian Journal for Science and Engineering*, 36(6):1091, 2011.

Deniz E., Orhan H. and Sokół J. Classes of analytic functions defined by a differential operator related to conic domains. *Ukrains kyi Matematychnyi Zhurnal*, 67(09):1217–1231, 2015.

Ding S. S., Ling Y. and Bao G. J. Some properties of a class of analytic functions. *Journal of Mathematical Analysis and Applications*, 195(1):71–81, 1995.

Doha E. H. The first and second kind Chebyshev coefficients of the moments for the general order derivative on an infinitely differentiable function. *International Journal of Computer Mathematics*, 51(1-2):21–35, 1994.

Duren P. L. Univalent functions. *Grundlehren der mathematischen Wissenschaften*, volume 259, Springer-Verlag, 1983.

Fekete M. and Szegö G. Eine bemerkung über ungerade schichte funktionen. *J. London Math. Soc.*, 8:85–89, 1933.

Frasin B. A. Family of analytic functions of complex order. *Acta Math. Acad. Paedagog. Nyházi.(NS)*, 22(2):179–191, 2006.

Frasin B. A. Convexity of integral operators of $p-$valent functions. *Math. Comput. Model*, 51:601–605, 2010.

Frasin B. A. and Aouf M. K. New subclasses of bi-univalent functions. *Appl. Math. Lett.*, 24(9):1569–1573, 2011.

Goodman A. W. *Univalent Functions*, volume I. Polygonal, Washington, 1983.

Goodman A. W. On uniformly convex functions. *Ann. Polon. Math.*, 56:87–92, 1991.

Goyal S. P. and Goswami P. Estimate for initial Maclaurin coefficients of bi-univalent functions for a class defined by fractional derivatives. *J. Egyptian Math. Soc.*, 20:179–182, 2012.

Grenander U. and Szegö G. *Toeplitz Forms and Their Applications*. University of California Press, Berkeley, 1958.

Hayami T. and Owa S. Coefficient bounds for bi-univalent functions. *Pan Amer. Math. J.*, 22(4):15–26, 2012.

Hummel J. The coefficient regions of starlike functions. *Pacific J. Math.*, 7:1381–1389, 1957.

Hummel J. Extremal problems in the class of starlike functions. *Proc. Amer. Math. Soc.*, 11:741–749, 1960.

Irmak H., Lee S. H. and Cho N. E. Some multivalently starlike functions with negative coefficients and their subclasses defined by using a differential operator. *Kyungpook Math. J.*, 37:43–51, 1997.

Kadioğlu E. On subclass of univalent functions with negative coefficients. *Applied Mathematics and Computation*, 146(2-3):351–358, 2003.

Kanas S. and Wisniowska A. Conic regions and $k-$uniform convexity. 105:327–336, 1999.

Keogh F. R. and Merkes E. P. A coefficient inequality for certain classes of analytic functions. *Proc. Amer. Math. Soc.*, 20:8–12, 1969.

Kim Y. C. and Srivastava H. M. Some subordination properties for spirallike functions. *Applied Mathematics and Computation*, 203(2):838–842, 2008.

Kim Y. J. and Merkes E. P. On an integral of powers of a spirallike function. *Kyungpook Math.*, 12:249–252, 1972.

Kumar S. S., Kumar V. and Ravichandran V. Estimates for the initial coefficients of bi-univalent functions. *arXiv preprint arXiv:1203.5480*, 2012.

Lewin M. On a coefficient problem for bi-univalent functions. *Proc. Amer. Math. Soc.*, 18:63–68, 1967.

Ma W. and Minda D. A unified treatment of some special classes of univalent functions. In *Proceedings of the Conference on Complex Analysis; International Press, Cambridge, Mass., USA*, pages 157–169, 1994.

Ma W. C. and Minda D. A unified treatment of some special classes of functions. In *Proceedings of the Conference on Complex Analysis, Tianjin*, pages 157–169, 1992.

Magesh N. and Yamini J. J. Coeffcient bounds for certain subclasses of bi-univalent functions. *Internat. Math. Forum*, 8(27):1337–1344, 2013.

Mason J. C. Chebyshev polynomials approximations for the l-membrane eigenvalue problem. *SIAM J. Appl. Math.*, 15:172–186, 1967.

Miller S. S. and Mocanu P. T. Second order differential inequalities in the complex plane,. *J. Math. Anal. Appl.*, 65:289–305, 1978.

Miller S. S., Mocanu P. T. and Reade M. O. Starlike integral operators. *Pacific J. Math.*, 79(1):157–168, 1978.

Moustafa A. O. A study on starlike and convex properties for hypergeometric functions. *Journal of Inequalities in Pure and Applied Mathematics*, 10(3):1–16, 2009. Article 87.

Murugusundaramoorthy G., Magesh N. and Prameela V. Coefficient bounds for certain subclasses of bi-univalent functions. *Abs. Appl. Anal.*, 3017:3, 2013. Article Id 573017.

Mustafa N. Fekete-Szegö problem for certain subclass of analytic and bi-univalent functions. *Journal of Scientific and Engineering Research*, 4(8):390–400, 2017.

Mustafa N. Upper bound estimate for the second Hankel determinant of a certain subclass of analytic and bi-univalent functions. *Journal of Scientific and Engineering Research*, 4(7):383–393, 2017.

Nasr M. A. and Aouf M. K. Starlike function of complex order. *J. Natur. Sci. Math.*, 25(1):1–12, 1985.

Netanyahu E. The minimal distance of the image boundary from the origin and the second coefficient of a univalent function in $|z| < 1$. *Arch. Rational Mech. Anal.*, 32:100–112, 1969.

Noonan J. W. and Thomas D. K. On the second Hankel determinant of areally mean p-valent functions. *Trans. Amer. Math. Soc.*, 223:337–346, 1976.

Orhan H., Deniz E. and Raducanu D. The fekete–Szegö problem for subclasses of analytic functions defined by a differential operator related to conic domains. *Comput. Math. Appl.*, 59(1):283–295, 2010.

Orhan H., Magesh N. and Yamini J. Bounds for the second Hankel determinant of certain bi-univalent functions. *Turkish J. Math.*, 40:65–678, 2016.

Oros G. I. and Oros G. A convexity property for an integral operator $\mathbb{F}_m$. *Stud. Univ. Babeş-Bolyai Math.*, 55(3):169–177, 2010.

Owa S. Some applications of the fractional calculus. *Research Notes in Mathematics*, 138:164–175, 1985.

Owa S. and Aouf M. K. On subclasses of univalent functions with negative coefficients. *Pusan Kyongnam Mathematical Journal*, 4:57–73, 1988.

Pescar V. and Owa S. Sufficient conditions for univalence of certain integral operators. *Indian J. Math.*, 42(3):347–35, 2000.

Pfaltzgraff J. A. Univalence of the integral of $(f'(z))^\lambda$. *Bulletin of the London Mathematical Society*, 7(3):254–256, 1975.

Pommerenke C. H. *Univalent Functions*. Vandenhoeck and Rupercht, Göttingen, 1975.

Porwal S. An application of a Poisson distribution series on certain analytic functions. *J. Complex Anal.*, 9841:1–3, 2014. Art. Id 35.

Porwal S. and Dixit K. K. An application of generalized Bessel functions on certain analytic functions. *Acta Universitatis Matthiae Belii. Series Mathematics*, pages 51–57, 2013.

Prema S. and Keerthi B. S. Coefficient bounds for certain subclasses of analytic functions. *J. Math. Anal.*, 4(1):22–27, 2013.

Rønning F. On starlike functions associated with parabolic regions. *Ann. Univ. Mariae Curie-Skłodowska Sect. A*, 45:117–122, 1991.

Silverman H. Univalent functions with negative coefficients. *Proc. Amer. Math. Soc.*, 51(1):106–116, 1975.

Srivastava H. M. Some inequalities and other results associated with certain subclasses of univalent and bi-univalent analytic functions. In *Nonlinear Analysis*, pages 607–630. Springer, 2012.

Srivastava H. M., Bulut S., Çağlar M. and Yağmur N. Coefficient estimates for a general subclass of analytic and bi-univalent functions. *Filomat*, 27(5):831–842, 2013.

Srivastava H. M., Mishra A. K. and Gochhayat P. Certain subclasses of analytic and bi-univalent functions. *Appl. Math. Lett.*, 23:1188–1192, 2010.

Srivastava H. M., Murugusundaramoorthy G. and Magesh N. On certain subclasses of bi-univalent functions associated with hohlov operator. *Global J. Math. Anal.*, 1(2):67–73, 2013.

Srivastava H. M. and Owa S. *Current Topics in Analytic Function Theory*. World Scientific, 1992.

Srivastava H. M., Owa S. and Chatterjea S. K. A note on certain classes of starlike functions. *Rend. Sem. Mat. Univ. Padova*, 77:115–124, 1987.

Taha T. S. *Topics in univalent funtion theory*. PhD thesis, King's College London (University of London), 1980.

Topkaya S. and Mustafa N. The general subclasses of the analytic functions and their various properties. *Asian Research Journal of Mathematics*, 7(4):1–12, 2017.

Wiatrowski P. The coefficients of a certain family of holomorphic functions. *Zeszyty Nauk. Uniw.Lodz. Nauki Mat. Pryrod. Ser.*, 3:75–85, 1971.

Wright E. M. On the coefficients of power series having exponential singularities. *J. London Math. Soc.*, 8:71–79, 1933.

Xu Q. H., Gui Y. C. and Srivastava H. M. Coefficient estimates for a certain subclass of analytic and bi-univalent functions. *Appl. Math. Lett.*, 25(6):990–994, 2012.

Xu Q. H., Xiao H. G. and Srivastava H. M. A certain general subclass of analytic and bi-univalent functions and associated coefficient estimate problems. *Appl. Math. Comput.*, 218(23):11461–11465, 2012.

Zaprawa P. Estimates of initial coefficients for bi-univalent functions. *Abstr. Appl. Anal.*, 3574:1–6, 2014. Article Id 80.

Zaprawa P. On the Fekete-Szegö problem for classes of bi-univalent functions. *Bulletin of the Belgian Mathematical Society - Simon Stevin*, 21:169–178, 2014.

# Chapter 11

---

# Free Stochastic Integrals for Weighted-Semicircular Motion Induced by Orthogonal Projections

**Ilwoo Cho**

*Saint Ambrose University, Department of Mathematics & Statistics,*
*518 W. Locust St. Davenport, Iowa, U. S. A.*

---

## 11.1 Introduction

In Cho (2018), we introduced how to construct *semicircular elements* from a given mutually orthogonal $|\mathbb{Z}|$-many *projections* in a $C^*$-*probability space*, motivated by the weighted-semicircularity and the corresponding semicircularity obtained from Cho (2017), Cho (2016), Cho (2018) and Cho and Jorgensen (2017). In this chapter, we study free calculus induced by the semicircular elements in the sense of Cho (2018). Our construction of semicircular elements is similar to, but different from the usual, or the traditional constructions of semicircular elements in *free probability theory* (e.g., Hiai and Petz (2000) and Radulescu (1994)); this construction of Cho (2018) is the generalization for the weighted-semicircularity and the semicircularity of Cho (2017), Cho (2016), Cho (2018) and Cho and Jorgensen (2017).

The main purposes of this chapter are (i) to apply the main results of Cho (2018) in free stochastic calculus on Banach ∗-algebras, (ii) to study how mutually free weighted-semicircular, or semicircular elements act under the settings of Cho (2018)], and (iii) to measure the actions among those elements under our free-stochastic-calculus models, by considering the free distributions of stochastic integrals for the elements.

### 11.1.1 Motivation and background

*p-adic* and *Adelic analysis* provide fundamental tools, and play important roles not only in *number theory* locally, respectively, globally, but also in *geometry* and *functional analysis* (e.g., Et and Colak (1995 )). So, we cannot help emphasizing their importances and applications in related fields (e.g., Cho (2017), Cho (2016), Cho (2018) and Cho and Jorgensen (2017)).

In Cho and Jorgensen (2017), the author and Jorgensen constructed-and-studied *weighted-semicircular elements* and semicircular elements induced from the *p*-adic analysis on the *p-adic number fields* $\mathbb{Q}_p$ for *primes p*, by using concepts and terminology from free probability theory. In Cho (2017), the author extended weighted-semicircularity and semicircularity of Cho and Jorgensen (2017) under *free product* of the structures of Cho and Jorgensen (2017) over primes and integers, and then established maximal *free weighted-semicircular family* and the corresponding *free semicircular family* in a certain free product Banach ∗-probability space. The free distributions of free reduced words in weighted-semicircular elements, or those in semicircular elements were computed-and-characterized there. As applications, the free stochastic integration in terms of free stochastic motions generated by weighted-semicircular elements of Cho (2017) was considered in Cho (2016); and the differences (or close-ness) between our weighted-semicircular laws of Cho (2017) and the semicircular law were studied in Cho (2018).

Motivated by the number-theoretic construction of weighted-semicircularity and semicircularity from Cho (2017), Cho (2016), Cho (2018) and Cho and Jorgensen (2017), we extended, or generalized our *p*-adic, or Adelic weighted-semicircularity and semicircularity, in Cho (2018). Weighted-semicircular elements and semicircular elements were constructed there from arbitrarily given mutually orthogonal $|\mathbb{Z}|$-many projections in a fixed $C^*$-probability space. And free-distributional data of free reduced words and free sums in such elements were studied there.

In this chapter, we use weighted-semicircularity and semicircularity from Cho (2018) to study free stochastic integrals, motivated by the main results of Cho (2016). We study free distributions of such integrals. For more about free probability theory, e.g., see Speicher (1998), Voiculescu (1997) and Voiculescu et al. (1992). Readers can study semicircular elements and their applications in Hiai and Petz (2000) and Radulescu (1994). Also, for free stochastic calculus, e.g., see Speicher (2001) and cited papers therein.

## 11.1.2   Overview

In this chapter, we study free probability on *free filterizations* in the sense of Cho (2018) (Also see below), in particular, free-distributional data obtained from our free weighted-semicircular, and semicircular families. Based on such data, free stochastic calculus models are established in terms of weighted-semicircular stochastic processes acting on free filterizations. And then free distributions of free stochastic integrals for our such processes are considered.

In Section 11.2, we briefly mention about background of our proceeding works.

In Section 11.3, certain Banach ∗-probability spaces are constructed from fixed mutually orthogonal $|\mathbb{Z}|$-many projections.

In Sections 11.4 and 11.5, on the free-probabilistic structures of Section 3, we construct-and-study weighted-semicircular elements and semicircular elements induced by projections.

In Section 11.6, by doing free product on our free-probabilistic structures of Section 11.5, we enlarge our weighted-semicircularity and semicircularity. i.e., free filterizations are constructed.

In Section 11.7, we establish $j$-th sectionized weighted-semicircular processes, called $j$-th *weighted-semicircular motions*, for all $j \in \mathbb{Z}$.

In Section 11.8, by fixing a $j$-th weighted-semicircular motions of Section 11.7, we study corresponding $j$-th *free stochastic integrals* of *simple adapted biprocesses*.

## 11.2   Preliminaries

For more about motivations, see Cho and Jorgensen (2017), Cho (2017), Cho (2016) and Cho (2018). Also, for constructions and properties of weighted-semicircularity and corresponding semicircularity induced by orthogonal projections, see Cho (2018). We use same concepts, and notations used in Cho (2018) which will be precisely introduced in text.

Readers can study fundamental *free probability theory* from Speicher (1998), Voiculescu (1997) and Voiculescu et al. (1992) (and the cited papers therein). *Free probability* is understood as the noncommutative operator-algebraic version of classical *probability theory* and *statistics*. The classical *independence* is replaced by the *freeness*, by replacing *measures* on sets to *linear functionals* on algebras. It has various applications not only in pure mathematics (e.g., Hiai and Petz (2000), Radulescu (1994) and Voiculescu (1997)), but also in related topics (e.g., Cho (2017), Cho (2016), Cho (2018) and Cho and Jorgensen (2017)). In particular, we will use combinatorial free probabilistic approach of *Speicher* (e.g., Speicher (1998)). *Free moments* and *free cumulants* of operators, *free probability spaces*, and *free product of algebras* will be considered without introducing detailed concepts and backgrounds.

Free stochastic calculus, and free-stochastic structures are well-introduced and studied in Speicher (2001).

## 11.3 On Banach *-algebras induced by orthogonal projections

Let $(B, \psi_B)$ be a topological *-*probability space* (e.g., $C^*$-probability space, or $W^*$-probability space, or Banach *-probability space), where $B$ is a *topological *-algebra* ($C^*$-algebra, resp., $W^*$-algebra, resp., Banach *-algebra), and $\psi_B$ is a *bounded linear functional on* $B$.

An operator $a$ of $B$ is said to be a *free random variable* whenever we regard it as an element of $(B, \psi_B)$. As usual, $a$ is said to be *self-adjoint*, if $a^* = a$ in $B$, where $a^*$ is the *adjoint of* $a$.

**Definition 11.3.1.** A self-adjoint free random variable $a$ is said to be weighted-semicircular in $(B, \psi_B)$ with its weight $t_0 \in \mathbb{C}^\times = \mathbb{C} \setminus \{0\}$ (or in short, $t_0$-semicircular), if $a$ satisfies the free-cumulant computation,

$$k_n^{\psi_B}(a, ..., a) = \begin{cases} k_2^{\psi_B}(a, a) = t_0 & \text{if } n = 2 \\ 0 & \text{otherwise,} \end{cases} \tag{11.3.1}$$

for all $n \in \mathbb{N}$, where $k_n^{\psi_B}(...)$ means the free cumulant on $B$ in terms of $\psi_B$ under the Möbius inversion of Speicher (1998).

If $t_0 = 1$ in (11.3.1), the 1-semicircular element $a$ is simply said to be semicircular in $(B, \psi_B)$,

By definition, a self-adjoint free random variable $a$ is *semicircular in* $(B, \psi_B)$, if and only if $a$ satisfies

$$k_n^{\psi_B}(a, ..., a) = \begin{cases} 1 & \text{if } n = 2 \\ 0 & \text{otherwise,} \end{cases} \tag{11.3.2}$$

for all $n \in \mathbb{N}$, if and only if it is 1-semicircular in the sense of (11.3.1).

By the *Möbius inversion of* Speicher (1998), one can characterize the weighted-semicircularity (11.3.1) as follows: $a$ is $t_0$-semicircular in $(B, \psi_B)$, if and only if

$$\psi(a^n) = \omega_n t_0^{\frac{n}{2}} c_{\frac{n}{2}}, \tag{11.3.3}$$

for all $n \in \mathbb{N}$, where

$$c_k = \frac{1}{k+1} \begin{pmatrix} 2k \\ k \end{pmatrix} = \frac{(2k)!}{k!(k+1)!}$$

are the *k-th Catalan numbers*, for all $k, n \in \mathbb{N}$, where

$$\omega_n \overset{def}{=} \begin{cases} 1 & \text{if } n \text{ is even} \\ 0 & \text{if } n \text{ is odd,} \end{cases}$$

for all $n \in \mathbb{N}$.

Similarly, a free random variable $a$ is semicircular in $(A, \psi)$, if and only if

$$\psi(a^n) = \omega_n c_{\frac{n}{2}}, \tag{3.4}$$

for all $n \in \mathbb{N}$, where $\omega_n$ are in the sense of (11.3.4).

So, one can use the $t_0$-semicircularity (11.3.1) (or the semicircularity (11.3.2)), and its characterization (11.3.3) (resp., (11.3.4)) alternatively, from below.

Recall that, if a free random variable $x \in (B, \psi_B)$ is *self-adjoint*, then the sequences

$$\{\psi_B(x^n)\}_{n=1}^{\infty}, \text{ and } \{k_n^{\psi_B}(x, ..., x)\}_{n=1}^{\infty}$$

provide equivalent free distributions of $x$.

Indeed, the *Möbius inversion* (of Speicher (1998)) satisfies

$$\psi(a^n) = \sum_{\pi \in NC(n)} \left( \prod_{V \in \pi} k_{|V|}(a, ..., a) \right),$$

and

$$k_n(a, ..., a) = \sum_{\pi \in NC(n)} \left( \prod_{V \in \theta} \psi(a^{|V|}) \right) \mu(\pi, 1_n),$$

where $NC(n)$ is the *lattice of all noncrossing partitions* over $\{1, ..., n\}$, and "$V \in \pi$" means "$V$ is a *block* of $\pi$," and where $\mu$ is the *Möbius functional* on the *incidence algebra* induced by $\{NC(n)\}_{n=1}^{\infty}$ (e.g., Speicher (1998)).

Now, let $(A, \psi)$ be a given $C^*$-probability space containing *projections* $q_j \in A$ in the sense that:

$$q_j^* = q_j = q_j^2 \text{ in } A,$$

for all $j \in \mathbb{Z}$, where $\mathbb{Z}$ is the set of all *integers*. Assume, for convenience, that

$$q_j \in \mathbb{C}^{\times} = \mathbb{C} \setminus \{0\}, \text{ for all } j \in \mathbb{Z}.$$

Moreover, assume that these projections $\{q_j\}_{j \in \mathbb{Z}}$ are *mutually orthogonal* from each other in $A$, in the sense that:

$$q_i q_j = \delta_{i,j} q_j \text{ in } A, \forall i, j \in \mathbb{Z}, \tag{11.3.5}$$

where $\delta$ is the *Kronecker delta*.

Now, fix the family $\{q_j\}_{j \in \mathbb{Z}}$ of mutually orthogonal projections of $A$, and we denote it by $\mathbf{Q}$,

$$\mathbf{Q} = \{q_j : j \in \mathbb{Z}\} \text{ in } A, \tag{11.3.6}$$

satisfying (11.3.5).

And let $Q$ be the $C^*$-subalgebra of $A$ generated by $\mathbf{Q}$ of (11.3.6),

$$Q \overset{def}{=} C^*(\mathbf{Q}) \subseteq A. \tag{11.3.7}$$

**Proposition 11.3.2.** *Let $Q$ be a $C^*$-subalgebra (11.3.7) of a given $C^*$-algebra $A$, generated by $\mathbf{Q}$ of (11.3.6). Then*

$$Q \overset{*-iso}{=} \underset{j \in \mathbb{Z}}{\oplus} (\mathbb{C} \cdot q_j) \overset{*-iso}{=} \mathbb{C}^{\oplus|\mathbb{Z}|}, \tag{3.8}$$

*in $A$.*

*Proof.* The proof of (11.3.8) is straightforward by the mutual orthogonality (11.3.5) of the generator set $\mathbf{Q}$ of $Q$ in $A$. $\qquad\square$

Define now linear functionals $\psi_j$ on $Q$ by
$$\psi_j(q_i) = \delta_{ij}\,\psi(q_j), \text{ for all } i \in \mathbb{Z}, \tag{11.3.9}$$
for all $j \in \mathbb{Z}$, where $\psi$ is the linear functional inducing our fixed $C^*$-probability space $(A, \psi)$. The linear functionals $\{\psi_j\}_{j\in\mathbb{Z}}$ are well-defined on $Q$ by the structure theorem (11.3.8) of $Q$.

If $q = \sum_{j\in\mathbb{Z}} t_j q_j \in Q$, then

$$q^k = \left(\sum_{j\in\mathbb{Z}} t_j q_j\right)^k \overset{\text{equi}}{=} \left(\bigoplus_{j\in\mathbb{Z}} t_j q_j\right)^k$$

$$= \bigoplus_{j\in\mathbb{Z}} t_j^k q_j \overset{\text{equi}}{=} \sum_{j\in\mathbb{Z}} t_j^k q_j,$$

where "$\overset{\text{equi}}{=}$" means "being equivalent to," and hence,

$$\psi_j(q^k) = \psi_j\left(t_j^k q_j\right) = t_j^k\,\psi(q_j),$$

on $Q$, by (11.3.8) and (11.3.9), for all $k \in \mathbb{N}$, for all $j \in \mathbb{Z}$.

**Definition 11.3.3.** The $C^*$-probability spaces $(Q, \psi_j)$ are called the $j$-th $C^*$-probability spaces of $Q$ in $(A, \psi)$, where $Q$ is the $C^*$-subalgebra (3.7) of $A$, and $\psi_j$ are in the sense of (11.3.9), for all $j \in \mathbb{Z}$.

Now, let's define a *bounded linear transformations* $c$ and $a$ "acting on the $C^*$-algebra $Q$" by linear morphisms satisfying
$$c(q_j) = q_{j+1}, \text{ and } a(q_j) = q_{j-1}, \tag{11.3.10}$$
for all $j \in \mathbb{Z}$.

**Definition 11.3.4.** We call these bounded linear transformations $c$ and $a$ of (11.3.10), the *creation*, respectively, the *annihilation* on $Q$.

Define now a new Banach-space operator $l$ on $Q$ by
$$l = c + a \text{ on } Q, \tag{11.3.11}$$
where $c$ and $a$ are the creation, and the annihilation on $Q$ in the sense of (11.3.10).

**Definition 11.3.5.** We call the linear transformation $l$ of (11.3.11), the *radial operator* on $Q$.

By the definition (11.3.11), one has

$$l\left(\sum_{j\in\mathbb{Z}} t_j\,q_j\right) = \sum_{j\in\mathbb{Z}} t_j\,(q_{j+1} + q_{j-1}).$$

Now, define a *Banach algebra* $\mathfrak{L}$ by

$$\mathfrak{L} \overset{def}{=} \overline{\mathbb{C}[\{l\}]}^{\|\cdot\|}, \tag{11.3.12}$$

generated by the radial operator $l$, equipped with the *operator norm,*

$$\|T\| = \sup\{\|Tq\|_Q : \|q\|_Q = 1\},$$

where $\|.\|_Q$ is the $C^*$-*norm on* $Q$, and where $\overline{X}^{\|\cdot\|}$ mean the operator-norm closures of sets $X$ in the *operator space* $B(Q)$ consisting of all bounded linear transformations acting on $Q$.

On the Banach algebra $\mathfrak{L}$ of (11.3.12), we define a unary operation $(*)$ by

$$\left(\sum_{n=0}^{\infty} t_n l^n\right)^* = \sum_{n=0}^{\infty} \overline{t_n}\, l^n \text{ in } \mathfrak{L}, \tag{11.3.13}$$

where $\overline{z}$ are the *conjugates of* $z$, for all $z \in \mathbb{C}$.

Then the operation (11.3.13) forms a well-defined *adjoint on* $\mathfrak{L}$, and hence this Banach algebra $\mathfrak{L}$ forms a *Banach $*$-algebra.*

**Definition 11.3.6.** We call the Banach $*$-algebra $\mathfrak{L}$ of (11.3.12), the *radial (Banach $*$-)algebra on* $Q$.

Now, let $\mathfrak{L}$ be the radial algebra on $Q$. Construct now the *tensor product Banach $*$-algebra,*

$$\mathfrak{L}_Q = \mathfrak{L} \otimes_{\mathbb{C}} Q. \tag{11.3.14}$$

Since $\mathfrak{L}$ is a Banach $*$-algebra, and $Q$ is a $C^*$-algebra, the tensor product $*$-algebra $\mathfrak{L}_Q$ of (11.3.14) is a well-determined Banach $*$-algebra under product topology.

**Definition 11.3.7.** We call the Banach $*$-algebra $\mathfrak{L}_Q$ of (3.14), the *radial projection (Banach $*$-)algebra on* $Q$.

---

## 11.4    Weighted-semicircular elements induced by Q

In this section, we construct weighted-semicircular elements induced by the family $\mathbf{Q}$ of mutually orthogonal projections in a fixed $C^*$-probability space $(A, \psi)$. Let $(Q, \psi_j)$ be $j$-th $C^*$-probability space of $Q$ in $(A, \psi)$, where $\psi_j$ are the linear functionals (3.9) for all $j \in \mathbb{Z}$, and let $\mathfrak{L}_Q$ be the radial projection algebra (3.14) on $Q$.

Remark that, if

$$u_j = l \otimes q_j \in \mathfrak{L}_Q, \text{ for all } j \in \mathbb{Z}, \tag{11.4.1}$$

then

$$u_j^n = (l \otimes q_j)^n = l^n \otimes q_j, \text{ for all } n \in \mathbb{N},$$

since $q_j^n = q_j$, for all $n \in \mathbb{N}$, for $j \in \mathbb{Z}$.

So, one can construct a linear functional $\varphi_j$ on the radial projection algebra $\mathfrak{L}_Q$ by a linear morphism satisfying that

$$\varphi_j \left( (l \otimes q_i)^n \right) \overset{def}{=} \psi_j \left( l^n(q_i) \right), \tag{11.4.2}$$

for all $n \in \mathbb{N}$, for all $i, j \in \mathbb{Z}$.

Note that such linear functionals $\varphi_j$ of (11.4.2) are well-defined by (11.3.8) and (11.3.14). We call the Banach $*$-probability spaces

$$\left( \mathfrak{L}_Q, \varphi_j \right), \text{ for all } j \in \mathbb{Z}, \tag{11.4.3}$$

the *j-th (Banach-$*$-)probability spaces on Q*.

Observe first that, if $c$ and $a$ are the creation, respectively, the annihilation on $Q$ in the sense of (11.3.10), then

$$ca = 1_Q = ac, \tag{11.4.4}$$

where $1_Q \in B(Q)$ is the identity operator on $Q$ satisfying $1_Q(q) = q$, for all $q \in Q$.

Indeed, for any $q_j \in \mathbf{Q}$ in $Q$,

$$ca(q_j) = c(a(q_j)) = c(q_{j-1}) = q_{j-1+1} = q_j,$$

and

$$ac(q_j) = a(c(q_j)) = a(q_{j+1}) = q_{j+1-1} = q_j,$$

for all $j \in \mathbb{Z}$. More generally, one has

$$c^n a^n = 1_Q = a^n c^n, \text{ for all } n \in \mathbb{N},$$

and

$$c^{n_1} a^{n_2} = a^{n_2} c^{n_1}, \text{ for all } n_1, n_2 \in \mathbb{N}, \tag{11.4.4$'$}$$

by (11.4.4).

Thus, by (11.4.4) and (11.4.4)$'$, one can get that

$$l^n = (c + a)^n = \sum_{k=0}^{n} \binom{n}{k} c^k a^{n-k}, \tag{11.4.5}$$

in $\mathfrak{L}$, for all $n \in \mathbb{N}$, where

$$\binom{n}{k} = \frac{n!}{k!(n-k!)}, \text{ for all } k \le n \in \mathbb{N}_0 = \mathbb{N} \cup \{0\}.$$

Note that, for any $n \in \mathbb{N}$,

$$l^{2n-1} = \sum_{k=0}^{2n-1} \binom{2n-1}{k} c^k a^{n-k}, \tag{11.4.6}$$

by (11.4.5). So, the formula (11.4.6) does not contain $1_Q$-terms by (11.4.4) and (4.4)$'$.

Note also that, for any $n \in \mathbb{N}$, one has

$$l^{2n} = \sum_{k=0}^{2n} \binom{2n}{k} c^k a^{n-k}$$

$$= \binom{2n}{n} c^n a^n + [\text{Rest terms}],$$

(11.4.7)

by (11.4.5).

**Proposition 11.4.1.** *Let $l$ be the radial operator generating the radial algebra $\mathfrak{L}$ on $Q$. Then*

*(11.4.8)*   $l^{2n-1}$ *does not contain* $1_Q$-*terms in* $\mathfrak{L}$,

*(11.4.9)*   $l^{2n}$ *contains* $\binom{2n}{n} \cdot 1_Q$ *in* $\mathfrak{L}$.

*Proof.* The statement (11.4.8) (resp., the statement (11.4.9)) is proven by (11.4.6) (resp., by (11.4.7)) with help of (11.4.4), (11.4.4)$'$ and (11.4.5).   $\square$

By (11.3.9) and (11.4.8), one can obtain that

$$\varphi_j \left( u_j^{2n-1} \right) = \psi_j \left( l^{2n-1}(q_j) \right) = 0,$$

(11.4.10)

for all $n \in \mathbb{N}$.

Similarly,

$$\varphi_j \left( u_j^{2n} \right) = \psi_j \left( l^{2n}(q_j) \right) = \psi_j \left( \binom{2n}{n} q_j + [\text{Rest teimrs}](q_j) \right)$$

by (11.4.7)

$$= \binom{2n}{n} \psi_j(q_j) = \binom{2n}{n} \psi(q_j),$$

(11.4.11)

by (11.3.9), for all $n \in \mathbb{N}$.

**Theorem 11.4.2.** *Fix $j \in \mathbb{Z}$, and let $u_j = l \otimes q_j$ be the corresponding gener-ating operator of the $j$-th probability space $(\mathfrak{L}_Q, \varphi_j)$. Then*

$$\varphi_j \left( u_j^n \right) = \omega_n \left( \left( \tfrac{n}{2} + 1 \right) \psi(q_j) \right) c_{\frac{n}{2}},$$

(11.4.12)

*where $\omega_n$ are in the sense of (3.3), and $c_k$ are the $k$-th Catalan numbers, for all $n \in \mathbb{N}$.*

*If $i \neq j$ in $\mathbb{Z}$, and if $u_i = l \otimes q_i \in (\mathfrak{L}_Q, \varphi_j)$, then*
$\varphi_j \left( u_i^n \right) = 0$, *for all $n \in \mathbb{N}$.*

*Proof.* Observe that

$$\varphi_j \left( u_j^{2n-1} \right) = 0,$$

by (11.4.10). And, consider that

$$\varphi_j \left( u_j^{2n} \right) = \binom{2n}{n} \psi(q_j) = \left( \tfrac{n+1}{n+1} \right) \binom{2n}{n} \psi(q_j)$$

$$= ((n+1)\psi(q_j)) \left( \tfrac{1}{n+1} \binom{2n}{n} \right)$$

$$= ((n+1)\psi(q_j)) c_n,$$

by (11.4.11), for all $n \in \mathbb{N}$.

Now, suppose $i \neq j$ in $\mathbb{Z}$, and $u_i$ is the $i$-th generating operator of $\mathfrak{L}_Q$. Then

$$\varphi_j(u_i^n) = 0, \text{ for all } n \in \mathbb{N},$$

by (11.3.9) and (11.4.2).                              □

Motivated by the free-distributional data (11.4.12) of the generating operator $u_j$ of $\mathfrak{L}_Q$, we define the following morphism,

$$E_{j,Q} : \mathfrak{L}_Q \to \mathfrak{L}_Q,$$

by a linear transformation satisfying that

$$E_{j,Q}(u_i^n) \overset{def}{=} \begin{cases} \dfrac{\psi(q_j)^{n-1}}{([\frac{n}{2}]+1)} u_j^n & \text{if } i = j \\[2ex] 0_{\mathfrak{L}_Q}, \text{ the zero operator of } \mathfrak{L}_Q & \text{otherwise,} \end{cases} \tag{11.4.13}$$

for all $n \in \mathbb{N}$, $i, j \in \mathbb{Z}$, where $[\frac{n}{2}]$ means the *minimal integer* greater than or equal to $\frac{n}{2}$, for example,

$$[\tfrac{3}{2}] = 2 = [\tfrac{4}{2}].$$

The morphisms $E_{j,Q}$ of (11.4.13) are well-defined linear transformations on $\mathfrak{L}_Q$ because of the construction (3.14) of $\mathfrak{L}_Q = \mathfrak{L} \otimes_{\mathbb{C}} Q$, and by the structure theorem (11.3.8) of $Q$.

Define now a new linear functional $\tau_j$ on $\mathfrak{L}_Q$ by

$$\tau_j \overset{def}{=} \varphi_j \circ E_{j,Q} \text{ on } \mathfrak{L}_Q, \forall j \in \mathbb{Z}, \tag{11.4.14}$$

where $\varphi_j$ are in the sense of (11.4.2), and $E_{j,Q}$ are in the sense of (11.4.13).

**Definition 11.4.3.** The well-defined Banach $*$-probability spaces

$$\mathfrak{L}_Q(j) \overset{denote}{=} (\mathfrak{L}_Q, \tau_j) \tag{11.4.15}$$

are called the $j$-th filtered (Banach-$*$-)probability spaces of the radial projection algebra $\mathfrak{L}_Q$ on $Q$, where $\tau_j$ are the linear functionals (4.14), for all $j \in \mathbb{Z}$.

On the $j$-th filtered probability space $\mathcal{L}_Q(j)$ of (11.4.15), One can obtain that

$$\tau_j\left(u_j^n\right) = \varphi_j\left(E_{j,Q}\left(u_j^n\right)\right)$$

$$= \varphi_j\left(\frac{\psi(q_j)^{n-1}}{([\frac{n}{2}]+1)}\left(u_j^n\right)\right) = \frac{\psi(q_j)^{n-1}}{([\frac{n}{2}]+1)}\varphi_j\left(u_j^n\right)$$

$$= \frac{\psi(q_j)^{n-1}}{([\frac{n}{2}]+1)}\omega_n\left(\left(\frac{n}{2}+1\right)\psi(q_j)\right)c_{\frac{n}{2}}$$

i.e.,

$$\tau_j\left(u_j^n\right) = \begin{cases} \psi(q_j)^n\,c_{\frac{n}{2}} & \text{if } n \text{ is even} \\ 0 & \text{if } n \text{ is odd,} \end{cases} \tag{11.4.16}$$

for all $n \in \mathbb{N}$, for $j \in \mathbb{Z}$.

**Theorem 11.4.4.** *Let $\mathcal{L}_Q(j) = (\mathcal{L}_Q, \tau_j)$ be the $j$-th filtered probability space of the radial projection algebra $\mathcal{L}_Q$ on $Q$, where $\tau_j$ are in the sense of (11.4.14), for an arbitrarily fixed $j \in \mathbb{Z}$. Then*

$$\tau_j\left(u_i^n\right) = \delta_{j,i}\,\omega_n\psi(q_j)^n\,c_{\frac{n}{2}}, \tag{11.4.17}$$

*for all $n \in \mathbb{N}$, for all $i \in \mathbb{Z}$, where $\omega_n$ are in the sense of (11.3.3).*

*Proof.* If $i = j$ in $\mathbb{Z}$, then the free momental data (11.4.17) holds true by (4.16), for all $n \in \mathbb{N}$.

If $i \neq j$ in $\mathbb{Z}$, then, by the very definition (11.4.13) of the $j$-th filterization $E_{j,Q}$, and also by the definition (11.4.2) of $\varphi_j$,

$$\tau_j\left(u_i^n\right) = 0.$$

Therefore, the above formula (11.4.17) holds true, for all $i, j \in \mathbb{Z}$. □

The following theorem is a direct consequence of the above free distribution (11.4.17).

**Theorem 11.4.5.** *Let $\mathcal{L}_Q(j)$ be the $j$-th filtered probability space (11.4.15) of $\mathcal{L}_Q$ for $j \in \mathbb{Z}$, and let $u_j = l \otimes q_j$ be the $j$-th generating operator of $\mathcal{L}_Q$. Then $u_j$ is $\psi(q_j)^2$-semicircular in $\mathcal{L}_Q(j)$.*

*Proof.* Observe that, for any fixed $j \in \mathbb{Z}$,

$$u_j^* = (l \otimes q_j)^* = l^* \otimes q_j^* = l \otimes q_j = u_j,$$

in $\mathcal{L}_Q$. So, the operator $u_j$ is self-adjoint in $\mathcal{L}_Q$, and hence, it is a self-adjoint free random variable in the $j$-th filtered probability space $\mathcal{L}_Q(j)$.

Thus, the $\psi(q_j)^2$ semicircularity of $u_j$ is guaranteed by the free-moment computation (11.4.17), by the characterization (11.3.3) of the weighted-semicircularity (11.3.1). □

Readers can check an alternative proof of the above theorem in Cho (2018). Indeed, in Cho (2018), based on the free-moment formula

$$\tau_j\left(u_j^n\right) = \omega_n\,\psi(q_j)^n\,c_{\frac{n}{2}},$$

from (11.4.16), we obtained the free-cumulant formula

$$k_n^j (u_j, \ ..., \ u_j) = \begin{cases} \psi(q_j)^2 & \text{if } n = 2 \\ 0 & \text{otherwise,} \end{cases} \qquad (11.4.18)$$

for all $n \in \mathbb{N}$, where $k_n^j(...)$ means the free cumulant on the radial projection algebra $\mathfrak{L}_Q$ in terms of $\tau_j$, for all $j \in \mathbb{Z}$. Thus, by the definition (11.3.1), the self-adjoint free random variables $u_j$ is $\psi(q_j)^2$-semicircular in $\mathfrak{L}_Q(j)$, for all $j \in \mathbb{Z}$.

---

## 11.5   Semicircular elements induced by Q

The main results of Section 11.4 show that, for any $j \in \mathbb{Z}$, the $j$-th generating operator $u_j = l \otimes q_j$ of $\mathfrak{L}_Q$ is $\psi(q_j)^2$-semicircular in the $j$-th filtered probability space $\mathfrak{L}_Q(j)$. (Remark that all other $i$-th generating operators $u_i$ of $\mathfrak{L}_Q(j)$ have zero free distribution, whenever $i \neq j$ in $\mathbb{Z}$.) i.e., we have

$$\tau_j \left( u_j^n \right) = \omega_n \psi(q_j)^n c_{\frac{n}{2}},$$

equivalently, $\qquad\qquad\qquad\qquad\qquad\qquad\qquad\qquad\qquad (11.5.1)$

$$k_n^j (u_j, \ ..., \ u_j) = \begin{cases} \psi(q_j)^2 & \text{if } n = 2 \\ 0 & \text{otherwise,} \end{cases}$$

for all $n \in \mathbb{N}$, by (11.4.17) and (11.4.18), respectively.

Let $U_j$ be a free random variable in the $j$-th filtered probability space $\mathfrak{L}_Q(j)$ induced by the $\psi(q_j)^2$-semicircular element $u_j$,

$$U_j = \tfrac{1}{\psi(q_j)} \, u_j \in \mathfrak{L}_Q(j), \qquad (11.5.2)$$

for $j \in \mathbb{Z}$. Remember that, in Section 11.3, we assumed that $\psi(q_j) \in \mathbb{C}^\times$, and hence, the operators $U_j$ of (5.2) are well-defined in the projection radial algebra $\mathfrak{L}_Q$, for all $j \in \mathbb{Z}$.

Then the operator $U_j$ of (11.5.2) is semicircular in $\mathfrak{L}_Q(j)$ under a certain additional condition.

**Theorem 11.5.1.** *Let* $U_j = \tfrac{1}{\psi(q_j)} \, u_j$ *be a free random variable (5.2) of the $j$-th filtered probability space $\mathfrak{L}_Q(j)$, for $j \in \mathbb{Z}$. If*
$$\psi(q_j) \in \mathbb{R}^\times = \mathbb{R} \setminus \{0\} \text{ in } \mathbb{C},$$
*then $U_j$ is semicircular in $\mathfrak{L}_Q(j)$.*

*Proof.* Since $\psi(q_j)$ is assumed to be a real number in $\mathbb{R}^\times$, this operator $U_j$ is self-adjoint in the radial projection algebra $\mathfrak{L}_Q$, by the self-adjointness of $\psi(q_j)^2$-semicircular element $u_j$ in $\mathfrak{L}_Q(j)$.

Observe now that

$$
\begin{aligned}
\tau_j \left( U_j^n \right) &= \tau_j \left( \tfrac{1}{\psi(q_j)^n} u_j^n \right) = \tfrac{1}{\psi(q_j)^n} \tau_j (u_j^n) \\
&= \tfrac{1}{\psi(q_j)^n} \left( \omega_n \psi(q_j)^n c_{\frac{n}{2}} \right) = \omega_n c_{\frac{n}{2}},
\end{aligned}
\tag{11.5.3}
$$

by the $\psi(q_j)^2$-semicircularity (5.1) of $u_j$, for all $n \in \mathbb{N}$, for $j \in \mathbb{Z}$.

Therefore, by the characterization (11.3.4) of the semicircularity (11.3.2), this self-adjoint operator $U_j$ is semicircular in $\mathfrak{L}_Q(j)$, for $j \in \mathbb{Z}$.  □

---

## 11.6   The free product Banach $*$-probability space $\underset{j \in \mathbb{Z}}{\star} \mathfrak{L}_Q(j)$

A family $\{a_n\}_{n \in \Lambda}$ in an arbitrary (topological or pure-algebraic) $*$-probability space $(B, \varphi)$ is said to be a *free family*, if all elements $a_n$ of the family are mutually free from each other in $(B, \varphi)$, where $\Lambda$ is a countable (finite or infinite) index set. For a free family $\{a_n\}_{n \in \Lambda}$, if every element $a_n$ is weighted-semicircular (or semicircular), then we call the free family, *free weighted-semicircular* (resp., *semicircular*) *family* in $(B, \varphi)$.

Recall that, for a fixed $C^*$-probability space $(A, \psi)$, if there exists a family $\{q_j\}_{j \in \mathbb{Z}}$ of mutually orthogonal projections, then one can construct $\psi(q_j)^2$-semicircular elements

$$
u_j = l \otimes q_j
\tag{11.6.1}
$$

in the $j$-th filtered probability spaces $\mathfrak{L}_Q(j) = (\mathfrak{L}_Q, \tau_j)$, for all $j \in \mathbb{Z}$, by (11.4.16) and (11.4.17).

Moreover, one can construct semicircular elements

$$
U_j = \tfrac{1}{\psi(q_j)} u_j \text{ in } \mathfrak{L}_Q(j),
\tag{11.6.2}
$$

whenever $\psi(q_j) \in \mathbb{R}^\times$ in $\mathbb{C}$, for $j \in \mathbb{Z}$, by (11.5.3).

**Assumption and Notation 11.5.1** In the rest of this chapter, we assume

$$
\psi(q_j) \in \mathbb{R}^\times \text{ in } \mathbb{C}, \text{ for all } j \in \mathbb{Z}.
$$

It means that, for any $\psi(q_j)^2$-semicircular elements $u_j$ of (11.6.1), one has their corresponding semicircular elements $U_j$ of (11.6.2), for "all" $j \in \mathbb{Z}$. □

Now, we will construct the *free product Banach ∗-probability space* $(\mathfrak{L}_Q(\mathbb{Z}),$ $\tau)$, by

$$(\mathfrak{L}_Q(\mathbb{Z}),\ \tau)\ \overset{def}{=}\ \underset{j\in\mathbb{Z}}{\star}\mathfrak{L}_Q(j)$$

$$=\left(\underset{j\in\mathbb{Z}}{\star}\mathfrak{L}_Q,\ \underset{j\in\mathbb{Z}}{\star}\tau_j\right),$$

satisfying (11.6.3)

$$\mathfrak{L}_Q(\mathbb{Z})=\underset{j\in\mathbb{Z}}{\star}\mathfrak{L}_Q\ \overset{\text{∗-iso}}{=}\ (\mathfrak{L}_Q)^{\star|\mathbb{Z}|}\,,\ \text{and}\ \tau=\underset{j\in\mathbb{Z}}{\star}\tau_j,$$

where $(\star)$ means the *free product* (over $\mathbb{C}$) in the sense of Speicher (1998), Voiculescu (1997) and Voiculescu et al. (1992).

**Definition 11.6.1.** The free product Banach ∗-probability space

$\mathfrak{L}_Q(\mathbb{Z})\ \overset{denote}{=}\ (\mathfrak{L}_Q(\mathbb{Z}),\ \tau)$

of (11.6.3) is called the free filterization of $Q$.

By the very construction (11.6.3) of the free filterization $(\mathfrak{L}_Q(\mathbb{Z}),\ \tau)$, we obtain the following proposition.

**Proposition 11.6.2.** *Let* $(\mathfrak{L}_Q(\mathbb{Z}),\ \tau)$ *be the free filterization (11.6.3) of* $Q$, *and let* $u_j$ *and* $U_j$ *be in the sense of (11.6.1) and (11.6.2), respectively.*

*(11.6.4) The family* $\{u_j\in\mathfrak{L}_Q(j)\}_{j\in\mathbb{Z}}$ *is a free weighted-semicircular family in* $\mathfrak{L}_Q(\mathbb{Z})$.

*(11.6.5) The family* $\{U_j\in\mathfrak{L}_Q(j)\}_{j\in\mathbb{Z}}$ *is a free semicircular family in* $\mathfrak{L}_Q(\mathbb{Z})$.

*Proof.* By the very definition (11.6.3) of the free filterization $\mathfrak{L}_Q(\mathbb{Z})$, all entries $u_j$ of the family in (6.4) are free from each other for all $j\in\mathbb{Z}$. Indeed, all $u_j$ are contained in mutually distinct free blocks $\mathfrak{L}_Q(j)$ of $\mathfrak{L}_Q(\mathbb{Z})$, for all $j\in\mathbb{Z}$. So, the family

$\{u_j\in\mathfrak{L}_Q(j)\}_{j\in\mathbb{Z}}$

forms a free family in $\mathfrak{L}_Q(\mathbb{Z})$. Since every $u_j$ is $\psi(q_j)^2$-semicircular in $\mathfrak{L}_Q(j)$, it is $\psi(q_j)^2$-semicircular in $(\mathfrak{L}_Q(\mathbb{Z}),\ \tau)$, too. Indeed,

$\tau\left(u_j^n\right)=\tau_j\left(u_j^n\right)=\omega_n\psi(q_j)^n c_{\frac{n}{2}}$, for all $n\in\mathbb{N}$,

for all $j\in\mathbb{Z}$.

Thus, this family in (11.6.4) is a free weighted-semicircular family in $\mathfrak{L}_Q(\mathbb{Z})$.

Similarly, one can conclude the family

$\{U_j\in\mathfrak{L}_Q(j)\}_{j\in\mathbb{Z}}$

is a free semicircular family in $\mathfrak{L}_Q(\mathbb{Z})$. So, the statement (11.6.5) holds. $\square$

*Remark* 11.6.3. Let $\mathfrak{L}_Q(\mathbb{Z})$ be the free filterization of $Q$, and let $\mathfrak{L}_Q(j)$ be free blocks of $\mathfrak{L}_Q(\mathbb{Z})$, as in (6.3), for all $j \in \mathbb{Z}$. For a fixed $j_1 \in \mathbb{Z}$, let's take the generating operators $u_{j_1}$ and $u_{j_2}$ of $\mathfrak{L}_Q(j_1)$, where $j_2 \neq j_1$ in $\mathbb{Z}$. Then, as free random variables in $\mathfrak{L}_Q(\mathbb{Z})$, one has that:
$$\tau\left(u_{j_1}^n\right) = \tau_{j_1}\left(u_{j_1}^n\right) = \omega_n \psi(q_{j_1})^n c_{\frac{n}{2}},$$
meanwhile,
$$\tau\left(u_{j_2}^n\right) = \tau_{j_1}\left(u_{j_2}^n\right) = 0,$$
by (11.4.17), for all $n \in \mathbb{N}$.

Similarly, if we take $u_{j_1}$ and $u_{j_2}$ in the different free block $\mathfrak{L}_Q(j_2)$ of $\mathfrak{L}_Q(\mathbb{Z})$, then
$$\tau\left(u_{j_1}^n\right) = \tau_{j_2}\left(u_{j_1}^n\right) = 0,$$
while
$$\tau\left(u_{j_2}^n\right) = \tau_{j_2}\left(u_{j_2}^n\right) = \omega_n \psi(q_{j_2})^n c_{\frac{n}{2}},$$
for all $n \in \mathbb{N}$.

So, under the same notations, $u_{j_1}$ is $\psi(q_{j_1})^2$-semicircular in $\mathfrak{L}_Q(\mathbb{Z})$, if and only if
$$u_{j_1} \in \mathfrak{L}_Q(j_1) \text{ in } \mathfrak{L}_Q(\mathbb{Z}).$$
If one picks $u_{j_1}$ in a free block $\mathfrak{L}_Q(j_2)$ of $\mathfrak{L}_Q(\mathbb{Z})$, where $j_2 \neq j_1$ in $\mathbb{Z}$, then it cannot be weighted-semicircular in $\mathfrak{L}_Q(\mathbb{Z})$.

So, from below, we need to be careful to check where $u_j$'s are picked in $\mathfrak{L}_Q(\mathbb{Z})$.

Motivated by the above remark, and by our filtering process, now, we focus on the following sub-structure of $\mathfrak{L}_Q(\mathbb{Z})$. Let

$$\mathcal{U} = \{u_j \in \mathfrak{L}_Q(j) : j \in \mathbb{Z}\},$$

and                                                          (11.6.6)

$$\mathcal{S} = \{U_j \in \mathfrak{L}_Q(j) : j \in \mathbb{Z}\}$$

be our free weighted-semicircular family (11.6.4), respectively, our free semicircular family (11.6.5) in $\mathfrak{L}_Q(\mathbb{Z})$. Construct the Banach $*$-subalgebra $\mathbb{L}_Q$ and $\mathbb{L}'_Q$ of $\mathfrak{L}_Q(\mathbb{Z})$ by
$$\mathbb{L}_Q = B^*\left(\mathcal{U}\right), \text{ and } \mathbb{L}'_Q = B^*\left(\mathcal{S}\right), \tag{6.7}$$
where $B^*(X)$ mean the Banach $*$-subalgebras of $\mathfrak{L}_Q(\mathbb{Z})$ generated by subsets $X$ of $\mathfrak{L}_Q(\mathbb{Z})$, and $\mathcal{U}$ and $\mathcal{S}$ are in the sense of (11.6.6).

**Theorem 11.6.4.** *Let $\mathbb{L}_Q$ and $\mathbb{L}'_Q$ be the Banach $*$-subalgebras (11.6.7) of the free filterization $\mathfrak{L}_Q(\mathbb{Z})$. Then*
$$\mathbb{L}_Q \overset{*\text{-}iso}{=} \mathbb{L}'_Q, \tag{11.6.8}$$
*where "$\overset{*\text{-}iso}{=}$" means "being Banach-$*$-isomorphic."*

*Moreover,*
$$\mathbb{L}_Q \overset{*\text{-}iso}{=} \underset{j \in \mathbb{Z}}{\star}\left(\overline{\mathbb{C}[\{u_j\}]}\right) \text{ in } \mathfrak{L}_Q(\mathbb{Z}), \tag{11.6.9}$$
*where*
$$\overline{\mathbb{C}[\{u_j\}]} \subset \mathfrak{L}_Q(j) \text{ in } \mathfrak{L}_Q(\mathbb{Z}), \text{ for all } j \in \mathbb{Z},$$
*where $\overline{X}$ are the Banach-norm-topology closures of subsets $X$ of $\mathfrak{L}_Q(\mathbb{Z})$.*

*Proof.* Let $\mathbb{L}_Q = B^*(\mathcal{U})$ be a Banach $*$-subalgebra of $\mathfrak{L}_Q(\mathbb{Z})$ as in (11.6.7), where $\mathcal{U}$ is the free weighted-semicircular family of (11.6.6). Then

$$\mathbb{L}_Q = B^*(\mathcal{U}) = \overline{\mathbb{C}[\mathcal{U}]}$$

$$= \overline{\mathbb{C}[\{u_j \in \mathfrak{L}_Q(j) : j \in \mathbb{Z}\}]}$$

$$= \overline{\underset{j\in\mathbb{Z}}{\star}(\mathbb{C}[\{u_j\}])}$$

by the freeness on $\mathcal{U}$

$$= \underset{j\in\mathbb{Z}}{\star}\overline{\mathbb{C}[\{u_j\}]}, \tag{11.6.10}$$

in $\mathfrak{L}_Q(\mathbb{Z})$. So, we obtain the isomorphism theorem (11.6.9) by (11.6.10). Also, one can get that

$$\mathbb{L}_Q = \underset{j\in\mathbb{Z}}{\star}\overline{\mathbb{C}[\{u_j\}]}$$

by (11.6.10)

$$= \underset{j\in\mathbb{Z}}{\star}\overline{\mathbb{C}[\{\psi(q_j)U_j\}]}$$

$$= \underset{j\in\mathbb{Z}}{\star}\overline{\mathbb{C}[\{U_j\}]} = \overline{\underset{j\in\mathbb{Z}}{\star}\mathbb{C}[\{U_j\}]}$$

$$= \overline{\mathbb{C}[\mathcal{S}]} = B^*(\mathcal{S}) = \mathbb{L}'_Q. \tag{11.6.11}$$

Therefore, the isomorphism theorem (11.6.8) holds by (11.6.11). □

So, from below, we use $\mathbb{L}_Q$ and $\mathbb{L}'_Q$ alternatively by (11.6.8), as a $*$-isomorphic Banach $*$-subalgebra (11.6.9) in $\mathfrak{L}_Q(\mathbb{Z})$.

**Definition 11.6.5.** Let $\mathbb{L}_Q$ be the Banach $*$-subalgebra (11.6.7) in the free filterization $(\mathfrak{L}_Q(\mathbb{Z}), \tau)$. Under the restricted linear functional $\tau = \tau\mid_{\mathbb{L}_Q}$ on $\mathbb{L}_Q$, the Banach $*$-probability space

$$\mathbb{L}_Q \overset{denote}{=} (\mathbb{L}_Q, \tau) \tag{11.6.12}$$

is called the (free-)semicircular filterization of $Q$ (in $\mathfrak{L}_Q(\mathbb{Z})$).

---

## 11.7 Weighted-semicircular processes on free filterizations

Let $\mathbb{L}_Q = (\mathbb{L}_Q, \tau)$ be the semicircular filterization (11.6.12) in the free filterization $\mathfrak{L}_Q(\mathbb{Z})$, and let

$$\mathbb{L}_Q(j) = \left(\overline{\mathbb{C}[\{u_j\}]}, \tau_j\right) \subset \mathfrak{L}_Q(j) \tag{7.1}$$

be the free blocks of the semicircular filterization $\mathbb{L}_Q$ of $Q$, and let

$$u_j = l \otimes q_j \in \mathbb{L}_Q(j),$$

for all $j \in \mathbb{Z}$. Then $u_j$ is $\psi(q_j)^2$-semicircular in $\mathbb{L}_Q$ (and hence, in $\mathcal{L}_Q(\mathbb{Z})$), for all $j \in \mathbb{Z}$, by (11.4.17) and (11.4.18). In other words, whenever $u_i$ are given in $\mathbb{L}_Q$, they are contained in the free blocks $\mathbb{L}_Q(i)$ of (11.7.1), for $i$, and hence, they are $\psi(q_i)^2$-semicircular in $\mathbb{L}_Q$, for all $i \in \mathbb{Z}$.

Let

$$\mathbb{R}_0^+ = [0, \infty) = \{t \in \mathbb{R} : t \geq 0\}$$

and let

$$M = L^\infty(\mathbb{R}_0^+),$$

the commutative *von Neumann algebra* consisting of all *essentially bounded* (usual) *Lebesgue measurable functions* on $\mathbb{R}_0^+$.

Define now an element $\chi_t$ of $M$ by

$$\chi_t(x) = \chi_{[0,t)}(x), \ \forall t, \ x \in \mathbb{R}_0^+, \tag{11.7.2}$$

where $\chi_{[0,t)}$ are the *characteristic functions* for the half-open intervals $[0, t)$ in $\mathbb{R}_0^+$.

Define now a *tensor product Banach *-algebra* $\mathbb{L}_Q(M)$ by

$$\mathbb{L}_Q(M) \overset{def}{=} M \otimes_{\mathbb{C}} \mathbb{L}_Q, \tag{11.7.3}$$

under product topology.

For the semicircular filterization $\mathbb{L}_Q$, and the Banach *-algebra $\mathbb{L}_Q(M)$ of (11.7.3), define a system of operators

$$\mathcal{X} = \{X_t : \mathbb{L}_Q \to \mathbb{L}_Q(M) \mid t \in \mathbb{R}_0^+\}, \tag{11.7.4}$$

where every element $X_t$ of the family $\mathcal{X}$ of (11.7.4) satisfies

$$X_t(T) = \chi_t \otimes T, \text{ for all } t \in \mathbb{R}_0^+,$$

for all $T \in \mathbb{L}_Q$, where $\chi_t$ are in the sense of (7.2) in $M = L^\infty(\mathbb{R}_0^+)$. For example, one has

$$X_t(u_k) = \chi_t \otimes u_k = \chi_t \otimes (l \otimes q_k),$$

for all $t \in \mathbb{R}_0^+$, and $k \in \mathbb{Z}$.

Define now a binary operation on the set $\mathcal{X}$ of (11.7.4) by

$$X_{t_1} X_{t_2} = X_{t_1+t_2}, \text{ for all } t_1, t_2 \in \mathbb{R}_0^+. \tag{11.7.5}$$

Then, this operation (11.7.5) is well-defined on $\mathcal{X}$, because $\mathbb{R}_0^+ = (\mathbb{R}_0^+, +)$ is a well-defined *monoid* (a *semigroup* with its operation-identity 0) under the usual addition $(+)$.

For instance, one can get that

$$X_{t_1} X_{t_2}(u_j) = X_{t_1+t_2}(u_j) = \chi_{t_1+t_2} \otimes u_j, \tag{11.7.6}$$

by (11.7.5), for all $t_1, t_2 \in \mathbb{R}_0^+$, for all $u_j \in \mathbb{L}_Q$ for $j \in \mathbb{Z}$.

Under the binary operation (11.7.5) satisfying (11.7.6), this system $\mathcal{X}$ of (11.7.4) forms a monoid, which is monoid-isomorphic to $(\mathbb{R}_0^+, +)$. Indeed, one can construct a bijective map

$$X_t \in \mathcal{X} \longmapsto t \in \mathbb{R}_0^+,$$

such that

$$X_{t_1} X_{t_2} = X_{t_1+t_2} \longmapsto t_1 + t_2,$$

for all $t, t_1, t_2 \in \mathbb{R}_0^+$.

Inductive to (11.7.5), one has that,

$$\prod_{k=1}^{m} X_{t_k} = X_{\Sigma_{k=1}^n t_k}, \qquad (11.7.6)'$$

on $\mathcal{X}$, for all $t_k \in \mathbb{R}_0^+$, $j \in \mathbb{Z}$, for $k = 1, ..., m$, for all $m \in \mathbb{N}$.

**Lemma 11.7.1.** *Let $\mathcal{X}$ be the monoid (11.7.4) under its binary operation (11.7.5), and let $u_j$ be the $j$-th generating operator of the semicircular filterization $\mathbb{L}_Q$, for $j \in \mathbb{Z}$. Then*

$(11.7.7)$   $X_t \left( u_j^n \right) = \chi_t \otimes u_j^n$ *in* $\mathbb{L}_Q(M)$,

$(11.7.8)$   $\left( \prod_{k=1}^{m} X_{t_k} \right) \left( u_j^n \right) = \chi_{\Sigma_{k=1}^m t_k} \otimes u_j^n$ *in* $\mathbb{L}_Q(M)$,

*for all $X_t, X_{t_k} \in \mathcal{X}$, for $k = 1, ..., m$, for all $m, n \in \mathbb{N}$.*

*Proof.* The proof of (11.7.7) is straightforward by (11.7.4), (11.7.5), (11.7.6) and (11.7.6)'. And the statement (11.7.8) is shown by (11.7.7).   □

Define a linear morphism,

$$F : \mathbb{L}_Q(M) \to M = L^\infty \left( \mathbb{R}_0^+ \right),$$

by a linear transformation satisfying that

$$F\left( f \otimes T \right) = \left( \tau\left( T \right) \right) f, \qquad (11.7.9)$$

for all $f \otimes T \in \mathbb{L}_Q(M)$, where $\tau$ is the linear functional (11.6.12) on the semicircular filterization $\mathbb{L}_Q$.

Define now a linear functional $\tau^M : \mathbb{L}_Q(M) \to \mathbb{C}$ by a linear transformation satisfying

$$
\begin{aligned}
\tau^M \left( f \otimes T \right) &= \int_0^\infty F\left( f \otimes T \right) dx \\
&= \left( \tau(T) \right) \left( \int_0^\infty f \, dx \right),
\end{aligned}
\qquad (11.7.10)
$$

for all $f \otimes T \in \mathbb{L}_Q(M)$, with $f \in M$, and $T \in \mathbb{L}_Q$, where $F$ is in the sense of (11.7.9).

The morphisms $\tau^M$ of (11.7.10) are indeed well-determined linear functionals on $\mathbb{L}_Q(M)$, providing Banach $*$-probability spaces

$$\left( \mathbb{L}_Q(M), \ \tau^M \right), \text{ for all } j \in \mathbb{Z}.$$

**Definition 11.7.2.** The Banach $*$-probability spaces

$$\mathbb{L}_Q(M) \overset{denote}{=} \left( \mathbb{L}_Q(M), \ \tau^M \right) \qquad (11.7.11)$$

are called the (free-stochastic) $\mathbb{L}_Q$-acting (Banach $*$-)probability space.

For example, if $X_t\left(u_j^n\right) = \chi_t \otimes u_j^n$ is a free random variable of the $\mathbb{L}_Q$-acting probability space $\mathbb{L}_Q(M)$ of (11.7.11), for $X_t \in \mathcal{X}$, $j \in \mathbb{Z}$, $n \in \mathbb{N}$, then

$$\tau^M\left(X_t(u_j^n)\right) = t\,\tau_j\left(u_j^n\right) = t\,\omega_n \psi(q_j)^n c_{\frac{n}{2}},$$

by (11.7.10) and by the $\psi(q_j)^2$-semicircular of $u_j$ in $\mathbb{L}_Q$ (for all $j \in \mathbb{Z}$) because

$$\int_0^\infty \chi_t\,dx = t, \text{ for all } t \in \mathbb{R}_0^+.$$

**Proposition 11.7.3.** *Let $\mathbb{L}_Q(M)$ be the $\mathbb{L}_Q$-acting probability space (11.7.11) for $j \in \mathbb{Z}$, and let*
$$X_t(u_j) \in \mathbb{L}_Q(M), \text{ for } u_j \in \mathcal{U},$$
*Then*
$$\tau^M\left(X_t(u_j^n)\right) = t\,\omega_n\,\psi(q_j)^n\,c_{\frac{n}{2}}, \tag{11.7.12}$$
*for all $n \in \mathbb{N}$, for all $t \in \mathbb{R}_0^+$, where $\omega_n$ are in the sense of (11.3.3).*

*Proof.* By (11.7.11), one has that
$$\tau^M\left((X_t(u_j))^n\right) = \tau^M\left(\chi_t \otimes u_j^n\right) = t\left(\omega_n \psi(q_j)^n c_{\frac{n}{2}}\right),$$
for all $n \in \mathbb{N}$. $\square$

The above free-moment formula (11.7.12) shows that we have a certain system,
$$X_j^{(n)} = \left\{X_t(u_j^n)\right\}_{t \in \mathbb{R}_0^+}, \tag{11.7.13}$$
determined both by weighted-semicircularity on $\mathbb{L}_Q$, and *time flow* $\mathbb{R}_0^+$ (or the monoid $\mathcal{X}$ of (11.7.4)), for each $j \in \mathbb{Z}$, for all $n \in \mathbb{N}$, in $\mathbb{L}_Q(M)$.

**Definition 11.7.4.** The free-stochastic process,
$$X_j \overset{denote}{=} X_j^{(2)} = \left\{X_t(u_j^2)\right\}_{t \in \mathbb{R}_0^+}, \tag{11.7.14}$$
is called the $j$-th weighted-semicircular (free-)motion in the $\mathbb{L}_Q$-acting probability space $\mathbb{L}_Q(M)$, for $j \in \mathbb{Z}$, where $X_j^{(2)}$ is in the sense of (11.7.13).

Let $X_j$ be the $j$-th weighted-semicircular motion (11.7.14) in $\mathbb{L}_Q(M)$, and let
$$X_t(u_j^2) = \chi_t \otimes u_j^2 \in X_j, \text{ for } t \in \mathbb{R}_0^+.$$

Then
$$\begin{aligned}\left(X_t(u_j^2)\right)^n &= \left(\chi_t \otimes u_j^2\right)^n = \chi_t^n \otimes u_j^{2n}\\ &= \chi_t \otimes u_j^{2n} = X_t(u_j^{2n}),\end{aligned} \tag{11.7.15}$$
and
$$\left(X_t(u_j^2)\right)^* = \chi_t^* \otimes \left(u_j^2\right)^* = X_t(u_j^2), \tag{11.7.16}$$
in $\mathbb{L}_Q(M)$, for all $n \in \mathbb{N}$.

By the self-adjointness (11.7.16) of $X_t(u_j)$ (for all $t \in \mathbb{R}_0^+$) in the $\mathbb{L}_Q$-acting probability space $\mathbb{L}_Q(M)$, one obtains the following free distribution of $X_t(u_j)$.

**Theorem 11.7.5.** *Let $X_j$ be the j-th weighted-semicircular motion (11.7.14) in the $\mathbb{L}_Q$-acting probability space $\mathbb{L}_Q(M)$, where $\mathbb{L}_Q$ is the semicircular filterization of $Q$, and $M = L^\infty(\mathbb{R}_0^+)$, for $j \in \mathbb{Z}$, and let $X_t(u_j^2) \in X_j$, for $t \in \mathbb{R}_0^+$. Then the free distribution of $X_t(u_j^2)$ is represented by the sequence,*

$$\left(t\psi(q_j)^{2n} c_n\right)_{n=1}^\infty, \tag{11.7.17}$$

*in $\mathbb{L}_Q(M)$, for all $n \in \mathbb{N}$.*

*Proof.* Note that, by (11.7.15),

$\left(X_t(u_j^2)\right)^n = X_t(u_j^{2n})$ in $\mathbb{L}_Q(M)$, $\forall n \in \mathbb{N}$.

Thus, one has

$$\tau^M\left(\left(X_t(u_j^2)\right)^n\right) = \tau^M\left(X_t(u_j^{2n})\right)$$

$$= \left(\int_0^t \chi_t \, dx\right)\left(\tau\left(u_j^{2n}\right)\right) = t\left(\tau_j\left(u_j^{2n}\right)\right)$$

$$= t\,\psi(q_j)^{2n}\, c_n,$$

for all $n \in \mathbb{N}$, by the $\psi(q_j)^2$-semicircularity of the generating operator $u_j$ of the semicircular filterization $\mathbb{L}_Q$ in $\mathbb{L}_Q(M)$.

By the self-adjointness (11.7.16) of $X_t(u_j^2)$ in $\mathbb{L}_Q(M)$, the free-moment sequence

$\left(t\psi(q_j)^{2n} c_n\right)_{n=1}^\infty$

represents the free distribution of the flows $X_t(u_j^2)$ in the motion $X_j$, for all $t \in \mathbb{R}_0^+$. $\qquad\square$

The above theorem characterizes the free distributions of flows $X_t(u_j^2)$ in the j-th weighted-semicircular motion $X_j$ as elements of the $\mathbb{L}_Q$-acting probability space $\mathbb{L}_Q(M)$, for all $j \in \mathbb{Z}$. One can realize that the free distributions (11.7.17) of the flows $X_t(u_j)$ are dictated by the weighted-semicircularity of $u_j$ in the semicircular filterization $\mathbb{L}_Q$.

---

## 11.8   Free stochastic integrals for $X_j$

Throughout this section, let

$$\mathbb{L}_Q(M) = \left(\mathbb{L}_Q(M), \tau^M\right)$$

be the corresponding $\mathbb{L}_Q$-acting probability space (11.7.11), where $\mathbb{L}_Q(M)$ is in the sense of (11.7.3), and $\tau^M$ is in the sense of (11.7.10).

Note that, by definition,

$$\mathbb{L}_Q(M) \overset{def}{=} M \otimes_{\mathbb{C}} \mathbb{L}_Q$$

where $\mathbb{L}_Q$ is our semicircular filterization of $Q$

$$= M \otimes_{\mathbb{C}} \left( \underset{j \in \mathbb{Z}}{\star} \mathbb{L}_j \right) \text{ with } \mathbb{L}_j = \overline{\mathbb{C}[\{u_j\}]}$$

$$\overset{\text{*-iso}}{=} \underset{j \in \mathbb{Z}}{\star} \left( M \otimes_{\mathbb{C}} \mathbb{L}_j \right). \tag{11.8.1}$$

**Notation** Let's denote the $M$-affiliated free blocks $M \otimes_{\mathbb{C}} \mathbb{L}_j$ of the $\mathbb{L}_Q$-acting probability space $\mathbb{L}_Q(M)$ in (11.8.1) by $\mathbb{L}_Q(j, M)$, i.e.,

$$\mathbb{L}_Q(j, M) \overset{denote}{=} M \otimes_{\mathbb{C}} \overline{\mathbb{C}[\{u_j\}]},$$

for all $j \in \mathbb{N}$. From the very construction, and by (11.8.1), $\{\mathbb{L}_Q(j, M)\}_{j \in \mathbb{Z}}$ are free from each other in $\mathbb{L}_Q(M)$. Sometimes, we will say that $\mathbb{L}_Q(j, M)$ are the $j$-th free blocks of $\mathbb{L}_Q(M)$, for all $j \in \mathbb{Z}$. $\square$

Of course, the free blocks $\mathbb{L}_Q(j, M)$ of $\mathbb{L}_Q(M)$ are equipped with their (restricted) linear functionals $\tau_j^M = \tau^M |_{\mathbb{L}_Q(j,M)}$, for all $j \in \mathbb{Z}$.

Fix $j \in \mathbb{Z}$, and let

$$X_j = \left\{ X_t(u_j^2) \right\}_{t \in \mathbb{R}_0^+} \in \mathbb{L}_Q(j, M), \tag{11.8.2}$$

be the $j$-th *weighted-semicircular motion* (11.7.14) in $\mathbb{L}_Q(M)$.

In (11.7.17), we characterizes the free distributions of flows $X_t(u_j^2)$ in the motion $X_j$ of (11.8.2) in $\mathbb{L}_Q(M)$:

$$\left( \tau^M \left( \left( X_t(u_j^2) \right)^n \right) \right)_{n=1}^{\infty} = \left( \tau^M \left( X_t(u_j^{2n}) \right) \right)_{n=1}^{\infty}$$

$$= \left( t\psi(q_j)^{2n} c_n \right)_{n=1}^{\infty}. \tag{11.8.2}'$$

In this section, for the fixed free stochastic motion $X_j$ of (11.8.2) satisfying (11.8.2)', we construct-and-study *free stochastic integrals* for $X_j$.

Define now the tensor product Banach *-algebra

$$\mathbb{L}_Q^{\otimes 2} = \mathbb{L}_Q \otimes_{\mathbb{C}} \mathbb{L}_Q^{op} \tag{11.8.3}$$

of the semicircular filterization $\mathbb{L}_Q$ of $Q$, where $\mathbb{L}_Q^{op}$ is the opposite algebra with opposite product,

$$(T_1, T_2) \in \mathbb{L}_Q^{op} \times \mathbb{L}_Q^{op} \longmapsto T_2 T_1 \in \mathbb{L}_Q^{op},$$

equipped with the same topology for $\mathbb{L}_Q$

And act the algebra $\mathbb{L}_Q^{\otimes 2}$ of (11.8.3) on the Banach *-algebra $\mathbb{L}_Q(M)$, by the $\mathbb{L}_Q$-$\mathbb{L}_Q$-*bimodule action*, denoted by #, from left and right, i.e.,

$$(T_1 \otimes T_2) \# (f \otimes T) = f \otimes T_1 T T_2, \tag{11.8.4}$$

under linearity, for all $f \otimes T \in \mathbb{L}_Q(M)$, for all $T_1 \otimes T_2 \in \mathbb{L}_Q^{\otimes 2}$.

For example, if we take

$$u_{j_1}^{n_1} \otimes u_{j_2}^{n_2} \in \mathbb{L}_Q^{\otimes 2},$$

for $j_1, j_2 \in \mathbb{Z}$, and $n_1, n_2 \in \mathbb{N}$, and if

$$\chi_t \otimes u_j^m \in \mathbb{L}_Q(M),$$

for $t \in \mathbb{R}_0^+$, $j \in \mathbb{Z}$, $m \in \mathbb{N}$, then

$$\left(u_{j_1}^{n_1} \otimes u_{j_2}^{n_2}\right) \# \left(\chi_t \otimes u_j^m\right) = \chi_t \otimes \left(u_{j_1}^{n_1} u_j^m u_{j_2}^{n_2}\right),$$

by (11.8.4), where the tensor factor $u_{j_1}^{n_1} u_j^m u_{j_2}^{n_2}$ in the right-hand side of the above expression means a free (non-reduced, in general,) word in $\mathcal{U}$, as free random variables of the semicircular filterization $\mathbb{L}_Q$, where $\mathcal{U}$ is the free weighted-semicircular family (11.6.6) of the free filterization $\mathfrak{L}_Q(\mathbb{Z})$.

**Definition 11.8.1.** An element $T$ of the Banach $*$-algebra $\mathbb{L}_Q^{\otimes 2}$ of (11.8.3) is said to be a simple adapted biprocess, if there exists $N \in \mathbb{N}$, such that
$$T = \sum_{k=1}^{N} s_k \left(u_{i_k}^{n_k} \otimes u_{j_k}^{m_k}\right), \tag{11.8.5}$$
with

$$s_k \in \mathbb{C}, \; n_k, \; m_k \in \mathbb{N}, \; i_k, \; j_k \in \mathbb{Z}.$$

In fact, the general definition of *simple adapted biprocesses* in *free stochastic calculus* is more complicated than our definition (11.8.5). Since "our" simple adapted biprocesses (11.8.5) satisfy the suitable conditions of the general definition of simple adapted biprocesses (e.g., Speicher (2001)), we use (11.8.5) as our formal definition without loss of generality under our settings.

Observe that, for $X_t\left(u_k\right)^m = \chi_t \otimes u_k^m \in \mathbb{L}_Q(M)$, for $t \in \mathbb{R}_0^+$, $k \in \mathbb{Z}$, $m \in \mathbb{N}$, if $T$ is a simple adapted biprocess (11.8.5) in $\mathbb{L}_Q^{\otimes 2}$, then
$$T \# \left(\chi_t \otimes u_k^m\right) = \sum_{k=1}^{N} \left(\chi_t \otimes u_{i_k}^{n_k} u_k^m u_{j_k}^{m_k}\right), \tag{11.8.6}$$
in $\mathbb{L}_Q(M)$, for all $t \in \mathbb{R}_0^+$ and $k \in \mathbb{Z}$, for $N \in \mathbb{N}$.

By (11.8.6), if $X_t\left(u_k\right)^m = \chi_t \otimes u_k^m \in \mathbb{L}_Q(M)$, for $t \in \mathbb{R}_0^+$, $k \in \mathbb{Z}$, $m \in \mathbb{N}$, and if $u_{i_1}^{n_1} \otimes u_{i_2}^{n_2} \in \mathbb{L}_Q^{\otimes 2}$ is a simple adapted biprocess for $i_1$, $i_2 \in \mathbb{Z}$, and $n_1$, $n_2 \in \mathbb{N}$, then
$$\left(u_{i_1}^{n_1} \otimes u_{i_2}^{n_2}\right) \# \left(\chi_t \otimes u_k^m\right) = \chi_t \otimes \left(u_{i_1}^{n_1} u_k^m u_{i_2}^{n_2}\right), \tag{11.8.6$'$}$$

**Definition 11.8.2.** Let $T$ be a simple adapted biprocess (11.8.5). Define the free stochastic integral $\int_0^t T dX_j$ of $T$ for the $j$-th weighted-semicircular motion $X_j$ of (11.8.1), by

$$\int_0^t T \, dX_j \overset{def}{=} \sum_{k=0}^{n-1} T \# \left(u_{t_{k+1},j}^2 - u_{t_k,j}^2\right), \tag{11.8.7}$$
whenever

$$t_0 = 0 < t_1 < t_2 < \dots < t_{n-1} < t_n = t,$$

for $t \in \mathbb{R}_0^+$, and $\tag{11.8.8}$

$$u_{t,j} = X_t(u_j^2) = \chi_t \otimes u_j^2 \in X_j, \; \text{for } j \in \mathbb{Z}.$$

We call the integral (11.8.7), the $j$-th weighted-semicircular stochastic integral of $T$ for $j \in \mathbb{Z}$ (in short, $j$-th w-s-integral of $T$), for $t \in \mathbb{R}_0^+$.

Observe the definition (11.8.7) more in detail;

$$\int_0^t T \, dX_j = \sum_{k=0}^{n-1} \left( \left( \sum_{l=1}^N s_l \left( u_{i_l}^{n_l} \otimes u_{j_l}^{m_l} \right) \right) \# \left( u_{t_{k+1},j}^2 - u_{t_k,j}^2 \right) \right)$$

$$= \sum_{k=0}^{n-1} \sum_{l=1}^N s_l \left( u_{i_l}^{n_l} \otimes u_{j_l}^{m_l} \right) \# \left( u_{t_{k+1},j}^2 - u_{t_k,j}^2 \right)$$

$$= \sum_{k=0}^{n-1} \sum_{l=1}^N s_l \left( u_{i_l}^{n_l} u_{t_{k+1},j}^2 u_{j_l}^{m_l} - u_{i_l}^{n_l} u_{t_k,j}^2 u_{j_l}^{m_l} \right)$$

$$= \sum_{k=0}^{n-1} \sum_{l=1}^N s_l \left( u_{i_l}^{n_l} \left( u_{t_{k+1},j}^2 - u_{t_k,j}^2 \right) u_{j_l}^{m_l} \right), \tag{11.8.9}$$

by (11.8.7), whenever the condition (11.8.8) holds, where the products of $u_j$ and $u_{t,j}$ in (11.8.9) mean the resulted action (11.8.4), and

$$u_{s,j}^2 = X_s \left( u_j^2 \right) \text{ in } X_j, \text{ for } s \in \mathbb{R}_0^+.$$

If we fix a basic simple adapted biprocess

$$u_{i_1,i_2}^{n_1,n_2} = u_{i_1}^{n_1} \otimes u_{i_2}^{n_2} \in \mathbb{L}_Q^{\otimes 2}, \tag{11.8.10}$$

for $i_1, i_2 \in \mathbb{Z}$, and $n_1, n_2 \in \mathbb{N}$, then one can obtain the corresponding $j$-th w-s integral of $u_{i_1,i_2}^{n_1,n_2}$,

$$\int_0^t u_{i_1,i_2}^{n_1,n_2} \, dX_j = \sum_{k=0}^{n-1} \left( u_{i_1}^{n_1} \left( u_{t_{k+1},j}^2 - u_{t_k,j}^2 \right) u_{i_2}^{n_2} \right),$$

by (11.8.6)′ and (11.8.9), whenever

$$0 = t_0 < t_1 < \dots < t_{n-1} < t_n = t,$$

for some $n \in \mathbb{N}$.

**Notation** Let $X_j = \{ u_{t,j}^2 \stackrel{denote}{=} X_t(u_j^2) \}_{t \in \mathbb{R}_0^+}$ be the $j$-th weighted-semicircular motion (11.7.14) for a fixed $j \in \mathbb{Z}$. We write

$$u_{j,t_1<t_2}^2 \stackrel{denote}{=} u_{t_2,j}^2 - u_{t_1,j}^2 \text{ in } \mathbb{L}_Q(M),$$

for all $t_1 < t_2 \in \mathbb{R}_0^+$. □

By using the above new notation, one can re-write (11.8.9) as follows; if $T$ is in the sense of (11.8.5), then

$$\int_0^t T \, dX_j = \sum_{k=0}^{n-1} \sum_{l=1}^N \left( u_{i_l}^{n_l} \left( u_{j,t_k<t_{k+1}}^2 \right) u_{j_l}^{m_l} \right),$$

whenever (11.8.8) holds.

And hence, the $j$-th w-s-integral of a basic simple adapted biprocess $u_{i_1,i_2}^{n_1,n_2}$ of (11.8.10) satisfies that

$$\int_0^t u_{i_1,i_2}^{n_1,n_2} \, dX_j = \sum_{k=0}^{n-1} u_{i_1}^{n_1} u_{j,t_k<t_{k+1}}^2 u_{i_2}^{n_2}, \tag{11.8.11}$$

under (11.8.8).

Remark that the $j$-th w-s integral (11.8.11) of $u_{i_1,i_2}^{n_1,n_2}$ can be understood as a free random variable in the $\mathbb{L}_Q$-acting probability space $\mathbb{L}_Q(M)$. It means that the $j$-th w-s-integral (11.8.11) has its own free-distributional data in $\left(\mathbb{L}_Q(M),\ \tau^M\right)$.

Observe that

$$\tau^M\left(\int_0^\infty u_{i_1,i_2}^{n_1,n_2} dX_j\right) = \tau_j^M\left(\sum_{k=0}^{n-1}\left(u_{i_1}^{n_1} u_{j,t_k<t_{k+1}}^2 u_{i_2}^{n_2}\right)\right)$$

by (11.8.1) and (11.8.11) under the condition (11.8.8)

$$= \sum_{k=0}^{n-1} \tau_j^M\left(\chi_{t_k<t_{k+1}} \otimes u_{i_1}^{n_1} u_j^2 u_{i_2}^{n_2}\right)$$

where

$$\chi_{t<s} = \chi_{[t,s)} \text{ in } M = L^\infty(\mathbb{R}_0^+),$$

for $t < s \in \mathbb{R}_0^+$, and hence, it goes to

$$= \sum_{k=0}^{n-1}\left(\int_0^\infty \chi_{[t_k,t_{k+1})}\ dx\right) \tau\left(u_{i_1}^{n_1} u_j^2 u_{i_2}^{n_2}\right). \tag{11.8.12}$$

In the formula (11.8.12), note that the product $u_{i_1}^{n_1} u_j^2 u_{i_2}^{n_2}$ are free (non-reduced, or reduced) words in the semicircular filterization $\mathbb{L}_Q$. For example, if

$$i_1 = j \neq i_2 \text{ in } \mathbb{Z},$$

then

$$u_{i_1}^{n_1} u_j^2 u_{i_2}^{n_2} = u_j^{n_1+2} u_{i_2}^{n_2},$$

as a free "reduced" word with its length-2 in $\mathbb{L}_Q$; and if

$$i_1 = j = i_2 \text{ in } \mathbb{Z},$$

then

$$u_{i_1}^{n_1} u_j^2 u_{i_2}^{n_2} = u_j^{n_1+n_2+2}$$

contained in the free block $\overline{\mathbb{C}[\{u_j\}]}$ of $\mathbb{L}_Q$ by (11.6.9), and hence, it becomes a free reduced word with its length-1 in $\mathbb{L}_Q$; and if

$$i_1 \neq j, \text{ and } j \neq i_2 \text{ in } \mathbb{Z},$$

equivalently, the triple $(i_1, j, i_2) \in \mathbb{Z}^3$ is *alternating* in $\mathbb{Z}$, then $u_{i_1}^{n_1} u_j^2 u_{i_2}^{n_2}$ is a free reduced word with its length-3 in $\mathbb{L}_Q$ (e.g., see Speicher (1998), Voiculescu (1997) and Voiculescu et al. (1992)).

**Proposition 11.8.3.** *Let* $u_{i_1,i_2}^{n_1,n_2} = u_{i_1}^{n_1} \otimes u_{i_2}^{n_2}$ *be a simple adapted biprocess* (11.8.10) *in* $\mathbb{L}_Q^{\otimes 2}$, *for* $i_1,\, i_2 \in \mathbb{Z}$, *and* $n_1,\, n_2 \in \mathbb{N}$. *And let* $\int_0^t \left(u_{i_1,i_2}^{n_1,n_2}\right) dX_j$ *be the $j$-th w-s-integral of* $u_{i_1,i_2}^{n_1,n_2}$ *in* $\mathbb{L}_Q(M)$ *for* $t \in \mathbb{R}_0^+$. *Suppose first that*

$$i_1 = j = i_2 \ \text{in} \ \mathbb{Z}.$$

*Then*
$$\tau^M\left(\int_0^t u_{j,j}^{n_1,n_2}\right) dX_j = t\omega_{n_1+n_2+2}\psi(q_j)^{n_1+n_2+2}c_{\frac{n_1+n_2+2}{2}}. \qquad (11.8.13)$$

*Assume now that either*

$$i_1 = j \neq i_2, \ \text{or} \ i_1 \neq j = i_2 \ \text{in} \ \mathbb{Z}.$$

*Then*
$$\tau^M\left(\int_0^t u_{j,i_2}^{n_1,n_2} dX_j\right) = t\omega_{n_1+2}\omega_{n_2}\psi(q_j)^{n_1+2}\psi(q_{i_2})^{n_2}c_{\frac{n_1+2}{2}}c_{\frac{n_2}{2}},$$

*respectively*                                  (11.8.14)

$$\tau^M\left(\int_0^t u_{i_1,j}^{n_1,n_2} dX_j\right) = t\omega_{n_1}\omega_{n_2+2}\psi(q_{i_1})^{n_1}\psi(q_j)^{n_2+2}c_{\frac{n_1}{2}}c_{\frac{n_2+2}{2}}.$$

*Finally, let* $(i_1,\, j,\, i_2) \in \mathbb{Z}^3$ *is mutually distinct in* $\mathbb{Z}$, *in the sense that:*

$$i_1 \neq j,\, j \neq i_2 \ \text{and} \ i_1 \neq i_2 \ \text{in} \ \mathbb{Z}.$$

*Then*
$$\tau^M\left(\int_0^t u_{i_1,i_2}^{n_1,n_2} dX_j\right) = t\,\omega_{n_1}\omega_{n_2}\psi(q_{i_1})^{n_1}\psi(q_j)^2\psi(q_{i_2})^{n_2}c_{\frac{n_1}{2}}c_{\frac{n_2}{2}}. \quad (11.8.15)$$

*Proof.* One can get that
$$\tau^M\left(\int_0^t u_{i_1,i_2}^{n_1,n_2} dX_j\right)$$

$$= \sum_{k=0}^{n-1}(t_{k+1} - t_k)\left(\tau\left(u_{i_1}^{n_1}u_j^2 u_{i_2}^{n_2}\right)\right)$$
by (11.8.12)
$$= \left(\tau\left(u_{i_1}^{n_1}u_j^2 u_{i_2}^{n_2}\right)\right)\left(\sum_{k=0}^{n-1}(t_{k+1} - t_k)\right)$$

$$= t\,\tau\left(u_{i_1}^{n_1}u_j^2 u_{i_2}^{n_2}\right), \qquad\qquad\qquad\qquad (11.8.16)$$
under (11.8.8).

Suppose first that $i_1 = j = i_2$ in $\mathbb{Z}$. Then, the free word $u_{i_1}^{n_1}u_j^2 u_{i_2}^{n_2}$ becomes the free reduced word $u_j^{n_1+n_2+2}$ with its length-1 in the semicircular filterization $\mathbb{L}_Q$. Thus we have that

$$\tau^M\left(\int_0^t u_{i_1,i_2}^{n_1,n_2} dX_j\right) = t\,\tau\left(u_j^{n_1+n_2+2}\right) = t\,\tau_j\left(u_j^{n_1+n_2+2}\right)$$

by (11.8.16)

$$= t\,\omega_{n_1+n_2+2}\psi(q_j)^{n_1+n_2+2}c_{\frac{n_1+n_2+2}{2}},$$

by the $\psi(q_j)^2$-semicircularity of $u_j \in \mathcal{U}$ in $\mathbb{L}_Q$, where $\mathcal{U}$ is the free weighted-semicircular family (11.6.6) generating $\mathbb{L}_Q$. Therefore, the formula (11.8.13) holds whenever $i_1 = j = i_2$ in $\mathbb{Z}$.

Assume now that $i_1 = j \neq i_2$ in $\mathbb{Z}$. Then the free word $u_{i_1}^{n_1}u_j^2u_{i_2}^{n_2}$ forms the free reduced word $u_j^{n_1+2}u_{i_2}^{n_2}$ with its length-2 in $\mathbb{L}_Q$. So, by (11.8.16), one can get that

$$\tau^M\left(\int_0^t u_{i_1,i_2}^{n_1,n_2}dX_j\right) = t\,\tau\left(u_j^{n_1+2}u_{i_2}^{n_2}\right)$$
$$= t\left(\tau_j\left(u_j^{n_1+2}\right)\right)\left(\tau_{i_2}\left(u_{i_2}^{n_2}\right)\right)$$

by the freeness of $u_j$ and $u_{i_2}$ in $\mathbb{L}_Q$ (See (11.6.4) or (11.6.9))

$$= t\left(\omega_{n_1+2}\psi(q_j)^{n_1+2}c_{\frac{n_1+2}{2}}\right)\left(\omega_{n_2}\psi(q_{i_2})^{n_2}c_{\frac{n_2}{2}}\right)$$

by the weighted-semicircularity of $u_j$ and $u_{i_2}$ in $\mathbb{L}_Q$

$$= t\omega_{n_1+2}\omega_{n_2}\,\psi(q_j)^{n_1+2}\psi(q_{i_2})^{n_2}c_{\frac{n_1+2}{2}}c_{\frac{n_2}{2}}.$$

Therefore, the first formula of (11.8.14) holds whenever $i_1 = j \neq i_2$ in $\mathbb{Z}$. Similarly, if $i_1 \neq j = i_2$ in $\mathbb{Z}$, then

$$\tau^M\left(\int_0^t u_{i_1,i_2}^{n_1,n_2}dX_j\right) = t\,\omega_{n_1}\omega_{n_2+2}\psi(q_{i_1})^{n_1}\psi(q_j)^{n_2+2}c_{\frac{n_1}{2}}c_{\frac{n_2+1}{2}}, \text{ vspace*-8pt}$$

and hence, the second formula of (11.8.14) holds, too.

Finally, let's assume that the entries of $(i_1, j, i_2) \in \mathbb{Z}^3$ is mutually distinct, i.e.,

$$i_1 \neq j,\ j \neq i_2,\ \text{and}\ i_1 \neq i_2,$$

in $\mathbb{Z}$. For convenience, we will say that $(i_1, j, i_2)$ is mutually distinct in $\mathbb{Z}$ from below. Then the free word $u_{i_1}^{n_1}u_j^2u_{i_2}^{n_2}$, itself, forms a free reduced word in $\mathbb{L}_Q$, moreover, the free generators $u_{i_1}$, $u_j$ and $u_{i_2}$ of this free reduced word are mutually free from each other. So, by (11.8.16),

$$\tau^M\left(\int_0^t u_{i_1,i_2}^{n_1,n_2}dX_j\right) = t\,\tau\left(u_{i_1}^{n_1}u_j^2u_{i_2}^{n_2}\right)$$

$$= t\left(\tau_{i_1}\left(u_{i_1}^{n_1}\right)\right)\left(\tau_j\left(u_j^2\right)\right)\left(\tau_{i_2}\left(u_{i_2}^{n_2}\right)\right)$$

$$= t\left(\omega_{n_1}\psi(q_{i_1})^{n_1}c_{\frac{n_1}{2}}\right)\left(\psi(q_j)^2c_1\right)\left(\omega_{n_2}\psi(q_{i_2})^{n_2}c_{\frac{n_2}{2}}\right)$$

$$= t\,\omega_{n_1}\omega_{n_2}\psi(q_{i_1})^{n_1}\psi(q_j)^2\psi(q_{i_2})^{n_2}c_{\frac{n_1}{2}}c_{\frac{n_2}{2}},$$

since

$$c_1 = \frac{1}{2} \begin{pmatrix} 2 \\ 1 \end{pmatrix} = \frac{2!}{2!1!} = 1.$$

Therefore, the formula (11.8.15) holds. □

**Assumption and Notation 11.8.1** (in short, **AN 11.8.1**, from below)
In the following text, if we write "a simple adapted biprocess $u_{i_1,i_2}^{n_1,n_2} \in \mathbb{L}_Q^{\otimes 2}$
follows (11.8.13), or (11.8.14), or (11.8.15)," then it means "

$$i_1 = j = i_2 \text{ in } \mathbb{Z},$$

respectively,

$$\text{either } i_1 \neq j = i_2, \text{ or } i_1 = j \neq i_2 \text{ in } \mathbb{Z},$$

respectively

$$(i_1, \ j, \ i_2) \in \mathbb{Z}^3 \text{ is mutually distinct in } \mathbb{Z}.$$

□

As we discussed about, one can realize the $j$-th w-s integrals

$$T = \int_0^t u_{i_1,i_2}^{n_1,n_2} dX_j \in \mathbb{L}_Q(M)$$

of basic simple adapted biprocesses $u_{i_1,i_2}^{n_1,n_2} \in \mathbb{L}_Q^{\otimes 2}$, as free random variables
of the $\mathbb{L}_Q$-acting probability space $\mathbb{L}_Q(M)$. So, it is meaningful to consider
free-distributional data of these free stochastic integrals.
By (11.8.19), one can obtain the following free-distributional data of
$\int_0^t u_{i_1,i_2}^{n_1,n_2} dX_j$ in $\mathbb{L}_Q(M)$.

**Theorem 11.8.4.** *Let* $T = \int_0^t u_{i_1,i_2}^{n_1,n_2} dX_j$ *be the $j$-th w-s integral of a simple
adapted biprocess* $u_{i_1,i_2}^{n_1,n_2} \in \mathbb{L}_Q^{\otimes n}$ *for the fixed $j$-th weighted-semicircular motion
$X_j$, for $j \in \mathbb{Z}$.*

*(11.8.17) If $u_{i_1,i_2}^{n_1,n_2}$ follows (11.8.13) under **AN 11.8.1**, then*

$$\tau^M(T) = t \, \omega_{n_1+n_2+2} \psi(q_j)^{n_1+n_2+2} c_{\frac{n_1+n_2+2}{2}} ;$$

*(11.8.18) if $u_{i_1,i_2}^{n_1,n_2}$ follows (11.8.14) under **AN 11.8.1**, then either*

$$\tau^M(T) = t\omega_{n_1+2}\omega_{n_2}\left(\psi(q_j)^{n_1+2}\psi(q_{i_2})^{n_2}c_{\frac{n_1+2}{2}}c_{\frac{n_2}{2}}\right),$$

*or*

$$\tau^M(T) = t\,\omega_{n_1}\omega_{n_2+2}\left(\psi(q_{i_1})^{n_1}\psi(q_j)^{n_2+2}c_{\frac{n_1}{2}}c_{\frac{n_2+2}{2}}\right);$$

(11.8.19) *if* $u_{i_1,i_2}^{n_1,n_2}$ *follows* (11.8.15) *under* **AN 11.8.1**, *then*

$$\tau^M(T) = t\,\omega_{n_1}\omega_{n_2}\psi(q_{i_1})^{n_1}\psi(q_j)^2\psi(q_{i_2})^{n_2}c_{\frac{n_1}{2}}c_{\frac{n_2}{2}},$$

*for all* $l \in \mathbb{N}$.
*Moreover, under the same hypotheses, one can get that*

$$\tau^M(T^*) = \tau^M(T). \qquad (11.8.20)$$

*Proof.* Suppose a simple adapted biprocess $u_{i_1,i_2}^{n_1,n_2}$ follows (11.8.13). Then the $j$-th w-s integral $T$ satisfies that

$$T = \sum_{k=0}^{n-1}\left(\chi_{t_k<t_{k+1}} \otimes u_j^{n_1+n_2+2}\right)$$

$$= \left(\sum_{k=0}^{n-1}\chi_{t_k<t_{k+1}}\right) \otimes u_j^{n_1+n_2+2}$$

under (11.8.8), for all $l \in \mathbb{N}$. So,

$$\tau^M(T) = t\tau\left(u_j^{n_1+n_2+2}\right) = t\tau_j\left(u_j^{n_1+n_2+2}\right)$$

$$= t\omega_{l(n_1+n_2+2)}\psi(q_j)^{n_1+n_2+2}c_{\frac{n_1+n_2+2}{2}},$$

for all $l \in \mathbb{N}$. Therefore, the statement (11.8.17) holds.

Now, assume that $u_{i_1,i_2}^{n_1,n_2}$ follows (11.8.14). Then, one has either

$$T = \sum_{k=0}^{n-1}\left(\chi_{t_k<t_{k+1}} \otimes u_j^{n_1+2}u_{i_2}^{n_2}\right),$$

*or*

$$T^l = \sum_{k=0}^{n-1}\left(\chi_{t_k<t_{k+1}} \otimes u_{i_1}^{n_1}u_j^{n_2+2}\right),$$

in $\mathbb{L}_Q(M)$, for all $l \in \mathbb{N}$. Thus, we obtain either

$$\tau^M \left(T^l\right) = t \, \tau \left(u_j^{n_1+2} u_{i_2}^{n_2}\right)$$

$$= t \, \tau \left(u_j^{n_1+2} u_{i_2}^{n_2}\right)$$

$$= t \left(\tau_j \left(u_j^{n_1+2}\right) \tau_{i_2} \left(u_{i_2}^{n_2}\right)\right)$$

$$= t \left(\omega_{n_1+2} \omega_{n_2} \psi(q_j)^{n_1+2} \psi(q_{i_2})^{n_2} c_{\frac{n_1+2}{2}} c_{\frac{n_2}{2}}\right)$$

$$= t \, \omega_{n_1+2} \omega_{n_2} \left(\psi(q_j)^{n_1+2} \psi(q_{i_2})^{n_2} c_{\frac{n_1+2}{2}} c_{\frac{n_2}{2}}\right), \qquad (11.8.21)$$

respectively,

$$\tau^M(T) = t \omega_{n_1} \omega_{n_2+2} \left(\psi(q_{i_1})^{n_1} \psi(q_j)^{n_2+2} c_{\frac{n_1}{2}} c_{\frac{n_2}{2}}\right),$$

in the similar arguments with (11.8.21). Therefore, the statement (11.8.18) holds true.

Finally, assume that $u_{i_1,i_2}^{n_1,n_2}$ follows (11.8.15). Then,

$$T = \sum_{k=0}^{n-1} \left(\chi_{t_k < t_{k+1}} \otimes u_{i_1}^{n_1} u_j^2 u_{i_2}^{n_2}\right)$$

in $\mathbb{L}_Q(M)$, for all $l \in \mathbb{N}$. So, one can get that

$$\tau^M \left(T\right) = t \, \tau \left(u_{i_1}^{n_1} u_j^2 u_{i_2}^{n_2}\right)$$

$$= t \left(\tau_{i_1} \left(u_{i_1}^{n_1}\right) \tau_j \left(u_j^2\right) \tau_{i_2} \left(u_{i_2}^{n_2}\right)\right)$$

$$= t \left(\omega_{n_1} \psi(q_{i_1})^{n_1} c_{\frac{n_1}{2}}\right) \left(\psi(q_j)^2 c_1\right) \left(\omega_{n_2} \psi(q_{i_2})^{n_2} c_{\frac{n_2}{2}}\right)$$

$$= t \, \omega_{n_1} \omega_{n_2} \left(\psi(q_{i_1})^{n_1} \psi(q_j)^2 \psi(q_{i_2})^{n_2} c_{\frac{n_1}{2}} c_{\frac{n_2}{2}}\right),$$

for all $l \in \mathbb{N}$. Therefore, the free-momental formula (11.8.19) holds.

Observe now that, if $u_{i_1,i_2}^{n_1,n_2}$ follows (11.8.13), then

$$T^* = \left( \sum_{k=0}^{n-1} \chi_{t_k < t_{k+1}} \otimes u_j^{n_1+n_2+2} \right)^* = T$$

in $\mathbb{L}_Q(M)$ under (11.8.8). Thus this self-adjointness of the $j$-th w-s integral $T$ let us have

$$\tau^M (T^*) = \tau^M(T).$$

Now, if $u_{i_1,i_2}^{n_1,n_2}$ follows (11.8.14), then the $j$-th w-s integral $T$ satisfies either

$$T^* = \left( \sum_{k=0}^{n-1} \chi_{t_k < t_{k+1}} \otimes u_j^{n_1+2} u_{i_2}^{n_2} \right)^*$$

$$= \sum_{k=0}^{n-1} \left( \chi_{t_k < t_{k+1}} \otimes u_{i_2}^{n_2} u_j^{n_1+2} \right),$$

or                                                                    (11.8.22)

$$T^* = \sum_{k=0}^{n-1} \left( \chi_{t_k < t_{k+1}} \otimes u_j^{n_2+2} u_{i_1}^{n_1} \right),$$

in $\mathbb{L}_Q(M)$. Therefore, one can conclude that

$$\tau^M (T^*) = \tau^M (T),$$

by (11.8.22).

Finally, suppose now that $u_{i_1,i_2}^{n_1,n_2}$ follows (11.8.15). Then

$$T^* = \sum_{k=0}^{n-1} \left( \chi_{t_k < t_{k+1}} \otimes u_{i_1}^{n_1} u_j^2 u_{i_2}^{n_2} \right)^*$$

$$= \sum_{k=0}^{n-1} \left( \chi_{t_k < t_{k+1}} \otimes u_{i_2}^{n_2} u_j^2 u_{i_1}^{n_1} \right)$$

in $\mathbb{L}_Q(M)$. So,

$$\tau^M (T^*) = \tau (T).$$

Therefore, the formula (11.8.20) holds true.                 $\square$

The free-moment computations (11.8.17), (11.8.18), (11.8.19) and (11.8.20) not only give the free-distributional data of $j$-th w-s integrals of "basic" simple adapted processes $u_{i_1,i_2}^{n_1,n_2}$, but also provide ways to compute free-distributional data of $j$-th w-s integrals of arbitrary simple adapted biprocesses in the $\mathbb{L}_Q$-acting probability space $\mathbb{L}_Q(M)$.

## 11.9 Glossary

**360 Degree Review:** Performance review that includes feedback from superiors, peers, subordinates, and clients.

**Abnormal Variation:** Changes in process performance that cannot be accounted for by typical day-to-day variation. Also referred to as non-random variation.

**Acceptable Quality Level (AQL):** The minimum number of parts that must comply with quality standards, usually stated as a percentage.

**Activity:** The tasks performed to change inputs into outputs.

**Adaptable:** An adaptable process is designed to maintain effectiveness and efficiency as requirements change. The process is deemed adaptable when there is agreement among suppliers, owners, and customers that the process will meet requirements throughout the strategic period.

## Bibliography

Blackadar B. Operator algebras, (2013) Published by Springer.

Cho I. Free Semicircular Families in Free Product Banach ∗-Algebras Induced by $p$-Adic Number Fields, Compl. Anal. Oper. Theo., 11, no. 3, (2017) 507 - 565.

Cho I. $p$-Adic Free Stochastic Integrals for $p$-Adic Weighted-Semicircular Motions Determined by Primes $p$, Libertas Math. (New Series), 36, no. 2, (2016) 65 - 110.

Cho I. Adelic analysis and functional analysis on the finite Adele ring, Opuscula Math., 38, no. 2, (2018) 139 - 185.

Cho I. Semicircular-Like Laws and the Semicircular Law Induced by Orthogonal Projections, Compl. Anal. Oper. Theo., (2018) To Appear.

Cho I. and Jorgensen P. E. T. Semicircular Elements Induced by $p$-Adic Number Fields, Opuscula Math., 35, no. 5, (2017) 665 - 703.

Hiai F. and Petz D. The Semicircular Law, Free Random Variables and Entropy, Math. Survey & Monographs, vol. 77, ISBN: 0-8218-4135-1, (2000) Published by Amer. Math. Soc..

Radulescu F. Random Matrices, Amalgamated Free Products and Subfactors of the $C^*$-Algebra of a Free Group of Nonsingular Index, Invent. Math., 115, (1994) 347 - 389.

Speicher R. Combinatorial Theory of the Free Product with Amalgamation and Operator-Valued Free Probability Theory, Amer. Math. Soc. Mem., vol 132, no. 627, (1998).

Speicher R. Free Calculus, arXiv:math / 0104004, (2001).

Voiculescu D. (Editor), Free Probability Theory, The Fields Institute Comm., vol 12, ISBN: 0-8218-0675-0, (1997).

Vladimirov V. S., Volovich I. V., and Zelenov E. I. $p$-Adic Analysis and Mathematical Physics, Ser. Soviet & East European Math., vol 1, ISBN: 978-981-02-0880-6, (1994) World Scientific.

Voiculescu D., Dykemma K., and Nica A. Free Random Variables, CRM Monograph Series, vol 1., (1992) Published by Amer. Math. Soc..

# Chapter 12

# On Difference Operators and Their Applications

**Pinakadhar Baliarsingh**

*Department of Mathematics, School of Applied Sciences, KIIT, Bhubaneswar, India*

**Hemen Dutta**

*Department of Mathematics, Gauhati University, Guwahati, India*

## 12.1 Introduction

In recent years, a number of extensive applications of summability theory in particular, theory of sequence spaces have been developed. The most crucial application is being used in the study of difference operators through sequence spaces and matrix theory. In 1981, the idea of difference sequence spaces based on difference operator with order one was introduced by Kizmaz Kızmaz (1981) and it was further generalized by Et and Colak Et and Colak (1995 ) to the case of an integral order $m$. The idea was further extended and studied by many researchers like Et and Basarır Et and Basarir (1997), Mursaleen Michael (1952) Altay and Başar Altay and Başar (2004), Başar and Aydın Basar and Aydand Aydın (2004), Ahmad and Mursaleen Ahmad (et al.), Bektaş et al. Bektas et al. (2004), Malkowsky et al. Malkowsky (et al.), and others. Recently, the operator is further generalized to the case of fractional order by Baliarsingh Baliarsingh (2013). The idea of difference operators has been extensively used to develop the study of sequence spaces through several standard methods involving topological structures, dual spaces, matrix transformations etc. (see Ahmad (et al.); Bektas et al. (2004); Et and Colak (1995 ); Kadak and Baliarsingh (2015); Malkowsky (et al.); Mursaleen and Noman (2010)) and different modern techniques involving approximations and spectrum of linear operators, fractional derivatives, matrix inversions etc. (see Altay and Başar (2004); Baliarsingh and Dutta (2015); Baliarsingh (2016); Baliarsingh and Nayak (2017); Dutta and Baliarsingh (2012, 2014); Mursaleen (et al.); Nayak et al. (2014 )).

In fact, the theory of difference sequence spaces has made a significant contribution in enveloping the theory of classical and fractional calculus. The theory of fractional calculus deals with the investigation of derivatives and integrations of a function with arbitrary orders. Fractional derivative provides an extensive knowledge for description of memory and hereditary properties of various materials and processes including certain natural and physical phenomena. The application of fractional derivatives becomes more apparent in modeling mechanical and electrical properties of real materials as well as in the description of rheological properties of rocks and in many other fields. Especially, the theory of fractional derivatives has been extensively used in the study of fractal theory, theory of control of dynamic systems, theory of viscoelasticity, electrochemistry, diffusion processes and many others (see Blutzer and Torvik (1996); Dreisigmeyer and Young (2003); Kilbas et al. (2006)).

The idea of fractional calculus has been frequently used in various theories involving the solution of diverse problems in mathematics, science, and engineering and in order to stimulate more interest in the subject and to show its utility, several definitions of fractional derivatives have been developed. The most popular definitions of fractional derivatives and integrations have been introduced by Riemann-Liouville and Grunwald-Letnikov which are stated below:

### 12.1.1   Preliminaries and definitions

(i) The left and right *Riemann-Lioville* fractional derivatives of order $\alpha$, respectively are given by

$$_aD_x^\alpha f(x) = \frac{1}{\Gamma(n-\alpha)} \frac{d^n}{dx^n} \int_a^x (x-t)^{n-\alpha-1} f(t)dt,$$

$$_xD_b^\alpha f(x) = \frac{(-1)^n}{\Gamma(n-\alpha)} \frac{d^n}{dx^n} \int_x^b (t-x)^{n-\alpha-1} f(t)dt.$$

(ii) The left and right *Riemann-Lioville* fractional integrals of order $\alpha$, respectively are given by

$$_aI_x^\alpha f(x) = \frac{1}{\Gamma(\alpha)} \int_a^x (x-t)^{\alpha-1} f(t)dt,$$

$$_xI_b^\alpha f(x) = \frac{1}{\Gamma(\alpha)} \int_x^b (t-x)^{\alpha-1} f(t)dt.$$

(iii) *Grunwald-Letnikov* fractional derivative of order $\alpha$, is given by

$$_aD_x^\alpha f(x) = \lim_{h \to 0} \frac{1}{h^\alpha} \sum_{j=0}^{\left[\frac{x-\alpha}{h}\right]} (-1)^j \binom{\alpha}{j} f(x - jh).$$

Later on Caputo reformulated the classical definition of the Riemann-Liouville fractional derivative and solved the fractional differential equations with certain initial conditions. The left and right fractional integrals of order $\alpha$, due to Caputo are given by

$$_a^cD_x^\alpha f(x) = \frac{(-1)^n}{\Gamma(n - \alpha)} \int_a^x (x - t)^{n-\alpha-1} f^{(n)}(t)dt,$$

$$_x^cD_b^\alpha f(x) = \frac{(-1)^n}{\Gamma(n - \alpha)} \int_x^b (t - x)^{n-\alpha-1} f^{(n)}(t)dt.$$

Recently, the Riemann-Liouville fractional derivatives of a function $f(t)$ have been modified by JumaricJumarie (2006,?) as

$$(f(t))^{(\alpha)} = \frac{1}{\Gamma(1 - \alpha)} \frac{d}{dt} \int_0^t (t - x)^{-\alpha}(f(x) - f(0))dx.$$

Using this definition, Jumarie gave two basic formulas regarding Leibniz and Chain rules of the fractional derivatives. Let $f(t)$ and $g(t)$ be two functions. Then their composition is denoted by $f(g(t))$ and the formulas due to Jumarie are as follows:

$$(f(t)g(t))^{(\alpha)} = (f(t))^{(\alpha)}g(t) + f(t)(g(t))^{(\alpha)},$$
$$(f(g(t)))^{(\alpha)} = f'_g(g(t))^{(\alpha)}.$$

Later on, LiuLiu (2015) has proved that these formulas are incorrect by taking two different counter examples and modified them as

$$(f(t)g(t))_J^{(\alpha)} = \sum_{j=0}^{\infty} \binom{\alpha}{j} f^j(t) g_{R-L}^{\alpha-j}(t) - \frac{f(0)g(0)}{t^\alpha \Gamma(1 - \alpha)},$$

$$(f(g(t)))_J^{(\alpha)} = \sum_{j=1}^{\infty} \binom{\alpha}{j} \frac{t^{j-\alpha}j!}{\Gamma(j - \alpha + 1)} \sum_{m=1}^{j} f^{(m)}(g) \sum \prod_{k=1}^{j} \frac{1}{P_k!} \left(\frac{g^{(k)}}{k!}\right)^{P_k}$$
$$+ \frac{f(g(t)) - f(g(0))}{t^\alpha \Gamma(1 - \alpha)},$$

where $\sum$ extends over all combinations of nonnegative integer values of $P_1, P_2, \ldots, P_n$ such that $\sum_{k=1}^{n} k P_k = n$ and $\sum_{k=1}^{n} P_k = m$ (see Liu (2015)).

The formulas and definitions of fractional calculus introduced by Leibniz have also been reformulated by Grunwald-Letnikov. Using special functions such as Euler gamma function, hyper geometric functions and Mittag-Leffler function, Kilbas, et al.Kilbas et al. (2006) reformulated the ideas on fractional derivatives introduced by Riemann-Liouville. More investigations on fractional calculus and its several applications to real world problems in several field of sciences such as electro chemistry, fluid mechanics, bio technology and applied mathematics are found in Blutzer and Torvik (1996); Dreisigmeyer and Young (2003); Jonsson and Yngvason (1995); Mainardi (1996); Podlubny (1999).

Initially, it was Kizmaz Kızmaz (1981) who introduced the idea of difference sequence space associated with basic sequences $\ell_\infty, c$ and $c_0$ by defining the difference operator $\Delta$ of order one, where

$$(\Delta x)_k = x_k - x_{k+1}, (k \in \mathbb{N}). \tag{12.1}$$

Later on, these sequence spaces have been generalized to the case of integral order $m$ by Et and Colak Et and Colak (1995 ) using operator $\Delta^m$ and

$$(\Delta^m x)_k = \sum_{i=0}^{m} (-1)^i \binom{m}{i} x_{k+i}, (k \in \mathbb{N}). \tag{12.2}$$

Recently, Baliarsingh Baliarsingh (2013) (see also Baliarsingh and Dutta (2015, 2014, 2015)) generalized the above difference operator by introducing fractional difference operator $\Delta^\alpha$, where

$$(\Delta^\alpha x)_k = \sum_{i=0}^{\infty} (-1)^i \frac{\Gamma(\alpha+1)}{i!\Gamma(\alpha-i+1)} x_{k+i}, (k \in \mathbb{N}), \tag{12.3}$$

where $\alpha > 0$. It is presumed that the infinite series defined in (12.3) is convergent. In particular, we observed that

- $\Delta^{\frac{1}{2}} x_k = x_k - \dfrac{1}{2}x_{k+1} - \dfrac{1}{8}x_{k+2} - \dfrac{1}{16}x_{k+3} - \dfrac{5}{128}x_{k+4} - \dfrac{7}{256}x_{k+5} - \dfrac{21}{1024}x_{k+6} + \dots$

- $\Delta^{\frac{1}{3}} x_k = x_k - \dfrac{1}{3}x_{k+1} - \dfrac{1}{9}x_{k+2} - \dfrac{5}{81}x_{k+3} - \dfrac{10}{243}x_{k+4} - \dfrac{22}{729}x_{k+5} - \dfrac{154}{6561}x_{k+6} + \dots$

- $\Delta^{\frac{2}{3}} x_k = x_k - \dfrac{2}{3}x_{k+1} - \dfrac{1}{9}x_{k+2} - \dfrac{4}{81}x_{k+3} - \dfrac{7}{243}x_{k+4} - \dfrac{14}{729}x_{k+5} - \dfrac{91}{6561}x_{k+6} + \dots$

- $\Delta^{\frac{1}{6}} x_k = x_k - \dfrac{1}{6}x_{k+1} - \dfrac{5}{72}x_{k+2} - \dfrac{55}{1296}x_{k+3} - \dfrac{935}{31104}x_{k+4} - \dfrac{21505}{933120}x_{k+5} + \dots$

It is noted that $\Delta^{-\alpha}$ is taken as the inverse operator of $\Delta^{\alpha}$, where

$$(\Delta^{-\alpha}x)_k = \sum_{i=0}^{\infty}(-1)^i \binom{-\alpha}{i} x_{k+i}, (k \in \mathbb{N}), \qquad (12.4)$$

Clearly, if we replace $\alpha$ as 1 in Eqn.(12.4), we obtain that $(\Delta^{-1}x)_k = \sum_{i=0}^{\infty} x_{k+i}, (k \in \mathbb{N})$. It is too difficult to study the convergence of the series defined in (12.4). For instance, we have the following example:

**Example 12.1.1.** Let $x = (x_k)$ be a constant sequence with $x_k = 1$ for all $k \in \mathbb{N}$. Although the sequence $x = (x_k)$ is convergent, but for a proper fraction $\alpha$, $(\Delta^{\alpha}x)_k \to 0$ as $k \to \infty$ whereas $(\Delta^{-\alpha}x)_k \to \infty$ as $k \to \infty$.

Now, for proper fractions $\alpha, \beta > 0$, we have the following results:

**Theorem 12.1.2** (Baliarsingh (2013); Baliarsingh and Dutta (2015)). **(i)** *The difference operator $\Delta^{\alpha}$ is a linear operator that represents an infinite triangular matrix satisfying $\|\Delta^{\alpha}\| = 2^{\alpha}$.*

**(ii)** $\Delta^{\alpha}.\Delta^{-\alpha} = \Delta^{-\alpha}.\Delta^{\alpha} = I.$

**(iii)** $\Delta^{\alpha}.\Delta^{\beta} = \Delta^{\beta}.\Delta^{\alpha} = \Delta^{\alpha+\beta},$
*where $I$ is the identity operator and $\|A\|$ represents supremum over $\ell_1$ norms of the rows of the matrix $A$.*

As already discussed, difference operators based on integral and non integral orders have been become a prominent area of research which provides major connections among different areas of mathematics. In this context, several difference operators have been introduced through sequence spaces and subsequently, applied in many areas of pure and applied mathematics such as summability theory (see Aasma et al. (2017); Dutta (2009); Dutta and Rhoades (2016)), approximation theory (see Mursaleen (et al.); Nayak et al. (2014 )), matrix and spectral theory (see Altay and Başar (2004); Baliarsingh and Dutta (2015); Dutta and Baliarsingh (2012, 2014); Kadak and Baliarsingh (2015)) and the theory of fractional calculus (see Baliarsingh (2016); Baliarsingh and Nayak (2017)). It is known that most of the results based on classical theory of calculus may not be applicable for the case of fractional calculus. As a result, these deviations and non uniform behaviors develop the dynamic nature of fractional calculus. Recently, using generalized fractional difference operator, some of these dynamic properties have been studied by Daliarsingh Baliarsingh (2016). Using the proposed difference operator due to Baliarsingh (2016); Baliarsingh and Nayak (2017), the primary goal of this chapter is to establish certain results on the dynamic nature of fractional calculus.

## 12.2 Fractional difference operator

Let $w$ be the space of all real valued sequences. By $\ell_\infty, c$ and $c_0$, we denote the spaces of all bounded, convergent and null sequences, respectively, normed by $\|x\|_\infty = \sup_k |x_k|$. Let $x = (x_k)$ be any sequence in $w$, and $h$ be a positive constant in $(0, 1]$. Then for real numbers $a, b$ and $c$, the generalized difference sequence via difference operator $\Delta_h^{a,b,c} : w \to w$ (see Baliarsingh (2016)) is defined by

$$(\Delta_h^{a,b,c} x)_k = \sum_{i=0}^{\infty} \frac{(-a)_i(-b)_i}{i!(-c)_i h^{a+b-c}} x_{k-i}, (k \in \mathbb{N}), \tag{12.5}$$

where $(\alpha)_k$ denotes the *Pochhammer* symbol or *shifted factorial* of a real number $\alpha$ which is being defined using familiar Euler gamma function as

$$(\alpha)_k = \begin{cases} 1, & (\alpha = 0 \text{ or } k = 0) \\ \dfrac{\Gamma(\alpha + k)}{\Gamma(\alpha)} = \alpha(\alpha+1)(\alpha+2)\dots(\alpha+k-1), & (k \in \mathbb{N}) \end{cases}$$

Note that the series defined in (12.5) need not be convergent for all positive $a, b, c$ and any sequence $x = (x_k)$. In fact, it converges for specific values of $a, b, c$ and a suitable choice of the sequence $x = (x_k)$, which may be presumed throughout the text. Eventually, the operator $\Delta_h^{a,b,c}$ represents a triangle as

$$(\Delta_h^{a,b,c})_{nk} = \begin{cases} 1, & (n = k) \\ (-1)^{n-k} \frac{a(a-1)\dots(a-(n-k-1))b(b-1)\dots(b-(n-k-1))}{(n-k)!c(c-1)\dots(c-(n-k-1))h^{a+b-c}}, & (0 \le k < n) \\ 0, & (k > n) \end{cases}$$

In particular, the difference operator $\Delta_h^{a,b,c}$ includes following special cases:

(i) The difference operator $\Delta^{(1)}$ for $a = 1, b = c$ and $h = 1$ (see Ahmad (et al.); Altay and Başar (2004)).

(ii) The difference operator $\Delta^{(m)}$ for $a = m \in \mathbb{N}_0, b = c$ and $h = 1$ (see Et and Colak (1995 ); Malkowsky (et al.)).

(iii) The difference operator $\Delta^{(\alpha)}$ for $a = \alpha \in \mathbb{R}$, (the set of all real numbers), $b = c$ and $h = 1$ (see Baliarsingh (2013); Baliarsingh and Dutta (2015)).

(iv) The difference operator $\Delta_\nu^r$ for $a = r, b = c, h = 1$ and $\nu = (1, 1, 1, \dots)$ (see Dutta and Baliarsingh (2012)).

**Theorem 12.2.1** (Baliarsingh and Nayak (2017); Gasper and Rahman (2004)). *The series defined in (12.5) converges absolutely if the following two conditions hold:*

**(I)** *For all $c > a + b$ with $c \notin \mathbb{N}$ and*

**(II)** *For all convergent sequence $x = (x_k)$ with $|x_k| \leq 1$.*

**Note:**  The above two conditions are sufficient rather than necessary for the convergence of (12.5).

We provide the proof of above remark by using the following examples:

**Example 12.2.2.**  Consider $c$ as a positive integer i.e., $c = 8, a = b = 2, h = 1$ and the sequence $x = (x_k)$ with $x_k = 1$ for all $k \in \mathbb{N}$. Then clearly, $c > a + b$, but $c \in \mathbb{N}$. It is noticed that although the condition **(I)** is not satisfying, the difference sequence

$$(\Delta_1^{2,2,8} x)_k = 1 - \frac{2.2}{8} + \frac{2.1.2.1}{8.7} = \frac{4}{7},$$

is convergent. Conversely, if we take a sequence $x = (x_k)$ with $x_k = k^3$ for all $k \in \mathbb{N}$, then it is clear that $|x_k| > 1$. Although the condition **(II)** is not satisfying, for $u = 4, b = c, h = 1$, we obtain

$$(\Delta_1^{4,b,b} x)_k = k^3 - 4(k-1)^3 + 6(k-2)^3 - 4(k-3)^3 + (k-4)^3 = 0,$$

which is convergent.

**Theorem 12.2.3.**  *For positive reals $a, b,$ and $c$ satisfying the condition* **(I)**, *the difference operator $\Delta_h^{a,b,c}$ is a linear operator and satisfies*

$$\|\Delta_h^{a,b,c}\| \leq \sup_m \left| \frac{\binom{b}{m}}{h^{a+b-c}\binom{c}{m}} \right| 2^a.$$

*Proof.* The proof of the linearity of $\Delta_h^{a,b,c}$ is simple, only to determine its upper bound. Now, using condition **(I)**, we may write

$$\|\Delta_h^{a,b,c}\| = \sup_n \sum_{k=0}^{n} |(\Delta_h^{a,b,c})_{nk}|$$

$$= \sup_n \sum_{k=0}^{n} \left| (-1)^{n-k} \frac{a(a-1)\dots(a-(n-k-1)b(b-1)\dots(b-(n-k-1))}{(n-k)!c(c-1)\dots(c-(n-k-1)h^{a+b-c}} \right|$$

$$= \lim_{n\to\infty} \sum_{k=0}^{n} \left| \frac{\binom{a}{n-k}\binom{b}{m_{lq}}}{\binom{c}{n-k}h^{a+b-c}} \right|$$

$$\leq \sup_m \left| \frac{\binom{b}{m}}{h^{a+b-c}\binom{c}{m}} \right| \left| \sum_{k=0}^{\infty} \binom{a}{k} \right| = \sup_m \left| \frac{\binom{b}{m}}{h^{a+b-c}\binom{c}{m}} \right| 2^a$$

The value $1/h^{a+b-c}$ is finite for all $h \in (0,1]$ under the condition **(I)**. This concludes the result.  □

Now, rewriting Eqn.(12.5) by using the notation $D(a, b, c, h, i)$, we have

$$(\Delta_h^{a,b,c} x)_k = \sum_{i=0}^{k} D(a, b, c, h, i) x_{k-i}, \, (k \in \mathbb{N}), \tag{12.6}$$

where,

$$D(a, b, c, h, i) = \frac{(-a)_i (-b)_i}{i! (-c)_i h^{a+b-c}}. \tag{12.7}$$

and we establish the following relations:

**Theorem 12.2.4.** *For positive reals $a, b$ and $c$ such that $b = c$, we have*

(i) *For all $a \neq i, i - 1$ and $i \geq 1$,*

$$D(a, b, c, h, i) = \frac{(-1)^i}{h^a} \binom{a}{i}$$

$$= \frac{a}{(a-i)h} D(a-1, b, c, h, i) = \frac{a-i+1}{ih} D(a, b, c, h, i-1).$$

(ii) $D(a, b, c, h, i) = D(a-1, b, c, h, i-1) + D(a-1, b, c, h, i).$

(iii) $D(-a, b, c, h, i) = D(-a, b, c, h, i-1) + h D(1-a, b, c, h, i).$

*Proof.* (i).

$$D(a, b, b, h, i) = \frac{(-a)_i}{i! h^a}$$

$$= \frac{-a(-a+1)(-a+2)\ldots(-a+i-1)}{i! h^a}$$

$$= (-1)^i \frac{a(a-1)(a-2)\ldots(a+i-1)(a-i)!}{i!(a-i)! h^a}$$

$$= \frac{(-1)^i}{h^a} \binom{a}{i}.$$

Again, we have

$$D(a, b, b, h, i) = \frac{-a(-a+1)(-a+2)\ldots(-a+i-1)}{i! h^a}$$

$$= \frac{-a(-a+1)(-a+2)\ldots(-a+i-1)(-a+i)}{i!(-a+i) h^a}$$

$$= \frac{a}{(a-i)h} D(a-1, b, c, h, i).$$

Also, it is seen that

$$D(a, b, b, h, i) = \frac{-a(-a+1)(-a+2)\ldots\ldots(-a+i-2)(-a+i-1)}{i(i-1)! h^a}$$

$$= \frac{a-i+1}{ih} D(a, b, c, h, i-1).$$

(ii). Taking the right hand side, we have

$$D(a-1, b, b, h, i-1) + D(a-1, b, b, h, i)$$

$$= \frac{(-1)^{i-1}}{h^a} \binom{a-1}{i-1} + \frac{(-1)^i}{h^a} \binom{a-1}{i}$$

$$= \frac{(-1)^i}{h^a} \left[ \binom{a-1}{i} - \binom{a-1}{i-1} \right]$$

$$= \frac{(-1)^i}{h^a} \binom{a}{i} = D(a, b, b, h, i).$$

(iii). Proof is similar to that of (ii). □

**Theorem 12.2.5.** *(i) For all $i \in \mathbb{N}_0$ and $i \neq a, b, c$,*

$$D(a, b, c, h, i+1) = \frac{(-a+i)(-b+i)}{(-c+i)(i+1)} D(a, b, c, h, i).$$

*(ii) For $i \geq 1$ and $i \neq a+1, b+1$,*

$$D(a, b, c, h, i) = \frac{(i-b-1)}{(i-c-1)} \left( 1 - \frac{a+1}{i} \right) D(a, b, c, h, i-1).$$

*(iii) For $i \geq 1$ and $c \neq -1$,*

$$D(a+1, b+1, c+1, h, i) = -\frac{(a+1)(b+1)}{ih(c+1)} D(a, b, c, h, i-1).$$

*(iv) For $i \geq 1$ and $c - a \neq 1$,*

$$D(a+1, b+1, c+1, h, i) = \frac{(a+1)}{h(c-a-1)} D(a, b+1, c+1, h, i)$$
$$- \frac{(c+2)}{h(c-a-1)} D(a+1, b+1, c+2, h, i).$$

*(v) For $i \geq 1$ and $a \neq b$,*

$$D(a+1, b+1, c+1, h, i) = \frac{(a+1)}{h(a-b)} D(a, b+1, c+1, h, i)$$
$$- \frac{(b+1)}{h(a-b)} D(a+1, b, c+1, h, i).$$

*(vi) For $i \geq 1$ and $c - a \neq 1$,*

$$D(-a, -b, -c, h, i) = \frac{a}{h(c-a-1)} D(-(a+1), -b, -c, h, i)$$
$$- \frac{(c-1)}{h(c-a-1)} D(-a, -b, -c+1, h, i).$$

*(vii)* For $i \geq 1$ and $a \neq b$,

$$D(-a, -b, -c, h, i) = \frac{a}{h(a-b)} D(-(a+1), -b, -c, h, i)$$

$$- \frac{b}{h(a-b)} D(-a, -(b+1), -c, h, i).$$

*(viii)* For $i \geq 1$, $c + i \neq 0$ and $b \neq 0$,

$$D(-a, -(b+1), -(c+1), h, i) = \frac{c(b+i)}{b(c+i)} D(-a, -b, -c, h, i).$$

*Proof.* The proofs are elementary verifications as discussed in Theorem 12.2.4, hence omitted. $\qquad\square$

---

## 12.3   Related sequence spaces

In the present section, using the difference operator defined in (12.5), the following classes of difference sequence spaces(see Baliarsingh (2016)) are defined:

$$X(\Delta_h^{a,b,c}) = \left\{ x = (x_k) \in w : \Delta_h^{a,b,c}(x) \in X \right\},$$

where $X = c, c_0$ and $\ell_\infty$. It is noticed that each element of the above classes can be obtained by taking the $\Delta_h^{a,b,c}$-transform of the sequence $x$.

**Theorem 12.3.1.** *Let $X = c, c_0, \ell_\infty$ and $a, b, c$ be positive reals satisfying condition **(I)**. Then the space $X(\Delta_h^{a,b,c})$ overlaps with $X$, but one does not contain another.*

*Proof.* We prove the theorem for the space $\ell_\infty$, and for others it follows the similar arguments.
Let us consider $x \in \ell_\infty$. Then there exists a constant $M$ such that

$$\sup_k |x_k| = M < \infty.$$

Now,

$$\sup_k \left| (\Delta_h^{a,b,c} x)_k \right| \leq \sum_{i=0}^{\infty} \left| \frac{(-a)_i (-b)_i}{i!(-c)_i h^{a+b-c}} x_{k-i} \right|$$

$$\leq \sum_{i=0}^{\infty} \left| \frac{(-a)_i (-b)_i}{i!(-c)_i h^{a+b-c}} \right| |x_{k-i}|$$

$$\leq \lambda M < \infty,$$

where $\lambda = \sup_m \left| \frac{\binom{b}{m}}{h^{a+b-c}\binom{c}{m}} \right| 2^a$, as suggested in previous section.

Let us also consider a sequence $x = (x_k)$ with $x_k = k$ for all $k \in \mathbb{N}$. Clearly $x \notin \ell_\infty$, but for $a = 2, b = c$ and $h = 1$, we observe that

$$(\Delta_1^{2,b,b} x)_k = k - 2(k+1) + k + 2 = 0.$$

As a result, we get $\Delta_1^{2,b,b} x \in \ell_\infty$. This completes the proof. $\square$

**Theorem 12.3.2.** *Let* $X = c, c_0, \ell_\infty$ *and* $a \geq 1$ *with* $b = c$, *the classes* $X(\Delta_h^{a,b,c})$ *are complete normed linear spaces, norm defined by*

$$\|x\|_{\Delta_h^{a,b,c}} = \sum_{i=0}^{[a]} |x_k| + \sup_k \left| \Delta_h^{a,b,c}(x_k) \right|. \tag{12.8}$$

*where* $[a]$ *indicates the integral part of* $a$.

*Proof.* To prove this theorem we refer Baliarsingh (2016). $\square$

**Theorem 12.3.3** (Baliarsingh (2016)). *Let* $X = c, c_0, \ell_\infty$ *and* $0 \leq a < 1$ *with* $b = c$, *the classes* $X(\Delta_h^{a,b,c})$ *are semi-normed spaces, semi-norm defined by*

$$g(x) = \sup_k \left| \Delta_h^{a,b,c}(x_k) \right|. \tag{12.9}$$

*Proof.* For $0 \leq a < 1$ with $b = c$, let us consider

$$g(x) = \sup_k \left| \sum_{i=0}^{\infty} \frac{(-a)_i}{i! h^a} x_{k-i} \right| = 0,$$

Then it does not necessarily imply that $x = \theta = (0, 0, 0, \dots)$. This follows from the following example:

Let us take $x = (1, 1, 1, \dots)$, a constant sequence, then it is clear that $x \neq \theta$, whereas $g(x) = 0$. $\square$

**Note:** It is noticed that if $a$ is an integer, $b = c$ and $h = 1$, then the norm defined in Theorem 12.3.2 reduces to the norm (see Et and Colak (1995 ); Kızmaz (1981); Malkowsky (et al.))

$$\|x\|_{\Delta_h^{a,b,c}} = \sum_{i=0}^{a} |x_k| + \sup_k \left| \Delta_h^{a,b,c}(x_k) \right|.$$

## 12.4    Application to fractional derivatives

In this section, we state some applications of the difference operator $\Delta_h^{a,b,c}$ in order to generalize the notion of fractional calculus.

Let $h \to 0$ and $f(x)$ be a differentiable(with fractional order) function. Associated to this function $f(x)$, define the sequence $f_h(.) = (f(x - kh))_{k \in \mathbb{N}_0}$, and the sequence spaces $\Delta_{h,x}^{a,b,c}(f_h(.))$ via the difference operator $\Delta_{h,x}^{a,b,c}$ as

$$\Delta_{h,x}^{a,b,c} f(x) = \sum_{i=0}^{\infty} \frac{(-a)_i(-b)_i}{i!(-c)_i h^{a+b-c}} f(x - ih). \tag{12.10}$$

Clearly, the difference operator defined in Eqn.(12.10) is a linear operator and it generalizes the concept of integral and fractional order derivatives. For instances,

- For $a = 1, 2, b = c$, we have $\Delta_{h,x}^{1,b,b} \equiv \dfrac{d}{dx}$, and $\Delta_{h,x}^{2,b,b} \equiv \left(\dfrac{d}{dx}\right)^2$,

  more specifically, $\Delta_{h,x}^{1,b,b} f(x) = \frac{f(x)-f(x-h)}{h}$

  and also  $\Delta_{h,x}^{2,b,b} f(x) = \frac{f(x)-2f(x-h)+f(x-2h)}{h^2}$,

- For $a = \alpha \in \mathbb{R}, b = c$, the operator $\Delta_{h,x}^{\alpha,b,b}$ reduces to fractional difference operator $\left(\dfrac{d}{dx}\right)^{\alpha}$, where

  $$\Delta_{h,x}^{\alpha,b,b} f(x) = \sum_{i=0}^{\infty} \frac{(-\alpha)_i}{i! h^{\alpha}} f(x - ih),$$

- For $\alpha \in \mathbb{R}(\notin \mathbb{N}), b = c$, the operator $\Delta_{h,x}^{-\alpha,b,b}$ reduces to fractional integro operator $\left(\dfrac{d}{dx}\right)^{-\alpha}$, where

  $$\Delta_{h,x}^{-\alpha,b,b} f(x) = \sum_{i=0}^{\infty} \frac{(\alpha)_i}{i! h^{-\alpha}} f(x - ih).$$

However, due to restriction of Euler gamma function, for $a = \alpha \in \mathbb{N}, b = c$, the operator $\Delta_{h,x}^{-\alpha,b,b}$ reduces to integral operator $\left(\dfrac{d}{dx}\right)^{-\alpha}$ by splitting the integer $\alpha$ to a sum of finite number of proper fractions.

**Theorem 12.4.1** (Baliarsingh (2016)). *If $\alpha, \beta > 0$, and $f(x)$ is a nonconstant function, then*

(i) The difference operator $\Delta_{h,x}^{\alpha,b,c}$ is a linear operator over $\mathbb{R}$.

(ii) $\Delta_{h,x}^{\alpha,b,b}\left(\Delta_{h,x}^{\beta,b,b}f(x)\right) = \Delta_{h,x}^{\alpha,b,b}\left(\sum_{i=0}^{\infty}\frac{(-\beta)_i}{i!h^{-\alpha}}f(x-ih)\right) = \Delta_{h,x}^{\alpha+\beta,b,b}f(x).$

(iii) $\Delta_{h,x}^{\alpha,b,b}\left(\Delta_{h,x}^{-\alpha,b,b}f(x)\right) = \Delta_{h,x}^{-\alpha,b,b}\left(\Delta_{h,x}^{\alpha,b,b}f(x)\right) = f(x).$

*Proof.* The proof of the theorem is found in Baliarsingh (2016). □

**Theorem 12.4.2** (Baliarsingh (2016)). *Let $\alpha$ and $0 \neq \beta \in \mathbb{R}$, then*

$$\Delta_{h,x}^{\alpha,b,b}x^{\beta} = x^{\beta-\alpha}\frac{\Gamma(\beta+1)}{\Gamma(\beta+1-\alpha)}.$$

*Proof.* From Eqn. (12.10), if $h \to 0$, it is seen that

$$\Delta_{h,x}^{\alpha,b,b}x^{\beta}$$

$$=\sum_{i=0}^{\infty}\frac{(-\alpha)_i(x-ih)^{\beta}}{i!h^{\alpha}}$$

$$=x^{\beta}h^{-\alpha}\left[1-\alpha\left(1-\frac{h}{x}\right)^{\beta}+\frac{\alpha(\alpha-1)}{2!}\left(1-\frac{2h}{x}\right)^{\beta}-\frac{\alpha(\alpha-1)(\alpha-2)}{3!}\right.$$
$$\left.\times\left(1-\frac{3h}{x}\right)^{\beta}+\dots\right]$$

$$=x^{\beta-\alpha}\delta^{-\alpha}\left[1-\alpha\left(1-\delta\right)^{\beta}+\frac{\alpha(\alpha-1)}{2!}\left(1-2\delta\right)^{\beta}-\frac{\alpha(\alpha-1)(\alpha-2)}{3!}\right.$$
$$\left.\times\left(1-3\delta\right)^{\beta}+\dots\right]$$

$$=x^{\beta-\alpha}\delta^{-\alpha}\left[1-\alpha\left(1-\beta\delta+\frac{\beta(\beta-1)}{2!}\delta^2-\frac{\beta(\beta-1)(\beta-2)}{3!}\delta^3+\dots\right)\right.$$

$$+\frac{\alpha(\alpha-1)}{2!}\left(1-2\delta+\frac{\beta(\beta-1)}{2!}(2\delta)^2-\frac{\beta(\beta-1)(\beta-2)}{3!}(2\delta)^3+\dots\right)$$

$$-\frac{\alpha(\alpha-1)(\alpha-2)}{3!}$$

$$\left.\times\left(1-3\delta+\frac{\beta(\beta-1)}{2!}(3\delta)^2-\frac{\beta(\beta-1)(\beta-2)}{3!}(3\delta)^3+\dots\right)+\dots\right]$$

$$=x^{\beta-\alpha}\left[\beta\delta^{1-\alpha}\left(\alpha-2\frac{\alpha(\alpha-1)}{2!}+3\frac{\alpha(\alpha-1)(\alpha-2)}{3!}-\dots\right)-\frac{\beta(\beta-1)}{2!}\delta^{2-\alpha}\right.$$

$$\times\left(\alpha-2^2\frac{\alpha(\alpha-1)}{2!}+3^2\frac{\alpha(\alpha-1)(\alpha-2)}{3!}-\dots\right)+\dots$$

$$+(-1)^{k-1}\frac{\beta(\beta-1)\dots(\beta-k+1)}{k!}\delta^{k-\alpha}$$

$$\left.\times\left(\alpha-2^k\frac{\alpha(\alpha-1)}{2!}+3^k\frac{\alpha(\alpha-1)(\alpha-2)}{3!}-\dots\right)\right]$$

$$=x^{\beta-\alpha}\frac{\beta(\beta-1)\dots(\beta-k+1)}{\Gamma(\alpha+1)}\Gamma(\alpha+1),$$

$$=x^{\beta-\alpha}\frac{\Gamma(\beta+1)}{\Gamma(\beta-\alpha+1)}.$$

It is noticed that $\delta^{k-\alpha}\to 0$, as $h\to 0$ for all $k>\alpha$, where $\delta=h/x$.

It is being suggested that the above formula is valid for $\beta\neq 0$. For $\beta=0$, $f(x)=x^\beta=1$, which is a constant function. As a result, we have $\Delta_{h,x}^{\alpha,b,b}x^\beta=0$, but using this formula, one may claim that $\Delta_{h,x}^{\alpha,b,b}x^\beta=x^{-\alpha}\frac{1}{\Gamma(1-\alpha)}$ which is a contradiction. $\square$

*Remark* 12.4.3. In particular, for any $\alpha\in\mathbb{R}$ and the constant function $f(x)=c$, we have

$$\frac{d^\alpha}{dt^\alpha}(c)=\begin{cases}0, & (\alpha=1,2,3,\dots)\\ c & (\alpha=0)\\ \frac{x^{-\alpha}c}{\Gamma(1-\alpha)}, & (\text{otherwise})\end{cases}.$$

**Example 12.4.4.** Now, we consider certain functions and determine their fractional derivatives.

- Let $f(x)=x$ and $\alpha=1/2$. Then

$$\Delta_{h,x}^{\alpha,b,b}f(x)=\frac{2}{\sqrt{\pi}}x^{1/2}.$$

- Let $f(x)=x^{1/2}$ and $\alpha=1/2$. Then

$$\Delta_{h,x}^{\alpha,b,b}f(x)=\sqrt{\pi}/2.$$

- Let $f(x) = x^2$ and $\alpha = 1/2$. Then

$$\Delta_{h,x}^{\alpha,b,b} f(x) = \frac{8}{3\sqrt{\pi}} x^{3/2}.$$

- Let $f(x) = x \sin x$ and $\alpha \in \mathbb{R}$. Then

$$\Delta_{h,x}^{\alpha,b,b} f(x) = \sum_{i=0}^{\infty} (-1)^i \frac{x^{2i+2-\alpha}}{\Gamma(2i+2-\alpha)}.$$

- Let $f(x) = x \cos x$ and $\alpha \in \mathbb{R}$. Then

$$\Delta_{h,x}^{\alpha,b,b} f(x) = \sum_{i=0}^{\infty} (-1)^i \frac{x^{2i+1-\alpha}}{\Gamma(2i+1-\alpha))}.$$

Basically, non integral calculus involves more complicated and comprehensive results due to their unusual and violating behaviors. Applying Theorem 12.4.2, we discuss following results on integral calculus which may violate for non integral cases:

*Remark 12.4.5. Fractional derivative of a periodic function of period $T$ need not preserve the periodicity $T$, even not be periodic also. For instance, we consider the function $f(x) = \cos x + 2$ which is periodic with period $2\pi$, whereas its $1/2$-derivative is found to be*

$$\Delta_{h,x}^{1/2,b,b}(\cos x + 2) = \sum_{i=0}^{\infty} (-1)^i \frac{x^{2i-1/2}}{\Gamma(2i+1-1/2))} + \frac{2x^{-1/2}}{\Gamma(1/2)}$$

$$= \frac{1}{\sqrt{x}} \left[ \sum_{i=0}^{\infty} (-1)^i \frac{x^{2i}}{\Gamma(2i+1/2))} + \frac{2}{\sqrt{\pi}} \right],$$

*which is not periodic. This is due to the fact that the fractional derivatives of a constant function are not necessarily zero or periodic.*

*Remark 12.4.6 (Baliarsingh (2016)). Let $f = f(x)$ and $g = g(x)$ be two functions. For any integer $n$,*

$$\Delta_{h,x}^{n,b,b}(fg) = \sum_{k=0}^{n} \binom{n}{k} \Delta_{h,x}^{k,b,b}(f) \Delta_{h,x}^{n-k,b,b}(g),$$

*but it does not hold for a proper fraction $\alpha$. For example, we take $f(x) = x^p$, $(p > 0)$ and $g(x) = x^q$, $(q > 0)$ and using Theorem 12.4.2, we have*

$$\Delta_{h,x}^{\alpha,b,b}(x^{p+q}) = x^{p+q-\alpha} \frac{\Gamma(p+q+1)}{\Gamma(p+q+1-\alpha)}$$

$$\neq \sum_{k=0}^{\infty} \frac{\Gamma(\alpha+1)}{k!\Gamma(\alpha-k+1)} \Delta_{h,x}^{\alpha,b,b}(f) \Delta_{h,x}^{\alpha-k,b,b}(g)$$

$$= \sum_{k=0}^{\infty} \frac{\Gamma(\alpha+1)}{k!\Gamma(\alpha-k+1)} \frac{\Gamma(p+1)}{\Gamma(p-\alpha+1)} \frac{\Gamma(q+1)}{\Gamma(q-\alpha+k+1)} x^{p+q-2\alpha+k}$$

*Remark 12.4.7 (Baliarsingh (2016)). Let $f = f(x)$ and $g = g(x)$ be two functions. For any integer $n$, the Chain rule*

$$\Delta_{h,x}^{n,b,b}(f(g(x)) = \Delta_{h,g(x)}^{n,b,b}(f(g)).\Delta_{h,x}^{1,b,b}g(x),$$

*is true but, it does not hold for a proper fraction $\alpha$. For example, we take $f(x)$ and $g(x)$ as considered in Remark 12.4.6 and using Theorem 12.4.2 on $f(g(x))$, we derive that*

$$\Delta_{h,x}^{\alpha,b,b}f(g(x)) = \Delta_{h,x}^{\alpha,b,b}(x^{pq}) = x^{pq-\alpha}\frac{\Gamma(pq+1)}{\Gamma(pq+1-\alpha)}. \qquad (12.11)$$

*But, applying Chain rule and Theorem 12.4.2, one can easily calculate*

$$\Delta_{h,g(x)}^{\alpha,b,b}f(g(x)).\Delta_{h,x}^{1,b,b}g(x) = \Delta_{h,x^q}^{\alpha,b,b}(x^{pq}).\Delta_{h,x}^{1,b,b}x^q$$

$$= \Delta_{h,t}^{\alpha,b,b}(t^p).\Delta_{h,x}^{1,b,b}x^q \quad (\text{where } t = x^q)$$

$$= t^{p-\alpha}\frac{\Gamma(p+1)}{\Gamma(p+1-\alpha)}x^{q-1}\frac{\Gamma(q+1)}{\Gamma(q+1-1)}$$

$$= x^{qp-q\alpha}\frac{\Gamma(p+1)}{\Gamma(p+1-\alpha)}.qx^{q-1}$$

$$= x^{qp-q\alpha+q-1}q\frac{\Gamma(p+1)}{\Gamma(p+1-\alpha)} \qquad (12.12)$$

*Combining Eqns. (12.11) and (12.12), we complete the proof.*

*Remark 12.4.8. Let us consider two functions as $f(x) = x^{1/6}$ and $g(x) = x^{1/2}$, then*

$$f(x)g(x) = x^{2/3}.$$

*For $\alpha = 5/3$, the left hand side of the formula due to Liu Liu (2015) is simplified as*

$$(f(x)g(x))^{(\alpha)} = (x^{2/3})^{(5/3)} = \frac{\frac{2}{3}\Gamma(\frac{2}{3})}{\Gamma(0)}x^{\frac{2}{3}-\frac{5}{3}} = 0. \qquad (12.13)$$

*Since $f(0) = g(0) = 0$ and on analyzing the right hand side, it is observed that*

$$\sum_{j=0}^{\infty} \binom{5/3}{j} (f(x))^{(j)} (g(x))^{(\frac{5}{3}-j)}$$

$$= \binom{5/3}{0} (x^{1/6})^{(0)} (x^{1/2})^{(\frac{5}{3})} + \binom{5/3}{1} (x^{1/6})^{(1)} (x^{1/2})^{(\frac{2}{3})}$$

$$+ \binom{5/3}{2} (x^{1/6})^{(2)} (x^{1/2})^{(-\frac{1}{3})}$$

$$+ \binom{5/3}{3} (x^{1/6})^{(3)} (x^{1/2})^{(-\frac{4}{3})} + \binom{5/3}{4} (x^{1/6})^{(4)} (x^{1/2})^{(-\frac{7}{3})} + \cdots$$

$$= \frac{\sqrt{\pi}}{2x} \left[ \frac{1}{\Gamma(-\frac{1}{6})} - \frac{5}{18\Gamma(\frac{5}{6})} - \frac{25}{324\Gamma(\frac{11}{6})} - \frac{275}{17496\Gamma(\frac{17}{6})} - \frac{4675}{314928\Gamma(\frac{23}{6})} \right.$$

$$\left. - \frac{150535}{5668704\Gamma(\frac{29}{6})} - \cdots \right]$$

$$= \Gamma_{5/3}(x) \neq 0.$$

*Now, calculating the value of $\Gamma_{5/3}(x)$ for all $t \in \mathbb{R}$ then it is observed that it takes the value zero asymptotically as $x \to \infty$ and takes non zero values as $x \to 0$. Therefore, the well known Leibniz formula involving fractional derivatives due to Liu is well posed for all values of $x$ except in the neighborhood of $0$.*

---

## 12.5  Geometrical interpretation

In this section, using Theorem 12.4.2 of the previous section we provide the geometrical interpretation of fractional derivatives of certain functions and mention their non uniform and unusual behaviors.

The fractional derivatives of a constant function need not be zero, whereas its derivatives in integral cases are zero. This idea suggests almost all dynamic behaviors of fractional calculus. From the following figure, it is mentioned that for any positive non integral real number $\alpha$, the $\alpha$-th derivatives of the constant function are asymptotically zero, but not exactly zero i.e, the corresponding curves are approaching to the line $f(x) - 0$ as $x \to \infty$. But for any negative values of $\alpha$, the graphs of the derivatives are gradually increased and approaching infinity as $x \to \infty$.

In Figure 12.1, we have taken fractional derivatives of different orders starting from $\alpha = 0$ to $1.6$ with step size $0.1$ for the function $f(x) = x$ and plotted them with respect to different values of $x$ ($0 \le x \le 5$). In addition to that we have also included its integration of order $\alpha = -0.1$ to $-1.6$ with same step size. In fact, the total sets of curves mentioned in Figure 12.2 are categorized into 4 different sets. The set of all green curves represent the fractional derivatives of $f(x) = x$ of order $\alpha$ with $0 < \alpha \leq 1$. They all are intersecting each other before they intersect the original curve $f(x) = x$. Another set of blue curves belongs to the family of fractional derivatives of the given function with $1 < \alpha \leq 1.6$. It is seen that all the curves are approaching the x-axis as we increase the value of $\alpha$. For $\alpha \to 0$, the curves of the subsequent derivatives are approaching the line $f(x) = x$ and for $\alpha = 0$ it exactly overlaps with the function itself which is shown by maroon color. The set of red curves represents fractional integrations of the given function with order $\alpha$ with $-1.6 \leq \alpha < 0$. It is also noticed that the curves of this family meet each other after intersecting the original function and for $\alpha < -1$, all the curves approach to the parabola $y = x^2/2$.

Furthermore, fractional derivatives of the functions $\cos x$ and $\sin x$ have been plotted for the order $\alpha$ varying from 0 to 1.3 in Figures 12.3 and 12.4 (see Baliarsingh (2016)), respectively. In Figures 12.5 and 12.6 , fractional derivatives of the function $f(x) = x\cos x$ and $f(x) = x\sin x$ have been mentioned for different orders from $-2$ to 2.

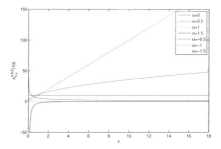

**FIGURE 12.1**: Fractional derivatives of the function $f(x) = 10$.

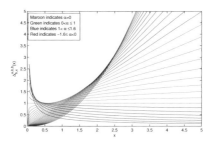

**FIGURE 12.2**: Fractional derivatives of the function $f(x) = x$ with order $\alpha$, where $-1.6 < \alpha \leq 1.6$.

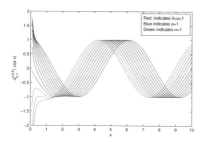

**FIGURE 12.3**: Fractional derivatives of the function $f(x) = \cos x$ with order $\alpha$, where $0 < \alpha \leq 1.3$.

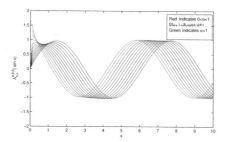

**FIGURE 12.4**: Fractional derivatives of the function $f(x) = \sin x$ with order $\alpha$, where $0 < \alpha \leq 1.3$.

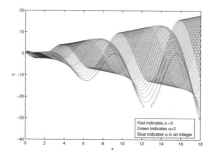

**FIGURE 12.5**: Fractional derivatives of the function $f(x) = x \cos x$ with order $\alpha$, where $-2 \leq \alpha \leq 2$.

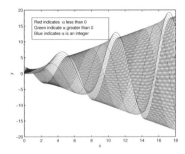

**FIGURE 12.6**: Fractional derivatives of the function $f(x) = x \sin x$ with order $\alpha$, where $-2 \leq \alpha \leq 2$.

---

# Bibliography

Aasma A., Dutta H., and Natarajan P. N. An introductory course in summability theory, John Wiley & Sons, Inc., 2017.

Ahmad Z. U., Mursaleen M., and Köthe-Toeplitz duals of some new sequence spaces and their matrix maps, Publ. Inst. Math. (Beograd) 42(56) (1987) 57–61.

Altay B. and Başar F. On the fine spectrum of the difference operator $\Delta$ on $c_0$ and $c$, Inform. Sci., 168(2004), 217–224.

Baliarsingh P. Some new difference sequence spaces of fractional order and their dual spaces, Appl. Math. Comput. 219(18) (2013) 9737–9742.

Baliarsingh P. and Dutta S. On the classes of fractional order difference sequence spaces and their matrix transformations, Appl. Math. Comput. 250 (2015), 665–674.

Baliarsingh P. and Dutta S. A note on paranormed difference sequence spaces of fractional order and their matrix transformations, J. Egypt. Math. Soc. 22(2), 249–253 (2014).

Baliarsingh P. and Dutta S. Unifying approach to the difference operators and their applications, Bol. Soc. Paran. Mat., 33(1) (2015) 49-57.

Baliarsingh P. and Dutta S. On an explicit formula for inverse of triangular matrices, J. Egypt. Math. Soc., 23 (2015) 297-302.

Baliarsingh P. On a fractional difference operator, Alexandria Eng. J., 55(2) (2016) 1811-1816.

Baliarsingh P. and Nayak L. A note on fractional difference operators, Alexandria Eng. J., (2017), dio.org/10.1016/j.aej.2017.02.022.

Başar F. and Aydın C. Some new difference sequence spaces, Appl. Math. Comput, 3(157) (2004) 677–693.

Bektas C. A., Et M. and Çolak R. Generalized difference sequence spaces and their dual spaces. J. Math. Anal. Appl., 2(292) (2004) 423–432.

Blutzer R. L. and Torvik P. J. On the fractional calculus model of viscoelastic behaviour, J. Rheol. 30 (1996) 133-135.

Dreisigmeyer W. D. and Young P. M. Nonconservative Lagrangian mechanics: a generalized function approach. J. Phys. A 36(30) (2003) 8297-8310.

Dutta H. Characterization of certain matrix classes involving generalized difference summability spaces, Appl. Sci., 11 (2009) 60–67.

Dutta H. and Rhoades B. E. (eds), Current topics in summability theory and applications, Springer, 2016.

Dutta S. and Baliarsingh P. On the fine spectra of the generalized rth difference operator $\Delta_\nu^r$ on the sequence space $\ell_1$, Appl. Math. Comput. 219 (18) (2012) 1776–1784.

Dutta S. and Baliarsingh P. On some Toeplitz matrices and their inversion, J. Egypt. Math. Soc., 22(3) (2014) 420–424.

Et M. and Colak R. On some generalized difference sequence spaces, Soochow J. Math. 21(4) (1995) 377–386.

Et M. and Basarir M. On some new generalized difference sequence spaces, Periodica Math. Hungar., 35(3), (1997) 169–175.

Gasper G. and Rahman M. Basic hypergeometric series, Cambridge University Press, 2004.

Jonsson T. and Yngvason J. Waves and distributions, World Scientific Publishing, 1995.

Jumarie G. On the representation of fractional Brownian motion as an integral with respect to $(dt)^\alpha$. Appl. Math. Lett. 18(7) (2005)739-48.

Jumarie G. New stochastic fractional models for Malthusian growth, the Poissonian birth process and optimal management of populations. Math. Comput. Modell. 44(3) (2006)231-54.

Kadak U. and Baliarsingh P. On certain Euler difference sequence spaces of fractional order and related dual properties, J. Nonlinear Sci. Appl. 8 (2015) 997–1004.

Kilbas A. A., Srivastava H. M., and Trujillo J. J. Theory and applications of fractional differential equations, North-Holland Mathematics Studies, 204, 2006.

Kızmaz H. On certain sequence spaces, Canad. Math. Bull., 24(2) (1981) 169-176.

Liu C. Counterexamples on Jumaries two basic fractional calculus formulae, Commun. Nonlinear Sci. Numer. Simulat., 22 (2015) 92–94.

Mainardi F. The fundamental solutions for the fractional diffusion-wave equation. Appl. Math. Lett. 9(6) (1996) 23-28.

Malkowsky E., Mursaleen M., and Suantai S. The dual spaces of sets of difference sequences of order $m$ and matrix transformations, Acta Math. Sin. (English Series) 23(3) (2007) 521–532.

Mursaleen M. Generalized spaces of difference sequences, J. Math. Anal. Appl. 203(3) (1996) 738–745.

Mursaleen M. and Noman A. K. Compactness by the Hausdorff measure of noncompactness, Nonlinear Anal. 73, (2010) 2541–2557.

Mursaleen M., Ansari K. J., and Asif Khan. Some approximation results by $(p, q)$-analogue of BernsteinStancu operators, Appl. Math. Comput. 264 (2015) 392-402.

Nayak L., Das G., and Ray B. K. An estimate of the rate of convergence of Fourier series in the generalized Hölder metric by deferred Cesàro mean, J. Math. Anal. Appl. 420 (2014) 563–575.

Podlubny I. Fractional differential equations, Academic Press, Inc. 1999.

# Chapter 13

## Composite Ostrowski and Trapezoid Type Inequalities for Riemann-Liouville Fractional Integrals of Functions with Bounded Variation

**Silvestru Sever Dragomir**

*Mathematics, College of Engineering & Science, Victoria University, Melbourne City, Australia & DST-NRF Centre of Excellence, in the Mathematical and Statistical Sciences, School of Computer Science and Applied Mathematics, University of the Witwatersrand, Johannesburg, South Africa*

## 13.1    Introduction

Let $f : [a, b] \to \mathbb{C}$ be a complex valued Lebesgue integrable function on the real interval $[a, b]$. The *Riemann-Liouville fractional integrals* are defined for $\alpha > 0$ by

$$J_{a+}^{\alpha} f(x) = \frac{1}{\Gamma(\alpha)} \int_a^x (x - t)^{\alpha-1} f(t) \, dt$$

for $a < x \leq b$ and

$$J_{b-}^{\alpha} f(x) = \frac{1}{\Gamma(\alpha)} \int_x^b (t - x)^{\alpha-1} f(t) \, dt$$

for $a \leq x < b$, where $\Gamma$ is the *Gamma function*. For $\alpha = 0$, they are defined as

$$J_{a+}^0 f(x) = J_{b-}^0 f(x) = f(x) \text{ for } x \in (a, b).$$

The following Ostrowski type inequalities for functions of bounded variation generalize the corresponding results for the Riemann integral obtained in

425

Dragomir (1999b), Dragomir (2001), Dragomir (2000) and have been established recently by the author in Dragomir (2017):

**Theorem 13.1.1.** *Let $f : [a, b] \to \mathbb{C}$ be a complex valued function of bounded variation on the real interval $[a, b]$.*

*(i) For any $x \in (a, b)$ we have*

$$\left| J_{a+}^{\alpha} f(x) + J_{b-}^{\alpha} f(x) - \frac{f(x)}{\Gamma(\alpha + 1)} \left[ (x - a)^{\alpha} + (b - x)^{\alpha} \right] \right| \tag{13.1}$$

$$\leq \frac{1}{\Gamma(\alpha)} \left[ \int_a^x (x - t)^{\alpha - 1} \bigvee_t^x (f) \, dt + \int_x^b (t - x)^{\alpha - 1} \bigvee_x^t (f) \, dt \right]$$

$$\leq \frac{1}{\Gamma(\alpha + 1)} \left[ (x - a)^{\alpha} \bigvee_a^x (f) + (b - x)^{\alpha} \bigvee_x^b (f) \right]$$

$$\leq \frac{1}{\Gamma(\alpha + 1)}$$

$$\times \begin{cases} \left[ \frac{1}{2}(b - a) + \left| x - \frac{a+b}{2} \right| \right]^{\alpha} \bigvee_a^b (f); \\[2mm] ((x - a)^{\alpha p} + (b - x)^{\alpha p})^{1/p} \left( (\bigvee_a^x (f))^q + \left( \bigvee_x^b (f) \right)^q \right)^{1/q} \\ \text{with } p, \ q > 1, \ \frac{1}{p} + \frac{1}{q} = 1; \\[2mm] \left[ \frac{1}{2} \bigvee_a^b (f) + \frac{1}{2} \left| \bigvee_a^x (f) - \bigvee_x^b (f) \right| \right] ((x - a)^{\alpha} + (b - x)^{\alpha}), \end{cases}$$

*and*

$$\left| J_{x+}^{\alpha} f(b) + J_{x-}^{\alpha} f(a) - \frac{f(x)}{\Gamma(\alpha + 1)} \left[ (x - a)^{\alpha} + (b - x)^{\alpha} \right] \right| \tag{13.2}$$

$$\leq \frac{1}{\Gamma(\alpha)} \left[ \int_x^b (b - t)^{\alpha - 1} \bigvee_x^t (f) \, dt + \int_a^x (t - a)^{\alpha - 1} \bigvee_t^x (f) \, dt \right]$$

$$\leq \frac{1}{\Gamma(\alpha + 1)} \left[ (x - a)^{\alpha} \bigvee_a^x (f) + (b - x)^{\alpha} \bigvee_x^b (f) \right]$$

$$\leq \frac{1}{\Gamma(\alpha + 1)}$$

$$\times \begin{cases} \left[ \frac{1}{2}(b - a) + \left| x - \frac{a+b}{2} \right| \right]^{\alpha} \bigvee_a^b (f); \\[2mm] ((x - a)^{\alpha p} + (b - x)^{\alpha p})^{1/p} \left( (\bigvee_a^x (f))^q + \left( \bigvee_x^b (f) \right)^q \right)^{1/q} \\ \text{with } p, \ q > 1, \ \frac{1}{p} + \frac{1}{q} = 1; \\[2mm] \left[ \frac{1}{2} \bigvee_a^b (f) + \frac{1}{2} \left| \bigvee_a^x (f) - \bigvee_x^b (f) \right| \right] ((x - a)^{\alpha} + (b - x)^{\alpha}). \end{cases}$$

*(ii) For any $x \in [a, b]$ we have*

$$\left| \frac{J_{a+}^{\alpha} f(b) + J_{b-}^{\alpha} f(a)}{2} - \frac{1}{\Gamma(\alpha+1)} f(x)(b-a)^{\alpha} \right| \qquad (13.3)$$

$$\leq \frac{1}{\Gamma(\alpha)} \int_a^x \frac{(b-t)^{\alpha-1} + (t-a)^{\alpha-1}}{2} \bigvee_t^x (f) \, dt$$

$$+ \frac{1}{\Gamma(\alpha)} \int_x^b \frac{(b-t)^{\alpha-1} + (t-a)^{\alpha-1}}{2} \bigvee_x^t (f) \, dt$$

$$\leq \frac{1}{2\Gamma(\alpha+1)} \left[(b-a)^{\alpha} + (x-a)^{\alpha} - (b-x)^{\alpha}\right] \bigvee_a^x (f)$$

$$+ \frac{1}{2\Gamma(\alpha+1)} \left[(b-a)^{\alpha} + (b-x)^{\alpha} - (x-a)^{\alpha}\right] \bigvee_x^b (f)$$

$$\leq \frac{1}{2\Gamma(\alpha+1)} \begin{cases} \left[(b-a)^{\alpha} + |(x-a)^{\alpha} - (b-x)^{\alpha}|\right] \bigvee_a^b (f), \\ (b-a)^{\alpha} \left[ \bigvee_a^b (f) + \left| \bigvee_a^x (f) - \bigvee_x^b (f) \right| \right]. \end{cases}$$

The following mid-point inequalities that can be derived from Theorem 13.1.1 are of interest as well:

$$\left| J_{a+}^{\alpha} f\left(\frac{a+b}{2}\right) + J_{b-}^{\alpha} f\left(\frac{a+b}{2}\right) - \frac{1}{2^{\alpha-1}\Gamma(\alpha+1)} f\left(\frac{a+b}{2}\right) \right| \qquad (13.4)$$

$$\leq \frac{1}{\Gamma(\alpha)} \int_a^{\frac{a+b}{2}} \left(\frac{a+b}{2} - t\right)^{\alpha-1} \bigvee_t^{\frac{a+b}{2}} (f) \, dt$$

$$+ \frac{1}{\Gamma(\alpha)} \int_{\frac{a+b}{2}}^b \left(t - \frac{a+b}{2}\right)^{\alpha-1} \bigvee_{\frac{a+b}{2}}^t (f) \, dt$$

$$\leq \frac{1}{2^{\alpha}\Gamma(\alpha+1)} (b-a)^{\alpha} \bigvee_a^b (f),$$

$$\left| J_{\frac{a+b}{2}+}^{\alpha} f(b) + J_{\frac{a+b}{2}-}^{\alpha} f(a) - \frac{1}{2^{\alpha-1}\Gamma(\alpha+1)} f\left(\frac{a+b}{2}\right) \right| \qquad (13.5)$$

$$\leq \frac{1}{\Gamma(\alpha)} \left[ \int_{\frac{a+b}{2}}^b (b-t)^{\alpha-1} \bigvee_{\frac{a+b}{2}}^t (f) \, dt + \int_a^{\frac{a+b}{2}} (t-a)^{\alpha-1} \bigvee_t^{\frac{a+b}{2}} (f) \, dt \right]$$

$$\leq \frac{1}{2^{\alpha}\Gamma(\alpha+1)} (b-a)^{\alpha} \bigvee_a^b (f)$$

and

$$\left| \frac{J_{a+}^{\alpha} f\left(b\right) + J_{b-}^{\alpha} f\left(a\right)}{2} - \frac{1}{\Gamma\left(\alpha + 1\right)} f\left(\frac{a+b}{2}\right)\left(b - a\right)^{\alpha} \right| \qquad (13.6)$$

$$\leq \frac{1}{\Gamma\left(\alpha\right)} \int_{a}^{\frac{a+b}{2}} \frac{\left(b - t\right)^{\alpha - 1} + \left(t - a\right)^{\alpha - 1}}{2} \bigvee_{t}^{\frac{a+b}{2}} \left(f\right) dt$$

$$+ \frac{1}{\Gamma\left(\alpha\right)} \int_{\frac{a+b}{2}}^{b} \frac{\left(b - t\right)^{\alpha - 1} + \left(t - a\right)^{\alpha - 1}}{2} \bigvee_{\frac{a+b}{2}}^{t} \left(f\right) dt$$

$$\leq \frac{1}{2\Gamma\left(\alpha + 1\right)} \left(b - a\right)^{\alpha} \bigvee_{a}^{b} \left(f\right).$$

For more related results see Aglić Aljinović (2014)-Dragomir (1999a) and Dragomir (2014)-Yue (2013).

Motivated by the above results, in this chapter we establish some composite Ostrowski and generalized trapezoid type inequalities for the Riemann-Liouville fractional integrals of functions of bounded variation. Applications for composite mid-point and trapezoid inequalities are provided as well. They generalize the know results holding for the classical Riemann integral.

---

## 13.2 Some identities

We have the following two parameter identity:

**Lemma 13.2.1.** *Let* $f : [a, b] \to \mathbb{C}$ *be a complex valued Lebesgue integrable function on the real interval* $[a, b]$ *and* $\lambda, \mu \in \mathbb{C}$.
*(i) If* $x \in (a, b)$, *then we have the representations*

$$J_{a+}^{\alpha} f\left(x\right) + J_{b-}^{\alpha} f\left(x\right) = \frac{1}{\Gamma\left(\alpha + 1\right)} \left[\lambda\left(x - a\right)^{\alpha} + \mu\left(b - x\right)^{\alpha}\right] \qquad (13.7)$$

$$+ \frac{1}{\Gamma\left(\alpha\right)} \int_{a}^{x} \left(x - t\right)^{\alpha - 1} \left[f\left(t\right) - \lambda\right] dt$$

$$+ \frac{1}{\Gamma\left(\alpha\right)} \int_{x}^{b} \left(t - x\right)^{\alpha - 1} \left[f\left(t\right) - \mu\right] dt.$$

*and*

$$J_{x-}^{\alpha} f(a) + J_{x+}^{\alpha} f(b) = \frac{1}{\Gamma(\alpha+1)} \left[ \lambda (x-a)^{\alpha} + \mu (b-x)^{\alpha} \right] \tag{13.8}$$

$$+ \frac{1}{\Gamma(\alpha)} \int_a^x (t-a)^{\alpha-1} \left[ f(t) - \lambda \right] dt$$

$$+ \frac{1}{\Gamma(\alpha)} \int_x^b (b-t)^{\alpha-1} \left[ f(t) - \mu \right] dt.$$

*(ii) We have the representation*

$$\frac{J_{a+}^{\alpha} f(b) + J_{b-}^{\alpha} f(a)}{2} = \frac{\lambda+\mu}{2} \frac{1}{\Gamma(\alpha+1)} (b-a)^{\alpha} \tag{13.9}$$

$$+ \frac{1}{2\Gamma(\alpha)} \left[ \int_a^b (b-t)^{\alpha-1} \left[ f(t) - \lambda \right] dt + \int_a^b (t-a)^{\alpha-1} \left[ f(t) - \mu \right] dt \right].$$

*Proof.* (i) We have that

$$\frac{1}{\Gamma(\alpha)} \int_a^x (x-t)^{\alpha-1} \left[ f(t) - \lambda \right] dt = J_{a+}^{\alpha} f(x) - \frac{\lambda}{\Gamma(\alpha+1)} (x-a)^{\alpha} \tag{13.10}$$

for $a < x \leq b$ and

$$\frac{1}{\Gamma(\alpha)} \int_x^b (t-x)^{\alpha-1} \left[ f(t) - \mu \right] dt = J_{b-}^{\alpha} f(x) - \frac{\mu}{\Gamma(\alpha+1)} (b-x)^{\alpha} \tag{13.11}$$

for $a \leq x < b$.

By adding these equalities for $x \in (a,b)$ we get the representation (13.7).
By the definition of fractional integrals we have

$$J_{x+}^{\alpha} f(b) = \frac{1}{\Gamma(\alpha)} \int_x^b (b-t)^{\alpha-1} f(t) dt$$

for $a \leq x < b$ and

$$J_{x-}^{\alpha} f(a) = \frac{1}{\Gamma(\alpha)} \int_a^x (t-a)^{\alpha-1} f(t) dt$$

for $a < x \leq b$.

Therefore

$$\frac{1}{\Gamma(\alpha)} \int_a^x (t-a)^{\alpha-1} \left[ f(t) - \lambda \right] dt = J_{x-}^{\alpha} f(a) - \frac{\lambda}{\Gamma(\alpha+1)} (x-a)^{\alpha}$$

*and*

$$\frac{1}{\Gamma(\alpha)} \int_x^b (b-t)^{\alpha-1} \left[ f(t) - \mu \right] dt = J_{x+}^{\alpha} f(b) - \frac{\mu}{\Gamma(\alpha+1)} (b-x)^{\alpha}.$$

By adding these equalities for $x \in (a, b)$ we get the representation (13.8).

From (13.10) for $x = b$ we have

$$\frac{1}{\Gamma(\alpha)} \int_a^b (b-t)^{\alpha-1} [f(t) - \lambda] \, dt = J_{a+}^\alpha f(b) - \frac{\lambda}{\Gamma(\alpha+1)} (b-a)^\alpha. \quad (13.12)$$

From (13.10) for $x = a$ we have

$$\frac{1}{\Gamma(\alpha)} \int_a^b (t-a)^{\alpha-1} [f(t) - \mu] \, dt = J_{b-}^\alpha f(a) - \frac{\mu}{\Gamma(\alpha+1)} (b-a)^\alpha. \quad (13.13)$$

If we add the equalities (13.12) and (13.17) and divide by 2 we get the equality (13.9). $\quad\square$

**Corollary 13.2.2.** *With the assumptions of Lemma 13.2.1 we have*

$$J_{a+}^\alpha f\left(\frac{a+b}{2}\right) + J_{b-}^\alpha f\left(\frac{a+b}{2}\right) = \frac{1}{2^\alpha \Gamma(\alpha+1)} (\lambda + \mu)(b-a)^\alpha \quad (13.14)$$

$$+ \frac{1}{\Gamma(\alpha)} \int_a^{\frac{a+b}{2}} \left(\frac{a+b}{2} - t\right)^{\alpha-1} [f(t) - \lambda] \, dt$$

$$+ \frac{1}{\Gamma(\alpha)} \int_{\frac{a+b}{2}}^b \left(t - \frac{a+b}{2}\right)^{\alpha-1} [f(t) - \mu] \, dt$$

*and*

$$J_{\frac{a+b}{2}-}^\alpha f(a) + J_{\frac{a+b}{2}+}^\alpha f(b) = \frac{1}{2^\alpha \Gamma(\alpha+1)} (\lambda + \mu)(b-a)^\alpha \quad (13.15)$$

$$+ \frac{1}{\Gamma(\alpha)} \int_a^{\frac{a+b}{2}} (t-a)^{\alpha-1} [f(t) - \lambda] \, dt$$

$$+ \frac{1}{\Gamma(\alpha)} \int_{\frac{a+b}{2}}^b (b-t)^{\alpha-1} [f(t) - \mu] \, dt.$$

**Corollary 13.2.3.** *With the assumptions of Lemma 13.2.1 we have*

$$\frac{1}{b-a} \int_a^b \frac{J_{a+}^\alpha f(x) + J_{b-}^\alpha f(x)}{2} \, dx \quad (13.16)$$

$$= \frac{1}{\Gamma(\alpha+2)} \left(\frac{\lambda+\mu}{2}\right)(b-a)^\alpha$$

$$+ \frac{1}{2\Gamma(\alpha)} \frac{1}{b-a} \int_a^b \left(\int_a^x (x-t)^{\alpha-1} [f(t) - \lambda] \, dt\right) dx$$

$$+ \frac{1}{2\Gamma(\alpha)} \frac{1}{b-a} \int_a^b \left(\int_x^b (t-x)^{\alpha-1} [f(t) - \mu] \, dt\right) dx$$

*and*

$$\frac{1}{b-a} \int_a^b \frac{J_{x-}^\alpha f(a) + J_{x+}^\alpha f(b)}{2} dx \qquad (13.17)$$

$$= \frac{1}{\Gamma(\alpha+2)} \left(\frac{\lambda+\mu}{2}\right) (b-a)^\alpha$$

$$+ \frac{1}{\Gamma(\alpha)} \frac{1}{b-a} \int_a^b \left(\int_a^x (t-a)^{\alpha-1} [f(t) - \lambda] dt\right) dx$$

$$+ \frac{1}{\Gamma(\alpha)} \frac{1}{b-a} \int_a^b \left(\int_x^b (b-t)^{\alpha-1} [f(t) - \mu] dt\right) dx.$$

The above inequalities (13.7), (13.8) and (13.9) have some particular cases of interest out of which we list the following.

1. If we take $\mu = \lambda$ in Lemma 13.2.1, then for $f \in L[a,b]$ and $x \in (a,b)$ we get the equalities

$$J_{a+}^\alpha f(x) + J_{b-}^\alpha f(x) - \frac{\lambda}{\Gamma(\alpha+1)} [(x-a)^\alpha + (b-x)^\alpha] \qquad (13.18)$$

$$+ \frac{1}{\Gamma(\alpha)} \int_a^x (x-t)^{\alpha-1} [f(t) - \lambda] dt$$

$$+ \frac{1}{\Gamma(\alpha)} \int_x^b (t-x)^{\alpha-1} [f(t) - \lambda] dt,$$

$$J_{x-}^\alpha f(a) + J_{x+}^\alpha f(b) = \frac{\lambda}{\Gamma(\alpha+1)} [(x-a)^\alpha + (b-x)^\alpha] \qquad (13.19)$$

$$+ \frac{1}{\Gamma(\alpha)} \int_a^x (t-a)^{\alpha-1} [f(t) - \lambda] dt$$

$$+ \frac{1}{\Gamma(\alpha)} \int_x^b (b-t)^{\alpha-1} [f(t) - \lambda] dt$$

*and*

$$\frac{J_{a+}^\alpha f(b) + J_{b-}^\alpha f(a)}{2} = \frac{\lambda}{\Gamma(\alpha+1)} (b-a)^\alpha \qquad (13.20)$$

$$+ \frac{1}{2\Gamma(\alpha)} \int_a^b \left[(b-t)^{\alpha-1} + (t-a)^{\alpha-1}\right] [f(t) - \lambda] dt$$

for $\lambda \in \mathbb{C}$.

If we take in (13.18)-(13.20) $\lambda = f(x)$, $x \in (a,b)$ then we get, see also

Dragomir (2017)

$$J_{a+}^{\alpha} f(x) + J_{b-}^{\alpha} f(x) = \frac{f(x)}{\Gamma(\alpha + 1)} \left[ (x - a)^{\alpha} + (b - x)^{\alpha} \right] \tag{13.21}$$

$$+ \frac{1}{\Gamma(\alpha)} \int_a^x (x - t)^{\alpha - 1} \left[ f(t) - f(x) \right] dt$$

$$+ \frac{1}{\Gamma(\alpha)} \int_x^b (t - x)^{\alpha - 1} \left[ f(t) - f(x) \right] dt,$$

$$J_{x-}^{\alpha} f(a) + J_{x+}^{\alpha} f(b) = \frac{f(x)}{\Gamma(\alpha + 1)} \left[ (x - a)^{\alpha} + (b - x)^{\alpha} \right] \tag{13.22}$$

$$+ \frac{1}{\Gamma(\alpha)} \int_a^x (t - a)^{\alpha - 1} \left[ f(t) - f(x) \right] dt$$

$$+ \frac{1}{\Gamma(\alpha)} \int_x^b (b - t)^{\alpha - 1} \left[ f(t) - f(x) \right] dt$$

and

$$\frac{J_{a+}^{\alpha} f(b) + J_{b-}^{\alpha} f(a)}{2} = \frac{f(x)}{\Gamma(\alpha + 1)} (b - a)^{\alpha} \tag{13.23}$$

$$+ \frac{1}{2\Gamma(\alpha)} \int_a^b \left[ (b - t)^{\alpha - 1} + (t - a)^{\alpha - 1} \right] \left[ f(t) - f(x) \right] dt.$$

In particular, we have the mid-point equalities, see also Dragomir (2017)

$$J_{a+}^{\alpha} f \left( \frac{a + b}{2} \right) + J_{b-}^{\alpha} f \left( \frac{a + b}{2} \right) \tag{13.24}$$

$$= \frac{1}{2^{\alpha - 1} \Gamma(\alpha + 1)} (b - a)^{\alpha} f \left( \frac{a + b}{2} \right)$$

$$+ \frac{1}{\Gamma(\alpha)} \int_a^{\frac{a+b}{2}} \left( \frac{a + b}{2} - t \right)^{\alpha - 1} \left[ f(t) - f \left( \frac{a + b}{2} \right) \right] dt$$

$$+ \frac{1}{\Gamma(\alpha)} \int_{\frac{a+b}{2}}^b \left( t - \frac{a + b}{2} \right)^{\alpha - 1} \left[ f(t) - f \left( \frac{a + b}{2} \right) \right] dt,$$

$$J_{\frac{a+b}{2}-}^{\alpha} f(a) + J_{\frac{a+b}{2}+}^{\alpha} f(b) = \frac{1}{2^{\alpha - 1} \Gamma(\alpha + 1)} (b - a)^{\alpha} f \left( \frac{a + b}{2} \right) \tag{13.25}$$

$$+ \frac{1}{\Gamma(\alpha)} \int_a^{\frac{a+b}{2}} (t - a)^{\alpha - 1} \left[ f(t) - f \left( \frac{a + b}{2} \right) \right] dt$$

$$+ \frac{1}{\Gamma(\alpha)} \int_{\frac{a+b}{2}}^b (b - t)^{\alpha - 1} \left[ f(t) - f \left( \frac{a + b}{2} \right) \right] dt$$

and

$$\frac{J_{a+}^{\alpha} f(b) + J_{b-}^{\alpha} f(a)}{2} \tag{13.26}$$

$$= \frac{1}{\Gamma(\alpha+1)} (b-a)^{\alpha} f\left(\frac{a+b}{2}\right)$$

$$+ \frac{1}{2\Gamma(\alpha)} \int_a^b \left[(b-t)^{\alpha-1} + (t-a)^{\alpha-1}\right] \left[f(t) - f\left(\frac{a+b}{2}\right)\right] dt.$$

If we take in (13.18)-(13.20) $\lambda = \frac{1}{b-a} \int_a^b f(s) \, ds$, $x \in (a,b)$ then we get

$$J_{a+}^{\alpha} f(x) + J_{b-}^{\alpha} f(x) = \frac{1}{\Gamma(\alpha+1)} \left[(x-a)^{\alpha} + (b-x)^{\alpha}\right] \frac{1}{b-a} \int_a^b f(s) \, ds$$

$$\tag{13.27}$$

$$+ \frac{1}{\Gamma(\alpha)} \int_a^x (x-t)^{\alpha-1} \left[f(t) - \frac{1}{b-a} \int_a^b f(s) \, ds\right] dt$$

$$+ \frac{1}{\Gamma(\alpha)} \int_x^b (t-x)^{\alpha-1} \left[f(t) - \frac{1}{b-a} \int_a^b f(s) \, ds\right] dt,$$

$$J_{x-}^{\alpha} f(a) + J_{x+}^{\alpha} f(b) = \frac{1}{\Gamma(\alpha+1)} \left[(x-a)^{\alpha} + (b-x)^{\alpha}\right] \frac{1}{b-a} \int_a^b f(s) \, ds$$

$$\tag{13.28}$$

$$+ \frac{1}{\Gamma(\alpha)} \int_a^x (t-a)^{\alpha-1} \left[f(t) - \frac{1}{b-a} \int_a^b f(s) \, ds\right] dt$$

$$+ \frac{1}{\Gamma(\alpha)} \int_x^b (b-t)^{\alpha-1} \left[f(t) - \frac{1}{b-a} \int_a^b f(s) \, ds\right] dt$$

and

$$\frac{J_{a+}^{\alpha} f(b) + J_{b-}^{\alpha} f(a)}{2} \tag{13.29}$$

$$= \frac{1}{\Gamma(\alpha+1)} (b-a)^{\alpha} \frac{1}{b-a} \int_a^b f(s) \, ds$$

$$+ \frac{1}{2\Gamma(\alpha)} \int_a^b \left[(b-t)^{\alpha-1} + (t-a)^{\alpha-1}\right] \left[f(t) - \frac{1}{b-a} \int_a^b f(s) \, ds\right] dt.$$

If we take in (13.18)-(13.20) $\lambda = \frac{1}{2}\left[f\left(a\right) + f\left(b\right)\right]$, $x \in \left(a, b\right)$ then we get

$$J_{a+}^{\alpha} f\left(x\right) + J_{b-}^{\alpha} f\left(x\right) = \frac{1}{\Gamma\left(\alpha + 1\right)}\left[\left(x - a\right)^{\alpha} + \left(b - x\right)^{\alpha}\right]\frac{1}{2}\left[f\left(a\right) + f\left(b\right)\right]$$

$$\tag{13.30}$$

$$+\frac{1}{\Gamma\left(\alpha\right)}\int_{a}^{x}\left(x - t\right)^{\alpha - 1}\left[f\left(t\right) - \frac{1}{2}\left[f\left(a\right) + f\left(b\right)\right]\right]dt$$

$$+\frac{1}{\Gamma\left(\alpha\right)}\int_{x}^{b}\left(t - x\right)^{\alpha - 1}\left[f\left(t\right) - \frac{1}{2}\left[f\left(a\right) + f\left(b\right)\right]\right]dt,$$

$$J_{x-}^{\alpha} f\left(a\right) + J_{x+}^{\alpha} f\left(b\right) = \frac{1}{\Gamma\left(\alpha + 1\right)}\left[\left(x - a\right)^{\alpha} + \left(b - x\right)^{\alpha}\right]\frac{1}{2}\left[f\left(a\right) + f\left(b\right)\right]$$

$$\tag{13.31}$$

$$+\frac{1}{\Gamma\left(\alpha\right)}\int_{a}^{x}\left(t - a\right)^{\alpha - 1}\left[f\left(t\right) - \frac{1}{2}\left[f\left(a\right) + f\left(b\right)\right]\right]dt$$

$$+\frac{1}{\Gamma\left(\alpha\right)}\int_{x}^{b}\left(b - t\right)^{\alpha - 1}\left[f\left(t\right) - \frac{1}{2}\left[f\left(a\right) + f\left(b\right)\right]\right]dt$$

and

$$\frac{J_{a+}^{\alpha} f\left(b\right) + J_{b-}^{\alpha} f\left(a\right)}{2} \tag{13.32}$$

$$= \frac{1}{\Gamma\left(\alpha + 1\right)}\left(b - a\right)^{\alpha}\frac{1}{2}\left[f\left(a\right) + f\left(b\right)\right]$$

$$+\frac{1}{2\Gamma\left(\alpha\right)}\int_{a}^{b}\left[\left(b - t\right)^{\alpha - 1} + \left(t - a\right)^{\alpha - 1}\right]\left[f\left(t\right) - \frac{1}{2}\left[f\left(a\right) + f\left(b\right)\right]\right]dt.$$

2. For $\mu \neq \lambda$ we can get other identities. For instance, if we take $\lambda = f\left(a\right)$ and $\mu = f\left(b\right)$ in (13.18)-(13.20) then we get for $x \in \left(a, b\right)$

$$J_{a+}^{\alpha} f\left(x\right) + J_{b-}^{\alpha} f\left(x\right) = \frac{1}{\Gamma\left(\alpha + 1\right)}\left[f\left(a\right)\left(x - a\right)^{\alpha} + f\left(b\right)\left(b - x\right)^{\alpha}\right] \quad (13.33)$$

$$+\frac{1}{\Gamma\left(\alpha\right)}\int_{a}^{x}\left(x - t\right)^{\alpha - 1}\left[f\left(t\right) - f\left(a\right)\right]dt$$

$$+\frac{1}{\Gamma\left(\alpha\right)}\int_{x}^{b}\left(t - x\right)^{\alpha - 1}\left[f\left(t\right) - f\left(b\right)\right]dt,$$

$$J_{x-}^{\alpha} f\left(a\right) + J_{x+}^{\alpha} f\left(b\right) = \frac{1}{\Gamma\left(\alpha + 1\right)}\left[f\left(a\right)\left(x - a\right)^{\alpha} + f\left(b\right)\left(b - x\right)^{\alpha}\right] \quad (13.34)$$

$$+\frac{1}{\Gamma\left(\alpha\right)}\int_{a}^{x}\left(t - a\right)^{\alpha - 1}\left[f\left(t\right) - f\left(a\right)\right]dt$$

$$+\frac{1}{\Gamma\left(\alpha\right)}\int_{x}^{b}\left(b - t\right)^{\alpha - 1}\left[f\left(t\right) - f\left(b\right)\right]dt.$$

and

$$\frac{J^{\alpha}_{a+}f(b) + J^{\alpha}_{b-}f(a)}{2} = \frac{f(a)+f(b)}{2}\frac{1}{\Gamma(\alpha+1)}(b-a)^{\alpha} \tag{13.35}$$

$$+ \frac{1}{2\Gamma(\alpha)}\int_a^b (b-t)^{\alpha-1}[f(t)-f(a)]\,dt$$

$$+ \frac{1}{2\Gamma(\alpha)} + \int_a^b (t-a)^{\alpha-1}[f(t)-f(b)]\,dt.$$

If we take $\lambda = \frac{f(a)+f(x)}{2}$ and $\mu = \frac{f(x)+f(b)}{2}$ in (13.18)-(13.20) then we get for $x \in (a,b)$ that

$$J^{\alpha}_{a+}f(x) + J^{\alpha}_{b-}f(x) = \frac{1}{2\Gamma(\alpha+1)}[(x-a)^{\alpha}f(a)+(b-x)^{\alpha}f(b)] \tag{13.36}$$

$$+ \frac{1}{2\Gamma(\alpha+1)}[(x-a)^{\alpha}+(b-x)^{\alpha}]f(x)$$

$$+ \frac{1}{\Gamma(\alpha)}\int_a^x (x-t)^{\alpha-1}\left[f(t)-\frac{f(a)+f(x)}{2}\right]\,dt$$

$$+ \frac{1}{\Gamma(\alpha)}\int_x^b (t-x)^{\alpha-1}\left[f(t)-\frac{f(x)+f(b)}{2}\right]\,dt,$$

$$J^{\alpha}_{x-}f(a) + J^{\alpha}_{x+}f(b) = \frac{1}{2\Gamma(\alpha+1)}[(x-a)^{\alpha}f(a)+(b-x)^{\alpha}f(b)] \tag{13.37}$$

$$+ \frac{1}{2\Gamma(\alpha+1)}[(x-a)^{\alpha}+(b-x)^{\alpha}]f(x)$$

$$+ \frac{1}{\Gamma(\alpha)}\int_a^x (t-a)^{\alpha-1}\left[f(t)-\frac{f(a)+f(x)}{2}\right]\,dt$$

$$+ \frac{1}{\Gamma(\alpha)}\int_x^b (b-t)^{\alpha-1}\left[f(t)-\frac{f(x)+f(b)}{2}\right]\,dt.$$

and

$$\frac{J^{\alpha}_{a+}f(b) + J^{\alpha}_{b-}f(a)}{2} \tag{13.38}$$

$$= \frac{1}{2}\frac{1}{\Gamma(\alpha+1)}(b-a)^{\alpha}\left[\frac{f(a)+f(b)}{2}+f(x)\right]$$

$$+ \frac{1}{2\Gamma(\alpha)}\int_a^b (b-t)^{\alpha-1}\left[f(t)-\frac{f(a)+f(x)}{2}\right]\,dt$$

$$+ \frac{1}{2\Gamma(\alpha)}\int_a^b (t-a)^{\alpha-1}\left[f(t)-\frac{f(x)+f(b)}{2}\right]\,dt.$$

We can also take $\lambda = f\left(\frac{a+x}{2}\right)$ and $\mu = f\left(\frac{x+b}{2}\right)$ or $\lambda = \frac{1}{x-a}\int_a^x f(s)\,ds$ and $\mu = \frac{1}{b-x}\int_x^b f(s)\,ds$ in (13.18)-(13.20) to get other similar equalities. The details are not presented here.

## 13.3    Inequalities for bounded function

Now, for $\phi$, $\Phi \in \mathbb{C}$ and $[a, b]$ an interval of real numbers, define the sets of complex-valued functions

$$\bar{U}_{[a,b]}(\phi, \Phi)$$
$$:= \left\{ f : [a, b] \to \mathbb{C} \mid \operatorname{Re}\left[ (\Phi - f(t)) \left( \overline{f(t)} - \bar{\phi} \right) \right] \geq 0 \text{ for almost every } t \in [a, b] \right\}$$

and

$$\bar{\Delta}_{[a,b]}(\phi, \Phi) := \left\{ f : [a, b] \to \mathbb{C} \mid \left| f(t) - \frac{\phi + \Phi}{2} \right| \leq \frac{1}{2} |\Phi - \phi| \text{ for a.e. } t \in [a, b] \right\}.$$

The following representation result may be stated.

**Proposition 13.3.1.** *For any $\phi$, $\Phi \in \mathbb{C}$, $\phi \neq \Phi$, we have that $\bar{U}_{[a,b]}(\phi, \Phi)$ and $\bar{\Delta}_{[a,b]}(\phi, \Phi)$ are nonempty, convex and closed sets and*

$$\bar{U}_{[a,b]}(\phi, \Phi) = \bar{\Delta}_{[a,b]}(\phi, \Phi). \tag{13.39}$$

*Proof.* We observe that for any $z \in \mathbb{C}$ we have the equivalence

$$\left| z - \frac{\phi + \Phi}{2} \right| \leq \frac{1}{2} |\Phi - \phi|$$

if and only if
$$\operatorname{Re}\left[ (\Phi - z)(\bar{z} - \phi) \right] \geq 0.$$

This follows by the equality

$$\frac{1}{4} |\Phi - \phi|^2 - \left| z - \frac{\phi + \Phi}{2} \right|^2 = \operatorname{Re}\left[ (\Phi - z)(\bar{z} - \phi) \right]$$

that holds for any $z \in \mathbb{C}$.

The equality (13.39) is thus a simple consequence of this fact.     □

On making use of the complex numbers field properties we can also state that:

**Corollary 13.3.2.** *For any $\phi, \Phi \in \mathbb{C}$, $\phi \neq \Phi$, we have that*

$$\bar{U}_{[a,b]}(\phi, \Phi) = \{ f : [a, b] \to \mathbb{C} \mid (\operatorname{Re}\Phi - \operatorname{Re} f(t))(\operatorname{Re} f(t) - \operatorname{Re}\phi) \tag{13.40}$$
$$+ (\operatorname{Im}\Phi - \operatorname{Im} f(t))(\operatorname{Im} f(t) - \operatorname{Im}\phi) \geq 0 \text{ for a.e. } t \in [a, b] \}.$$

Now, if we assume that $\operatorname{Re}(\Phi) \geq \operatorname{Re}(\phi)$ and $\operatorname{Im}(\Phi) \geq \operatorname{Im}(\phi)$, then we can define the following set of functions as well:

$$\bar{S}_{[a,b]}(\phi, \Phi) := \{f : [a,b] \to \mathbb{C} \mid \operatorname{Re}(\Phi) \geq \operatorname{Re} f(t) \geq \operatorname{Re}(\phi) \qquad (13.41)$$
$$\text{and } \operatorname{Im}(\Phi) \geq \operatorname{Im} f(t) \geq \operatorname{Im}(\phi) \text{ for a.e. } t \in [a,b]\}.$$

One can easily observe that $\bar{S}_{[a,b]}(\phi, \Phi)$ is closed, convex and

$$\emptyset \neq \bar{S}_{[a,b]}(\phi, \Phi) \subseteq \bar{U}_{[a,b]}(\phi, \Phi). \qquad (13.42)$$

We have

**Theorem 13.3.3.** *Let* $f : [a,b] \to \mathbb{C}$ *be a complex valued Lebesgue integrable function on the real interval* $[a,b]$ *and* $\phi, \Phi \in \mathbb{C}, \phi \neq \Phi$ *such that* $f \in \bar{\Delta}_{[a,b]}(\phi, \Phi)$.
*(i) For any* $x \in (a,b)$ *we have*

$$\left| J_{a+}^{\alpha} f(x) + J_{b-}^{\alpha} f(x) - \frac{1}{\Gamma(\alpha+1)} [(x-a)^{\alpha} + (b-x)^{\alpha}] \frac{\phi+\Phi}{2} \right| \qquad (13.43)$$
$$\leq \frac{1}{\Gamma(\alpha+1)} |\Phi - \phi| \left[ \frac{(x-a)^{\alpha} + (b-x)^{\alpha}}{2} \right]$$

*and, in particular*

$$\left| J_{a+}^{\alpha} f\left(\frac{a+b}{2}\right) + J_{b-}^{\alpha} f\left(\frac{a+b}{2}\right) - \frac{1}{2^{\alpha-1}\Gamma(\alpha+1)} (b-a)^{\alpha} \frac{\phi+\Phi}{2} \right| \qquad (13.44)$$
$$\leq \frac{1}{2^{\alpha-1}\Gamma(\alpha+1)} |\Phi - \phi| (b-a)^{\alpha}.$$

*(ii) For any* $x \in (a,b)$ *we have*

$$\left| J_{x-}^{\alpha} f(a) + J_{x+}^{\alpha} f(b) - \frac{1}{\Gamma(\alpha+1)} [(x-a)^{\alpha} + (b-x)^{\alpha}] \frac{\phi+\Phi}{2} \right| \qquad (13.45)$$
$$\leq \frac{1}{\Gamma(\alpha+1)} |\Phi - \phi| \left[ \frac{(x-a)^{\alpha} + (b-x)^{\alpha}}{2} \right],$$

*and, in particular*

$$\left| J_{\frac{a+b}{2}-}^{\alpha} f(a) + J_{\frac{a+b}{2}+}^{\alpha} f(b) - \frac{1}{2^{\alpha-1}\Gamma(\alpha+1)} (b-a)^{\alpha} \frac{\phi+\Phi}{2} \right| \qquad (13.46)$$
$$\leq \frac{1}{2^{\alpha-1}\Gamma(\alpha+1)} |\Phi - \phi| (b-a)^{\alpha},$$

*(iii) We have*

$$\left| \frac{J_{a+}^{\alpha} f(b) + J_{b-}^{\alpha} f(a)}{2} - \frac{1}{\Gamma(\alpha+1)} (b-a)^{\alpha} \frac{\phi+\Phi}{2} \right| \qquad (13.47)$$
$$\leq \frac{1}{2\Gamma(\alpha+1)} |\Phi - \phi| (b-a)^{\alpha}.$$

*Proof.* (i) From the identity (13.18) we have for $x \in (a, b)$ that

$$J_{a+}^{\alpha} f(x) + J_{b-}^{\alpha} f(x) - \frac{1}{\Gamma(\alpha+1)} \left[ (x-a)^{\alpha} + (b-x)^{\alpha} \right] \frac{\phi + \Phi}{2} \qquad (13.48)$$

$$= \frac{1}{\Gamma(\alpha)} \int_a^x (x-t)^{\alpha-1} \left[ f(t) - \frac{\phi + \Phi}{2} \right] dt$$

$$+ \frac{1}{\Gamma(\alpha)} \int_x^b (t-x)^{\alpha-1} \left[ f(t) - \frac{\phi + \Phi}{2} \right] dt.$$

If $f \in \bar{\Delta}_{[a,b]}(\phi, \Phi)$, then by taking the modulus in (13.48)

$$\left| J_{a+}^{\alpha} f(x) + J_{b-}^{\alpha} f(x) - \frac{1}{\Gamma(\alpha+1)} \left[ (x-a)^{\alpha} + (b-x)^{\alpha} \right] \frac{\phi + \Phi}{2} \right|$$

$$\leq \frac{1}{\Gamma(\alpha)} \left| \int_a^x (x-t)^{\alpha-1} \left[ f(t) - \frac{\phi + \Phi}{2} \right] dt \right|$$

$$+ \frac{1}{\Gamma(\alpha)} \left| \int_x^b (t-x)^{\alpha-1} \left[ f(t) - \frac{\phi + \Phi}{2} \right] dt \right|$$

$$\leq \frac{1}{\Gamma(\alpha)} \int_a^x (x-t)^{\alpha-1} \left| f(t) - \frac{\phi + \Phi}{2} \right| dt$$

$$+ \frac{1}{\Gamma(\alpha)} \int_x^b (t-x)^{\alpha-1} \left| f(t) - \frac{\phi + \Phi}{2} \right| dt$$

$$\leq \frac{1}{2\Gamma(\alpha)} |\Phi - \phi| \left[ \int_a^x (x-t)^{\alpha-1} dt + \int_x^b (t-x)^{\alpha-1} dt \right],$$

$$= \frac{1}{2\alpha\Gamma(\alpha)} |\Phi - \phi| \left[ (x-a)^{\alpha} + (b-x)^{\alpha} \right]$$

$$= \frac{1}{\Gamma(\alpha+1)} |\Phi - \phi| \left[ \frac{(x-a)^{\alpha} + (b-x)^{\alpha}}{2} \right],$$

which proves (13.43).

(ii) From the identity (13.19) we have for $x \in (a, b)$ that

$$J_{x-}^{\alpha} f(a) + J_{x+}^{\alpha} f(b) - \frac{1}{\Gamma(\alpha+1)} \left[ (x-a)^{\alpha} + (b-x)^{\alpha} \right] \frac{\phi + \Phi}{2}$$

$$= \frac{1}{\Gamma(\alpha)} \int_a^x (t-a)^{\alpha-1} \left[ f(t) - \frac{\phi + \Phi}{2} \right] dt$$

$$+ \frac{1}{\Gamma(\alpha)} \int_x^b (b-t)^{\alpha-1} \left[ f(t) - \frac{\phi + \Phi}{2} \right] dt.$$

Now, by employing a similar argument as in (i) we deduce the desired result (13.45).

(iii) From the identity (13.20) we have

$$\frac{J_{a+}^{\alpha} f(b) + J_{b-}^{\alpha} f(a)}{2} - \frac{1}{\Gamma(\alpha+1)} (b-a)^{\alpha} \frac{\phi + \Phi}{2} \qquad (13.49)$$

$$= \frac{1}{\Gamma(\alpha)} \int_{a}^{b} \frac{(b-t)^{\alpha-1} + (t-a)^{\alpha-1}}{2} \left[ f(t) - \frac{\phi + \Phi}{2} \right] dt.$$

If $f \in \bar{\Delta}_{[a,b]}(\phi, \Phi)$, then by taking the modulus in (13.49) we get

$$\left| \frac{J_{a+}^{\alpha} f(b) + J_{b-}^{\alpha} f(a)}{2} - \frac{1}{\Gamma(\alpha+1)} (b-a)^{\alpha} \frac{\phi + \Phi}{2} \right|$$

$$\leq \frac{1}{\Gamma(\alpha)} \int_{a}^{b} \frac{(b-t)^{\alpha-1} + (t-a)^{\alpha-1}}{2} \left| f(t) - \frac{\phi + \Phi}{2} \right| dt$$

$$\leq \frac{1}{2\Gamma(\alpha)} |\Phi - \phi| \int_{a}^{b} \frac{(b-t)^{\alpha-1} + (t-a)^{\alpha-1}}{2} dt$$

$$= \frac{1}{2\Gamma(\alpha+1)} |\Phi - \phi| (b-a)^{\alpha},$$

which proves the inequality (13.47). $\qquad \square$

**Corollary 13.3.4.** *With the assumptions of Theorem 13.3.3 we have*

$$\left| \frac{1}{b-a} \int_{a}^{b} \frac{J_{a+}^{\alpha} f(x) + J_{b-}^{\alpha} f(x)}{2} dx - \frac{1}{\Gamma(\alpha+2)} \frac{\phi + \Phi}{2} (b-a)^{\alpha} \right| \qquad (13.50)$$

$$\leq \frac{1}{2\Gamma(\alpha+2)} |\Phi - \phi| (b-a)^{\alpha}$$

*and*

$$\left| \frac{1}{b-a} \int_{a}^{b} \frac{J_{x-}^{\alpha} f(a) + J_{x+}^{\alpha} f(b)}{2} dx - \frac{1}{\Gamma(\alpha+2)} \frac{\phi + \Phi}{2} (b-a)^{\alpha} \right| \qquad (13.51)$$

$$\leq \frac{1}{2\Gamma(\alpha+2)} |\Phi - \phi| (b-a)^{\alpha}.$$

*Remark* 13.3.5. If the function $f : [a,b] \to \mathbb{R}$ is measurable and there exist the constants $m$, $M$ such that $m \leq f(t) \leq M$ for a.e. $t \in [a,b]$, then for any $x \in (a,b)$ we have by (13.43) and (13.44) that

$$\left| J_{a+}^{\alpha} f(x) + J_{b-}^{\alpha} f(x) - \frac{1}{\Gamma(\alpha+1)} [(x-a)^{\alpha} + (b-x)^{\alpha}] \frac{m+M}{2} \right| \qquad (13.52)$$

$$\leq \frac{1}{\Gamma(\alpha+1)} (M-m) \left[ \frac{(x-a)^{\alpha} + (b-x)^{\alpha}}{2} \right]$$

and, in particular

$$\left| J_{a+}^{\alpha} f \left( \frac{a+b}{2} \right) + J_{b-}^{\alpha} f \left( \frac{a+b}{2} \right) - \frac{1}{2^{\alpha-1}\Gamma(\alpha+1)} (b-a)^{\alpha} \frac{m+M}{2} \right| \quad (13.53)$$

$$\leq \frac{1}{2^{\alpha-1}\Gamma(\alpha+1)} (M-m)(b-a)^{\alpha}.$$

By (13.45) and (13.46) we have that

$$\left| J_{x-}^{\alpha} f(a) + J_{x+}^{\alpha} f(b) - \frac{1}{\Gamma(\alpha+1)} [(x-a)^{\alpha} + (b-x)^{\alpha}] \frac{m+M}{2} \right| \quad (13.54)$$

$$\leq \frac{1}{\Gamma(\alpha+1)} (M-m) \left[ \frac{(x-a)^{\alpha} + (b-x)^{\alpha}}{2} \right],$$

and, in particular

$$\left| J_{\frac{a+b}{2}-}^{\alpha} f(a) + J_{\frac{a+b}{2}+}^{\alpha} f(b) - \frac{1}{2^{\alpha-1}\Gamma(\alpha+1)} (b-a)^{\alpha} \frac{m+M}{2} \right| \quad (13.55)$$

$$\leq \frac{1}{2^{\alpha-1}\Gamma(\alpha+1)} (M-m)(b-a)^{\alpha}.$$

From (13.47) we have

$$\left| \frac{J_{a+}^{\alpha} f(b) + J_{b-}^{\alpha} f(a)}{2} - \frac{1}{\Gamma(\alpha+1)} (b-a)^{\alpha} \frac{m+M}{2} \right| \quad (13.56)$$

$$\leq \frac{1}{2\Gamma(\alpha+1)} (M-m)(b-a)^{\alpha}.$$

## 13.4 Composite inequalities for functions of bounded variation

Let $f : [a, b] \to \mathbb{C}$ be a Lebesgue integrable function on $[a, b]$. For $\gamma \in [0, 1]$ and $x \in (a, b)$ we have from (13.7)-(13.9) for $\lambda = (1 - \gamma) f(a) + \gamma f(x)$ and $\mu = (1 - \gamma) f(x) + \gamma f(b)$ the following 3-point representations

$$J_{a+}^{\alpha} f(x) + J_{b-}^{\alpha} f(x) \tag{13.57}$$
$$= \frac{1}{\Gamma(\alpha + 1)} \left[ (1 - \gamma)(x - a)^{\alpha} f(a) + \gamma (b - x)^{\alpha} f(b) \right]$$
$$+ \frac{1}{\Gamma(\alpha + 1)} \left[ \gamma (x - a)^{\alpha} + (1 - \gamma)(b - x)^{\alpha} \right] f(x)$$
$$+ \frac{1}{\Gamma(\alpha)} \int_a^x (x - t)^{\alpha - 1} \left[ f(t) - (1 - \gamma) f(a) - \gamma f(x) \right] dt$$
$$+ \frac{1}{\Gamma(\alpha)} \int_x^b (t - x)^{\alpha - 1} \left[ f(t) - (1 - \gamma) f(x) - \gamma f(b) \right] dt,$$

$$J_{x-}^{\alpha} f(a) + J_{x+}^{\alpha} f(b) \tag{13.58}$$
$$= \frac{1}{\Gamma(\alpha + 1)} \left[ (1 - \gamma)(x - a)^{\alpha} f(a) + \gamma (b - x)^{\alpha} f(b) \right]$$
$$+ \frac{1}{\Gamma(\alpha + 1)} \left[ \gamma (x - a)^{\alpha} + (1 - \gamma)(b - x)^{\alpha} \right] f(x)$$
$$+ \frac{1}{\Gamma(\alpha)} \int_a^x (t - a)^{\alpha - 1} \left[ f(t) - (1 - \gamma) f(a) - \gamma f(x) \right] dt$$
$$+ \frac{1}{\Gamma(\alpha)} \int_x^b (b - t)^{\alpha - 1} \left[ f(t) - (1 - \gamma) f(x) - \gamma f(b) \right] dt.$$

and

$$\frac{J_{a+}^{\alpha} f(b) + J_{b-}^{\alpha} f(a)}{2} \tag{13.59}$$
$$= \frac{1}{2} \left[ (1 - \gamma) f(a) + \gamma f(b) + f(x) \right] \frac{1}{\Gamma(\alpha + 1)} (b - a)^{\alpha}$$
$$+ \frac{1}{2\Gamma(\alpha)} \int_a^b (b - t)^{\alpha - 1} \left[ f(t) - (1 - \gamma) f(a) - \gamma f(x) \right] dt$$
$$+ \frac{1}{2\Gamma(\alpha)} \int_a^b (t - a)^{\alpha - 1} \left[ f(t) - (1 - \gamma) f(x) - \gamma f(b) \right] dt.$$

**Theorem 13.4.1.** *Assume that $f : [a, b] \to \mathbb{C}$ is a function of bounded variation on $[a, b]$ and $\gamma \in [0, 1]$*

*(i) If $x \in (a, b)$, then*

$$\left| J_{a+}^{\alpha} f(x) + J_{b-}^{\alpha} f(x) - \frac{1}{\Gamma(\alpha+1)} \left[ (1-\gamma)(x-a)^{\alpha} f(a) + \gamma (b-x)^{\alpha} f(b) \right] \right.$$

$$\left. - \frac{1}{\Gamma(\alpha+1)} \left[ \gamma (x-a)^{\alpha} + (1-\gamma)(b-x)^{\alpha} \right] f(x) \right| \qquad (13.60)$$

$$\leq \frac{1}{\Gamma(\alpha)} \int_a^x (x-t)^{\alpha-1} \left[ (1-\gamma) \bigvee_a^t (f) + \gamma \bigvee_t^x (f) \right] dt$$

$$+ \frac{1}{\Gamma(\alpha)} \int_x^b (t-x)^{\alpha-1} \left[ (1-\gamma) \bigvee_x^t (f) + \gamma \bigvee_t^b (f) \right] dt$$

$$\leq \frac{1}{\Gamma(\alpha+1)} \max\{1-\gamma, \gamma\} \left[ (x-a)^{\alpha} \bigvee_a^x (f) + (b-x)^{\alpha} \bigvee_x^b (f) \right]$$

$$\leq \frac{1}{\Gamma(\alpha+1)} \max\{1-\gamma, \gamma\}$$

$$\times \begin{cases} \left[ \frac{1}{2}(b-a) + \left| x - \frac{a+b}{2} \right| \right]^{\alpha} \bigvee_a^b (f); \\[2mm] \left( (x-a)^{\alpha p} + (b-x)^{\alpha p} \right)^{1/p} \left( \left( \bigvee_a^x (f) \right)^q + \left( \bigvee_x^b (f) \right)^q \right)^{1/q} \\ \text{with } p, \ q > 1, \ \frac{1}{p} + \frac{1}{q} = 1; \\[2mm] \left[ \frac{1}{2} \bigvee_a^b (f) + \frac{1}{2} \left| \bigvee_a^x (f) - \bigvee_x^b (f) \right| \right] \left( (x-a)^{\alpha} + (b-x)^{\alpha} \right), \end{cases}$$

*and*

$$\left| J_{x-}^{\alpha} f(a) + J_{x+}^{\alpha} f(b) - \frac{1}{\Gamma(\alpha+1)} \left[ (1-\gamma)(x-a)^{\alpha} f(a) + \gamma (b-x)^{\alpha} f(b) \right] \right.$$

$$(13.61)$$

$$\left. - \frac{1}{\Gamma(\alpha+1)} \left[ \gamma (x-a)^{\alpha} + (1-\gamma)(b-x)^{\alpha} \right] f(x) \right|$$

$$\leq \frac{1}{\Gamma(\alpha)} \int_a^x (t-a)^{\alpha-1} \left[ (1-\gamma) \bigvee_a^t (f) + \gamma \bigvee_t^x (f) \right] dt$$

$$+ \frac{1}{\Gamma(\alpha)} \int_x^b (b-t)^{\alpha-1} \left[ (1-\gamma) \bigvee_x^t (f) + \gamma \bigvee_t^b (f) \right] dt$$

$$\leq \frac{1}{\Gamma(\alpha+1)} \max\{1-\gamma,\gamma\} \left[(x-a)^\alpha \bigvee_a^x (f) + (b-x)^\alpha \bigvee_x^b (f)\right]$$

$$\leq \frac{1}{\Gamma(\alpha+1)} \max\{1-\gamma,\gamma\}$$

$$\times \begin{cases} \left[\frac{1}{2}(b-a) + \left|x - \frac{a+b}{2}\right|\right]^\alpha \bigvee_a^b (f); \\[3mm] ((x-a)^{\alpha p} + (b-x)^{\alpha p})^{1/p} \left(\left(\bigvee_a^x (f)\right)^q + \left(\bigvee_x^b (f)\right)^q\right)^{1/q} \\ \text{with } p,\ q > 1,\ \frac{1}{p} + \frac{1}{q} = 1; \\[3mm] \left[\frac{1}{2}\bigvee_a^b (f) + \frac{1}{2}\left|\bigvee_a^x (f) - \bigvee_x^b (f)\right|\right]((x-a)^\alpha + (b-x)^\alpha). \end{cases}$$

*(ii)* If $x \in [a,b]$, then

$$\left|\frac{J_{a+}^\alpha f(b) + J_{b-}^\alpha f(a)}{2} - \frac{1}{2}\left[(1-\gamma)f(a) + \gamma f(b) + f(x)\right]\frac{1}{\Gamma(\alpha+1)}(b-a)^\alpha\right|$$

$$\tag{13.62}$$

$$\leq \frac{1}{2\Gamma(\alpha)} \int_a^b (b-t)^{\alpha-1} \left[(1-\gamma)\bigvee_a^t (f) + \gamma\bigvee_t^x (f)\right] dt$$

$$+ \frac{1}{2\Gamma(\alpha)} \int_a^b (t-a)^{\alpha-1} \left[(1-\gamma)\bigvee_x^t (f) + \gamma\bigvee_t^b (f)\right] dt$$

$$\leq \frac{1}{2\Gamma(\alpha+1)}(b-a)^\alpha \max\{1-\gamma,\gamma\} \bigvee_a^b (f).$$

*Proof.* (i) Since $f$ is of bounded variation, then we have for $a \leq t \leq x$ that

$$|f(t) - (1-\gamma)f(a) - \gamma f(x)| \tag{13.63}$$

$$= |(1-\gamma)f(t) + \gamma f(t) - (1-\gamma)f(a) - \gamma f(x)|$$

$$= |(1-\gamma)[f(t) - f(a)] + \gamma[f(t) - f(x)]|$$

$$\leq (1-\gamma)|f(t) - f(a)| + \gamma|f(x) - f(t)|$$

$$\leq (1-\gamma)\bigvee_a^t (f) + \gamma\bigvee_t^x (f) \leq \max\{1-\gamma,\gamma\}\bigvee_a^x (f)$$

and for $x \leq t \leq b$ that

$$|f(t) - (1-\gamma)f(x) - \gamma f(b)| \tag{13.64}$$

$$= |(1-\gamma)f(t) + \gamma f(t) - (1-\gamma)f(x) - \gamma f(b)|$$

$$= |(1-\gamma)[f(t) - f(x)] + \gamma[f(t) - f(b)]|$$

$$\leq (1-\gamma)|f(t) - f(x)| + \gamma|f(b) - f(t)|$$

$$\leq (1-\gamma)\bigvee_x^t (f) + \gamma\bigvee_t^b (f) \leq \max\{1-\gamma,\gamma\}\bigvee_x^b (f).$$

Using (13.57) for $x \in (a,b)$ we have

$$\left| J_{a+}^{\alpha} f(x) + J_{b-}^{\alpha} f(x) - \frac{1}{\Gamma(\alpha+1)} \left[ (1-\gamma)(x-a)^{\alpha} f(a) + \gamma(b-x)^{\alpha} f(b) \right] \right.$$
$$\left. - \frac{1}{\Gamma(\alpha+1)} \left[ \gamma(x-a)^{\alpha} + (1-\gamma)(b-x)^{\alpha} \right] f(x) \right|$$

$$\leq \frac{1}{\Gamma(\alpha)} \int_{a}^{x} (x-t)^{\alpha-1} |f(t) - (1-\gamma)f(a) - \gamma f(x)| \, dt$$

$$+ \frac{1}{\Gamma(\alpha)} \int_{x}^{b} (t-x)^{\alpha-1} |f(t) - (1-\gamma)f(x) - \gamma f(b)| \, dt$$

$$\leq \frac{1}{\Gamma(\alpha)} \int_{a}^{x} (x-t)^{\alpha-1} \left[ (1-\gamma) \bigvee_{a}^{t} (f) + \gamma \bigvee_{t}^{x} (f) \right] dt$$

$$+ \frac{1}{\Gamma(\alpha)} \int_{x}^{b} (t-x)^{\alpha-1} \left[ (1-\gamma) \bigvee_{x}^{t} (f) + \gamma \bigvee_{t}^{b} (f) \right] dt$$

$$\leq \max\{1-\gamma,\gamma\} \frac{1}{\Gamma(\alpha)} \left[ \bigvee_{a}^{x} (f) \int_{a}^{x} (x-t)^{\alpha-1} \, dt + \bigvee_{x}^{b} (f) \int_{x}^{b} (t-x)^{\alpha-1} \, dt \right]$$

$$= \max\{1-\gamma,\gamma\} \frac{1}{\Gamma(\alpha+1)} \left[ (x-a)^{\alpha} \bigvee_{a}^{x} (f) + (b-x)^{\alpha} \bigvee_{x}^{b} (f) \right],$$

which prove the first two inequalities in (13.60).

The last part is obvious by making use of the elementary Hölder type inequalities for positive real numbers $c$, $d$, $m$, $n \geq 0$

$$mc + nd \leq \begin{cases} \max\{m,n\}(c+d); \\[2mm] (m^p + n^p)^{1/p}(c^q + d^q)^{1/q} \text{ with } p, \ q > 1, \ \frac{1}{p} + \frac{1}{q} = 1. \end{cases}$$

The inequality (13.61) follows in a similar way and we omit the details.

(ii) By (13.59) we get

$$\left| \frac{J_{a+}^{\alpha} f(b) + J_{b-}^{\alpha} f(a)}{2} - \frac{1}{2} \left[ (1-\gamma)f(a) + \gamma f(b) + f(x) \right] \frac{1}{\Gamma(\alpha+1)} (b-a)^{\alpha} \right|$$

$$\leq \frac{1}{2\Gamma(\alpha)} \int_{a}^{b} (b-t)^{\alpha-1} |f(t) - (1-\gamma)f(a) - \gamma f(x)| \, dt$$

$$+ \frac{1}{2\Gamma(\alpha)} \int_{a}^{b} (t-a)^{\alpha-1} |f(t) - (1-\gamma)f(x) - \gamma f(b)| \, dt$$

$$\leq \frac{1}{2\Gamma(\alpha)} \int_{a}^{b} (b-t)^{\alpha-1} \left[ (1-\gamma) \bigvee_{a}^{t} (f) + \gamma \bigvee_{t}^{x} (f) \right] dt$$

$$+ \frac{1}{2\Gamma(\alpha)} \int_{a}^{b} (t-a)^{\alpha-1} \left[ (1-\gamma) \bigvee_{x}^{t} (f) + \gamma \bigvee_{t}^{b} (f) \right] dt$$

$$\leq \max\left\{1-\gamma,\gamma\right\} \bigvee_a^x (f) \, \frac{1}{2\Gamma(\alpha)} \int_a^b (b-t)^{\alpha-1} \, dt$$

$$+ \max\left\{1-\gamma,\gamma\right\} \bigvee_x^b (f) \, \frac{1}{2\Gamma(\alpha)} \int_a^b (t-a)^{\alpha-1} \, dt$$

$$= \max\left\{1-\gamma,\gamma\right\} \frac{1}{2\Gamma(\alpha+1)} (b-a)^\alpha \bigvee_a^b (f),$$

which proves (13.62). $\square$

**Corollary 13.4.2.** *With the assumptions of Theorem 13.4.1, we have*

$$\left| J_{a+}^\alpha f\left(\frac{a+b}{2}\right) + J_{b-}^\alpha f\left(\frac{a+b}{2}\right) \right. \tag{13.65}$$

$$- \frac{1}{2^\alpha \Gamma(\alpha+1)} \left[(1-\gamma) f(a) + \gamma f(b)\right] (b-a)^\alpha$$

$$\left. - \frac{1}{2^\alpha \Gamma(\alpha+1)} f\left(\frac{a+b}{2}\right) (b-a)^\alpha \right|$$

$$\leq \frac{1}{\Gamma(\alpha)} \int_a^{\frac{a+b}{2}} \left(\frac{a+b}{2}-t\right)^{\alpha-1} \left[(1-\gamma) \bigvee_a^t (f) + \gamma \bigvee_t^{\frac{a+b}{2}} (f)\right] dt$$

$$+ \frac{1}{\Gamma(\alpha)} \int_{\frac{a+b}{2}}^b \left(t-\frac{a+b}{2}\right)^{\alpha-1} \left[(1-\gamma) \bigvee_{\frac{a+b}{2}}^t (f) + \gamma \bigvee_t^b (f)\right] dt$$

$$\leq \frac{1}{2^\alpha \Gamma(\alpha+1)} \max\left\{1-\gamma,\gamma\right\} \bigvee_a^b (f) (b-a)^\alpha$$

$$\left| J_{\frac{a+b}{2}-}^\alpha f(a) + J_{\frac{a+b}{2}+}^\alpha f(b) - \frac{1}{2^\alpha \Gamma(\alpha+1)} \left[(1-\gamma) f(a) + \gamma f(b)\right] (b-a)^\alpha \right.$$

$$\tag{13.66}$$

$$\left. - \frac{1}{2^\alpha \Gamma(\alpha+1)} f\left(\frac{a+b}{2}\right) (b-a)^\alpha \right|$$

$$\leq \frac{1}{\Gamma(\alpha)} \int_a^{\frac{a+b}{2}} (t-a)^{\alpha-1} \left[(1-\gamma) \bigvee_a^t (f) + \gamma \bigvee_t^{\frac{a+b}{2}} (f)\right] dt$$

$$+ \frac{1}{\Gamma(\alpha)} \int_{\frac{a+b}{2}}^b (b-t)^{\alpha-1} \left[(1-\gamma) \bigvee_{\frac{a+b}{2}}^t (f) + \gamma \bigvee_t^b (f)\right] dt$$

$$\leq \frac{1}{2^\alpha \Gamma(\alpha+1)} \max\left\{1-\gamma,\gamma\right\} \bigvee_a^b (f) (b-a)^\alpha$$

*and*

$$\left| \frac{J_{a+}^{\alpha}f(b) + J_{b-}^{\alpha}f(a)}{2} \right. \tag{13.67}$$

$$\left. - \frac{1}{2}\left[(1-\gamma)f(a) + \gamma f(b) + f\left(\frac{a+b}{2}\right)\right] \frac{1}{\Gamma(\alpha+1)}(b-a)^{\alpha} \right|$$

$$\leq \frac{1}{2\Gamma(\alpha)}\int_{a}^{b}(b-t)^{\alpha-1}\left[(1-\gamma)\bigvee_{a}^{t}(f) + \gamma\bigvee_{t}^{\frac{a+b}{2}}(f)\right]dt$$

$$+ \frac{1}{2\Gamma(\alpha)}\int_{a}^{b}(t-a)^{\alpha-1}\left[(1-\gamma)\bigvee_{\frac{a+b}{2}}^{t}(f) + \gamma\bigvee_{t}^{b}(f)\right]dt$$

$$\leq \frac{1}{2\Gamma(\alpha+1)}(b-a)^{\alpha}\max\{1-\gamma,\gamma\}\bigvee_{a}^{b}(f).$$

*Remark* 13.4.3. If we take $\gamma = \frac{1}{2}$ in Corollary 13.4.2, then we get the following composite mid-point and trapezoid inequalities

$$\left| J_{a+}^{\alpha}f\left(\frac{a+b}{2}\right) + J_{b-}^{\alpha}f\left(\frac{a+b}{2}\right) - \frac{1}{2^{\alpha}\Gamma(\alpha+1)}\frac{f(a)+f(b)}{2}(b-a)^{\alpha} \right.$$

$$\tag{13.68}$$

$$\left. - \frac{1}{2^{\alpha}\Gamma(\alpha+1)}f\left(\frac{a+b}{2}\right)(b-a)^{\alpha} \right| \leq \frac{1}{2^{\alpha+1}\Gamma(\alpha+1)}\bigvee_{a}^{b}(f)(b-a)^{\alpha}$$

$$\left| J_{\frac{a+b}{2}-}^{\alpha}f(a) + J_{\frac{a+b}{2}+}^{\alpha}f(b) - \frac{1}{2^{\alpha}\Gamma(\alpha+1)}\frac{f(a)+f(b)}{2}(b-a)^{\alpha} \right. \tag{13.69}$$

$$\left. - \frac{1}{2^{\alpha}\Gamma(\alpha+1)}f\left(\frac{a+b}{2}\right)(b-a)^{\alpha} \right| \leq \frac{1}{2^{\alpha+1}\Gamma(\alpha+1)}\bigvee_{a}^{b}(f)(b-a)^{\alpha}$$

*and*

$$\left| \frac{J_{a+}^{\alpha}f(b) + J_{b-}^{\alpha}f(a)}{2} - \frac{1}{2}\left[\frac{f(a)+f(b)}{2} + f\left(\frac{a+b}{2}\right)\right] \frac{1}{\Gamma(\alpha+1)}(b-a)^{\alpha} \right|$$

$$\tag{13.70}$$

$$\leq \frac{1}{4\Gamma(\alpha+1)}(b-a)^{\alpha}\bigvee_{a}^{b}(f).$$

# Bibliography

Aglić Aljinović A. Montgomery identity and Ostrowski type inequalities for Riemann-Liouville fractional integral . *J. Math.*, (Art. ID 503195):1–6, 2014.

Dragomir S. S. The Ostrowski's integral inequality for Lipschitzian mappings and applications. *Comput. Math. Appl.*, (38):33–37, 1999a.

Dragomir S. S. The Ostrowski integral inequality for mappings of bounded variation . *Bull. Austral. Math. Soc.*, (60):495–508, 1999b.

Dragomir S. S. On the midpoint quadrature formula for mappings with bounded variation and applications . *Kragujevac J. Math.*, (22):22, 2000.

Dragomir S. S. On the Ostrowski's integral inequality for mappings with bounded variation and applications. *Math. Ineq. Appl.*, (4):59–66, 2001

Dragomir S. S. Refinements of the Ostrowski inequality in terms of the cumulative variation and applications. *Analysis (Berlin)*, (34):223–240, 2014.

Dragomir S. S. Ostrowski type inequalities for Riemann-Liouville fractional integrals of bounded variation, Holder and Lipschitzian functions. *RGMIA Res. Rep. Coll.*, (20):Art. 48, 2017.

Yue H. Ostrowski inequality for fractional integrals and related fractional inequalities. *Transylv. J. Math. Mech.*, (5):85–89, 2013.

# Chapter 14

# $kn - 2$ Scalar Matrix and Its Functional Equations by Mathematical Modeling

**Pasupathi Narasimman**

*Department of Mathematics, Thiruvalluvar University College of Arts and Science, Kariyampatti, Tirupattur, Tamilnadu, India*

**Hemen Dutta**

*Department of Mathematics, Gauhati University, Guwahati, India*

## 14.1 Introduction

For the past two decades, functional equation and its stability have been a rapidly developing area and a number of research papers are being published on the same. Inspite of the growing interest, all these publications lack a proper methodology and go only by a trial and error method to form a various type of functional equations. Very recently, the research articles Narasimman and Amuda (2016); Narasimman et al. (2016) introduced a proper method to

449

form an additive functional equation using $CSM$ and $RSM$ matrices and the famous additive Cauchy functional equation and logical functional equation have also been modeled using identity matrix and logical matrix. As a continuation of the same this chapter elucidates a possible method to model various types of functional equations like additive, quadratic and mixed type of additive and quadratic using eigenvalues and eigenvectors of newly established $kn - 2$ scalar matrices with suitable numerical examples.

The study of functional equations is a contemporary area of mathematics that provides a powerful approach to working with important concepts and relationships in analysis and algebra such as symmetry, linearity and equivalence. Although the systematic study of such equations is a relatively recent area of mathematical study, they have been considered earlier in various forms by mathematicians such as Euler in the $18^{th}$ century and Cauchy in the $19^{th}$ century.

The theory of functional equations is a growing branch of mathematics which has contributed a lot to the development of the strong tools in today's mathematics. Many new applied problems and theories have motivated functional equations to develop new approaches and methods. D'Alembert, Euler, Gauss, Cauchy, Abel, Weierstrass, Darboux and Hilbert are among the great mathematicians who have been concerned with functional equations and methods of solving them.

There have been many researchers studied the solution and stability of different types of functional equations such as additive, quadratic, cubic, quartic and mixed type of additive-quadratic, quadratic-cubic and so on. However, studies over the origin and formation of such functional equations are not convincing as they are only structured based trial and error method. This fact has influenced a number of authors to undergo a formal study on modeling a functional equations.

This chapter is organized as follows: In Section 14.1, we give introduction and discuss the preliminaries and definitions of functional equations. In Sections 14.2 and 14.3, authors introduce and discuss the new type of $kn - 2$ scalar matrix and its various possible model matrices. In Section 14.4, authors propose a new method to model additive, quadratic and mixed type functional equations through eigenvalues and eigenvectors of $kn - 2$ matrices. Also, authors present some numerical examples in Section 14.5. In Section 14.6, authors model the most famous logical functional equations. Finally, Section 14.7 ends with conclusion and scope for further study.

### 14.1.1    Preliminaries and definitions

A Hungarian Mathematician J. Aczel Aczel (1966), an excellent specialist in functional equations, defines the functional equation as follows:

**Definition 14.1.1. Functional Equation:**
Functional equations are equations in which both sides are terms con-

structed from the finite number of unknown functions and a finite number of independent variables.

**Example 14.1.2.** *(i) Cauchy Additive Functional Equation*

$$f(x + y) = f(x) + f(y).$$

*(ii) Lee and Park Quadratic Functional Equation*

$$f(x + y) + f(x - y) = 2f(x) + 2f(y).$$

*(iii) Abbas Najati Cubic Functional Equation*

$$f(2x + y) + f(2x - y) = 8f(x) + 2f(y).$$

*(iv) J.M.Rassias Quartic Functional Equation*

$$3f(x + 3y) + f(3x - y) = 15f(x + y) + 15f(x - y) + 80f(y).$$

**Definition 14.1.3. Solutions of Functional Equation:**

A solution of a functional equation is a function which satisfies the equation.

**Example 14.1.4.** *(i) Cauchy Functional Equations*

$$f(x + y) = f(x) + f(y), \tag{14.1}$$
$$f(x + y) = f(x)f(y),$$
$$f(xy) = f(x) + f(y),$$

*have solutions* $f(x) = kx$, $f(x) = e^x$, $f(x) = lnx$, *respectively.*

*(ii)* $f(x) = cx + a$ *is the solution of the Jensen functional equation*

$$f\left(\frac{x + y}{2}\right) = \frac{f(x) + f(y)}{2}.$$

The functional equation (14.1) is the most famous among the functional equations. It is often called the additive Cauchy functional equation in honor of A. L. Cauchy. Here we note only that it appeared as early as 1791 (A. M. Legendre) and 1809 (C. F. Gauss). It was, however, A. L. Cauchy who described all continuous solutions in 1821; for references, see Kuczma (1985) and Ivan (2009). The properties of this functional equation are frequently applied to the development of theories of other functional equations. Moreover, the properties of the additive Cauchy equation are powerful tools in almost every field of natural and social sciences.

In 2011, K. Ravi et al.Ravi et al. (2011) introduced the following additive functional equation

$$f(2x + y) + f(x - 2y) = 3f(x) - f(y) \tag{14.2}$$

and they investigated generalized Hyers-Ulam stability theorem in uniform and nonuniform version in a fuzzy sense.

Consider the functional equation

$$f(x + y) + f(x - y) = 2f(x) + 2f(y). \tag{14.3}$$

The function $f(x) = x^2$ is the solution of the functional equation (14.3). Hence it is called the quadratic functional equation or Euler - Lagrange functional equation Rassias (1992) and every solution of the quadratic functional equation (14.3) is called quadratic function Czerwik (2002, 2003); Jung (1998, 1999). The functional equation (14.3) is a familiar equation and this equation was dealt by many authors such as F. Skof Skof (1983), P. W. Cholewa Cholewa (1984), S. Czerwik Czerwik (1992) and J. M. Rassias Rassias (1982, 1984, 1985, 2001a, 2002, 2005).

In 2001, J. M. Rassias Rassias (2001b) introduced the the pioneering cubic functional equation

$$f(x + 2y) - 3f(x + y) + 3f(x) - f(x - y) = 6f(y) \tag{14.4}$$

and established the solution of the Ulam stability problem for these cubic mappings. It is easy to show that the function $f(x) = x^3$ satisfies the functional equation(14.4), which is called a cubic functional equation and every solution of the cubic functional equation is said to be a cubic mapping.

In 2005, K. W. Jun and H. M. Kim Jun and Kim (2002) obtained the general solution of a generalized quadratic and additive type functional equation of the form

$$f(x + ay) + af(x - y) = f(x - ay) + af(x + y) \tag{14.5}$$

for any integer $a$ with $a \neq -1, 0, 1$.

A. Najati and M. B. Moghimi Najati and Moghimi (2008) dealt the functional equation

$$f(2x + y) + f(2x - y) = f(x + y) + f(x - y) + 2f(2x) - 2f(x) \tag{14.6}$$

which is derived from quadratic and additive functions and established the general solution of equation (14.6) and investigated the Hyers-Ulam-Rassias stability for equation (14.6).

In 2017, P. Narasimman and A. Bodaghi Narasimman and Bodaghi (2017) obtained the general solution and investigated the generalized Hyers-Ulam-Rassias stability for the new mixed type additive and cubic functional equation

$$3f(x + 3y) - f(3x + y) \tag{14.7}$$
$$= 12[f(x + y) + f(x - y)] - 16[f(x) + f(y)] + 12f(2y) - 4f(2x).$$

Very recently, using the fixed point method, T. Z. Xu et al. Xu et al.

(2010) investigated the generalized Hyers-Ulam stability of the general mixed additive-quadratic-cubic-quartic functional equation

$$f(x + ny) + f(x - ny) = n^2 f(x + y) + n^2 f(x - y) + 2(1 - n^2)f(x) \quad (14.8)$$
$$+ \frac{n^4 - n^2}{12}[f(2y) + f(-2y) - 4f(y) - 4f(-y)]$$

for fixed integers $n$ with $n \neq 0, 1$ in multi-Banach spaces. It is easy to show that the function $f(x) = x^4 + x^3 + x^2 + x$ satisfies the functional equation (14.8), which is called a mixed type functional equation.

The stability problem of various functional equations and mappings such as the Cauchy equation, the Jensen equation, the quadratic equation, the cubic equation, the quartic equation and various versions on more general domains and ranges have been investigated by a number of authors Jun and Lee (2001); Kannappan (1995); Ravi and Arunkumar (2006); Ravi et al. (2007, 2009a,b).

## 14.2 Main results

In this section, authors define the new $kn - 2$ scalar matrices and introduce a new method using their eigenvalues and eigenvectors to produce a various type of functional equations such as additive, quadratic and mixed type of additive and quadratic with suitable numerical examples.

**Definition 14.2.1.** $kn - 2$ **Scalar Matrix:**
A $kn - 2$ scalar matrix is a second order square matrix with its second row elements are $k$ times of its first row elements. Also, its main diagonal elements are same, where $k$ is any non-zero integer.

**Example 14.2.2.** *A matrix $A = \begin{pmatrix} kn & n \\ k^2 n & kn \end{pmatrix}$ be the $kn - 2$ matrix.*

**Numerical Example 14.2.3.**

(i) $\begin{pmatrix} -6 & 3 \\ 12 & -6 \end{pmatrix}$ *for $n = 3$ and $k = -2$.*

(ii) $\begin{pmatrix} 1 & 1 \\ 1 & 1 \end{pmatrix}$ *for $n = 1$ and $k = 1$.*

**Theorem 14.2.4.** *The eigenvalues of any $kn - 2$ scalar matrix of order 2 are $0$ and $2kn$.*

*Proof.* Characteristic equation of the matrix $A$ is $\|A - \lambda I\| = 0$,

$$\begin{vmatrix} kn - \lambda & n \\ k^2 n & kn - \lambda \end{vmatrix} = 0, \text{ which imply } \lambda = 0 \text{ and } \lambda = 2kn. \qquad \square$$

**Numerical Example 14.2.5.** *The* $kn - 2$ *matrices and their eigenvalues are*

(i) $\begin{pmatrix} -6 & 3 \\ 12 & -6 \end{pmatrix}$; $\lambda_1 = 0$ *and* $\lambda_2 = -12$.

(ii) $\begin{pmatrix} 1 & 1 \\ 1 & 1 \end{pmatrix}$; $\lambda_1 = 0$ *and* $\lambda_2 = 2$.

**Theorem 14.2.6.** *The eigenvector of* $kn - 2$ *matrix when* $\lambda = 0$ *is always second diagonals with the following condition.*

*In the second diagonal elements of a matrix, second element is divided by* $k$ *and any one of these elements has to be multiplied by negative sign. That is the eigenvector is*

$$X_1 = \begin{pmatrix} -a_{12} \\ \frac{a_{21}}{k} \end{pmatrix} \text{ or } \begin{pmatrix} a_{12} \\ -\frac{a_{21}}{k} \end{pmatrix}.$$

*Proof.* Let a matrix $A = \begin{pmatrix} kn & n \\ k^2 n & kn \end{pmatrix}$ be the $kn - 2$ matrix. When $\lambda = 0$ the eigenvector $(A - \lambda I)X = 0$ implies $knx_1 + nx_2 = 0, k^2 nx_1 + knx_2 = 0$ and solving the equations, we get the possible eigenvectors are

$$X_1 = \begin{pmatrix} -n \\ kn \end{pmatrix} (or) \begin{pmatrix} n \\ -kn \end{pmatrix} (or) \begin{pmatrix} -1 \\ k \end{pmatrix} (or) \begin{pmatrix} 1 \\ -k \end{pmatrix}.$$

$\square$

**Theorem 14.2.7.** *The eigenvector of* $kn - 2$ *matrix when* $\lambda = 2kn$ *is always sum of first and second row of a matrix, respectively or always* $\begin{pmatrix} 1 \\ k \end{pmatrix}$. *That is the eigenvectors are*

$$X = \begin{pmatrix} x_1 \\ x_2 \end{pmatrix} = \begin{pmatrix} s_1 \\ s_2 \end{pmatrix} (or) \begin{pmatrix} -s_1 \\ -s_2 \end{pmatrix} (or) \begin{pmatrix} 1 \\ k \end{pmatrix} (or) \begin{pmatrix} -1 \\ -k \end{pmatrix},$$

*where* $s_1 = a_{11} + a_{12}, s_2 = a_{21} + a_{22}$ *in* $kn - 2$ *matrix.*

*Proof.* Let a matrix $A = \begin{pmatrix} kn & n \\ k^2 n & kn \end{pmatrix}$ be the $kn - 2$ matrix. When $\lambda = 2kn$ the eigenvector $(A - \lambda I)X = 0$ implies $-knx_1 + nx_2 = 0, k^2 nx_1 - knx_2 = 0$

and solving the equations, we get the possible eigenvectors are

$$X = \begin{pmatrix} x_1 \\ x_2 \end{pmatrix} = \begin{pmatrix} kn + n \\ k^2 n + kn \end{pmatrix} \ (or) \ \begin{pmatrix} -(kn + n) \\ -(k^2 n + kn) \end{pmatrix}$$

$$(or) \ \begin{pmatrix} 1 \\ k \end{pmatrix} \ (or) \ \begin{pmatrix} -1 \\ -k \end{pmatrix}.$$

□

**Numerical Example 14.2.8.** *Consider a* $kn - 2$ *matrix for* $n = 1$ *and* $k = -5$ *is* $\begin{pmatrix} -5 & 1 \\ 25 & -5 \end{pmatrix}$. *When* $\lambda_1 = 0$, *the eigenvector* $X_1$ *is*

$$X_1 = \begin{pmatrix} 1 \\ 5 \end{pmatrix} (or) \begin{pmatrix} -1 \\ -5 \end{pmatrix}.$$

*When* $\lambda_2 = -10$, *the eigenvector* $X_2$ *is*

$$X_2 = \begin{pmatrix} -4 \\ 20 \end{pmatrix} (or) \begin{pmatrix} 4 \\ -20 \end{pmatrix} (or) \begin{pmatrix} -1 \\ 5 \end{pmatrix} (or) \begin{pmatrix} 1 \\ -5 \end{pmatrix}.$$

---

## 14.3 Possible model matrices of $kn - 2$ matrix

In this section, we discuss about the possible model matrices of $kn - 2$ matrix using eigenvectors $X_1$, $X_2$ and their corresponding eigenvalues $\lambda_1$ and $\lambda_2$.

The possible model matrices for $kn - 2$ matrix $A = \begin{pmatrix} a_{11} & a_{12} \\ a_{21} & a_{22} \end{pmatrix}$ are $M_1 = [\alpha X_1, \beta X_2]$ and $M_2 = [\beta X_2, \alpha X_1]$, where $\alpha, \beta \in \mathbb{R}$.

Using $M_1$ and different possible values of $\alpha, \beta$, we arrive the following model matrices $M = \begin{pmatrix} m_{11} & m_{12} \\ m_{21} & m_{22} \end{pmatrix}$.

$$M = \begin{pmatrix} -n & kn+n \\ kn & k^2n+kn \end{pmatrix}, \quad M = \begin{pmatrix} -n & -(kn+n) \\ kn & -(k^2n+kn) \end{pmatrix}, \quad M = \begin{pmatrix} -n & 1 \\ kn & k \end{pmatrix},$$

$$M = \begin{pmatrix} -n & -1 \\ kn & -k \end{pmatrix}, \quad M = \begin{pmatrix} n & kn+n \\ -kn & k^2n+kn \end{pmatrix}, \quad M = \begin{pmatrix} n & -(kn+n) \\ -kn & -(k^2n+kn) \end{pmatrix},$$

$$M = \begin{pmatrix} n & 1 \\ -kn & k \end{pmatrix}, \quad M = \begin{pmatrix} n & -1 \\ -kn & -k \end{pmatrix}, \quad M = \begin{pmatrix} -1 & kn+n \\ k & k^2n+kn \end{pmatrix},$$

$$(14.9)$$

$$M = \begin{pmatrix} -1 & -(kn+n) \\ k & -(k^2n+kn) \end{pmatrix}, \quad M = \begin{pmatrix} -1 & 1 \\ k & k \end{pmatrix}, \quad M = \begin{pmatrix} -1 & -1 \\ k & -k \end{pmatrix},$$

$$M = \begin{pmatrix} 1 & -(kn+n) \\ -k & -(k^2n+kn) \end{pmatrix}, \quad M = \begin{pmatrix} 1 & 1 \\ -k & k \end{pmatrix}, \quad M = \begin{pmatrix} 1 & -1 \\ -k & -k \end{pmatrix},$$

$$M = \begin{pmatrix} 1 & kn+n \\ -k & k^2n+kn \end{pmatrix}.$$

Using $M_2$ and different possible values of $\alpha, \beta$, we arrive the following model matrices $M = \begin{pmatrix} m_{11} & m_{12} \\ m_{21} & m_{22} \end{pmatrix}$.

$$M = \begin{pmatrix} kn+n & -n \\ k^2n+kn & kn \end{pmatrix}, \quad M = \begin{pmatrix} -(kn+n) & -n \\ -(k^2n+kn) & kn \end{pmatrix}, \quad M = \begin{pmatrix} 1 & -n \\ k & kn \end{pmatrix},$$

$$M = \begin{pmatrix} -1 & -n \\ -k & kn \end{pmatrix}, \quad M = \begin{pmatrix} kn+n & n \\ k^2n+kn & -kn \end{pmatrix}, \quad M = \begin{pmatrix} -(kn+n) & n \\ -(k^2n+kn) & -kn \end{pmatrix},$$

$$M = \begin{pmatrix} 1 & n \\ k & -kn \end{pmatrix}, \quad M = \begin{pmatrix} -1 & n \\ -k & -kn \end{pmatrix}, \quad M = \begin{pmatrix} kn+n & -1 \\ k^2n+kn & k \end{pmatrix},$$

$$(14.10)$$

$$M = \begin{pmatrix} -(kn+n) & -1 \\ -(k^2n+kn) & k \end{pmatrix}, \quad M = \begin{pmatrix} 1 & -1 \\ k & k \end{pmatrix}, \quad M = \begin{pmatrix} -1 & -1 \\ -k & k \end{pmatrix},$$

$$M = \begin{pmatrix} -(kn+n) & 1 \\ -(k^2n+kn) & -k \end{pmatrix}, \quad M = \begin{pmatrix} 1 & 1 \\ k & -k \end{pmatrix}, \quad M = \begin{pmatrix} -1 & 1 \\ -k & -k \end{pmatrix},$$

$$M = \begin{pmatrix} kn+n & 1 \\ k^2n+kn & -k \end{pmatrix}.$$

**Definition 14.3.1. Additive Matrix:** A matrix is said to be additive matrix, if it leads to the additive functional equations.

**Definition 14.3.2. Quadratic Matrix:** A matrix is said to be quadratic matrix, if it leads to the quadratic functional equations.

**Definition 14.3.3. Mixed Matrix:** A matrix is said to be mixed matrix, if it leads to the Mixed type functional equations like additive-quadratic, quadratic-cubic, cubic-quartic, additive-quadratic-cubic, and so on.

---

## 14.4 New method for modeling functional equations

In this section, we introduce new methods of model additive, quadratic and mixed type functional equations by using the eigenvalues and eigenvectors of $kn - 2$ matrix.

### 14.4.1 New method for modeling additive functional equations

Let $X$ and $Y$ be vector spaces. Define a mapping $f : X \to Y$. Now, using $A$ and $M_1$ or $M_2$ we may model the following additive functional equation

$$k \left[ f(a_{11}x + m_{11}y) + f(a_{12}x + m_{12}y) \right] \tag{14.11}$$
$$= f(a_{21}x + m_{21}y) + f(a_{22}x + m_{22}y) + \alpha f \left( (\lambda_1 - \lambda_2)y \right)$$

for all $x, y \in X$, $\alpha \in \mathbb{R}$, $\lambda_1$ and $\lambda_2$ are the eigenvalues of $A$.

It is easy to show that the function $f(x) = x$ satisfies the functional equation (14.11), which is called additive functional equation. Hence, the $kn - 2$ matrices are additive matrices, since it gives additive functional equations.

### 14.4.2 New method for modeling quadratic functional equations

Let $X$ and $Y$ be vector spaces. Define a mapping $f : X \to Y$. Now, using $A$ and $M_1$ we may model the following quadratic functional equation

$$k^2 \left[ f(a_{11}x + m_{11}y) + f(a_{12}x + m_{12}y) \right] \tag{14.12}$$
$$= f(a_{21}x + m_{21}y) + f(a_{22}x + m_{22}y) + (-\alpha)kf \left( (\lambda_1 - \lambda_2)\sqrt{xy} \right)$$

for all $x, y \in X$, $\alpha \in \mathbb{R}$, $\lambda_1$ and $\lambda_2$ are the eigenvalues of $A$. Now, using $A$ and $M_2$ we may model the following quadratic functional equation

$$k^2 \left[ f(a_{11}x + m_{11}y) + f(a_{12}x + m_{12}y) \right] \tag{14.13}$$
$$= f(a_{21}x + m_{21}y) + f(a_{22}x + m_{22}y) + (-\alpha)f \left( (\lambda_1 - \lambda_2)\sqrt{xy} \right)$$

for all $x, y \in X$, $\alpha \in \mathbb{R}$, $\lambda_1$ and $\lambda_2$ are the eigenvalues of $A$.

It is easy to show that the function $f(x) = x^2$ satisfies the functional equations (14.12) and (14.13), which is called quadratic functional equation. Hence, the $kn - 2$ matrices are quadratic matrices, since it gives quadratic functional equations.

### 14.4.3 New method for modeling mixed type functional equations

Let $X$ and $Y$ be vector spaces. Define a mapping $f : X \to Y$. Now, using $A$ and $M_1$ we may model the following mixed additive-quadratic functional equation

$$f(k)\left[f(a_{11}x + m_{11}y) + f(a_{12}x + m_{12}y)\right] \qquad (14.14)$$
$$= f(a_{21}x + m_{21}y) + f(a_{22}x + m_{22}y)$$
$$+ \frac{(-\alpha)k}{2}\left[f(2(\lambda_1 - \lambda_2)\sqrt{xy}) - 2f((\lambda_1 - \lambda_2)\sqrt{xy})\right]$$
$$+ \frac{\alpha}{2}\left[4f((\lambda_1 - \lambda_2)y) - f(2(\lambda_1 - \lambda_2)y)\right]$$

for all $x, y \in X$, $\alpha \in \mathbb{R}$, $\lambda_1$ and $\lambda_2$ are the eigenvalues of $A$. Now, using $A$ and $M_2$ we may model the following mixed additive-quadratic functional equation

$$f(k)\left[f(a_{11}x + m_{11}y) + f(a_{12}x + m_{12}y)\right] \qquad (14.15)$$
$$= f(a_{21}x + m_{21}y) + f(a_{22}x + m_{22}y)$$
$$+ \frac{(-\alpha)}{2}\left[f(2(\lambda_1 - \lambda_2)\sqrt{xy}) - 2f((\lambda_1 - \lambda_2)\sqrt{xy})\right]$$
$$+ \frac{\alpha}{2}\left[4f((\lambda_1 - \lambda_2)y) - f(2(\lambda_1 - \lambda_2)y)\right]$$

for all $x, y \in X$, $\alpha \in \mathbb{R}$, $\lambda_1$ and $\lambda_2$ are the eigenvalues of $A$.

It is easy to show that the function $f(x) = x + x^2$ satisfies the functional equations (14.14) and (14.15), which is called mixed additive-quadratic functional equation. Hence, the $kn - 2$ matrices are mixed type matrices, since it gives mixed type functional equations.

---

## 14.5 Numerical example

In this section, authors give numerical examples for modeling the additive, quadratic and mixed type additive-quadratic functional equations using the proposed new method.

Let $B = \begin{pmatrix} 8 & -2 \\ -32 & 8 \end{pmatrix}$ for $n = -2$ and $k = -4$ be the $kn - 2$ matrix.

By Theorem (14.2.4), the eigenvalues are 0 and 16. By Theorems (14.2.6) and (14.2.7), the eigenvectors are

$$X_1 = \begin{pmatrix} 2 \\ 8 \end{pmatrix} \text{ and } X_2 = \begin{pmatrix} 6 \\ -24 \end{pmatrix}.$$

Now, using (14.9) and $M_1 = [\alpha X_1, \beta X_2]$, we arrive the following model matrices of $B$.

$$N_1 = \begin{pmatrix} 2 & 6 \\ 8 & -24 \end{pmatrix} \text{ for } \alpha = 1, \ \beta = 1,$$

$$N_2 = \begin{pmatrix} 2 & -6 \\ 8 & 24 \end{pmatrix} \text{ for } \alpha = 1, \ \beta = -1,$$

$$N_3 = \begin{pmatrix} 2 & 1 \\ 8 & -4 \end{pmatrix} \text{ for } \alpha = 1, \ \beta = \frac{1}{6},$$

$$N_4 = \begin{pmatrix} 2 & -1 \\ 8 & 4 \end{pmatrix} \text{ for } \alpha - 1, \ \beta - \frac{-1}{6},$$

$$N_5 = \begin{pmatrix} -2 & 6 \\ -8 & -24 \end{pmatrix} \text{ for } \alpha = -1, \ \beta = 1,$$

$$N_6 = \begin{pmatrix} -2 & -6 \\ -8 & 24 \end{pmatrix} \text{ for } \alpha = -1, \ \beta = -1.$$

$$N_7 = \begin{pmatrix} -2 & 1 \\ -8 & -4 \end{pmatrix} \text{ for } \alpha = -1, \ \beta = \frac{1}{6}.$$

$$N_8 = \begin{pmatrix} -2 & -1 \\ -8 & 4 \end{pmatrix} \text{ for } \alpha = -1, \ \beta = -\frac{1}{6}.$$

$$N_9 = \begin{pmatrix} -1 & 6 \\ -4 & -24 \end{pmatrix} \text{ for } \alpha = -\frac{1}{2}, \ \beta = 1.$$

$$N_{10} = \begin{pmatrix} -1 & -6 \\ -4 & 24 \end{pmatrix} \text{ for } \alpha = -\frac{1}{2}, \ \beta = -1.$$

$$N_{11} = \begin{pmatrix} -1 & 1 \\ -4 & -4 \end{pmatrix} \text{ for } \alpha = -\frac{1}{2}, \ \beta = \frac{1}{6}.$$

$$N_{12} = \begin{pmatrix} -1 & -1 \\ -4 & 4 \end{pmatrix} \text{ for } \alpha = -\frac{1}{2}, \ \beta - -\frac{1}{6}.$$

$$N_{13} = \begin{pmatrix} 1 & -6 \\ 4 & 24 \end{pmatrix} \text{ for } \alpha = \frac{1}{2}, \ \beta = -1.$$

$$N_{14} = \begin{pmatrix} 1 & 1 \\ 4 & -4 \end{pmatrix} \text{ for } \alpha = \frac{1}{2}, \ \beta = \frac{1}{6}.$$

$$N_{15} = \begin{pmatrix} 1 & -1 \\ 4 & 4 \end{pmatrix} \quad \text{for} \quad \alpha = \tfrac{1}{2}, \ \beta = -\tfrac{1}{6}.$$

$$N_{16} = \begin{pmatrix} 1 & 6 \\ 4 & -24 \end{pmatrix} \quad \text{for} \quad \alpha = \tfrac{1}{2}, \ \beta = 1.$$

Now, by using (14.10) and $M_2 = [\beta X_2, \alpha X_1]$, we can arrive the following model matrices $B$.

$$N_{17} = \begin{pmatrix} 6 & 2 \\ -24 & 8 \end{pmatrix} \quad \text{for} \quad \alpha = 1, \ \beta = 1,$$

$$N_{18} = \begin{pmatrix} -6 & 2 \\ 24 & 8 \end{pmatrix} \quad \text{for} \quad \alpha = 1, \ \beta = -1,$$

$$N_{19} = \begin{pmatrix} 1 & 2 \\ -4 & 8 \end{pmatrix} \quad \text{for} \quad \alpha = 1, \ \beta = \tfrac{1}{6},$$

$$N_{20} = \begin{pmatrix} -1 & 2 \\ 4 & 8 \end{pmatrix} \quad \text{for} \quad \alpha = 1, \ \beta = \tfrac{-1}{6},$$

$$N_{21} = \begin{pmatrix} 6 & -2 \\ -24 & -8 \end{pmatrix} \quad \text{for} \quad \alpha = -1, \ \beta = 1,$$

$$N_{22} = \begin{pmatrix} -6 & -2 \\ 24 & -8 \end{pmatrix} \quad \text{for} \quad \alpha = -1, \ \beta = -1.$$

$$N_{23} = \begin{pmatrix} 1 & -2 \\ -4 & -8 \end{pmatrix} \quad \text{for} \quad \alpha = -1, \ \beta = \tfrac{1}{6}.$$

$$N_{24} = \begin{pmatrix} -1 & -2 \\ 4 & -8 \end{pmatrix} \quad \text{for} \quad \alpha = -1, \ \beta = -\tfrac{1}{6}.$$

$$N_{25} = \begin{pmatrix} 6 & -1 \\ -24 & -4 \end{pmatrix} \quad \text{for} \quad \alpha = -\tfrac{1}{2}, \ \beta = 1.$$

$$N_{26} = \begin{pmatrix} -6 & -1 \\ 24 & -4 \end{pmatrix} \quad \text{for} \quad \alpha = -\tfrac{1}{2}, \ \beta = -1.$$

$$N_{27} = \begin{pmatrix} 1 & -1 \\ -4 & -4 \end{pmatrix} \quad \text{for} \quad \alpha = -\tfrac{1}{2}, \ \beta = \tfrac{1}{6}.$$

$$N_{28} = \begin{pmatrix} -1 & -1 \\ 4 & -4 \end{pmatrix} \quad \text{for} \quad \alpha = -\tfrac{1}{2}, \ \beta = -\tfrac{1}{6}.$$

$$N_{29} = \begin{pmatrix} -6 & 1 \\ 24 & 4 \end{pmatrix} \quad \text{for} \quad \alpha = \tfrac{1}{2}, \ \beta = -1.$$

$$N_{30} = \begin{pmatrix} 1 & 1 \\ -4 & 4 \end{pmatrix} \quad \text{for} \quad \alpha = \tfrac{1}{2}, \ \beta = \tfrac{1}{6}.$$

$$N_{31} = \begin{pmatrix} -1 & 1 \\ 4 & 4 \end{pmatrix} \quad \text{for} \quad \alpha = \tfrac{1}{2}, \ \beta = -\tfrac{1}{6}.$$

$$N_{32} = \begin{pmatrix} 6 & 1 \\ -24 & 4 \end{pmatrix} \quad \text{for} \quad \alpha = \tfrac{1}{2}, \ \beta = 1.$$

Now, using $kn - 2$ scalar matrix $B$ and its model matrices, we produce the following functional equations by applying (14.11), (14.12), (14.13), (14.14) and (14.15).

Now, from the matrices $B$ and $N_1$ by applying (14.11), we model the additive functional equation

$$-4\left[f(8x + 2y) + f(-2x + 6y)\right] \tag{14.16}$$
$$= f(-32x + 8y) + f(8x - 24y) + f(-16y)$$

for all $x, y \in X$. By applying (14.12), we get the following quadratic functional equation

$$16\left[f(8x + 2y) + f(-2x + 6y)\right] \tag{14.17}$$
$$= f(-32x + 8y) + f(8x - 24y) + 4f(-16\sqrt{xy})$$

for all $x, y \in X$. By applying (14.14), we get the following additive-quadratic functional equation

$$f(-4)\left[f(8x + 2y) + f(-2x + 6y)\right] \tag{14.18}$$
$$= f(-32x + 8y) + f(8x - 24y) + \frac{4}{2}\left[f(2(-16)\sqrt{xy}) - 2f(-16\sqrt{xy})\right]$$
$$+ \frac{1}{2}\left[4f(-16y) - f(2(-16)y)\right]$$

for all $x, y \in X$.

Now, From the matrices $B$ and $N_9$ by applying (14.11), we arrive the following additive functional equation

$$-4\left[f(8x - y) + f(-2x + 6y)\right] \tag{14.19}$$
$$= f(-32x - 4y) + f(8x - 24y) - \frac{1}{2}f(-16y)$$

for all $x, y \in X$. By applying (14.12), we get the following quadratic functional equation

$$16\left[f(8x - y) + f(-2x + 6y)\right] \tag{14.20}$$
$$= f(-32x - 4y) + f(8x - 24y) - \frac{4}{2}f(-16\sqrt{xy})$$

for all $x, y \in X$. By applying (14.14), we get the following additive-quadratic functional equation

$$f(-4)\left[f(8x - y) + f(-2x + 6y)\right] \tag{14.21}$$
$$= f(-32x - 4y) + f(8x - 24y) - \frac{4}{2(2)}\left[f(2(-16)\sqrt{xy}) - 2f(-16\sqrt{xy})\right]$$
$$- \frac{1}{2(2)}\left[4f(-16y) - f(2(-16)y)\right]$$

for all $x, y \in X$.

Now, From the matrices $B$ and $N_{21}$, by applying (14.11), we arrive the following additive functional equation

$$-4\left[f(8x+6y)+f(-2x-2y)\right] \tag{14.22}$$
$$= f(-32x-24y)+f(8x-8y)-f(-16y)$$

for all $x, y \in X$. By applying (14.13), we get the following quadratic functional equation

$$16\left[f(8x+6y)+f(-2x-2y)\right] \tag{14.23}$$
$$= f(-32x-24y)+f(8x-8y)+f(-16\sqrt{xy})$$

for all $x, y \in X$. By applying (14.15), we get the following additive-quadratic functional equation

$$f(-4)\left[f(8x+6y)+f(-2x-2y)\right] \tag{14.24}$$
$$= f(-32x-24y)+f(8x-8y)+\frac{1}{2}\left[f(2(-16)\sqrt{xy})-2f(-16\sqrt{xy})\right]$$
$$-\frac{1}{2}\left[4f(-16y)-f(2(-16)y)\right]$$

for all $x, y \in X$.

Now, From the matrices $B$ and $N_{31}$, by applying (14.11), we have arrived the following additive functional equation

$$-4\left[f(8x-y)+f(-2x+y)\right] \tag{14.25}$$
$$= f(-32x+4y)+f(8x+4y)+\frac{1}{2}f(-16y)$$

for all $x, y \in X$. By applying (14.13), we get the following quadratic functional equation

$$16\left[f(8x-y)+f(-2x+y)\right] \tag{14.26}$$
$$= f(-32x+4y)+f(8x+4y)-\frac{1}{2}f(-16\sqrt{xy})$$

for all $x, y \in X$. By applying (14.15), we get the following additive-quadratic functional equation

$$f(-4)\left[f(8x-y)+f(-2x+y)\right] \tag{14.27}$$
$$= f(-32x+4y)+f(8x+4y)-\frac{1}{2(2)}\left[f(2(-16)\sqrt{xy})-2f(-16\sqrt{xy})\right]$$
$$+\frac{1}{2(2)}\left[4f(-16y)-f(2(-16)y)\right]$$

for all $x, y \in X$.

Similarly, from the other model matrices of $B$ we can produce more additive, quadratic and mixed additive-quadratic functional equations.

## 14.6 Modeling the functional equations corresponding to logical matrix

In this section, authors model additive, quadratic and mixed type functional equations corresponding to logical matrix.

**Definition 14.6.1. Logical matrix:** A matrix whose entries are all either 0 or 1 is known as logical matrix.

**Definition 14.6.2. Logical Functional Equation:** A functional equation corresponding to logical matrix obtained by using its model matrix is known as logical functional equation.

**Example 14.6.3.** $L = \begin{pmatrix} 1 & 1 \\ 1 & 1 \end{pmatrix}$ *is a well known logical matrix. This logical matrix is also known as* $kn - 2$ *matrix for* $n = 1$ *and* $k = 1$. *By Theorem (14.2.4), the eigenvalues are 0 and 2 and by the Theorems (14.2.6) and (14.2.7), the eigenvectors are*

$$X_1 = \begin{pmatrix} -1 \\ 1 \end{pmatrix} \text{ and } X_2 = \begin{pmatrix} 2 \\ 2 \end{pmatrix}.$$

Now, using (14.9) by $M_1 = [\alpha X_1, \beta X_2]$ we can arrive the following model matrices of $L$.

$$L_1 = \begin{pmatrix} -1 & 2 \\ 1 & 2 \end{pmatrix} \text{ for } \alpha = 1, \ \beta = 1.$$

$$L_2 = \begin{pmatrix} -1 & -2 \\ 1 & -2 \end{pmatrix} \text{ for } \alpha = 1, \ \beta = -1.$$

$$L_3 = \begin{pmatrix} -1 & 1 \\ 1 & 1 \end{pmatrix} \text{ for } \alpha = 1, \ \beta = \tfrac{1}{2}.$$

$$L_4 = \begin{pmatrix} -1 & -1 \\ 1 & -1 \end{pmatrix} \text{ for } \alpha = 1, \ \beta = -\tfrac{1}{2}.$$

$$L_5 = \begin{pmatrix} 1 & 2 \\ -1 & 2 \end{pmatrix} \text{ for } \alpha = -1, \ \beta = 1.$$

$$L_6 = \begin{pmatrix} 1 & -2 \\ -1 & -2 \end{pmatrix} \text{ for } \alpha = -1, \ \beta = -1.$$

$$L_7 = \begin{pmatrix} 1 & 1 \\ -1 & 1 \end{pmatrix} \text{ for } \alpha = -1, \ \beta = \tfrac{1}{2}.$$

$$L_8 = \begin{pmatrix} 1 & -1 \\ -1 & -1 \end{pmatrix} \text{ for } \alpha = -1, \ \beta = -\tfrac{1}{2}.$$

Now, using (14.10) and by $M_2 = [\beta X_2, \alpha X_1]$, we arrive the following model matrices of $L$.

$$L_9 = \begin{pmatrix} 2 & -1 \\ 2 & 1 \end{pmatrix} \quad \text{for} \quad \alpha = 1, \ \beta = 1.$$

$$L_{10} = \begin{pmatrix} -2 & -1 \\ -2 & 1 \end{pmatrix} \quad \text{for} \quad \alpha = 1, \ \beta = -1.$$

$$L_{11} = \begin{pmatrix} 1 & -1 \\ 1 & 1 \end{pmatrix} \quad \text{for} \quad \alpha = 1, \ \beta = \tfrac{1}{2}.$$

$$L_{12} = \begin{pmatrix} -1 & -1 \\ -1 & 1 \end{pmatrix} \quad \text{for} \quad \alpha = 1, \ \beta = -\tfrac{1}{2}.$$

$$L_{13} = \begin{pmatrix} 2 & 1 \\ 2 & -1 \end{pmatrix} \quad \text{for} \quad \alpha = -1, \ \beta = 1.$$

$$L_{14} = \begin{pmatrix} -2 & 1 \\ -2 & -1 \end{pmatrix} \quad \text{for} \quad \alpha = -1, \ \beta = -1.$$

$$L_{15} = \begin{pmatrix} 1 & 1 \\ 1 & -1 \end{pmatrix} \quad \text{for} \quad \alpha = -1, \ \beta = \tfrac{1}{2}.$$

$$L_{16} = \begin{pmatrix} -1 & 1 \\ -1 & -1 \end{pmatrix} \quad \text{for} \quad \alpha = -1, \ \beta = -\tfrac{1}{2}.$$

Now, using $L$ and $L_1$ by applying (14.11) we can model the following additive functional equation

$$f(x - y) + f(x + 2y) = f(x + y) + f(x + 2y) + f(-2y) \tag{14.28}$$

for all $x, y \in X$, which implies,

$$f(x - y) - f(x + y) = f(-2y) \tag{14.29}$$

for all $x, y \in X$. Now, by applying (14.12) we can model the following quadratic functional equation

$$f(x - y) + f(x + 2y) = f(x + y) + f(x + 2y) - f(-2\sqrt{xy}) \tag{14.30}$$

for all $x, y \in X$, which implies,

$$f(x - y) - f(x + y) = -f(-2\sqrt{xy}) \tag{14.31}$$

for all $x, y \in X$. Now, by applying (14.14) we can model the following additive-quadratic functional equation

$$f(1) [f(x - y) + f(x + 2y)] \tag{14.32}$$

$$= f(x + y) + f(x + 2y) - \frac{1}{2} [f(2(-2)\sqrt{xy}) - 2f(-2\sqrt{xy})]$$

$$+ \frac{1}{2} [4f(-2y) - f(2(-2)y)]$$

for all $x, y \in X$.

Now, using $L$ and $L_8$ by applying (14.11) we can model the following additive functional equation

$$f(x+y) + f(x-y) = f(x-y) + f(x-y) - f(-2y) \qquad (14.33)$$

for all $x, y \in X$, which implies,

$$f(x+y) - f(x-y) = -f(-2y) \qquad (14.34)$$

for all $x, y \in X$. Now, by applying (14.12) we can model the following quadratic functional equation

$$f(x+y) + f(x-y) = f(x-y) + f(x-y) + f(-2\sqrt{xy}) \qquad (14.35)$$

for all $x, y \in X$, which implies,

$$f(x+y) - f(x-y) = f(-2\sqrt{xy}) \qquad (14.36)$$

for all $x, y \in X$. Now, by applying (14.14) we can model the following additive-quadratic functional equation

$$f(1) \left[ f(x+y) + f(x-y) \right] \qquad (14.37)$$
$$= f(x-y) + f(x-y) + \frac{1}{2} \left[ f(2(-2)\sqrt{xy}) - 2f(-2\sqrt{xy}) \right]$$
$$- \frac{1}{2} \left[ 4f(-2y) - f(2(-2)y) \right]$$

for all $x, y \in X$.

Similarly, from the other model matrices of $L$ we can produce more additive, quadratic and mixed additive-quadratic functional equations. Here after, we call these functional equations are logical functional equations.

## 14.7 Conclusion

In this way, the new type of $kn - 2$ matrix is introduced, defined and its eigenvalues and eigenvectors have been discussed and explained. We introduced a very first method to model the additive, quadratic and mixed type functional equations through eigenvalues and eigenvectors of matrices.

This is the first attempt to model additive, quadratic and mixed type functional equations by using eigenvalues and eigenvectors of matrices. A famous logical functional equation has been modeled using logical matrix of order 2 with the proposed new method.

## 14.7.1   Scope for further study

This study can be furthered in the future with Hyers- Ulam- Rassias stability as well as Ulam-J. M. Rassias product-sum stability of those functional equations generated by the introduced method and also this could be extended to higher order matrices. The proposed problem is open and can lead to contributions from other researchers and take the stability of functional equations field to the next level.

---

# Bibliography

Aczel J. *Lectures on functional equations and their applications.* Academic Press, New York, 1966.

Cholewa P. W. Remarks on the stability of functional equations. *Aequationes Math.*, 27:76–86, 1984.

Czerwik S. On the stability of the quadratic mapping in normed spaces. *Abh. Math. Sem. Univ. Hamburg*, 62:59–64, 1992.

Czerwik S. *Functional equations and inequalities in several variables.* World Scientific Publishing Company, New Jersey, London, Singapore and Hong Kong, 2002.

Czerwik S. *Stability of functional equations of Ulam - Hyers - Rassias type.* Hadronic Press, Palm Harbor, Florida, 2003.

Ivan N. Pexider's functional equation. *Jan Vilém Pexider (1874–1914). (English), Matfyzpress*, 2009:51–56, 2009.

Jun K. and Lee Y. On the hyers-ulam-rassias stability of a pexiderized quadratic inequality. *Math. Inequal, Appl.*, 4:93–118, 2001.

Jun K. W. and Kim H. M. The generalized hyers-ulam-rassias stability of a cubic functional equation. *J. Math. Anal. Appl.*, 274:867–878, 2002.

Jung S. M. On the hyers-ulam stability of the functional equations that have the quadratic property. *J. Math. Anal. Appl.*, 222:126 – 137, 1998.

Jung S. M. On the hyers-ulam-rassias stability of a quadratic functional equation. *J. Math. Anal. Appl.*, 232:384 – 393, 1999.

Kannappan P. Quadratic functional equation and inner product spaces. *Results Math.*, 27:368 – 372, 1995.

Kuczma M. *An introduction to the theory of functional equations and inequalities.* Panstwowe wydawnictwo naukowe, Krakow, 1985.

Najati A. and Moghimi M. B. Stability of a functional equation deriving from quadratic and additive functions in quasi-banach spaces. *J.Math.Anal.Appl.*, 337:399–415, 2008.

Narasimman P. and Amuda R. A new method to modelling the additive functional equations. *Appl. Math. Inf. Sci.*, 10(3):1047–1051, 2016.

Narasimman P. and Bodaghi A. Solution and stability of a mixed type functional equation. *Filomat*, 31:5:1229–1239, 2017.

Narasimman P., Ravi K., and Pinelas S. Modelling the additive functional equations through rsm matrices. *Journal of Advances in Mathematics*, 12(10):6714–6719, 2016.

Rassias J. M. On approximation of approximately linear mappings by linear mapping. *J.Funct. Anal.*, 46(1):126–130, 1982.

Rassias J. M. On approximation of approximately linear mappings by linear mappings. *Bull.Sci. Math.*, 108(4):445–446, 1984.

Rassias J. M. On a new approximation of approximately linear mapping by linear mappings. *Discuss. Math.*, 7:193–196, 1985.

Rassias J. M. On the stability of the euler-lagrange functional equation. *Chinese J. Math.*, 20:185–190, 1992.

Rassias J. M. Hyers-ulam stability of the quadratic functional equation in several variables. *J. Ind. Math. Soc.*, 68:65–73, 2001a.

Rassias J. M. Solution of the ulam stability problem for the cubic mapping. *Glasnik Mathematica*, 36(56):63–72, 2001b.

Rassias J. M. Solution of the ulam stability problem for an euler type quadratic functional equation. *Southeast Asian Bull.Math.*, 26(1):101–112, 2002.

Rassias J. M. On the general quadratic functional equation. *Bol.Soc. Mat. Mexicana*, 11:259 – 268, 2005.

Ravi K. and Arunkumar M. On the generalized hyers - ulam -rassias stability of a quadratic functional equations. *International Journal of Pure and Applied Mathematics*, 28(1):85–94, 2006.

Ravi K., Arunkumar M., and Narasimman P. Fuzzy stability of an additive functional equation. *International Journal of Mathematics and Statistics*, 9(A 11):88–105, Autumn 2011.

Ravi K., Arunkumar M., and Rassias J. M. On the ulam stability for the orthogonally general euler-lagrange type functional equation. *International Journal of Mathematics and statistics*, 7(Fe07):143–156, 2007.

Ravi K., Kodandan R., and Narasimman P. Ulam stability of a quadratic functional equation. *International Journal of Pure and Applied Mathematics*, 51(1):87–101, 2009a.

Ravi K., Narasimman P., and Kishore Kumar R. Generalized hyers- ulam-rassias stability and j. m. rassias stability of a quadratic functional equation. *International Journal of Mathematical Sciences and Engineering Applications*, 3(2):79–94, 2009b.

Skof F. Local properties and approximations of operators. *Rend. Sem. Mat. Fis Milano*, 53:113–129, 1983.

Xu T., Rassias J. M., and Xu W. Generalized ulam-hyers stability of a general mixed aqcq-functional equation in multi-banach spaces: a fixed point approach. *European Journal of Pure and Applied Mathematics*, 3(6):1032–1047, 2010.

# Chapter 15

# Renorming $c_0$ and Fixed Point Property

**Veysel Nezir**

*Kafkas University, Faculty of Science and Letters, Department of Mathematics, Kars, Turkey*

**Nizami Mustafa**

*Kafkas University, Faculty of Science and Letters, Department of Mathematics, Kars, Turkey*

**Hemen Dutta**

*Gauhati University, Department of Mathematics, Guwahati, India*

## 15.1 Introduction and preliminaries

We begin by giving a brief introduction to metric fixed point theory.

It can be said that researches on fixed point theory started in 1912 by L.E.J. Brouwer's Brouwer (1911) result: for $n \in \mathbb{N}$, for $C$ equal to the closed unit ball of $\mathbb{R}^n$, every norm-to-norm continuous map $f : C \longrightarrow C$ has a fixed point. His result was later extended to every compact convex subset of $\mathbb{R}^n$

469

and in 1930, Schauder Schauder (1930) extended to the same result to every Banach space. It was seen that the class of continuous maps was very large and fixed point theory was worked on smaller class of mappings. In 1922, the well known principle was introduced by Banach and he demonstrated so called Banach Contraction theorem Banach (1922): If $(X, d)$ is a complete metric space, and $f : X \longrightarrow X$ is a strict contraction, then $f$ has a unique fixed point in $X$. Then, easily the following corollary was obtained in terms of Banach spaces: for a nonempty, closed, bounded, and convex subset $C$ of a Banach space $(X, \| \cdot \|)$, if $T : C \longrightarrow C$ is a strict contraction for the metric $d = d_{\| \cdot \|}$ generated by the norm, then $T$ has a fixed point.

Then, 1965 was very efficient year for the fixed point theory. Indeed, firstly, Browder Browder (1965a) gave a more balanced theorem: [★] [ For every closed, bounded, convex (non-empty) subset $E$ of a Hilbert space $(X, \langle \cdot, \cdot \rangle)$ with associated norm $\| \cdot \|$, every nonexpansive mappings $U : E \to E$, $U$ has a fixed point in $E$ [Here, $U$ is nonexpansive means that $\|Ux - Uy\| \leq \|x - y\|$, for all $x, y \in E$] ]. Later in 1965, Browder Browder (1965b) and Göhde Göhde (1965) each (independently) generalized the previous theorem [★] to all uniformly convex Banach spaces $(X, \| \cdot \|)$; for example, $X = L^p$, $1 < p < \infty$, with its usual norm $\| \cdot \|_p$.

Later in 1965, Kirk Kirk (1965) generalized theorem [★] of Browder to all reflexive Banach spaces $X$ with so-called "normal structure": those spaces such that all non-singleton closed, bounded, and convex sets have a greater diameter than radius. This is a very large class of spaces. Spaces $(X, \| \cdot \|)$ with the property of Browder [★] became known as spaces with "the fixed point property for nonexpansive mappings". We often abbreviate this and write FPP (n.e.). We also note that the sequence spaces $(c_0, \| \cdot \|_\infty)$ and $(\ell^1, \| \cdot \|_1)$ are both nonreflexive and do not have the FPP (n.e.).

Returning to Kirk's theorem, we may ask if further generalizations are possible. Even after 51 years, for a long time, it was unclear whether or not all Banach spaces with the property FPP (n.e.) were reflexive. This and related questions have been and still are central themes in metric fixed point theory.

For example, a Banach space $X$ $(X^*)$is said to have the weak fixed point property (w-FPP (n.e.)) (the weak* fixed point property (w*-FPP (n.e.))) if every nonexpansive mapping on every nonempty weak compact convex set $C$ in $X$ ($C$ in $X^*$, respectively) has a fixed point and for a long time, it had remained open whether or not every Banach space possessed w-FPP until Alspach Alspach (1981) provided an example of a fixed point free nonexpansive mapping on a weakly compact subset of Banach space $X = (L^1[0, 1], \| \cdot \|_1)$. But contrary, short-time later, Maurey showed $(c_0, \| \cdot \|_\infty)$, the Banach space of real-valued sequences that converge to zero, with the absolute supremum norm $\| \cdot \|_\infty$ and reflexive subspaces of $L^1[0, 1]$ do have w-FPP using ultrafilter techniques.

There have been many strategies to check whether or not a Banach space $X$ $(X^*)$ possesses w-FPP (w*-FPP, respectively) or FPP and different type Banach spaces have been the subject of many papers in the last 50 or so years interms of possessing or not possessing w-FPP or FPP and some researches tried to show the connection between reflexivity and FPP; and some like the editor of this book, Hemen Dutta with his joint studies such as Gunduz and Dutta (2017); Kir et al. (2017) worked to find fixed point property oriented results via iterative approximations for different type of mappings when there is FPP in the reflexive Banach spaces.

Conversely to Maurey's result, Downling and Lennard Dowling and Lennard (1997) showed every nonreflexive subspace of $L^1[0,1]$ fails the fixed point property.

Before that result, Carothers et al. Carothers et al. (1996) in 1996 established the analogous result for nonreflexive subspaces of the Lorentz function space $L_{w,1}(0,\infty)$. Even earlier, in 1991, Carothers et al. Carothers et al. (1991) showed the Lorentz Space $L_{w,1}(\mu)$ enjoys the weak-star fixed point property if $(X, \Sigma, \mu)$ is a $\sigma$-fine measure space; that is, if $C$ is a weak-star compact convex subset of $L_{w,1}(\mu)$, then, every nonexpansive mapping on $C$ has a fixed point.

Due to the researches on fixed point theory in $(c_0, \|\cdot\|)_\infty$ and theorems about $c_0$, we are interested to understand more about Banach spaces $(X, \|\cdot\|)$ that contain subspaces isomorphic to $c_0$. Equivalently, we are interested in Banach spaces that contain "$c_0$-summing basic sequences". A sequence $(x_n)_{n\in\mathbb{N}}$ in a Banach space $(X, \|\cdot\|)$ is a $c_0$-summing basic sequence if there exist constants $0 < A \leq B < \infty$ such that for all $t \in c_{00}$,

$$A \sup_{n\in\mathbb{N}} \left| \sum_{k=n}^\infty t_k \right| \leq \left\| \sum_{n=1}^\infty t_n x_n \right\| \leq B \sup_{n\in\mathbb{N}} \left| \sum_{k=n}^\infty t_k \right| .$$

Reflexive Banach spaces $(X, \|\cdot\|)$ (for example, $L^p$, $1 < p < \infty$, and Hilbert spaces) do not contain $c_0$-summing basic sequences. On the other hand, many non-reflexive Banach spaces do. Lorentz-Marcinkiewicz spaces $l^0_{w,\infty}$ discussed in the Ph.D. thesis of the first author Nezir (2012), written under supervisor of Chris Lennard, are of this type.

In 1979, Goebel and Kuczumow Goebel and Kuczumow (1979) constructed very irregular closed, bounded, convex, non-weak*-compact subsets $K$ of $l^1$, and proved that such $K$ have FPP(n.e.). Then, in Everest's Ph.D. thesis Everest (2013), under supervision of Lennard, Everest found new and larger class of sets that have the fixed point property for nonexpansive mappings in $l^1$ by using Goebel and Kuczumow ideas.

We recall that Lin Lin (2008) gave the first example of a non-reflexive Banach space $(X, \|\cdot\|)$ with the fixed point property for nonexpansive mappings and showed this fact for $(\ell^1, \|\cdot\|_1)$ with the equivalent norm $\|\|\cdot\|\|^*$ given by

$$\|\|x\|\|^* = \sup_{k\in\mathbb{N}} \frac{8^k}{1+8^k} \sum_{n=k}^\infty |x_n|, \text{ for all } x = (x_n)_{n\in\mathbb{N}} \in \ell^1 .$$

We wonder $(c_0, \|\cdot\|_\infty)$ analogue of P. K. Lin's work. That is, can we renorm $c_0$ to have the fixed point property for nonexpansive mappings or not? While this is a famous open question, $c_0$ analogue of Goebel and Kuczumow's theory (with an equivalent norm of course) has also great importance since it would be the first step to find a candidate equivalent norm to work on $c_0$ analogue of P. K. Lin's work.

In contrast to Goebel and Kuczumow's result for $l^1$, Dowling, Lennard and Turett Dowling et al. (1998) showed that any closed infinite dimensional subspace of $(c_0, \|\cdot\|_\infty)$ also fails FPP(n.e.). Also, in 2004, they Dowling et al. (2004) showed that every non-weakly compact, closed, bounded, convex (c.b.c.) subset $K$ of $(c_0, \|\cdot\|_\infty)$ fails FPP for $\|\cdot\|_\infty$-nonexpansive mappings. Thus, to think about $c_0$ analogue of Goebel and Kuczumow's work, firstly, we have to consider it with an equivalent norm for $c_0$. That is, we can work on a question "do there exist any renormings of $c_0$ and a nonempty closed, bounded and convex subset $C$ so that every nonexpansive mapping has fixed point property?"

In this chapter, we give positive answer for this question when the mapping is also affine. To this end, we construct two different equivalent norms, one looking like P. K. Lin's norm, with interesting results. Our results are very related with the first author's work in the joint paper with Lennard that we can recall as below:

In 2011, Lennard and Nezir Lennard and Nezir (2011) show that the closed convex hull of any asymptotically isometric (ai) $c_0$-summing basic sequence $(x_n)_{n\in\mathbb{N}}$ in a Banach space, $E := \overline{co}(\{x_n : n \in \mathbb{N}\})$, fails the fixed point property for affine nonexpansive mappings.

In their paper, first of all, they work on some specific ai $c_0$-summing basic sequences in $c_0$.

For example, they fix $b \in (0,1)$ and define the sequence $(f_n)_{n\in\mathbb{N}}$ in $c_0$ by setting $f_1 := b\,e_1$, $f_2 := b\,e_2$, and $f_n := e_n$ for every $n \geq 3$ where $(e_n)_{n\in\mathbb{N}}$ is defined to be 1 in its $n^{\text{th}}$ coordinate, and 0 in all other coordinates such that for both $(c_0, \|\cdot\|_\infty)$ and $(\ell^1, \|\cdot\|_1)$, the sequence $(e_n)_{n\in\mathbb{N}}$ is an unconditional basis.

Next, they define the closed, bounded, convex subset $E = E_b$ of $c_0$ by

$$E := \left\{ \sum_{n=1}^{\infty} t_n\, f_n : 1 = t_1 \geq t_2 \geq \cdots \geq t_n \downarrow_n 0 \right\}.$$

Then, they define the sequence $(\eta_n)_{n\in\mathbb{N}}$ in $E$ in the following way: $\eta_1 := f_1$ and $\eta_n := f_1 + \cdots + f_n$ for every $n \geq 2$. Note that

$$E = \left\{ \sum_{n=1}^{\infty} \alpha_n\, \eta_n : \text{ each } \alpha_n \geq 0 \text{ and } \sum_{n=1}^{\infty} \alpha_n = 1 \right\}.$$

Then, the following theorem is given as a result in their work:

**Theorem 15.1.1.** *Assume $b \in (0,1)$. Then the closed convex hull of the sequence $(\eta_n)_{n\in\mathbb{N}}$, $E = \overline{co}(\{\eta_n : n \in \mathbb{N}\})$ is such that there exists a fixed point free affine $\|\cdot\|_\infty$-nonexpansive mapping $U : E \longrightarrow E$.*

Note that it can be easily seen that the sequence $(\eta_n)_{n\in\mathbb{N}}$ is an ai $c_0$-summing basic sequence.

Next, they generalize their result more and prove the following theorems.

**Theorem 15.1.2.** *Assume $\overrightarrow{b} = (b_n)_{n\in\mathbb{N}}$ is any increasing sequence in $(0, 1]$ with $b_n \uparrow_n 1$. Let $(f_n)_{n\in\mathbb{N}}$ be a sequence in $c_0$ by the following way: $f_n := b_n\, e_n$, for every $n \in \mathbb{N}$. Next, take $E = E_{\overrightarrow{b}}$ of $c_0$ such that*

$$E := \left\{ \sum_{n=1}^{\infty} t_n\, f_n : 1 = t_1 \geq t_2 \geq \cdots \geq t_n \downarrow_n 0 \right\}.$$

*Then, there exists a fixed point free affine $\|\cdot\|_\infty$-nonexpansive mapping $U : E \longrightarrow E$.*

**Theorem 15.1.3.** *Assume $\overrightarrow{b} = (b_n)_{n\in\mathbb{N}}$ is any sequence in $(0, \infty)$ converging to some $\kappa > 0$. Define the sequence $(f_n)_{n\in\mathbb{N}}$ in $c_0$ given by $f_n := b_n\, e_n$ for every $n \in \mathbb{N}$. Next, using this sequence define a subset $E = E_{\overrightarrow{b}}$ of $c_0$ by*

$$E := \left\{ \sum_{n=1}^{\infty} t_n\, f_n : 1 = t_1 \geq t_2 \geq \cdots \geq t_n \downarrow_n 0 \right\}.$$

*Then, there exists a fixed point free affine $\|\cdot\|_\infty$-nonexpansive mapping $U : E \longrightarrow E$.*

**Theorem 15.1.4.** *Assume $\overrightarrow{b} = (b_n)_{n\in\mathbb{N}}$ is any bounded sequence in $(0, \infty)$. Define the sequence $(f_n)_{n\in\mathbb{N}}$ in $c_0$ given by $f_n := b_n\, e_n$, for every $n \in \mathbb{N}$. Next, using this sequence define a subset $E = E_{\overrightarrow{b}}$ of $c_0$ by*

$$E := \left\{ \sum_{n=1}^{\infty} t_n\, f_n : 1 = t_1 \geq t_2 \geq \cdots \geq t_n \downarrow_n 0 \right\}.$$

*Then, there exists a fixed point free affine $\|\cdot\|_\infty$-nonexpansive mapping $U : E \longrightarrow E$.*

More importantly, they give their main result as below:

**Theorem 15.1.5.** *Let $L \in (0, \infty)$. Banach space containing an L-scaled asymptotically isometric $c_0$-summing basic sequence $(x_n)_{n\in\mathbb{N}}$ fails the fixed point property for affine nonexpansive mappings since when the closed convex hull of the sequence $(x_n)_{n\in\mathbb{N}}$, $E := \overline{co}(\{x_n : n \in \mathbb{N}\})$ is taken, then there exists a fixed point free affine contractive mapping $U : E \longrightarrow E$.*

In this study, we construct two different equivalent norms on $c_0$ and aim to show the above mentioned sets have the fixed point property for affine nonexpansive mappings respect to the norms constructed. We see that for one of the equivalent norms we develop, $c_0$ does not contain any asymptotically isometric copy of $c_0$ and the set $E$ used in Theorem 15.1.1 has FPP for affine nonexpansive mappings while they all have FPP for affine nonexpansive mappings when we use the other renorming. That means second renorming allows larger class of non-weakly compact closed, bounded and convex subsets of $c_0$ to have FPP for affine nonexpansive mappings. In fact, for our second renorming, we see more is possible. We study closed convex hull of some $c_0$-summing basic sequences given in the first author's Ph.D. thesis Nezir (2012) under supervision of Lennard. We should note that for those sets, we gave the following results there:

Assume $(\gamma_n)_{n \in \mathbb{N}}$ is a sequence in $(0, \infty)$ for which there exists $\Gamma > 0$ with $[\Gamma \leq \gamma_N$, for all $N \in \mathbb{N}]$ and $\sigma := \sum_{n=2}^{\infty} |\gamma_n - \gamma_{n-1}| < \infty$ ; and suppose $(b_n)_{n \in \mathbb{N}}$ is a sequence in $(0, \infty)$ converging to some $\lambda \in (0, \infty)$. Define the sequence by $\eta_n := \gamma_n (b_1 e_1 + b_2 e_2 + b_3 e_3 + b_4 e_4 + .... + b_n e_n)$, for all $n \in \mathbb{N}$. Also suppose that $(\eta_n)_{n \in \mathbb{N}}$ satisfies a lower $c_0$-summing estimate. That is, suppose $\exists K \in (0, \infty)$ s.t. $\forall \alpha = (\alpha_n)_{n \in \mathbb{N}} \in c_{00}$, $K \sup_{n \geq 1} \left| \sum_{j=n}^{\infty} \alpha_j \right| \leq \left\| \sum_{j=1}^{\infty} \alpha_j \eta_j \right\|$ . Then, $(\eta_n)_{n \in \mathbb{N}}$ is an $L$-scaled asymptotically isometric $c_0$-summing basic sequence. Moreover, on the closed convex hull of $(\eta_n)_{n \in \mathbb{N}}$, $E := \overline{co}(\{\eta_n : n \in \mathbb{N}\})$, there exists a fixed point free affine $\| \cdot \|_\infty$-contractive mapping $U : E \longrightarrow E$.

Then, acknowledging these sets $E$'s too, we conclude that for our second renorming with the equivalent norm $|||\cdot|||$ given below on $c_0$, the closed convex hull of the sequence $(\mu_n (b_1 e_1 + b_2 e_2 + b_3 e_3 + b_4 e_4 + .... + b_n e_n))_{n \in \mathbb{N}}$ has FPP for affine $|||\cdot|||$-nonexpansive mappings where $(\mu_n)_{n \in \mathbb{N}}$ and $(b_n)_{n \in \mathbb{N}}$ are any sequence in $(0, \infty)$.

For $x = (\xi_k)_k \in c_0$, define $|||x|||$ by

$$|||x||| := \lim_{p \to \infty} \sup_{k \in \mathbb{N}} \gamma_k \left( \sum_{j=k}^{\infty} \frac{|\xi_j|^p}{j^2} \right)^{\frac{1}{p}} \quad \text{where } \gamma_k \uparrow_k 1 \text{ and } \gamma_k \text{ is strictly increasing.}$$

We should also note that when we work on these equivalent norms, we are motivated firstly by Goebel and Kuczumow's work Goebel and Kuczumow (1979) and secondly by Ph.D. thesis of Everest Everest (2013), written under the supervision of Lennard, and finally by the significant work of Lin Lin (2008). Thus, the general idea of the main result comes from the paper by Goebel and Kuczumow Goebel and Kuczumow (1979) and constructing the sets, we follow Nezir's previous paper written jointly with Lennard Lennard and Nezir (2011). We believe that the results given in the chapter are of some interests for specialists since this work introduces a class of equivalent norms on $c_0$ with important properties that may generate stages and ideas of the study directed towards the problem, $c_0$-analogue of Lin's theorem about $\ell^1$, if $c_0$ can be renormed to have the fixed point property for nonexpansive mappings. These will be first results of the chapter by the following two sections and the later section will generalize the former results to the whole space but we begin with some preliminaries.

We need to note that most work in this chapter is inspired by and derived from the following studies: "Nezir, V. Renorming $c_0$ and affine fixed point property, submitted", "Nezir, V. and Mustafa, N. $c_0$ can be renormed to have the fixed point property for affine nonexpansive mappings, to appear in Filomat", "Nezir, V., Oran, S. and Dutta, H. A large class of nonweakly compact subsets in $c_0$ with fixed point property for affine nonexpansive mappings when $c_0$ is renormed", "Ateş, T. Families of equivalent norms on $c_0$ and fixed point property, Master's thesis, Kafkas University, in preparation" and Güven (2017); Nezir (2017); Oran (2017).

## 15.1.1 Preliminaries

**Definition 15.1.6.** Let $K$ be a non-empty closed, bounded, convex subset of a Banach space $(X, \|\cdot\|)$. Let $T : K \longrightarrow K$ be a mapping.

1. We say $T$ is *affine* if

   for all $\lambda \in [0, 1]$, for all $x, y \in K$, $T\big((1 - \lambda)x + \lambda y\big) = (1 - \lambda)T(x) + \lambda T(y)$ .

2. We say $T$ is *nonexpansive* if

   $\|T(x) - T(y)\| \leq \|x - y\|$ , *for all* $x, y \in K$.

   Also, we say that $K$ has the *fixed point property for nonexpansive mappings*
   [FPP(n.e.)] if for all nonexpansive mappings $T : K \longrightarrow K$, there exists $z \in K$ with $T(z) = z$.

Let $(X, \|\cdot\|)$ be a Banach space and $E \subseteq X$. We will denote the closed, convex hull of $E$ by $\overline{\mathrm{co}}(E)$. As usual, $(c_0, \|\cdot\|_\infty)$ is given by $c_0 := \left\{ x = (x_n)_{n\in\mathbb{N}} : \text{each } x_n \in \mathbb{R} \text{ and } \lim_{n\to\infty} x_n = 0 \right\}$ . Further, $\|x\|_\infty := \sup_{n\in\mathbb{N}} |x_n|$, for all $x = (x_n)_{n\in\mathbb{N}} \in c_0$; and $(\ell^1, \|\cdot\|_1)$ is defined by

$$\ell^1 := \left\{ x = (x_n)_{n\in\mathbb{N}} : \text{each } x_n \in \mathbb{R} \text{ and } \|x\|_1 := \sum_{n=1}^{\infty} |x_n| < \infty \right\} .$$

Let $n \in \mathbb{N}$. The scalar sequence $e_n$ is the well-known canonical basis of $c_0$. Moreover, $c_{00}$ is the space of all scalar sequences that have only finitely many non-zero terms.

We recall now the definition of an *asymptotically isometric $c_0$-summing basic sequence* in a Banach space $(X, \|\cdot\|)$, from Lennard and Nezir (2011).

**Definition 15.1.7.** Consider $(X, \|\cdot\|)$ is a Banach space and $(x_n)_{n\in\mathbb{N}}$ is a sequence in $X$ that satisfies the following condition; then, we say $(x_n)_{n\in\mathbb{N}}$ is an *asymptotically isometric (ai) $c_0$-summing basic sequence* in $(X, \|\cdot\|)$: There exists a null sequence $(\varepsilon_n)_{n\in\mathbb{N}}$ in $[0, \infty)$ such that for all sequences $(t_n)_{n\in\mathbb{N}} \in c_{00}$,

$$\sup_{n\geq 1} \left(\frac{1}{1 + \varepsilon_n}\right) \left|\sum_{j=n}^{\infty} t_j\right| \leq \left\|\sum_{j=1}^{\infty} t_j x_j\right\| \leq \sup_{n\geq 1}(1 + \varepsilon_n) \left|\sum_{j=n}^{\infty} t_j\right| .$$

Note that in the above definition we may replace $c_{00}$ by $\ell^1$. Also, if $L > 0$ such that the sequence $(z_n/L)_{n \in \mathbb{N}}$ is an *asymptotically isometric $c_0$-summing basic sequence*, we will call the sequence $(z_n)_{n \in \mathbb{N}}$ an *L-scaled asymptotically isometric $c_0$-summing basic sequence* in $(X, \| \cdot \|)$.

**Definition 15.1.8.** Dowling et al. (2001) We call a Banach space $(X, \| \cdot \|)$ contains an asymptotically isometric (ai) copy of $c_0$ if there exist a sequence $(x_n)_n$ in $X$ and a null sequence $(\varepsilon_n)_n$ in $(0, 1)$ so that

$$\sup_n (1 - \varepsilon_n)|a_n| \leq \left\| \sum_{n=1}^{\infty} a_n x_n \right\| \leq \sup_n |a_n| ,$$

for all $(a_n)_n \in c_0$.

**Theorem 15.1.9.** *Dowling et al. (2001) If a Banach space $(X, \| \cdot \|)$ contains an ai copy of $c_0$, then $X$ fails FPP(n.e.).*

**Theorem 15.1.10.** *Lennard and Nezir (2011) Let $L \in (0, \infty)$. If a Banach space contains an L-scaled asymptotically isometric $c_0$-summing basic sequence $(x_n)_{n \in \mathbb{N}}$, then $E := \overline{co}(\{x_n : n \in \mathbb{N}\})$ fails the fixed point property for affine nonexpansive mappings. Indeed, more is true. There exists an affine contractive mapping $U : E \longrightarrow E$ that is fixed point free.*

One can easily obtain the following partial analogue result to (Goebel and Kuczumow, 1979, Lemma 1) by the following lemma.

**Lemma 15.1.11.** *Let $(x_n)_n$ be a bounded sequence in a Banach space $(X, \| \cdot \|)$. Consider a function $s : X \to [0, \infty)$ given by*

$$s(y) = \limsup_m \left\| \frac{1}{m} \sum_{k=1}^{m} x_k - y \right\| , \quad \forall y \in X .$$

*Then, if $X$ has weak Banach-Saks property and $x \in X$ is the weak limit of the sequence $(x_n)_n$, then there exists a subsequence $(x_{n_k})_k$ whose norm limit is $x$ such that if $s$ is redefined via this subsequence, we have $s(x) = 0$ and $s(y) = \|y - x\|$ , $\forall y \in X$ and for any equivalent norm $\|\cdot\|$ on $X$.*
*Thus, since $c_0$ has weak Banach-Saks property Núñez (1989), the above can be applied.*

---

## 15.2 A renorming of $c_0$ and a large class of non-weakly compact, closed, bounded, and convex subsets with FPP for affine nonexpansive mappings

In this section, we will construct an equivalent norm to the usual norm on $c_0$ and obtain interesting properties in terms of fixed point theory.

**Definition 15.2.1.** Let $\alpha \in \mathbb{R}$. For $x = (\xi_k)_k \in c_0$, define

$$\|x\| = \|x\|_\infty + \sup_{j \in \mathbb{N}} \sum_{k=1}^{\infty} Q_k \left|\xi_k - \alpha\xi_j\right| \text{ where } \sum_{k=1}^{\infty} Q_k = 1, \ Q_k \downarrow_k 0$$
$$\text{and } Q_k > Q_{k+1}, \ \forall k \in \mathbb{N}.$$

Then, as we show below, it can be seen that for any $\alpha$, $\|.\|$ is an equivalent norm on $c_0$.

First of all, it is clear that for any $x \in c_0$, $\|x\| \geq \|x\|_\infty$. Next, for $j \in \mathbb{N}$,

$$\sum_{\substack{k=1 \\ j \neq k}}^{\infty} Q_k \left|\xi_k - \alpha\xi_j\right| + Q_j \left|1 - \alpha\right| \left|\xi_j\right| \leq \sup_{\substack{k \in \mathbb{N} \\ j \neq k}} \left|\xi_k - \alpha\xi_j\right| + \left|1 - \alpha\right| \left|\xi_j\right|$$

Then, using the arguments in the paper Santos et al. (2014), choosing $N \in \mathbb{N}$ so that $\|x\|_\infty = |\xi_N|$ we can say that

1. If $0 < \alpha \leq 1$ then

$$\|x\| \leq (3 - \alpha) \|x\|_\infty + \alpha \sup_k \left\{ |\xi_k| \ : \ sgn\left(\zeta_k\right) = -sgn\left(\xi_N\right) \right\}.$$

   Hence, $\|x\| \leq (3 - \alpha) \|x\|_\infty + \alpha \|x\|_\infty = 3 \|x\|_\infty$.

2. If $\alpha > 1$ then

$$\|x\| \leq \alpha \|x\|_\infty + \alpha \|x\|_\infty + \sup_k \left\{ |\xi_k| \ : \ sgn\left(\xi_k\right) = -sgn\left(\xi_N\right) \right\}$$
$$\leq (2\alpha + 1) \|x\|_\infty.$$

3. If $\alpha < 0$ then

$$\|x\| \leq \|x\|_\infty + \sup_{\substack{k \in \mathbb{N} \\ j \neq k}} \left|\xi_k - \alpha\xi_j\right| + \left|1 - \alpha\right| \left|\xi_j\right|$$
$$\leq (3 - 2\alpha) \|x\|_\infty.$$

4. If $\alpha = 0$ then

$$\|x\| := \|x\|_\infty + \sum_{k=1}^{\infty} Q_k \left|\xi_k\right| \leq 2\|x\|_\infty.$$

Thus, for any $\alpha$, $\|.\|$ is an equivalent norm on $c_0$.

**Theorem 15.2.2.** *Consider the equivalent norm on $c_0$ given in Definition 15.2.1. Then, if $\alpha = 0$ or if $Q_1 > \frac{2|\alpha|}{1+2|\alpha|}$ when $|\alpha| > 1$, then $(c_0, \|.\|)$ does not contain an asymptotically isometric copy of $c_0$.*

*Proof. Case 1: $\alpha = 0$.*

This case can be considered as trivial since the norm is a generalization of the norm in (Dowling et al., 1998, Example 5) and so we just need to use the same idea as in there.

*Case 2:* $|\alpha| > 1$ and $Q_1 > \frac{2|\alpha|}{1+2|\alpha|}$.

By contradiction, assume $(c_0, \|.\|)$ does contain an asymptotically isometric copy of $c_0$. Then, $2|\alpha| > 2 > 1$ and so for the following equivalent norm $\|.\|^{\sim}$, $(c_0, \|.\|^{\sim})$ would also contain an asymptotically isometric copy of $c_0$.

$$\|x\|^{\sim} = \|x\|_{\infty} + \sup_{j \in \mathbb{N}} \sum_{k=1}^{\infty} Q_k \left|\xi_k - 2\alpha \xi_j\right| \text{ for } x = (\xi_k)_k \in c_0$$

$$\text{where } \sum_{k=1}^{\infty} Q_k = 1, \ Q_k \downarrow_k 0 \text{ and } Q_k > Q_{k+1}, \ \forall k \in \mathbb{N}.$$

Then, there exists a null sequence $(\varepsilon_n)_n$ in $(0,1)$ and a sequence $(x_n)_n$ in $c_0$ such that

$$\heartsuit \left[ \begin{array}{c} \text{for every } n \in \mathbb{N} \text{ and every choice of scalars } t_1, t_2, \ldots, t_n, \\ \text{it follows that } \max_{1 \leq k \leq n} (1-\varepsilon_k)|t_k| \leq \left\|\sum_{k=1}^{n} t_k x_k\right\|^{\sim} \leq \max_{1 \leq k \leq n} |t_k|. \end{array} \right]$$

Without loss of generality we can assume that the sequence $(x_n)_n$ converges pointwise to 0. Actually, from the right hand side inequality in $\heartsuit$ it follows that $(x_n)_n$ does converge weakly to 0.

For each $n \in \mathbb{N}$, let $x_n = (\xi_j^n)_j$.

Note that, for every $x \in c_0$, we had showed above that there exists $L > 1$ such that $\|x\|_{\infty} \geq \left\|\frac{x}{L}\right\|$ and similar result can be obtained for the norm $\|.\|^{\sim}$ too. Now, without loss of generality, by passing to a subsequence if necessary, we may assume there exists $s \in \mathbb{N}$ such that $\|x_s\|_{\infty} > \frac{1}{2|\alpha|-1}$. We can do this since for $L > 1$, the sequence $(x_n)_n$ can be replaced with $(\frac{x_n}{L})_n$ so that the condition $\heartsuit$ is satisfied for null sequence $(\varepsilon_n)_n$ in $(0,1)$ and so there exists $s \in \mathbb{N}$ such that $\varepsilon_s < 1 - \frac{1}{2|\alpha|-1}$ and $\|x_s\|_{\infty} \geq \left\|\frac{x_s}{L}\right\| > 1 - \varepsilon_s > \frac{1}{2|\alpha|-1}$.

Now, there exists $r \in \mathbb{N}$ s.t. $\xi_r^s \neq 0$ and, as previously, since $x_s \in c_0$, there exists $N^{(s)} \in \mathbb{N}$ such that $\|x_s\|_{\infty} = |\xi_{N^{(s)}}^s| \geq |\xi_r^s|$. Hence, take $p = \min\{r \mid |\xi_r^s| = |\xi_{N^{(s)}}|\}$.

Now, let $\delta = \left(Q_1 - \frac{2|\alpha|}{1+2|\alpha|}\right) \frac{|\alpha|+2|\alpha|^2}{4|\alpha|^3+6|\alpha|^2+5|\alpha|+1}$.

Now, choose $N_1 \geq p$ so that $\sum_{k=1+N_1}^{\infty} Q_k < (1+2|\alpha|)\delta$. Choose $N_2 \in \mathbb{N}$ so that $\varepsilon_n < \min\left\{1 - \frac{1}{2|\alpha|-1}, \ \delta\right\}$ for all $n \geq \max\{s, N_2\}$. Choose $M \geq \max\{s, N_2\}$ so that $|\xi_j^n| < \frac{(1+2|\alpha|)\delta}{4|\alpha|}$ for $j = 1, 2, \ldots, N_1$ and for all $n \geq M$. Note that $1 \geq \|x_s\|^{\sim}$ and $1 \geq \|x_n\|^{\sim}$ and so $1 \geq |\xi_j^s|$ and $1 \geq |\xi_j^n|$ for all $j \in \mathbb{N}$.

Therefore, for each $n \geq M$,

$$
\begin{aligned}
\|x_n\|_\infty &\leq \|x_n\|^{\sim} \\
&\leq (1 + 2|\alpha|) \|x_n\|_\infty + \sum_{k=1}^{N_1} Q_k |\xi_k^n| + \sum_{k=1+N_1}^{\infty} Q_k |\xi_k^n| \\
&< (1 + 2|\alpha|) \|x_n\|_\infty + \frac{(1+2|\alpha|)\,\delta}{4|\alpha|} \sum_{k=1}^{N_1} Q_k + \sum_{k=1+N_1}^{\infty} Q_k.
\end{aligned}
$$

Thus, for each $n \geq M$,

$$
\|x_n\|^{\sim} < (1 + 2|\alpha|) \|x_n\|_\infty + \frac{(1+2|\alpha|)^2 \delta}{2|\alpha|}.
$$

By the triangle inequality $\|x_n\|_\infty \leq \frac{1}{2}\|x_n + x_s\|_\infty + \frac{1}{2}\|x_n - x_s\|_\infty$ and so either $\|x_n + x_s\|_\infty \geq \|x_n\|_\infty$ or $\|x_n - x_s\|_\infty \geq \|x_n\|_\infty$.
If $\|x_n + x_s\|_\infty \geq \|x_n\|_\infty$ then we have

$$
\begin{aligned}
1 = \max\{1,1\} &\geq \|x_s + x_n\|^{\sim} \\
&= \|x_s + x_n\|_\infty + \sup_{j \in \mathbb{N}} \sum_{k=1}^{\infty} Q_k \left| \xi_k^s + \xi_k^n - 2\alpha \left(\xi_j^s + \xi_j^n\right) \right| \\
&\geq \|x_s + x_n\|_\infty + Q_1(2|\alpha| - 1)|\xi_p^s| - 2Q_1|\alpha||\xi_p^n| - Q_1|\xi_1^n| \\
&\geq \|x_n\|_\infty + Q_1(2|\alpha| - 1)|\xi_p^s| - 2Q_1|\alpha||\xi_p^n| - Q_1|\xi_1^n| \\
&> \frac{1}{1+2|\alpha|}\|x_n\|^{\sim} - \frac{(1+2|\alpha|)\delta}{2|\alpha|} + Q_1(2|\alpha| - 1)|\xi_p^s| \\
&\quad - 2Q_1|\alpha|\left(|\xi_p^n| + |\xi_1^n|\right) \\
&> \frac{1}{1+2|\alpha|}\|x_n\|^{\sim} - \frac{(1+2|\alpha|)\delta}{2|\alpha|} + Q_1 - 2|\alpha|\left(|\xi_p^n| + |\xi_1^n|\right) \\
&> \frac{1}{1+2|\alpha|}(1 - \varepsilon_n) - \frac{(1+2|\alpha|)\delta}{2|\alpha|} + Q_1 - 2|\alpha|\left(|\xi_p^n| + |\xi_1^n|\right) \\
&> \frac{1}{1+2|\alpha|}(1 - \delta) - \frac{(1+2|\alpha|)\delta}{2|\alpha|} + Q_1 - 2|\alpha|\left(|\xi_p^n| + |\xi_1^n|\right) \\
&> 1 + \frac{\delta}{2|\alpha|}
\end{aligned}
$$

which is not possible (contradiction). Similarly we arrive at a contradiction if we assume that $\|x_n - x_s\|_\infty \geq \|x_n\|_\infty$. $\qquad\square$

### 15.2.1 A large class of non-weakly compact, c.b.c. sets in $c_0$ having FPP for affine $\|\cdot\|$-nonexpansive mappings

**Example 15.2.3.** *Fix $b \in (0,1)$. We define the sequence $(f_n)_{n \in \mathbb{N}}$ in $c_0$ by setting $f_1 := b\,e_1$ and $f_n := e_n$, for all integers $n \geq 2$.*

*Let us define the sequence $(\eta_n)_{n \in \mathbb{N}}$ in $E$ in the following way. Let $\eta_1 := f_1$ and $\eta_n := f_1 + \cdots + f_n$, for all integers $n \geq 2$. Next, define the closed, bounded, convex subset $E = E_b$ of $c_0$ by*

$$E := \left\{ \sum_{n=1}^{\infty} \alpha_n \eta_n : \text{ each } \alpha_n \geq 0 \text{ and } \sum_{n=1}^{\infty} \alpha_n = 1 \right\}.$$

**Theorem 15.2.4.** *There exist $b \in (0, 1)$ and $\alpha > 1$ such that the set $E$ defined as in the example above has the fixed point property for $\|.\|$-nonexpansive affine mappings when $Q_1 > \frac{2\alpha}{1+2\alpha}$.*

*Proof.* First of all, recall that we need the condition $Q_1 > \frac{2\alpha}{1+2\alpha}$ to eliminate the possibility of having an a.i. copy of $c_0$ inside. Let $b \in (0, 1)$, $Q_1 < \frac{2b}{1+b}$ and $\alpha < 2Q_1$.

We note that if $x \in E$, then there exists a sequence of scalars $(\alpha_n)_n$ with $\alpha_n \geq 0, \forall n \in \mathbb{N}$ satisfying $\sum_{n=1}^{\infty} \alpha_n = 1$ and $x = \sum_{n=1}^{\infty} \alpha_n \eta_n$. Hence, $\exists q \in \mathbb{N}$ such that $\alpha_q \geq \frac{1}{2^{q+1}}$.

Then inspired by this fact, we will consider two different sets:

$$E_1 := \left\{ \sum_{n=1}^{\infty} \alpha_n \eta_n : \text{ each } \alpha_n \geq 0, \; \alpha_1 > 0 \text{ and } \sum_{n=1}^{\infty} \alpha_n = 1 \right\}$$

and

$$E_2 := \left\{ \sum_{n=1}^{\infty} \alpha_n \eta_n : \text{ each } \alpha_n \geq 0, \; \alpha_1 = 0 \text{ and } \sum_{n=1}^{\infty} \alpha_n = 1 \right\}$$

noting $E = E_1 \cup E_2$. Now, provided that $Q_2 < \frac{1}{1+2\alpha}$, choose $p \in \mathbb{N}$ so that $Q_2 \geq \left( 1 - \frac{1}{2^{p+1}} \right) \frac{1}{3b}$ and

$$E_1 = E_p := \left\{ \sum_{n=1}^{\infty} \alpha_n \eta_n : \text{ each } \alpha_n \geq 0, \; \alpha_1 \geq \frac{1}{2^{p+1}} \text{ and } \sum_{n=1}^{\infty} \alpha_n = 1 \right\} \quad (\clubsuit)$$

and next choose $b \in (0, 1)$ so that $\frac{1-b^2}{b(3b^2-6b+7)} < \frac{1}{2^{p+2}}$. Then, we obtain that $b > \frac{3+\sqrt{20}}{11} > \frac{\sqrt{5}-1}{2}$ which will help us for further steps. ($\clubsuit\clubsuit$)

**Case A:** We will be working on the set $E_1 = E_p$.

Let $T : E_p \to E_p$ be an affine nonexpansive mapping. Then, there exists a sequence $\left( x^{(n)} \right)_{n \in \mathbb{N}} \in E_p$ such that $\left\| T x^{(n)} - x^{(n)} \right\| \underset{n}{\to} 0$ and so $\left\| T x^{(n)} - x^{(n)} \right\|_{\infty} \underset{n}{\to} 0$. Without loss of generality, passing to a subsequence if necessary, there exists $z \in c_0$ such that $x^{(n)}$ converges to $z$ in weak topology. Then, by Lemma 15.4.7, we can define a function $s : c_0 \to [0, \infty)$ by

$$s(y) = \limsup_{m} \left\| \frac{1}{m} \sum_{k=1}^{m} x^{(k)} - y \right\|, \quad \forall y \in c_0.$$

Then,
$$s(y) = \|y - z\|, \ \forall y \in c_0.$$

Define
$$W := \overline{E_p}^{\sigma(l^\infty, l^1)} = \left\{ \sum_{n=1}^\infty \alpha_n \eta_n : \text{ each } \alpha_n \geq 0, \ \alpha_1 \geq \frac{1}{2^{p+1}} \text{ and } \sum_{n=1}^\infty \alpha_n \leq 1 \right\}$$

**Case A.1:** $z \in E_p$.

Then, we have $s(Tz) = \|Tz - z\|$.

Also,
$$
\begin{aligned}
s(Tz) &= \limsup_m \left\| Tz - \frac{1}{m} \sum_{k=1}^m x^{(k)} \right\| \\
&\leq \limsup_m \left\| Tz - T\left( \frac{1}{m} \sum_{k=1}^m x^{(k)} \right) \right\| \\
&\quad + \limsup_m \left\| \frac{1}{m} \sum_{k=1}^m x^{(k)} - T\left( \frac{1}{m} \sum_{k=1}^m x^{(k)} \right) \right\|.
\end{aligned}
$$

Then, since $T$ is affine,
$$
\begin{aligned}
s(Tz) &\leq \limsup_m \left\| Tz - T\left( \frac{1}{m} \sum_{k=1}^m x^{(k)} \right) \right\| + \limsup_m \left\| \frac{1}{m} \sum_{k=1}^m x^{(k)} - \frac{1}{m} \sum_{k=1}^m Tx^{(k)} \right\| \\
&\leq \limsup_m \left\| z - \frac{1}{m} \sum_{k=1}^m x^{(k)} \right\| \\
&= s(z).
\end{aligned}
$$

Therefore, $\|z - Tz\| \leq 0$ and so $Tz = z$.

**Case A.2:** $z \in W \setminus E_p$.

Then, $z$ is of the form
$$\sum_{n=1}^\infty \gamma_n \eta_n$$

such that
$$\gamma_n \geq 0, \ \forall n \in \mathbb{N}, \ \gamma_1 \geq \frac{1}{2^{p+1}} \text{ and } \sum_{n=1}^\infty \gamma_n < 1.$$

Define $\delta := 1 - \sum_{n=1}^\infty \gamma_n$ and next define $h_\lambda := (\gamma_1 + \lambda\delta)\eta_1 + (\gamma_2 + (1 - \lambda)\delta)\eta_2 + \sum_{n=3}^\infty \gamma_n \eta_n$.

We want $h_\lambda$ to be in $E_p$, so we restrict values of $\lambda$ to be in $\left[ -\frac{\gamma_1}{\delta}, \frac{\gamma_2}{\delta} + 1 \right]$, then

$$\|h_\lambda - z\| = \max \begin{cases} \begin{aligned} &\max\{b\delta,\ (1-\lambda)\delta\} + Q_1|1-\alpha|b\delta \\ &+Q_2|(1-\lambda)\delta - \alpha b\delta| \\ &+(1-Q_1-Q_2)\alpha b\delta, \end{aligned} \\[2mm] \begin{aligned} &\max\{b\delta,\ (1-\lambda)\delta\} + Q_1|b\delta - \alpha(1-\lambda)\delta| \\ &+Q_2|1-\alpha||1-\lambda|\delta \\ &+(1-Q_1-Q_2)\alpha|1-\lambda|\delta, \end{aligned} \\[2mm] \max\{b\delta,\ (1-\lambda)\delta\} + Q_1 b\delta + Q_2|1-\lambda|\delta \end{cases}.$$

Then using the hypothesis and considering solely the case $\lambda > 1 - \frac{b}{\alpha}$, we would have $b > 1 - \lambda$ and $\alpha b > 1 - \lambda$ and so

$$\begin{aligned} \|h_\lambda - z\| &= b\delta + Q_1|1-\alpha|b\delta + Q_2|(1-\lambda)\delta - \alpha b\delta| + (1-Q_1-Q_2)\alpha b\delta \\ &= (1+\alpha-Q_1)\delta b - Q_2(1-\lambda)\delta. \end{aligned}$$

Define $\Gamma := \min\limits_{\lambda \in \left[1-\frac{b}{\alpha},\ \frac{\gamma_2}{\delta}+1\right]} \|h_\lambda - z\|$. Then, $\Gamma = (1+\alpha-Q_1-Q_2\frac{1}{\alpha})\delta b$.

Also, define for each $j \in \mathbb{N}$,

$$\|x\|_{(j)} = \|x\|_\infty + \sum_{k=1}^\infty Q_k \left|\xi_k - \alpha\xi_j\right|. \text{ Then, } \|x\| = \sup_{j\in\mathbb{N}} \|x\|_{(j)}.$$

Fix $y \in E_p$ of the form $\sum_{n=1}^\infty t_n\eta_n$ such that $t_n \geq 0,\ \forall n \in \mathbb{N},\ t_1 \geq \frac{1}{2^{p+1}}$ and $\sum_{n=1}^\infty t_n = 1$.

Then, using the definition of the equivalent norm, we obtain that

$$\begin{aligned} \|y-z\|_{(1)} \geq\ & |\gamma_1 + \delta - t_1 + (t_1-\gamma_1)b| \\ &+Q_1\left|(t_1-\gamma_1)b + \sum_{k=2}^\infty(t_k-\gamma_k) - \alpha\sum_{k=2}^\infty(t_k-\gamma_k) - \alpha(t_1-\gamma_1)b\right| \\ &+Q_2\left|\sum_{k=2}^\infty(t_k-\gamma_k) - \alpha\sum_{k=2}^\infty(t_k-\gamma_k) - \alpha(t_1-\gamma_1)b\right| \\ &+Q_3\left|\sum_{k=3}^\infty(t_k-\gamma_k) - \alpha\sum_{k=2}^\infty(t_k-\gamma_k) - \alpha(t_1-\gamma_1)b\right| \\ &+Q_4\left|\sum_{k=4}^\infty(t_k-\gamma_k) - \alpha\sum_{k=2}^\infty(t_k-\gamma_k) - \alpha(t_1-\gamma_1)b\right| \\ &+\dots, \end{aligned}$$

$$\|y - z\|_{(2)} \geq |\gamma_1 + \delta - t_1 + (t_1 - \gamma_1)b|$$

$$+Q_1 \left| (t_1 - \gamma_1)b + \sum_{k=2}^{\infty}(t_k - \gamma_k) - \alpha \sum_{k=2}^{\infty}(t_k - \gamma_k) \right|$$

$$+Q_2 \left| \sum_{k=2}^{\infty}(t_k - \gamma_k) - \alpha \sum_{k=2}^{\infty}(t_k - \gamma_k) \right|$$

$$+Q_3 \left| \sum_{k=3}^{\infty}(t_k - \gamma_k) - \alpha \sum_{k=2}^{\infty}(t_k - \gamma_k) \right|$$

$$+Q_4 \left| \sum_{k=4}^{\infty}(t_k - \gamma_k) - \alpha \sum_{k=2}^{\infty}(t_k - \gamma_k) \right|$$

$$+ \dots .$$

Hence,

$$\|y - z\| \geq \|y - z\|_{(1)} \geq |\delta + (\gamma_1 - t_1)(1 - b)|$$

$$+ \left| \begin{array}{c} Q_1(t_1 - \gamma_1)b + Q_1 \sum\limits_{k=2}^{\infty}(t_k - \gamma_k) + Q_2 \sum\limits_{k=2}^{\infty}(t_k - \gamma_k) \\ +Q_3 \sum\limits_{k=3}^{\infty}(t_k - \gamma_k) + Q_4 \sum\limits_{k=4}^{\infty}(t_k - \gamma_k) + \cdots \\ -\alpha \sum\limits_{j=1}^{\infty}Q_j \left[ \sum\limits_{k=2}^{\infty}(t_k - \gamma_k) + (t_1 - \gamma_1)b \right] \end{array} \right|$$

$$\geq |\delta + (\gamma_1 - t_1)(1 - b)|(1 + \alpha - Q_1) - (1 - Q_1)|\delta - (t_1 - \gamma_1)|$$
$$-(1 - Q_1 - Q_2)(2 - \delta - t_1 - \gamma_1).$$

**Subcase A.2.1:** Let $\delta + \gamma_1 - t_1 \geq 0$.
Then,

$$\|y - z\| - \Gamma \geq [\delta + (\gamma_1 - t_1)(1 - b)](1 + \alpha - Q_1) - (1 - Q_1)[\delta + (\gamma_1 - t_1)]$$

$$-(1 - Q_1 - Q_2)(2 - \delta - t_1 - \gamma_1) - (1 + \alpha - Q_1 - Q_2\frac{1}{\alpha})b\delta$$

Here we should note that $2 - \delta - t_1 - \gamma_1 \geq 0$ and $1 - Q_1 - Q_2 > 0$.

**Subcase A.2.1.1:** Let $\delta < t_1$ and assume $\alpha \leq \frac{(1 - Q_1)(1+b)}{1-b}$.
Then, $2(1 - Q_1) - (1 - b)(1 + \alpha - Q_1) \geq 0$ and so

$$\|y - z\| - \Gamma \geq \delta(1 - b)(1 + \alpha - Q_1) + \gamma_1(1 - b)(1 + \alpha - Q_1) - 2(1 - Q_1)$$

$$+2Q_2 \left( 1 + \frac{b}{2\alpha} \right) + [2(1 - Q_1) - (1 - b)(1 + \alpha - Q_1)]t_1$$

$$\geq 2Q_2\frac{2\alpha + b}{2\alpha} - 2(1 - Q_1) \left( 1 - \frac{1}{2^{p+1}} \right)$$

$$\geq \frac{2b + 4}{1 + 2\alpha} \left[ Q_2 \left( 1 - \frac{1}{2^{p+1}} \right) \frac{1}{3b} \right]$$

$$\geq 0.$$

due to choice of $p$ from (♣) at the beginning of the proof.

**Subcase A.2.1.2:** Let $\delta \geq t_1$.

Assume that $\alpha \geq \dfrac{-(b^2-8b+7)+\sqrt{(b^2-8b+7)^2+2^{p+5}(1-b)(1+b)^2}}{4(1-b^2)}$.

Then, noting (♣♣) at the beginning of the proof, we have

$$
\begin{aligned}
\|y-z\| - \Gamma \;\geq\; & (\delta - t_1)(1-b)(1+\alpha - Q_1) + (1+\alpha - Q_1)\gamma_1(1-b) \\
& + 2(1-Q_1)t_1 - 2(1-Q_1) \\
\geq\; & \frac{(1-b)(3-b)}{2^{p+1}(1+b)} + \alpha \frac{1-b}{2^{p+1}} - \frac{2}{1+2\alpha} \\
\geq\; & \frac{1}{1+2\alpha}\left[2(1-b^2)\alpha^2 + (b^2 - 8b + 7)\alpha - 2^{p+2}(1+b)\right] \\
\geq\; & 0
\end{aligned}
$$

**Subcase A.2.2:** Let $\delta + \gamma_1 - t_1 < 0$. Then, $\delta < |t_1 - \gamma_1|$.

Consider the assumption for $\alpha$ in Subcase A.2.1 and those in (♣♣). Then, $\alpha > \frac{1}{b}$.

Now, since we have

$$
\begin{aligned}
\|y-z\| \;\geq\; & \frac{\|y-z\|_{(1)} + \|y-z\|_{(2)}}{2} \\
\geq\; & |\gamma_1 + \delta - t_1 + (t_1 - \gamma_1)b| + \frac{\alpha|t_1 - \gamma_1|b}{2}.
\end{aligned}
$$

Hence,

$$
\begin{aligned}
\|y-z\| - \Gamma \;\geq\; & \left(\frac{\alpha b}{2} - (1-b)\right)|t_1 - \gamma_1| - (b(1+\alpha - Q_1) - 1)\delta \\
\geq\; & \left[\frac{\alpha b}{2} - (1-b) - b(1+\alpha - Q_1) + 1\right]|t_1 - \gamma_1| \\
\geq\; & 0.
\end{aligned}
$$

In conclusion, from all cases, we see that there exist $b \in (0,1)$ and $\dfrac{-(b^2-8b+7)+\sqrt{(b^2-8b+7)^2+2^{p+5}(1-b)(1+b)^2}}{4(1-b^2)} \leq \alpha \leq \dfrac{(1-Q_1)(1+b)}{1-b}$ such that if $\lambda$ is chosen to be in $\left[1 - \frac{b}{\alpha}, \frac{\gamma_2}{\delta} + 1\right]$, for any $y \in E_p$ and for $z \in W \setminus E_p$, $\|y-z\| \geq \Gamma$ where

$$
\Gamma := \min_{\lambda \in \left[1 - \frac{b}{\alpha}, \frac{\gamma_2}{\delta} + 1\right]} \|h_\lambda - z\|.
$$

Then, there exists unique $h_{\lambda_0}$ with $\lambda_0 \in \left[1 - \frac{b}{\alpha}, \frac{\gamma_2}{\delta} + 1\right]$ such that $\|h_{\lambda_0} - z\|$ is minimizer of $\Gamma$.

Now, define a subset in our set by

$$
\Lambda := \{y \;:\; \|y-z\| \leq \Gamma\}.
$$

Note that $\Lambda \subseteq E_p$ is a nonempty compact convex subset such that for any

$h \in \Lambda$,

$$
\begin{aligned}
s\,(Th) \;&=\; \limsup_{m}\left\|Th - \frac{1}{m}\sum_{k=1}^{m}x^{(k)}\right\| \\[2mm]
&\leq\; \limsup_{m}\left\|Th - T\left(\frac{1}{m}\sum_{k=1}^{m}x^{(k)}\right)\right\| \\[2mm]
&\quad +\limsup_{m}\left\|\frac{1}{m}\sum_{k=1}^{m}x^{(k)} - T\left(\frac{1}{m}\sum_{k=1}^{m}x^{(k)}\right)\right\| \\[1mm]
&\quad \text{(since } T \text{ is affine)} \\[2mm]
&=\; \limsup_{m}\left\|Th - T\left(\frac{1}{m}\sum_{k=1}^{m}x^{(k)}\right)\right\| \\[2mm]
&\quad +\limsup_{m}\left\|\frac{1}{m}\sum_{k=1}^{m}x^{(k)} - \frac{1}{m}\sum_{k=1}^{m}Tx^{(k)}\right\| \\[2mm]
&\leq\; \limsup_{m}\left\|h - \frac{1}{m}\sum_{k=1}^{m}x^{(k)}\right\| \\[2mm]
&=\; s(h).
\end{aligned}
$$

Also, $s(Th) = \|z - Th\|$ and $s(h) = \|z - h\|$. Hence,

$$
\|z - Th\| \leq \|z - h\| \;\;\Longrightarrow\;\; \|z - Th\| = \|z - h\|
$$
$$
\Longrightarrow\;\; Th \in \Lambda.
$$

Therefore, $T(\Lambda) \subseteq \Lambda$ and since $T$ is continuous, Brouwer's Fixed Point Theorem Brouwer (1911) tells us that $T$ has a fixed point such that $h = h_{\lambda_0}$ is the unique minimizer of $\|y - z\| \; : \; y \in E_p$ and $Th = h$.

Hence, $E_p$ has FPP (n.e.) as desired.

**Case B:** We will be working on the set $E_2$.

Let $T : E_2 \to E_2$ be an affine nonexpansive mapping. Then, there exists a sequence $\left(x^{(n)}\right)_{n \in \mathbb{N}} \in E_2$ such that $\left\|Tx^{(n)} - x^{(n)}\right\| \underset{n}{\to} 0$ and so $\left\|Tx^{(n)} - x^{(n)}\right\|_{\infty} \underset{n}{\to} 0$. Without loss of generality, passing to a subsequence if necessary, there exists $z \in c_0$ such that $x^{(n)}$ converges to $z$ in weak topology. Then, by Lemma 15.4.7, we can define a function $s : c_0 \to [0, \infty)$ by

$$
s\,(y) = \limsup_{m}\left\|\frac{1}{m}\sum_{k=1}^{m}x^{(k)} - y\right\|, \quad \forall y \in c_0\,.
$$

Then,

$$
s\,(y) = \|y - z\|, \quad \forall y \in c_0.
$$

Define

$$
W^{\sim} := \overline{E_2}^{\,\sigma(l^{\infty},\,l^{1})} = \left\{\sum_{n=1}^{\infty}\alpha_n\,\eta_n \;:\; \text{each } \alpha_n \geq 0,\; \alpha_1 = 0 \text{ and } \sum_{n=1}^{\infty}\alpha_n \leq 1\right\}
$$

**Case B.1:** This case is similar to Case A.1.

**Case B.2:** $z \in W^{\sim} \setminus E_2$. This case is same as Case A.2 up to calculation of $\|y - z\|$.

Then, $z$ is of the form

$$\sum_{n=1}^{\infty} \gamma_n \eta_n$$

such that

$$\gamma_n \geq 0, \ \forall n \in \mathbb{N}, \ \gamma_1 = 0 \text{ and } \sum_{n=1}^{\infty} \gamma_n < 1.$$

Define $\delta := 1 - \sum_{n=1}^{\infty} \gamma_n$ and next define $h_\lambda := (\gamma_1 + \lambda \delta)\eta_1 + (\gamma_2 + (1 - \lambda)\delta)\eta_2 + \sum_{n=3}^{\infty} \gamma_n \eta_n$. We want $h_\lambda$ to be in $E_2$, so we restrict values of $\lambda$ to be in $\left[ -\frac{\gamma_1}{\delta}, \frac{\gamma_2}{\delta} + 1 \right]$, then define $\Gamma := \min_{\lambda \in \left[ 1 - \frac{b}{\alpha}, \frac{\gamma_2}{\delta} + 1 \right]} \|h_\lambda - z\|$. Then, $\Gamma = (1 + \alpha - Q_1 - Q_2 \frac{1}{\alpha})\delta b$.

Fix $y \in E_2$ of the form $\sum_{n=1}^{\infty} t_n \eta_n$ such that $t_n \geq 0, \ \forall n \in \mathbb{N}, \ t_1 = 0$ and $\sum_{n=1}^{\infty} t_n = 1$.

Assume $\alpha \geq \frac{2 - b + \sqrt{(2+b)^2 - 8b^2}}{4b}$. Then, as in Subcase A.2.2, (considering $\|x\|_{(1)} = \|x\|_{(2)}, \forall x \in E_2$)

$$\|y - z\| \geq \frac{\|y - z\|_{(2)} + \|y - z\|_{(3)}}{2}$$

$$\geq \delta + \frac{\alpha|t_2 - \gamma_2|}{2}$$

Hence,

$$\|y - z\| - \Gamma \geq \left[ 1 + Q_1 b + Q_2 \frac{b}{\alpha} - b - b\alpha \right] \delta$$

$$\geq (1 + 2\alpha + b - b\alpha - 2b\alpha^2) \frac{\delta}{1 + 2\alpha}$$

$$\geq 0$$

Here we should note that if $\frac{-(b^2 - 8b + 7) + \sqrt{(b^2 - 8b + 7)^2 + 2^{p+5}(1-b)(1+b)^2}}{4(1-b^2)} \leq \alpha \leq \frac{(1-Q_1)(1+b)}{1-b}$, then $\alpha \geq \frac{2 - b + \sqrt{(2+b)^2 - 8b^2}}{4b}$ when $b \leq \frac{14}{15}$.

Then, we conclude the remaining parts of the Case B same as Case A.

In conclusion, sets $E_1$ and $E_2$ have the fixed point property for affine $\| \cdot \|$-nonexpansive mappings for some $\alpha > 1$ and $b \in (0, 1)$. Then, if $T : E \longrightarrow E$ is an affine $\| \cdot \|$-nonexpansive mapping, if $T(E) \subseteq E_1$, then we consider the restricted mapping defined from $E_1$ to $E_1$ and say that there exists a fixed point in $E_1$ and so in $E$ and if $T(E) \subseteq E_2$, then we consider the restricted mapping defined from $E_2$ to $E_2$ and say that there exists a fixed point in $E_2$ and so in $E$; thus the set $E$ has the fixed point property for affine $\| \cdot \|$-nonexpansive mappings as desired.    □

Now, we consider another set.

**Example 15.2.5.** *Fix $b \in (0,1)$. We define the sequence $(f_n)_{n \in \mathbb{N}}$ in $c_0$ by setting $f_1 := b\,e_1$, $f_2 := b\,e_2$ and $f_n := e_n$, for all integers $n \geq 3$.*

*Let us define the sequence $(\eta_n)_{n \in \mathbb{N}}$ in $E$ in the following way. Let $\eta_1 := f_1$ and $\eta_n := f_1 + \cdots + f_n$, for all integers $n \geq 2$. Next, define the closed, bounded, convex subset $E = E_b$ of $c_0$ by*

$$E := \left\{ \sum_{n=1}^{\infty} \alpha_n \, \eta_n : \text{ each } \alpha_n \geq 0 \text{ and } \sum_{n-1}^{\infty} \alpha_n = 1 \right\}.$$

**Theorem 15.2.6.** *There exist $b \in (0, 1)$ and $\alpha > 1$ such that the set $E$ defined as in the example above has the fixed point property for $\|.\|$-nonexpansive affine mappings when $Q_1 > \frac{2\alpha}{1+2\alpha}$.*

*Proof.* As in the previous theorem, first of all, recall that we need the condition $Q_1 > \frac{2\alpha}{1+2\alpha}$ to eliminate the possibility of having an a.i. copy of $c_0$ inside. Let $b \in (0,1)$ and $Q_1 < \frac{2b}{1+b}$.

Here, we will consider three different sets:

$$E_1 := \left\{ \sum_{n=1}^{\infty} \alpha_n \, \eta_n : \text{ each } \alpha_n \geq 0, \; \alpha_1 > 0 \text{ and } \sum_{n=1}^{\infty} \alpha_n = 1 \right\},$$

$$E_2 := \left\{ \sum_{n=1}^{\infty} \alpha_n \, \eta_n : \text{ each } \alpha_n \geq 0, \; \alpha_1 = 0, \; \alpha_2 > 0 \text{ and } \sum_{n=1}^{\infty} \alpha_n = 1 \right\}$$

and

$$E_3 := \left\{ \sum_{n=1}^{\infty} \alpha_n \, \eta_n : \text{ each } \alpha_n \geq 0, \; \alpha_1 = 0, \; \alpha_2 = 0 \text{ and } \sum_{n=1}^{\infty} \alpha_n = 1 \right\}$$

noting $E = E_1 \cup E_2 \cup E_3$. Now, provided that $Q_2 < \frac{1}{1+2\alpha}$, choose $p \in \mathbb{N}$ so that $Q_2 \geq \left(1 - \frac{1}{2^{p+1}}\right) \frac{1}{3b}$ and

$$E_2 = E_p := \left\{ \sum_{n=1}^{\infty} \alpha_n \, \eta_n : \begin{array}{l} \text{each } \alpha_n \geq 0, \; \alpha_1 = 0, \\ \alpha_2 \geq \frac{1}{2^{p+1}} \text{ and } \sum_{n=1}^{\infty} \alpha_n = 1 \end{array} \right\},$$

$$E_1 = E_{p'} := \left\{ \sum_{n=1}^{\infty} \alpha_n \, \eta_n : \text{ each } \alpha_n \geq 0, \; \alpha_1 \geq \frac{1}{2^{p+1}} \text{ and } \sum_{n=1}^{\infty} \alpha_n = 1 \right\}$$

(♣♣♣) and next choose $b \in (0,1)$ so that $\frac{1-b^2}{b(3b^2-6b+7)} < \frac{1}{2^{p+2}}$. Then, we obtain that $b > \frac{3+\sqrt{20}}{11} > \frac{\sqrt{5}-1}{2}$ which will help us for further steps. (♣♣♣♣)

**Case A:** We will be working on the set $E_2 = E_p$ similar to Case A of Theorem 15.2.4.

Define

$$
\begin{aligned}
W : &= \overline{E_2}^{\sigma(l^\infty, l^1)} \\
&= \left\{ \sum_{n=1}^\infty \alpha_n \, \eta_n : \text{ each } \alpha_n \geq 0, \ \alpha_1 = 0, \ \alpha_2 \geq \frac{1}{2^{p+1}} \text{ and } \sum_{n=1}^\infty \alpha_n \leq 1 \right\}
\end{aligned}
$$

**Case A.1:** $z \in E_2$. Then, this case will be same as Case A.1 of Theorem 15.2.4.

**Case A.2:** $z \in W \setminus E_2$. This case will be also similar to Case A.2 of Theorem 15.2.4 but the following changes will occur:

First of all, assume $\alpha < 2Q_1 + 2Q_2$.

$z$ is of the form

$$
\sum_{n=1}^\infty \gamma_n \eta_n
$$

such that

$$
\gamma_n \geq 0, \ \forall n \in \mathbb{N}, \ \gamma_1 = 0,
$$

$$
\gamma_2 \geq \frac{1}{2^{p+1}} \text{ and } \sum_{n=1}^\infty \gamma_n < 1.
$$

Define $\delta := 1 - \sum_{n=1}^\infty \gamma_n$ and next define

$$
h_\lambda := (\gamma_1 + \lambda\delta)\eta_1 + (\gamma_2 + (1-\lambda)\delta)\eta_2 + \sum_{n=3}^\infty \gamma_n\eta_n.
$$

We want $h_\lambda$ to be in $E_2$, so we restrict values of $\lambda$ to be in $\left[-\frac{\gamma_1}{\delta}, \frac{\gamma_2}{\delta} + 1\right]$.

Fix $y \in E_2$ of the form $\sum_{n=1}^\infty t_n \eta_n$ such that $t_n \geq 0, \ \forall n \in \mathbb{N}, \ t_1 = 0, \ t_2 \geq \frac{1}{2^{p+1}}$ and $\sum_{n=1}^\infty t_n = 1$. Then,

$$
\|h_\lambda - z\| = \max \left\{
\begin{array}{l}
b\delta + Q_1|1 - \alpha|b\delta \\
+Q_2|(1-\lambda)b\delta - \alpha b\delta| \\
+(1 - Q_1 - Q_2)\alpha b\delta, \\[1em]
b\delta + Q_1|b\delta - \alpha(1-\lambda)b\delta| \\
+Q_2|1 - \alpha||1 - \lambda|b\delta \\
+(1 - Q_1 - Q_2)\alpha|1 - \lambda|b\delta, \\[1em]
b\delta + Q_1 b\delta + Q_2|1 - \lambda|b\delta
\end{array}
\right\}.
$$

Define

$$
\Gamma := \min_{\lambda \in \left[-\frac{\gamma_1}{\delta}, \frac{\gamma_2}{\delta}+1\right]} \|h_\lambda - z\|.
$$

Hence, $\Gamma = (1 + \alpha - Q_1 - Q_2)b\delta$ in the interval $[0, 1]$.

Then, replace each $t_j$ with $t_{j+1}$ and $\gamma_j$ with $\gamma_{j+1}$ in the Case A of Theorem 15.2.4 and use similar subcases to complete the Case A since $\|y - z\| - (1 + \alpha - Q_1 - Q_2)b\delta \geq \|y - z\| - (1 + \alpha - Q_1 - Q_2\frac{1}{\alpha})b\delta$ which yields $\|y - z\| - (1 + \alpha - Q_1 - Q_2)b\delta \geq 0$.

**Case B:** We will be working on the set $E_1$ similar to Case A of Theorem 15.2.4.

Define

$$W^{\sim} := \overline{E_1}^{\sigma(l^\infty, l^1)} = \left\{ \sum_{n=1}^{\infty} \alpha_n \eta_n : \text{ each } \alpha_n \geq 0, \ \alpha_1 \geq \frac{1}{2^{p+1}} \text{ and } \sum_{n=1}^{\infty} \alpha_n \leq 1 \right\}$$

**Case B.1:** $z \in E_1$. Then, this case will be same as Case A.1 of Theorem 15.2.4.

**Case B.2:** $z \in W^{\sim} \setminus E_1$. This case will be also similar to Case A.2 but the following changes will occur:

$z$ is of the form $\sum_{n=1}^{\infty} \gamma_n \eta_n$ such that $\gamma_n \geq 0$, $\forall n \in \mathbb{N}$, $\gamma_1 \geq \frac{1}{2^{p+1}}$ and $\sum_{n=1}^{\infty} \gamma_n < 1$. Define $\delta := 1 - \sum_{n=1}^{\infty} \gamma_n$ and next define $h_\lambda := (\gamma_1 + \lambda\delta)\eta_1 + (\gamma_2 + (1 - \lambda)\delta)\eta_2 + \sum_{n=3}^{\infty} \gamma_n \eta_n$. We want $h_\lambda$ to be in $E_{p'}$, so we restrict values of $\lambda$ to be in $\left[-\frac{\gamma_1}{\delta}, \frac{\gamma_2}{\delta} + 1\right]$.

Fix $y \in E_{p'}$ of the form $\sum_{n=1}^{\infty} t_n \eta_n$ such that $t_n \geq 0$, $\forall n \in \mathbb{N}$, $t_1 \geq \frac{1}{2^{p+1}}$ and $\sum_{n=1}^{\infty} t_n = 1$.

Then, using the definition of the equivalent norm, we obtain that

$$\|y - z\|_{(1)} \geq \left| \gamma_1 + \gamma_2 + \delta - t_1 - t_2 + (t_1 - \gamma_1)b + (t_2 - \gamma_2)b \right.$$
$$+ \left| \begin{array}{l} Q_1(t_1 - \gamma_1)b + Q_1(t_2 - \gamma_2)b + Q_2(t_2 - \gamma_2)b \\ + Q_1 \sum_{k=3}^{\infty}(t_k - \gamma_k) + Q_2 \sum_{k=3}^{\infty}(t_k - \gamma_k) \\ + Q_3 \sum_{k=3}^{\infty}(t_k - \gamma_k) + Q_4 \sum_{k=4}^{\infty}(t_k - \gamma_k) + \cdots \\ - \alpha \sum_{j=1}^{\infty} Q_j \left[ \sum_{k=3}^{\infty}(t_k - \gamma_k) + (t_1 - \gamma_1)b + (t_2 - \gamma_2)b \right] \end{array} \right|$$
$$= \left| \gamma_1 + \gamma_2 + \delta - t_1 - t_2 + (t_1 - \gamma_1)b + (t_2 - \gamma_2)b \right.$$
$$+ \left| \begin{array}{l} (Q_1 - \alpha)(\delta + (\gamma_1 - t_1 + \gamma_2 - t_2)(1 - b)) + Q_2(t_2 - \gamma_2)b \\ + Q_2 \sum_{k=3}^{\infty}(t_k - \gamma_k) + Q_3 \sum_{k=3}^{\infty}(t_k - \gamma_k) + Q_4 \sum_{k=4}^{\infty}(t_k - \gamma_k) + \cdots \end{array} \right|,$$

$$\|y - z\|_{(2)} \geq |\gamma_1 + \gamma_2 + \delta - t_1 - t_2 + (t_1 - \gamma_1)b + (t_2 - \gamma_2)b|$$

$$+ \left| \begin{array}{l} Q_1(t_1 - \gamma_1)b + (Q_1 + Q_2)(t_2 - \gamma_2)b + Q_1 \sum_{k=3}^{\infty} (t_k - \gamma_k) \\ + Q_2 \sum_{k=3}^{\infty} (t_k - \gamma_k) + Q_3 \sum_{k=3}^{\infty} (t_k - \gamma_k) + Q_4 \sum_{k=4}^{\infty} (t_k - \gamma_k) + \cdots \\ - \alpha \sum_{j=1}^{\infty} Q_j \left[ \sum_{k=3}^{\infty} (t_k - \gamma_k) + (t_2 - \gamma_2)b \right] \end{array} \right|$$

$$= |\gamma_1 + \gamma_2 + \delta - t_1 - t_2 + (t_1 - \gamma_1)b + (t_2 - \gamma_2)b|$$

$$+ \left| \begin{array}{l} (Q_1 - \alpha)(\delta + (\gamma_1 - t_1 + \gamma_2 - t_2)(1 - b)) + \alpha(t_1 - \gamma_1)b \\ + Q_2(t_2 - \gamma_2)b + Q_2 \sum_{k=3}^{\infty} (t_k - \gamma_k) + Q_3 \sum_{k=3}^{\infty} (t_k - \gamma_k) \\ + Q_4 \sum_{k=4}^{\infty} (t_k - \gamma_k) + \cdots \end{array} \right|$$

and similarly

$$\|y - z\|_{(3)} \geq |\gamma_1 + \gamma_2 + \delta - t_1 - t_2 + (t_1 - \gamma_1)b + (t_2 - \gamma_2)b|$$

$$+ \left| \begin{array}{l} (Q_1 - \alpha)(\delta + (\gamma_1 - t_1 + \gamma_2 - t_2)(1 - b)) + \alpha(t_1 - \gamma_1)b \\ + \alpha(t_2 - \gamma_2)b + Q_2(t_2 - \gamma_2)b + Q_2 \sum_{k=3}^{\infty} (t_k - \gamma_k) \\ + Q_3 \sum_{k=3}^{\infty} (t_k - \gamma_k) + Q_4 \sum_{k=4}^{\infty} (t_k - \gamma_k) + \cdots \end{array} \right|$$

Hence,

$$\|y - z\| \geq \|y - z\|_{(1)}$$

$$\geq |\delta + (\gamma_1 - t_1 + \gamma_2 - t_2)(1 - b)|(1 + \alpha - Q_1) - Q_2|t_2 - \gamma_2|b$$

$$- \sum_{k=2}^{\infty} Q_k \left| \sum_{j=3}^{\infty} (t_j - \gamma_j) \right| - \sum_{k=4}^{\infty} Q_k |t_3 - \gamma_3|$$

$$- \sum_{k=5}^{\infty} Q_k |t_4 - \gamma_4| - \sum_{k=6}^{\infty} Q_k |t_5 - \gamma_5| - \cdots$$

$$\geq |\delta + (\gamma_1 - t_1 + \gamma_2 - t_2)(1 - b)|(1 + \alpha - Q_1) - Q_2|t_2 - \gamma_2|b$$

$$- (1 - Q_1)|\delta + (\gamma_1 - t_1 + \gamma_2 - t_2)| - \sum_{k=4}^{\infty} Q_k |t_3 - \gamma_3|$$

$$- \sum_{k=4}^{\infty} Q_k |t_4 - \gamma_4| - \sum_{k=4}^{\infty} Q_k |t_5 - \gamma_5| - \cdots.$$

Thus,

$$
\begin{aligned}
\|y - z\| &\geq |\delta + (\gamma_1 - t_1 + \gamma_2 - t_2)(1 - b)|(1 + \alpha - Q_1) - Q_2|t_2 - \gamma_2|b \\
&\quad -(1 - Q_1)|\delta + (\gamma_1 - t_1 + \gamma_2 - t_2)| - \sum_{k=4}^{\infty} Q_k \sum_{j=3}^{\infty} |t_j - \gamma_j| \\
&\geq |\delta + (\gamma_1 - t_1 + \gamma_2 - t_2)(1 - b)|(1 + \alpha - Q_1) - Q_2|t_2 - \gamma_2|b \\
&\quad -(1 - Q_1)|\delta + (\gamma_1 - t_1 + \gamma_2 - t_2)| \\
&\quad -(1 - Q_1 - Q_2 - Q_3)\sum_{j=3}^{\infty} |t_j - \gamma_j| \\
&\geq |\delta + (\gamma_1 - t_1 + \gamma_2 - t_2)(1 - b)|(1 + \alpha - Q_1) \\
&\quad -(1 - Q_1)|\delta + (\gamma_1 - t_1 + \gamma_2 - t_2)| \\
&\quad -(1 - Q_1 - Q_3)(2 - \delta - t_1 - t_2 - \gamma_1 - \gamma_2) \\
&\quad +Q_2(2 - \delta - t_1 - \gamma_1).
\end{aligned}
$$

**Subcase B.2.1:** Let $\delta + \gamma_1 - t_1 + \gamma_2 - t_2 > 0$.
Then,

$$
\begin{aligned}
\|y - z\| - \Gamma &\geq [\delta + (\gamma_1 - t_1 + \gamma_2 - t_2)(1 - b)](1 + \alpha - Q_1) \\
&\quad -(1 - Q_1)[\delta + (\gamma_1 - t_1 + \gamma_2 - t_2)(1 - b)] \\
&\quad -(1 - Q_1 - Q_3)(2 - \delta - t_1 - t_2 - \gamma_1 - \gamma_2) \\
&\quad +Q_2(2 - \delta - t_1 - \gamma_1) - (1 + \alpha - Q_1 - Q_2)b\delta.
\end{aligned}
$$

Here we should note that $2 - \delta - t_1 - \gamma_1 - t_2 - \gamma_2 \geq 0$ and $1 - Q_1 - Q_2 > 0$.

**Subcase B.2.1.1:** Let $\delta < t_1 + t_2$ and assume $\alpha \leq \frac{(1 - Q_1)(1 + b)}{1 - b}$.

Then, we will obtain the same method as the proof of Subcase A.2.1.1 in the previous theorem. Indeed, firstly, $2(1 - Q_1) - (1 - b)(1 + \alpha - Q_1) \geq 0$ and so

$$
\begin{aligned}
\|y - z\| - \Gamma &\geq \delta(1 - b)(1 + \alpha - Q_1) + \gamma_1(1 - b)(1 + \alpha - Q_1) - 2(1 - Q_1) \\
&\quad +2Q_2\left(1 + \frac{b}{2\alpha}\right) + [2(1 - Q_1) - (1 - b)(1 + \alpha - Q_1)]t_1 \\
&\geq 2Q_2\frac{2\alpha + b}{2\alpha} - 2(1 - Q_1)\left(1 - \frac{1}{2^{p+1}}\right) \\
&\geq \frac{2b + 4}{1 + 2\alpha}\left[Q_2 - \left(1 - \frac{1}{2^{p+1}}\right)\frac{1}{3b}\right] \\
&\geq 0.
\end{aligned}
$$

due to choice of $p$ from (♣♣♣) at the beginning of the proof.

**Subcase B.2.1.2:** Let $\delta \geq t_1 + t_2$. Then, this case is also completed by the same method as in Subcase A.2.1.2 in the previous theorem.

Assume that $\alpha \geq \frac{-(b^2 - 8b + 7) + \sqrt{(b^2 - 8b + 7)^2 + 2^{p+5}(1 - b)(1 + b)^2}}{4(1 - b^2)}$.

Then, noting (♣♣♣) at the beginning of the proof, we have

$$
\begin{aligned}
\|y - z\| - \Gamma \;&\geq\; (\delta - t_1 - t_2)(1 - b)(1 + \alpha - Q_1) + (1 + \alpha - Q_1)\gamma_1(1 - b) \\
&\quad + 2(1 - Q_1)t_1 - 2(1 - Q_1) \\
&\geq\; \left(\frac{1 - b}{1 + b} + \alpha\right)\frac{1 - b}{2^{p+1}} + \frac{1 - b}{2^p(1 + b)} - \frac{2}{1 + 2\alpha} \\
&=\; \frac{(1 - b)(3 - b)}{2^{p+1}(1 + b)} + \alpha\frac{1 - b}{2^{p+1}} - \frac{2}{1 + 2\alpha} \\
&\geq\; \frac{1}{1 + 2\alpha}\left[2(1 - b^2)\alpha^2 + (b^2 - 8b + 7)\alpha - 2^{p+2}(1 + b)\right] \\
&\geq\; 0
\end{aligned}
$$

**Subcase B.2.2:** Let $\delta < t_1 - \gamma_1 + t_2 - \gamma_2$ and so $\delta < |t_1 - \gamma_1 + t_2 - \gamma_2|$.

Consider the assumption for $\alpha$ in Subcase B.2.1 and those in (♣♣♣). Then, $\alpha > \frac{1}{b}$.

Now, since we have

$$
\begin{aligned}
\|y - z\| \;&\geq\; \frac{\|y - z\|_{(1)} + \|y - z\|_{(3)}}{2} \\
&\geq\; |\delta - (1 - b)(t_1 - \gamma_1 + t_2 - \gamma_2)| + \frac{\alpha|t_1 - \gamma_1 + t_2 - \gamma_2|b}{2}.
\end{aligned}
$$

Hence,

$$
\begin{aligned}
\|y - z\| - \Gamma \;&\geq\; \left(\frac{\alpha b}{2} - (1 - b)\right)|t_1 - \gamma_1 + t_2 - \gamma_2| \\
&\quad - (b(1 + \alpha - Q_1 - Q_2) - 1)\delta \\
&\geq\; \left[\frac{\alpha b}{2} - b(\alpha - Q_1 - Q_2)\right]|t_1 - \gamma_1 + t_2 - \gamma_2| \\
&\geq\; 0.
\end{aligned}
$$

In conclusion, from all cases, we see that there exist $b \in (0, 1)$ and $\frac{-(b^2 - 8b + 7) + \sqrt{(b^2 - 8b + 7)^2 + 2^{p+5}(1 - b)(1 + b)^2}}{4(1 - b^2)} \leq \alpha \leq \frac{(1 - Q_1)(1 + b)}{1 - b}$ such that if $\lambda$ is chosen to be in $\left[1 - \frac{b}{\alpha}, \frac{\gamma_2}{\delta} + 1\right]$, for any $y \in E_{p'}$ and for $z \in W^{\sim} \setminus E_{p'}$, $\|y - z\| \geq \Gamma$ where

$$
\Gamma := \min_{\lambda \in \left[1 - \frac{b}{\alpha}, \frac{\gamma_2}{\delta} + 1\right]} \|h_\lambda - z\|.
$$

Then, there exists unique $h_{\lambda_0}$ with $\lambda_0 \in \left[1 - \frac{b}{\alpha}, \frac{\gamma_2}{\delta} + 1\right]$ such that $\|h_{\lambda_0} - z\|$ is minimizer of $\Gamma$.

Now, define a subset in our set by

$$
\Lambda := \{y \,:\, \|y - z\| \leq \Gamma\}.
$$

Note that $\Lambda \subseteq E_{p'}$ is a nonempty compact convex subset such that for any

$h \in \Lambda$,

$$
\begin{aligned}
s\,(Th) &= \limsup_m \left\| Th - \frac{1}{m} \sum_{k=1}^{m} x^{(k)} \right\| \\
&\leq \limsup_m \left\| h - \frac{1}{m} \sum_{k=1}^{m} x^{(k)} \right\| \\
&= s(h).
\end{aligned}
$$

Also, $s(Th) = \|z - Th\|$ and $s(h) = \|z - h\|$. Hence,

$$
\begin{aligned}
\|z - Th\| \leq \|z - h\| &\implies \|z - Th\| = \|z - h\| \\
&\implies Th \in \Lambda.
\end{aligned}
$$

Therefore, $T(\Lambda) \subseteq \Lambda$ and since $T$ is continuous, Brouwer's Fixed Point Theorem Brouwer (1911) tells us that $T$ has a fixed point such that $h = h_{\lambda_0}$ is the unique minimizer of $\|y - z\| \; : \; y \in E_{p'}$ and $Th = h$.

Hence, $E_{p'}$ has FPP (n.e.) as desired.

**Case C:** We will be working on the set $E_3$.

Define

$$
W^{\sim\sim} := \overline{E_3}^{\,\sigma(l^\infty,\, l^1)} = \left\{ \sum_{n=1}^{\infty} \alpha_n \, \eta_n : \begin{array}{l} \text{each } \alpha_n \geq 0, \; \alpha_1 = 0, \\ \alpha_2 = 0 \text{ and } \sum_{n=1}^{\infty} \alpha_n \leq 1 \end{array} \right\}
$$

Then, replace each $t_j$ with $t_{j+1}$ and $\gamma_j$ with $\gamma_{j+1}$ in the Case B of Theorem 15.2.4 and use similar subcases together with the same assumptions for $\alpha$ for the set $W^\sim$ there to complete Case C.

Then, finally we conclude our proof similarly to the last part of the proof of Theorem 15.2.4. □

**Example 15.2.7.** *Fix $b \in (0,1)$. We define the sequence $(f_n)_{n \in \mathbb{N}}$ in $c_0$ by setting $f_1 := b\,e_1$, $f_2 := b\,e_2$, $f_3 := b\,e_3$ and $f_n := e_n$, for all integers $n \geq 4$.*

*Let us define the sequence $(\eta_n)_{n \in \mathbb{N}}$ in $E$ in the following way. Let $\eta_1 := f_1$ and $\eta_n := f_1 + \cdots + f_n$, for all integers $n \geq 2$. Next, define the closed, bounded, convex subset $E = E_b$ of $c_0$ by*

$$
E := \left\{ \sum_{n=1}^{\infty} \alpha_n \, \eta_n : \; \text{each } \alpha_n \geq 0 \text{ and } \sum_{n=1}^{\infty} \alpha_n = 1 \right\}.
$$

**Theorem 15.2.8.** *There exist $b \in (0,\,1)$ and $\alpha > 1$ such that the set $E$ defined as in the example above has the fixed point property for $\|.\|$-nonexpansive affine mappings when $Q_1 > \frac{2\alpha}{1+2\alpha}$.*

*Proof.* As in the previous theorems, first of all, recall that we need the condition $Q_1 > \frac{2\alpha}{1+2\alpha}$ to eliminate the possibility of having an a.i. copy of $c_0$ inside. Let $b \in (0,1)$, $Q_1 < \frac{2b}{1+b}$ and $\alpha < 2Q_1 + 2Q_2 + 2Q_3$.

We will use the same method as in the previous theorem and using same conditions for $\alpha$ and $b$, we will consider four different sets:

$$E_1 := \left\{ \sum_{n=1}^{\infty} \alpha_n \, \eta_n : \text{ each } \alpha_n \geq 0, \ \alpha_1 > 0 \text{ and } \sum_{n=1}^{\infty} \alpha_n = 1 \right\},$$

$$E_2 := \left\{ \sum_{n=1}^{\infty} \alpha_n \, \eta_n : \text{ each } \alpha_n \geq 0, \ \alpha_1 = 0, \ \alpha_2 > 0 \text{ and } \sum_{n=1}^{\infty} \alpha_n = 1 \right\}$$

and

$$E_3 := \left\{ \sum_{n=1}^{\infty} \alpha_n \, \eta_n : \text{ each } \alpha_n \geq 0, \ \alpha_1 = 0, \ \alpha_2 = 0 \text{ and } \sum_{n=1}^{\infty} \alpha_n = 1 \right\}$$

$$E_4 := \left\{ \sum_{n=1}^{\infty} \alpha_n \, \eta_n : \begin{array}{l} \text{each } \alpha_n \geq 0, \ \alpha_1 = 0, \ \alpha_2 = 0, \\ \alpha_3 = 0 \text{ and } \sum_{n=1}^{\infty} \alpha_n = 1 \end{array} \right\}$$

noting $E = E_1 \cup E_2 \cup E_3 \cup E_4$. Now, provided that $Q_2 < \frac{1}{1+2\alpha}$, choose $p \in \mathbb{N}$ so that $Q_2 \geq \left(1 - \frac{1}{2^{p+1}}\right) \frac{1}{3b}$ and

$$E_3 = E_{p''} := \left\{ \sum_{n=1}^{\infty} \alpha_n \, \eta_n : \begin{array}{l} \text{each } \alpha_n \geq 0, \ \alpha_1 = 0, \ \alpha_2 = 0, \\ \alpha_3 \geq \frac{1}{2^{p+1}} \text{ and } \sum_{n=1}^{\infty} \alpha_n = 1 \end{array} \right\},$$

$$E_2 = E_p := \left\{ \sum_{n=1}^{\infty} \alpha_n \, \eta_n : \begin{array}{l} \text{each } \alpha_n \geq 0, \ \alpha_1 = 0, \\ \alpha_2 \geq \frac{1}{2^{p+1}} \text{ and } \sum_{n=1}^{\infty} \alpha_n = 1 \end{array} \right\},$$

$$E_1 = E_{p'} := \left\{ \sum_{n=1}^{\infty} \alpha_n \, \eta_n : \text{ each } \alpha_n \geq 0, \ \alpha_1 \geq \frac{1}{2^{p+1}} \text{ and } \sum_{n=1}^{\infty} \alpha_n = 1 \right\}$$

(♣♣♣♣♣) and next choose $b \in (0,1)$ so that $\frac{1-b^2}{b(3b^2-6b+7)} < \frac{1}{2^{p+2}}$. Then, we obtain that $b > \frac{3+\sqrt{20}}{11} > \frac{\sqrt{5}-1}{2}$ which will help us for further steps. (♣♣♣♣♣♣)

Then, we will repeat the same argument as in the previous theorem but there will be the following change in the second cases of Theorem 15.2.6's proof for each set $E_i$ which will help us apply the same proof as the proof of the previous theorem.

Let $i \in \{1, 2, 3\}$ and $z \in \overline{E_i}^{\,\sigma(l^{\infty}, \, l^1)} \setminus E_i$.

Define $\delta := 1 - \sum_{n=1}^{\infty} \gamma_n$ and next define

$$h_\lambda := (\gamma_1 + \tfrac{\lambda}{2}\delta)\eta_1 + (\gamma_2 + \tfrac{\lambda}{2}\delta)\delta)\eta_2 + (\gamma_3 + (1-\lambda)\delta)\eta_3 + \sum_{n=4}^{\infty} \gamma_n \eta_n.$$

We want $h_\lambda$ to be in $E_i$, so we restrict values of $\lambda$ to be in $\left[-\frac{\gamma_1}{\delta}, 1\right]$. Then, one can see that technical calculations show that

$$
\|h_\lambda - z\| = \begin{cases} (1-\lambda)b\delta \\ \quad +(\alpha - Q_1 - Q_2(1-\frac{\lambda}{2}) - Q_3(1-\lambda))b\delta & \text{if } \lambda < 0 \\[2ex] \delta b \\ \quad +(\alpha - Q_1 - Q_2(1-\frac{\lambda}{2}) - Q_3(1-\lambda)b\delta & \text{if } \lambda \in [0, 1) \end{cases}
$$

Define

$$
\Gamma := \min_{\lambda \in \left[-\frac{\gamma_1}{\delta}, 1\right]} \|h_\lambda - z\|.
$$

Hence, $\Gamma = (1 + \alpha - Q_1 - Q_2 - Q_3)b\delta$. Then, imitating the proof of the previous theorem, we conclude our proof. $\qquad\square$

Now, we generalize our sets and our theorems.

**Example 15.2.9.** *Fix $b \in (0, 1)$ and $N \in \mathbb{N}$. We define the sequence $(f_n)_{n \in \mathbb{N}}$ in $c_0$ by setting $f_n := b\,e_n$ for $n \leq N$ and $f_n := e_n$, for all integers $n \geq N+1$.*

*Let us define the sequence $(\eta_n)_{n \in \mathbb{N}}$ in $E$ in the following way. Let $\eta_1 := f_1$ and $\eta_n := f_1 + \cdots + f_n$, for all integers $n \geq 2$. Next, define the closed, bounded, convex subset $E = E_b$ of $c_0$ by*

$$
E := \left\{ \sum_{n=1}^{\infty} \alpha_n \eta_n : \text{ each } \alpha_n \geq 0 \text{ and } \sum_{n=1}^{\infty} \alpha_n = 1 \right\}.
$$

**Theorem 15.2.10.** *There exist $b \in (0, 1)$ and $\alpha > 1$ such that the set $E$ defined as in the example above has the fixed point property for $\|.\|$-nonexpansive affine mappings when $Q_1 > \frac{2\alpha}{1+2\alpha}$.*

*Proof.* We repeat the arguments as in Theorem 15.2.6 and constructing $N+1$ different sets inductively, we will have the Case 1 of each set exactly same as the one in Theorem 15.2.4. For the Case 2, we will make similar assumptions to those in previous theorems again but only the following change will occur: Assume $\alpha < 2Q_1 + 2Q_2 + 2Q_3 + \cdots + 2Q_N$ and write

$$
h_\lambda := \left(\gamma_1 + \frac{\lambda}{N-1}\delta\right)\eta_1 + \left(\gamma_2 + \frac{\lambda}{N-1}\delta\right)\eta_2 + \cdots + \left(\gamma_{N-1} + \frac{\lambda}{N-1}\delta\right)\eta_{N-1}
$$

$$
+ (\gamma_N + (1-\lambda)\delta)\eta_N + \sum_{n=N+1}^{\infty} \gamma_n \eta_n.
$$

Then, define

$$
\Gamma := \min_{\lambda \in \left[-\frac{\gamma_1}{\delta}, 1\right]} \|h_\lambda - z\|
$$

and one can see that we get $\Gamma = (1 + \alpha - Q_1 - Q_2 - Q_3 - \cdots - Q_N)b\delta$.

Then, the proof is completed similarly to the proofs of Theorem 15.2.4 and Theorem 15.2.6. $\qquad\square$

## 15.3　Another renorming of $c_0$ allowing larger class of non-weakly compact, closed, bounded, and convex subsets to have FPP for affine nonexpansive mappings

We will consider the following norm.

**Definition 15.3.1.** For $x = (\xi_k)_k \in c_0$, define

$$\||x\|| : \; = \; \lim_{p \to \infty} \sup_{k \in \mathbb{N}} \gamma_k \left( \sum_{j=k}^{\infty} \frac{|\xi_j|^p}{j^2} \right)^{\frac{1}{p}} \quad \text{where } \gamma_k \uparrow_k 1, \; \gamma_k \text{ is strictly increasing.}$$

We can admit easily that the above expression is indeed an equivalent norm on $c_0$ due to the following facts:

**Lemma 15.3.2.** *For any* $x = (\xi_i)_{i \in \mathbb{N}} \in c_0$,

$$\|x\|_\infty = \lim_{p \to \infty} \left( \sum_{k=1}^{\infty} \frac{|\xi_k|^p}{k^2} \right)^{\frac{1}{p}}$$
$$= \lim_{p \to \infty} \left( \sum_{k=1}^{\infty} \frac{|\xi_k|^p}{2^k} \right)^{\frac{1}{p}}$$

*Proof.* Let $x = (\xi_i)_{i \in \mathbb{N}} \in c_0$. We will consider $x \neq (0, 0, \cdots)$ otherwise proof of the claim is clear.

Then,

$$\exists N \in \mathbb{N} \; \ni \; \|x\|_\infty = \sup_{k \in \mathbb{N}} |\xi_k| = \max_{k \in \mathbb{N}} |\xi_k| = |\xi_N|.$$

Due to power mean inequalities formula Hardy et al. (1952),

$$\|x\|_\infty \; = \; \max_{k \leq N} |\xi_k| = \max \{ |\xi_1|, |\xi_2|, \cdots, |\xi_N| \}$$

$$= \; \lim_{p \to \infty} \left( \frac{|\xi_1|^p + |\xi_2|^p + \cdots + |\xi_N|^p}{N} \right)^{\frac{1}{p}} \quad = \; \lim_{p \to \infty} \left( \sum_{k=1}^{N} \frac{|\xi_k|^p}{N} \right)^{\frac{1}{p}}.$$

Also, due to weighted power mean inequalities formula Hardy et al. (1952),

$$\|x\|_\infty \; = \; \max_{k \leq N} |\xi_k| = \max \{ |\xi_1|, |\xi_2|, \cdots, |\xi_N| \}$$

$$= \; \lim_{p \to \infty} \left( \frac{|\xi_1|^p + |\xi_2|^p + \cdots + |\xi_N|^p}{2^N} \right)^{\frac{1}{p}} \quad = \; \lim_{p \to \infty} \left( \sum_{k=1}^{N} \frac{|\xi_k|^p}{2^N} \right)^{\frac{1}{p}}.$$

**Claim 2.**

$$\|x\|_\infty = \lim_{p\to\infty}\left(\sum_{k=1}^\infty \frac{|\xi_k|^p}{k^2}\right)^{\frac{1}{p}}$$

$$= \lim_{p\to\infty}\left(\sum_{k=1}^\infty \frac{|\xi_k|^p}{2^k}\right)^{\frac{1}{p}}$$

Indeed,

$$\|x\|_\infty \le \lim_{p\to\infty}\left(\sum_{k=1}^N \frac{|\xi_k|^p}{2^k}\right)^{\frac{1}{p}}$$

$$\le \lim_{p\to\infty}\left(\sum_{k=1}^\infty \frac{|\xi_k|^p}{k^2}\right)^{\frac{1}{p}}.$$

On the other hand, $\exists s \in \mathbb{N}$ such that $|\xi_k| < \frac{1}{k^2}$, $\forall k \ge s$.
Thus,

$$\lim_{p\to\infty}\left(\sum_{k=1}^\infty \frac{|\xi_k|^p}{k^2}\right)^{\frac{1}{p}} = \lim_{p\to\infty}\left(\sum_{k=1}^{s-1} \frac{|\xi_k|^p}{k^2} + \sum_{k=s}^\infty \frac{|\xi_k|^p}{k^2}\right)^{\frac{1}{p}}$$

$$\le \lim_{p\to\infty}\left(\sum_{k=1}^{s-1} \frac{|\xi_k|^p}{k^2} + \frac{|\xi_s|^p}{s^2} + \int_s^\infty \frac{|\xi_k|^p}{k^2}\,dk\right)^{\frac{1}{p}}$$

$$= \lim_{p\to\infty}\left(\sum_{k=1}^{s} \frac{|\xi_k|^p}{k^2} + \int_s^\infty \frac{|\xi_k|^p}{k^2}\,dk\right)^{\frac{1}{p}}.$$

Hence,

$$\lim_{p\to\infty}\left(\sum_{k=1}^\infty \frac{|\xi_k|^p}{k^2}\right)^{\frac{1}{p}} \le \lim_{p\to\infty}\left(\sum_{k=1}^{s} \frac{|\xi_k|^p}{k^2} + \int_s^\infty \frac{1}{k^{2p+2}}\,dk\right)^{\frac{1}{p}}$$

$$\le \lim_{p\to\infty}\left(|\xi_N|^p \sum_{k=1}^{s} \frac{1}{k^2} - \frac{1}{(2p+1)\,s^{2p+1}}\right)^{\frac{1}{p}}$$

$$\le \lim_{p\to\infty}\left(|\xi_N|^p\left[1 + \int_1^s \frac{1}{k^2}\,dk\right] - \frac{1}{(2p+1)\,s^{2p+1}}\right)^{\frac{1}{p}}$$

$$= |\xi_N| = \|x\|_\infty.$$

Then, it is easy to see that our norm is equivalent to the usual norm on $c_0$. $\square$

**Example 15.3.3.** *We will be considering the closed convex hull of summing basis. That is, we define the sequence $(f_n)_{n\in\mathbb{N}}$ in $c_0$ by setting $f_n := e_n$, for every $n \ge 1$. Next, define the closed, bounded, convex subset $E = E_b$ of $c_0$ by*

$$E := \left\{\sum_{n=1}^\infty t_n f_n : 1 = t_1 \ge t_2 \ge \cdots \ge t_n \downarrow_n 0\right\}.$$

Let us define the sequence $(\eta_n)_{n \in \mathbb{N}}$ in $E$ in the following way. Let $\eta_1 := f_1$ and $\eta_n := f_1 + \cdots + f_n$, for every $n \geq 2$. It is straightforward to check that

$$E := \left\{ \sum_{n=1}^{\infty} \alpha_n \eta_n : \text{ each } \alpha_n \geq 0 \text{ and } \sum_{n=1}^{\infty} \alpha_n = 1 \right\}.$$

Then, it is well known that $E$ is the closed convex hull of $(\eta_n)_{n \in \mathbb{N}}$ such that right shift mapping is fixed point free affine $\|\cdot\|_\infty$-nonexpansive mapping.

**Theorem 15.3.4.** *The set $E$ defined as in the example above has the fixed point property for $\|\|\cdot\|\|$-nonexpansive affine mappings where the equivalent norm $\|\|\cdot\|\|$ on $c_0$ is given as in Definition 15.3.1.*

*Proof.* Let $T : E \to E$ be an affine nonexpansive mapping. Then, there exists a sequence $\left(x^{(n)}\right)_{n \in \mathbb{N}} \in E$ such that $\|\|Tx^{(n)} - x^{(n)}\|\| \xrightarrow[n]{} 0$ and so $\|Tx^{(n)} - x^{(n)}\|_\infty \xrightarrow[n]{} 0$. Without loss of generality, passing to a subsequence if necessary, there exists $z \in c_0$ such that $x^{(n)}$ converges to $z$ in weak topology. Then, by Lemma 15.4.7, we can define a function $s : c_0 \to [0, \infty)$ by

$$s(y) = \limsup_m \left\|\left\| \frac{1}{m} \sum_{k=1}^{m} x^{(k)} - y \right\|\right\| , \quad \forall y \in c_0 .$$

Then,

$$s(y) = \|\|y - z\|\| , \quad \forall y \in c_0.$$

Define

$$W := \overline{E}^{\sigma(l^\infty, l^1)} = \left\{ \sum_{n=1}^{\infty} \alpha_n \eta_n : \text{ each } \alpha_n \geq 0 \text{ and } \sum_{n=1}^{\infty} \alpha_n \leq 1 \right\}$$

**Case 1:** $z \in E$.
Then, we have $s(Tz) = \|\|Tz - z\|\|$.
Also,

$$s(Tz) \leq \limsup_m \left\|\left\| Tz - T\left( \frac{1}{m} \sum_{k=1}^{m} x^{(k)} \right) \right\|\right\|$$

$$+ \limsup_m \left\|\left\| \frac{1}{m} \sum_{k=1}^{m} x^{(k)} - T\left( \frac{1}{m} \sum_{k=1}^{m} x^{(k)} \right) \right\|\right\|.$$

But since $T$ is affine,

$$s(Tz) \leq \limsup_m \left\|\left\| Tz - T\left( \frac{1}{m} \sum_{k=1}^{m} x^{(k)} \right) \right\|\right\|$$

$$+ \limsup_m \left\|\left\| \frac{1}{m} \sum_{k=1}^{m} x^{(k)} - \frac{1}{m} \sum_{k=1}^{m} Tx^{(k)} \right\|\right\|.$$

Hence,

$$s(Tz) \leq \limsup_{m} \left|\left|\left| z - \frac{1}{m} \sum_{k=1}^{m} x^{(k)} \right|\right|\right|$$

$$= s(z).$$

Therefore, $|||z - Tz||| \leq 0$ and so $Tz = z$.

**Case 2:** $z \in W \setminus E$.

Then, $z$ is of the form $\sum_{n=1}^{\infty} \sigma_n \eta_n$ such that $\sum_{n=1}^{\infty} \sigma_n < 1$.

Define $\delta := 1 - \sum_{n=1}^{\infty} \sigma_n$ and define

$$h_\lambda := (\sigma_1 + \lambda\delta)\eta_1 + (\sigma_2 + (1 - \lambda)\delta)\eta_2 + \sum_{n=3}^{\infty} \sigma_n \eta_n.$$

We want $h_\lambda$ to be in $E$, so we restrict values of $\lambda$ to be in $\left[ -\frac{\sigma_1}{\delta}, \frac{\sigma_2}{\delta} + 1 \right]$, then

$$|||h_\lambda - z||| = |||\lambda\delta e_1 + (1 - \lambda)\delta(e_1 + e_2)|||$$

$$= \lim_{p \to \infty} \sup \left\{ \gamma_1 \left[ \delta^p + \frac{|1 - \lambda|^p \delta^p}{4} \right]^{\frac{1}{p}}, \frac{\gamma_2 |1 - \lambda|\delta}{4} \right\}$$

$$= \max \left\{ \gamma_1 |1 - \lambda|\delta, \frac{\gamma_2 |1 - \lambda|\delta}{4} \right\}.$$

Define

$$\Gamma := \min_{\lambda \in \left[ -\frac{\gamma_1}{\delta}, \frac{\gamma_2}{\delta} + 1 \right]} |||h_\lambda - z|||.$$

Hence, $\Gamma = 0$ and so there exists unique $h_{\lambda_0}$ with $\lambda_0 \in \left[ -\frac{\gamma_1}{\delta}, \frac{\gamma_2}{\delta} + 1 \right]$ such that $|||h_{\lambda_0} - z|||$ minimizer of $\Gamma$.

Now, define a subset in our set by

$$\Lambda := \{ y : |||y - z||| \leq \Gamma \}.$$

Note that $\Lambda \subseteq E$ is a nonempty compact convex subset such that for any $h \in \Lambda$,

$$s(Th) \leq \limsup_{m} \left|\left|\left| Th - T\left( \frac{1}{m} \sum_{k=1}^{m} x^{(k)} \right) \right|\right|\right|$$

$$+ \limsup_{m} \left|\left|\left| \frac{1}{m} \sum_{k=1}^{m} x^{(k)} - T\left( \frac{1}{m} \sum_{k=1}^{m} x^{(k)} \right) \right|\right|\right|.$$

Since $T$ is affine

$$
\begin{aligned}
s\left(Th\right) \;\leq\; & \limsup_{m} \left\lVert\!\left\lVert\!\left\lVert Th - T\left(\frac{1}{m}\sum_{k=1}^{m} x^{(k)}\right)\right\rVert\!\right\rVert\!\right\rVert \\
& + \limsup_{m} \left\lVert\!\left\lVert\!\left\lVert \frac{1}{m}\sum_{k=1}^{m} x^{(k)} - \frac{1}{m}\sum_{k=1}^{m} Tx^{(k)}\right\rVert\!\right\rVert\!\right\rVert \\
\leq\; & \limsup_{m} \left\lVert\!\left\lVert\!\left\lVert h - \frac{1}{m}\sum_{k=1}^{m} x^{(k)}\right\rVert\!\right\rVert\!\right\rVert \\
=\; & s(h).
\end{aligned}
$$

Also, $s(Th) = \lVert\!\lVert\!\lVert z - Th\rVert\!\rVert\!\rVert$ and $s(h) = \lVert\!\lVert\!\lVert z - h\rVert\!\rVert\!\rVert$. Hence, (due to uniqueness of minimum value) $\lVert\!\lVert\!\lVert z - Th\rVert\!\rVert\!\rVert \leq \lVert\!\lVert\!\lVert z - h\rVert\!\rVert\!\rVert \implies \lVert\!\lVert\!\lVert z - Th\rVert\!\rVert\!\rVert = \lVert\!\lVert\!\lVert z - h\rVert\!\rVert\!\rVert \implies Th \in \Lambda$.

Therefore, $T(\Lambda) \subseteq \Lambda$ and since $T$ is continuous, Brouwer's Fixed Point Theorem Brouwer (1911) tells us that $T$ has a fixed point such that $h = h_{\lambda_0}$ is the unique minimizer of $\lVert\!\lVert\!\lVert y - z\rVert\!\rVert\!\rVert : y \in E$ and $Th = h$.

Hence, $E$ has FPP (n.e.) as desired. $\qquad\square$

**Theorem 15.3.5.** *Fix $b \in (0,1)$ and define the sequence $(f_n)_{n\in\mathbb{N}}$ in $c_0$ by the following way: $f_1 := be_1$, $f_2 := be_2$, and $f_n := e_n$, for every $n \geq 3$ where $(e_n)_{n\in\mathbb{N}}$ is defined to be 1 in its $n$th coordinate, and 0 in all other coordinates such that for both $(c_0, \lVert\cdot\rVert_\infty)$ and $(\ell^1, \lVert\cdot\rVert_1)$, the sequence $(e_n)_{n\in\mathbb{N}}$ is an unconditional basis.*

*Next, define the closed, bounded, convex subset $E = E_b$ of $c_0$ by*

$$
E := \left\{ \sum_{n=1}^{\infty} t_n f_n : 1 = t_1 \geq t_2 \geq \cdots \geq t_n \downarrow_n 0 \right\}.
$$

*Then, define the sequence $(\eta_n)_{n\in\mathbb{N}}$ in $E$ in the following way: $\eta_1 := f_1$ and $\eta_n := f_1 + \cdots + f_n$, for every $n \geq 2$. Note that $E$ is the closed convex hull the sequence $(\eta_n)_{n\in\mathbb{N}}$. Then, the set $E$ has the fixed point property for $\lVert\!\lVert\cdot\rVert\!\rVert$-nonexpansive affine mappings where the equivalent norm $\lVert\!\lVert\cdot\rVert\!\rVert$ on $c_0$ is given as in Definition 15.3.1.*

*Proof.* We will use exactly the same method as the proof of Theorem 15.3.4 but just consider the following statements for the case 2 in the proof of Theorem 15.3.4.

$$
\begin{aligned}
\lVert\!\lVert\!\lVert h_\lambda - z\rVert\!\rVert\!\rVert &= \lVert\!\lVert\!\lVert \lambda\delta\eta_1 + (1-\lambda)\delta\eta_2 \rVert\!\rVert\!\rVert \\
&= \lVert\!\lVert\!\lVert \lambda\delta be_1 + (1-\lambda)\delta(be_1 + be_2)\rVert\!\rVert\!\rVert \\
&= b \lim_{p\to\infty} \sup \left\{ \gamma_1\left[\delta^p + \frac{|1-\lambda|^p \delta^p}{4}\right]^{\frac{1}{p}}, \frac{\gamma_2|1-\lambda|\delta}{4}\right\} \\
&= b \max\left\{ \gamma_1|1-\lambda|\delta, \frac{\gamma_2|1-\lambda|\delta}{4}\right\}.
\end{aligned}
$$

Define

$$\Gamma := \min_{\lambda \in \left[-\frac{\gamma_1}{\delta}, \frac{\gamma_2}{\delta}+1\right]} \||h_\lambda - z\||.$$

Hence, $\Gamma = 0$. $\qquad\square$

Now, we generalize our result more and give the following theorems.

**Theorem 15.3.6.** *Assume $\vec{b} = (b_n)_{n \in \mathbb{N}}$ is any increasing sequence in $(0, 1]$ with $b_n \uparrow_n 1$. Define the sequence $(f_n)_{n \in \mathbb{N}}$ in $c_0$ by the following way: $f_n := b_n e_n$, for every $n \in \mathbb{N}$. Next, take $E = E_{\vec{b}}$ of $c_0$ such that*

$$E := \left\{ \sum_{n=1}^{\infty} t_n f_n : 1 = t_1 \geq t_2 \geq \cdots \geq t_n \downarrow_n 0 \right\}.$$

*Then, the set $E$ has the fixed point property for $\||\cdot\||$ noncxpansive affine mappings where the equivalent norm $\||\cdot\||$ on $c_0$ is given as in Definition 15.3.1.*

*Proof.* We will use exactly same method as the proof of Theorem 15.3.4 but just consider the following statements for the case 2 in the proof of Theorem 15.3.4.

$$
\begin{aligned}
\||h_\lambda - z\|| &= \||\lambda \delta \eta_1 + (1 - \lambda)\delta \eta_2\|| \\
&= \||\lambda \delta b_1 e_1 + (1 - \lambda)\delta(b_1 e_1 + b_2 e_2)\|| \\
&= \lim_{p \to \infty} \sup \left\{ \gamma_1 \left[ b_1{}^p \delta^p + \frac{b_2{}^p |1 - \lambda|^p \delta^p}{4} \right]^{\frac{1}{p}}, \frac{\gamma_2 b_2 |1 - \lambda|\delta}{4} \right\} \\
&= b_2 \lim_{p \to \infty} \sup \left\{ \gamma_1 \left[ \left(\frac{b_1}{b_2}\right)^p \delta^p + \frac{|1 - \lambda|^p \delta^p}{4} \right]^{\frac{1}{p}}, \frac{\gamma_2 |1 - \lambda|\delta}{4} \right\} \\
&\leq b_2 \max \left\{ \gamma_1 |1 - \lambda|\delta, \frac{\gamma_2 |1 - \lambda|\delta}{4} \right\}.
\end{aligned}
$$

$$\text{Define } \Gamma := \min_{\lambda \in \left[-\frac{\gamma_1}{\delta}, \frac{\gamma_2}{\delta}+1\right]} \||h_\lambda - z\||.$$

Hence, $\Gamma = 0$ since $\Gamma \leq \min\limits_{\lambda \in \left[-\frac{\gamma_1}{\delta}, \frac{\gamma_2}{\delta}+1\right]} b_2 \max \left\{ \gamma_1 |1 - \lambda|\delta, \frac{\gamma_2 |1 - \lambda|\delta}{4} \right\} = 0.$

$\qquad\square$

**Theorem 15.3.7.** *Assume $\vec{b} = (b_n)_{n \in \mathbb{N}}$ is any sequence in $(0, \infty)$ converging to some $\kappa > 0$. Define the sequence $(f_n)_{n \in \mathbb{N}}$ in $c_0$ by the following way: $f_n := b_n e_n$, for every $n \in \mathbb{N}$. Next, take $E = E_{\vec{b}}$ of $c_0$ such that*

$$E := \left\{ \sum_{n=1}^{\infty} t_n f_n : 1 = t_1 \geq t_2 \geq \cdots \geq t_n \downarrow_n 0 \right\}.$$

*Then, the set $E$ has the fixed point property for $\||\cdot\||$-nonexpansive affine mappings where the equivalent norm $\||\cdot\||$ on $c_0$ is given as in Definition 15.3.1.*

*Proof.* We will use exactly same method as the proof of Theorem 15.3.4 but just consider the following statements for the case 2 in the proof of Theorem 15.3.4.

$$
\begin{aligned}
\||h_\lambda - z\|| &= \||\lambda\delta\eta_1 + (1-\lambda)\delta\eta_2\|| \\
&= \||\lambda\delta b_1 e_1 + (1-\lambda)\delta(b_1 e_1 + b_2 e_2)\|| \\
&= \limsup_{p\to\infty} \left\{ \gamma_1 \left[ b_1{}^p \delta^p + \frac{b_2{}^p |1-\lambda|^p \delta^p}{4} \right]^{\frac{1}{p}}, \; \frac{\gamma_2 b_2 |1-\lambda|\delta}{4} \right\} \\
&= \max\{b_1, b_2\} \limsup_{p\to\infty} \left\{ \gamma_1 \left[ \left( \frac{b_1}{\max\{b_1,b_2\}} \right)^p \delta^p \right. \right. \\
&\qquad \left. \left. + \left( \frac{b_2}{\max\{b_1,b_2\}} \right)^p \frac{|1-\lambda|^p \delta^p}{4} \right]^{\frac{1}{p}}, \right. \\
&\qquad \left. \frac{b_2 \gamma_2 |1-\lambda|\delta}{4\max\{b_1,b_2\}} \right\} \\
&\le \max\{b_1, b_2\} \max \left\{ \gamma_1 |1-\lambda|\delta, \; \frac{\gamma_2 |1-\lambda|\delta}{4} \right\}.
\end{aligned}
$$

$$
\text{Define } \Gamma := \min_{\lambda\in\left[-\frac{\gamma_1}{\delta}, \frac{\gamma_2}{\delta}+1\right]} \||h_\lambda - z\||.
$$

Hence, $\Gamma = 0$ since

$$
\Gamma \le \min_{\lambda\in\left[-\frac{\gamma_1}{\delta}, \frac{\gamma_2}{\delta}+1\right]} \max\{b_1, b_2\} \max \left\{ \gamma_1 |1-\lambda|\delta, \; \frac{\gamma_2 |1-\lambda|\delta}{4} \right\} = 0.
$$

$\square$

Then, the following corollaries are immediate by the proof of Theorem 15.3.7.

**Corollary 15.3.8.** *Assume* $\overrightarrow{b} = (b_n)_{n\in\mathbb{N}}$ *is any bounded sequence in* $(0,\infty)$. *Define the sequence* $(f_n)_{n\in\mathbb{N}}$ *in* $c_0$ *by the following way:* $f_n := b_n e_n$, *for every* $n \in \mathbb{N}$. *Next, take* $E = E_{\overrightarrow{b}}$ *of* $c_0$ *such that*

$$
E := \left\{ \sum_{n=1}^{\infty} t_n f_n : 1 = t_1 \ge t_2 \ge \cdots \ge t_n \downarrow_n 0 \right\}.
$$

*Then, the set* $E$ *has the fixed point property for* $\||\cdot\||$-*nonexpansive affine mappings where the equivalent norm* $\||\cdot\||$ *on* $c_0$ *is given as in Definition 15.3.1.*

Then, we give our most generalized results by the following theorem and further corollaries whose proofs are straightforward via the proof of Theorem 15.3.7.

**Theorem 15.3.9.** *Let $\overrightarrow{b} = (b_n)_{n \in \mathbb{N}}$ be any sequence in $(0, \infty)$. We define the sequence $(f_n)_{n \in \mathbb{N}}$ in $c_0$ by setting $f_n := b_n e_n$, for all $n \in \mathbb{N}$. Next, define the closed, bounded, convex subset $E = E_{\overrightarrow{b}}$ of $c_0$ by*

$$E := \left\{ \sum_{n=1}^{\infty} t_n f_n : 1 = t_1 \geq t_2 \geq \cdots \geq t_n \downarrow_n 0 \right\} \ . \ \textit{Then, the set } E \textit{ has the}$$

*fixed point property for $|||\cdot|||$-nonexpansive affine mappings where the equivalent norm $|||\cdot|||$ on $c_0$ is given as in Definition 15.3.1.*

**Corollary 15.3.10.** *Assume $(\mu_n)_{n \in \mathbb{N}}$ is a sequence in $(0, \infty)$ for which there exists $\Gamma > 0$ with $[\Gamma \leq \mu_N$, for every $N \in \mathbb{N}]$ and $\sigma := \sum_{n=2}^{\infty} |\mu_n - \mu_{n-1}| < \infty$ ; and assume $(b_n)_{n \in \mathbb{N}}$ is a sequence in $(0, \infty)$ converging to some $\lambda \in (0, \infty)$. Define the sequence $(\eta_n)_{n \in \mathbb{N}}$ by setting $\eta_n := \mu_n(b_1 e_1 + b_2 e_2 + b_3 e_3 + b_4 e_4 + \dots + b_n e_n)$, for all $n \in \mathbb{N}$. Also suppose that $(\eta_n)_{n \in \mathbb{N}}$ satisfies a lower $c_0$-summing estimate. That is, suppose $\exists K \in (0, \infty)$ s.t. $\forall \alpha = (\alpha_n)_{n \in \mathbb{N}} \in c_{00}$, $K \sup_{n \geq 1} \left| \sum_{j=n}^{\infty} \alpha_j \right| \leq \left\| \sum_{j=1}^{\infty} \alpha_j \eta_j \right\|$ . Consider the closed convex hull of $(\eta_n)_{n \in \mathbb{N}}$, $E := \overline{co}(\{\eta_n : n \subset \mathbb{N}\})$.*

*Then, the set $E$ has the fixed point property for $|||\cdot|||$-nonexpansive affine mappings where the equivalent norm $|||\cdot|||$ on $c_0$ is given as in Definition 15.3.1.*

**Corollary 15.3.11.** *Let $(\mu_n)_{n \in \mathbb{N}}$ be a sequence in $(0, \infty)$ and let $(b_n)_{n \in \mathbb{N}}$ be any sequence in $(0, \infty)$. Define the sequence $(\eta_n)_{n \in \mathbb{N}}$ by setting $\eta_n := \mu_n(b_1 e_1 + b_2 e_2 + b_3 e_3 + b_4 e_4 + \dots + b_n e_n)$, for all $n \in \mathbb{N}$.*

*Let $E$ be the closed convex hull of $(\eta_n)_{n \in \mathbb{N}}$, $E := \overline{co}(\{\eta_n : n \in \mathbb{N}\})$.*

*Then, the set $E$ has the fixed point property for $|||\cdot|||$-nonexpansive affine mappings where the equivalent norm $|||\cdot|||$ on $c_0$ is given as in Definition 15.3.1.*

---

## 15.4 $c_0$ can be renormed to have the fixed point property for affine nonexpansive mappings

We need to note that this and next section derive from and use similar methods of the first and second author's recent accepted paper entitled "$c_0$ can be renormed to have the fixed point property for affine nonexpansive mappings" which is to be published by Filomat journal. However, here and next section we use a different renorming that yields similar results to those of the paper mentioned. Firstly, we would like to give some details and a literature review on the subject.

The sequence spaces $(c_0, \| \cdot \|_\infty)$ and $(l^1, \| \cdot \|_1)$ are both nonreflexive and possess the w-FPP(n.e.), but do not have the FPP(n.e.). Researchers do not know whether or not every reflexive Banach space $(X, \| \cdot \|)$ has the fixed point property for nonexpansive maps but this and related questions have been and still are central themes in metric fixed point theory.

For example, Dowling and Lennard Dowling and Lennard (1997) showed that every nonreflexive subspace of $L^1[0,1]$ fails the fixed point property and Domínguez Benavides Domínguez Benavides (2009) proved that given a reflexive Banach space $(X, \|\cdot\|)$, there exists an equivalent norm $\|\cdot\|_\sim$ on $X$ such that $(X, \|\cdot\|_\sim)$ has the fixed point property for nonexpansive mappings. In 2014, motivated by work in Nezir Nezir (2012), Lennard and Nezir Lennard and Nezir (2014) used the above-described theorem of Domínguez Benavides and the Strong James' Distortion Theorems (Dowling et al., 2000, Theorem 8), to prove the following theorem: If a Banach space is a Banach lattice, or has an unconditional basis, or is a symmetrically normed ideal of operators on an infinite-dimensional Hilbert space, then it is reflexive if and only if it has an equivalent norm that has the fixed point property for *cascading nonexpansive mappings*. Moreover, in Lennard and Nezir (2017), Lennard and Nezir proved that $l^1$ cannot be equivalently renormed to have the FPP for semi-strongly asymptotically nonexpansive maps. Moreover, by a similar proof to that of Theorem 10 of Dowling, Lennard and Turett Dowling et al. (2000), they showed that $c_0$ cannot be equivalently renormed to have the FPP for strongly asymptotically nonexpansive maps. From this, we conclude that if $(X, \|\cdot\|)$ is a non-reflexive Banach lattice, then $(X, \|\cdot\|)$ fails the fixed point property for $\|\cdot\|$-semi-strongly asymptotically nonexpansive mappings. They strengthened this result: If a Banach space is a Banach lattice then it is reflexive if and only if it has the fixed point property for *affine semi-strongly asymptotically nonexpansive mappings*.

In 2008, P.K. Lin Lin (2008) showed that $\ell^1$ can be renormed to have FPP with the equivalent norm $\|\cdot\|$ given by

$$\|x\| = \sup_{k \in \mathbb{N}} \frac{8^k}{1 + 8^k} \sum_{n=k}^\infty |x_n|, \text{ for all } x = (x_n)_{n \in \mathbb{N}} \in \ell^1 \ .$$

The analogue of P. K. Lin's work for $(c_0, \|\cdot\|_\infty)$ is still an open question; that is, it is still unknown whether or not there exists any renorming of $c_0$ such that it can have the FPP (n.e) with respect to that equivalent norm.

Moreover, in the Ph.D. Thesis of Carlos Hernández written under supervision of Maria Japón Pineda (Hernández-Linares, 2011, Theorem 4.2.1) and in their recent paper (Hernández-Linares and Japón, 2012, Theorem 3.2) , they invented an equivalent norm on $L^1$ that has the FPP for all closed bounded sets and all affine nonexpansive mappings. This partially extends Lin's $\ell^1$ theorem to $L^1$.

In the next section, we show that Banach space $c_0$ is also in the same category as $L^1$; that is, $c_0$ can be renormed to have the fixed point property for affine nonexpansive mappings. Next, we generalize it in our last section by showing that if $\rho(\cdot)$ is an equivalent norm to the usual norm on $c_0$ such that

$$\limsup_n \rho\left(\frac{1}{n}\sum_{m=1}^{n} x_m + x\right) = \limsup_n \rho\left(\frac{1}{n}\sum_{m=1}^{n} x_m\right) + \rho(x)$$

for every weakly null sequence $(x_n)_n$ and for all $x \in c_0$, then for every $\lambda > 0$, $c_0$ with the norm $\|\cdot\|_\rho = \rho(\cdot) + \lambda\,\|\!|\cdot|\!\|$ has the FPP for affine $\|\cdot\|_\rho$-nonexpansive self-mappings.

In this section, proofs of our theorems and lemmas are inspired by the proofs of theorems and lemmas given by Helga Fetter and Berta Gamboa de Buen Fetter and de Buen (2009) such that they extend P.K. Lin's work Lin (2008). We implement our ideas and get our desired result with the help of their work.

**Lemma 15.4.1.** *Fetter and de Buen (2009) Let $(X, \|\cdot\|)$ be a Banach space and let $C \subset X$ be a closed convex nonempty (ccne) subset. Assume $T : C \to C$ a fixed point free nonexpansive mapping. Then, there exists a ccne $T$-invariant set $D$ and a $> 0$ such that for every ccne $T$-invariant set $D' \subset D$, $\mathrm{diam}\, D' > a$.*

**Lemma 15.4.2.** *Let $(X, \|\cdot\|)$ be a Banach space and let $C \subset X$ be a ccne subset. Assume $T : C \to C$ an affine nonexpansive mapping and $(x_n)_n \subset C$ be an approximate fixed point sequence (afps). Consider a function $\Phi : C \to [0, \infty)$ given by*

$$\Phi(y) = \limsup_m \left\|\frac{1}{m}\sum_{n=1}^{m} x_n - y\right\| \,, \quad \forall y \in C \,.$$

*If $d > \inf_{x \in C} \Phi(x)$ and $D = \{x \in C : \Phi(x) \le d\}$, then $D$ is a ccne $T$-invariant set with $\mathrm{diam}\, D \le 2d$.*

**Lemma 15.4.3.** *Let $(X, \|\cdot\|)$ be a Banach space and let $C \subset X$ be a ccne subset. Assume $T : C \to C$ an affine nonexpansive fixed point free mapping. Let $D \subset C$ and $a > 0$ be as in lemma 15.4.1. Then*

$$\inf\left\{\limsup_m \left\|\frac{1}{m}\sum_{n=1}^{m} x_n - y\right\| : (x_n)_n \subset D \,,\ (x_n)_n \text{ is an afps} \,,\ y \in D\right\} \ge \frac{a}{2}.$$

*Proof.* Assume by contradiction that

$$\inf\left\{\limsup_m \left\|\frac{1}{m}\sum_{n=1}^{m} x_n - y\right\| : (x_n)_n \subset D \,,\ (x_n)_n \text{ is an afps} \,,\ y \in D\right\} < \frac{a}{2}.$$

Then, there exists an afps $(x_n)_n \subset D$ such that

$$D' = \left\{y \in D : \limsup_m \left\|\frac{1}{m}\sum_{n=1}^{m} x_n - y\right\| \le \frac{a}{2}\right\} \ne \emptyset.$$

By Lemma 15.4.2, $D'$ is ccne, $T$-invariant and $\mathrm{diam}\, D' \le a$ which is a contradiction. $\qquad\square$

**Lemma 15.4.4.** *Let $(X, \| \cdot \|)$ be a Banach space and let $C \subset X$ be a ccne subset. Assume $T : C \to C$ an affine nonexpansive fixed point free mapping. Let $D \subset C$ and $a > 0$ be as in Lemma 15.4.1 so that $D$ is a $T$-invariant ccne set such that if $D' \subset D$ is a $T$-invariant ccne set then $\operatorname{diam} D' > a$. Then,*

$$\inf_{z \in X} \left\{ \limsup_m \left\| \frac{1}{m} \sum_{n=1}^m x_n - z \right\| : (x_n)_n \subset D, \ (x_n)_n \text{ is an afps} \right\} \geq \frac{a}{4}.$$

*Proof.* Using Lemma 15.4.3, for every $s \in \mathbb{N}$, for every afps $(x_k)_k \subset D$ and for any $z \in X$,

$$\frac{a}{2} \leq \limsup_m \left\| \frac{1}{m} \sum_{n=1}^m x_n - \frac{1}{s} \sum_{r=1}^s x_r \right\| \leq \limsup_m \left\| \frac{1}{m} \sum_{n=1}^m x_n - z \right\|$$
$$+ \left\| z - \frac{1}{s} \sum_{r=1}^s x_r \right\|.$$

Thus,

$$\frac{a}{2} \leq \limsup_s \limsup_m \left\| \frac{1}{m} \sum_{n=1}^m x_n - \frac{1}{s} \sum_{r=1}^s x_r \right\| \leq \limsup_m \left\| \frac{1}{m} \sum_{n=1}^m x_n - z \right\|$$
$$+ \limsup_s \left\| z - \frac{1}{s} \sum_{r=1}^s x_r \right\|$$
$$= 2 \limsup_m \left\| \frac{1}{m} \sum_{n=1}^m x_n - z \right\|.$$

$\square$

**Corollary 15.4.5.** *Let $(X, \| \cdot \|)$ be a Banach space and let $C \subset X$ be a ccne subset. Assume that $T : C \to C$ is an affine nonexpansive fixed point free mapping and that for every $T$-invariant ccne set $D \subset C$, let $a > 0$ be as in Lemma 15.4.1 and $\operatorname{diam} D > a$. Then,*

$$\inf \left\{ \limsup_m \left\| \frac{1}{m} \sum_{n=1}^m x_n - x \right\| : (x_n)_n \subset C, \ (x_n)_n \text{ is an afps, } x_n \xrightarrow[w^*]{} x \right\} \geq \frac{a}{4}.$$

Now we give another lemma but from now on, unless it is stated as a different norm on $c_0$, the norm $\||\cdot\||$ on $c_0$ given in section 15.3 will be used.

**Lemma 15.4.6.** *Assume that $(x_n)_n \subset (c_0, \| \cdot \|_\infty)$ and $x_n \xrightarrow{w} 0$. Then, there exists a subsequence $(x_{n_k})_k$ of $(x_n)_n$ and a sequence $(u_k)_k$ with $u_k = \sum_{i=m_k+1}^{m_{k+1}} a_i e_i$ where $(e_i)_i$ is the canonical basis and $(a_i)_i \subset \mathbb{R}$ such that*

$$\lim_{j \to \infty} \left\| \frac{1}{j} \sum_{k=1}^j x_{n_k} - u_j \right\|_\infty = \lim_{j \to \infty} \left\| \left\| \frac{1}{j} \sum_{k=1}^j x_{n_k} - u_j \right\| \right\| = 0.$$

The proof of the above is straightforward by the proof of the Bessaga-Pełczyński Selection Principle (Diestel, 1984, p.46), Bessaga and Pełczyński (1958) and by passing to a subsequence of $(x_n)_n$ that is equivalent to the block basic sequence $(u_k)_k$ and is essentially disjointly supported.

One can easily obtain the following partial analogue result to (Goebel and Kuczumow, 1979, Lemma 1) by the following lemma.

**Lemma 15.4.7.** *Let $(x_n)_n$ be a bounded sequence in a Banach space $(X, \|\cdot\|)$. Consider a function $s : X \to [0, \infty)$ given by*

$$s(y) = \limsup_m \left\| \frac{1}{m} \sum_{k=1}^{m} x_k - y \right\| \ , \ \forall y \in X .$$

*Then, if $X$ has weak Banach-Saks property and $x \in X$ is the weak limit of the sequence $(x_n)_n$, then there exists a subsequence $(x_{n_k})_k$ whose norm limit is $x$ such that if $s$ is redefined via this subsequence, we have $s(x) = 0$ and $s(y) = \|y - x\|$ , $\forall y \in X$ and for any equivalent norm $\|\cdot\|$ on $X$.*

*Thus, since $c_0$ has weak Banach-Saks property Núñez (1989), the above can be applied.*

**Lemma 15.4.8.** *Assume that $(x_n)_n \subset (c_0, \|\cdot\|_\infty)$, $x_n \xrightarrow{w} x$ and that*

$$\lim_{j \to \infty} \left\| \left\| \frac{1}{j} \sum_{n=1}^{j} x_n - x \right\| \right\| = a \text{ exists. Then, } \lim_{j \to \infty} \left\| \frac{1}{j} \sum_{n=1}^{j} x_n - x \right\|_\infty = a.$$

*Proof.* Scaling $\gamma_n$, we may assume that $\gamma_k \uparrow_k 1$, $\gamma_k$ is strictly increasing and we can redefine the equivalent norm according to this change. We may also suppose without loss of generality that $x_n \xrightarrow{w} 0$. Let $(x_{n_k})_k$ be a subsequence of $(x_n)_n$ such that $\lim_{j \to \infty} \left\| \frac{1}{j} \sum_{k=1}^{j} x_{n_k} \right\|_\infty$ exists. By Lemma 15.4.6, we may assume that there is a sequence $(u_k)_k$ with $u_k = \sum_{i=m_k+1}^{m_{k+1}} a_i e_i$ where $(e_i)_i$ is the canonical basis and $(a_i)_i \subset \mathbb{R}$ such that

$$\lim_{j \to \infty} \left\| \frac{1}{j} \sum_{k=1}^{j} x_{n_k} - u_j \right\|_\infty = \lim_{j \to \infty} \left\| \left\| \frac{1}{j} \sum_{k=1}^{j} x_{n_k} - u_j \right\| \right\| = 0.$$

Define $y_k = x_{n_k}$ for every $k \in \mathbb{N}$. Then,

$$\lim_{j \to \infty} \left\| \left\| \frac{1}{j} \sum_{k=1}^{j} y_k \right\| \right\| = \lim_{j \to \infty} \|\|u_j\|\| = a \text{ and } \lim_{j \to \infty} \left\| \frac{1}{j} \sum_{k=1}^{j} y_k \right\|_\infty = \lim_{j \to \infty} \|u_j\|_\infty.$$

By the definition of the equivalent norm $\|\|\cdot\|\|$, there exists $l \in \mathbb{N}$ such that

$$\|\|u_j\|\| = \gamma_l \lim_{p \to \infty} \left( \sum_{i=l}^{m_{j+1}} \frac{|a_i|^p}{i^2} \right)^{\frac{1}{p}} \text{ where } a_i = 0 \text{ in case } i \leq m_j.$$

Since $\gamma_n < \gamma_{n+1}$, if $n \leq m_j$, we have that

$$
\begin{aligned}
\gamma_n \lim_{p \to \infty} \left( \sum_{i=n}^{m_{j+1}} \frac{|a_i|^p}{i^2} \right)^{\frac{1}{p}} &= \gamma_n \lim_{p \to \infty} \left( \sum_{i=m_j+1}^{m_{j+1}} \frac{|a_i|^p}{i^2} \right)^{\frac{1}{p}} \\
&\leq \gamma_{n+1} \lim_{p \to \infty} \left( \sum_{i=m_j+1}^{m_{j+1}} \frac{|a_i|^p}{i^2} \right)^{\frac{1}{p}} \\
&= \gamma_{n+1} \lim_{p \to \infty} \left( \sum_{i=n+1}^{m_{j+1}} \frac{|a_i|^p}{i^2} \right)^{\frac{1}{p}}
\end{aligned}
$$

and $m_j + 1 \leq l \leq m_{j+1}$. Also,

$$
\gamma_l \lim_{p \to \infty} \left( \sum_{i=l}^{m_{j+1}} \frac{|a_i|^p}{i^2} \right)^{\frac{1}{p}} \geq \gamma_{m_j+1} \lim_{p \to \infty} \left( \sum_{i=m_j+1}^{m_{j+1}} \frac{|a_i|^p}{i^2} \right)^{\frac{1}{p}}
$$

and so

$$
\frac{\gamma_l - \gamma_{m_j+1}}{\gamma_{m_j+1}} \lim_{p \to \infty} \left( \sum_{i=l}^{m_{j+1}} \frac{|a_i|^p}{i^2} \right)^{\frac{1}{p}} \geq \lim_{p \to \infty} \left( \sum_{i=m_j+1}^{l-1} \frac{|a_i|^p}{i^2} \right)^{\frac{1}{p}}.
$$

Hence, we obtain that

$$
\begin{aligned}
\left| \|u_j\|_\infty - \|\|u_j\|\| \right| &= \sup_{m_j+1 \leq i \leq m_{j+1}} |a_i| - \gamma_l \lim_{p \to \infty} \left( \sum_{i=l}^{m_{j+1}} \frac{|a_i|^p}{i^2} \right)^{\frac{1}{p}} \\
&= \lim_{p \to \infty} \left( \sum_{i=m_j+1}^{m_{j+1}} \frac{|a_i|^p}{i^2} \right)^{\frac{1}{p}} - \gamma_l \lim_{p \to \infty} \left( \sum_{i=l}^{m_{j+1}} \frac{|a_i|^p}{i^2} \right)^{\frac{1}{p}} \\
&\leq (1 - \gamma_l) \lim_{p \to \infty} \left( \sum_{i=l}^{m_{j+1}} \frac{|a_i|^p}{i^2} \right)^{\frac{1}{p}} + \lim_{p \to \infty} \left( \sum_{i=m_j+1}^{l-1} \frac{|a_i|^p}{i^2} \right)^{\frac{1}{p}} \\
&\leq \left( 1 + \frac{\gamma_l - \gamma_{m_j+1}}{\gamma_{m_j+1}} - \gamma_l \right) \lim_{p \to \infty} \left( \sum_{i=l}^{m_{j+1}} \frac{|a_i|^p}{i^2} \right)^{\frac{1}{p}} \\
&= \left( \frac{\gamma_l}{\gamma_{m_j+1}} - \gamma_l \right) \lim_{p \to \infty} \left( \sum_{i=l}^{m_{j+1}} \frac{|a_i|^p}{i^2} \right)^{\frac{1}{p}}.
\end{aligned}
$$

Then, taking the limit as $j \to \infty$, we get that $\lim_{j \to \infty} \|u_j\|_\infty - \|\|u_j\|\| = 0$ and so

$$
\lim_{j \to \infty} \left\| \left\| \frac{1}{j} \sum_{k=1}^{j} y_k \right\| \right\| = \lim_{j \to \infty} \|\|u_j\|\| = \lim_{j \to \infty} \|u_j\|_\infty = \lim_{j \to \infty} \left\| \frac{1}{j} \sum_{k=1}^{j} y_k \right\|_\infty = a.
$$

Since every subsequence of $\frac{1}{j}\sum\limits_{k=1}^{j}x_k$ has in turn a subsequence such that

$$\lim_{j\to\infty}\left\|\frac{1}{j}\sum_{k=1}^{j}x_{n_k}\right\|_{\infty}=a,$$ we get the desired result. Also, the reciprocal can be proved in the same way. $\qquad\square$

**Lemma 15.4.9.** *Assume that $(x_n)_n\subset(c_0,\|\cdot\|_\infty)$, $x_n\xrightarrow{w}x$ and that*

$$\lim_{s\to\infty}\lim_{j\to\infty}\left\|\left\|\frac{1}{j}\sum_{n=1}^{j}x_n-\frac{1}{s}\sum_{n=1}^{s}x_n\right\|\right\|=a \text{ and } \lim_{j\to\infty}\left\|\left\|\frac{1}{j}\sum_{n=1}^{j}x_n\right\|\right\|$$

*exist. Then, there exists a subsequence $(y_n)_n$ of $(x_n)_n$ such that*

$$\lim_{s\to\infty}\lim_{j\to\infty}\left\|\left\|\frac{1}{j}\sum_{n=1}^{j}y_n-\frac{1}{s}\sum_{n=1}^{s}y_n\right\|\right\|=a=2\lim_{j\to\infty}\left\|\left\|\frac{1}{j}\sum_{n=1}^{j}x_n\right\|\right\|.$$

*Proof.* Similarly to the proof of Lemma 15.4.8, scaling $\gamma_n$, we may assume that $\gamma_k\uparrow_k 1$, $\gamma_k$ is strictly increasing and we can redefine the equivalent norm according to this change and we can let $(x_{n_k})_k$ be a subsequence of $(x_n)_n$ such that $\lim\limits_{j\to\infty}\left\|\frac{1}{j}\sum\limits_{k=1}^{j}x_{n_k}\right\|_{\infty}$ exists. Furthermore, by Lemma 15.4.6, we may assume that there is a sequence $(u_k)_k$ with $u_k=\sum\limits_{i=m_k+1}^{m_{k+1}}a_ie_i$ where $(e_i)_i$ is the canonical basis and $(a_i)_i\subset\mathbb{R}$ such that

$$\lim_{j\to\infty}\left\|\frac{1}{j}\sum_{k=1}^{j}x_{n_k}-u_j\right\|_{\infty}=\lim_{j\to\infty}\left\|\left\|\frac{1}{j}\sum_{k=1}^{j}x_{n_k}-u_j\right\|\right\|=0.$$

We may also assume, by passing a subsequence that

$$\lim_{s\to\infty}\lim_{j\to\infty}\left\|\left\|\frac{1}{j}\sum_{n=1}^{j}x_n-\frac{1}{s}\sum_{n=1}^{s}x_n\right\|\right\| \text{ and } \lim_{s\to\infty}\lim_{j\to\infty}\|\|u_j-u_s\|\|$$

exist. Define $y_k=x_{n_k}$ for every $k\in\mathbb{N}$. Then, for every $j,s\in\mathbb{N}$,

$$\|\|u_j-u_s\|\| - \left\|\left\|\frac{1}{j}\sum_{n=1}^{j}y_n-u_j\right\|\right\|$$
$$- \left\|\left\|\frac{1}{s}\sum_{n=1}^{s}y_n-u_s\right\|\right\|$$
$$\leq \left\|\left\|\frac{1}{j}\sum_{n=1}^{j}y_n-\frac{1}{s}\sum_{n=1}^{s}y_n\right\|\right\|$$
$$\leq \|\|u_j-u_s\|\|+\left\|\left\|\frac{1}{j}\sum_{n=1}^{j}y_n-u_j\right\|\right\|+\left\|\left\|\frac{1}{s}\sum_{n=1}^{s}y_n-u_s\right\|\right\|.$$

and

$$
\begin{aligned}
\|u_j - u_s\|_\infty &- \left\| \frac{1}{j} \sum_{n=1}^{j} y_n - u_j \right\|_\infty \\
&- \left\| \frac{1}{s} \sum_{n=1}^{s} y_n - u_s \right\|_\infty \\
&\leq \left\| \frac{1}{j} \sum_{n=1}^{j} y_n - \frac{1}{s} \sum_{n=1}^{s} y_n \right\|_\infty \\
&\leq \|u_j - u_s\|_\infty + \left\| \frac{1}{j} \sum_{n=1}^{j} y_n - u_j \right\|_\infty + \left\| \frac{1}{s} \sum_{n=1}^{s} y_n - u_s \right\|_\infty .
\end{aligned}
$$

Hence, $\lim\limits_{s\to\infty} \lim\limits_{j\to\infty} \|\|u_j - u_s\|\| = a$. Assume that $s > j$. By the definition of the equivalent norm $\|\| \cdot \|\|$, there exists $l \in \mathbb{N}$ with $m_j + 1 \leq l \leq m_{s+1}$ such that

$$
\|\|u_j - u_s\|\| = \gamma_l \lim_{p\to\infty} \left( \sum_{i=l}^{m_{s+1}} \frac{|a_i'|^p}{i^2} \right)^{\frac{1}{p}} \quad \text{where } a_i' = \begin{cases} 0 & \text{if } i \leq m_j \text{ or } m_{j+1} < i < m_s \\ a_i & \text{if } m_j + 1 \leq i \leq m_{j+1} \\ & \text{or } m_s + 1 \leq i \leq m_{s+1} \end{cases}
$$

Since $\gamma_n < \gamma_{n+1}$, if $n \leq m_j$, we have that

$$
\gamma_n \lim_{p\to\infty} \left( \sum_{i=n}^{m_{s+1}} \frac{|a_i'|^p}{i^2} \right)^{\frac{1}{p}} = \gamma_n \lim_{p\to\infty} \left( \sum_{i=m_j+1}^{m_{s+1}} \frac{|a_i'|^p}{i^2} \right)^{\frac{1}{p}} \leq \gamma_{n+1} \lim_{p\to\infty} \left( \sum_{i=m_j+1}^{m_{s+1}} \frac{|a_i'|^p}{i^2} \right)^{\frac{1}{p}}
$$

$$
= \gamma_{n+1} \lim_{p\to\infty} \left( \sum_{i=n+1}^{m_{s+1}} \frac{|a_i|^p}{i^2} \right)^{\frac{1}{p}} .
$$

Hence we obtain that

$$
\gamma_l \lim_{p\to\infty} \left( \sum_{i=l}^{m_{s+1}} \frac{|a_i'|^p}{i^2} \right)^{\frac{1}{p}} \geq \gamma_{m_j+1} \lim_{p\to\infty} \left( \sum_{i=m_j+1}^{m_{s+1}} \frac{|a_i'|^p}{i^2} \right)^{\frac{1}{p}}
$$

and so

$$
\frac{\gamma_l - \gamma_{m_j+1}}{\gamma_{m_j+1}} \lim_{p\to\infty} \left( \sum_{i=l}^{m_{j+1}} \frac{|a_i|^p}{i^2} \right)^{\frac{1}{p}} \geq \lim_{p\to\infty} \left( \sum_{i=m_j+1}^{l-1} \frac{|a_i'|^p}{i^2} \right)^{\frac{1}{p}} .
$$

Thus, we have

$$\left| \|u_j - u_s\|_\infty - \|\|u_j - u_s\|\| \right| = \sup_{m_j+1 \le i \le m_{s+1}} |a_i'| - \gamma_l \lim_{p\to\infty} \left( \sum_{i=l}^{m_{s+1}} \frac{|a_i'|^p}{i^2} \right)^{\frac{1}{p}}$$

$$= \lim_{p\to\infty} \left( \sum_{i=m_j+1}^{m_{s+1}} \frac{|a_i'|^p}{i^2} \right)^{\frac{1}{p}} - \gamma_l \lim_{p\to\infty} \left( \sum_{i=l}^{m_{s+1}} \frac{|a_i'|^p}{i^2} \right)^{\frac{1}{p}}$$

$$\le (1 - \gamma_l) \lim_{p\to\infty} \left( \sum_{i=l}^{m_{s+1}} \frac{|a_i'|^p}{i^2} \right)^{\frac{1}{p}} + \lim_{p\to\infty} \left( \sum_{i=m_j+1}^{l-1} \frac{|a_i'|^p}{i^2} \right)^{\frac{1}{p}}.$$

Therefore,

$$\left| \|u_j - u_s\|_\infty - \|\|u_j - u_s\|\| \right| \le \left( 1 + \frac{\gamma_l - \gamma_{m_j+1}}{\gamma_{m_j+1}} - \gamma_l \right) \lim_{p\to\infty} \left( \sum_{i=l}^{m_{s+1}} \frac{|a_i'|^p}{i^2} \right)^{\frac{1}{p}}$$

$$= \left( \frac{\gamma_l}{\gamma_{m_j+1}} - \gamma_l \right) \lim_{p\to\infty} \left( \sum_{i=l}^{m_{s+1}} \frac{|a_i'|^p}{i^2} \right)^{\frac{1}{p}}.$$

Assume that $\|u_j\|_\infty \le d$ (since the sequence is bounded). Then,

$$0 \le \left| \|u_j - u_s\|_\infty - \|\|u_j - u_s\|\| \right| \le \left( \frac{\gamma_l}{\gamma_{m_j+1}} - \gamma_l \right) 2d \text{ taking the limit as } j \to \infty$$

and as $s \to \infty$ next, we get that

$\lim_{s\to\infty} \lim_{j\to\infty} \left| \|u_j - u_s\|_\infty - \|\|u_j - u_s\|\| \right| = 0$ and so using Lemma 15.4.8 and Lemma 15.4.7 (and considering we had passed to a subsequence already) we obtain that

$$a = \lim_{s\to\infty} \lim_{j\to\infty} \|\|u_j - u_s\|\| = \lim_{s\to\infty} \lim_{j\to\infty} \|u_j - u_s\|_\infty$$

$$= \lim_{s\to\infty} \lim_{j\to\infty} \left\| \frac{1}{j} \sum_{k=1}^{j} x_k - \frac{1}{s} \sum_{k=1}^{s} x_k \right\|_\infty$$

$$= \lim_{j\to\infty} \left\| \frac{1}{j} \sum_{k=1}^{j} x_k \right\|_\infty + \lim_{s\to\infty} \left\| \frac{1}{s} \sum_{k=1}^{s} x_k \right\|_\infty$$

$$= 2 \lim_{j\to\infty} \left\| \frac{1}{j} \sum_{k=1}^{j} x_k \right\|_\infty = 2 \lim_{j\to\infty} \left\| \left\| \frac{1}{j} \sum_{k=1}^{j} x_k \right\| \right\|.$$

$\square$

**Lemma 15.4.10.** *Assume that $C \subset (c_0, \| \cdot \|_\infty)$ is a ccne set, $T : C \to C$*

*is an affine nonexpansive fixed point free mapping,* $(x_n)_n \subset C$ *is an afps,* $x_n \underset{w}{\longrightarrow} u$, *such that*

$$\lim_{j \to \infty} \left\| \frac{1}{j} \sum_{n=1}^{j} x_n - u \right\|_{\infty} \text{ exists, } \Phi : C \to [0, \infty) \text{ given by}$$

$$\Phi(x) = \limsup_{m} \left\| \frac{1}{m} \sum_{n=1}^{m} x_n - x \right\|_{\infty} \text{ and } D = \left\{ x : \ \Phi(x) \leq \frac{d}{4} \right\} \neq \emptyset$$

*and so* $\{x : \ \Phi(x) \leq d\} \neq \emptyset$.

*Assume further that* $(y_n)_n \subset D$ *and* $y_n \underset{w}{\longrightarrow} y$. *Then,*

$$\limsup_{m} \left\| \frac{1}{m} \sum_{n=1}^{m} y_n - y \right\|_{\infty} \leq d - \lim_{m \to \infty} \left\| \frac{1}{m} \sum_{n=1}^{m} x_n - u \right\|_{\infty}.$$

*Proof.* For every $s \in \mathbb{N}$,

$$
\begin{aligned}
\frac{d}{4} &\geq \limsup_{m} \left\| \frac{1}{m} \sum_{n=1}^{m} x_n - y_s \right\|_{\infty} = \limsup_{m} \left\| \frac{1}{m} \sum_{n=1}^{m} x_n - u + u - y_s \right\|_{\infty} \\
&\geq \frac{1}{2} \lim_{m \to \infty} \left\| \frac{1}{m} \sum_{n=1}^{m} x_n - u \right\|_{\infty} + \frac{1}{2} \| y_s - y + y - u \|_{\infty}.
\end{aligned}
$$

Therefore,

$$
\begin{aligned}
\frac{d}{4} &\geq \limsup_{s} \limsup_{m} \frac{1}{s} \sum_{k=1}^{s} \left\| \frac{1}{m} \sum_{n=1}^{m} x_n - y_k \right\|_{\infty} \\
&\geq \limsup_{s} \limsup_{m} \left\| \frac{1}{m} \sum_{n=1}^{m} x_n - \frac{1}{s} \sum_{n=1}^{s} y_n \right\|_{\infty} \\
&\geq \frac{1}{4} \lim_{m \to \infty} \left\| \frac{1}{m} \sum_{n=1}^{m} x_n - u \right\|_{\infty} + \frac{1}{4} \| y - u \|_{\infty} + \frac{1}{4} \lim_{s \to \infty} \left\| \frac{1}{s} \sum_{n=1}^{s} y_n - y \right\|_{\infty}.
\end{aligned}
$$

Thus,

$$\limsup_{m} \left\| \frac{1}{m} \sum_{n=1}^{m} y_n - y \right\|_{\infty} \leq d - \lim_{m \to \infty} \left\| \frac{1}{m} \sum_{n=1}^{m} x_n - u \right\|_{\infty} - \| y - u \|_{\infty}.$$

$\square$

Now, we prove the corresponding lemma in $(c_0, |||\cdot|||)$.

**Lemma 15.4.11.** *Assume that $C \subset (c_0, \|\cdot\|_\infty)$ is a ccne set,$T : C \to C$ is an affine nonexpansive fixed point free mapping, $(x_n)_n \subset C$ is an afps, $x_n \xrightarrow{w} u$, such that*

$$\lim_{j\to\infty} \left\|\left\| \frac{1}{j} \sum_{n=1}^{j} x_n - u \right\|\right\| \text{ exists, } (u_n)_n \text{ is as in lemma 15.4.6,}$$

$$\lim_{j\to\infty} \left\|\left\| \frac{1}{j} \sum_{n=1}^{j} x_n - u - u_j \right\|\right\| = 0, \ \Phi : C \to [0,\infty) \text{ given by}$$

$$\Phi(x) = \limsup_m \left\|\left\| \frac{1}{m} \sum_{n=1}^{m} x_n - x \right\|\right\| \text{ and } D = \{x : \ \Phi(x) \le d\} \neq \emptyset.$$

*Assume further that $(y_n)_n \subset D$, $y_n \xrightarrow{w} y$ and $\lim_{n\to\infty} \left\|\left\| \frac{1}{n} \sum_{k=1}^{n} y_k - y \right\|\right\|$ exists. Then,*

$$\lim_m \left\|\left\| \frac{1}{m} \sum_{n=1}^{m} y_n - y \right\|\right\| \le d - \lim_{m\to\infty} \left\|\left\| \frac{1}{m} \sum_{n=1}^{m} x_n - u \right\|\right\|.$$

*Proof.* Assume $\varepsilon > 0$, $k$ such that $\|\|y - P_k y\|\| < \varepsilon$ and $\|\|u - P_k u\|\| < \varepsilon$ where $P_k$ is the natural projection. By passing to a subsequence of $(y_n)_n$ we may also assume that there is a sequence $(v_n)_n$ with $v_n = \sum_{i=r_n+1}^{r_{n+1}} b_i e_i$ and $\lim_{n\to\infty} \left\|\left\| \frac{1}{n} \sum_{k=1}^{n} y_k - y - v_n \right\|\right\|$. Let $N > k$ be such that for $n > N$,

$$\left\|\left\| \frac{1}{n} \sum_{k=1}^{n} x_k - u - u_n \right\|\right\| < \varepsilon \text{ and } \left\|\left\| \frac{1}{n} \sum_{k=1}^{n} y_k - y - v_n \right\|\right\| < \varepsilon. \text{ Then, if } m_n > r_{j+1} > N, \text{ then}$$

$$\limsup_n \|\|u_n + P_k u - v_j - P_k y_k\|\|$$

$$\le \limsup_n \left( \begin{array}{c} \left\|\left\| \frac{1}{n} \sum_{k=1}^{n} x_n - y_j \right\|\right\| + \left\|\left\| \frac{1}{n} \sum_{k=1}^{n} x_n - u - u_n \right\|\right\| \\ + \|\|y_j - y - v_j\|\| + \|\|u - P_k u\|\| \\ + \|\|y - P_k y\|\| \end{array} \right)$$

$$\le d + 4\varepsilon.$$

Then, if $r_{j+1} \ge k$ and $m_n + 1 > r_{j+1}$ and $r_j + 1 \le s_j \le r_{j+1}$ is such that $\|\|v_j\|\| = \gamma_{s_j} P_{s_j} \|v_j\|_\infty$, we get $\|\|u_n + P_k u - v_j - P_k y_k\|\| \ge \gamma_{s_j} (\|v_j\|_\infty + \|u_n\|_\infty) = \|\|v_j\|\| + \|\|u_n\|\|$. Hence, since $u_n \xrightarrow{w} 0$, using Lemma 15.4.8, $d + 4\varepsilon \ge \|\|v_j\|\| + \gamma_{s_j} \lim_n \|u_n\|_\infty = \|\|v_j\|\| + \gamma_{s_j} \lim_n \|\|u_n\|\|$ and by passing to the limit as $j \to \infty$, we have $d + 4\varepsilon \ge \lim_j \|\|v_j\|\| + \lim_n \|\|u_n\|\| = \lim_j \left\|\left\| \frac{1}{j} \sum_{k=1}^{j} y_k - y \right\|\right\| + \lim_n \left\|\left\| \frac{1}{n} \sum_{m=1}^{n} x_m - u \right\|\right\|$ and this concludes the proof. $\square$

**Theorem 15.4.12.** *Assume that $C \subset (c_0, \| \cdot \|_\infty)$ is a ccne set, $T : C \to C$ is an affine nonexpansive fixed point free mapping, $(x_n)_n \subset C$ is an afps, $x_n \underset{w}{\to} 0$ and that*

$$D = \left\{ x : \limsup_m \left\| \frac{1}{m} \sum_{n=1}^m x_n - x \right\|_\infty \le d \right\} \text{ is assumed not empty.}$$

*If*

$$c = \inf \left\{ \limsup_m \left\| \frac{1}{m} \sum_{n=1}^m y_n - y \right\| : (y_n)_n \subset D, (y_n)_n \text{ is an afps with } y_n \underset{w}{\to} y \right\},$$

*then*

$$\inf \left\{ \limsup_m \left\| \frac{1}{m} \sum_{n=1}^m y_n - z \right\| : z \in D, (y_n)_n \subset D \text{ is an afps with } y_n \underset{w}{\to} y \right\} \ge 2c.$$

*Proof.* By contradiction, assume that there exists $z \in D$ and $(y_n)_n \subset D$ is an afps with $y_n \underset{w}{\to} y$ such that $a = \lim\limits_{m \to \infty} \left\| \frac{1}{m} \sum_{n=1}^m y_n - z \right\| < 2c$. Then, by the hypothesis, $\limsup\limits_m \left\| \frac{1}{m} \sum_{n=1}^m y_n - y \right\| \ge c$ and

$$z \in D_1 = \left\{ u \in D : \limsup_m \left\| \frac{1}{m} \sum_{n=1}^m y_n - u \right\| \le a \right\} \ne \emptyset$$

and $D_1$ is a ccne which is $T$-invariant by Lemma 15.4.1. Let $(u_n)_n \subset D_1$ be an afps which converges weakly to $u$. Then, by Lemma 15.4.11,

$$\limsup_m \left\| \frac{1}{m} \sum_{n=1}^m u_n - u \right\| \le a - \lim_{m \to \infty} \left\| \frac{1}{m} \sum_{n=1}^m y_n - y \right\| < 2c - c = c$$

which is a contradiction. $\qquad\square$

**Theorem 15.4.13.** *Consider the equivalent norm $\|\|\cdot\|\|$ to the usual norm $\| \cdot \|_\infty$ of $c_0$ given below.*
*For $x = (\xi_k)_k \in c_0$, define*

$$\|\|x\|\| := \lim_{p \to \infty} \sup_{k \in \mathbb{N}} \gamma_k \left( \sum_{j=k}^\infty \frac{|\xi_j|^p}{j^2} \right)^{\frac{1}{p}}$$

*where $\gamma_k \uparrow_k 3$, $\gamma_k$ is strictly increasing with $\gamma_k > 2$, $\forall k \in \mathbb{N}$.*

*Then, $(c_0, \|\|\cdot\|\|)$ has the fixed point property for affine $\|\|\cdot\|\|$-nonexpansive self-mappings.*

*Proof.*

Define $\||x\||_k := \gamma_k \lim_{p \to \infty} \left( \sum_{j=k}^{\infty} \frac{|\xi_j|^p}{j^2} \right)^{\frac{1}{p}}$, $\forall k \in \mathbb{N}$.

Then, $\||x\|| = \sup_{k \in \mathbb{N}} \||x\||_k$. By contradiction, assume that $C \subset (c_0, \| \cdot \|_\infty)$ is a ccne set, $T : C \to C$ is an affine nonexpansive fixed point free mapping and that $C$ satisfies the hypothesis of Corollary 15.4.5. Thus,

$$b = \inf \left\{ \limsup_m \left\|\left\| \frac{1}{m} \sum_{n=1}^{m} y_n - y \right\|\right\| : (y_n)_n \subset C \text{ is an afps with } y_n \xrightarrow{w} y \right\} > 0.$$

Let $\alpha = 4\gamma_1 - 8$, $\varepsilon_1 = \frac{\alpha}{12}$, assume that $(x_n)_n \subset C$ is an afps with $x_n \xrightarrow{w} x_0$ and that

$$b \leq \lim_n \left\|\left\| \frac{1}{n} \sum_{i=1}^{n} x_i - x_0 \right\|\right\| < (1 + \varepsilon_1)b,$$

Without loss of generality we may assume $x_0 = 0$ and by Lemma 15.4.9, taking an appropriate subsequence, $\lim_{n \to \infty} \left\|\left\| \frac{1}{n} \sum_{i=1}^{n} x_i \right\|\right\|$ exists and

$$\lim_{m \to \infty} \lim_{n \to \infty} \left\|\left\| \frac{1}{m} \sum_{i=1}^{m} x_i - \frac{1}{n} \sum_{i=1}^{n} x_i \right\|\right\| = 2 \lim_{n \to \infty} \left\|\left\| \frac{1}{n} \sum_{i=1}^{n} x_i \right\|\right\| < 2(1 + \varepsilon_1)b.$$

Let

$$D = \left\{ z \in C : \limsup_n \left\|\left\| \frac{1}{n} \sum_{i=1}^{n} x_i - z \right\|\right\| \leq 2(1 + \varepsilon_1)b \right\}.$$

Then by the above, $D$ is a ccne set. Now, let

$$c = \inf \left\{ \limsup_m \left\|\left\| \frac{1}{m} \sum_{n=1}^{m} y_n - y \right\|\right\| : (y_n)_n \subset D \text{ is an afps with } y_n \xrightarrow{w} y \right\}.$$

Note that $c \geq b$ and that if $(y_n)_n \subset D$ is an afps with $y_n \xrightarrow{w} y$ and $\lim_{n \to \infty} \left\|\left\| \frac{1}{n} \sum_{i=1}^{n} y_i - y \right\|\right\|$ exists, by Lemma 15.4.8, $b \leq c \leq \lim_{n \to \infty} \left\|\left\| \frac{1}{n} \sum_{i=1}^{n} y_i - y \right\|\right\| =$

$$\lim_{n\to\infty}\left\|\frac{1}{n}\sum_{i=1}^{n}y_i-y\right\|_{\infty} \quad\text{and}\quad b\le\lim_{n\to\infty}\left\|\left\|\frac{1}{n}\sum_{i=1}^{n}x_i\right\|\right\|=\lim_{n\to\infty}\left\|\frac{1}{n}\sum_{i=1}^{n}x_i\right\|_{\infty}\quad\text{and so}$$

$$
\begin{aligned}
\frac{1}{\gamma_1}\|\|y\|\|_1 &= \|y\|_\infty = \lim_{n\to\infty}\left\|\frac{1}{n}\sum_{i=1}^{n}x_i\right\|_\infty + \|y\|_\infty - \lim_{n\to\infty}\left\|\frac{1}{n}\sum_{i=1}^{n}x_i\right\|_\infty \qquad (15.1)\\[2mm]
&\le \lim_{n\to\infty}\left(2\left\|\frac{1}{n}\sum_{i=1}^{n}x_i-y\right\|_\infty - \left\|\frac{1}{n}\sum_{i=1}^{n}x_i\right\|_\infty\right)\\[2mm]
&\le \frac{4}{\gamma_1}\lim_{n\to\infty}\lim_{m\to\infty}\left\|\left\|\frac{1}{n}\sum_{i=1}^{n}x_i-\frac{1}{m}\sum_{i=1}^{m}y_i\right\|\right\| - \lim_{n\to\infty}\left\|\frac{1}{n}\sum_{i=1}^{n}x_i\right\|_\infty\\[2mm]
&\quad -2\lim_{m\to\infty}\left\|\frac{1}{m}\sum_{i=1}^{m}y_i-y\right\|_\infty\\[2mm]
&\le \frac{8}{\gamma_1}(1+\varepsilon_1)b-3b=b\left(\frac{8}{\gamma_1}-3+\frac{8}{\gamma_1}\varepsilon_1\right).
\end{aligned}
$$

Now let $x\in D$ satisfy

$$\|\|x\|\|_1\le\|\|x\|\|\le(1+\varepsilon_1)b\le(1+\varepsilon_1)c. \qquad (15.2)$$

Such an element exists since $\lim_{n\to\infty}\left\|\left\|\frac{1}{n}\sum_{i=1}^{n}x_i\right\|\right\|<(1+\varepsilon_1)b$ and so for $n$ large

enough $\left\|\left\|\frac{1}{n}\sum_{i=1}^{n}x_i\right\|\right\|<(1+\varepsilon_1)b$.

Let $\varepsilon_2=\frac{\alpha}{3(8+\alpha)}$ and $m$ large enough so that

$$\frac{1}{\gamma_1}\|\|(I-P_m)x\|\|_1=\|(I-P_m)x\|_\infty<c\varepsilon_2. \qquad (15.3)$$

Let $\varepsilon_3=\frac{\alpha\Phi_m}{4(8+\alpha)(1-\Phi_m)}$ where $\Phi_m=1-\gamma_m$ and let $(y_n)_n\subset D$ be an afps with $y_n\underset{w}{\to}y$ and $\lim_{n\to\infty}\left\|\left\|\frac{1}{n}\sum_{i=1}^{n}y_i-y\right\|\right\|$ exists such that

$$
\begin{aligned}
\frac{1}{\gamma_1}\lim_{n\to\infty}\left\|\left\|\frac{1}{n}\sum_{i=1}^{n}y_i-y\right\|\right\|_1 &= \lim_{n\to\infty}\left\|\frac{1}{n}\sum_{i=1}^{n}y_i-y\right\|_\infty=\lim_{n\to\infty}\left\|\left\|\frac{1}{n}\sum_{i=1}^{n}y_i-y\right\|\right\|\\[2mm]
&< (1+\varepsilon_3)c.
\end{aligned}
$$

Thus, there exists $N\in\mathbb{N}$ such that for every $k\ge N$,

$$\frac{1}{\gamma_1}\left\|\frac{1}{k}\sum_{i=1}^{k}y_i-y\right\|_\infty\le(1+\varepsilon_3)c. \qquad (15.4)$$

Now let $\lambda=\frac{\Phi_m}{2(1-\Phi_m)}$ and set $z=\lambda x+(1-\lambda)\frac{1}{N}\sum_{i=1}^{N}y_i\in D$ (due to convexity of $D$) and pick $j>N$.

Now note that if $k > m$, then $\|\|u\|\|_k \leq \frac{\gamma_k}{\gamma_1}\|\|u\|\|_1$, $\forall u \in c_0$ and so using the inequalities 15.1, 15.3 and 15.4, we have

$$
\left\|\left\|\frac{1}{j}\sum_{i=1}^{j} y_i - z\right\|\right\|_k = \left\|\left\|(I - P_m)\frac{1}{j}\sum_{i=1}^{j} y_i - z\right\|\right\|_k
$$

$$
= \left\|\left\|(I - P_m)\frac{1}{j}\sum_{i=1}^{j} y_i - y + y - z\right\|\right\|_k
$$

$$
\leq \left\|\left\|\frac{1}{j}\sum_{i=1}^{j} y_i - y\right\|\right\|_k + (1 - \lambda)\left\|\left\|\frac{1}{N}\sum_{i=1}^{N} y_i - y\right\|\right\|_k
$$

$$
+ \lambda \left[ \begin{array}{c} \|\|(I - P_m)x\|\|_k \\ + \\ \|\|y\|\|_k \end{array} \right].
$$

Thus,

$$
\left\|\left\|\frac{1}{j}\sum_{i=1}^{j} y_i - z\right\|\right\|_k \leq \left[\gamma_k(1 + \varepsilon_3)(2 - \lambda) + \lambda\gamma_k\varepsilon_2 + \lambda\gamma_k\left(\frac{8}{\gamma_1} - 3 + \frac{8}{\gamma_1}\varepsilon_1\right)\right]c
$$

$$
\leq \left[2 + 2\varepsilon_3 + \lambda\left(\varepsilon_2 - 4 + \frac{8}{\gamma_1} + \frac{8}{\gamma_1}\varepsilon_1 - \varepsilon_3\right)\right]c
$$

$$
\leq \left[2 - \frac{\alpha\Phi_m^2}{8(8 + \alpha)(1 - \Phi_m)^2}\right]c.
$$

But if $k \leq m$, then

$$
\left\|\left\|\frac{1}{j}\sum_{i=1}^{j} y_i - z\right\|\right\|_k \leq \frac{\gamma_k}{\gamma_1}\left\|\left\|\frac{1}{j}\sum_{i=1}^{j} y_i - y + y - z\right\|\right\|_1
$$

$$
\leq \gamma_m \left[\frac{\lambda}{\gamma_1}(\|\|x\|\|_1 + \|\|y\|\|_1) + \frac{1}{\gamma_1}\left(\begin{array}{c}\left\|\left\|\frac{1}{j}\sum_{i=1}^{j} y_i - y\right\|\right\|_1 \\ + \\ (1 - \lambda)\left\|\left\|\frac{1}{N}\sum_{i=1}^{N} y_i - y\right\|\right\|_1\end{array}\right)\right]
$$

$$
\leq \left[2 - 2\Phi_m + (1 - \Phi_m)\left\{\lambda\left(\frac{9}{\gamma_1} - 3 + \frac{9}{\gamma_1}\varepsilon_1\right) + 2\varepsilon_3\right\}\right]c
$$

$$
\leq \left[2 - \frac{\Phi_m(20 + 3\alpha)}{2(8 + \alpha)}\right]c.
$$

Hence,

$$
\lim_j \left\|\left\|\frac{1}{j}\sum_{i=1}^{j} y_i - z\right\|\right\|
$$

$$
\leq \max\left\{\left[2 - \frac{\alpha\Phi_m^2}{8(8 + \alpha)(1 - \Phi_m)^2}\right]c, \left[2 - \frac{\Phi_m(20 + 3\alpha)}{2(8 + \alpha)}\right]c\right\} < 2c
$$

contradicting with Theorem 15.4.12.      □

## 15.5 Family of equivalent norms on $c_0$ with FPP for affine nonexpansive mappings

Firstly, we would like to remind that this section is also derived from and uses similar methods of the first and second author's recent accepted paper entitled "$c_0$ can be renormed to have the fixed point property for affine non-expansive mappings" which is to be published by Filomat journal. However, we take a diferent renorming to construct our family.

In this section, we generalize our result from the previous section and we would like to note that the proof of our theorem below is inspired by the proofs of theorems and lemmas given by Carlos A. Hernández-Linares, María A. Japón and Enrique Llorens-Fuster Hernández-Linares et al. (2012) such that they extend P.K. Lin's work Lin (2008). We implement our ideas and get our desired result with the help of their work. In our theorem, the reader will notice that Lemma 15.4.7 shows that our hypothesis in our theorem below is reasonable. Now, let us see our theorem with its proof and a remark concluding our chapter.

**Theorem 15.5.1.** *If $\rho(\cdot)$ is an equivalent norm to the usual norm on $c_0$ such that*

$$\limsup_n \rho\left(\frac{1}{n}\sum_{m=1}^n x_m + x\right) = \limsup_n \rho\left(\frac{1}{n}\sum_{m=1}^n x_m\right) + \rho(x)$$

*for every weakly null sequence $(x_n)_n$ and for all $x \in c_0$, then for every $\lambda > 0$, $c_0$ with the norm $\|\cdot\|_\rho = \rho(\cdot) + \lambda\,\|\|\cdot\|\|$ has the FPP for affine $\|\cdot\|_\rho$-nonexpansive self-mappings.*

*Proof.* First of all, let us define for any $k \in \mathbb{N}$ and for any $x = (\xi_i)_{i\in\mathbb{N}} \in c_0$,

$$S_k(x) := \lim_{p\to\infty}\left(\sum_{j=k}^\infty \frac{|\xi_j|^p}{j^2}\right)^{\frac{1}{p}} \quad \text{and} \quad \rho_k(x) := \rho(x) + \lambda\gamma_k S_k(x). \tag{15.5}$$

Note that it is clear that for every $x \in c_0$, $\|\|x\|\| = \sup_k S_k(x)$ and due to Lemma 15.4.7, for every weakly null sequence $(x_n)_n$, there exists a subsequence $(x_{n_l})_l$ such that for every $x \in c_0$ and for all $k \in \mathbb{N}$,

$$\limsup_j \rho_k\left(\frac{1}{j}\sum_{l=1}^j x_{n_l} + x\right) = \limsup_j \rho_k\left(\frac{1}{j}\sum_{l=1}^j x_{n_l}\right) + \rho_k(x). \tag{15.6}$$

Now, by contradiction, assume that $\left(c_0, \|\cdot\|_\rho\right)$ fails the fixed point property

for affine $\| \cdot \|_\rho$-nonexpansive self-mappings. Let $T$ and $D$ be as in Corollary 15.4.5 and $\tau$ be the weak $(\sigma(l^\infty, l^1))$-topology in $c_0$ that we will denote by $w$.
Define

$$M := \inf \left\{ \limsup_m \left\| \frac{1}{m} \sum_{n=1}^m x_n - x \right\|_\rho : (x_n)_n \subset D , \ (x_n)_n \text{ is an afps}, \ x_n \underset{w}{\to} x \right\}.$$

Note that $M > 0$. Assume that $A(D)$ is the set of all $(x_n)_n \subset D$ such that $(x_n)_n$ is an afps converging weakly $(w)$ to some $x \in c_0$ and such that

$$\lim_j \rho \left( \frac{1}{j} \sum_{k=1}^j x_k - u \right), \ \lim_j \left\| \frac{1}{j} \sum_{k=1}^j x_k - u \right\| \text{ and } \lim_j S_k \left( \frac{1}{j} \sum_{k=1}^j x_k - u \right) \text{ exist for}$$

any $u \in c_0$ and for all $k \in \mathbb{N}$.

Now, due to separability of $c_0$, we can say that

$$M := \inf \left\{ \lim_m \left\| \frac{1}{m} \sum_{n=1}^m x_n - x \right\|_\rho : (x_n)_n \subset A(D) , \ x_n \underset{w}{\to} x \right\}.$$

Without loss of generality we can assume that $M = 1$ and from the equivalence of the norms, we can obtain that

$$c := \inf \{ \|\|x\|\| : \rho(x) = \lambda \} > 0 \text{ and } d := \inf \left\{ \|\|x\|\| : \|x\|_\rho = \lambda \right\} > 0. \ (15.7)$$

Choose $\delta_1 > 0$ such that $\frac{1+\delta_1}{1+c} + 2\delta_1 < 1$ an afps $(x_n)_n \subset A(D)$ such that weak-$\lim_n x_n = x$ and

$$\lim_m \left\| \frac{1}{m} \sum_{n=1}^m x_n - x \right\|_\rho < 1 + \delta_1. \text{ Without loss of generality we can assume that}$$

$x = 0$. Now, note that $K := \left\{ z \in D : \lim_m \left\| \frac{1}{m} \sum_{n=1}^m x_n - z \right\|_\rho \leq 2 + 2\delta_1 \right\}$ is a nonempty closed, bounded, convex and $T$-invariant subset and there exists $n_0 \in \mathbb{N}$ such that for every $n \geq n_0$, $x_n \in K$.

Define

$$Q := \inf \left\{ \lim_m \left\| \frac{1}{m} \sum_{n=1}^m y_n - y \right\|_\rho : (y_n)_n \subset K \cap A(D) , \ y_n \underset{w}{\to} y \right\}$$

and note that

$$1 \leq Q \leq \lim_m \left\| \frac{1}{m} \sum_{n=1}^m x_n \right\|_\rho \leq 1 + \delta_1. \tag{15.8}$$

Define $A(K) := \{ (y_n)_n \subset A(D) : y_n \in K, \forall n \in \mathbb{N} \}$ and pick an arbitrary afps $(y_n)_n \subset A(K)$ with $y_n \underset{w}{\to} y$. Then, for every $k \in \mathbb{N}$, without loss of generality, by passing to the appropriate subsequence of $(x_n)_n$ if necessary, we have

$$2 + 2\delta_1 \geq \limsup_s \lim_m \left\| \frac{1}{m} \sum_{n=1}^{m} x_n - \frac{1}{s} \sum_{n=1}^{s} y_n \right\|_\rho$$

$$\geq \limsup_s \lim_m \rho_k \left( \frac{1}{m} \sum_{n=1}^{m} x_n - \frac{1}{s} \sum_{n=1}^{s} y_n \right)$$

$$= \limsup_s \left[ \limsup_m \rho_k \left( \frac{1}{m} \sum_{n=1}^{m} x_n \right) + \rho_k \left( \frac{1}{s} \sum_{n=1}^{s} y_n \right) \right] \quad \text{by (15.6)}$$

$$= \limsup_m \rho_k \left( \frac{1}{m} \sum_{n=1}^{m} x_n \right) + \limsup_s \rho_k \left( \frac{1}{s} \sum_{n=1}^{s} y_n - y \right) + \rho_k(y) \quad \text{by (15.6)}$$

Hence,

$$2 + 2\delta_1 \geq \lim_m \rho \left( \frac{1}{m} \sum_{n=1}^{m} x_n \right) + \lambda \gamma_k \lim_m S_k \left( \frac{1}{m} \sum_{n=1}^{m} x_n \right) + \lim_s \rho \left( \frac{1}{s} \sum_{n=1}^{s} y_n - y \right)$$

$$+ \lambda \gamma_k \lim_s S_k \left( \frac{1}{s} \sum_{n=1}^{s} y_n - y \right) + \rho_k(y)$$

$$= \lim_m \rho \left( \frac{1}{m} \sum_{n=1}^{m} x_n \right) + \lambda \gamma_k \lim_m \left\| \frac{1}{m} \sum_{n=1}^{m} x_n \right\| + \lim_s \rho \left( \frac{1}{s} \sum_{n=1}^{s} y_n - y \right)$$

$$+ \lambda \gamma_k \lim_s \left\| \frac{1}{s} \sum_{n=1}^{s} y_n - y \right\| + \rho_k(y)$$

$$\geq \gamma_k \left[ \lim_m \left\| \frac{1}{m} \sum_{n=1}^{m} x_n \right\|_\rho + \lim_s \left\| \frac{1}{s} \sum_{n=1}^{s} y_n - y \right\|_\rho \right] + \rho_k(y)$$

$$\geq 2\gamma_k + \rho_k(y).$$

Thus, if $(y_n)_n \subset A(K)$ is an afps with $y_n \xrightarrow{w} y$, we have

$$\rho_k(y) \leq 2(1 - \gamma_k) + 2\delta_1 < 2 + 2\delta_1. \tag{15.9}$$

Now, we choose $s$ such that $\frac{1+\delta_1}{1+c} < 2\delta_1 < s < 1$ and note that from (15.7), for all $u \in c_0$, $\rho(u) \leq \frac{\|u\|_\rho}{1+c}$. Thus,

$$\lim_m \rho \left( \frac{1}{m} \sum_{n=1}^{m} x_n \right) \leq \frac{\lim_m \left\| \frac{1}{m} \sum_{n=1}^{m} x_n \right\|_\rho}{1+c} < \frac{1+\delta_1}{1+c}$$

and so there exists $n_0 \in \mathbb{N}$ such that

$$\rho \left( \frac{1}{n_0} \sum_{n=1}^{n_0} x_n \right) < \frac{\lim_m \left\| \frac{1}{m} \sum_{n=1}^{m} x_n \right\|_\rho}{1+c} < \frac{1+\delta_1}{1+c}$$

But recalling we took $x = 0$,

$$\limsup_k \rho_k \left( \frac{1}{n_0} \sum_{n=1}^{n_0} x_n \right) = \rho \left( \frac{1}{n_0} \sum_{n=1}^{n_0} x_n \right) + \lambda \limsup_k \gamma_k S_k \left( \frac{1}{n_0} \sum_{n=1}^{n_0} x_n \right)$$

$$= \rho \left( \frac{1}{n_0} \sum_{n=1}^{n_0} x_n \right).$$

Therefore, there exists $k_0 \in \mathbb{N}$ such that for all $k \geq k_0$,

$$\rho_k \left( \frac{1}{n_0} \sum_{n=1}^{n_0} x_n \right) < \frac{1 + \delta_1}{1 + c} \tag{15.10}$$

and

$$q_k := \frac{1 + \delta_1}{1 + c} + 2(1 - \gamma_k) + 2\delta_1 < s < 1 \leq Q. \tag{15.11}$$

Since $K$ is a bounded set, $\exists H > 0$ such that $\rho(x) < H$ for all $x \in K$. Thus,

$$\rho_k \left( \frac{1}{n_0} \sum_{n=1}^{n_0} x_n \right) < H, \forall k \in \mathbb{N}. \tag{15.12}$$

Now, let us define $s_0 := 1 - M(1 - \gamma_{k_0})$. Note that $s_0 < 1$. Define also

$$h := H + 2 + 2\delta_1 \text{ and note that by (15.8)}, h > Q > s_0 Q \tag{15.13}$$

Now, choose $\alpha \in (0,1)$ that satisfies $\alpha < \frac{2Q(1-s_0)}{h - s_0 Q}$; hence, $(2 - \alpha)Q + \alpha s = 2Q - \alpha(Q - s) < 2Q$ and
$(2 - \alpha)s_0 Q + \alpha h = 2s_0 Q + \alpha(h - s_0 Q) < 2s_0 Q + 2Q(1 - s_0) = 2Q$.
Thus, we can find $\delta_2 > 0$ such that

$$(2 - \alpha)(Q + \delta_2) + \alpha s < 2Q \tag{15.14}$$

and

$$(2 - \alpha)s_0(Q + \delta_2) + \alpha h < 2Q. \tag{15.15}$$

Note that for

$$W := \max\left\{(2 - \alpha)(Q + \delta_2) + \alpha s, \; (2 - \alpha)s_0(Q + \delta_2) + \alpha h\right\}, W < 2Q. \tag{15.16}$$

Now, let $(y_n)_n \subset A(K)$ be an afps with $y_n \xrightarrow{w} y$ and $\lim_m \left\| \frac{1}{m} \sum_{n=1}^{m} y_n - y \right\|_\rho < Q + \delta_2$. Then, $\exists N_0 \in \mathbb{N}$ such that $\forall m \geq N_0$,

$$\left\| \frac{1}{m} \sum_{n=1}^{m} y_n - y \right\|_\rho < Q + \delta_2. \tag{15.17}$$

Also,

$$\lim_m \rho_k \left( \frac{1}{m} \sum_{n=1}^m y_n - y \right) = \lim_m \rho \left( \frac{1}{m} \sum_{n=1}^m y_n - y \right) + \lambda \gamma_k \lim_m S_k \left( \frac{1}{m} \sum_{n=1}^m y_n - y \right)$$

$$= \lim_m \rho \left( \frac{1}{m} \sum_{n=1}^m y_n - y \right) + \lambda \gamma_k \lim_m \left\| \left\| \frac{1}{m} \sum_{n=1}^m y_n - y \right\| \right\|$$

$$= \lim_m \left\| \frac{1}{m} \sum_{n=1}^m y_n - y \right\|_\rho - (1 - \gamma_k) \lambda \lim_m \left\| \left\| \frac{1}{m} \sum_{n=1}^m y_n - y \right\| \right\|$$

$$\leq [1 - (1 - \gamma_k)d] \lim_m \left\| \frac{1}{m} \sum_{n=1}^m y_n - y \right\|_\rho$$

by definition of $d$ from (15.7)

$$< [1 - (1 - \gamma_k)d] (Q + \delta_2)$$

and we can find $N_1 \geq N_0$ such that for all $m \geq N_1$ and for all $k \leq k_0$

$$\rho_k \left( \frac{1}{m} \sum_{n=1}^m y_n - y \right) < [1 - (1 - \gamma_k)d] (Q + \delta_2) \leq s_0(Q + \delta_2). \qquad (15.18)$$

Now, define $z_0 := \alpha \frac{1}{n_0} \sum_{n=1}^{n_0} x_n + (1 - \alpha) \frac{1}{N_1} \sum_{n=1}^{N_1} y_n$ and note that since $K$ is convex, $z_0 \in K$. Now, we show that $\lim_m \left\| \frac{1}{m} \sum_{n=1}^m y_n - z_0 \right\|_\rho \leq W$ and to prove that we will observe for all $k \in \mathbb{N}$ and for all $m \geq N_1$, $\rho_k \left( \frac{1}{m} \sum_{n=1}^m y_n - z_0 \right) \leq W$.

First, we need to take the equation

$$\frac{1}{m} \sum_{n=1}^m y_n - z_0 = \frac{1}{m} \sum_{n=1}^m y_n - y - (1 - \alpha) \left( \frac{1}{N_1} \sum_{n=1}^{N_1} y_n - y \right) - \alpha \left( \frac{1}{n_0} \sum_{n=1}^{n_0} x_n - y \right)$$

into consideration.

We will have two cases to see this for $m \geq N_1$.

**Case 3.** $k > k_0$ :

$$\rho_k \left( \frac{1}{m} \sum_{n=1}^m y_n - z_0 \right) \leq \rho_k \left( \frac{1}{m} \sum_{n=1}^m y_n - y \right) + (1 - \alpha) \rho_k \left( \frac{1}{N_1} \sum_{n=1}^{N_1} y_n - y \right)$$

$$+ \alpha \left[ \rho_k \left( \frac{1}{n_0} \sum_{n=1}^{n_0} x_n \right) + \rho_k (y) \right]$$

$$\leq \left\| \frac{1}{m} \sum_{n=1}^m y_n - y \right\|_\rho + (1 - \alpha) \left\| \frac{1}{N_1} \sum_{n=1}^{N_1} y_n - y \right\|_\rho$$

$$+ \alpha \left[ \rho_k \left( \frac{1}{n_0} \sum_{n=1}^{n_0} x_n \right) + \rho_k (y) \right]$$

*Then,*

$$\rho_k\left(\frac{1}{m}\sum_{n=1}^{m}y_n - z_0\right) < (2-\alpha)(Q+\delta_2) + \alpha\left[\rho_k\left(\frac{1}{n_0}\sum_{n=1}^{n_0}x_n\right) + \rho_k(y)\right] \text{ by (15.17)}$$

$$< (2-\alpha)(Q+\delta_2) + \alpha\left[\frac{1+\delta_1}{1+c} + 2(1-\gamma_k) + 2\delta_1\right]$$

$$\text{by (15.9) and (15.10)}$$

$$= (2-\alpha)(Q+\delta_2) + \alpha q_k < (2-\alpha)(Q+\delta_2) + \alpha s \text{ by (15.11)}$$

$$\leq W \text{ from (15.16).}$$

**Case 4.** $k \leq k_0$ :

$$\rho_k\left(\frac{1}{m}\sum_{n=1}^{m}y_n - z_0\right) \leq \rho_k\left(\frac{1}{m}\sum_{n=1}^{m}y_n - y\right) + (1-\alpha)\rho_k\left(\frac{1}{N_1}\sum_{n=1}^{N_1}y_n - y\right)$$

$$+\alpha\left[\rho_k\left(\frac{1}{n_0}\sum_{n=1}^{n_0}x_n\right) + \rho_k(y)\right]$$

$$\leq s_0(2-\alpha)(Q+\delta_2) + \alpha\left[\rho_k\left(\frac{1}{n_0}\sum_{n=1}^{n_0}x_n\right) + \rho_k(y)\right] \text{ by (15.18)}$$

$$< s_0(2-\alpha)(Q+\delta_2) + \alpha\left[H + 2 + 2\delta_1\right] \text{ by (15.12) and (15.9)}$$

$$\leq s_0(2-\alpha)(Q+\delta_2) + \alpha h \text{ by (15.13)}$$

$$\leq W \text{ from (15.16).}$$

*Then,* $\rho_k\left(\frac{1}{m}\sum_{n=1}^{m}y_n - z_0\right) \leq W$ for all $k \in \mathbb{N}$ and for all $m \geq N_1$.

*Thus,* $\left\|\frac{1}{m}\sum_{n=1}^{m}y_n - y\right\|_\rho \leq W$ for all $m \geq N_1$ and so

$$\limsup_m \left\|\frac{1}{m}\sum_{n=1}^{m}y_n - y\right\|_\rho \leq W.$$

Now, we can say that there exists a nonempty subset

$K_0 := \left\{z \in K : \limsup_m \left\|\frac{1}{m}\sum_{n=1}^{m}y_n - z\right\|_\rho \leq W\right\}$ and an afps $(u_n)_n \subset K_0 \cap$

$A(D)$ with $u_n \xrightarrow{w} u \in c_0$. Then, for every $k \in \mathbb{N}$,

$$W > \limsup_s \limsup_m \left\|\frac{1}{m}\sum_{n=1}^{m}y_n - \frac{1}{s}\sum_{n=1}^{s}u_n\right\|_\rho$$

$$\geq \limsup_s \limsup_m \rho_k\left(\frac{1}{m}\sum_{n=1}^{m}y_n - \frac{1}{s}\sum_{n=1}^{s}u_n\right)$$

$$= \lim_m \rho_k\left(\frac{1}{m}\sum_{n=1}^{m}y_n - y\right) + \lim_s \rho_k\left(\frac{1}{s}\sum_{n=1}^{s}u_n - u\right) + \rho_k(y-u) \text{ by (15.6).}$$

Hence,

$$W \geq \lim_m \rho\left(\frac{1}{m}\sum_{n=1}^{m} y_n - y\right) + \lambda\gamma_k \lim_m \left|\left|\left|\frac{1}{m}\sum_{n=1}^{m} y_n - y\right|\right|\right|$$

$$+ \lim_s \rho\left(\frac{1}{s}\sum_{n=1}^{s} u_n - u\right) + \lambda\gamma_k \lim_s \left|\left|\left|\frac{1}{s}\sum_{n=1}^{s} u_n - u\right|\right|\right|.$$

Thus, by taking limits as $k$ approaches to infinity, we obtain that

$$W \geq \lim_m \left|\left|\frac{1}{m}\sum_{n=1}^{m} y_n - y\right|\right|_\rho + \lim_s \left|\left|\frac{1}{s}\sum_{n=1}^{s} u_n - u\right|\right|_\rho \geq Q + Q = 2Q$$

which contradicts the definition of $W$ by (15.16). Therefore, $\left(c_0, \|\cdot\|_\rho\right)$ has the fixed point property for affine $\|\cdot\|_\rho$-nonexpansive self-mappings. $\qquad\square$

*Remark* 15.5.2. Since $\|x\|_\infty = \lim\limits_{p\to\infty}\left(\sum\limits_{k=1}^{\infty}\frac{|\xi_k|^p}{2^k}\right)^{\frac{1}{p}} = \lim\limits_{p\to\infty}\left(\sum\limits_{k=1}^{\infty}\frac{|\xi_k|^p}{k^2}\right)^{\frac{1}{p}}$, $\forall x = (\xi_i)_{i\in\mathbb{N}} \in c_0$, it is easy to see our results have been derived from the below results obtained in the recent study of the first and second author as we mentioned at the beginning of the section:

For $x = (\xi_k)_k \in c_0$, define

$$|||x|||^\sim := \lim_{p\to\infty}\sup_{k\in\mathbb{N}}\gamma_k\left(\sum_{j=k}^{\infty}\frac{|\xi_j|^p}{2^j}\right)^{\frac{1}{p}}$$

where $\gamma_k \uparrow_k 3$, $\gamma_k$ is strictly increasing with $\gamma_k > 2$, $\forall k \in \mathbb{N}$,

then $(c_0, |||\cdot|||^\sim)$ has the fixed point property for affine $|||\cdot|||^\sim$-nonexpansive self-mappings and if $\rho^\sim(\cdot)$ is an equivalent norm to the usual norm on $c_0$ such that

$$\limsup_n \rho^\sim\left(\frac{1}{n}\sum_{m=1}^{n} x_m + x\right) = \limsup_n \rho^\sim\left(\frac{1}{n}\sum_{m=1}^{n} x_m\right) + \rho^\sim(x)$$

for every weakly null sequence $(x_n)_n$ and for all $x \in c_0$, then for every $\lambda > 0$, $c_0$ with the norm $\|\cdot\|_{\rho^\sim} = \rho^\sim(\cdot) + \lambda|||\cdot|||^\sim$ has the FPP for affine $\|\cdot\|_{\rho^\sim}$-nonexpansive self-mappings.

# Bibliography

Alspach D. E. A fixed point free nonexpansive map. *Proceedings of the American Mathematical Society*, 82(3):423–424, 1981.

Banach S. Sur les opérations dans les ensembles abstraits et leur application aux équations intégrales. *Fund. Math*, 3(1):133–181, 1922.

Bessaga C. and Pełczyński A. On bases and unconditional convergence of series in Banach spaces. *Studia Mathematica*, 17(2):151–164, 1958.

Brouwer L. E. J. Über abbildung von mannigfaltigkeiten. *Mathematische Annalen*, 71(1):97–115, 1911.

Browder F. E. Fixed-point theorems for noncompact mappings in Hilbert space. *Proc. Nat. Acad. Sci. U.S.A.*, 53(6):1272–1276, 1965a.

Browder F. E. Fixed-point theorems for noncompact mappings in Hilbert space. *Proc. Nat. Acad. Sci. U.S.A.*, 54(4):1041–1044, 1965b.

Carothers N. L., Dilworth S. J., and Lennard C. J. On a localization of the ukk property and the fixed point property in $L_{w,1}$. *Lecture notes in Pure and Applied Mathematics*, pages 111–124, 1996.

Carothers N. L., Dilworth S. J., Lennard C. J., and Trautman D. A. A fixed point property for the Lorentz space $L_{p,1}(\mu)$. *Indiana University Mathematics Journal*, 40(1):345, 1991.

Diestel J. *Sequences and Series in Banach Spaces*, volume 92 of *Graduate Texts in Mathematics*. Springer, New York, 1984.

Domínguez Benavides T. A renorming of some nonseparable Banach spaces with the Fixed Point Property. *J. Math. Anal. Appl.*, 350(2):525–530, 2009.

Dowling P. N. and Lennard C. J. Every nonreflexive subspace of $L_1[0, 1]$ fails the fixed point property. *Proc. Am. Math. Soc.*, 125(2):443–446, 1997.

Dowling P. N., Lennard C. J., and Turett B. Asymptotically isometric copies of $c_0$ in Banach spaces. *J. Math. Anal. Appl.*, 219:377–391, 1998.

Dowling P. N., Lennard C. J., and Turett B. Some fixed point results in $l^1$ and $c_0$. *Nonlinear Analysis*, 39:929–936, 2000.

Dowling P. N., Lennard C. J., and Turett B., 2001. Renormings of $l^1$ and $c_0$ and fixed point properties. In *Handbook of Metric Fixed Point Theory*, pages 269–297. Springer Netherlands.

Dowling P. N., Lennard C. J., and Turett B. Weak compactness is equivalent to the fixed point property in $c_0$. *Proc. Amer. Math. Soc.*, 132(6):1659–1666, 2004.

Everest T., 2013. *Fixed points of nonexpansive maps on closed, bounded, convex sets in $l^1$*. PhD thesis, University of Pittsburgh.

Fetter H. and de Buen B. G. Banach spaces with a basis that are hereditarily asymptotically isometric to $l^1$ and the fixed point property. *Nonlinear Analysis*, 71:4598–4608, 2009.

Goebel K. and Kuczumow T., 1979. Irregular convex sets with fixed-point property for nonexpansive mappings. In *Colloquium Mathematicae*, volume 2, pages 259–264.

Göhde D. Zum Prinzip der kontraktiven Abbildung. *Math. Nachr.*, 30:251–258, 1965.

Gunduz B. and Dutta H. On the convergence of an iteration process for nonself totally asymptotically i-quasi-nonexpansive mappings. *Proceedings of the Jangjeon Mathematical Society*, 20(3):377–389, 2017.

Güven A., 2017. Characterizing nonweakly compact subsets of $c_0$ with fixed point property for affine nonexpansive mappings when $c_0$ is renormed. Master's thesis, Kafkas University.

Hardy G. H., Littlewood J. E., and Pólya G. *Inequalities*. Cambridge University Press, 1952.

Hernández-Linares C. A., 2011. *Propiedad de punto fijo, normas equivalentes y espacios de funciones no-conmutativos*. PhD thesis, Universidad de Sevilla.

Hernández-Linares C. A. and Japón M. A. Renormings and fixed point property in non commutative $l_1$ spaces ii: Affine mappings. *Nonlinear Analysis*, 75(13):5357–5361, 2012.

Hernández-Linares C. A., Japón M. A., and Llorens-Fuster E. On the structure of the set of equivalent norms on $l^1$ with the fixed point property. *Journal of Mathematical Analysis and Applications*, 387(2):645–654, 2012.

Kir M., Akturk M. A., Dutta H., and Yolacan E. An extension of rational contractive condition in partially ordered rectangular metric spaces. *Miskolc Mathematical Notes*, 18(2):873–888, 2017.

Kirk W. A. A fixed point theorem for mappings which do not increase distances. *The American Mathematical Monthly*, 72(9):1004–1006, 1965.

Lennard C. and Nezir V. The closed, convex hull of every ai $c_0$-summing basic sequence fails the fpp for affine nonexpansive mappings. *J. Math. Anal. Appl.*, 381:678–688, 2011.

Lennard C. and Nezir V. Reflexivity is equivalent to the perturbed fixed point property for cascading nonexpansive maps in Banach lattices. *Nonlinear Analysis: Theory, Methods & Applications*, 95:414–420, 2014.

Lennard C. and Nezir V. Semi-strongly asymptotically non-expansive mappings and their applications on fixed point theory. *Haccettepe Journal of Mathematics and Statistics*, 46(4):613–620, 2017.

Lin P. K. There is an equivalent norm on $\ell_1$ that has the fixed point property. *Nonlinear Analysis*, 68:2303–2308, 2008.

Nezir V., 2012. *Fixed point properties for $c_0$-like spaces*. PhD thesis, University of Pittsburgh.

Nezir V. A new look to the usual norm of $c_0$ and candidates to renormings of $c_0$ with fixed point property. *Kafkas Üniversitesi Fen Bilimleri Enstitüsü Dergisi*, 10(2):85–102, 2017.

Núñez C. Characterization of Banach spaces of continuous vector valued functions with the weak Banach-Saks property. *Illinois Journal of Mathematics*, 33:27–41, 1989.

Oran S., 2017. Fixed point property for affine nonexpansive mappings on a very large class of nonweakly compact subsets in $c_0$ respect to an equivalent norm. Master's thesis, Kafkas University.

Santos F. E. C., Fetter H., Gamboa de Buen B., and Núñez-Medina F. Directionally bounded sets in c0 with equivalent norms. *Journal of Mathematical Analysis and Applications*, 419(2):727–737, 2014.

Schauder J. Der fixpunktsatz in funktionalraümen. *Studia Mathematica*, 2(1):171–180, 1930.

# Chapter 16

# Steinhaus Type Theorems over Valued Fields: A Survey

**P.N. Natarajan**

*Old No. 2/3, New No. 3/3, Second Main Road, R.A. Puram, Chennai, India & Formerly Head of the Department of Mathematics, Ramakrishna Mission Vivekananda College, Chennai, India. E-mail: pinnangudinatarajan@gmail.com*

## 16.1   Introduction

In this chapter, we present a brief survey of the Steinhaus type theorems over complete, non-trivially valued fields proved so far. By a complete, non-trivially valued $K$, we mean $K = \mathbb{R}$ (the field of real numbers) or $\mathbb{C}$ (the field of complex numbers) or a complete, non-trivially valued, non-archimedean field. For classical summability theory, standard references are Cooke (1950), Hardy (1949), Maddox (1970), while, for summability theory over non-archimedean fields, a standard reference is Natarajan (2015). Basic knowledge of Analysis - both classical and non-archimdean - is assumed, standard references for non-archimedean analysis being Bachman (1964), Narici et al. (1971). With this very brief introduction, we shall recall some basic concepts and theorems in summability theory.

## 16.2 Basic concepts and definitions in summability theory

Throughout this chapter, $K = \mathbb{R}$ or $\mathbb{C}$ or a complete, non-trivially valued, non-archimedean field. We shall explicitly mention which field is chosen depending on the context.

Given an infinite matrix $A = (a_{nk})$, $a_{nk} \in K$, $n, k = 0, 1, 2, \ldots$ and a sequence $x = \{x_k\}$, $x_k \in K$, $k = 0, 1, 2, \ldots$, by the $A$-transform of $x = \{x_k\}$, we mean the sequence $Ax = \{(Ax)_n\}$,

$$(Ax)_n = \sum_{k=0}^{\infty} a_{nk} x_k, \quad n = 0, 1, 2, \ldots,$$

where, we suppose that the series on the right converge. If $\lim_{n \to \infty} (Ax)_n = t$, we say that $x = \{x_k\}$ is summable $A$ or $A$-summable to $t$. If $\lim_{n \to \infty} (Ax)_n = t$, whenever, $\lim_{k \to \infty} x_k = s$, we say that $A$ is convergence-preserving or conservative. If, further,

$$\lim_{n \to \infty} (Ax)_n = \lim_{k \to \infty} x_k,$$

we say that $A$ is a regular or a Toeplitz matrix. The following results, characterizing a conservative and a regular matrix, in terms of the entries of the matrix, are well-known.

**Theorem 16.2.1** (see Cooke (1950), Hardy (1949), Maddox (1970)). *When $K = \mathbb{R}$ or $\mathbb{C}$, $A = (a_{nk})$ is conservative if and only if*

$$\left. \begin{array}{l} (i) \; \sup_{n \geq 0} \sum_{k=0}^{\infty} |a_{nk}| < \infty; \\[2mm] (ii) \; \lim_{n \to \infty} a_{nk} = \delta_k \; \text{exists}, k = 0, 1, 2, \ldots; \\[2mm] \text{and} \\[2mm] (iii) \; \lim_{n \to \infty} \sum_{k=0}^{\infty} a_{nk} = \delta \; \text{exists}. \end{array} \right\} \quad (16.1)$$

*Further, $A$ is regular if and only if (16.1) holds with $\delta_k = 0$, $k = 0, 1, 2, \ldots$ and $\delta = 1$.*

**Theorem 16.2.2** (see Monna (1963), Natarajan (2015)). *When $K$ is a complete, non-trivially valued, non-archimedean field, $A = (a_{nk})$ is conservative* *if and only if*

$$
\left.
\begin{aligned}
&(i) \sup_{n,k} |a_{nk}| < \infty; \\
&(ii) \lim_{n \to \infty} a_{nk} = \delta_k \ exists \ , k = 0, 1, 2, \ldots; \\
&and \\
&(iii) \lim_{n \to \infty} \sum_{k=0}^{\infty} a_{nk} = \delta \ exists.
\end{aligned}
\right\}
\tag{16.2}
$$

*Further, $A$ is regular if and only if (16.2) holds, with $\delta_k = 0$, $k = 0, 1, 2, \ldots$ and $\delta = 1$.*

*Remark* 16.2.3. In the context of Theorem 16.2.2, it is worth noting that the absence of an analogue in non-archimedean analysis for the signum function in classical analysis possibly made Monna Monna (1963) use functional analytic tools like the analogue of Uniform Boundedness Principle to prove Theorem 16.2.2. However, Natarajan Natarajan (1991) proved Theorem 16.2.2 without using functional analytic tools.

Let $X, Y$ be sequence spaces and $A = (a_{nk})$ be an infinite matrix. If $\{(Ax)_n\} \in Y$, whenever $x = \{x_k\} \in X$, we write

$$
A \in (X, Y).
$$

So, if $A$ is conservative, $A \in (c, c)$, where $c$ is the space of all convergent sequences in $K$. If $A$ is regular, we write

$$
A \in (c, c; P),
$$

$P$ denoting "preservation of limit", i.e., $\lim_{n \to \infty} (Ax)_n = \lim_{k \to \infty} x_k$, $x = \{x_k\} \in c$.

$A = (a_{nk})$ is called a Schur matrix if $\{(Ax)_n\} \in c$, whenever $x = \{x_k\} \in \ell_\infty$, where $\ell_\infty$ is the space of all bounded sequences in $K$, i.e., $A$ is a Schur matrix if

$$
A \in (\ell_\infty, c).
$$

For $x = \{x_k\} \in \ell_\infty$, define

$$
\|x\| = \sup_{k \geq 0} |x_k|. \tag{16.3}
$$

When $K = \mathbb{R}$ or $\mathbb{C}$, $\ell_\infty$ is a Banach space with respect to the norm defined by (16.3); when $K$ is a complete, non-trivially valued, non-archimedean field, $K$ is a non-archimedean Banach space with respect to the norm defined by (16.3). Note that $c$ is a closed subspace of $\ell_\infty$, in both cases, under the norm (16.3).

The following results are again well-known (see Cooke (1950), Hardy (1949), Maddox (1970) for $K = \mathbb{R}$ or $\mathbb{C}$; Natarajan (1978), Natarajan (2015) when $K$ is a complete, non-trivially valued, non-archimedean field).

**Theorem 16.2.4.** *When* $K = \mathbb{R}$ *or* $\mathbb{C}$, $A = (a_{nk})$ *is a Schur matrix, i.e.,* $A \in (\ell_\infty, c)$ *if and only if*

$$
\left.
\begin{array}{l}
(i) \ \lim_{n \to \infty} a_{nk} = \delta_k \text{ exists }, k = 0, 1, 2, \ldots; \\[2mm]
\text{and} \\[4mm]
(ii) \ \sum_{k=0}^{\infty} |a_{nk}| \text{ converges uniformly with respect to} \\[2mm]
\quad n = 0, 1, 2, \ldots.
\end{array}
\right\} \qquad (16.4)
$$

**Theorem 16.2.5.** *When* $K$ *is a complete, non-trivially valued, non-archimedean field,* $A \in (\ell_\infty, c)$ *if and only if*

$$
\left.
\begin{array}{l}
(i) a_{nk} \to 0, k \to \infty, n = 0, 1, 2, \ldots; \\[2mm]
\text{and} \\[2mm]
(ii) \sup_{k \geq 0} |a_{n+1,k} - a_{nk}| \to 0, n \to \infty.
\end{array}
\right\} \qquad (16.5)
$$

Using Theorems 16.2.1 - 16.2.5, we have the following important result.

**Theorem 16.2.6** (Steinhaus). *An infinite matrix* $A = (a_{nk})$ *cannot be both a regular and a Schur matrix, i.e., given a regular matrix* $A$, *there exists a bounded, divergent sequence which is not $A$-summable. In other words, symbolically,*

$$
(c, c; P) \cap (\ell_\infty, c) = \phi.
$$

Let $X, Y$ be sequence spaces such that there is some notion of limit or sum in $X, Y$. Then, we denote by $(X, Y; P)$ that subclass of $(X, Y)$ consisting of infinite matrices which preserve this limit or sum. If there is a notion of limit in $X$ and a notion of sum in $Y$ or vice versa, $(X, Y, ; P')$ denotes that subclass of $(X, Y)$ consisting of infinite matrices which preserve the limit in $X$ and the sum in $Y$ or vice versa. With this understanding, if $X, Y, Z$ are sequence spaces such that $X \subset Z$, then results of the type

$$
(X, Y; P) \cap (Z, Y) = \phi
$$

and

$$
(X, Y; P') \cap (Z, Y) = \phi
$$

are called "Steinhaus type" theorems.

When $K$ is a complete, non-trivially valued field, we present a survey of various Steinhaus type theorems proved so far.

## 16.3     A Steinhaus type theorem involving the sequence spaces $\Lambda_r$, $r \geq 1$ being integer

In the context of Steinhaus theorem, we introduce the sequence space $\Lambda_r$, $r = 1, 2, \ldots$. The sequence space $\Lambda_r$, $r \geq 1$ being a fixed integer, is defined as the set of all $x = \{x_k\} \in \ell_\infty$ such that

$$|x_{k+r} - x_k| \to 0, k \to \infty.$$

Note that $\Lambda_r$ is a closed subspace of $\ell_\infty$ with respect to the norm defined by (16.3).

When $K = \mathbb{R}$ or $\mathbb{C}$, the following result, improving Steinhaus theorem (Theorem 16.2.6), is proved here (see Natarajan (1987), Natarajan (2015)). We give a constructive proof.

**Theorem 16.3.1.** *When $K = \mathbb{R}$ or $\mathbb{C}$,*

$$(c, c; P) \cap \left( \Lambda_r - \bigcup_{i=1}^{r-1} \Lambda_i, c \right) = \phi.$$

*Proof.* Let $A = (a_{nk})$ be a regular matrix. We can now choose two sequences of positive integers $\{n(m)\}$, $\{k(m)\}$ such that if $m = 2p$, $n(m) > n(m-1)$, $k(m) > k(m-1) + (2m-5)r$, then

$$\sum_{k=0}^{k(m-1)+(2m-5)r} |a_{n(m),k}| < \frac{1}{16},$$

$$\sum_{k=k(m)+1}^{\infty} |a_{n(m),k}| < \frac{1}{16};$$

and if

$$m = 2p+1, n(m) > n(m-1), k(m) > k(m-1) + (m-2)r,$$

then

$$\sum_{k=0}^{k(m-1)+(m-2)r} |a_{n(m),k}| < \frac{1}{16},$$

$$\sum_{k=k(m-1)+(m-2)r+1}^{k(m)} |a_{n(m),k}| > \frac{7}{8},$$

and

$$\sum_{k=k(m)+1}^{\infty} |a_{n(m),k}| < \frac{1}{16}.$$

Define the sequence $x = \{x_k\}$ as follows:

if $k(2p-1) < k \le k(2p)$, then

$$
x_k = \begin{cases}
\frac{2p-2}{2p-1}, & \text{if } k = k(2p-1) + 1; \\[4pt]
1, & \text{if } k(2p-1) + 1 < k \le k(2p-1) + r; \\[4pt]
\frac{2p-3}{2p-1}, & \text{if } k = k(2p-1) + r + 1; \\[4pt]
1, & \text{if } k(2p-1) + r + 1 < k \le k(2p-1) + 2r; \\[4pt]
\vdots & \\[4pt]
1, & \text{if } k(2p-1) + (2p-4)r + 1 < k \le k(2p-1) + (2p-3)r; \\[4pt]
\frac{1}{2p-1}, & \text{if } k = k(2p-1) + (2p-3)r + 1; \\[4pt]
\frac{2p-2}{2p-1}, & \text{if } k(2p-1) + (2p-3)r + 1 < k \le k(2p-1) + (2p-2)r; \\[4pt]
0, & \text{if } k = k(2p-1) + (2p-2)r + 1; \\[4pt]
\frac{2p-3}{2p-1}, & \text{if } k(2p-1) + (2p-2)r + 1 < k \le k(2p-1) + (2p-1)r; \\[4pt]
0, & \text{if } k = k(2p-1) + (2p-1)r + 1; \\[4pt]
\vdots & \\[4pt]
\frac{1}{2p-1}, & \text{if } k(2p-1) + (4p-6)r + 1 < k \le k(2p-1) + (4p-5)r; \\[4pt]
0, & \text{if } k(2p-1) + (4p-5)r < k \le k(2p),
\end{cases}
$$

and

if $k(2p) < k \le k(2p+1)$, then

$$
x_k = \begin{cases}
\frac{1}{2p}, & \text{if } k(2p) < k \le k(2p) + r; \\[4pt]
\frac{2}{2p}, & \text{if } k(2p) + r < k \le k(2p) + 2r; \\[4pt]
\vdots & \\[4pt]
\frac{2p-1}{2p}, & \text{if } k(2p) + (2p-2)r < k \le k(2p) + (2p-1)r; \\[4pt]
1, & \text{if } k(2p) + (2p-1)r < k \le k(2p+1).
\end{cases}
$$

We note that

if $k(2p-1) < k \le k(2p)$,

$$
|x_{k+r} - x_k| < \frac{1}{2p-1};
$$

while,

if $k(2p) < k \le k(2p+1)$,

$$
|x_{k+r} - x_k| < \frac{1}{2p}.
$$

Thus, $|x_{k+r} - x_k| \to 0$, $k \to \infty$, showing that $x = \{x_k\} \in \Lambda_r$. However,

$$|x_{k+1} - x_k| = \frac{2p - 2}{2p - 1}, \quad \text{if } k = k(2p - 1) + (2p - 3)r, p = 1, 2, \ldots$$

so that

$$|x_{k+1} - x_k| \nrightarrow 0, k \to \infty,$$

and consequently

$$x \notin \Lambda_1.$$

In a similar manner, we can prove that

$$x \notin \Lambda_i, i = 2, 3, \ldots, (r - 1).$$

Thus

$$x = \{x_k\} \in \Lambda_r - \bigcup_{i=1}^{r-1} \Lambda_i.$$

Further,

$$\left. \begin{array}{l} |(Ax)_{n(2p)}| < \frac{1}{16} + \frac{1}{16} = \frac{1}{8}, \\ |(Ax)_{n(2p+1)}| > \frac{7}{8} - \frac{1}{16} - \frac{1}{16} = \frac{3}{4} \end{array} \right\}, p = 1, 2, \ldots,$$

which shows that $\{(Ax)_n\} \notin c$, completing the proof of the theorem. □

In the context of Theorem 16.3.1, Theorem 16.3.2 below indicates a deviation in the non-archimedean case from the classical case.

**Theorem 16.3.2.** *If $K$ is a complete, non-trivially valued, non-archimedean field,*

$$(c, c; P) \cap (\Lambda_r, c) = \phi,$$

$r \geq 1$ *being a fixed integer.*

*Proof.* We will show that there exists a regular matrix with entries in $Q_p$, the $p$-adic field for a prime $p$, which sums all sequences in $\Lambda_r$. Consider the infinite matrix $A = (a_{nk})$, $a_{nk} \in Q_p$, $n, k = 0, 1, 2, \ldots$, where

$$\left. \begin{array}{ll} a_{nk} & = \frac{1}{r}, \quad k = n, n + 1, \ldots, n + r - 1; \\ & = 0, \quad \text{otherwise} \end{array} \right\}, n = 0, 1, 2, \ldots.$$

Now,

$$\begin{aligned} (Ax)_{n+1} - (Ax)_n &= \frac{\begin{array}{c} (x_{n+1} + x_{n+2} + \cdots + x_{n+r}) \\ -(x_n + x_{n+1} + \cdots + x_{n+r-1}) \end{array}}{r} \\ &= \frac{x_{n+r} - x_n}{r} \\ &\to 0, n \to \infty, \end{aligned}$$

if $x = \{x_k\} \in \Lambda_r$. Thus $\{(Ax)_n\}$ is a Cauchy sequence in $Q_p$, which is complete. So $\{(Ax)_n\} \in c$, i.e., $A$ sums all sequences in $\Lambda_r$. It is clear that $A$ is regular, completing the proof. □

## 16.4　On theorems of Steinhaus type proved by Maddox

In this section, we need the following sequence spaces:

$$\ell_1 = \left\{ x = \{x_k\} : \sum_{k=0}^{\infty} |x_k| < \infty \right\};$$

$$\gamma = \left\{ x = \{x_k\} : \sum_{k=0}^{\infty} x_k \text{ converges} \right\};$$

$$\gamma_\infty = \left\{ x = \{x_k\} : \{s_n\} \in \ell_\infty, s_n = \sum_{k=0}^{n} x_k, n = 0, 1, 2, \dots \right\}.$$

Maddox Maddox (1967) proved the following Steinhaus type result:

**Theorem 16.4.1.** *When* $K = \mathbb{R}$ *or* $\mathbb{C}$,

$$(\gamma, \gamma; P) \cap (\gamma_\infty, \gamma) = \phi,$$

*where,* $(\gamma, \gamma; P)$ *denotes the set of all infinite matrices* $A \in (\gamma, \gamma)$ *such that*

$$\sum_{n=0}^{\infty} (Ax)_n = \sum_{k=0}^{\infty} x_k, \quad x = \{x_k\} \in \gamma.$$

Maddox Maddox (1967) also observed that

$$(\ell_1, \ell_1; P) \cap (\gamma, \ell_1) \neq \phi,$$

where again, $(\ell_1, \ell_1; P)$ is the set of all infinite matrices $A \in (\ell_1, \ell_1)$ such that

$$\sum_{n=0}^{\infty} (Ax)_n = \sum_{k=0}^{\infty} x_k, \quad x = \{x_k\} \in \ell_1.$$

We write $A = (a_{nk}) \in (\ell_1, \ell_1; P)'$ if $A \in (\ell_1, \ell_1, P)$ with the additional condition

$$a_{nk} \to 0, \ k \to \infty, \ n = 0, 1, 2, \dots.$$

Natarajan Natarajan (1997) proved that given an $(\ell_1, \ell_1; P)'$ matrix $A$, there exists a sequence $x = \{x_k\} \in \gamma$, whose $A$-transform is not in $\ell_1$, i.e.,

$$(\ell_1, \ell_1; P)' \cap (\gamma, \ell_1) = \phi.$$

To prove the above result, we need a lemma, the proof of which is modelled on that of Fridy's Fridy (1970).

It is well-known that $A = (a_{nk}) \in (\ell_1, \ell_1)$ (see Mears (1937)) if and only if

$$\sup_{k \geq 0} \sum_{n=0}^{\infty} |a_{nk}| < \infty.$$

Further, $A \in (\ell_1, \ell_1; P)$ if and only if $A \in (\ell_1, \ell_1)$ and

$$\sum_{n=0}^{\infty} a_{nk} = 1, \ k = 0, 1, 2, \ldots.$$

**Lemma 16.4.2.** *If $A = (a_{nk}) \in (\ell_1, \ell_1)$ with $a_{nk} \to 0$, $k \to \infty$, $n = 0, 1, 2, \ldots$ and*

$$\overline{\lim_{k \to \infty}} \left| \sum_{n=0}^{\infty} a_{nk} \right| > 0,$$

*then there exists a sequence $x = \{x_k\} \in \gamma$ such that $\{(Ax)_n\} \notin \ell_1$.*

*Proof.* By hypothesis, for some $\epsilon > 0$, there exists a strictly increasing sequence $\{k(i)\}$ of positive integers such that

$$\left| \sum_{n=0}^{\infty} a_{n,k(i)} \right| \geq 2\epsilon, \ i = 1, 2, \ldots.$$

In particular,

$$\left| \sum_{n=0}^{\infty} a_{n,k(1)} \right| \geq 2\epsilon.$$

We then choose a positive integer $n(1)$ such that

$$\sum_{n > n(1)} |a_{n,k(1)}| < \min \left( \frac{1}{2}, \frac{\epsilon}{2} \right),$$

this being possible since $\sum_{n=0}^{\infty} |a_{nk}| < \infty$, $k = 0, 1, 2, \ldots$, using the fact that $A \in (\ell_1, \ell_1)$. It now follows that

$$\left| \sum_{n=0}^{n(1)} a_{n,k(1)} \right| > \epsilon.$$

In general, having chosen $k(j)$, $n(j)$, $j \leq m - 1$, choose a positive integer $k(m) > k(m-1)$ such that

$$\left| \sum_{n=0}^{\infty} a_{n,k(m)} \right| \geq 2\epsilon,$$

$$\sum_{n=0}^{n(m-1)} |a_{n,k(m)}| < \min \left( \frac{1}{2}, \frac{\epsilon}{2} \right),$$

and then choose a positive integer $n(m) > n(m-1)$ such that

$$\sum_{n > n(m)} |a_{n,k(m)}| < \min\left(\frac{1}{2^m}, \frac{\epsilon}{2}\right),$$

so that

$$\left| \sum_{n=n(m-1)+1}^{n(m)} a_{n,k(m)} \right| > 2\epsilon - \frac{\epsilon}{2} - \frac{\epsilon}{2}$$

$$= \epsilon.$$

Let the sequence $x = \{x_k\}$ be defined by

$$x_k = \frac{(-1)^{i+1}}{i}, \quad \text{if } k = k(i);$$
$$= 0, \quad \text{if } k \neq k(i), \ i = 1, 2, \ldots.$$

It is clear that $x = \{x_k\} \in \gamma$. Defining $n(0) = 0$, we have,

$$\sum_{n=0}^{n(N)} |(Ax)_n| \geq \sum_{m=1}^{N} \sum_{n=n(m-1)+1}^{n(m)} |(Ax)_n|$$

$$= \sum_{m=1}^{N} \sum_{n=n(m-1)+1}^{n(m)} \left| \sum_{i=1}^{\infty} a_{n,k(i)} x_{k(i)} \right|$$

$$= \sum_{m=1}^{N} \sum_{n=n(m-1)+1}^{n(m)} \left| \sum_{i=1}^{\infty} a_{n,k(i)} \frac{(-1)^{i+1}}{i} \right|$$

$$\geq \sum_{m=1}^{N} \sum_{n=n(m-1)+1}^{n(m)} \left\{ \left| \frac{(-1)^{m+1}}{m} a_{n,k(m)} \right| \right.$$

$$\left. - \sum_{\substack{i=1 \\ i \neq m}}^{\infty} \left| \frac{(-1)^{i+1}}{i} a_{n,k(i)} \right| \right\}$$

$$= \sum_{m=1}^{N} \sum_{n=n(m-1)+1}^{n(m)} \left\{ \frac{1}{m} |a_{n,k(m)}| \right.$$

$$\left. - \sum_{\substack{i=1 \\ i \neq m}}^{\infty} \frac{1}{i} |a_{n,k(i)}| \right\}$$

$$> \epsilon \sum_{m=1}^{N} \frac{1}{m} - \sum_{m=1}^{N} \sum_{n=n(m-1)+1}^{n(m)} \sum_{\substack{i=1 \\ i \neq m}}^{\infty} |a_{n,k(i)}|$$

$$\left(\text{since } \frac{1}{i} \leq 1\right).$$

Now,

$$\sum_{m=1}^{\infty} \sum_{n=n(m-1)+1}^{n(m)} \sum_{i<m} |a_{n,k(i)}| \leq \sum_{m=1}^{\infty} \frac{1}{2^m}$$

$$= 1.$$

Similarly, we can prove that

$$\sum_{m=1}^{\infty} \sum_{n=n(m-1)+1}^{n(m)} \sum_{i>m} |a_{n,k(i)}| \leq \frac{1}{2}.$$

Consequently

$$\sum_{n=0}^{n(N)} |(Ax)_n| > \epsilon \sum_{m=1}^{N} \frac{1}{m} - \frac{3}{2}.$$

Since $\sum_{m=1}^{\infty} \frac{1}{m}$ diverges, $\{(Ax)_n\} \notin \ell_1$, completing the proof of the lemma. $\square$

We now prove the following.

**Theorem 16.4.3.** *When $K = \mathbb{R}$ or $\mathbb{C}$,*

$$(\ell_1, \ell_1; P)' \cap (\gamma, \ell_1) = \phi.$$

*Proof.* Let $A = (a_{nk}) \in (\ell_1, \ell_1; P)' \cap (\gamma, \ell_1)$. Since $\sum_{n=0}^{\infty} a_{nk} = 1$, $k = 0, 1, 2, \ldots$,

$$\varlimsup_{k \to \infty} \left| \sum_{n=0}^{\infty} a_{nk} \right| = 1 > 0.$$

In view of Lemma 16.4.2, there exists a sequence $x = \{x_k\} \in \gamma$ such that $\{(Ax)_n\} \notin \ell_1$, which contradicts our assumption, completing the proof of the theorem. $\square$

*Remark* 16.4.4. The condition

$$a_{nk} \to 0, \ k \to \infty, \ n = 0, 1, 2, \ldots$$

is essential for the validity of Theorem 16.4.3. If the condition is dropped,

Theorem 16.4.3 fails to hold, as the following example illustrates: The infinite matrix

$$A = (a_{nk}) = \begin{bmatrix} 1 & 1 & 1 & \cdots \\ 0 & 0 & 0 & \cdots \\ 0 & 0 & 0 & \cdots \\ \cdots & \cdots & \cdots & \cdots \end{bmatrix}$$

is in $(\ell_1, \ell_1; P) \cap (\gamma, \ell_1)$ and $a_{0k} \not\to 0$, $k \to \infty$.

---

## 16.5   On a theorem of Steinhaus type of Fridy over valued fields

When $K = \mathbb{R}$ or $\mathbb{C}$, it was proved by Fridy Fridy (1970) that

$$(\ell_1, \ell_1; P) \cap (\ell_\alpha, \ell_1) = \phi, \ \alpha > 1,$$

where the sequence space $\ell_\alpha$ is defined by

$$\ell_\alpha = \left\{ x = \{x_k\} : \sum_{k=0}^{\infty} |x_k|^\alpha < \infty \right\}, \alpha \geq 1.$$

Defining, for $x = \{x_k\} \in \ell_\alpha$, $\alpha \geq 1$,

$$\|x\| = \left( \sum_{k=0}^{\infty} |x_k|^\alpha \right)^{\frac{1}{\alpha}},$$

$\ell_\alpha$ is a Banach space, whatever be $K$, with respect to the norm defined above.

The result of Fridy, as such, fails to hold when $K$ is a complete, non trivially valued, non-archimedean field as the following example illustrates:

Let $K = \mathbb{Q}_3$, the 3-adic field and $A$ be the infinite matrix defined by

$$A \equiv (a_{nk}) = \begin{bmatrix} \frac{1}{4} & \frac{1}{4} & \frac{1}{4} & \cdots \\ \frac{1}{4}\left(\frac{3}{4}\right) & \frac{1}{4}\left(\frac{3}{4}\right) & \frac{1}{4}\left(\frac{3}{4}\right) & \cdots \\ \frac{1}{4}\left(\frac{3}{4}\right)^2 & \frac{1}{4}\left(\frac{3}{4}\right)^2 & \frac{1}{4}\left(\frac{3}{4}\right)^2 & \cdots \\ \cdots & \cdots & \cdots & \cdots \end{bmatrix},$$

i.e., $a_{nk} = \frac{1}{4}\left(\frac{3}{4}\right)^n$, $k = 0, 1, 2, \ldots; n = 0, 1, 2, \ldots$.

Now,

$$\sup_{k \geq 0} \sum_{n=0}^{\infty} |a_{nk}|_3 = \frac{1}{1 - \rho} < \infty, \text{ where } \rho = |3|_3$$

and

$$\sum_{n=0}^{\infty} a_{nk} = 1, k = 0, 1, 2, \ldots,$$

so that $A \in (\ell_1, \ell_1; P)$. For $\alpha > 1$, if $x = \{x_k\} \in \ell_\alpha \subset c_0$,

$$\sum_{n=0}^{\infty} |(Ax)_n|_3 = \sum_{n=0}^{\infty} \left| \sum_{k=0}^{\infty} a_{nk} x_k \right|_3$$

$$= \sum_{n=0}^{\infty} \left| \frac{1}{4} \left( \frac{3}{4} \right)^n \right|_3 \left| \sum_{k=0}^{\infty} x_k \right|_3$$

$$= \left| \sum_{k=0}^{\infty} x_k \right|_3 \frac{1}{1-\rho} < \infty,$$

so that $A \in (\ell_\alpha, \ell_1)$ too. Thus

$$(\ell_1, \ell_1; P) \cap (\ell_\alpha, \ell_1) \neq \phi,$$

when $K$ is a complete, non-trivially valued, non-archimedean field.

However, the following is an attempt to salvage Fridy's result in a general form in the non-archimedean set up.

When $K = \mathbb{R}$ or $\mathbb{C}$, a complete characterization of the matrix class $(\ell_\alpha, \ell_\beta)$, $\alpha, \beta \geq 2$, does not seem to be available in the literature. A result of Koskela Koskela (1978) in this direction characterizes only non-negative matrices in $(\ell_\alpha, \ell_\beta)$, when $\alpha \geq \beta > 1$. A known simple sufficient condition (Maddox (1970), p. 174, Theorem 9) for a matrix to belong to $(\ell_\alpha, \ell_\alpha)$ is

$$A \in (\ell_\infty, \ell_\infty) \cap (\ell_1, \ell_1).$$

Sufficient conditions or necessary conditions for $A \in (\ell_\alpha, \ell_\beta)$, when $K = \mathbb{R}$ or $\mathbb{C}$, are available in literature (see e.g. Stieglitz and Tietz (1977)). Necessary and sufficient conditions for $A \in (\ell_1, \ell_1)$ are well-known and is due to Mears Mears (1937) (for alternate proofs, see Knopp and Lorentz Knopp and Lorentz (1949), Fridy Fridy (1969)). However, when $K$ is a complete, non-trivially valued, non-archimedean field, Natarajan (Natarajan (1986), Natarajan (2015)) obtained a complete characterization of $(\ell_\alpha, \ell_\alpha)$, $\alpha > 0$.

**Theorem 16.5.1.** *When $K$ is a complete, non-trivially valued, non-archimedean field, $A = (a_{nk}) \in (\ell_\alpha, \ell_\alpha)$, $\alpha > 0$, if and only if*

$$\sup_{k \geq 0} \sum_{n=0}^{\infty} |a_{nk}|^\alpha < \infty. \tag{16.6}$$

In this case, as usual, we write $A = (a_{nk}) \in (\ell_\alpha, \ell_\alpha; P)$ if $A \in (\ell_\alpha, \ell_\alpha)$ and $\sum_{n=0}^{\infty} (Ax)_n = \sum_{k=0}^{\infty} x_k$, $x = \{x_k\} \in \ell_\alpha$, noting that since $\ell_\alpha \subset c_0$, whenever $x = \{x_k\} \in \ell_\alpha$, $\{x_k\} \in c_0$ and so $\sum_{k=0}^{\infty} x_k$ converges in the non-archimedean case, where $c_0$ denotes the non-archimedean Banach space of all null sequences in $K$ under the norm defined by (16.3). It is easy to prove that $A = (a_{nk}) \in (\ell_\alpha, \ell_\alpha; P)$ if $A \in (\ell_\alpha, \ell_\alpha)$ and

$$\sum_{n=0}^{\infty} a_{nk} = 1, \ k = 0, 1, 2, \dots.$$

We write $A = (a_{nk}) \in (\ell_\alpha, \ell_\alpha; P)'$ if $A \in (\ell_\alpha, \ell_\alpha; P)$ and

$$a_{nk} \to 0, k \to \infty, n = 0, 1, 2, \dots.$$

We need the following lemma.

**Lemma 16.5.2.** *Let* $A = (a_{nk}) \in (\ell_\alpha, \ell_\alpha)$ *such that*

$$a_{nk} \to 0, k \to \infty, n = 0, 1, 2, \dots$$

*and*

$$\overline{\lim_{k \to \infty}} \sum_{n=0}^{\infty} |a_{nk}|^\alpha > 0.$$

*Then there exists a sequence* $x = \{x_k\} \in \ell_\beta$, $\beta > \alpha$ *such that* $\{(Ax)_n\} \notin \ell_\alpha$.

*Proof.* By hypothesis, for some $\epsilon > 0$, there exists a subsequence $\{k(i)\}$ of positive integers such that

$$\sum_{n=0}^{\infty} |a_{n,k(i)}|^\alpha \geq 2\epsilon, i = 1, 2, \dots.$$

In particular,

$$\sum_{n=0}^{\infty} |a_{n,k(1)}|^\alpha \geq 2\epsilon.$$

We then choose a positive integer $n(1)$ such that

$$\sum_{n > n(1)} |a_{n,k(1)}|^\alpha < \min\left(2^{-1}, \frac{\epsilon}{2}\right),$$

this being possible, since $A \in (\ell_\alpha, \ell_\alpha)$ and so $\sum_{n=0}^{\infty} |a_{n,k}|^\alpha < \infty$, $k = 0, 1, 2, \dots$.
It now follows that

$$\sum_{n=0}^{n(1)} |a_{n,k(1)}|^\alpha > \epsilon.$$

More generally, having chosen the positive integers $k(j)$, $n(j)$, $j \leq m - 1$, choose a positive integer $k(m)$ such that $k(m) > k(m - 1)$,

$$\sum_{n=0}^{\infty} |a_{n,k(m)}|^\alpha \geq 2\epsilon,$$

$$\sum_{n=0}^{n(m-1)} |a_{n,k(m)}|^\alpha < \min\left(2^{-m}, \frac{\epsilon}{2}\right),$$

and then choose a positive integer $n(m)$ such that $n(m) > n(m-1)$,

$$\sum_{n>n(m)} |a_{n,k(m)}|^\alpha < \min\left(2^{-m}, \frac{\epsilon}{2}\right),$$

so that

$$\sum_{n=n(m-1)+1}^{n(m)} |a_{n,k(m)}|^\alpha > 2\epsilon - \frac{\epsilon}{2} - \frac{\epsilon}{2}$$

$$= \epsilon.$$

Since $K$ is non-trivially valued, there exists $\pi \in K$ such that $0 < \rho = |\pi| < 1$. For each $i = 1, 2, \ldots$, choose a non-negative integer $\lambda(i)$ such that

$$\rho^{\lambda(i)+1} \le \frac{1}{i^{\frac{1}{\alpha}}} < \rho^{\lambda(i)}.$$

Define the sequence $x = \{x_k\}$ by

$$\left.\begin{array}{ll} x_k & = \pi^{\lambda(i)}, \quad \text{if } k = k(i); \\ & = 0, \qquad \text{if } k \ne k(i), \end{array}\right\} i = 1, 2, \ldots.$$

Now,

$$\sum_{k=0}^{\infty} |x_k|^\beta = \sum_{i=1}^{\infty} |x_{k(i)}|^\beta$$

$$= \sum_{i=1}^{\infty} \rho^{\beta\lambda(i)}$$

$$\le \frac{1}{\rho^\beta} \sum_{i=1}^{\infty} \frac{1}{i^{\frac{\beta}{\alpha}}}$$

$$< \infty, \quad \text{since } \beta > \alpha,$$

while,

$$\sum_{k=0}^{\infty} |x_k|^\alpha = \sum_{i=1}^{\infty} |x_{k(i)}|^\alpha$$

$$= \sum_{i=1}^{\infty} \rho^{\alpha\lambda(i)}$$

$$> \sum_{i=1}^{\infty} \frac{1}{i}$$

$$= \infty,$$

so that $x = \{x_k\} \in \ell_\beta - \ell_\alpha$. Defining $n(0) = 0$, we have,

$$\sum_{n=0}^{n(N)} |(Ax)_n|^\alpha \geq \sum_{m=1}^{N} \sum_{n=n(m-1)+1}^{n(m)} |(Ax)_n|^\alpha$$

$$= \sum_{m=1}^{N} \sum_{n=n(m-1)+1}^{n(m)} \left| \sum_{i=1}^{\infty} a_{n,k(i)} x_{k(i)} \right|^\alpha$$

$$= \sum_{m=1}^{N} \sum_{n=n(m-1)+1}^{n(m)} \left| \sum_{i=1}^{\infty} a_{n,k(i)} \pi^{\lambda(i)} \right|^\alpha$$

$$\geq \sum_{m=1}^{N} \sum_{n=n(m-1)+1}^{n(m)} \left\{ |a_{n,k(m)}|^\alpha \rho^{\alpha\lambda(m)} \right.$$

$$\left. - \sum_{i\neq m} |a_{n,k(i)}|^\alpha \rho^{\alpha\lambda(i)} \right\},$$

using the fact $|a + b|^\alpha \geq ||a|^\alpha - |b|^\alpha|, \alpha > 0$,
the valuation being non-archimedean

$$\geq \sum_{m=1}^{N} \sum_{n=n(m-1)+1}^{n(m)} \left\{ |a_{n,k(m)}|^\alpha m^{-1} \right.$$

$$\left. - \frac{1}{\rho^\alpha} \sum_{i\neq m} |a_{n,k(i)}|^\alpha \right\},$$

since $\rho^{\alpha(\lambda(i)+1)} \leq \frac{1}{i} \leq 1$ so that $\rho^{\alpha\lambda(i)} \leq \frac{1}{\rho^\alpha}$

$$> \sum_{m=1}^{N} \epsilon m^{-1} - \frac{1}{\rho^\alpha} \sum_{m=1}^{N} \sum_{n=n(m-1)+1}^{n(m)} \sum_{i\neq m} |a_{n,k(i)}|^\alpha.$$

We note that

$$\sum_{m=1}^{N} \sum_{n=n(m-1)+1}^{n(m)} \sum_{i<m} |a_{n,k(i)}|^\alpha < \sum_{m=1}^{\infty} \frac{1}{2^m}$$

$$= 1.$$

Similarly, it can be shown that

$$\sum_{m=1}^{N} \sum_{n=n(m-1)+1}^{n(m)} \sum_{i>m} |a_{n,k(i)}|^\alpha < \frac{1}{2}.$$

Thus

$$\sum_{n=0}^{n(N)} |(Ax)_n|^\alpha > \epsilon \sum_{m=1}^{N} \frac{1}{m} - \frac{3}{2}.$$

Since $\sum_{m=1}^{\infty} \frac{1}{m} = \infty$, it follows that $\{(Ax)_n\} \notin \ell_\alpha$, completing the proof of the lemma. $\square$

We now have the following result.

**Theorem 16.5.3.** *When $K$ is a complete, non-trivially valued, non-archimedean field,*

$$(\ell_\alpha, \ell_\alpha; P)' \cap (\ell_\beta, \ell_\alpha) = \phi, \beta > \alpha.$$

*Proof.* Suppose $A = (a_{nk}) \in (\ell_\alpha, \ell_\alpha; P)' \cap (\ell_\beta, \ell_\alpha)$. Then

$$\sum_{n=0}^{\infty} |a_{nk}|^\alpha \geq \left| \sum_{n=0}^{\infty} a_{nk} \right|^\alpha = 1, k = 0, 1, 2, \ldots,$$

since the valuation of $K$ is non-archimedean and so $|a + b|^\alpha \leq |a|^\alpha + |b|^\alpha$, $\alpha > 0$. Thus

$$\overline{\lim_{k \to \infty}} \sum_{n=0}^{\infty} |a_{nk}|^\alpha \geq 1 > 0.$$

In view of Lemma 16.5.2, there exists $x = \{x_k\} \in \ell_\beta$ such that $\{(Ax)_n\} \notin \ell_\alpha$, which contradicts our assumption. The proof of the theorem is now complete. $\square$

*Remark 16.5.4.* We can prove that the condition

$$a_{nk} \to 0, k \to \infty, n = 0, 1, 2, \ldots$$

is indispensable, in the sense that if we remove the condition,

$$(\ell_\alpha, \ell_\alpha; P) \cap (\ell_\beta, \ell_\alpha) \neq \phi, \beta > \alpha > 1.$$

However, the following result holds in the non-archimedean set up.

**Theorem 16.5.5.** *When $K$ is a complete, non-trivially valued, non-archimedean field,*

$$(\ell_\alpha, \ell_\alpha; P) \cap (\ell_\infty, \ell_\alpha) = \phi, \alpha > 0.$$

*Proof.* Suppose $A \in (\ell_\alpha, \ell_\alpha; P) \cap (\ell_\infty, \ell_\alpha)$, $\alpha > 0$. Then $A \in (\ell_\alpha, \ell_\alpha; P)' \cap (\ell_\beta, \ell_\alpha)$, $\beta > \alpha > 0$, which is a contradiction, establishing the theorem. $\square$

## 16.6   Some Steinhaus type theorems of Natarajan over valued fields

For the contents of this section, one can refer to Natarajan (1996).
$(\ell_1, c; P')$ denotes the class of all infinite matrices $A \in (\ell_1, c)$ such that

$$\lim_{n \to \infty} (Ax)_n = \sum_{k=0}^{\infty} x_k, x = \{x_k\} \in \ell_1.$$

When $K = \mathbb{R}$ or $\mathbb{C}$, it is known (Stieglitz and Tietz, 1977, p.4,17), that $A = (a_{nk}) \in (\ell_1, c)$ if and only if

$$\left. \begin{array}{l} (i) \sup_{n,k} |a_{nk}| < \infty; \\[2mm] and \\[2mm] (ii) \lim_{n \to \infty} a_{nk} = \delta_k \text{ exists}, k = 0, 1, 2, \ldots \end{array} \right\} \qquad (16.7)$$

We now prove the following result.

**Theorem 16.6.1.** *When $K = \mathbb{R}$ or $\mathbb{C}$, $A = (a_{nk}) \in (\ell_1, c; P')$ if and only if (16.7) holds with $\delta_k = 1$, $k = 0, 1, 2, \ldots$.*

*Proof.* Let $A = (a_{nk}) \in (\ell_1, c; P')$. Let $e_k$ be the sequence in which 1 occurs in the $k$th place and 0 elsewhere, $k = 0, 1, 2, \ldots$, i.e.,

$$e_k = \{x_i^{(k)}\}_{i=1}^{\infty},$$

where

$$\begin{aligned} x_i^{(k)} &= 1, \quad \text{if } i = k; \\ &= 0, \quad \text{otherwise.} \end{aligned}$$

Then $e_k \in \ell_1$, $k = 0, 1, 2, \ldots$, $\displaystyle\sum_{i=0}^{\infty} x_i^{(k)} = 1$ and $(Ae_k)_n = a_{nk}$ so that $\lim_{n \to \infty} a_{nk} = 1$, i.e., $\delta_k = 1$, $k = 0, 1, 2, \ldots$. Conversely, let (16.7) hold with $\delta_k = 1$, $k = 0, 1, 2, \ldots$. Let $x = \{x_k\} \in \ell_1$.

In view of (16.7)(i), $\displaystyle\sum_{k=0}^{\infty} a_{nk} x_k$ converges, $n = 0, 1, 2, \ldots$. Now,

$$(Ax)_n = \sum_{k=0}^{\infty} a_{nk} x_k$$

$$= \sum_{k=0}^{\infty} (a_{nk} - 1) x_k + \sum_{k=0}^{\infty} x_k,$$

this being true since $\sum\limits_{k=0}^{\infty} a_{nk}x_k$ and $\sum\limits_{k=0}^{\infty} x_k$ both converge. Since $\sum\limits_{k=0}^{\infty} |x_k| < \infty$, given $\epsilon > 0$, there exists a positive integer $N$ such that

$$\sum_{k>N} |x_k| < \frac{\epsilon}{2A}, \tag{16.8}$$

where $A = \sup\limits_{n,k} |a_{nk} - 1|$. Since $\lim\limits_{n \to \infty} a_{nk} = 1$, $k = 0, 1, 2, \ldots, N$, we can choose a positive integer $N' > N$ such that

$$|a_{nk} - 1| < \frac{\epsilon}{2(N+1)M}, n \geq N', k = 0, 1, 2, \ldots, N, \tag{16.9}$$

where $M > 0$ is such that $|x_k| \leq M$, $k = 0, 1, 2, \ldots$. Now, for $n \geq N'$,

$$\left| \sum_{k=0}^{\infty} (a_{nk} - 1)x_k \right| = \left| \sum_{k=0}^{N} (a_{nk} - 1)x_k + \sum_{k>N} (a_{nk} - 1)x_k \right|$$

$$\leq \sum_{k=0}^{N} |a_{nk} - 1||x_k| + \sum_{k>N} |a_{nk} - 1||x_k|$$

$$< (N+1)\frac{\epsilon}{2(N+1)M}M + A\frac{\epsilon}{2A},$$

$$\text{in view of } (16.8), (16.9)$$

$$= \epsilon,$$

so that

$$\lim_{n \to \infty} \sum_{k=0}^{\infty} (a_{nk} - 1)x_k = 0$$

and consequently

$$\lim_{n \to \infty} (Ax)_n = \sum_{k=0}^{\infty} x_k.$$

Thus $A \in (\ell_1, c; P')$, completing the proof of the theorem. $\square$

**Theorem 16.6.2.** *When $K = \mathbb{R}$ or $\mathbb{C}$,*

$$(\ell_1, c; P') \cap (\ell_p, c) = \phi, p > 1$$

*Proof.* Let $A = (a_{nk}) \in (\ell_1, c; P') \cap (\ell_p, c)$, $p > 1$. It is known (Stieglitz and Tietz, 1977, p.4,16) that $A \in (\ell_p, c)$, $p > 1$, if and only if $(16.7)(ii)$ holds and

$$\sup_{n \geq 0} \sum_{k=0}^{\infty} |a_{nk}|^q < \infty, \tag{16.10}$$

where $\frac{1}{p} + \frac{1}{q} = 1$. Using (16.10), it follows that

$$\sum_{k=0}^{\infty} |\delta_k|^q < \infty,$$

which contradicts the fact the $\delta_k = 1$, $k = 0, 1, 2, \ldots$, since $A \in (\ell_1, c; P')$ and so $\sum_{k=0}^{\infty} |\delta_k|^q$ diverges. This establishes the theorem. $\square$

*Remark* 16.6.3. Since $(\ell_\infty, c) \subset (c, c) \subset (c_0, c) \subset (\ell_p, c)$, $p > 1$, we have,

$$(\ell_1, c; P') \cap (X, c) = \phi$$

when $X = \ell_\infty, c, c_0, l_p, p > 1$.

In the remaining part of the present section, $K$ is a complete, non-trivially valued, non-archimedean field. In this case too, it can be proved that $A = (a_{nk}) \in (\ell_1, c)$ if and only if (16.7) holds. Theorem 16.6.1 continues to hold in this case too.

In the non-archimedean set up, we have the following theorem.

**Theorem 16.6.4.** *When $K$ is a complete, non-trivially valued, non-archimedean field,*

$$(\ell_1, c; P') \cap (\ell_\infty, c) = \phi.$$

*Proof.* Let $A = (a_{nk}) \in (\ell_1, c; P') \cap (\ell_\infty, c)$. Then

$$\lim_{n \to \infty} \sup_{k \geq 0} |a_{nk} - 1| = 0$$

(see Natarajan (1978), Theorem 2). So for any $\epsilon$, $0 < \epsilon < 1$, there exists a positive integer $N$ such that

$$|a_{nk} - 1| < \epsilon, n \geq N, k = 0, 1, 2, \ldots.$$

In particular,

$$|a_{N,k} - 1| < \epsilon, k = 0, 1, 2, \ldots.$$

Thus,

$$\lim_{k \to \infty} |a_{N,k} - 1| \leq \epsilon,$$

$$i.e., \ |0 - 1| \leq \epsilon,$$

since $A \in (\ell_\infty, c)$, $\lim_{k \to \infty} a_{nk} = 0$, $n = 0, 1, 2, \ldots$, by Theorem 2 of Natarajan (1978),

$$i.e., \ 1 \leq \epsilon,$$

a contradiction on the choice of $\epsilon$, completing the proof. $\square$

*Remark* 16.6.5. However,

$$(\ell_1, c; P') \cap (c, c) \neq \phi,$$

when $K$ is a complete, non-trivially valued, non-archimedean field, as the following example illustrates: Consider the infinite matrix

$$A \equiv (a_{nk}) = \begin{bmatrix} 1 & 0 & 0 & 0 & 0 & 0 & \dots \\ 1 & -1 & 0 & 0 & 0 & 0 & \dots \\ 1 & 1 & -2 & 0 & 0 & 0 & \dots \\ 1 & 1 & 1 & -3 & 0 & 0 & \dots \\ 1 & 1 & 1 & 1 & -4 & 0 & \dots \\ \dots & \dots & \dots & \dots & \dots & \dots & \dots \end{bmatrix},$$

$$i.e., \ a_{nk} = 1, \text{ if } k \leq n-1;$$
$$= -(n-1), \text{ if } k = n;$$
$$= 0, \text{ otherwise.}$$

Then $\sup_{n,k} |a_{nk}| \leq 1 < \infty$, $\lim_{n\to\infty} a_{nk} = 1$, $k = 0, 1, 2, \dots$ and $\lim_{n\to\infty} \sum_{k=0}^{\infty} a_{nk} = 0$ so that $A \in (\ell_1, c; P') \cap (c, c)$ (see (16.1), proving our claim. Since $(c, c) \subset (c_0, c) \subset (\ell_p, c)$, $p > 1$, it follows that

$$(\ell_1, c; P') \cap (X, c) \neq \phi,$$

when $X = c, c_0, \ell_p$, $p > 1$. This indicates a violent departure from the case $K = \mathbb{R}$ or $\mathbb{C}$, when $K$ is a complete, non-trivially valued, non-archimedean field.

In the non-archimedean case, $(c_0, c; P')$ denotes the class of all infinite matrices $A \in (c_0, c)$ such that

$$\lim_{n\to\infty} (Ax)_n = \sum_{k=0}^{\infty} x_k, x = \{x_k\} \in c_0.$$

In this context, we recall that $\sum_{k=0}^{\infty} x_k$ converges if and only if $\{x_k\} \in c_0$. We can easily prove the following result.

*Remark* 16.6.6. When $K$ is a complete, non-trivially valued, non-archimedean field,

$$(c_0, c; P') = (\ell_1, c; P').$$

*Remark* 16.6.7. In the context of what has been discussed in this section, we recall that $\ell_p$, $p \geq 1$, $c_0, c, \ell_\infty$ are all linear spaces with respect to coordinate-wise addition and scalar multiplication irrespective of how $K$ is chosen. When $K = \mathbb{R}$ or $\mathbb{C}$, $c_0, c, \ell_\infty$ are Banach spaces with respect to the norm defined by (16.3), while they are non-archimedean Banach spaces under the above norm when $K$ is a complete, non-trivially valued, non-archimedean field.

Whatever be $K$, $\ell_p$, $p \geq 1$, is a Banach space with respect to the norm

$$\|x\| = \left( \sum_{k=0}^{\infty} |x_k|^p \right)^{\frac{1}{p}}, x = \{x_k\} \in \ell_p.$$

Whatever be $K$, if $A = (a_{nk}) \in (\ell_1, c; P')$, then $A$ is bounded and $\|A\| = \sup_{n,k} |a_{nk}|$. However, $(\ell_1, c; P')$ is not a subspace of $BL(\ell_1, c)$, i.e., the space of all bounded linear mappings of $\ell_1$ into $c$, since $\lim_{n \to \infty} 2a_{nk} = 2$, $k = 0, 1, 2, \ldots$ and consequently $2A \notin (\ell_1, c; P')$, when $A \in (\ell_1, c; P')$.

## 16.7   Some more Steinhaus type theorems over valued fields I

For the contents of this section, the reader can refer to Natarajan (1999). $(\ell_1, \gamma; P)$ denotes the class of all infinite matrices $A = (a_{nk}) \in (\ell_1, \gamma)$ such that

$$\sum_{n=0}^{\infty} (Ax)_n = \sum_{k=0}^{\infty} x_k, x = \{x_k\} \in \ell_1.$$

When $K = \mathbb{R}$ or $\mathbb{C}$, it is known (see Stieglitz and Tietz (1977), p.7,48) that $A = (a_{nk}) \in (\ell_1, \gamma)$ if and only if

$$\left.
\begin{array}{l}
(i) \sup_{m,k} \left| \sum_{n=0}^{m} a_{nk} \right| < \infty; \\
\\
and \\
\\
(ii) \sum_{n=0}^{\infty} a_{nk} \text{ converges}, k = 0, 1, 2, \ldots
\end{array}
\right\} \qquad (16.11)$$

We now prove the following result when $K = \mathbb{R}$ or $\mathbb{C}$.

**Theorem 16.7.1.** *When $K = \mathbb{R}$ or $\mathbb{C}$, a matrix $A = (a_{nk}) \in (\ell_1, \gamma; P)$ if and only if (16.11) holds and*

$$\sum_{n=0}^{\infty} a_{nk} = 1, k = 0, 1, 2, \ldots. \qquad (16.12)$$

*Proof.* Let $A \in (\ell_1, \gamma; P)$. So $A \in (\ell_1, \gamma)$ and thus (16.11) holds. For $k = 0, 1, 2, \ldots$, let $e_k = \{0, \ldots, 0, 1, 0, \ldots\}$, 1 occurring in the $k$th place and 0 elsewhere. Then $e_k \in \ell_1$, $k = 0, 1, 2, \ldots$ and so $\sum_{n=0}^{\infty} (Ae_k)_n = 1$, i.e., $\sum_{n=0}^{\infty} a_{nk} = 1$, $k = 0, 1, 2, \ldots$ so that (16.12) holds.

Conversely, let (16.11) and (16.12) hold. So $A \in (\ell_1, \gamma)$. Let $B = (b_{mk})$, where

$$b_{mk} = \sum_{n=0}^{m} a_{nk}, m, k = 0, 1, 2, \ldots.$$

Using (16.11)(i) and (16.12), we have,

$$\sup_{m,k} |b_{mk}| < \infty$$

and

$$\lim_{m \to \infty} b_{mk} = 1, k = 0, 1, 2, \ldots.$$

Using Theorem 16.6.1,

$$B \in (\ell_1, c; P').$$

Let, now, $x = \{x_k\} \in \ell_1$. So

$$\lim_{m \to \infty} \sum_{k=0}^{\infty} b_{mk} x_k \text{ exists and is equal to } \sum_{k=0}^{\infty} x_k,$$

$$i.e., \lim_{m \to \infty} \sum_{k=0}^{\infty} \left( \sum_{n=0}^{m} a_{nk} \right) x_k = \sum_{k=0}^{\infty} x_k,$$

$$i.e., \lim_{m \to \infty} \sum_{n=0}^{m} \left( \sum_{k=0}^{\infty} a_{nk} x_k \right) = \sum_{k=0}^{\infty} x_k,$$

$$i.e., \sum_{n=0}^{\infty} \left( \sum_{k=0}^{\infty} a_{nk} x_k \right) = \sum_{k=0}^{\infty} x_k,$$

$$i.e., \sum_{n=0}^{\infty} (Ax)_n = \sum_{k=0}^{\infty} x_k,$$

$$i.e., A \in (\ell_1, \gamma; P),$$

completing the proof of the theorem. $\square$

When $K = \mathbb{R}$ or $\mathbb{C}$, we recall that Maddox Maddox (1967) proved that

$$(\gamma, \gamma; P) \cap (\gamma_\infty, \gamma) = \phi.$$

In this context, it is worth noting that the identity matrix, i.e., the matrix $I = (i_{nk})$, where $i_{nk} = 1$, if $k = n$ and $i_{nk} = 0$, if $k \neq n$, is in $(\ell_1, \gamma; P) \cap (\gamma_\infty, \gamma)$ so that

$$(\ell_1, \gamma; P) \cap (\gamma_\infty, \gamma) \neq \phi.$$

Since $(\gamma, \gamma) \supset (\gamma_\infty, \gamma)$, it follows that

$$(\ell_1, \gamma; P) \cap (\gamma, \gamma) \neq \phi.$$

We note that $(\gamma, \gamma; P) \subset (\ell_1, \gamma; P)$ and $(c_0, \gamma) \subset (\gamma, \gamma)$. Having "enlarged" the class $(\gamma, \gamma; P)$ to $(\ell_1, \gamma; P)$, we would like to "contract" the class $(\gamma, \gamma)$ to $(c_0, \gamma)$ and attempt a Steinhaus type theorem involving the classes $(\ell_1, \gamma; p)$ and $(c_0, \gamma)$.

**Theorem 16.7.2.** *When $K = \mathbb{R}$ or $\mathbb{C}$, $(\ell_1, \gamma; P) \cap (c_0, \gamma) = \phi$.*

*Proof.* Let $A = (a_{nk}) \in (\ell_1, \gamma; P) \cap (c_0, \gamma)$. Since $A \in (c_0, \gamma)$,

$$\sup_{m \geq 0} \sum_{k=0}^{\infty} \left| \sum_{n=0}^{m} a_{nk} \right| \leq M < \infty \qquad (16.13)$$

(see Stieglitz and Tietz (1977), p.6,43). Now, for $L = 0, 1, 2, \ldots$, $m = 0, 1, 2, \ldots$,

$$\sum_{k=0}^{L} \left| \sum_{n=0}^{m} a_{nk} \right| \leq \sum_{k=0}^{\infty} \left| \sum_{n=0}^{m} a_{nk} \right|$$
$$\leq M.$$

Taking limit as $m \to \infty$, we have,

$$\sum_{k=0}^{L} \left| \sum_{n=0}^{\infty} a_{nk} \right| \leq M, L = 0, 1, 2, \ldots.$$

Taking limit as $L \to \infty$, we get,

$$\sum_{k=0}^{\infty} \left| \sum_{n=0}^{\infty} a_{nk} \right| \leq M,$$

which is a contradiction, since $\sum_{n=0}^{\infty} a_{nk} = 1$, $k = 0, 1, 2, \ldots$, using (16.12). This proves the theorem. $\square$

**Corollary 16.7.3.** *Since $c_0 \subset c \subset \ell_\infty$, $(\ell_\infty, \gamma) \subset (c, \gamma) \subset (c_0, \gamma)$ so that*

$$(\ell_1, \gamma; P) \cap (X, \gamma) = \phi,$$

*when $X = c_0, c, \ell_\infty$.*

Hereafter, in this section, let us suppose that $K$ is a complete, non-trivially valued, non-archimedean field. In this case, we note that

$$\gamma = c_0 \quad \text{and} \quad \gamma_\infty = \ell_\infty.$$

It is easy to prove the following results.

**Theorem 16.7.4.** $(\ell_1, \gamma) = (\ell_1, c_0) = (c_0, c_0)$. *A matrix* $A = (a_{nk}) \in (\ell_1, c_0)$ *if and only if*

$$\left.\begin{array}{l} (i)\, \sup_{n,k} |a_{nk}| < \infty; \\[4pt] and \\[4pt] (ii)\, \lim_{n \to \infty} a_{nk} = 0,\, k = 0, 1, 2, \ldots . \end{array}\right\} \qquad (16.14)$$

**Theorem 16.7.5.** $(\ell_1, \gamma; P) = (\ell_1, c_0; P) = (c_0, c_0; P) = (\gamma, \gamma; P)$. $A = (a_{nk}) \in (\ell_1, c_0; P)$ *if and only (16.12) and (16.14) hold.*

**Theorem 16.7.6.** *A matrix* $A = (a_{nk}) \in (c, c_0)$ *if and only if (16.14) holds and*

$$\lim_{n \to \infty} \sum_{k=0}^{\infty} a_{nk} = 0. \qquad (16.15)$$

Theorem 16.7.2 fails to hold when $K$ is a complete, non-trivially valued, non-archimedean field, since $(\ell_1, c_0) = (c_0, c_0)$ in this case. We also have,

$$(\ell_1, c_0; P) \cap (c, c_0) \neq \phi,$$

as the following example illustrates. Consider the infinite matrix

$$A \equiv (a_{nk}) = \begin{bmatrix} 1 & -1 & 0 & 0 & 0 & 0 & \ldots \\ 0 & 2 & -2 & 0 & 0 & 0 & \ldots \\ 0 & 0 & 3 & -3 & 0 & 0 & \ldots \\ 0 & 0 & 0 & 4 & -4 & 0 & \ldots \\ \ldots & \ldots & \ldots & \ldots & \ldots & \ldots & \ldots \end{bmatrix},$$

$$\begin{array}{ll} i.e., & a_{nk} = n+1, \text{if } k = n; \\ & = -(n+1), \text{if } k = n+1; \\ & = 0, \text{otherwise.} \end{array}$$

Then (16.12), (16.14), (16.15) hold so that $A \in (\ell_1, c_0; P) \cap (c, c_0)$. These remarks point out significant departure from the case $K = \mathbb{R}$ or $\mathbb{C}$.

The following lemma is needed in the sequel.

**Lemma 16.7.7.** *The following statements are equivalent:*

*(a)* $A = (a_{nk}) \in (\ell_\infty, c_0)$;

*(b)*

$$\left.\begin{array}{l} (i)\, \lim_{k \to \infty} a_{nk} = 0,\, n = 0, 1, 2, \ldots; \\[4pt] and \\[4pt] (ii)\, \lim_{n \to \infty} \sup_{k \geq 0} |a_{nk}| = 0. \end{array}\right\} \qquad (16.16)$$

*(c)*

$$
\left.
\begin{aligned}
&(i)(16.14)(ii)\,holds;\\
&and\\
&(ii)\ \lim_{k\to\infty}\ \sup_{n\geq0}|a_{nk}| = 0.
\end{aligned}
\right\}
\qquad (16.17)
$$

*Proof.* For the proof of "(a) is equivalent to (b)", see Natarajan (1978). We now prove that (b) and (c) are equivalent. Let (b) hold. For every fixed $k = 0, 1, 2, \ldots,$

$$
|a_{nk}| \leq \sup_{k'\geq0}|a_{n,k'}|
$$

$$
\to 0, n \to \infty, \text{using } (16.16)(ii),
$$

so that $(16.17)(i)$ holds. Again, using $(16.16)(ii)$, given $\epsilon > 0$, we can choose a positive integer $N$ such that

$$
\sup_{k\geq0}|a_{nk}| < \epsilon, n > N. \qquad (16.18)
$$

In view of $(16.16)(i)$, for $n = 0, 1, 2, \ldots, N$, we can find a positive integer $L$ such that

$$
|a_{nk}| < \epsilon, k > L. \qquad (16.19)
$$

$(16.18)$ and $(16.19)$ imply that

$$
|a_{nk}| < \epsilon, k > L, n = 0, 1, 2, \ldots,
$$
$$
i.e., \sup_{n\geq0}|a_{nk}| \leq \epsilon, k > L,
$$
$$
i.e.,\ \lim_{k\to\infty}\ \sup_{n\geq0}|a_{nk}| = 0,
$$

so that $(16.17)(ii)$ holds. In other words, (b) implies (c). Similarly we can prove that (c) implies (b). Thus (b) and (c) are equivalent, completing the proof of the lemma. □

We can now prove the following Steinhaus type result.

**Theorem 16.7.8.** $(\ell_1, c_0; P) \cap (\ell_\infty, c_0) = \phi$, *when* $K$ *is a complete, non-trivially valued, non-archimedean field.*

*Proof.* Let $A = (a_{nk}) \in (\ell_1, c_0; P) \cap (\ell_\infty, c_0)$. Then

$$
\sum_{n=0}^{\infty} a_{nk} = 1, k = 0, 1, 2, \ldots.
$$

So,

$$1 = \left| \sum_{n=0}^{\infty} a_{nk} \right|$$

$$\leq \sup_{n \geq 0} |a_{nk}|$$

$$\rightarrow 0, k \rightarrow \infty,$$

using Lemma 16.7.7. This is a contradiction establishing the theorem. $\square$

Using Theorem 16.7.5 and Theorem 16.7.8, we have,

**Corollary 16.7.9.** $(c_0, c_0; P) \cap (\ell_\infty, c_0) = \phi.$

## 16.8 Some more Steinhaus type theorems over valued fields II

The reader can refer to Natarajan (2008) for the contents of this section. First, as usual, we discuss the case $K = \mathbb{R}$ or $\mathbb{C}$. When $K = \mathbb{R}$ or $\mathbb{C}$, the following results are well-known (see Stieglitz and Tietz (1977), p.4,12 & 14). $A = (a_{nk}) \in (c_0, c)$ if and only if

$$\left. \begin{array}{l} (i) \sup_{n \geq 0} \sum_{k=0}^{\infty} |a_{nk}| < \infty; \\ and \\ (ii) \lim_{n \to \infty} a_{nk} = \delta_k \text{ exists, } k = 0, 1, 2, \ldots. \end{array} \right\} \quad (16.20)$$

$A = (a_{nk}) \in (\gamma, c; P')$ if and only if

$$\sup_{n \geq 0} \sum_{k=0}^{\infty} |a_{nk} - a_{n,k+1}| < \infty; \quad (16.21)$$

and (16.20) (ii) holds with $\delta_k = 1$, $k = 0, 1, 2, \ldots$, where, we recall that $(\gamma, c; P')$ is the subclass of $(\gamma, c)$ such that

$$\lim_{n \to \infty} (Ax)_n = \sum_{k=0}^{\infty} x_k, x = \{x_k\} \in \gamma.$$

We now have

**Theorem 16.8.1.** *When* $K = \mathbb{R}$ *or* $\mathbb{C}$,

$$(\gamma, c; P') \cap (c_0, c) = \phi.$$

*Proof.* Let $A = (a_{nk}) \in (\gamma, c; P') \cap (c_0, c)$. Using (16.20), it follows that $\sum_{k=0}^{\infty} |\delta_k|$ converges, which is a contradiction, since $\delta_k = 1$, $k = 0, 1, 2, \ldots$, completing the proof. $\qquad \square$

**Corollary 16.8.2.** *Since $c_0 \subset c \subset \ell_\infty$, it follows that $(\ell_\infty, c) \subset (c, c) \subset (c_0, c)$ so that*
$$(\gamma, c; P') \cap (X, c) = \phi,$$
*when $X = c_0, c, \ell_\infty$.*

It is known (see Stieglitz and Tietz (1977), p.7,48) that $A = (a_{nk}) \in (\ell_1, \gamma)$ if and only if

$$\left.\begin{array}{l} (i) \sup_{m,k} \left| \sum_{n=0}^{m} a_{nk} \right| < \infty; \\[2mm] \text{and} \\[4mm] (ii) \sum_{n=0}^{\infty} a_{nk} \text{ converges}, k = 0, 1, 2, \ldots. \end{array}\right\} \qquad (16.22)$$

It is easy to prove that $A \in (\ell_1, \gamma; P)$ if and only if (16.22) holds and

$$\sum_{n=0}^{\infty} a_{nk} = 1, k = 0, 1, 2, \ldots,$$

where, $(\ell_1, \gamma; P)$ denotes the subclass of $(\ell_1, \gamma)$ such that

$$\sum_{n=0}^{\infty} (Ax)_n = \sum_{k=0}^{\infty} x_k, x = \{x_k\} \in \ell_1.$$

It is also known (see Stieglitz and Tietz (1977), p.7,47) that $A = (a_{nk}) \in (\ell_p, \gamma)$, $p > 1$, if and only if (16.22) (ii) holds and

$$\sup_{m \geq 0} \sum_{k=0}^{\infty} \left| \sum_{n=0}^{\infty} a_{nk} \right|^q < \infty, \qquad (16.23)$$

where $\frac{1}{p} + \frac{1}{q} = 1$.

We now have the following Steinhaus type theorem.

**Theorem 16.8.3.** *When $K = \mathbb{R}$ or $\mathbb{C}$,*
$$(\ell_1, \gamma; P) \cap (\ell_p, \gamma) = \phi, p > 1.$$

*Proof.* Let $A = (a_{nk}) \in (\ell_1, \gamma, P) \cap (\ell_p, \gamma)$, $p > 1$. It now follows that

$$\sum_{k=0}^{\infty} \left| \sum_{n=0}^{\infty} a_{nk} \right|^q < \infty,$$

a contradiction, since $\sum_{n=0}^{\infty} a_{nk} = 1$, $k = 0, 1, 2, \ldots$, completing the proof. $\qquad \square$

**Corollary 16.8.4.** *Since $\ell_p \subset c_0 \subset c \subset \ell_\infty$, $p > 1$, it follows that $(\ell_\infty, \gamma) \subset (c, \gamma) \subset (c_0, \gamma) \subset (\ell_p, \gamma)$ so that*

$$(\ell_1, \gamma; P) \cap (X, \gamma) = \phi,$$

*for $X = \ell_p, c_0, c, \ell_\infty$.*

It is known (see Stieglitz and Tietz (1977), p.6,43) that $A = (a_{nk}) \in (c_0, \gamma)$ if and only if (16.22) (ii) holds and

$$\sup_{m \geq 0} \sum_{k=0}^{\infty} \left| \sum_{n=0}^{m} a_{nk} \right| < \infty. \tag{16.24}$$

$A = (a_{nk}) \in (\gamma, \gamma)$ (see Stieglitz and Tietz (1977), p.7,45) if and only if (16.22)(ii) holds and

$$\sup_{m \geq 0} \sum_{k=0}^{\infty} \left| \sum_{n=0}^{m} (a_{nk} - a_{n,k-1}) \right| < \infty. \tag{16.25}$$

$A \in (\gamma, \gamma; P)$ if and only if (16.25) holds and $\displaystyle\sum_{n=0}^{\infty} a_{nk} = 1$, $k = 0, 1, 2, \ldots$,

where, $(\gamma, \gamma; P)$ denotes the subclass of $(\gamma, \gamma)$ such that

$$\sum_{n=0}^{\infty} (Ax)_n = \sum_{k=0}^{\infty} x_k, x = \{x_k\} \in \gamma.$$

We now have

**Theorem 16.8.5.** *When $K = \mathbb{R}$ or $\mathbb{C}$,*

$$(\gamma, \gamma; P) \cap (c_0, \gamma) = \phi.$$

*Proof.* Let $A = (a_{nk}) \in (\gamma, \gamma; P) \cap (c_0, \gamma)$. Using (16.24), it follows that

$$\sum_{k=0}^{\infty} \left| \sum_{n=0}^{\infty} a_{nk} \right| < \infty,$$

a contradiction, since $\displaystyle\sum_{n=0}^{\infty} a_{nk} = 1$, $k = 0, 1, 2, \ldots$. $\qquad\square$

**Corollary 16.8.6.** *We have*

$$(\gamma, \gamma; P) \cap (X, \gamma) = \phi,$$

*for $X = c_0, c, \ell_\infty$.*

It is known that $A = (a_{nk}) \in (c_0, \ell_1)$ (see Stieglitz and Tietz (1977), p.8,72) if and only if

$$\sup_N \sum_{k=0}^{\infty} \left| \sum_{n \in N} a_{nk} \right| < \infty, \tag{16.26}$$

where $N$ is a subset of $\mathbb{N}_0 = \{0, 1, 2, \dots\}$. It is also known that $A \in (\gamma, \ell_1)$ (see Stieglitz and Tietz (1977), p.9,74) if and only if

$$\sup_{N,K'} \left| \sum_{n \in N} \sum_{k \in K'} (a_{nk} - a_{n,k-1}) \right| < \infty, \tag{16.27}$$

where $N, K'$ are subsets of $\mathbb{N}_0 = \{0, 1, 2, \dots\}$.

One can prove that $A \in (\gamma, \ell_1; P)$ if and only if (16.27) holds and

$$\sum_{n=0}^{\infty} a_{nk} = 1, k = 0, 1, 2, \dots,$$

where, as usual, $(\gamma, \ell_1; P)$ denotes the subclass of $(\gamma, \ell_1)$ with

$$\sum_{n=0}^{\infty} (Ax)_n = \sum_{k=0}^{\infty} x_k, x = \{x_k\} \in \gamma.$$

We are now in a position to prove the following Steinhaus type result.

**Theorem 16.8.7.** *When $K = \mathbb{R}$ or $\mathbb{C}$,*

$$(\gamma, \ell_1; P) \cap (c_0, \ell_1) = \phi.$$

*Proof.* If $A = (a_{nk}) \in (\gamma, \ell_1; P) \cap (c_0, \ell_1)$, using (16.26), we get,

$$\sum_{k=0}^{\infty} \left| \sum_{n=0}^{\infty} a_{nk} \right| < \infty,$$

which is contradiction, since $\sum_{n=0}^{\infty} a_{nk} = 1, k = 0, 1, \dots$, completing the proof.

$\square$

**Corollary 16.8.8.**

$$(\gamma, \ell_1; P) \cap (X, \ell_1) = \phi,$$

*for $X = c_0, c, \ell_\infty$.*

*Remark* 16.8.9. $(c, \gamma; P')$ denotes the subclass of $(c, \gamma)$ with

$$\sum_{n=0}^{\infty} (Ax)_n = \lim_{k \to \infty} x_k, x = \{x_k\} \in c.$$

In the context of Steinhaus type theorems, we note that

$$(c, \gamma; P') \cap (\ell_\infty, \gamma) \neq \phi,$$

when $K = \mathbb{R}$ or $\mathbb{C}$.

In the remaining part of the present section, $K$ is a complete, non-trivially valued, non-archimedean field.

The following observations are worth recording in the context of Theorem 16.8.1. In this case, $\gamma = c_0$ and so

$$(c_0, c; P') \cap (c_0, c) = (c_0, c; P') \neq \phi.$$

We know that $A = (a_{nk}) \in (c_0, c)$ (see Natarajan (1996)) if and only if (16.2) (i) - (ii) hold. $A \in (c_0, c; P')$ if and only if $A \in (c_0, c)$ with $\delta_k = 1$, $k = 0, 1, 2, \ldots$. We will now prove that

$$(c_0, c; P') \cap (c, c) \neq \phi.$$

Consider the following matrix

$$A \equiv (a_{nk}) = \begin{bmatrix} 1 & -1 & 0 & 0 & 0 & 0 & \cdots \\ 1 & 1 & -2 & 0 & 0 & 0 & \cdots \\ 1 & 1 & 1 & -3 & 0 & 0 & \cdots \\ 1 & 1 & 1 & 1 & -4 & 0 & \cdots \\ \cdots & \cdots & \cdots & \cdots & \cdots & \cdots & \cdots \end{bmatrix}.$$

The reader can easily verify that $A \in (c_0, c; P') \cap (c, c)$, proving our claim. However, the following result is true.

**Theorem 16.8.10.** *When $K$ is a complete, non-trivially valued, non-archimedean field,*

$$(c_0, c; P') \cap (\ell_\infty, c) = \phi.$$

*Proof.* Let $A = (a_{nk}) \in (c_0, c; P') \cap (\ell_\infty, c)$. Since $A \in (\ell_\infty, c)$, $\sum\limits_{k=0}^{\infty} \delta_k$ converges, a contradiction, since $\delta_k = 1$, $k = 0, 1, 2, \ldots$ in this case, completing the proof. $\square$

In view of Remark 3.4 of Natarajan (1999),

$$(\ell_1, c_0; P) \cap (X, c_0) \neq \phi,$$

for $X = \ell_p, c_0, c$, while

$$(\ell_1, c_0; P) \cap (\ell_\infty, c_0) = \phi$$

(see Natarajan (1999), Theorem 3.6). Noting that, in this case,

$$(c_0, c_0; P) = (\ell_1, c_0; P),$$

we can write the above results with $(\ell_1, c_0; P)$ replaced by $(c_0, c_0; P)$.

We also note that

$$(c_0, \ell_1; P) \cap (c_0, \ell_1) = (c_0, \ell_1; P) \neq \phi$$

(compare with Theorem 16.8.7). However, we have the following:

**Theorem 16.8.11.** *When $K$ is complete, non-trivially valued, non-archimedean field,*

$$(c_0, \ell_1; P) \cap (c, \ell_1) = \phi.$$

*Proof.* Let $A = (a_{nk}) \in (c_0, \ell_1; P) \cap (c, \ell_1)$. Since $A \in (c_0, \ell_1; P)$,

$$\sum_{n=0}^{\infty} a_{nk} = 1, k = 0, 1, 2, \ldots.$$

Since $A \in (c, \ell)$,

$$\sum_{n=0}^{\infty} \left( \sum_{k=0}^{\infty} a_{nk} \right) \text{ converges.}$$

In view of the fact that convergence is equivalent to unconditional convergence in the non-archimedean set up (see Van Rooij and Schikhof (1971), p.133),

$$\sum_{k=0}^{\infty} \left( \sum_{n=0}^{\infty} a_{nk} \right) \text{ converges,}$$

which is a contradiction, since $\sum_{n=0}^{\infty} a_{nk} = 1$, $k = 0, 1, 2, \ldots$, completing the proof. $\qquad \square$

**Corollary 16.8.12.**

$$(c_0, \ell_1; P) \cap (X, \ell_1) = \phi,$$

*for $X = c, \ell_\infty$.*

*Remark* 16.8.13. Analogous to the class $(c, \gamma; P')$ when $K = \mathbb{R}$ or $\mathbb{C}$, we may be tempted to consider the class $(c, c_0; P')$, when $K$ is complete, non-trivially valued, non-archimedean field. However, we note that

$$(c, c_0; P') = \phi.$$

For, if $A = (a_{nk}) \in (c, c_0; P')$, we have,

$$\sum_{n=0}^{\infty} a_{nk} = 0, k = 0, 1, 2, \ldots \tag{16.28}$$

and

$$\sum_{n=0}^{\infty} \left( \sum_{k=0}^{\infty} a_{nk} \right) = 1. \tag{16.29}$$

(16.28) and (16.29) are contradictory, for,

$$
\begin{aligned}
1 &= \sum_{n=0}^{\infty} \left( \sum_{k=0}^{\infty} a_{nk} \right) \\
&= \sum_{k=0}^{\infty} \left( \sum_{n=0}^{\infty} a_{nk} \right), \text{ since convergence is equivalent} \\
&\qquad \text{to unconditional convergence} \\
&\qquad\quad \text{(see Van Rooij and Schikhof (1971), p. 133)} \\
&= 0,
\end{aligned}
$$

which is absurd, proving our claim.

---

# Bibliography

Bachman G. *Introduction to p-adic numbers and valuation theory.* Academic Press, 1964.

Cooke R. G. *Infinite matrices and sequence spaces.* Macmillan, 1950.

Fridy J. A. A note on absolute summability. *Proc. Amer. Math. Soc.*, 20:285–286, 1969.

Fridy J. A. Properties of absolute summability matrices. *Proc. Amer. Math. Soc.*, 24:583–585, 1970.

Hardy G. H. *Divergent Series.* Oxford, 1949.

Knopp K. and Lorentz G. G. Beiträge Zur absoluten Limitiernng. *Arch. Math.*, 2:10–16, 1949.

Koskela M. A characterization of non-negative matrix operators on $\ell^p$ to $\ell^q$ with $\infty > p \geq q > 1$. *Pacific J. Math.*, 75:165–169, 1978.

Maddox I. J. On theorems of Steinhaus type. *J. London Math. Soc.*, 42:239–244, 1967

Maddox I. J. *Elements of Functional Analysis.* Cambridge University Press, 1970.

Mears F. M. Absolute regularity and the Nörlund mean. *Ann. of Math.*, 38:594–601, 1937.

Monna A. F. Sur le théorème de Banach-Steinhaus. *Indag. Math.*, 25:121–131, 1963.

Narici L., Beckenstein E., and Bachman G. *Functional analysis and valuation theory*. Marcel Dekkar, 1971.

Natarajan P. N. The Steinhaus theorem for Toeplitz matrices in non-archimedean fields. *Comment. Math. Prace Mat.*, 20:417–422, 1978.

Natarajan P. N. Characterization of a class of infinite matrices with applications. *Bull. Austral. Math. Soc.*, 34:161–175, 1986.

Natarajan P. N. A Steinhaus type theorm. *Proc. Amer. Math. Soc.*, 99:559–562, 1987.

Natarajan P. N. Criterion for regular matrices in non-archimedean fields. *J. Ramanujan Math. Soc.*, 6:185–195, 1991.

Natarajan P. N. Some Steinhaus type theorems over valued fields. *Ann. Math. Blaise Pascal*, 3:183–188, 1996.

Natarajan P. N. A theorem of Steinhaus type. *J. Analysis*, 5:139–143, 1997.

Natarajan P. N. Some more Steinhaus type theorems over valued fields. *Ann. Math. Blaise Pascal*, 6:47–54, 1999.

Natarajan P. N. Some more Steinhaus type theorems over valued fields II. *Comm. Math. Analysis*, 5:1–7, 2008.

Natarajan P. N. *An Introduction to Ultrametric Summability Theory*. Springer, 2015, Second edition.

Stieglitz M. and Tietz H. Matrix transformationen von Folgenräumen Eine Ergebnisubersicht. *Math. Z.*, 154:1–16, 1977.

Van Rooij A. C. M. and Schikhof W. H. Non-archimedean analysis. *Nieuw Arch. Wisk.*, 29:120–160, 1971.

# Chapter 17

## The Interplay Between Topological Algebras Theory and Algebras of Holomorphic Functions

**Alberto Saracco**

*Dipartimento di Scienze Matematiche, Fisiche e Informatiche, Università degli Studi di Parma, Italy*

## 17.1 Introduction

There are several interactions between the general theory of commutative topological algebras over $\mathbb{C}$ with unity and that of the holomorphic functions of one or several complex variables.

This chapter is intended to be a review chapter of classical results in this subjects and a list of very difficult and long-standing open problems. A big reference section will provide the reader material to dwell into the proofs of theorems and understand more deeply the subjects.

This chapter will be divided in two parts.

In the first will be reviewed known results on the general theory of Banach algebras, Frechet algebras, $\mathcal{L}B$ and $\mathcal{L}F$ algebras, as well as the open problems in the field.

In the second part will be examined *concrete* algebras of holomorphic functions, again reviewing known results and open problems.

The chapter will mainly focus on the interplay between the two theories, showing how each one can be considered either an object of study per se or a mean to understand better the other subject.

## 17.2 Topological algebras

A *topological algebra* $A$ over a topological field $\mathbb{K}$ is a topological space endowed with continuous operations

$$+ : A \times A \to A \,,$$

$$\cdot : \mathbb{K} \times A \to A \,,$$

$$\star : A \times A \to A \,,$$

such that $(A, +, \cdot)$ is a vector space over $\mathbb{K}$ and $(A, +, \star)$ is an algebra. If the inner product $\star$ is commutative, we will call $A$ a *commutative (topological) algebra*. We will usually omit the adjective topological in the sequel. $A$ is said commutative algebra with unit if there is $\mathbf{1} \in A$ such that $\mathbf{1} \star a = a \star \mathbf{1} = a$ for each $a \in A$.

Topological algebras were introduced in 1931 by David van Dantzig in his doctoral theses van Dantzig (1931) and they were extensively studied by Izrail Gelfand, Mark Naĭmark and Georgi Šilov starting from the Forties.

Since we are interested in algebras of holomorphic functions, usually $\mathbb{K} = \mathbb{C}$, with the Euclidean topology. Moreover we will usually denote $a \star b$ simply by $ab$.

## 17.2.1   Banach algebras

A commutative algebra $B$ over $\mathbb{C}$ with unit $\mathbf{1}$ is said to be a *Banach algebra* if it is a Banach space with a norm $\| \cdot \|$ such that

$$\| \mathbf{1} \| = 1, \quad \| xy \| \leq \| x \| \| y \|,$$

for all $x, y \in B$.

Algebras of continuous $\mathbb{C}$-valued bounded functions on a set $D$, containing constant functions, endowed with the supremum norm, are Banach algebras.

Banach algebras of $\mathbb{C}$-valued functions are such that for all elements $b$ of the algebra

$$\| b^2 \| = \| b \|^2 . \tag{17.1}$$

A Banach algebra $B$ satisfying the above condition is called an *uniform algebra*. Actually this condition is very restrictive and the only uniform algebras are algebras of $\mathbb{C}$-valued functions.

From now on, $B$ will always denote a Banach algebra.

The *spectrum* of an element $b \in B$ is the subset of $\mathbb{C}$

$$\sigma(b) = \{\lambda \in \mathbb{C} \ : \ \lambda \cdot \mathbf{1} - b \text{ is not invertible}\}$$

For every $b \in B$, the spectrum $\sigma(b)$ is a non empty compact subset of $\mathbb{C}$. This implies that the only Banach field (up to an isometric isomorphism) is $\mathbb{C}$ (Gelfand-Mazur theorem).

## 17.2.1.1   Banach algebras: characters and spectrum

A *character* of $B$ is a homomorphism of algebras $\chi : B \to \mathbb{C}$. Every character $\chi$ of a Banach algebra is continuous. In particular the norm of every character is bounded by 1. If the character $\chi$ is not trivial (i.e. does not send $B$ in 0) than since $\chi(\mathbf{1}) = 1 \| \chi \| = 1$.

The *spectrum* of $B$ is the set $\mathfrak{M}(B)$ of all non trivial characters of $B$. If $B^*$ denotes the dual space of $B$ then

$$\mathfrak{M}(B) \subset \{b^* \in B^* \ : \ \| b^* \| \leq 1\},$$

thus the spectrum, with the topology induced by the weak-* topology on $B^*$ (called the *Gelfand topology*), is compact, thanks to Banach-Alaoglu.

$\mathfrak{M}(B)$ equipped with the Gelfand topology is a compact Hausdorff space

For every $b \in B$, the map $\hat{b} : \mathfrak{M}(B) \to \mathbb{C}$ defined by

$$\hat{b}(\chi) = \chi(b)$$

is called the *Gelfand transform* of $b$. By $\hat{B}$ we denote the set of all Gelfand transforms of $B$.

The Gelfand topology is the weakest topology on the spectrum $\mathfrak{M}(B)$ that makes every Gelfand transform a continuous map.

The map $\Gamma : B \to \mathcal{C}^0(\mathfrak{M}(B))$ sending an element of $B$ to its Gelfand transform is continuous (the space of continuous $\mathbb{C}$-valued functions on the spectrum being endowed with the sup norm). If $B$ is a uniform algebra, see (17.1), then this map makes $\hat{B}$ a function algebra on $\mathfrak{M}(B)$ isomorphic to $B$.

### 17.2.1.2   Banach algebras: maximal spectrum

The closure of any proper ideal of a Banach algebra $B$ is a proper ideal. Hence maximal ideals are closed. The set of all maximal ideals of $B$, $\Omega(B)$, is called the *maximal spectrum* of $B$.

There is a natural bijection between the spectrum and the maximal spectrum of an algebra, sending a non trivial character to its kernel:

$$T : \ \mathfrak{M}(B) \ \to \ \Omega(B), \ \varphi \mapsto \ \mathrm{Ker}\,\varphi.$$

From now on, by $\mathfrak{M}(B)$ we will denote both the spectrum and the maximal spectrum of $B$, via the above identification.

The spectrum of a Banach algebra is a non-empty compact space.

### 17.2.1.3   Banach algebras: boundaries

A closed subset $E$ of the spectrum $\mathfrak{M}(B)$ is called a *boundary* for $B$ if for every $b \in B$, its Gelfand trasform $\hat{b}$ attains its maximum on $E$:

$$\max_E |\hat{b}| \ = \ \max_{\mathfrak{M}(B)} |\hat{b}| \, .$$

$\mathfrak{M}(B)$ is obviously a boundary for $B$.

If $B$ is an algebra of functions on a compact $K$, then we can see each point $x \in K$ as a point of the spectrum (as the character of evaluation at $x$: $\varphi_x : b \to b(x)$), and obviously $K$ is a boundary for $B$.

The intersection of all boundaries of $B$ is itself a boundary, called the *Šilov boundary* of $B$. We will denote by $\gamma B$ the Šilov boundary of $B$.

### 17.2.1.4   Banach algebras: analytical properties of the spectrum

Many properties of the spectrum of a Banach algebra resemble basic properties of a domain $D$ of $\mathbb{C}^n$, where the holomorphic functions on the domain $D$ are the equivalent of the elements of the Banach algebra. These properties are thus usually named *analytic properties of the spectrum*.

Let $B$ be a Banach algebra and $\mathfrak{M}(B)$ its spectrum. A character $\varphi \in \mathfrak{M}(B)$ is called a *peak point* for the Banach algebra $B$ if there is an element $b \in B$ such that $\varphi(b) = 1$ and $|\psi(b)| < 1$ for all $\psi \in \mathfrak{M}(B)$, $\psi \neq \varphi$. Notice that the set of peak points for $B$ is a subset of the Šilov boundary of $B$, since given a peak point $\varphi \in \mathfrak{M}(B)$, $\hat{b}$ attains its maximum only in $\varphi$.

A character $\varphi \in \mathfrak{M}(B)$ is called a *local peak point* for the Banach algebra $B$ if there is a neighbourhood $U \subset \mathfrak{M}(B)$ of $\varphi$ and an element $b \in B$ such that $\varphi(b) = 1$ and $|\psi(b)| < 1$ for all $\psi \in U$, $\psi \neq \varphi$.

If $A$ is a uniform Banach algebra (i.e. an algebra of functions) then for any $U \subset \mathfrak{M}(A)$ not touching the Šilov boundary of $A$ $U \cap \gamma A = \emptyset$ we define the algebra $A(U)$ as the closure of $A|_{\overline{U}}$ in the sup norm on $\overline{U}$. Hugo Rossi in 1959 Rossi (1960) proved the following

**Theorem 17.2.1.** *Let $A$ be a uniform Banach algebra, and $U \subset \mathfrak{M}(A) \setminus \gamma A$. Any point $x \in U$ is not a local peak point for $A(\overline{U})$.*

This theorem, called *maximum modulus principle* asserts that for uniform algebras points out of the Šilov boundary are not even local peak points. This closely resembles the maximum modulus principle for holomorphic functions (non-constant holomorphic functions have no local maxima).

Rossi's maximum modulus principle strongly suggests something holomorphic is going on. Thus many attempts to put on (a subdomain of) the spectrum of a uniform Banach algebra a complex structure making the elements of the algebra holomorphic functions have been made. The maximum modulus principle made reasonable to conjecture that the spectrum of a uniform Banach algebra, except for the Šilov boundary, must in some sense have an underlying analytic structure.

**Analytic structure of the spectrum: a negative result.** In 1963 Stolzenberg, in a paper with the evocative title *A hull without anaytic structure* Stolzenberg (1963), gave a negative result.

Let $K \subset \mathbb{C}^n$ be a compact set and $\mathcal{Q}(K)$ be the closure of polynomials in $\mathcal{C}^0(K)$. Stolzenberg considers a relatively weak notion of analytic structure: given a point $z \in \mathfrak{M}(\mathcal{Q}(K))$, $(S_z, f_z)$ is said to be an *analytic structure through* $z$ if

1. there is $N \in \mathbb{Z}^+$ such that $S_z \subset \mathbb{C}^N$ is a connected analytic subset containing the origin;

2. $f_z : S_z \to \mathbb{C}^n$ is a non-constant holomorphic map such that $f_z(0) = z$ and $f_z(S_z) \subset \mathfrak{M}(\mathcal{Q}(K))$.

Any subset of the spectrum $\mathfrak{M}(\mathcal{Q}(K))$ admitting such an analytic structure satisfies the maximum modulus principle.

Then Stolzenberg constructs a set so bad that nowhere satisfies such a definition and uses it to construct a counterexample: let $P = \overline{\Delta} \times \overline{\Delta}$ be the closed bidisc of $\mathbb{C}^2$ and $\{z_j\}_{j \in \mathbb{N}} \subset \Delta \setminus \{0\}$ a countable dense sequence of points. Depending on $z_j$, Stolzenberg constructs a sequence of varieties in the bidisc $P$. The varieties $V_j$, as $j$ goes to infinity, are more and more wavy so that their limit $V$ is so bad that nowhere admits an analytic structure. Considering $K = V \cap bP$, one has $\mathfrak{M}(\mathcal{Q}(K)) \supset V$ and no subset of $\mathfrak{M}(\mathcal{Q}(K))$ admits an analytic structure.

**Analytic structure of the spectrum: positive results.** One year later, in 1964, Andrew Gleason Gleason (1964) proved a positive results for points of the spectrum corresponding to finitely generated ideals of the algebra $B$.

More precisely, let us as usual denote both the spectrum and the maximal spectrum by $\mathfrak{M}(B)$, so that $m \in \mathfrak{M}(B)$ will mean both a maximal ideal of $B$ and a character on $B$.

**Theorem 17.2.2** (Gleason, 1964). *Let $B$ be a Banach algebra and $m_0 \in \mathfrak{M}(B)$ be a maximal ideal of $B$ finitely generated by $b_1, \ldots, b_n \in B$. Then we can define the map $\varphi : \mathfrak{M}(B) \to \mathbb{C}^n$ by*

$$\varphi(m) = (\hat{b}_1(m), \cdots, \hat{b}_n(m)) \,.$$

*In this way $\varphi(m_0) = 0$. There is a neighbourhood of $0 \in \mathbb{C}^n$ such that*

1. *$\varphi|_{\varphi^{-1}(U)}$ is a homeomorphism onto a closed analytic subset of $U$;*

2. *the Gelfand transforms of elements of $B$ are holomorphic functions on $\varphi^{-1}(U)$.*

*Moreover any maximal ideal $m \in \varphi^{-1}(U)$ is generated by $a_1(m), \ldots, a_n(m)$, where $a_j(m) = a_j - \hat{a}_j(m) \cdot \mathbf{1}$.*

The key point in the proof of Gleason's theorem is a holomorphic version of the implicit function theorem which holds for Banach spaces.

By Gleason's theorem, if all maximal ideals of $B$ are finitely generated, then the spectrum $\mathfrak{M}(B)$ is a compact complex space on which the holomorphic functions in $\hat{B}$ separate points. Thus $\mathfrak{M}(B)$ must be a finite set and $B$ is a semi-local ring.

Finiteness properties of a topological algebra and structure properties of its spectrum have interesting links. Some of these were investigated by Artur Vaz Ferreira and Giuseppe Tomassini in Ferreira and Tomassini (1978).

Gleason's theorem was later generalized by Andrew Browder in 1971 Browder (1971):

**Theorem 17.2.3** (Browder, 1971). *Let $B$ be a Banach algebra and $m \in \mathfrak{M}(B)$ a maximal ideal of $B$. If*

$$\dim_{\mathbb{C}} m/m^2 = n < +\infty, \qquad (17.2)$$

*then there is a neighbourhood $U \ni m$ in $\mathfrak{M}(B)$, an analytic subset $V$ of the unit polydisc of $\mathbb{C}^n$ and a surjective homeomorphism $\tau : U \to V$ such that $\hat{b} \circ \tau$ is holomorphic on $V$, for all $b \in B$.*

Clearly, if $m$ is finitely generated as in the hypothesis of Gleason's theorem, then the hypothesis of Browder's theorem are fulfilled and near $m$ the spectrum is analytic. Gleason's theorem has thus a stronger hypothesis, but gives something more, i.e. the fact that near $m$ also the other maximal ideals are finitely generated, while Browder's theorem does not state nothing similar.

The finiteness hypothesis in Browder's theorem can be restated in terms of derivations. As usual, let $B$ be a Banach algebra and $\varphi \in \mathfrak{M}(B)$ a point of its spectrum. A *derivation* of $B$ at $\varphi$ is a linear functional $\delta : B \to \mathbb{C}$ satisfying the Leibniz rule, i.e. for all $a, b \in B$

$$\delta(ab) = \delta(a)\hat{b}(\varphi) + \hat{a}(\varphi)\delta(b).$$

Considering $m = \mathrm{Ker}\,\varphi$ (i.e. $m$ is the maximal ideal of $B$ corresponding to $\varphi$), we have that a linear $\delta : B \to \mathbb{C}$ is a derivation if and only if $\mathrm{Ker}\,\delta$ contains the unit $\mathbf{1}$ of the algebra and the ideal $m^2$. Thus the complex vector space of derivations at $\varphi$ (at $m$) $\mathcal{D}_\varphi = \mathcal{D}_m$ is isomorphic to $m/m^2$.

Thus the hypothesis (17.2) of Browder's theorem may be restated as

$$\dim_{\mathbb{C}} \mathcal{D}_m = n < +\infty.$$

### 17.2.1.5 $C^*$-algebras

A special kind of Banach algebras is that of $C^*$-algebras.

A $*$-*(Banach) algebra* $B$ is a Banach algebra (over $\mathbb{C}$) endowed with an involution $* : B \to B$ (we denote $*(b)$ by $b^*$) such that $\forall a, b \in B$, $\forall \lambda \in \mathbb{C}$

($*$-1) $(a + b)^* = a^* + b^*$;

($*$-2) $(\lambda a)^* = \bar{\lambda} a^*$;

($*$-3) $(ab)^* = b^* a^*$;

($*$-4) $(a^*)^* = a$.

If moreover

$(C^*)$ $\| a^* a \| = \| a \|^2$,

then $B$ is said to be a *(Banach) $C^*$-algebra*.

As a consequence of the $C^*$ condition, elements of $C^*$-algebras satisfy $\| a^* \| = \| a \|$ and commutative $C^*$-algebras are uniform algebras.

A classical example of $C^*$-algebra is that of $\mathbb{C}$-valued functions on a compact $K$, with the involution $*$ given by the conjugation.

Izrail Gelfand and Mark Naĭmark Gelfand and Neumark (1943) proved that a commutative $*$-algebra $B$ with unit is isometrically isomorphic to the algebra of continuous functions on its spectrum $C^0(\mathfrak{M}(B))$.

A Banach algebra $B$ of holomorphic functions can be made a $*$-algebra, by defining the involution $*$ in the following way

$$f^*(z) = \overline{f(\bar{z})} \quad \forall f \in B,$$

but cannot be made a $C^*$-algebra.

Since here we are mostly interested in algebras of holomorphic functions, we do not indulge in exploring $C^*$-algebras.

## 17.2.2 Fréchet algebras

A commutative algebra $F$ over $\mathbb{C}$ with unit $\mathbf{1}$ is said to be a *Fréchet algebra* if it is a Fréchet space with a family $\{p_n\}_{n\in\mathbb{N}}$ of submultiplicative seminorms, i.e. such that

$$p_n(xy) \le p_n(x)p_n(y),$$

for all $x, y \in F$.

We may assume (up to modifying the system of seminorms with an equivalent one) that, for all $n \in \mathbb{N}$,

  i) $p_n \le p_{n+1}$;

  ii) $p_n(\mathbf{1}) = 1$.

Actually giving a Fréchet algebra $F$ is equivalent to giving a projective system of Banach algebras

$$\{B_n, \psi_n : B_{n+1} \to B_n\}_{n\in\mathbb{N}},$$

the Fréchet algebra being the projective limit of the system $F = \varprojlim B_n$. For a precise definition of projective limit of a projective system, refer to Schaefer (1966).

### 17.2.2.1 Fréchet algebras: characters and spectrum

A *character* of $F$ is a homomorphism of algebras $\chi : F \to \mathbb{C}$. The set of all characters of $F$ is denoted by $\mathfrak{C}(F)$ It is not known whether or not all characters $\chi$ of a Fréchet algebra are continuous.

The *spectrum* of $F$ is the set $\mathfrak{M}(F)$ of all non trivial continuous characters of $F$. If $F^*$ denotes the dual space of $F$, then

$$\mathfrak{M}(F) \subset \{b^* \in F^* : \| b^* \| \le 1\}.$$

The spectrum, with the topology induced by the weak-* topology on $F^*$ (called the *Gelfand topology*), thanks to the Banach-Alaoglu theorem and to the fact that a Fréchet algebra is the projective limit of a projective system of Banach algebras

$$\{B_n, \psi_n : B_{n+1} \to B_n\}_{n\in\mathbb{N}},$$

admits a compact exhaustion by $\mathfrak{M}(B_n)$.

This implies, using the Gelfand-Mazur theorem, that a Fréchet field is isometric to $\mathbb{C}$, and hence Banach.

The fact that a Fréchet algebra $F$ is a projective limit of Banach algebras can be used to prove that the spectrum is dense in the set of all characters:

$$\overline{\mathfrak{M}(F)} = \mathfrak{C}(F).$$

**Definition 17.2.4.** Let $A$ be an algebra of function over a topological space $X$. We say that $X$ is *A-convex* if for every compact $K \subset X$, its $A$-hull

$$X_A = \{x \in X \mid |f(x)| \leq \max_{z \in K} |f(z)|\},$$

is compact, too.

As a consequense of Banach-Steinhaus theorem and the fact that Fréchet algebras are projective limits of Banach algebras, it can be proved that

**Theorem 17.2.5.** *The spectrum $\mathfrak{M}(F)$ of a Fréchet algebra $F$ is $F$-convex.*

Obviously this goes with the canonical Gelfand identification of $f \in F$ with its Gelfand transform $\hat{f}$. For a proof of this theorem, refer to (Della Sala et al., 2006, Theorems 7.32, 12.24).

### 17.2.2.2   Fréchet algebras: Michael's problem

**Open problem**. The problem whether or not all characters $\chi$ of a Fréchet algebra are continuous was first posed by Ernest Michael in 1952 (see Michael (1952)). Many attempts to solve the problem were made, based on methods of topological algebras theory.

*A positive result.* A first partial result was obtained in 1958 by Richard Arens: if the Fréchet algebra $F$ is finitely rationally generated, i.e. there are a finite numer of generators $x_1, \ldots, x_k \in F$ such that the subalgebra of elements of the form

$$\frac{P(x_1, \ldots, x_k)}{Q(x_1, \ldots, x_k)},$$

where $P$ and $Q$ are $F$-valued polynomials ($Q$ is of course required to have an inverse in $F$) is dense in $F$, then all characters of $F$ are continuous.

*Reducing Michael's problem to a specific algebra.* In 1975 Dennis Clayton Clayton (1975) reduced Michael's problem to solving it for a particular algebra. Let $l^\infty$ be the space of bounded complex sequences and $\mathcal{P}$ the algebra of functions $p : l^\infty \to \mathbb{C}$ which are polynomials in (a finite number of) the coordinate functions. Introducing on $\mathcal{P}$ the semi-norms

$$p_n(p) = \sup\{|p(z)| \; : \| z \|_\infty \leq n\}$$

consider $A$, the completion of $\mathcal{P}$ with respect to these semi-norms. $A$ is then a Fréchet algebra. Clayton proves that if there is a Fréchet algebra $B$ and $\varphi : B \to \mathbb{C}$ is a discontinuous homomorphism, then there is a discontinuous homomorphism $\psi : A \to \mathbb{C}$ (i.e. the answer to Michael's problem is negative if and only if the algebra $A$ is a counterexample). Thus Michael's problem reduces to a problem about the algebra $A$.

Martin Schottenloher in 1981 Schottenloher (1981) modified Clayton's construction to reduce Michael's problem to the same problem for a specific algebra of holomorphic functions, and similar reductions were found more recently also by Jorge Mujica Mujica (1986, 2016).

Even with these huge simplifications of the problem, still no answer was found.

*Holomorphic dynamics approach to Michael's problem.* A totally different approach to the problem was opened in 1986, when Peter Dixon and Jean Esterle Dixon and Esterle (1986) proved that

**Proposition 17.2.6.** *Suppose there is a Fréchet algebra $F$ with a discontinuous character. Then, for every sequence $\{s_n\}_{n\in\mathbb{N}}$ and any projective system*

$$\cdots \longrightarrow \mathbb{C}^{s_{n+1}} \xrightarrow{F_n} \mathbb{C}^{s_n} \longrightarrow \cdots ,$$

*where all $F_n$ are holomorphic, $\varprojlim \mathbb{C}^{s_n} \neq \emptyset$.*

This result reduced proving Michael's problem to finding an example where the above projective limit is empty. In particular one gets that all characters of a Fréchet algebra are continuous if there exists a sequence of holomorphic maps $F_k : \mathbb{C}^n \to \mathbb{C}^n$ such that

$$A = \bigcap_{k\in\mathbb{N}} F_0 \circ \cdots \circ F_k(\mathbb{C}^n) = \emptyset. \qquad (17.3)$$

The best part of this result is that now both the positive and negative answer to the Michael's problem can be given just by producing an example.

The bad part is that the example seems not easy at all to find.

Obviously if one of the $F_k$ is constant, then the intersection is not empty. If $n = 1$, by Picard's theorem, each non constant holomorphic entire function assumes all values but at most one. Hence each $F_0 \circ \cdots \circ F_k(\mathbb{C}^n)$ is a dense open subdomain of $\mathbb{C}$, hence by Baire's theorem also $A$ is a dense open set. The example has to be found in several variables.

If $n > 1$, there are domains $\Omega \subsetneq \mathbb{C}^n$ biholomorphic to $\mathbb{C}^n$ not dense in the whole of $\mathbb{C}^n$. These domains (called *Fatou-Bieberach domains*) could provide a way to construct the needed example.

Fatou-Bieberach domains can also be used to construct the so called *long* $\mathbb{C}^n$, which are complex manifolds of dimension $n$ exhausted by an increasing sequence of domains each biholomorphic to $\mathbb{C}^n$. Long $\mathbb{C}^n$ can have unexpected properties (e.g. they can even have no non-constant holomorphic or plurisubharmonic functions, see Boc Thaler and Forstnerič (2016)) and might play a role in the solution of the Michael's problem.

## 17.2.3   $\mathcal{L}B$ algebras and $\mathcal{L}F$ algebras

Just like one can obtain Fréchet algebras starting from a projective family of Banach algebras, one can produce new kinds of algebras starting from a directed family of Banach or Fréchet algebras.

Since, differently from Fréchet algebras, $\mathcal{L}B$ algebras and $\mathcal{L}F$ algebras do not have a nice presentation different from the one as limit, we give some details of the definition of direct limit. For a more precise definition of direct limits of directed system, refer to Schaefer (1966).

A topological vector space $E$ is said to be a *locally convex space* if the origin $O \in E$ has a basis of open convex neighbourhoods whose intersection is $\{O\}$. Notice that Banach and Fréchet algebras are locally convex spaces.

Let $E$ be a vector space, and an increasing sequence $E_n$ of subspaces of $E$ (i.e. such that $E_n \subset E_{n+1}$), each endowed with a topology $\tau_n$ such that $E_n$ is a a locally convex space and the topolgy $\tau_{n+1}$ induces $\tau_n$, and such that

$$E \;=\; \bigcup_{n \in \mathbb{N}} E_n \,.$$

$E$, endowed with the finest topology $\tau$ such that

1. $(E, \tau)$ is a locally convex space;

2. the inclusions $(E_n, \tau_n) \to (E, \tau)$ are continuous,

is called the *direct limit* of the directed system $(E_n, \tau_n)$.

An algebra which is a direct limit of Banach algebras is called a $\mathcal{L}B$ algebra, while an algebra which is a direct limit of Fréchet algebras is called a $\mathcal{L}F$ algebra.

A direct limit of complete locally convex spaces is a complete locally convex space, hence $\mathcal{L}B$ algebras and $\mathcal{L}F$ algebras are complete.

The spectrum of a $\mathcal{L}B$ algebra or $\mathcal{L}F$ algebra $A$ is defined similarly to what has been done for Banach and Fréchet algebras:

$$\mathfrak{M}(A) \;=\; \{\varphi : A \to \mathbb{C} \,|\, \varphi \text{ is a non trivial continuous homomorphism of algebras}\} \,.$$

Due to the fact that projective limits and direct limits are in duality, if $A$ is the direct limit of the (Banach or Fréchet) algebras $A_n$, then its spectrum $\mathfrak{M}(A)$ is the projective limit of the spectra $\mathfrak{M}(A_n)$.

---

## 17.3   Algebras of holomorphic functions

Let D be a domain (i.e. an open connected set) in $\mathbb{C}^n$ or more generally in a complex manifold or a complex space $X$.

By $\mathcal{O}(D)$ we denote the algebra of holomorphic functions on $D$.

In this section, we will present several algebras of holomorphic functions, i.e. subalgebras of $\mathcal{O}(D)$, which fall in one of the families introduced in the previous section. These concrete algebras serve both as an example and a stimulus to study the general theory as well as a way to suggest new interesting general problems.

### 17.3.1 The algebra of holomorphic functions $\mathcal{O}(X)$

We start our review with the algebra of holomorphic functions on a complex manifold (or space) $X$.

#### 17.3.1.1 Stein spaces and Stein manifolds

When we are interested in studying such an algebra, it is natural to consider only a certain class of complex manifolds (complex spaces), namely that of Stein manifolds (Stein spaces).

Let us show a few examples suggesting that we may want to impose some conditions on the space $X$.

Let $X$ be complex torus $\mathbb{T}^n = \mathbb{C}^n/\mathbb{Z}^n$ or a complex projective space $\mathbb{CP}^n$, being $X$ compact, all holomorphic functions on $X$ are bounded. Since bounded holomorphic entire functions are constant, it easily follows that the only holomorphic functions on such an $X$ are the constant functions, and the algebra $\mathcal{O}(X)$ is simply $\mathbb{C}$.

A slightly less trivial class of examples could be that of spaces $X$ where holomorphic functions do not separate points, i.e. there exists a couple of points $x \neq y \in X$ such that $f(x) = f(y)$ for all functions $f \in \mathcal{O}(X)$. Considering the quotient space $Y = X/\sim$, where $x \sim y$ iff $f(x) = f(y)$ for all functions $f \in \mathcal{O}(X)$, then $Y$ is a complex space such that $\mathcal{O}(Y) = \mathcal{O}(X)$ and where holomorphic functions separate points of $Y$.

Another (less immediate) condition we may want to ask is that of $X$ being holomorphically convex, i.e. such that for any compact $K \Subset X$ its holomorphic hull

$$\hat{K}_{\mathcal{O}(X)} = \{x \in X \mid |f(x)| \leq \max_{z \in K} |f(z)|, \ \forall f \in \mathcal{O}(X)\}$$

is also compact. This is exactly the notion of $A$-convexity given in much more generality in Definition 17.2.4, when $A = \mathcal{O}(X)$. Such a condition prevents the *Hartogs phenomenon* to happen, i.e. $X$ is the natural space for considering the algebra of functions $\mathcal{O}(X)$ and is not a subspace of a bigger complex space $\widetilde{X}$ with the same algebra. The Hartogs phenomenon is characteristic of several complex variables and does not happen in one complex variable, and was discovered by Fritz Hartogs in 1906 Hartogs (1906).

Thus, if we are interested in studying the algebra of holomorphic functions on a complex space $X$, we may limit ourselves to spaces such that

($S_1$) holomorphic functions separate points of $X$;

($S_2$) $X$ is holomorphically convex.

A complex manifold (resp. complex space) $X$ satisfying ($S_1$) and ($S_2$) is called a *Stein manifold* (resp. *Stein space*). A subdomain $D$ of a Stein space $X$ is said to be *locally Stein* if for every point $x \in bD$ there is a neighbourhood $V \ni x$ such that $V \cap D$ is Stein.

For a Stein manifold $X$ of dimension $n$ it holds a result very similar to the Whitney embedding theorem for real manifolds: they can be properly embedded in $\mathbb{C}^{2n+1}$.

For a review on Stein manifolds and spaces, we refer the reader to (Forstnerič, 2011, Ch. 2, 3) and (Della Sala et al., 2006, Ch. 11).

### 17.3.1.2  The spectrum of the algebra of holomorphic functions on a Stein space

Let $X$ be an $n$-dimensional complex space. Denote by

$$\delta : X \to \mathfrak{M}(\mathcal{O}(X))$$

the natural map which associates $x \in X$ to the maximal ideal

$$M_x = \{f \in \mathcal{O}(X) \mid f(x) = 0\}.$$

$\mathcal{O}(X)$ is a Fréchet algebra when endowed with the sup-seminorms on an increasing exhaustion of compacts.

A Stein space (manifold) is completely determined by its algebra of holomorphic function, as proved in great generality by Constantin Bănică and Octavian Stănăşilă in 1969 Bănică and Stănăşilă (1969).

**Theorem 17.3.1.** *A complex space $X$ is Stein iff $\delta$ is a homeomorphism. Moreover, for every complex space $Y$*

$$Hol(Y, X) \simeq Hom_{\mathcal{C}^0}(\mathcal{O}(X), \mathcal{O}(Y)).$$

*In particular, two Stein spaces are biholomorphic iff their Fréchet algebras of holomorphic functions are isomorphic.*

### 17.3.1.3  The envelope of holomorphy problem

As we discussed above, there are some complex spaces $X$ which are not the natural domains for holomorphic functions, i.e. Hartogs phenomenon happens and all holomorphic functions on $X$ extend (uniquely thanks to the identity principle) to a bigger domain. This does not happen if $X$ is Stein.

Hence, a natural problem arises. An *envelope* (or *hull*) *of holomorphy* of a complex space (or manifold) $X$ is a Stein space $\hat{X}$ and a holomorphic open embedding $j : X \to \hat{X}$ such that $j^* : \mathcal{O}(\hat{X}) \to \mathcal{O}(X)$ is an isomorphism of Fréchet algebras, i.e. all holomorphic functions on $X$ extend uniquely to the Stein space $\hat{X}$.

Since $X \subset \hat{X}$, and $\hat{X}$ is required to be Stein, an obvious necessary condition for $X$ to have an envelope of holomorphy is to have holomorphic functions that separate its points.

Notice that, thanks to the previous theorem, if an envelope of holomorphy exists, it is unique up to biholomorphisms.

Moreover notice that, even if $X$ is a complex manifold, its envelope of holomorphy may be a Stein space, and not a manifold. Indeed, just consider a Stein space $\hat{X}$ (with singular set $Z \neq \emptyset$). Then $X = \hat{X} \setminus Z$ is a complex manifold, which has the Stein space $\hat{X}$ as envelope of holomorphy, where $j : X = \hat{X} \setminus Z \to \hat{X}$ is the inclusion.

Thanks to Theorem 17.3.1, if a complex space $X$ admits an envelope of holomorphy $\hat{X}$, then its envelope —being a Stein space— is completely determined up to biholomorphisms by its algebra of holomorphic functions $\mathcal{O}(\hat{X}) \simeq \mathcal{O}(X)$. Hence, the natural candidate for $\hat{X}$ is the spectrum $\mathfrak{M}(\mathcal{O}(X))$. Thus the envelope of holomorphy problem may be restated in the following way.

**Envelope of holomorphy problem**: let $X$ be a complex space such that $\mathcal{O}(X)$ separates points of $X$. Give the spectrum $\mathfrak{M}(\mathcal{O}(X))$ a complex structure such that

1. the natural map $\delta : X \to \mathfrak{M}(\mathcal{O}(X))$ is an open holomorphic embedding;

2. all the Gelfand transforms $\hat{f}$, $f \in \mathcal{O}(X)$, are holomorphic functions on the spectrum.

We remark that $\mathfrak{M}(\mathcal{O}(X))$ is a Stein space thanks to condition (1) and the fact that $\mathfrak{M}(\mathcal{O}(X))$ is $\mathcal{O}(\hat{X})$-convex, as it follows from the general Theorem 17.2.5 on Fréchet algebras.

Using this characterization of the envelope of holomorphy problem, one usually says that $\mathcal{O}(X)$ is a *Stein algebra* iff $\mathcal{O}(X) \simeq \mathcal{O}(\hat{X})$ for a Stein space $\hat{X}$ or —equivalently— if $\mathfrak{M}(\mathcal{O}(X))$ can be given a complex structure satisfying (1) and (2) above.

**The envelope of holomorphy problem: positive results.** The envelope of holomorphy problem has a positive answer for a huge class of domains. Henri Cartan, Peter Thullen and Kiyoshi Oka gave the positive answer for all Riemann domains $X$ over $\mathbb{C}^n$, i.e. domains $X$ with a holomorphic projection $\pi : X \to \mathbb{C}^n$ which is locally a biholomorphism. For a presentation of the proof, we refer to the paper by Hugo Rossi Rossi (1963). The classical proofs by Cartan, Thullen and Oka use classical arguments of the theory of several complex variables. Their results were also obtained using a different strategy of proof, which used abstract theory of algebras of functions —namely, interpolating semi-norms, by Erret Bishop in 1963 Bishop (1963).

The approach of Bishop solves more generally the problem of the envelope of holomorphy for Riemann domains $D$ relative to a subalgebra $\mathcal{A} \subset \mathcal{O}(D)$.

The same is true for Riemann domains $X$ over a Stein manifold $Y$, i.e. domains $X$ with a holomorphic projection $\pi : X \to Y$ which is locally a biholomorphism. In this greater generality the result was proven in 1960 by Ferdinand Docquier and Hans Grauert Docquier and Grauert (1960).

In particular, all domains $D \subset \mathbb{C}^n$ admit an envelope of holomorphy (since subdomains are a trivial examples of Riemann domains). It is nevertheless worth noticing that if $X \subset Y$, where $Y$ is a Stein manifold, its envelope of holomorphy needs not to be contained in $Y$. This is false even when $Y = \mathbb{C}^n$. Henri Cartan pointed out a counterexample for $Y = \mathbb{C}^2$. Obviously there is no counterexample for $Y = \mathbb{C}$, since all domains in $\mathbb{C}$ are holomorphically convex, hence Stein.

**The envelope of holomorphy problem: negative results.** To construct a domain for which the envelope of holomorphy does not exist is quite a complex and technical task. A first example was found by Hans Grauert in 1963 Grauert (1963). A clear exposition of the example, which is a domain with a smooth boundary, almost everywhere strongly Levi-convex, can be found in Huckleberry (1970).

As we've already noted, subdomains of Stein spaces always admit an envelope of holomorphy. This is no longer the case for subdomains of Stein spaces, as proved by Mihnea Colţoiu and Klas Diederich in 2000 Coltoiu and Diederich (2000). Let $X$ be a Stein space and $D \subset X$ an open set. $D$ is said to verify the *strong hypersection condition* (SHSC) if for every open set $V \supset D$ and every closed complex hypersurface $H \subset V$ then $H \cap D$ is Stein.

**Theorem 17.3.2.** *Let $D$ be a relatively compact open set of a Stein space $X$ satisfying one of the following conditions:*

- *$D$ is locally Stein;*

- *$D$ is an increasing union of Stein spaces;*

- *$D$ satisfies SHSC.*

*Then either $D$ is Stein or it does not have an envelope of holomorphy.*

In Coltoiu and Diederich (2000) it is constructed a Stein space $X$ of pure dimension 3 and a closed analytic surface $A \subset X$ such that $D = X \setminus A$ satisfies SHSC and is not Stein, hence does not have an envelope of holomorphy.

It is worth citing the fact that the other two hypotheses in the above theorem are linked with some open problems, namely the Levi problem and the union problem.

**Open problem**: *Levi problem.* Is a locally Stein open subset of a Stein space Stein?

The answer is known to be positive for $X = \mathbb{C}^n$ (Fifties: Oka Oka (1950), Bremermann Bremermann (1954), Norguet Norguet (1954)), $X$ a Stein manifold (1960: Docquier and Grauert Docquier and Grauert (1960)), or even $X$ a Stein space with isolated singularities (1964: Andreotti and Narasimhan Andreotti and Narasimhan (1964)). For a survey of the Levi problem, refer to Siu (1978).

**Open problem**: *union problem.* Is an increasing union of Stein subdomains $D$ of a Stein space $X$ Stein?

The answer is known to be positive for $X = \mathbb{C}^n$ (1939: Behnke and Stein

Behnke and Stein (1939)) or $X$ a Stein manifold (1960: Docquier and Grauert Docquier and Grauert (1960)). The general problem is unsolved even for isolated singularities. If $X$ is a normal Stein space, $D$ is known to be a domain of holomorphy, i.e. for each point $x \in bD$ there is a function unbounded near $x$. Stein spaces are domain of holomorphy, but the converse fails to be true if the dimension of the space is at least 3.

*Other notions of envelope.* The notion of envelope is linked to the choice of an algebra of functions. It is worth noticing that in complex geometry, this notion can be generalized to other families of analytic objects, as divisors, principal divisors or analytic subsets. Since this survey is on algebras of functions, we do not indulge into these more general notions.

While all domains in $\mathbb{C}^n$ are Stein and hence coincide with their envelopes, if we consider small algebras things may get trickier. Consider e.g.

$$D = \mathbb{C} \setminus \{z \in R \,|\, z \leq 0\},$$

and the algebra $\mathcal{A}$ generated by the holomorphic function $\log z$. Then $D$ is not $\mathcal{A}$-convex and its $\mathcal{A}$-envelope is the infinite layer Riemann spiral projecting over $\mathbb{C} \setminus \{0\}$.

## 17.3.2  The algebra of bounded holomorphic functions $H^\infty(D)$

We denote by $H^\infty(D)$ the algebra of bounded holomorphic functions on $D$. $H^\infty(D)$ is a Banach algebra when endowed with the sup-norm:

$$\| f \|_\infty = \sup_{z \in D} |f(z)|.$$

The notation $H^\infty(D)$ is due to the fact that $H^\infty(D) = \mathcal{O}(D) \cap L^\infty(D)$ and the norm on $H^\infty(D)$ is just the restriction of the sup-norm on the Banach algebra $L^\infty(D)$.

The space $H^\infty(D)$ was firstly considered in the case $D = \Delta$, the unit disk of $\mathbb{C}$. In this situation, also the spaces $H^p(\Delta)$, $0 < p \leq \infty$ are defined. The spaces $H^p(\Delta)$ are usually known as Hardy spaces of the disk. For a nice and complete survey on these spaces, refer to the classical book of Peter Duren Duren (1970).

Let $0 < p < \infty$. A holomorphic function $f$ on $\Delta$ is said to be in $H^p(\Delta)$ if

$$\|f\|_p^p = \sup_{0<r<1} \int_0^1 |f(re^{2\pi i\theta})| \, d\theta < \infty$$

For $1 \leq p \leq \infty$, $H^p(\Delta)$ endowed with the norm $\| \cdot \|_p$ is a Banach algebra, and for $p = 2$ the Banach norm is induced by a Hermitian product, turning $H^2(\Delta)$ in a complex Hilbert space of functions.

Assume $H^\infty(D)$ separates the points of $D$, i.e. given any two points $z \neq w \in D$ there is a function $f \in H^\infty(D)$ such that $f(z) \neq f(w)$.

Then there is a natural embedding $\iota$ of the domain $D$ into the (maximal) spectrum $\mathfrak{M}(H^\infty(D))$ given by

$$\iota(z) \;=\; \{f \in H^\infty(D) \mid f(z) = 0\}\,,$$

i.e. each point $z$ of the domain is sent to the maximal ideal of functions vanishing in $z$.

Obviouly $\iota(D)$ is not all of the spectrum (e.g. the maximal ideals of functions vanishing at a point of the boundary are not in the image), also for the compactness of the spectrum.

Consider the set

$$\mathfrak{M}(H^\infty(D)) \setminus \overline{\iota(D)}\,,$$

called the *corona*.

The *corona conjecture* states that the corona is empty, i.e. the domain $D$ naturally embeds densely in the spectrum of $\mathfrak{M}(H^\infty(D))$.

We remark that the corona conjecture has an analytic equivalent form:

**Theorem 17.3.3.** *Assume bounded holomorphic functions separate points of $D$.*

*$\iota(D)$ is dense in $\mathfrak{M}(H^\infty(D))$ if and only if for all $f_1, \ldots, f_n \in H^\infty(D)$ such that*

$$\sum_{i=1}^{n} |f_i(z)|^2 \geq \delta > 0$$

*for all $z \in D$, there exists $g_1, \ldots, g_n \in H^\infty(D)$ such that*

$$\sum_{i=1}^{n} f_i(z) g_i(z) \;\equiv\; 1 \quad \forall z \in D.$$

The corona conjecture has a positive answer for the unit disk $\Delta \subset \mathbb{C}$ (first proof due to Lennart Carleson in 1962 Carleson (1962)), while the answer is negative in general (for a counterexample, see e.g. the one by Nessim Sibony Sibony (1987)) and is still an open problem for simple domains as the unit ball and unit polidisk in $\mathbb{C}^n$, $n > 1$.

In the disk the corona conjecture is deeply linked with the study of Carleson measures for $H^p(\Delta)$, i.e. measures $\mu$ on $\Delta$ such that the inclusion $H^p(\Delta) \; \hookrightarrow L^r(\mu)$ is continuous.

A class of algebras of functions strongly linked to the Hardy algebras is that of Bergmann algebras of functions:

$$\mathcal{A}^p(D) \;=\; \mathcal{O}(D) \cap L^p(D)\,,$$

considered as a subalgebra of $L^p(D)$. In Bergmann algebras a central problem is that of the characterization of Carleson measures, while there are no problems directly linked to the theory of topological algebras, hence we do not analyze them here.

To have a broader review on the corona problem and on Carleson measures of Hardy and Bergmann spaces, refer to Douglas et al. (2014) and Saracco (2017).

### 17.3.3  The algebras of holomorphic functions continuous (or more) up to the boundary $A^k(D)$

Let $D \subset \mathbb{C}^n$ be a domain. For $k \in \mathbb{N} \cup \{\infty\}$, we denote by $A^k(D) = \mathcal{C}^k(\overline{D}) \cap \mathcal{O}(D)$, i.e. the algebra of functions holomorphic on $D$ and of class $\mathcal{C}^k$ up to the boundary. Since $A^k(D)$ is a closed subalgebra of $\mathcal{C}^k(\overline{D})$, it is a Banach algebra if $D$ is bounded and $k \in \mathbb{N}$, when endowed with the norm

$$\| f \|_{A^k(D)} = \sup_{j \leq k, z \in \overline{D}} |f^{(j)}(z)|.$$

In case either $D$ is unbounded or $k = \infty$ (or both), then $A^k(D)$ is a Fréchet algebra, endowed with the seminorms ($1 \leq \ell \in \mathbb{N}$, $z_0 \in \overline{D}$ fixed)

$$p_\ell(f) = \| f_{|B_\ell(z_0) \cap D} \|_{A^\ell(B_\ell \cap D)},$$

where $B_\ell(z_0) = \{z \in \mathbb{C}^n \mid |z - z_0| < \ell\}$.

Convexity properties of the domain $D$ (plain convexity, or Levi-convexity at the points of the boundary) result in interesting properties of the spectrum of $A^k(D)$.

**Theorem 17.3.4.** *Let $D \Subset \mathbb{C}^n$ be a convex bounded domain (or a convex domain with smooth strongly Levi-convex boundary. Then*

$$\mathfrak{M}(A^k(D)) = \overline{D},$$

*for all $k \in \mathbb{N} \cup \{\infty\}$.*

The main ingredient in the proof is that, under the hypothesis, functions in $A^k(D)$ can be approximated by holomorphic functions in a system of Stein neighbourhoods whose intersection is $\overline{D}$. Since for any Stein manifold $X$ closed ideals of the algebra of holomorphic functions correspond to the points of $X$, it follows the thesis of the theorem.

A different way of proof, based on the $\mathcal{C}^\infty$-regularity for the $\bar{\partial}$-problem on $\overline{D}$ (result due to Joseph Kohn Kohn (1979)), allows the domain $D$ to have smooth Levi-convex boundary, i.e. to drop the strong Levi-convexity of the boundary. This was estabilished by Monique Hakim and Nessim Sibony in 1980 Hakim and Sibony (1980) and further generalized by Giuseppe Tomassini in 1983 Tomassini (1983) to the case when the domain $D$ is no long required to be bounded.

All these results hold (with the same proof) to domains $D$ in a complex manifold $X$ such that there exists a non-constant subharmonic function defined on a neighbourhood of $bD$.

**Theorem 17.3.5.** *Let $X$ be a complex manifold and $D \subset X$ a domain such that there exists a non-constant subharmonic function defined on a neighbourhood of $bD$. Then*

$$\mathfrak{M}(A^k(D)) = \overline{D},$$

*for all $k \in \mathbb{N} \cup \{\infty\}$.*

### 17.3.3.1  Šilov boundary of $A^k(D)$

If $D \Subset \mathbb{C}^n$ is a bounded domain and $k \in \mathbb{N}$, then $A^k(D)$ is a Banach algebra whose spectrum is $\mathfrak{M}(A^k(D)) = \overline{D}$.

Due to the maximum principle for holomorphic functions, the topological boundary of $\overline{D}$ is a boundary for the algebra $A^k(D)$. Actually, the name *boundary* for Banach algebras derives exactly from this remark. Hence the Šilov boundary of the algebra is a subset of the topological boundary of $\overline{D}$:

$$\gamma\mathfrak{M}(A^k(D)) \subset b\overline{D}.$$

Depending on $D$, however, the Šilov boundary may or may not coincide with the whole topological boundary.

If $D$ is strictly convex, then for every point of the boundary $z_0$ there is a (complex) hyperplane $\{L = 0\}$ through $z_0$ touching $\overline{D}$ only in $z_0$. Thus $f = 1/L$ is a function holomorphic on $\mathbb{C}^n \setminus \{z_0\}$. A small deformation of this function gives a function holomorphic in $D$ and even $\mathcal{C}^\infty$ up to the boundary, which attains its maximum at $z_0$. Hence in this case the Šilov boundary coincides with the topological boundary.

The above argument may be carried on also if $D$ is supposed to be strongly pseudoconvex, or even for other notions of convexity. As it should be clear, hypotheses of convexity on the domain, imply several nice properties on algebras of holomorphic functions (see Andersson et al. (2004)).

On the opposite side of the spectrum, where the boundary has big flat parts, let us suppose $D$ is a product domain, i.e.

$$D = D_1 \times \cdots \times D_h,$$

where $D_j \subset \mathbb{C}^{n_j}$ and $n_1 + \cdots + n_h = n$.

Then the Šilov boundary of the algebra $A^k(D)$ is the Cartesian product of the Šilov boundaries of the algebras $A^k(D_j)$, hence it is strictly contained in the topological boundary of $\overline{D}$.

It is worth noticing that, while the real dimension of the topological boundary is $2n - 1$ (if the topological boundary is a manifold), the real dimension of the Šilov boundary (again, if it is a manifold) is lower, for product domains.

In the particular case of a polidisk

$$D = \Delta_1 \times \cdots \times \Delta_n,$$

$\Delta_k \subset \mathbb{C}$ being disks, the Šilov boundary of $A^k(D)$ is exactly the product of the topological boundaries of the disks, i.e. a real torus of dimension $n$.

### 17.3.3.2 Finitely generated maximal ideals

Let now $D \Subset \mathbb{C}^n$ be a bounded domain with a (sufficiently) smooth Levi-convex boundary, $n > 1$. Let moreover $k \in \mathbb{N}$, so that $A^k(D)$ is a Banach algebra. As already noted, $\mathfrak{M}(A^k(D)) = \bar{D}$.

*Remark* 17.3.6. In this situation, Gleason's theorem (theorem 17.2.2) implies that finitely generated ideals have a structure of a complex space. This implies that points of $bD \subset \mathfrak{M}(A^k(D))$ are not finitely generated maximal ideals. Indeed, if $z \in bD$ was a finitely generated maximal ideal, there would exist a neighbourhood $z \in U \subset \bar{D}$ with a complex space structure. In particular $U \cap bD$ would locally disconnect $U$, which is a contradiction.

It remains to determine whether the points $a \in D \subset \mathfrak{M}(A^k(D))$ are finitely generated maximal ideals or not. This is known as the *Gleason problem*.

Given a point $a = (a_1, \ldots, a_n) \in D \Subset \mathbb{C}^n$, a natural candidate set of generators of the maximal ideal $M_a$ is given by the coordinate functions $z_j - a_j$, $j = 1, \ldots, n$. They actually are a set of generators, provided the boundary $bD$ is of class at least $\mathcal{C}^{k+3}$ and $D$ is convex, as proved by Gennadĭ Henkin in 1971 Henkin (1971).

**Theorem 17.3.7.** *Let $D \Subset \mathbb{C}^n$ be a bounded convex domain with boundary $bD$ is of class at least $\mathcal{C}^{k+3}$ ($k \in \mathbb{N}$) and $a = (a_1, \ldots, a_n) \in D$. Then the maximal ideal $M_a$ of $A^k(D)$ is finitely generated by the coordinate functions $z_j - a_j$, $j = 1, \ldots, n$.*

The quite elementary proof explicitly constructs, for $f \in M_a$, functions $h_j \in A^k(D)$, $j = 1, \ldots, n$, via an integral on the segment joining $a$ and a generic point $z \in \bar{D}$, such that

$$f(z) = \sum_{j=1}^{n} h_j(z)(z_j - a_j).$$

It is worth noticing that since $A^\infty(D)$ is not a Banach algebra (not even when $D$ is bounded), remark 17.3.6 does not apply in this case. It is indeed possible, by the very same proof, to prove the stronger

**Theorem 17.3.8.** *Let $D \subset \mathbb{C}^n$ be a (non necessarily bounded) convex domain with boundary $bD$ of class $\mathcal{C}^\infty$ and $a = (a_1, \ldots, a_n) \in D$. Then the maximal ideal $M_a$ of $A^\infty(D)$ is finitely generated by the coordinate functions $z_j - a_j$, $j = 1, \ldots, n$.*

Dropping the convexity property in favour of strong Levi-convexity at all points of the boundary (even if a greater smoothness of the boundary is required, mainly for technical reasons), both results remain true, although the proof is no longer constructive and heavily uses sheaf theory. The following theorems were established by Giuseppe Tomassini in 1983 Tomassini (1983).

**Theorem 17.3.9.** *Let $D \Subset \mathbb{C}^n$ be a bounded domain with strongly Levi-convex boundary $bD$ of class $\mathcal{C}^\infty$ and $a = (a_1, \ldots, a_n) \in D$. Then the maximal ideal $M_a$ of $A^k(D)$ is finitely generated by the coordinate functions $z_j - a_j$, $j = 1, \ldots, n$.*

**Theorem 17.3.10.** *Let $D \subset \mathbb{C}^n$ be a (non necessarily bounded) domain with strongly Levi-convex boundary $bD$ of class $\mathcal{C}^\infty$ and $a = (a_1, \ldots, a_n) \in D$. Then the maximal ideal $M_a$ of $A^\infty(D)$ is finitely generated by the coordinate functions $z_j - a_j$, $j = 1, \ldots, n$.*

### 17.3.4 The algebra of functions holomorphic in a neighbourhood of a compact $\mathcal{O}(K)$

Let $X$ be a complex space (usually a Stein space or even $\mathbb{C}^n$) and $K \subset X$ a closed set (possibly a compact). Consider the algebra $\mathcal{O}(K)$ of germs of holomorphic functions, i.e. the algebra of functions holomorphic on an open neighbourhood of $K$. Considering a family of open neighbourhoods $U_k$ of $K$, each one contained in the next, and whose intersection is $K$, $\mathcal{O}(K)$ is easily seen to be an $\mathcal{L}F$-algebra, being the direct limit of the Fréchet algebras $\mathcal{O}(U_k)$.

If $K$ is the closure of a bounded domain in $\mathbb{C}^n$ or in a Stein space $X$, the structure of the spectrum $\mathfrak{M}(\mathcal{O}(K))$ (which for obvious reasons is called the *holomorphic envelope* of $K$) was studied by Reese Harvey and Raymond Wells at the end of the Sixties Harvey and Wells (1969); Wells (1968).

Regularity hypotheses on the set $K$ give interesting algebraic properties of the algebra $\mathcal{O}(K)$. Indeed

**Theorem 17.3.11.** *If $K$ is a compact subset of a connected normal space $X$, then $\mathcal{O}(K)$ is an integrally closed ring.*

**Theorem 17.3.12** (Frisch 1965 Frisch (1965), Siu 1969 Siu (1969)). *If $K$ is a semianalytic compact set which has a fundamental system of Stein neighbourhoods in a complex space $X$, then $\mathcal{O}(K)$ is a noetherian ring.*

**Theorem 17.3.13** (Dales 1974 Dales (1974)). *If $K$ is a contractible semianalytic compact set which has a fundamental system of Stein neighbourhoods in a locally factorial complex space $X$, then $\mathcal{O}(K)$ is a factorial ring.*

### 17.3.5 The algebra of holomorphic functions with polynomial growth $\mathcal{P}(D)$

Let $X$ be a complex space endowed with a metric, and $D \subsetneq X$ a domain. A function $f : D \to \mathbb{C}$ of class $\mathcal{C}^\infty$ is said to have a *polynomial growth* if for every $K$ compact subset of $X$ there exists $m \in \mathbb{N}$ such that

$$\sup_{z \in K \cap D} d(z, bD)^m |f(z)|,$$

$d(z, bD)$ being the distance of $z$ from the boundary of $D$. The same definition may of course be given also for differential forms.

The name polynomial growth is due to the fact that, if $X = \mathbb{CP}^n$ with the Fubini-Study metric and $D = \mathbb{C}^n$, then the holomorphic functions with polynomial growth are exactly the polynomials.

The holomorphic functions with polinomial growth on $D$ form an algebra, denoted with $\mathcal{P}(D)$. Note that actually the algebra depends not only on $D$ but also on the ambient space $X$ and its metric. We do not stress these in the name, as the only result we cite here is in $\mathbb{C}^n$ with the Euclidean metric.

Let $X = \mathbb{C}^n$, with the euclidean metric, and $D \subsetneq \mathbb{C}^n$ be a domain with a smooth strongly Levi-convex boundary. In this situation, Paolo Cerrone and Giuseppe Tomassini proved in 1984 Cerrone and Tomassini (1984) some theorems of finiteness for ideals of functions of polynomial growth, leading to an interesting result on the $\bar{\partial}$-problem.

**Theorem 17.3.14.** *Let $D \subsetneq \mathbb{C}^n$ be a domain with a smooth strongly Levi-convex boundary and $\eta$ a $\mathcal{C}^\infty$-smooth $(p, q)$-form with polynomial growth. Then the equation*

$$\bar{\partial}\mu = \eta$$

*has a solution $\mu$, $\mathcal{C}^\infty$-smooth $(p, q - 1)$-form with polynomial growth.*

---

# Bibliography

Andersson M., Passare M., and Sigurdsson R. *Complex convexity and analytic functionals*, volume 225 of *Progress in Mathematics*. Birkhäuser Verlag, Basel, 2004.

Andreotti A. and Narasimhan R. Oka's Heftungslemma and the Levi problem for complex spaces. *Trans. Amer. Math. Soc.*, 111:345–366, 1964.

Behnke H. and Stein K. Konvergente Folgen von Regularitätsbereichen und die Meromorphiekonvexität. *Math. Ann.*, 116(1):204–216, 1939.

Bishop E. Holomorphic completions, analytic continuation, and the interpolation of semi-norms. *Ann. of Math. (2)*, 78:468–500, 1963.

Boc Thaler L. and Forstnerič F. A long $\mathbb{C}^2$ without holomorphic functions. *Anal. PDE*, 9(8):2031–2050, 2016.

Bremermann H. J. über die äquivalenz der pseudokonvexen Gebiete und der Holomorphiegebiete im Raum von $n$ komplexen Veränderlichen. *Math. Ann.*, 128:63–91, 1954.

Browder A. Point derivations and analytic structure in the spectrum of a Banach algebra. *J. Functional Analysis*, 7:156–164, 1971.

Bănică C. and Stănăşilă O. A result on section algebras over complex spaces. *Atti Accad. Naz. Lincei Rend. Cl. Sci. Fis. Mat. Natur. (8)*, 47:233–235 (1970), 1969.

Carleson L. Interpolations by bounded analytic functions and the corona problem. *Ann. of Math. (2)*, 76:547–559, 1962.

Cerrone P. and Tomassini G. Sur l'algèbre des fonctions holomorphes à croissance polynomiale. *Bull. Sci. Math. (2)*, 108(4):393–407, 1984.

Clayton D. A reduction of the continuous homomorphism problem for *F*-algebras. *Rocky Mountain J. Math.*, 5:337–344, 1975.

Coltoiu M. and Diederich K. On Levi's problem on complex spaces and envelopes of holomorphy. *Math. Ann.*, 316(1):185–199, 2000.

Dales H. G. The ring of holomorphic functions on a Stein compact set as a unique factorization domain. *Proc. Amer. Math. Soc.*, 44:88–92, 1974.

Della Sala G., Saracco A., Simioniuc A., and Tomassini G. *Lectures on complex analysis and analytic geometry*, volume 3 of *Appunti. Scuola Normale Superiore di Pisa (Nuova Serie) [Lecture Notes. Scuola Normale Superiore di Pisa (New Series)]*. Edizioni della Normale, Pisa, 2006.

Dixon P. G. and Esterle J. Michael's problem and the Poincaré-Fatou-Bieberbach phenomenon. *Bull. Amer. Math. Soc. (N.S.)*, 15(2):127–187, 1986.

Docquier F. and Grauert H. Levisches Problem und Rungescher Satz für Teilgebiete Steinscher Mannigfaltigkeiten. *Math. Ann.*, 140:94–123, 1960.

Douglas R. G., Krantz S. G., Sawyer E. T., Treil S., and Wicks B. D., 2014. A history of the corona problem. In *The corona problem*, volume 72 of *Fields Inst. Commun.*, pages 1–29. Springer, New York.

Duren P. L. *Theory of $H^p$ spaces*. Pure and Applied Mathematics, Vol. 38. Academic Press, New York-London, 1970.

Ferreira A. V. and Tomassini G. Finiteness properties of topological algebras. I. *Ann. Scuola Norm. Sup. Pisa Cl. Sci. (4)*, 5(3):471–488, 1978.

Forstnerič F. *Stein manifolds and holomorphic mappings*, volume 56 of *Ergebnisse der Mathematik und ihrer Grenzgebiete. 3. Folge. A Series of Modern Surveys in Mathematics [Results in Mathematics and Related Areas. 3rd Series. A Series of Modern Surveys in Mathematics]*. Springer, Heidelberg, 2011. The homotopy principle in complex analysis.

Frisch J. Fonctions analytiques sur un ensemble semi-analytique. *C. R. Acad. Sci. Paris*, 260:2974–2976, 1965.

Gelfand I. and Neumark M. On the imbedding of normed rings into the ring of operators in Hilbert space. *Rec. Math. [Mat. Sbornik] N.S.*, 12(54):197–213, 1943.

Gleason A. M. Finitely generated ideals in Banach algebras. *J. Math. Mech.*, 13:125–132, 1964.

Grauert H. Bemerkenswerte pseudokonvexe Mannigfaltigkeiten. *Math. Z.*, 81:377–391, 1963.

Hakim M. and Sibony N. Spectre de $A(\bar{\Omega})$ pour des domaines bornés faiblement pseudoconvexes réguliers. *J. Funct. Anal.*, 37(2):127–135, 1980.

Hartogs F. Zur Theorie der analytischen Funktionen mehrerer unabhängiger Veränderlichen, insbesondere über die Darstellung derselben durch Reihen, welche nach Potenzen einer Veränderlichen fortschreiten. *Math. Ann.*, 62(1):1–88, 1906.

Harvey R. and Wells, Jr. R. O. Compact holomorphically convex subsets of a Stein manifold. *Trans. Amer. Math. Soc.*, 136:509–516, 1969.

Henkin G. M. The approximation of functions in pseudo-convex domains and a theorem of Z. L. Leïbenzon. *Bull. Acad. Polon. Sci. Sér. Sci. Math. Astronom. Phys.*, 19:37–42, 1971.

Huckleberry A. T. Some famous examples in the study of holomorphic function algebras. *Seminario E. E. Levi*, 1970.

Kohn J. J. Subellipticity of the $\bar{\partial}$-Neumann problem on pseudo-convex domains: sufficient conditions. *Acta Math.*, 142(1-2):79–122, 1979.

Michael E. A. Locally multiplicatively-convex topological algebras. *Mem. Amer. Math. Soc.,*, No. 11:79, 1952.

Mujica J. *Complex analysis in Banach spaces*, volume 120 of *North-Holland Mathematics Studies*. North-Holland Publishing Co., Amsterdam, 1986. Holomorphic functions and domains of holomorphy in finite and infinite dimensions, Notas de Matemática [Mathematical Notes], 107.

Mujica J. Algebras of holomorphic functions and the Michael problem. *Rev. R. Acad. Cienc. Exactas Fís. Nat. Ser. A Math. RACSAM*, 110(1):1–6, 2016.

Norguet F. Sur les domaines d'holomorphie des fonctions uniformes de plusieurs variables complexes. (Passage du local au global.). *Bull. Soc. Math. France*, 82:137–159, 1954.

Oka K. Sur les fonctions analytiques de plusieurs variables. VII. Sur quelques notions arithmétiques. *Bull. Soc. Math. France*, 78:1–27, 1950.

Rossi H. The local maximum modulus principle. *Ann. of Math. (2)*, 72:1–11, 1960.

Rossi H. On envelopes of holomorphy. *Comm. Pure Appl. Math.*, 16:9–17, 1963.

Saracco A., 2017. The corona problem, carleson measures and applications. In *Mathematical Analysis and Applications - selected topics*. Wiley.

Schaefer H. H. *Topological vector spaces*. The Macmillan Co., New York; Collier-Macmillan Ltd., London, 1966.

Schottenloher M. Michael problem and algebras of holomorphic functions. *Arch. Math. (Basel)*, 37(3):241–247, 1981.

Sibony N. Problème de la couronne pour des domaines pseudoconvexes à bord lisse. *Ann. of Math. (2)*, 126(3):675–682, 1987.

Siu Y.-t. Noetherianness of rings of holomorphic functions on Stein compact series. *Proc. Amer. Math. Soc.*, 21:483–489, 1969.

Siu Y. T. Pseudoconvexity and the problem of Levi. *Bull. Amer. Math. Soc.*, 84(4):481–512, 1978.

Stolzenberg G. A hull with no analytic structure. *J. Math. Mech.*, 12:103–111, 1963.

Tomassini G. Sur les algèbres $A^0(\bar{D})$ et $A^\infty(\bar{D})$ d'un domaine pseudoconvexe non borné. *Ann. Scuola Norm. Sup. Pisa Cl. Sci. (4)*, 10(2):243–256, 1983.

van Dantzig D. *Studien over topologische algebra*, 1931. Thesis (Ph.D.) – University of Groningen.

Wells, Jr. R. O. Holomorphic hulls and holomorphic convexity of differentiable submanifolds. *Trans. Amer. Math. Soc.*, 132:245–262, 1968.